CHRISTOPHER ISHERWOOD

THE SIXTIES

DIARIES, VOLUME TWO:
1960–1969

Books by Christopher Isherwood

NOVELS
All the Conspirators
The Memorial
Mr. Norris Changes Trains
Goodbye to Berlin
Prater Violet
The World in the Evening
Down There on a Visit
A Single Man
A Meeting by the River

AUTOBIOGRAPHY & DIARIES
Lions and Shadows
Kathleen and Frank
Christopher and His Kind
My Guru and His Disciple
October (*with Don Bachardy*)
Volume One: 1939–1960
Lost Years: A Memoir: 1945–1951

BIOGRAPHY
Ramakrishna and His Disciples

PLAYS (*with W.H.Auden*)
The Dog Beneath the Skin
The Ascent of F6
On the Frontier

TRAVEL
Journey to a War (*with W.H.Auden*)
The Condor and the Cows

COLLECTIONS
Exhumations
Where Joy Resides

THE SIXTIES

DIARIES, VOLUME TWO:

1960–1969

CHRISTOPHER ISHERWOOD

Edited and Introduced by Katherine Bucknell
Foreword by Christopher Hitchens

HARPER

An Imprint of HarperCollins*Publishers*
www.harpercollins.com

HarperCollins books may be purchased for educational, business, or sales promotional use. For information, please write: Special Markets Department, HarperCollins Publishers, 10 East 53rd Street, New York, NY 10022.

Originally published in Great Britain in 2010 by Chatto & Windus.

FIRST U.S. EDITION

Library of Congress Control Number: 97005501

ISBN: 978-0-06-118019-4

10 11 12 13 14   OFF/RRD   10 9 8 7 6 5 4 3 2 1

Contents

Foreword

Why and when did we cease as a culture to divide time into reigns or epochs ("Colonial," "Georgian," and so forth) and begin to do so by decades? Very few decades really possess an identity, let alone an identity that "fits" the precise ten-year interlude. Thus, there were hardly any "forties" or "seventies", whereas there really *were*, with a definitive definite article, "the thirties" and "the sixties". And in both of these, albeit in different ways, Christopher Isherwood played an observant and a participant role.

Decades are nonetheless ragged: the thirties probably start with the 1929 financial crash and end with the German invasion of France in 1940. The sixties proper don't seriously begin until the Cuba crisis and then the Kennedy assassination, but they are still going on, in some ways, well into the mid-1970s. An emblematic book of the latter decade was *Voices from the Crowd*, a collection of essays against the bomb that came out in 1964 but had been provoked by the events of two years earlier. Among the contributors were Bertrand Russell, Philip Toynbee, John Osborne, Alan Sillitoe, and James Kirkup. One of them, Ray Gosling—then considered a literate voice of "the teenagers"—was very struck by the novel Christopher Isherwood had brought out that year: *Down There on a Visit*, and in particular by Christopher's recorded reaction to the Munich crisis of 1938:

> E.M. went back to the country by a late afternoon train. Keeping up my mood of celebration, I had supper with B. at the flat. Since I was there last, B. has brought a big mirror and hung it in the bedroom. We drank whisky and then had sex in front of it. "Like actors in a blue movie," B. said, "except that we're both much more attractive."

But there was something cruel and tragic and desperate about the way we made love; as though we were fighting naked to the death. There was a sort of rage in both of us—perhaps simply rage that we are trapped here in September 1938—which we vented on each other. It wasn't innocent fun, like the old times in Germany—and yet, just because it wasn't—it was fiercely exciting. We satisfied each other absolutely, without the smallest sentiment, like a pair of animals.

Having revered Isherwood as a radical oppositionist of the 1930s, the angry young Ray Gosling writing his piece—entitled "No Such Zone"—in 1964 felt that there was something rather escapist about this reaction. (He perhaps underestimated, as Isherwood never did, the usefulness of Eros as a means of warding off Thanatos.) Anyway, here is what Isherwood was writing on October 23, 1962, at the height of the crisis over the Cuban missiles and when the threat of actual annihilation seemed even more immediate than it had two dozen years previously. This time he was at the gym in California:

If we are to be fried alive, it seems funny to be working out; and yet that's precisely what one must do in a crisis, as I learned long ago, in 1938. I have also been prodded into getting on with both my novel and the Ramakrishna book today, and I have watered all the indoor plants. Now I must write to Frank Wiley and Glenn Porter, before I go to have supper with Gavin.

Exceptional in point of its dating, this is otherwise very nearly a "typical" Isherwood sixties diary entry. (Though the type and style of "workout," one is compelled to note, has altered or at any rate evolved since 1938.) But the themes are constant: a persistent register of anxiety about the outside world combined with a sort of fatalistic distancing from same, a permanent conscience about being behindhand with work, and a second-to-none commitment to friendship and socializing that forces one to wonder how he ever got any work done at all.

Of course this summary of mine does not include the consistent, ever-renewing love and concern that Isherwood felt for his companion Don Bachardy, but that phenomenon is imbricated in and with every page of this diary, even when it is not explicitly so.

Of the various types of "sixties" that were on offer—the political, the psychedelic, the black and ethnic or "identity" movements, the sexual, the newly uncensored musical and showbiz—Isherwood contrived to be a sort of quizzical Zelig at all of them. And yet, if you are a certain kind of British reader, you will not fail to notice that beneath all this hedonism and experiment there still remains a somewhat austere and self-reproaching English public-school man of the kind he'd sworn to escape,[1] forever piously reproving his own backslidings, vowing to do more manly exercises—even when these involve the telling of *japam* rosary-beads—and (to annex a line of Auden's) swearing to "concentrate more on my work." In similar key, there are endless regrets about wasted time and especially about evenings squandered in drink and drunkenness. It's often difficult to tell how hard on himself he's being here, since unlike Byron he never itemizes his booze intake. On the sole occasion when I met him, at Marguerite Lamkin's in Chester Square in the late 1970s, he sat with Don under the David Hockney painting of the two of them and appeared very lean and lucid. (That meeting led to a bewitching drawing by Don of my guest James Fenton. Incidentally, after the famous line that introduces his Berlin stories, it's charming to notice how Isherwood observes that Hockney always carries a camera.)

But then who else was around to notice that Aldous Huxley, who died on the same day in November 1963 as the assassination in Dallas, was being given regular doses of LSD to sweeten or to soften his end? Who else might have had a conversation with Mick Jagger, under the auspices of Tony Richardson, in the Australian outback, and elicited from him the gossip that the Beatles had abandoned the Maharishi after the guru had made a pass at one of them? (I wonder which one, don't you?) And who else was still matter-of-factly saying "Jewboy" or "nigger," depending entirely on how he happened to be feeling? Who else felt practically nothing at the murder of Dr. Martin Luther King, refused to sign any petitions about Vietnam, and apparently didn't even notice the Soviet invasion of Czechoslovakia? This is an idiosyncratic, unillusioned tour of the sixties that has few if any rivals.

The dead-pan and matter-of-fact humor is also rather distanced, as though seen through a lens. Don reads in the paper that "Norman Mailer" has stabbed his wife and thinks he's seeing the words "Arthur Miller," which cause him to feel that Marilyn

---

[1] Even the passage cited above from *Down There on a Visit* concludes by saying: "Whatever happens, I mean to work a lot on the China book. And I'll start doing my exercises again. For the first time this year." Isherwood was also working throughout the sixties on the genealogical and other research for what became *Kathleen and Frank*, his oblique homage to the Victorian and Edwardian—not the "eighties and nineties and noughties"—values of his parents and other forebears.

Monroe has been unfairly deprived of a mention: surely we are witnessing the birth of celebrity culture? Gore Vidal rings up and says: "Mole? Toad." Even the famous Swami is not always treated with unmixed reverence, at one point scattering sacred Ganges water over his devotees "vigorously, as if he were ridding a room of flies with DDT."

If one could follow just two Isherwoodian threads through the labyrinth of this decade they would be (apart from the devotion to Don and the amazing willingness to put up with the Swami, and the slight weirdness of that "green flash" that he keeps on seeing at sunset out to sea) the agony of creative collaboration and the distinct but related hell of solitary literary effort. It is astonishing, for someone like myself who took such pleasure in the final production of *Cabaret*, to read of how bleak and sour were the original discussions with Auden and Chester Kallman, and how unpromising was the whole original scheme and many of its successive stages. Surely the idea of a Berlin musical was "a natural." Ah, but nothing of that sort does come "naturally," and Isherwood was probably wise to understand that one only lives once but frets and worries enough for several lifetimes. His best maxim, taken from that other great English public-school and Cambridge queer "Morgan" Forster, was, "Get on with your own work: behave as if you were immortal." These industriously maintained diaries, written at a time when many people were mistaking work for play and vice versa, and taking their own desires as realities, are at once a vindication of that Forsterian injunction and an illustration of its limitations.

Christopher Hitchens
Washington, D.C.
May 30, 2009

Introduction

Christopher Isherwood had been pioneering the cultural trends of the 1960s ever since the 1930s. When Gerald Heard and Aldous Huxley despaired of Europe's future and took their pacifist vision to California in 1937, Isherwood soon followed them, and, emulating them at first, experimented during the 1940s and the 1950s with mysticism, Eastern religion, psychedelic drugs, and sexual freedom. As the black-and-white, buttoned-up Establishment of the post-war period was gradually overrun in the sixties by the Technicolor warmth of pop culture and youth on the march, he continued to lead the way in doing his own thing. He wanted not only to write well but also to live well. Yeats once argued that, "The intellect of man is forced to choose/Perfection of the life, or of the work";[1] Isherwood's lifelong friend W. H. Auden retorted that "perfection is possible in neither";[2] but Isherwood never ceased trying for perfection in both. With great determination in the face of social disapproval and emotional difficulty, he forged a notably unconventional and, eventually, deeply happy personal life. At the heart of this volume are the intertwining stories of his continuing devotion to his Indian guru Swami Prabhavananda and his intimate and complex relationship with the American portrait painter Don Bachardy, who was thirty years his junior. If the 1960s was the decade of rebellious youth, the decade of the generation gap, Isherwood was living right on the gap. This diary begins on his fifty-sixth birthday, when Bachardy was only twenty-six and desperately trying to grow up. In a sense, Isherwood had to grow up all over again with him, and this pulled him all the more tightly into the central impulse of the time.

[1] "The Choice," *The Winding Stair and Other Poems* (1933).
[2] "Writing," *The Dyer's Hand* (1962), p. 19.

These pages are thick with novel writing, script writing, college teaching, and Isherwood's myriad friendships with the creative stars who shaped the sixties—Francis Bacon, Richard Burton, Leslie Caron, Julie Harris, David Hockney, Jennifer Jones, Hope Lange, Somerset Maugham, John Osborne, Vanessa Redgrave, Tony Richardson, David Selznick, Igor Stravinsky, Gore Vidal, Tennessee Williams, and many others. His psychological insight often takes us right underneath the skin of his subjects, and in the background he unfolds, week by week, a concisely referenced sketch of the period. He records the mounting anxieties of the Cold War in Laos, Berlin, and Cuba, the end of the colonial age presaged by the Algerian war for independence, the space flight of Yuri Gagarin, the Kennedy–Nixon election, the eruption of assassinations and the burning of America's inner cities, the Vietnam War and the anti-war movement, the coming of Diggers, Hippies, Flower Children, Timothy Leary, Mick Jagger and Marianne Faithfull, the Summer of Love, the walk on the moon, and the changing fashions—for pointed winkle-picker shoes, minis, maxis, moustaches, Afros, the illustrations of Bouché, and the costume designs of Beaton.

Isherwood began the new decade by completing his seventh novel, *Down There on a Visit*, about four earlier phases of his life when he was a tourist among the marginalized—eccentrics, neurotics, defective lovers, refugees—indulging himself in a long deliberation about possible modes of living. His title reflects a debt to Hans Castorp, the tubercular hero of Thomas Mann's *The Magic Mountain*, who keeps saying on his arrival at the Sanatorium Berghof, "I am only up here on a visit." Castorp stays for seven years, enchanted by his spiritual as well as by his physical condition. By 1960, Isherwood had lived with Bachardy for seven years, and with *Down There on a Visit*, he wrote himself out of possible alternative lives into the orderly and productive calm of his healthy present reality. He was a successful middle-aged writer, well-connected, widely admired, settled in his own house in Santa Monica with a young partner he adored, looked up to in his community as a part-time professor and literary personality. His geographical and spiritual wanderings were behind him. Since 1939, he had been a regular temple-goer at his local Vedanta Society, the Hindu congregation led by his guru Swami Prabhavananda. Isherwood was committed to his path. That year, he worked with Charles Laughton on a play about Socrates, and he taught at Los Angeles State College and at the University of California at Santa Barbara. Fellow writers like

Auden and Truman Capote were to tell him *Down There on a Visit* was the best book he had ever written. Over the next decade, he would write two more novels and then turn away altogether from invention and fantasy to autobiography, writing only about real life.

But a longlasting storm was about to break; Don Bachardy was preparing to make a bid for independence. In January 1961, he moved to London to study painting at the Slade. Although Isherwood joined him a few months later, their relationship entered a period of strain that was to evolve dramatically into repeating and intensifying crises. Over the next few years, Bachardy had debut exhibitions in London, New York, Los Angeles, San Francisco, and elsewhere. He was courted on several levels by various different kinds of admirers, fell in and out of love, struggled to find his way forward as an artist, and felt more and more trapped by Isherwood's self-confidence, Isherwood's fame, Isherwood's bossiness, Isherwood's years.

On June 10, 1961, in London, Isherwood records in his diary that Bachardy continues to seem "a sort of magic boy" as he had done since 1953: "I still feel that about him now and then. Yesterday evening, for example ... he absolutely sparkled like a diamond. He seemed a creature of another kind, altogether." But ten days later, Isherwood recognizes that while Bachardy needs bolstering as he prepares to launch his first-ever gallery show, he is constantly at risk of being sidelined by Isherwood's presence. When Auden sat for Bachardy—for a work later acquired by the National Portrait Gallery—Auden talked over his head:

> Right now Don is drawing Wystan, who keeps talking to me as I write: Falstaff and Don Quixote are the only satisfying saints in literature, etc. etc. ... I think [Don] would like me to go away for quite a bit of the time between now and his show, when he needs my moral support. It's the old story: he can't have any friends of his own as long as I'm around, because, even if he finds them, they take more interest in me as soon as we meet.[1]

Bachardy could never fully participate in the lifelong conversation between these boyhood friends, however fond or well-disposed Auden may have felt toward him; yet he was riveted at the margin of the scene by the opportunity to witness and to portray Auden's celebrated talent and extraordinary face. It was the same with

[1] June 20, 1961.

many of Isherwood's friends, and since Bachardy couldn't risk sharing his own friends, he had to learn to hide them, an investment in duplicity with which he gradually became more and more uncomfortable. In New York six months later, for Bachardy's second gallery debut, he and Isherwood were both made miserable by the cold, by the city's hectic pace, by tight hotel quarters; Bachardy slammed a taxi door in Isherwood's face, breaking the skin. Isherwood returned to California alone.

Nevertheless, he knew that Bachardy remained the center of his life. He loved their house on Adelaide Drive and enjoyed being there alone for a while, but "the whole affair," owning property, the routine of work and play, "would still have no reason to exist without him. He is the ultimate reason why it's worthwhile bothering at all."[1] He was to write this sort of thing in his diary time and again in the years to come. Thus, Isherwood faced the greatest challenge of his personal life: to love Bachardy for Bachardy's sake rather than for his own. This was the test of his maturity, and, in due course, he was to draw upon all his resources to meet it—his religion, his friends, his teaching, and his work.

When Bachardy returned from New York, Isherwood saw in him, "a reserve. He doesn't seem so childishly open as before."[2] He also saw how hard it would be for Bachardy to go on painting now that the external goals of Slade course work and the first shows were behind him. They discussed creating a studio in the house so that he could do this in privacy. Some of the tension between them was sexual, although Isherwood is initially reticent about this in his diary. He had been the first to claim the right to have other partners, and he owed Bachardy the same freedom, but the practice caused them both considerable anguish as they struggled to find the terms on which it was possible in a relationship as intimate as theirs. Each wished to control what the other knew about him, but neither found it easy to settle on knowing only what the other wished to share. They were possessive and intuitive, and both drew their own conclusions with penetrating accuracy. As they grew older, Isherwood was to have fewer partners and Bachardy more; the changing dynamic between them called for continual and, for Isherwood, perhaps unexpected adjustments.

Isherwood was Bachardy's mentor and a father figure as well as his lover; like any child trying to break free from a parent, Bachardy still needed someone he could depend on, so even as he

[1] Feb. 12, 1962.
[2] Feb. 23, 1962.

tried to establish his own autonomous identity, he clung to the old bond. In April 1962, Isherwood wrote:

> ... Don made another of his declarations of independence. He has got to have a studio of his own, here at the house, and his own telephone, and his own money and his own friends.... And he quite realizes that he has to do nearly all of the getting himself. He only asked of me that I shall understand. Well, I do—and I sincerely believe that things would be much better if he could achieve all these objectives. The trouble is, some of them are really opposed to other deeper wishes, or perhaps one should rather say fears, in his nature. For example, he would do much better to have a studio away from the house altogether.... [H]e says jokingly that he wants to keep an eye on me. And I suspect that this isn't entirely a joke. He is afraid of leaving me *too* much alone. He doesn't want *my* independence.[1]

In fact, Isherwood understood Bachardy so well that he sometimes left him no room to discover who he was for himself. This was an especially excruciating feature of the trap Bachardy felt he was in, and he was often at pains to reverse the power structure implied by the vast difference in their ages. The diary records constantly shifting chemistry between them. They had no established code to follow, not only because they were homosexual, and not only because of the age difference, but also because there never can be a code between two individuals who are continually seeking a more complete fulfilment of self and of vocation. One or the other of them was always trying something new; neither possessed a nature that was easily—if ever—satisfied. And so their relationship followed an ambivalent, wayward path as each felt by turns that it was supporting or holding him back, satisfying his appetites or denying them; they drew closer and apart, closer and apart. A diary entry for June 1962 records, "after the party, drunk, Don told me he wants me to go away to San Francisco and leave him alone all summer...." But the very same entry introduces Bachardy's wish to be initiated by Swami Prabhavananda. A week later, they went together to Vedanta Place so he could learn how to meditate, and indeed on December 18 that year, Bachardy became Swami's disciple. This reaffirmed the depth of his devotion to Isherwood as his model in life and created a new, public bond between them: a shared form and place of worship.

[1] Apr. 16, 1962.

In June 1962, they turned their garage into the talked-of studio, and Isherwood overcame his fear of material expenditure so they could improve the house as well. The construction produced moments of intense and precarious joy. Over a few days at the end of the month, he described its progress:

> Don and I lay on the deck, which still has no railing and seems as insecure as a flying carpet, with the wind blowing up between the floorboards and the whole Canyon floating in the air around you....

> The workmen have now put up the trellis over the deck, casting a barred shadow. Don is in raptures. The framing of the view gives him exquisite pleasure and now he keeps saying how happy he is here and how happy he is with me. And so, of course, I am happy too....[1]

But the new domestic arrangements and the mantra were not enough. Things fell apart again in early August, and Isherwood left for Laguna Beach to stay with Swami, much as he used to do when he was unhappy with Bill Caskey in 1950. Almost immediately, Bachardy prevailed upon him to return home, over Swami's strong objections. Isherwood's tone, as he records episodes of screaming and anger, grows grimmer each time the episodes recur, although his underlying convictions do not change. On his fifty-eighth birthday, he writes: "Do I hate Don? Only the selfish part of me hates him, for rocking the boat. When I go beyond that, I feel real compassion, because he is suffering terribly. I still don't know if he really wants to leave me, or what. And I don't think he knows."[2] By September, Isherwood was considering that he ought to move out for a few months because he was older, surer, stronger, and he sensed that he was undermining Bachardy's efforts to grow up:

> Not to do this is to force *him* to go away, and this is wrong because he is the one who didn't feel really at home in this house, and now that he has his own studio he should be free to enjoy it.
>
> Then why don't I go away? Because it is such a lot of fuss and I don't want to leave *my* home and above all my books. I want to stay here and get on with my work, in my own tempo....

[1] June 25 and 27, 1962.
[2] Aug. 26, 1962.

Aren't I bad for him, now, under any circumstances? Probably. He only needs me in his weakness, not his strength; and he hates me for supporting his weakness.[1]

From the heart of this dark period, Isherwood produced *A Single Man*, a novel that articulates his anxieties about living alone and which is, in a sense, his own bid for freedom—freedom from grief over lost love, freedom to reveal to conventional readers the gay "monster" he had so long been obliged to hide in his published work, freedom from the demands of the ego and the limitations of individual identity. He first conceived of the book as a novel about an English woman. But Bachardy, even as they approached the nadir of their relations together, offered the crucial insight that Isherwood should write about himself: "this morning we went on the beach and discussed *The Englishwoman*, and Don, after hearing all my difficulties with it, made a really brilliant simple suggestion, namely that it ought to be *The Englishman*—that is, me. This is very far-reaching. . . ."[2]

The novel became centered in the daily routine of Isherwood's contemporary life in California; but the technique derives from Bloomsbury, from the novels of Forster and especially Virginia Woolf, splicing together the British and American literary traditions. It is modelled on *Mrs. Dalloway*, which Isherwood unreservedly praised that summer as: "one of the most truly beautiful novels or prose poems or whatever that I have ever read. It is prose written with absolute pitch, a perfect ear. You could perform it with instruments. Could I write a book like that and keep within the nature of my own style? I'd love to try."[3] Exactness of "pitch" affords subtle discrimination among sensations, enabling the author to explore the inchoate area between social existence and creaturely unconscious; Isherwood was increasingly drawn to this rich inner world both in his diaries and in almost all of his later work. When he finished reading *Mrs. Dalloway* just before his birthday, he wrote:

Woolf's use of the reverie is quite different from Joyce's stream of consciousness. Beside her Joyce seems tricky and vulgar and cheap, as she herself thought. Woolf's kind of reverie is less "realistic" but far more convincing and moving. It can convey

[1] Sept. 10, 1962.
[2] Sept. 18, 1962.
[3] Aug. 22, 1962.

tremendous and varied emotion. Joyce's emotional range is very small.[1]

Isherwood's early work is sometimes criticized for having an emotionally bland and undeveloped narrator. In fact, this was a deliberate strategy for concealing the narrator's homosexuality. Now, as the unspeakable homosexual elbowed his way to the center of *A Single Man*, Isherwood had found the technique to reveal the repressed feelings of such a character in all their complexity. The narrative is subtle, exact, unafraid, and powerful. Even fifty years later, the rage lurking behind the cultivated façade of the middle-aged literature professor called George frightens straight readers; civilized human beings hide this kind of anger from one another in order to be able to get along. Bachardy recognized the quality of the book right away:

> Yesterday, I showed Don the first twenty-eight pages of this second draft of my new novel. He was far more impressed, even, than I had hoped. He made me feel that I have found a new approach altogether; that, as he put it, the writing itself is so interesting from page to page that you don't even care what is going to happen. That's marvellous and a great incentive to go on with the work, because I feel that Don has a better *nose* than anyone I know. He sniffs out the least artifice or fudging. He was on his way out after reading it, and then he came back and embraced me and said, "I'm so proud of Old Dub."[2]

And, nearly a year later, it was Bachardy who came up with the title.[3] Isherwood felt that the book "spoke the truth,"[4] and, over the years, he referred to it with growing confidence, as his "masterpiece."[5]

Through the rest of 1962 and the start of 1963, the relationship between Isherwood and Bachardy continued its tumultuous course. In November, Bachardy wanted to separate for a few months, but Isherwood still refused to uproot himself: "If he wants out, then he must be the one to get out.... Most of the freedom Don is looking for could actually be achieved right here, living with me. He

---

[1] Aug. 26, 1962.
[2] Nov. 29, 1962.
[3] See Aug. 2, 1963.
[4] Sept. 7, 1964.
[5] See Oct. 31, 1963 and Nov. 23, 1964.

doesn't realize that yet. Okay, he can find it somewhere outside and then come back."[1]

What Bachardy found outside was a fairly serious love affair, and he introduced his lover openly at home, pushing Isherwood to acknowledge and to condone his behavior, or perhaps to somehow share in his pleasure or validate his choice. In July, Bachardy had told Isherwood that "he wished we could speak frankly about *everything* that we did." Isherwood had warned "this wasn't desirable" and noted in his diary Bachardy's humorous and defiant reply, "But I get to know almost everything you do, anyway."[2] In fact, this was Bachardy's way of warning Isherwood—the reverse would also have to be the case. He knew a great deal about Isherwood's earlier life and loves, best described by the cliché "the stuff of legend"; modelling himself as he did on Isherwood, he, too, wanted a legendary love life, and he wanted Isherwood to know about it. He sometimes felt he had to compete with all Isherwood's past partners as well as the optimistic boys still crowding around; so his affairs were partly conducted in self-defense, as a counterbalancing act.[3] For his part, Isherwood was prepared to blind himself to things he did not want to know about Bachardy, even if Bachardy was determined he should find out. They quarrelled about the lover a few days before Christmas; the day after Christmas, Isherwood wrote:

[These] are not things I want to dwell on yet. Maybe all will work out for the best—but I don't know that, and I don't even want to think it. When I suffer, I suffer as stupidly as an animal. It altogether stops me working. I am ashamed of such weakness....

Christmas (which I seem to hate more every year) was placid and almost joyous by comparison.... Don and I lay on the beach and talked affectionately. I think he would love it if he could discuss *everything* with me. But, alas, I am neither the Buddha nor completely senile. I have my limits. I cannot help minding. When I finally stop minding I also stop caring. Then I don't give a shit.[4]

He struggled to weather the affair, admitting to his pain and yet trying to dismiss it: "Am getting into a flap about the ... situation.

[1] Nov. 27, 1962.
[2] July 10, 1962.
[3] Conversation with me, Oct. 2006.
[4] Dec. 26, 1962.

Last night I had two if not three dreams about them.... And meanwhile Don—no doubt because of this—remains unusually sweet and affectionate. I ought to be grateful really. Oh—idiocy."[1] That winter, the younger lovers spent more and more time together, and Isherwood feared that an alternative domestic intimacy was building up in Bachardy's life: "... Don took him some of our plates; admittedly, not ones we use any more. I am wildly miserable, but only in spurts. What I am miserable about is the feeling that Don is gradually slipping away from me."[2] He was resigned to the fact that there was "no question, here, of finding any kind of solution on the personal level. I can only find a solution through prayer and japam."[3]

During this painful phase, Bachardy chose to tell Isherwood that he believed the bond between them was a mystical one. In his diary, Isherwood mentions a "sudden revelation" from Bachardy "about the Bowles experience in Tangier"; but he professes that the revelation left him feeling puzzled.[4] In October 1955, when they had taken hashish with Paul Bowles and his painter friend Ahmed Yacoubi, Bachardy experienced an episode of near-madness during which he sensed a plot to incapacitate Isherwood so that Yacoubi could force sex on Bachardy while Bowles watched. Alternating with the paranoia was a blissful recognition of his love for Isherwood, his need for Isherwood, and Isherwood's unconditional commitment to him. They left Bowles's apartment abruptly, but the spiral of ecstasy and madness continued into the small hours. When he later read about the *kundalini*—the spiritual energy which, when awakened, rises from the base of the spine through the seven *chakras*, or centers of consciousness located in the spinal canal and cerebrum, until it illuminates the brain—Bachardy recognized that he had had a mystical experience in Tangier. As he recognized this, the experience became vivid to him all over again. He didn't tell anyone because he was overwhelmed by the experience at the time that it occurred, and later, when he came to understand it, he thought it would sound presumptuous. He also knew that Swami disapproved of achieving mystical experiences through the use of drugs.[5] By confiding in Isherwood now, he seemed to wish to reassure him that the bond between them could not be broken by ordinary love affairs.

[1] Feb. 6, 1963.
[2] Feb. 9, 1963.
[3] Feb. 9, 1963.
[4] Jan. 3, 1963.
[5] Letter to me, Apr. 26, 2008.

Even if he professed to be puzzled by Bachardy's confidence, Isherwood continued to tell himself that the affair was a good thing for Bachardy, and just as he began to feel that it was therefore a good thing for himself, he discovered in March that Bachardy had begun a new romance. At last, Isherwood planned to move out for a while, mostly because he had found arrangements which suited him. He was reluctant to say much in his diary, remarking only that his relationship with Bachardy might end by summer or "might equally well lead to a much better relationship."[1] In mid-April, he settled in a borrowed house in San Francisco, where he concerned himself with his "psychological convalescence." He wrote, "Oh, I did so need to be alone! Now I am resolved to get on with my work, I mean my own work; and to exercise—I am hatefully fat.... Oh yes, I am happy to be here...."[2] Within two weeks, his thoughts turned to Bachardy, but he kept his resolve to leave him alone:

> Am starting to think a lot about Don, miss him, wish he would write. But I won't pester him. Why does he seem so unique, irreplaceable? Because I've trained him to be, and myself to believe that he is? Yes, partly. But saying that proves nothing; the deed is done and the feelings I feel are perfectly genuine. ... At least I have proved to myself that I can still live alone and function. In some respects I have never felt so truly on the beam.[3]

It was Bachardy who was having a terrible time. Isherwood copied into his diary part of a letter from him, "'Fits of doubt and gloom keep descending.... I don't want you to worry about me. I must do this alone. I must get through by myself. And I try hard to love you instead of just needing you.'" On this, Isherwood commented, "Well, of course I am terribly worried. I am even losing my confidence that this will end all right—though I wrote him a reassuring letter."[4]

After some uncertainty about whether Bachardy might join him in San Francisco, Isherwood drove home for Bachardy's twenty-ninth birthday on May 18. But the day was a fiasco:

> Yesterday, I rushed downtown ... and bought him a ring with an Australian sapphire, dark blue. This morning at breakfast he

[1] March 20, 1963.
[2] Apr. 14, 1963.
[3] Apr. 26, 1963.
[4] May 3, 1963.

shed tears, said he couldn't accept it. Our relationship is impossible for him. I am too possessive. He can't face the idea of having me around for another ten years or more, using up his life.

I said I absolutely agree with him. If it won't work, it must stop. Now he has gone out.... I cried a bit. Then drank coffee, felt a lot better, and began figuring. Don should start by getting a studio away from this place, where he can stay whenever he wants to. Also, he should go to a psychiatrist. (That was his idea.) And we must start thinking about selling this house.[1]

Perhaps Bachardy's protracted revolt against Isherwood was a factor in Isherwood's own revolt, which was building up to a climax during this same period, against Swami Prabhavananda. The diaries show that Isherwood invested more and more time and conviction in Swami and his teachings as Bachardy tested to its limits his relationship with Isherwood. Ever the skeptic, Isherwood questioned in the most practical sense whatever Swami taught him, seeking a balance that could work for him as a devotee living outside the monastery in his own household. In February 1961, he had written:

And what's left, if Don goes out of my life? Swami and Ramakrishna: yes. As much—more so—than ever. My japam has been getting more and more mechanical. But when I told Swami this, he didn't seem worried. He assured me that I will get the fruits of it sometime or other; and I really believe this. The only thing that sometimes disturbs me a little about his teaching is the idea that we—all of us who have "come to" Ramakrishna—are anyhow "saved," i.e. assured of not being reborn. This disturbs me because the idea seems too easily optimistic. But then—who am I to talk? Swami says it, and I do honestly believe that he somehow *knows*.[2]

In a long-running show of duty, Isherwood was completing the first draft of his biography *Ramakrishna and His Disciples* alongside his final draft of *A Single Man*. And he reluctantly agreed to travel with Swami to the Ramakrishna Math, or monastery, in India at the end of 1963 to help celebrate the hundredth anniversary of the birth of Swami Vivekananda. But he was swept by waves of defiance, manifested in physical illness:

[1] May 18, 1963.
[2] Feb. 12, 1962.

I still have this thing in my throat. And, psychosomatically, it gets worse every Wednesday when I have to read to the family up at Vedanta Place. A *passionate* psychosomatic revolt is brewing against the Indian trip ... I will not surrender my will; be made to do anything I don't like.[1]

In fact, all through 1962 and 1963, the period of his worst troubles with Bachardy and his most painful bouts of jealousy, Isherwood had a recurrent sore throat. In the summer of 1963, he twice records in his diary his intuition that the sore throat was linked to writing about Ramakrishna's death from throat cancer.[2] And he sometimes feared he himself had throat or jaw cancer. But the episodes of illness and the cancer anxiety had started earlier, and indeed, Isherwood had had trouble with sore throats long before. In *Lions and Shadows: An Education in the Twenties*, Hugh Weston, the youthful Auden character, announces that tonsillitis "means you've been telling lies!" At that period, the lies were essentially about homosexuality; the Isherwood character concludes that his life and his writing are "sham," so he leaves medical school with its conventional cures and travels to Berlin, where he can indulge his sexuality without guilt, supported by the theories of the American psychologist, Homer Lane: "Every disease, Lane had taught, is in itself a cure—if we know how to take it. There is only one sin: disobedience to the inner law of our own nature."[3]

In the 1960s, as in the 1920s and 1930s, happiness and good health continued to be proof to Isherwood of right living. Illness resulted from dishonesty, from being out of harmony with one's true self. Ramakrishna and Swami Prabhavananda had replaced Homer Lane and his disciple John Layard (who taught Lane's theories to Auden and Isherwood), and Isherwood had progressed to seeking spiritual liberation through their version of self-knowledge. According to one diary entry, what he admired most about Ramakrishna was his honesty, although he notes that Ramakrishna's honesty is not transparent to everyone, because it sometimes takes an exaggerated form, which Isherwood identifies as camp:

When Swami used to teach me that purity is telling the truth I used to think that this was, if anything, a rather convenient

[1] Oct. 31, 1963.
[2] Aug. 9 and 20, 1963.
[3] Chptr. VII.

belief for me to have, because it meant that I didn't have to be pure but only to refrain from lying about my impurity. Well, that's the minimum or negative interpretation. But, thinking about it in relation to Ramakrishna, I saw this: that the greatness of Ramakrishna is not expressed by the fact that he was under all circumstances "pure." No. And even if he was pure, that didn't mean he wasn't capable of anything. You always feel that about him—there was nothing that he might not have done—except one thing—tell a lie.

... It's funny that I, who am steeped in sex up to the eyebrows, can see quite clearly what Ramakrishna's kind of purity is capable of, and that most people just can't. I suppose it's having been around Swami so much *and* understanding camp. I am privileged; far more than I realize, most of the time.[1]

In *The World in the Evening*, Isherwood's character Charles Kennedy explains, "You can't camp about something you don't take seriously. You're not making fun of it; you're making fun out of it. You're expressing what's basically serious to you in terms of fun and artifice and elegance."[2]

When relations were bad with Bachardy, Isherwood didn't like to write in his diary at all, preferring silence to the risk of prevarication or of articulating indelibly a situation he hoped might improve. In June 1963, struggling to cope with the current lover constantly around the house, he records: "Diary keeping at this time seems definitely counterindicated.... Part of Don wants to run me right off the range and wreck our home beyond repair; part wants to keep on and see how things work out."[3]

Isherwood, too, was waiting to see how things would work out, and he was barely coping. He could not feel content with his longstanding refusal to come more closely under Swami's tutelage if his life at home as a householder devotee was the failure that his misery suggested it must be. And he was in no position to proclaim his shaky beliefs to anyone else, or even to read aloud to them from holy texts. And so his voice deserted him in the temple, and at home his diary-writing pen fell silent. The ménage à trois with Bachardy's lover certainly wasn't working: "Have now definitely said I don't want to have to meet [him] any more. I should never have done so in the first place. That kind of thing is messy and was

---

[1] Oct. 16, 1962.
[2] Prt. II, chptr. 3.
[3] June 2, 1963.

messy in the days of Lord Byron, and always will be messy. Unless one simply doesn't give a shit."[1] He continued to be racked by jealousy: "Jealousy: Not what they do together sexually. But the thought of their waking in the morning, little pats and squeezes, jokes, talk through the open doorway of the bathroom. For that one could kill."[2] A few months later, he had to tell Bachardy again and more fully just how he felt, because nothing had changed.

Airing his feelings strengthened Isherwood, but however much he suffered, he was little interested in advice. From Swami, he wanted information about his spirit and some understanding of what was going to happen to him when he died; he did not want a set of rules on how to behave. As he told Gerald Heard:

> I don't go to Swami for ethics, but for spiritual reassurance. "Does God really exist? Can you promise me he does?" Not, "Ought I, ought I not to act in the following way?" I feel this so strongly that I can quite imagine doing something of which I know Swami disapproves—but which I believe to be right, for me—and then going and telling him about it. That simply isn't very important. Advice on how to act—my goodness, if you want that, you can get it from a best friend, a doctor, a bank manager.[3]

However devoted Isherwood was to Swami, he often felt cramped and frowned upon by the congregation. He had many individual friends in it, but, for Isherwood, whenever individuals gathered into a group, they were transformed—into a crowd, a mob, something alien and impenetrable with which he could have no individual rapport, no private conversation. Groups imply a norm; Isherwood was temperamentally disposed to deviate from any norm, to make an exception of himself. He was enormously uncomfortable with the group trip to India. He made his speeches in front of the crowds at the eternal sessions of the Parliament of Religions, ate the mass meals seated on the floor in the halls and under the vast temporary canopies, but one day, exhausted by a traveller's tummy, he was suddenly revolted by what he had been saying and by the way he had been conforming to what others expected of him. He felt an urgent need to express who he really was:

[1] Sept. 19, 1963.
[2] Nov. 1, 1963.
[3] Nov. 22, 1962.

Just before going to bed, I started to get the gripes and shits. I shivered a lot and couldn't sleep all night. Lying awake in the dark, I was swept by gusts of furious resentment—against India, against being pushed around, even against Swami himself. I resolved to tell him that I refuse ever again to appear in the temple or anywhere else and talk about God. Part of this resolve is quite valid; I *do* think that when I give these God lectures it is Sunday religion in the worst sense. As long as I quite unashamedly get drunk, have sex, and write books like *A Single Man*, I simply cannot appear before people as a sort of lay minister. The inevitable result must be that my ordinary life becomes divided and untruthful. Or rather, in the end, the only truth left is in my drunkenness, my sex, and my art, not in my religion. For me religion must be quite private as far as I'm publicly concerned. I can still write about it *informatively*, but I must not appear before people on a platform as a living witness and example.[1]

Like Bachardy forcing Isherwood to acknowledge his lovers, Isherwood insisted that Swami recognize his whole personality, all his inclinations and all his loyalties. How could he accept Swami's assurance that he was saved unless Swami knew exactly who he was? In the diary, Isherwood tells how he built up to a confrontation with Swami as to a climactic moment in a play. "I realized I was going to make a scene and I needed time to rehearse it." He had to exaggerate his feelings in order to bring home to Swami just how strongly he felt. "Some instinct told me that this ultimatum must be drastic or it would make no impression at all." It was, in this sense, camp, and although it was generated by deeply serious feelings, it was also very funny:

"... the Ramakrishna Math is coming between me and God. I can't belong to any kind of institution. Because I'm not respectable—...

"I can't stand up on Sundays in nice clothes and talk about God. I feel like a prostitute. I've felt like that after all of these meetings of the parliament, when I've spoken.... I knew this was going to happen. I should never have agreed to come to India. After I promised you I'd come, I used to wake up every morning, feeling awful—

"... the first time I prostrated before you, that was a great

[1]Dec. 31, 1963.

moment in my life. It really meant something tremendous to me, to want to bow down before another human being. And here I've been making pranams to everyone.... And it's just taking all the significance out of doing it—"

... I felt that everybody knew a scene was taking place. I felt that I was acting hysterically. Indeed, I couldn't have looked Swami in the eye while I was saying all this. But I didn't have to, because I was wearing ... dark glasses....[1]

Exaggerating his feelings was a way of making them seem justified. It concealed his self-consciousness and his guilt about failing to live up to what Swami hoped for from him. And it freed him from any further constraints on his behavior. By Swami's lights, it would always remain possible for Isherwood to become the saint who could sit on the dais and give the lecture without needing to lie about or conceal an unacceptable personal life. Well into the 1970s, Swami occasionally teased Isherwood about the possibility of returning to the monastery, and Isherwood several times records this in the diary. But Isherwood had decided when he left the monastery during World War II that he could never follow such a strict and narrow path. He shaped the conclusion to *Prater Violet* around this decision, and he reaffirmed it at several cruxes later in his life. He moved among many worlds, pursued many relationships, and explored many imaginary alternatives in his fiction. As a writer, and simply as a human being, he wished to remain available to all varieties of experience.

As Isherwood must have known he would, Swami met the premeditated tantrum with unconditional love, despite bewilderment and wounded feelings: "Swami had barely understood a word. He was quite dismayed. 'I don't want to lose you, Chris,' he said. I told him there was absolutely no question of that. That I loved him as much as ever. That this had nothing to do with him. But still he didn't understand. He looked at me with hurt brown eyes ..."[2] The dialogue might have been spoken between lovers, for instance, between Isherwood and Bachardy. But then Swami himself fell ill, as Isherwood reported in the diary:

Swami ... in bed with a cough; very rumpled and sad.... The country dust is blamed; but I got a strong impression (later confirmed by Prema) that the sickness has a lot to do with me. This

[1] Jan. 2, 1964.
[2] Jan. 2, 1963.

is perhaps the only respect in which Swami can be described as sly; he is absolutely capable of getting sick to make you feel guilty, though I doubt if he realizes this—and it is purely instinctive.[1]

Swami's body now loaned itself to the playacting, building their confrontation up to melodrama and reducing it to comedy at the same time. His illness, like Isherwood's illness, is a bodily manifestation of camp—the psyche's exaggerated, theatrical account of its distress.

Underneath the playacting was something in which both Isherwood and Swami wholeheartedly believed and which was far more important to either of them than who would win this immediate power struggle. They loved one another, and beyond—or above this—they loved Ramakrishna and believed in the possibility of spiritual liberation. Their egos battled, but on a higher level, they were at one. In the end, it didn't matter whether Isherwood made the speeches or the pranams. Such actions occur only in the "as if" world of maya, the cosmic illusion of material reality which veils Brahman. Susan Sontag once wrote, "Camp sees everything in quotes."[2] In a sense, *maya* itself is camp—it is the "as if" world—all in quotes. The dynamic at work between Swami and Isherwood was also at work between Isherwood and Bachardy: they both believed in their relationship, their love, over and above any relationships with others; their egos battled, but they were at one. This is what Bachardy realized and confided in Isherwood when he told him the bond between them was a mystical one that could not be broken by other love affairs. And Isherwood was to offer Bachardy the same unconditional love that Swami offered him, the same freedom to do almost entirely as he pleased, whatever suffering it caused, rather than break this bond between them.

Through the mid-1960s, Isherwood and Bachardy lived apart a great deal of the time, with Bachardy in New York or London for long spells. But their relationship survived. In his diaries Isherwood from time to time remarks upon the sense in which their day-to-day life together was camp, a symbolic enactment of something sacred and hidden, something veiled in the safety and humor of exaggeration, yet made evident by it. It was a world for which no

[1] Jan. 3, 1963.
[2] "Notes on 'Camp'," *Partisan Review*, vol. 31, no. 4, 1964, pp. 515–530; rpt. *Against Interpretation* (1966).

words existed, but its speechless, creaturely innocence and warmth was partly embodied in the identities they adopted for themselves as animals: Isherwood a stubborn, hardworking old horse, Bachardy a skittish, needy kitten of irresistible softness and with sharp claws. As Bachardy departed at the beginning of 1965, after spending the Christmas period in Santa Monica, Isherwood records, "I told him that this short time together has been the best I have ever had with him. He said, 'Lately I've been thinking that the Animals haven't seen anything yet; they still haven't had their golden age.' I said, 'They'd better hurry.'"[1]

Meanwhile, Isherwood occupied himself with other friends and with money-making film jobs. Through the mid-1960s, he worked for Tony Richardson on scripts for *The Loved One, Reflections in a Golden Eye*, and *The Sailor from Gibraltar*. Each of Richardson's projects generated a new and complicated domestic ménage, with family, friends, and co-workers crammed into a rented star's house in Los Angeles, a remote farm, a yacht, or, when breaking from work, a villa in the South of France. Isherwood was fascinated by these households and by Richardson's many partners of both sexes, although he never allowed himself to be fully drawn into the circle. His diaries observe how Richardson's obsessive genius for manipulation produced plays and films of psychological intensity and sensual revelation, and how destructive this genius could be when set loose upon friends and acquaintances. Wary though he was of Richardson, Isherwood always accepted his offers of work, just as he had done earlier with Charles Laughton when Laughton was aging, mortally ill, surrounded by a retinue of young male chauffeurs, masseuses, bed companions, and a wife who wished to demonstrate that she was more important than any of them.

A third gifted Englishman, David Hockney, settled in Los Angeles in 1964. His paintings of light-struck swimming pools, palm trees, and beautiful young male bodies reveal clearly enough some of the things which drew him, just as they drew the others. Hockney was young enough to have grown up in the privation of wartime England, as well as its cold, dark climate. There was money in America, in the form of patronage and teaching posts, for example; and plenty—of just about everything—meant that projects could be accomplished quickly. Even in 1961, when he was living in England with Bachardy, Isherwood had been struck there by, "The utter fatalistic patience of everyone when a line has

[1] Jan. 7, 1965.

to be formed or a train or a bus waited for.... You feel the wartime mentality still very strongly here...."[1] Hockney loved to work, and he was ambitious. For him, as for Isherwood, the hedonism and glamor of southern California were a subject as well as a way of life, and he retained a strong degree of analytical detachment. Even during the trips they occasionally made together, Isherwood describes Hockney carrying a camera. Isherwood grew to love and admire him without qualification, for his energy and his impulse to experiment, and for his natural, unstinting generosity.

In 1965, Isherwood began teaching again, this time at UCLA where he was Regents' Professor and, in 1966, visiting professor. He had always relished the animal spirits at large on the Californian campuses; he felt invigorated by his students, and he spent large amounts of time reading and commenting on their work. In almost every class he taught, at least one talented young man was writing about his homosexual yearnings and handing his work to Isherwood as a step toward coming out. But Isherwood's colleagues tended to be conservative, and they brought out his toothed hatred of bourgeois married life. After one long evening spent among professors and their wives early in the decade, he wrote:

> There weren't enough martinis, there wasn't enough food, and there were too many guests. I don't think heterosexual parties are workable, anyhow, just as conversation groups.... And, oh dear, the academic atmosphere with its prissy caution! ...
>
> Sure, I am prejudiced, but I feel always more strongly how ignoble marriage usually is. How it drags down and shackles and degrades.... The squalid little shop, the little business premises you have to open, and the deadly social pattern which is then imposed on you—of dragging some dowdy little frump of a woman all around with you, wherever you go, for the next forty years. Not to mention the kids. It is a miserable compromise for the man, and he is apt to punish the woman for having blackmailed him into it.[2]

Isherwood had close friendships with women writers, artists, designers, and film stars, but he was less comfortable with women who chose a domestic role over a career or a serious personal occupation. Not only was he distressed by the unequal enslavement to financial necessity, but also he sensed in housewives a repressed

[1] Apr. 19, 1961.
[2] May 13, 1962.

bitterness. They seemed unable to avoid turning sacrifices made for their husbands' professional success into longterm silent accusations, such as his mother might have lodged against him: that men failed to recognize or care how much women were denying themselves. Committed as he was to the private life and to the inner life he felt it should nourish, Isherwood didn't believe it was necessary for either party to remain personally unfulfilled.

There were many contrasts to square faculty get-togethers. He still enjoyed the well-protected gay party scene in Hollywood; even though he had already found the boy with whom he wanted to spend his life, he sometimes attended playwright Jerry Lawrence's all-male evenings peopled by good-looking young would-be actors. The diaries also wryly report on many star parties. And at the height of the sixties, he describes a gallery opening for Bachardy which was successful to the point of hysteria, with actors, directors, playwrights, and monks cramming in, and art work flying out:

> Anne Baxter started the buying. She rushed across the room into Jo [Masselink]'s arms screaming, with a kind of tearful triumph, "I've bought two!" Vidya was there, viewing the scene with the amused world-weariness of a swami about to depart forever into the depths of India ... and Elsa Lanchester looking almost ladylike in a dark dress, gracious and bitchy-grand; and Jennifer Selznick in white, about to leave alone to drive to Big Sur ... and Dan[a] Woodbury quite drunk, saying it was a shame Rex [Evans] didn't exhibit Don's nudes of him, and then taking a fancy to Jim [Charlton] and leaving with him; and Gerald Heard and Michael [Barrie], bitchily arriving dead on time ... and old King Vidor being encouraged by his wife to paint again; and John Houseman, a little worried because he liked Don's work so much, almost more than he felt he should; and Cukor sly but friendly, planning a memorial supper for Maugham ... and Bill Inge terribly depressed about his life, sitting glum like a bankrupt on a couch....[1]

Two evenings earlier, he and Bachardy had spent the evening at home with Allen Ginsberg and a few others:

> Everybody got high, and Ginsberg recorded our conversation and chanted Hindu chants, and [Peter] Orlovsky took off his woollen cap and let his long greasy hair fall over his shoulders and

[1] Jan. 8, 1966.

kept asking me if I ever had raped anyone, and the boy Stephen [Bornstein] unrolled a picture scroll he had made, under the influence of something or other, to illustrate the Bardo Thodol.[1]

However willingly he explored the trends of the time, part of Isherwood always stood back, sometimes mocking, sometimes soberly assessing. In response to a request to endorse the Vietnam Summer antiwar project in 1967, he wrote in his diary:

> ... the whole Vietnam antiwar movement is something I must keep away from ... as a pacifist I must deny the rightness of every war, even the most apparently righteous ones. This war is too obviously unrighteous—indeed it is even politically deplorable.... Therefore objection to this war is primarily a political objection.... I believe Aldous would have agreed with me. And Gerald Heard.[2]

The painful episode with Swami in India was to lead to another novel, Isherwood's last, which he began writing in 1965, *A Meeting by the River*. It is a story of two brothers, a good one who becomes a Hindu monk and a bad one who tries to prevent him. The two brothers are modelled on various real life people, but both are, in a sense, also Isherwood. The "meeting" of the title is a meeting with himself, an exploration of his spiritual convictions and his human attachments embodied in two opposed character types. The bad brother, Patrick, walks away from the encounter with a sharpened appetite for the duplicitous life he was already leading; the good brother, Oliver, is illuminated by a vision of his late swami, which reassures him that both he and his brother are included in the swami's love. Isherwood wrote in his diary when he was drafting the book:

> ... the main action of the book is temptation—the temptation of any saint by any satan ...
> The key line is when Oliver says that he was inviting Patrick to come and judge the swami. He has to have Patrick's okay. He doesn't ever get it of course. What he does get is a spiritual intervention by the swami himself, proving to him that Patrick

[1]Jan. 8, 1966.
[2]Jul. 19, 1967. Heard was still alive but increasingly unwell from an ongoing series of strokes.

"belongs" whether he likes it or not, knows it or not. And this, in its turn, is sort of campily confirmed by Patrick's taking the dust of Oliver's feet.[1]

The formal show of respect to Oliver, who is now a swami himself, is a Hollywood gesture—extreme, slightly embarrassing; but the ritual act of devotion also expresses a true and innocent emotion struggling to life in the arch-villain Patrick.

While he was still working on *A Meeting by the River*, Isherwood also began a book about his parents, a new kind of autobiography, which he eventually called *Kathleen and Frank*. In 1966, he travelled to Austria where he worked on a Christmas T.V. special about the song "Silent Night," and he combined this with another trip to England, partly to review family papers that he wanted to use for the memoir. The memoir is the first in the trilogy of personal histories, or what he also called personal mythologies, which begins with the courtship and marriage of his parents during the reign of Queen Victoria and his father's death in World War I, moves on, in *Christopher and His Kind*, to thirties Berlin and life on the run from the Nazis with his first serious lover, Heinz Neddermeyer, then concludes, in *My Guru and His Disciple*, with an account of his religious conversion in southern California and his life as a follower of Ramakrishna. Many authors turn to memoir in middle age, and perhaps this was the natural progression for Isherwood, but it is a striking coincidence that he turned away from fiction once and for all and became newly interested in the facts about who he was and how he came to be that way just as Broadway attempted to assign him permanently to a sexually neutral destiny as "Herr Issyvoo," a stage figure based vaguely on the invented narrator in his own Berlin stories.

Isherwood had attempted a Berlin musical with Auden and Chester Kallman, but he had nothing to do with Kander and Ebb's *Cabaret*, which opened on Broadway in November 1966, and he was never able to like it. Bachardy went to New York without him to see it and to attend, on November 28, Truman Capote's Black and White Ball. Isherwood was delighted to be allowed to stay home in Santa Monica. But his diaries show his satisfaction when *Cabaret* proved to be a hit, even quoting from reviews. In fact, *Cabaret* changed Isherwood's life. It provided him with significant income, boosted in 1972 by proceeds from the film, and it made him, willy-nilly, a celebrity. The musical won eight Tony

---

[1] Jan. 4, 1966.

Awards, including Best Musical and Best Director, and it was a hit all over again in London when it opened in 1968 with Judi Dench in her first-ever singing role. Later, the film made Liza Minnelli a super star. On February 28, 1972, she was on the covers of both *Time* and *Newsweek* dressed as Sally Bowles; that March, the film won eight Academy Awards, including Best Director, Best Actress, and Best Supporting Actor (Joel Gray). "Herr Issyvoo" is still the "role" for which Isherwood is most widely recognized. But "Herr Issyvoo" had never been the real Christopher Isherwood. It was to be quite a task to reclaim his identity for himself.

Early on, Isherwood had an insight that *Kathleen and Frank* was "not about my father and my mother, it's about me. I mean, it is like an archaeological excavation. I dig into myself and find my father and my mother in me. I find all the figures of the past *inside* me, not outside."[1] But the more he discovered in their letters and diaries about what his parents actually thought and experienced, the more absorbed he became by them. It was lack of information about his father that had led Isherwood to devise in adolescence an imaginary father who fulfilled his own needs but left him at odds with the real world in which he must live:

> ... I really didn't know my father at all ... the myth about him was created for my own private reasons—i.e., that I needed an anti-heroic hero to oppose to the official hero figure erected by the patriots of the period, who were my deadly enemies.... [C]ertain aspects of my father had to be suppressed, because they were disconcertingly square; e.g. his references in his letters to "real men" etc.[2]

He now had materials that enabled him to pick apart his youthful myth, and so better understand himself as its maker. And he entirely rediscovered his mother, the figure who in his youth represented for him everything against which he wished to rebel. At one time, he had feared he would be swallowed up in her grief and her longing for the past, now he regretted his unkindness in not asking to read her diaries while she was still alive: "There all the while, in the drawers of her desk, lay the rows of little volumes of her masterpiece."[3] He explored with compassion every nuance of her relationship with her selfish and demanding parents who nearly

[1] Nov. 30, 1966.
[2] Jan. 22, 1967.
[3] *Kathleen and Frank* (*K&F*), chptr. 1.

prevented her from marrying and having a life of her own. And he recognized in his grandmother, Emily Greene Machell Smith, "a great psychosomatic virtuoso who could produce high fevers, large swellings and mysterious rashes within the hour; her ailments were roles into which she threw herself with abandon."[1] His own subtle and neurotic temperament beautifully fitted into the family portrait, and so did the all-absorbing mutual fascination he shared with Bachardy. Moreover, the Victorian atmosphere of tasselled drapery and ferns which was the setting for his grandmother's magnificent camp—a full-time activity for members of a newly rich class very much at its leisure—chimed revealingly with the vestiges of India-under-the-Raj that still clung to Swami's more earthly self.

Towards the end of 1967, Isherwood began writing a play of *A Meeting by the River* with Jim Bridges. He also began adapting for the stage Bernard Shaw's story "The Adventures of the Black Girl in Her Search for God," which led to a fiery production in the age of Black Power. Then, in 1968, around the time that Hockney began working on his celebrated double portrait of Isherwood and Bachardy, Bachardy began to work professionally as Isherwood's co-writer, first on the dramatization of *A Meeting by the River*, then on an adaptation of *I, Claudius* for Tony Richardson. They used the job to justify a trip in July and August 1969 to Tahiti and Australia, where Richardson was filming *Ned Kelly* starring Mick Jagger as the outlaw. Jagger, two years after the notorious Redlands drug bust, had recently been rearrested for possession of marijuana and was in Australia by permission of the judge who had agreed to delay his trial for the filming; Marianne Faithfull, rearrested with him, had marked her arrival in Sydney by swallowing a suicidal dose of barbiturates as the airplane landed. She was in the hospital in a coma, from which she luckily recovered. Of his own arrival on the set Isherwood wrote:

> ... Tony Richardson, looking like the Duke of Wellington, in a kind of Inverness mackintosh cape; we embraced in front of the whole crew and the actors, including Mick Jagger. It was such an improbable encounter, after these thousands of miles, like Stanley and Livingstone, rather. Mick Jagger, very pale, quiet, good-tempered, full of fun, ugly-beautiful, a bit like Beatrix Lehmann; he has the air of a castaway, someone saved from a wreck, but not in the least dismayed by it.[2]

[1] *K&F*, chptr. 1.
[2] Sept. 5, 1969.

The ranch house where they worked with Richardson on the script was heavily guarded to keep away gangs of students who had vowed to kidnap Jagger. "[T]here were ten policemen sitting up in the kitchen all night, waiting for the students who never showed. Incidentally, without knowing it, they were guarding a pot party which was going on in the living room!"[1] By December, the Isherwood–Bachardy script for *I, Claudius* had been dropped, though Richardson seemed to regret it. He wrote from London "that he wished we had been with him, implying that, in that case, we might have worked together."[2] In fact, Isherwood and Bachardy themselves had probably not spent enough time working on the script together; Bachardy had been preoccupied with a new and rather serious boyfriend. As collaborators, they were to have more success with later projects, though not for Richardson.

Despite separations and set-backs, Isherwood valued more and more the privilege denied his parents of spending his life with Bachardy over the long term. On his sixty-third birthday, he admonished himself not to feel guilty about his happiness but instead to understand it as the very evidence he perpetually sought that he was living in the right way. Happiness was not a distraction from spiritual intentions, but the path towards self-understanding and perpetual bliss:

> My life with Don seems, as of this minute and indeed of the past couple of months in general, to be in a marvellous phase of love, intimacy, mutual trust, tenderness, affection, fun, everything. We have plenty of money and more to come, presumably, very soon from *Cabaret*.... My health is good.... And I am very lucky to have work to occupy me for many many months ahead. What is bad, as of now, is my apparent spiritual condition.... I do "keep the line open" and try, throughout the day, to make acts of recollection. I am of course terribly uneasy about my "worldly" happiness; fearing to lose it and yet knowing that of course it will be necessary to lose it before I can find *ananda*. (Having said this, I suddenly ask myself.... How *can* love be profane if it really is love? In my own case, hasn't my relation with Don now become my true means of enlightenment?)[3]

Throughout the 1960s, Isherwood continued to fight the spiritual

[1] Sept. 5, 1969.
[2] Dec. 10, 1969.
[3] Aug. 29, 1967.

dryness that had worried him ever since Swami initiated him in 1939. He hungered to experience his belief as an emotion, not just an idea. Early in 1968, Swami fell gravely ill and was put into intensive care; in April, when he had recovered, he told Isherwood he had expected to die. If this was the camp of brinksmanship, the very real possibility of losing Swami had an enormous effect on Isherwood, intensifying his love for him and also his faith. Swami reported that he had seen his own guru, Brahmananda, coming towards him twice during his illness; he told Isherwood he had decided that if he recovered he would meditate more, and he told him that he had lost all personal desires. Isherwood writes, "his face seemed to shine with love and lack of anxiety. I thought to myself, I am in the presence of a saint...."[1] Over the following year, Isherwood noted that Swami could now "convey, as almost never before to the same degree, an absolute spiritual guarantee: *this thing is true*."[2] Thereafter, Isherwood focused on Swami more and more as Swami grew older and frailer. Ever since the death of his father in World War I, Isherwood had had an enormous curiosity about death; he wanted to find out, in the most literal and specific sense against his own needs in the future, what would happen to Swami when Swami died.

A different and greater address to his spiritual dryness was made by the youth and energy of Bachardy. If Swami, an old man, was his teacher, Bachardy, a young man, was the lesson set. By the end of the decade, Isherwood had indeed come to see his spiritual path as being made available to him *through* Bachardy. "[Don] wrote such a wonderful letter yesterday, and I realize more than ever that this is IT. Not just an individual. Or just a relationship, but THE WAY. The way through to everything else."[3] Thus, the conflict between his private emotional life and his spiritual life, which had reached its crisis during his journey to India, was resolved—his love for Bachardy and his devotion to Swami and Ramakrishna were one and the same.

[1] Apr. 11, 1968.
[2] May 31, 1969.
[3] Apr. 26, 1968.

Textual Note

American style and spelling are used throughout this book because Isherwood himself gradually adopted them. English spellings mostly disappeared from his diaries by the end of his first decade in California, although he sometimes reverted to them, for instance when staying at length in England. I have altered anomalies in keeping with the general trend; however, I have retained idiosyncrasies of phrasing and spelling which have a phonetic impact in order that his characteristically Anglo-American voice might resound in the writing. And I have let stand some English spellings that are accepted in America; Isherwood had no reason to change these.

I have made some very minor alterations silently, such as standardizing passages which Isherwood quotes from his own published books, from other published authors, and from letters. I have standardized punctuation for most dialogue and quotations, for obvious typos (which are rare), and very occasionally to ease the reader's progress. I have usually retained Isherwood's characteristic use of the semi-colon followed by an incomplete clause. I have spelled out many abbreviations, including names, for which Isherwood sometimes used only initials, because I believe he himself would have spelled these out for publication. Also, I have corrected the spellings of many names because he typically checked and corrected them himself. Square brackets mark emendations of any substance or interest and these are often explained in a footnote. Square brackets also mark information I have added to the text for clarity, such as surnames or parts of titles shortened by Isherwood. And square brackets indicate where I have removed or altered material in order to protect the privacy of individuals still living.

This book includes some footnotes written by Isherwood himself, in particular in the diary he kept in London from April to October 1961. Had Isherwood himself prepared the diary for publication, he would almost certainly have incorporated such material into the text, rewriting as necessary. I have not attempted to do this on his behalf. I have occasionally added, in square brackets, to his notes.

Readers will find supplemental information provided in several ways. Footnotes explain passing historical references, identify people who appear only once, offer translations of foreign passages, gloss slang, explain allusions to Isherwood's or other people's works in progress, give references to books of clear significance to Isherwood, sometimes provide information essential for making sense of jokes or witticisms, and so forth. For people, events, terms, organizations, and other things which appear more than once or which were of long-term importance to Isherwood, and for explanations too long to fit conveniently into a footnote, I have provided a glossary at the end of the volume. The glossary gives general biographical information about many of Isherwood's friends and acquaintances and offers details of particular relevance to Isherwood and to what he recorded in his diaries. A few very famous people—for instance, Katharine Hepburn or Mick Jagger—do not appear in the glossary because although Isherwood may have met them more than once, he knew them or at least wrote about them essentially in their capacity as celebrities. Others who were intimate friends—Igor Stravinsky or Aldous Huxley—are included even though their main achievements will be familiar to many readers. This kind of information is now easily available on the internet, but a reader of this diary should be able to find what he or she immediately wishes to know and to get a feel for what Isherwood himself or his contemporaries may have known, without putting the book down and turning to a computer. Isherwood has audiences of widely varied ages and cultural backgrounds, and I have aimed to make his diaries accessible to all of them. Where he himself fully introduces someone, I have avoided duplicating his work, and readers may need to use the index to refer back to figures introduced early in the text who sometimes appear much later. Hindu terminology is also explained in the glossary in accordance, generally, with the way the terms are used in Vedanta.

In any book of this size, there are many details which do not fit systematically into even the most flexible of structures, but I hope that my arrangement of the supplemental materials will be consistent enough that readers can find what help they want.

Acknowledgements

Don Bachardy has shown extraordinary patience in waiting for me to edit this volume of diaries, and all the while he has continued to answer endless questions, to share with me his astonishing knowledge of the movies, and above all to unfold in long conversations his understanding of and love for Christopher Isherwood. Needless to say, these would be different diaries without Don, and certainly I could never have completed my task as editor without his help and encouragement. I thank him for continuing to trust in me over the years, for the excitement we have shared about this material, and for his friendship.

Through the Hollywood Vedanta Society, I was introduced to the sunny energy of Pravrajika Vrajaprana, a Californian nun of the Ramakrishna Order and an open-hearted scholar. She has spent many hours teaching me about Vedanta, clarifying terminology, and setting Isherwood's practices and beliefs in a broader context. She has even hunted through this volume for mistakes in the footnotes and glossary. Any that remain are mine, not hers. I am extremely grateful to her, to Eduardo Acebo, to the late Peter Schneider, and to the many other nuns and monks at the Vedanta Society of Southern California who have generously answered a wide range of questions.

I have had research assistance from Douglas Murray, Anne Totterdell, Gosia Lawik, Christopher Hurley, and many other members of the staff at the London Library, and I thank each of them for their skill and tenacity. Christopher Phipps also helped with research before going on to create the detailed index without which this diary would be a disappointment to many readers.

Others who have answered myriad questions along the way, and with whom I have shared often delightful exchanges, include

Kathy and Jeff Allinson, Paul Barber, the late Thomas Braun, Charlotte Brown of the UCLA University Archives, Patrice Chaplin, Robert Craft, Tom Devine, Jane Faulkner of the UCSB Library, Robin French, Ronald Frost, P.N. Furbank, Grey Gowrie, Don Graham, Stephen Graham, Richard Grigg, John Gross, Pat Hardwick of the UCLA History Project, Nicky Haslam, John Heilpern, Nancy Hereford of the Center Theater Group in Los Angeles, Sue Hodson of the Huntington Library in San Marino, California, Samuel Hynes, Shayna Ingram of the UCSB English Department, Evelyn Jacomb, Frank Kermode, Vijay Khan, Robert Maguire, Lucy Maguire, Edward Mendelson, Breon Mitchell, Anthea Morton-Saner, Araceli Navarro, Axel Neubohn, John Julius Norwich, Richard W. Oram, Peter Parker, Geraldine Parsons, Christopher Pennington, Jan Pieńkowski, Sherrill Pinney, John Rechy, Andreas Reyneke, John Ridland, Andrew and Polly Robison, John Sandbrook of the UCLA vice-chancellor's office, David Segal, Michael Sragow, Walter Starcke, Rupert Strachwitz, Geoffrey Strachan, Hugh Thomas, Daniel Topolski, Annu Trivedi of the Nehru Memorial Museum and Library in Delhi, Roy Turner, Swami Tyagananda, Hugo Vickers, Bettina von Hase, Grace Wherry, Edmund White, Swami Yogeshananda, and John Zeigel.

I would like to thank my agent Stephanie Cabot for her thoughtful advice and continuing support, and I would also like to thank Caroline Dawnay for setting me up in this project years ago. For their forbearance as well as for their enthusiasm, I would like to thank Isherwood's most recent publishers, Alison Samuel, Jonathan Burnham, Clara Farmer, and Daniel Halpern. And to their colleagues, Lizzie Dipple, Dr. Anthony Hippisley, Rowena Skelton-Wallace, Amanda Telfer, Alison Tulett, and Terry Karten, thank you for your hawk eyes, your nerve, and your stamina.

A few trusted friends generously read and commented on parts of this book. Thank you Al Alvarez, Richard Davenport-Hines, Isabel Fonseca, John Fuller, Bob Maguire, Bobby Maguire, Robert McCrum, Blake Morrison, and Erik Tarloff.

I am also more than grateful to Jackie Edgar, Vilma Catbagan, Felisberta Rodrigues, Charlie Watson, Katrina Johnston, Elizabeth Jones, and Susan Mellett, for clearing the way to my desk. To my family, for permanent safe haven, joyful distraction, and leaving me alone over many, many long hours, I can never offer enough thanks.

The Sixties
1960–1969

August 27, 1960–December 31, 1969[1]

August 27. My birthday evening at Hope Lange's was cozy and quite pleasant. Just Hope, Glenn Ford, David Lange and a friend of his, Don [Bachardy] and me. Glenn seems to be around all the time now, but, we think, merely in loco parentis. He makes a big show of devotion to Hope—saying for instance that he has to learn polo for his part in *The Four Horsemen of the Apocalypse,* and that he's sure to get hurt, especially if Hope is watching the shooting of the polo scenes, thus causing him to show off and get too daring. Somehow, though, one doesn't believe in this. Again, Glenn hugged me when we said goodnight—and this, too, didn't altogether convince. You felt it wasn't *him.* Is it how he thinks Hope's bohemian friends should behave? Is he trying to get himself elected an honorary queer?

(It's strange, typing this diary. It seems much less intimate than handwriting, and already in that first paragraph I notice a primness. I suppose typing makes me instinctively try harder. But I'll get used to it.)[2]

Mr. Gardner and his brother arrived this morning to begin painting the house on the outside. But now we find that the existing coat of paint on the house is calcimine, and that if we had painted it with the paint Don bought yesterday it would all have started to peel off in a few days. So poor Don has to go clear down to Western Avenue to change the paint.

[1] Isherwood often began a fresh notebook of diaries near his birthday, August 26. He titled this one "August 26, 1960–October 16, 1962" with a note: "(A separate volume covers April–October 1961—a visit to England)."
[2] Isherwood had worsening arthritis in his right thumb, so in this entry he gave up writing by hand and began to type his diaries on loose sheets which he clipped into a binder.

Yesterday and again today I have been sketching an opening passage to serve as a frame for the four episodes of my novel.[1] I feel that I must start with myself, and at the present time—otherwise there will be no perspective—but just how to relate myself to my characters, I don't know. Because, after all, it is my characters who matter most in the stories, not me. As long as the characters come to life, I have achieved my purpose; in a book of this sort, philosophy doesn't greatly matter. I do see that it would be fatal to be too pat—to base the four narratives, for instance, on four reflections in a mirror, or some such crap. That would cheapen the whole effect. Nor, I think, must I suggest that I'm deliberately setting out on a Proustian time-safari; that'd give me a tiresome air of self-consciousness.

No—I see something different, even as I write this. Something much simpler. Some kind of a brief introduction and description of myself today—showing somehow, as it were, geological-psychological strata which correspond to the periods of my four episodes and which reveal the influence on me of the characters in them. Here's something really difficult but fascinating to work out, perhaps calling for a new technique of behavior description.

August 29. The day before yesterday, I ran into Michael Hall and Scott Schubach on the beach. So I asked them up to the house for drinks. (We had plenty!) They told a marvellous story about Scott. He has always been a bit ashamed of his Jewish background and especially his childhood, which was spent in a slum neighborhood of New York. He even had a block against remembering any of it. So, when he went through analysis (including lysergic acid) he decided he must face up to all of this. And so he and Michael paid a visit to the apartment house where he and his mother used to live. They found the apartment was now inhabited by a family of Puerto Ricans; when Michael explained to them in Spanish why Scott wanted to look around their place they were deeply touched and most hospitable. So Scott and Michael came inside, and at once Scott was violently moved: he remembered everything; it all came back to him—how the rooms had looked when he was little and the view from the window, and how his mother's voice had sounded, calling down to him in the street, and even how the grain of the woodwork had felt to his hands as a child. He wept. This was one of the greatest experiences of his life. The Puerto

---

[1] *Down There on a Visit* (1962), begun in 1955 and now nearing its final form though still titled *The Lost*.

Ricans wept, too.... And then Scott and Michael went to visit Scott's mother in her present home, and told her where they had been—and she questioned them, and found that they'd been in the wrong apartment!

Yesterday I had lunch up at Malibu Colony[1] with Doris and Len Kaufman. Doris was in a very strange mood. She kept saying, about Len, "He sassed me this morning—he hasn't got long to live, and he knows it." At first, the rest of us laughed at this. But Doris wasn't *quite* joking. You saw in her a feminine tyrannical determination not to let the male get away with *anything*, because, if the female once does, she's sunk.

September 1. Spent yesterday and this morning reading "The Beach of Falesá," the screenplay on it by Dylan Thomas, and the revised screenplay by Jan Read. There's certainly a great deal there already, but I think I can improve it.[2] Above all, it is a chance to air one of my favorite theories, that the truly evil man is the one who only pretends to believe in evil.

Incidentally, I'd always thought the [Robert Louis] Stevenson story was called "The Beach at Falesá." Somehow, I like *at* much better than *of.* Why? Partly, I suppose, because *at* dissociates the beach from Falesá itself—thereby suggesting that there is something peculiarly significant and sinister about the beach. But that's not the whole of the reason why I like *at. At* somehow goes very deep into my subconscious fantasy.

Talking of beaches, I've been going to the beach at Santa Monica Canyon quite regularly, lately, and swimming. Not because I terribly want to—I always have to do it alone, or almost always. But I want a tan, and I want to catch a little of what's left of the summer, which I've spent mostly indoors, writing this novel.

On the retaining wall below the road at Inspiration Point, somebody has painted in huge red letters, UNI IS PIMP, with an arrow coming in from the right to call extra attention to the inscription. Most mysterious. Who or what is UNI? Why IS PIMP rather than IS A PIMP? (This suggests a foreigner—a Mexican, maybe.) And what a strange accusation for nowadays, surely? It sounds so old-fashioned.

Don in very low spirits and inclined to vent them on me— because he has had to draw these "bad" drawings to be used on

---

[1] A gated community on the beach.
[2] For a film for Richard Burton; see Glossary under Burton.

the stage in *A Taste of Honey...*[1] I'm afraid I got irritated with him, because he moaned about it so much. I do hate that. Except of course when *I'm* sad—then I *demand* sympathy and am furious when I don't get it.

September 8. Maybe as a reaction from finishing the novel, maybe because of a change in the weather, I am having a wretched attack of arthritis, gout, or whatever—my thumbs both sore, the left hand *very*, and pain all up the left arm into the shoulder, so that last night I had to take a painkiller pill. I'm still dazed now, but I had a good sleep.

Yesterday evening, we gave a dinner party; Dorothy [Miller] spent the night here and cooked for us. We had Iris Tree, Ivan Moffat, Gavin Lambert, Wolfgang Reinhardt, Lesley Blanch and a Mademoiselle Yvonne Petranant who is the French Consul here and who has somehow so intimidated Lesley that she begged us to let her bring her. Romain Gary has already gone to France, and Lesley is most unwillingly preparing to follow him; she has now become one of southern California's most passionate lovers. She's really a most hysterical woman.

The party started quite well. Wolfgang told us about the experiments which are being made in Russia and elsewhere in producing prolonged sleep—a sort of hibernation which may last weeks or months and from which you get up quite refreshed and renewed. "Sleep is only in its infancy!" said Iris with her chuckle. And then they—chiefly Ivan—started an elaborate fantasy about future times in which people will have themselves put to sleep for five hundred or a thousand years at a stretch. And how one'll avoid waking up at the same time as some terrible bore. And how one'll try to figure out the best way of spacing out one's eighty years—etc. etc. It went on so persistently that it became boring. And then Wolfgang, who really is *diseased*, I suspect, and cannot bear to listen to any talk which isn't thoroughly negative and alarmist, began talking about dreams and how they've discovered that there are dreams that can kill you if you have a weak heart; they give you such a shock that you wake up and die. (This really horrified Iris.) And he said that probably our dream experiences are far more terrible than anything in our waking life and that maybe patients who have been given sedation suffer more horribly than those who endure the physical pain.

[1] Props for Tony Richardson's New York production of Shelagh Delaney's play.

And from this he went on to talk about the RAND Corporation;[1] how the experts say it doesn't matter if there is an atomic war because about eighty million will survive, which is quite sufficient. Of course, the people who survive will be the ones with money, because to survive you have to build a shelter and stay in it for three weeks. And when you get your shelter built, you should go to at least three different contractors, so nobody will know what it is you're building; because if the word gets around that you have a shelter, you'll be mobbed at the first emergency. For the same reason, you ought to have a submachine gun to kill people who try to force their way in.

This led to talk about the different nations and their policies. Wolfgang is violently anti-USA and pro-Russia. According to him, the Americans are the real warmongers; this he deduces because of the RAND Corporation. But then when Mlle. Petranant began speaking up for France, he said that French people were so scared of their own police that he had been asked by French friends to mail letters for them outside the country. At this, Petranant, who is a most unappetizing blonde lesbian type, threw back her head and laughed savagely. But Wolfgang stuck to his accusation, and said that the letters were about the Algerian situation. "Oh *well*—" said Petranant, "*Algeria*—that's different. That's an *internal* problem."[2]

Why does one entertain people one doesn't like? The only really relaxed part of the evening was right at the end, when we had Gavin alone. He is leaving shortly for New York, to work on the movie of *Vanity Fair*, then going on to Rome for the shooting of *The Roman Spring of Mrs. Stone*. We talked about the performance of *Taste of Honey*, all agreeing that it's really a nothing play.

Don finished reading "Paul"[3] yesterday. He seems very excited by it; and thinks it's so shocking that maybe it even won't be published. He is such a tremendous moral and emotional support to me, now, and so I'm correspondingly upset when he turns on me and is vicious and ugly. And yet—all that is on the surface. And again and again I have to remind myself that the whole art of life

---

[1] First-ever think tank. The *R*esearch *AN*d Development Institute, created by the U.S. Army Air Forces at the end of W.W. II to advise them on aircrafts, rockets, satellites, and other new technology; based in Santa Monica, California.

[2] The Algerians had been fighting for independence since 1954, attracting growing public support in France where activists refused to fight against them and even secretly assisted them. In February 1960, President de Gaulle promised self-determination for Algeria, but the war continued until 1962.

[3] Last section of *Down There on a Visit*.

is to lean on people, to involve oneself with them quite fearlessly and yet—when the props are kicked away—remain leaning, as it were, on empty air. Like levitation.

All kinds of boringly sensational tales about the split-up of Marguerite and Rory [Harrity]. [...] Rory has now gone home to his mother, like an old-fashioned bride.

After the opening of *Taste of Honey* the night before last, we had drinks with Glenn Ford and Hope in two grand, depressingly empty night spots, The Traders and Le Petit Jean. This was to say goodbye to Glenn, who's leaving for Europe. He asked "if he might" write to me, and said, "I want us to be friends—for reasons you don't even know about." His behavior is truly a mystery—I wish it were a more thrilling one. But there is something very nice about him.

September 10. I'm suffering from acute nervous laziness—the kind which is caused by having too much to do. Also, the weather is tropically steamy. The ocean yesterday was so strange, shining silver and quite smooth and streamy, like a vast river. Went down on the beach today; horrible, dirty, crowded and the water full of rocks.

Here are some of the things I have to do:

Think seriously about my opening lecture at Santa Barbara.[1] Get my earnings and expenses properly listed for the income tax. (I have taken on this new accountant of Jo and Ben [Masselink]'s, Ken Hogan, who is handsome and seems nice but is still going to charge "a minimum" of $150, which Jo says is a hundred dollars more than he charges *them!*) Finish chapter 10 of the Ramakrishna book,[2] (*plus* a promotion letter for the magazine which Prema wants me to write.[3]) Get started on the revision of my novel.

Nothing from Laughton, and my God I certainly do not want to get all involved with him yet.[4] As for the "Beach of Falesá" project, I think it will bog down in a financial deadlock; [Jim] Geller wants to ask for $10,000 down, which Hugh French will

---

[1] In May, Isherwood had accepted a new job teaching English at the University of California at Santa Barbara (UCSB); he was to begin September 22.
[2] *Ramakrishna and His Disciples* (1965); see Glossary for Ramakrishna and for other Hindu names and terms.
[3] The magazine of the Vedanta Society of Southern California, *Vedanta and the West*, in which Isherwood's Ramakrishna biography was appearing in installments; it was edited by Prema Chaitanya.
[4] Charles Laughton had taken a break from the play they were writing about Socrates while he did some television work and underwent a gall bladder operation in mid-August.

never in his life pay.[1]

I long for the fall and its beautiful sane weather and its empty beaches.

Last night we saw *The Prodigal*, this play by Jack Richardson, at UCLA.[2] It is arty and more than somewhat Frenchified, with the usual wa-wa talk—I parodied it to Don as, "No, Prince—it is not the birds who fear the sea; it is the sea which fears the birds." Just the same, it is entertaining and has a good idea.

Another inscription, in the evil-smelling tunnel under the coast highway, reveals that it is not just UNI but UNIHI that is a pimp—also shit. I suppose this means University High School. But the word "pimp" is still mysterious.

September 17. Well, I have finished off chapter 10 of the Ramakrishna book, *and* an open letter Prema wanted me to write, appealing to the readers to renew their subscriptions to the magazine, in order to read (his) "Student's Notebook." Oh, the ghastly coyness of the draft of the letter which Prema made to guide me! This kind of a chore is more difficult for me than any other sort of writing.

Also, I have brought our income tax accounts up to date. And I have to admit grudgingly that, once you have things listed in the way Mr. Hogan suggested, it is far simpler to keep track.

Laughton came down, three days ago. It was the first time he'd visited 147 Adelaide[3] since his operation. Poor thing, he still seems terribly shaky, and so old. Like a punch-drunk old fighter groggily declaring that he'll make a comeback, but not *quite* believing it himself. It was curious, how impressed he was when Geller told him that Hurok[4] believes the Socrates project is really box office, and will back it. I suppose I shouldn't be surprised, but I am. I can't help expecting that Laughton should be ready, at his time of life and with his fortune, to try something he wants to do, regardless of the money. Laughton brought a pair of young men with him; the one who chauffeurs him and a masseur. Their shameless grins

---

[1] Geller was then Isherwood's film agent; French, representing Richard Burton, was trying to make a package deal of the project. Later French became Isherwood's film agent. See Glossary.
[2] I.e., the University of California at Los Angeles. Richardson's adaptation of the Orestes myth was nominated for several drama awards in 1959–1960.
[3] Laughton's spare house next door to Isherwood and Bachardy.
[4] Solomon Hurok (1888–1974), Russian-born impresario who produced classical music, ballet and theatrical events under his rubric, "Sol Hurok presents ...."

and ever-so-slightly cautious familiarity. Courtiers. The masseur started right in with *jokes*. "Have you heard the latest? They've sent a rocket to the moon with a colored man in it. The headline's in the paper: THE JIG'S UP." We cackled away, and old Charles watched us, his head sunk into his shoulders. There was a faint smile on his face, as though he were being tickled. After a while, he said in his deep hoarse voice, "You're bloody funny, aren't you?" (He uses "bloody" on principle, you feel; it is part of his public performance of being British.)

I must be very careful not to let the next months slip through my fingers. It would be easy to do so. For, most likely, the work at Santa Barbara won't be so difficult and yet it could easily fill the rest of my week with mild fussing. I must get on with the revision of my novel. Would it be too much to try to have it done by the New Year? That sounds frantic—well, we'll see—

First, I must think seriously about my first lecture. The second is more or less set already, because it will be the same as the one I gave at USC last spring.[1] The third one is perhaps the most difficult—"The Nerve of the Novel"—but that's a long way off.[2]

Rory Harrity is back with Marguerite. There has been no communiqué issued and no one dares go see them and find out what the score is. I simply couldn't care less.

Mr. Gardner is painting the garage this weekend: that'll be more or less the end of our home decoration for the year. We look very handsome now, on the hillside, seen from the street below; a proper subtropical palazzo, with our blue shutters in the big window and our fringed white shades in the bathroom and our yellow slat-blinds in my workroom. As we walk along Maybery Road, Don points up at our dazzling white frontage and says, "Just look—*there*—that's where the animals live—!"

Ronny [Frost], the monk from Trabuco who is "on loan" to Hollywood while [John Markovich] is away seeing his family, came down and drove me into Beverly Hills yesterday, because the Simca is still being fixed. Ronny was just starting to be a concert pianist before he became a monk; a Texan boy with a pretty soft face and hair, a sort of Van Cliburn.[3] I daresay he had a sex problem. Anyhow, here he is, and terribly anxious to be reassured. His piano playing unsettles him. At Trabuco he practises, and this

[1] "How I Write a Novel," delivered May 5, 1960 at the University of Southern California.
[2] "What Is the Nerve of the Interest in a Novel"; see Glossary under Lectures 1960.
[3] The American pianist (b. 1934), also a southerner.

is obviously his great joy in life. But then, from time to time, he told me, he gets invitations to play at concerts; and he knows he mustn't, but nevertheless, he feels terrible. After all, he *has* been practising—

I realized that Ronny wanted to ask me about my time up at Vedanta Place: how come I went there, and why I left. I tried to tell him about it in a reassuring way—pointing out that I didn't start out specifically as a monk, that it all grew out of the Gita translation project, and that when I decided to leave, I did so quite gradually, that there was no dramatic break, that I remained in constant touch with Swami,[1] etc., etc. "So really," Ronny said, "it's just the same now as if you'd stayed there?" But I couldn't let him think that, so I owned that there had been a "jazzy" (the *words* I sometimes pick!) period right after I left, and that, indeed, people had often come to Swami and told him I was going to the dogs—and that Swami had charmingly shut them up. So then I got the conversation off on to Swami and how marvellously he had changed since I'd known him—and Sarada too—and Krishna—all proving that the spiritual life *did* work.... I hope Ronny was satisfied. Just when I was warming to the theme, we reached Beverly Hills.

September 20. Yesterday I started work on the revision of "Waldemar"[2]—the Munich crisis episode in London which I originally called "The Others." I now realize that there's a good deal wrong with it—at least, at the beginning. For now the question of my own state of mind becomes important: it has to be clearly defined so that it can later be contrasted with my state of mind in 1940, in the final episode. At present, I seem to be way off the mark.

The day after tomorrow, I start at Santa Barbara, and of course I have mild stage fright about this, although I know it won't actually be nearly such an ordeal as my first day at L.A. State,[3] and anyhow, I have no lecture the first week.

We have been much involved with the cast of *Taste of Honey*;

---

[1] Isherwood and his guru, Swami Prabhavananda, began translating the Bhagavad Gita in October 1942; it was published in August 1944. Isherwood tells about living as a monk at Vedanta Place in *Diaries: Volume One 1939–1960* (*D.1*). See also Glossary under Prabhavananda and Vedanta Place.
[2] The third section of *Down There on a Visit*, set during August and September 1938.
[3] Isherwood's first university post, at Los Angeles State College, from September 22, 1959 to June 1960.

said goodbye to Mary [Ure] and Joan [Plowright] on Sunday night; now they're both in New York.[1] As I was driving Joan back to Tony [Richardson]'s house, a cop gave me a speeding ticket and as it was within the Sawtelle[2] grounds I have to go clear downtown and settle it.

Last night we had Nigel Davenport, Billy Dee Williams and Andrew Ray to supper, with Hope Lange. This was a great success. Hope got quite drunk. Nigel, who is very intelligent, took *Vedanta for the Western World*[3] off with him. I gave Billy Dee a copy of *The World in the Evening*. When he'd had too many drinks he got on the Negro question and was a little tiresome.[4] Andrew got nicer and nicer. He is a great boxing fan. You feel in him the sort of mad towheaded recklessness which I associate with RAF pilots. A very strange boy. He couldn't be anything but English.

Don now has a truly admirable set of drawings of all five of the cast. I am very proud of him. And one day last week he made seventy dollars![5] His maximum so far.

Department of sweetness and shit: this beginning of a letter from Walter Starcke in Tokyo. "I had always heard how hard the Japanese were to know, but when I arrived and found such giving-ness, such affection, and such ease, the rug was really pulled out from under me and I felt adrift in a way I had not expected—not with treasures just outside of reach, but rather choaked (sic) with treasure closer than touch."

September 21. This morning, I went downtown to settle my speeding ticket. It was just a formality, after all, aside from the tiresome drive; a nice judge in a small room all by himself fined me fifteen dollars and that was that. He told me that the Sawtelle hospital grounds have to be strictly patrolled even at night because crazy

[1] Angela Lansbury, Plowright, Nigel Davenport, Andrew Ray, and Billy Dee Williams were to open on Broadway on October 4; Ure was not in the play but was sharing Tony Richardson's West L.A. house.
[2] A veterans' hospital on Sawtelle Boulevard.
[3] The 1945 collection of essays edited and introduced by Isherwood. Davenport (b. 1928), an Oxford-educated British actor, played supporting and character roles on the London stage, in films—including *Look Back in Anger* (1959), *A High Wind in Jamaica* (1965), *A Man for All Seasons* (1966), *Chariots of Fire* (1981)—and on T.V.
[4] Williams (b. 1937) is black; he was raised in Harlem, was a child actor and entertainer, and later became known in the made-for-T.V. film "Brian's Song" (1971), in two films opposite Diana Ross—*Lady Sings the Blues* (1972) and *Mahogany* (1975)—and others.
[5] Drawing portraits and fashion illustrations.

patients hide in the bushes ready to throw themselves in front of cars. Hence also the strict speeding regulation.

At the Pickwick Bookshop later I ran into Wilbur Flam[1] and we had quite a long talk. Admittedly, his marriage is chiefly kept going by his and Bertha's[2] interest in their children—how wonderful it is when they first walk, talk, etc. He hinted at nostalgia and restlessness,[3] but said nevertheless that he and Bertha never bore each other. We agreed to meet again and discuss all of these problems more fully. Oh God—how glad I am that I'm me and not him!

Lunch at Vedanta Place with Swami, and a Hindu publisher whose name I already forget, and Mr. Watumull, the Honolulu clothing manufacturer. The publisher had silver hair and a black coat and was a bit like an ugly Nehru. He talked and talked; telling one interesting thing—that Warren Hastings,[4] in giving permission for the publication of the first translation of the Gita into English, said (in effect), "This book will continue to have an influence upon the English for many years after they have all left India." Mr. Watumull was more likeable, however; he reminded me of Morgan [Forster].

This afternoon, I have been trying to analyze the psychology of Chris in the four episodes of my novel. The results are quite fairly encouraging. It does add up to something, and make a pattern. I don't want it to make too much of one. But I think there's more work to be done on "Paul" from this point of view.

Tomorrow, Santa Barbara. It really *is* quite an adventure; and I'm tense and excited at the prospect of it. The last few days, my pain, elusively in the intestines, has recurred. Will try to ignore it.

September 28 [Wednesday]. Damn it, my work is on the skids again! Since Santa Barbara, I've just futzed around and really done nothing to my novel, and tomorrow off I go up there again. Well, I've got to pull myself together, starting Saturday.

Tomorrow will be my first lecture up there and of course I've got a certain amount of stage fright about it. But I do believe I have the materials for a good and amusing lecture. As for last week, it was quite pleasant, though I don't feel that I did more than scratch the shell of shyness-aggression which some of my seminar students

[1] Not his real name.
[2] Not her real name.
[3] Before marrying, Flam had homosexual affairs.
[4] British colonial administrator (1732–1818); Governor of Bengal (1772), first Governor General of India (1773–1785).

were wearing. However, there was a pleasant drunken evening with Douwe Stuurman and the Warshaws. I really like Howard. And when I asked him if I might show him some of the work of a young artist I knew, he answered, "Was that the same one who drew Vera Stravinsky?" I started defensively, "Yes, but—" meaning to tell him how terrible the reproduction was. But Howard said he thought it was excellent and very interesting.[1] Then, the next morning, a boy named Frank Wiley[2] came who wanted to get into my seminar and he showed me part of a novel he's written. It's all about the Santa Barbara campus and exclusively (so far) a homosexual love-story! But, oh, so rambling and long-winded!

Now—since yesterday, really—there has been a dramatic development: Don is almost certainly going to New York to supervise the framing of his drawings of the *Taste of Honey* cast; they are to be exhibited in the foyer or outside the theater and he is to get $250 for them *if* the play is a hit! I don't want to go; much as I shall miss Don and much as I should like to see Olivier in *Becket*.[3] I cannot rush around as I'm involved in all this work. I must try to stay very calm. And, also—though Don is scared at the prospect—I know it will be wonderful for him to have this triumph, however big or small it turns out to be, alone.

Yesterday afternoon, at Tom Wright's, we met John Rechy, who wrote "The Fabulous Wedding of Miss Destiny." I liked him. He lives downtown, in the midst of his "world," and dresses exactly like a Pershing Square hustler; shirt open to the navel with sleeves rolled to the armpits, skintight jeans, a Christopher medal. He is rather charming. Not at all aggressive or sulky.

Early this morning, a dream.

Hard to describe its setting. There were a lot of people—Don not among them—in a small town or village; I can't be more precise about the architecture. What we were all doing there, I don't know. The action started when one of us, a man, went mad—not noisy but deadly berserk. He had a tommy gun and he was going to kill as many of us as he could. He protected himself from us by forcing a group of women to stand around him as a screen, so he couldn't be shot at. The women were wearing print dresses rather like pioneer women of the covered wagon period.

For a while we all scattered and were scared, awaiting the attack of the madman. At least, I was scared; but it didn't occur to me

[1] Warshaw was a painter (see Glossary); Bachardy's drawing of Vera Stravinsky was reproduced by photostat.
[2] Not his real name.
[3] English translation of Jean Anouilh's *Becket, ou l'Honneur de Dieu* (1959).

to run away altogether. Maybe it was somehow not possible. I kept on the outskirts of the crowd, moving around, with others, to various places where we could take cover when the shooting started, but always deciding that each place was unsuitable because it had no proper exit or way of quick escape.

It had seemed that the crowd was quite disorganized; but suddenly I realized that a part of it had gotten together and formed a clear plan of resistance. And just as I realized this, some gates opened in the wall of a building on the other side of the square, and the madman came out, protected by his screen of women. But the opposing force went to meet him, and they also were surrounded and screened by women. Only, I now saw that the "women" were men dressed in women's clothes and that they carried guns. The two groups advanced upon each other and mingled; there was no struggle of course, because everybody except the madman was on the same side.

A terribly tense pause. Then a shot. The madman had been shot. He was dead. And everybody was congratulating the woman who had done it. *She* was a real woman; not dressed like the others but wearing a black evening gown. She was handsome and blonde, and I knew she was a lesbian. She accepted congratulations with a harsh laugh and said something, probably ironic, about being "an old member of the shooting club." I was hostile to her. I was the only person in the crowd who disapproved of the shooting of the madman. She understood this and made some cutting remark about my being "a silly little man."

The dead madman was lying there. His head looked more like a big square block of ice which is starting to melt. The features were already becoming indistinct. I wanted to pray for him. I knelt to do this, feeling somewhat embarrassed because there were people all around and I thought they would think I was showing off. As a matter of fact, I don't think they were paying much attention to me. As I knelt, the floor collapsed under me—it was a house floor, although we were out of doors; but I only sank through about a foot onto another floor which was firm. So I went ahead and said my prayer, asking Ramakrishna to protect the madman. And then I woke up.

This was a nightmare, in that I was badly scared. The curious thing was, however, that I didn't wake up in the midst of my fear, as one usually does, but quite a while after it had passed.

Have just been talking to Charles Laughton on the phone. Terry [Jenkins] is arriving back here next Wednesday, and on Saturday he and Charles will fly to Japan! I am still anxious about Charles, for he seems still very shaky and depressed by his illness. He says the doctor told him that his relapse was far more serious than the operation itself. And he is so *desperate* to get well.

October 2. No work done. Largely because of hangovers but also outside interruptions. More about these in a moment.

Don is in New York. He phoned this morning from Julie [Harris]'s, where he's staying. He still hasn't been able to see the producer and get the exhibition of his drawings outside the theater definitely agreed on. But he has met Cecil Beaton, who loves his. drawings and is going to recommend him to *Harper's Bazaar.* So he's delighted and feels the trip was worthwhile even if the other thing falls through.

Yesterday I had a phone call from Charles Laughton next door, to say he has had two violent attacks in which he tried to kill himself. They were both in the Curson Road house, and somehow connected with Elsa.[1] (She thinks he is trying to ruin the beginning of her tour! And she claims that she is shattered. Really, the fuss these vain old hams make! What a temperament *I* could have thrown over my first lecture at Santa Barbara last Thursday! As a matter of fact it was a truly smashing success.) So I told Charles I felt the attacks were disguised mystical experiences. ("Oh, how *wonderfully* tactful of you!" Don exclaimed, when I told him this morning.) And I certainly did please and reassure Charles, who now says that he was trying to reach infinity. Anyhow, to protect him from doing himself violence, he has two male nurses and Bill Phipps, who actually sleeps with him in the same bed. Now Charles feels fine and lolls around the house, whispering so as not to be overheard by the male nurses, whom he is already trying to get rid of. And now he has a new worry: he thinks Elsa may be preparing to have him certified. I assured him that this would be impossible under the circumstances. But I suspect Bill Phipps is an alarmist [...]. He has told Charles that Elsa said she wished he was dead as he had nothing left to live for. Even if she did say this, there was no mortal need to repeat it.

Then, also yesterday, a boy named Erik Kaln[2] came to see me. He had been at Tom Wright's the other day, with a good-looking boy named Bill Small.[3] After Don and I had left, this Bill Small got very drunk, kissed everybody in the room, then became violent

[1] Lanchester, his wife. The Laughtons' main residence was on Curson *Avenue* in Hollywood.
[2] Not his real name.
[3] Not his real name.

in the car as they were driving away and yelled and wanted to kill himself. Since, he has been perfectly all right and has written an incredibly gooey article for the [paper] on which he works [...]. Erik Kaln is Bill's roommate, and he had called me saying he wanted advice on how to handle Bill, with whom he's going to Europe in a short while. But actually he talked almost entirely about himself and how he was suffering—until I told him he was a monster and was manipulating the whole situation. This he took well and we got along splendidly and laughed a lot. He is a blond Jewboy of twenty-two, with a bottle nose and rather wonderful green eyes.

Tomorrow Jill Macklem is coming with her husband, and later John Rechy, so that day will be shot, too. Well, hell—these are all people who had to be seen sometime. Presumably Charles will get out of my hair as soon as Terry arrives.

October 3. More about Santa Barbara. It was very funny to see how sincerely relieved and somewhat surprised Chancellor Gould was that my lecture was such a hit. Later, I got drunk at the house of a nice man named Geo Dangerfield and fell over a barbecue bowl outside on the beach in the dark and hurt my shin. Frances Warshaw put Mercurochrome (?) on it and it won't wash off.

With Bart Johnson[1] to see Elsa Lanchester last night, in Royce Hall.[2] The trouble is, she isn't quite first-rate. She fusses too much with her hands and she is scared of the audience; and she's often dirty in the wrong way. The advertising says that she has "a world." She doesn't. There is no magic in any of this. Maybe because it's so unspontaneous. Charles says that she has to learn every word she says on the stage—all the asides, everything—by heart. "She couldn't even say, 'Hello, Santa Barbara,'" says Charles, "because if she learnt that line, she'd have to say it in Stockton and Miami as well—all over the country." Wicked old Charles was half pleased that I didn't really like the show. At the same time, he was delighted because it appears to have been a smash hit. Or rather, Elsa thinks it was a smash hit.

Bart Johnson, nicey-nice in a suit, was terrified of Charles, who ignored him. We ate warmed-up stew and the gravy was burnt and shreds of meat got into our teeth.

A rather ridiculous fuss with Glade Bachardy about the *Examiner*.[3]

---

[1] Not his real name.
[2] *Elsa Lanchester—Herself*, at UCLA.
[3] William Randolph Hearst's first Los Angeles paper, a New York-style tabloid.

While I was at Santa Barbara, just before Don left, Glade entered some contest which necessitated her getting a new subscriber to the *Examiner*. So she gave my name. It was Don's fault, of course—he should have known how strongly I'd object. The idea of having this paper around is obscene to me; and I hate the little boys who throw it all over the garden. So I called her today and told her I wouldn't take it. And she immediately got tearful, like a child who has been told it mustn't do something. So then I have had all the bother of having to call the local distribution office and tell them to send the paper elsewhere. I refuse to feel the least guilty about this. Why should one pander eternally to the swinish reactionary attitudes of women like Glade and my mother? They have to be told that the paper is utter filth and that decent people won't have it around. And, on top of that I shall have to pay for the subscription.

October 10. And now I've missed some really important days— notably Don's triumphant return from New York—and it's too late to describe them properly. Actually, this was relatively speaking the greatest triumph Don will ever have in his life, perhaps—because it was the first and because it's doubtful if the praise of any two people will ever again mean quite as much to him as Beaton's and Bouché's[1] did. Now he's about to return to New York again—on a much more dubious enterprise; designing posters for Tennessee [Williams]'s *Period of Adjustment* and the play Julie will be in, *The Little Moon of Alban.*[2] Don doesn't really know how to do this, and maybe it will be a flop; but we agree that it's still better for him to go and make the attempt than not. The worst of it is, yesterday and today he has had a really cruel attack of tonsillitis. This evening he says it's getting better, and I only hope this isn't grim autosuggestion which will lose its power once he's on the plane. Well—let's hope for the best—

I, too, have a tiresome ailment. The fall I had over the barbecue bowl at Geo Dangerfield's house caused a hemorrhage under the skin of my ankle, and it has remained very sensitive all this time. Today Jack Lewis x-rayed it. He still isn't sure that the bone may not have been cracked. And, if it *is* cracked, I shall have to wear a cast. And that will mean I can't drive myself to Santa Barbara—unless I can borrow a car with an automatic gearshift. Right now,

[1] R. R. Bouché (1906–1963), Czech-born portrait artist and fashion illustrator whose work often appeared in *Vogue.*
[2] By James Costigan; it opened December 1 but lasted only two weeks. Both poster designs were based on Bachardy's drawings of cast members.

Lewis has put an elastic bandage on me, which feels quite good, but I don't notice any improvement.

Charles Laughton and Terry left today for Japan. Terry seems as placid as ever, though I think I detected a very faint uneasiness about Japanese food.

I have managed to do a little, a very little work on "Waldemar." I feel oppressed by the various lectures and talks which are ahead of me. The week after this one will be particularly tough: my lecture on "The Nerve of the Novel," which is probably the most difficult of the whole lot. A possible appearance on local T.V.; God knows what I'll say. And then, next day, a luncheon speech on "Writing—A Profession or a Way of Life?"[1] Here I hope to get in some spiteful digs at the Books of the Month and suggest that "the crowd is the real beast."

October 12. What rat-racing! As soon as Don recovered from his bad throat, he went into an emotional spin caused by his anxiety about the work to be done for Tennessee. And then he was mad at me for "aggressively" helping him when he didn't ask to be helped. And then he had a quarrel with Glade about this fucking *Herald-Express*[2] problem and said terrible things to her. Hasn't told me what they were, yet. As for the *Herald*, I arranged for it to be delivered to Jim Charlton at his office, and then called him and told him what I'd done. So that's taken care of.

Nevertheless, today, I have finished the opening and very difficult section of "Waldemar" which announces all the themes.

Don has now more or less decided to leave on Friday morning and go straight to Wilmington, Delaware, where *Period of Adjustment* is opening. This is much later than Tennessee wanted, but no doubt they will fix up something.

Lewis says that, now the X-ray photos are dry, he still can't see any signs of a fracture. But he still threatens me with a cast if the swelling doesn't go down soon.

Charles Laughton seems to have told Dorothy Miller quite a lot about his problems with Elsa. He is such a baby.

At last the sun is setting right into the ocean again, beyond Point

---

[1] Evidently, to the Channel City Club in Santa Barbara on October 21. Isherwood was to reuse this lecture, or the title anyway, on February 10, 1965 at UCLA.

[2] *The Herald and Express* was Hearst's evening paper; the *Examiner*, mentioned October 3, appeared in the morning. The *Examiner* merged with the *Herald*, but not until 1962; possibly Isherwood didn't know (or care) enough about the papers to distinguish between them.

Dume. Yesterday evening I was watching it and I distinctly saw the green flash,[1] very bright and localized, like the explosion of a bomb. This is the second time I've ever seen it. The first was with Caskey when we were living in South Laguna in 1951, and that time our experience was slightly suspect, because we were both drinking very strong martinis.

October 17. Don is in the East and won't be back at the earliest till the end of the week. He seems to have had a very reassuring talk with Tennessee, whom he now feels is really fond of him. (But Don will need to be reassured about this later, as he always does.) Tennessee had also said that he regards his friendship with me as one of the greatest friendships of his life.

There is a hot wind and the colors are sharp; this is glorious weather. But the wind is giving me shooting nerve pains in my buttocks and thighs. I have been worrying somewhat about my very heavy schedule this week at Santa Barbara, but now I've more or less figured out what I shall talk about in my two lectures; and the T.V. show will have to take care of itself.[2]

Also, there has been far too much drinking. This is bloating me and dulling me altogether and I'm up above 150 lbs. I have got to stop it and get on with my novel.

I miss Don. Without him I feel "restless and uneasy" and I worry about how he's getting along. Without him, my life is just a big bore. I *could* live alone, I guess; but then everything would have to be reconstructed.

October 18. This morning I fixed myself a Prairie Oyster, because I couldn't be bothered to eat breakfast; I wanted to get started on work. This was nostalgic. Thoughts of John [Layard] and Berlin.[3]

Last night was [a] perfect little gem of boredom. I drove all the way to Highland Park to see Del Huserik and his wife.[4] Why? Because he intimidates me and makes me feel guilty for not taking part in his aggressive Quaker projects. Del is as nervous as a witch.

---

[1] An optical phenomenon caused by the refraction of light through the atmosphere as the sun drops below the horizon; most easily seen in clear air across an unobstructed view like the ocean.

[2] With six students, at KEYT in Santa Barbara, October 20.

[3] In *Goodbye to Berlin*, Sally Bowles "practically live[s] on" Prairie Oysters—raw eggs mixed with Worcester sauce—and she offers one to the Christopher Isherwood character when he visits her flat in the Kurfürstendamm. See Glossary for Layard, a Berlin friend.

[4] In his day-to-day diary, Isherwood noted: "Marge?"

He wouldn't sit down and talk, which would have interested me, because his political opinions are his own. No, he had to play me jazz, Weill, Wagner, tiny little snippets of things which he then immediately switched off in favor of something else. And meanwhile his wife served her best supper; a symbolic act, because she didn't really want to have me there, only the *idea* that I was in the house as their guest. She's a sharp-faced discontented girl [and you wonder whether she] will give trouble later, like an unreliable make of car—a Simca, in fact. Mine has been through all kinds of trouble lately; now the lights have gone out on the instrument panel. Mr. Mead counsels patience: "If I may venture to suggest, Sir," "If you'll pardon me for remarking ..." etc.

How good to be quite quite sober this morning! I drank only a couple of glasses of wine with dinner. Nothing else. Tony Richardson was quite irritated about this; he is an absolutely incurable mischief-maker in tiny ways. He wanted me to stay and get drunk with them—Tom Wright and John Rechy were there—and then, having cancelled the Huseriks at the last moment—have supper *alone* with Frank Moore[1] while he went out to some engagement. What irritated me was that, of course, I would have liked to do this. But much more because of the Huseriks than because of Frank.

October 23. Don phoned yesterday to say that he must stay east until the end of the coming week, at the least. His designs for the Tennessee Williams and Julie Harris advertisements have been accepted, and that's what matters. And of course it's obvious that this may lead on to other jobs, and he really should stay there as long as it seems necessary. Still, I miss him more and more. Each day I feel it just slightly more.

Have just returned from a lunch at the de Grunwalds'[2] for Terry Rattigan, and indirectly for Angus Wilson. Oh dear God how I loathe lunches and the run of Hollywood people you meet at them! Besides which, of course, I didn't get to talk to Terry (who looked rather bloated) or to Angus, who is roly-poly and quite sweet. From Angus I get a rather depressing whiff of the London critical deadlock—everybody's fangs locked in someone else's back. But Angus himself is really understanding and sweet.

Told *everybody* about Don's success. That was really the only

---

[1] A pleasant, dark-haired gay friend in his mid-thirties; he kept an apartment in New York.
[2] Russian-born, English-educated producer and writer Anatole de Grunwald (1910–1967) and his wife.

satisfaction of having gone there. But David Selznick, as usual, was interesting. He'll still vote for Nixon, but admits that he's only fifty-one percent for him. He thinks inflation will come much faster with Kennedy. He also says that Kennedy is anti-Semitic.[1]

Well, at least I feel good about one thing; I got some work done on "Waldemar" today. I have been so bad lately, getting hangovers, and I'm fat and pouchy in consequence.

October 27. It's just eight in the morning, a beautiful one, and I plan to get off to an earlier start, so I can arrive punctually at 11 a.m. at the college and thereby frustrate Douwe, who always greets me with his faintly bitchy smile because I'm late. (Late for *what*, one may well ask.) Douwe is amazingly bitchy underneath, and full of old-maidish resentments. I don't dislike him for this, or I won't until he turns against *me*; indeed, I find it rather fascinating. It fascinated me the other night—last week, at that awful party at the Hoffmans' with the arty-method puffed-shit group of actors—when he suddenly exploded against Christopher Fry[2] and said he was "evil" and that he felt sick to his stomach, just being in the room with him.

I've still heard nothing more from Don. I now really do miss him terribly, more and more every day. Perhaps this is the difference between different sorts of relationships; there are people you miss instantly after parting, and then gradually less and less; and there are a few—very few in a lifetime—you become slowly and then increasingly aware of missing, at first it's discomfort and then misery and then agony, like being deprived of oxygen. Of course, I wouldn't be in the agony stage unless I thought we were being separated for a long time or forever; but it is getting *very* unpleasant.

Tony Richardson's *Sanctuary* was shown in the projection room yesterday. Parts of it were very impressive; but I do think they have missed the point at the end. The end ought to be about the execution; not the getting-together of these two boring little underlings, Temple and her husband. That's the story of the two

[1] The fourth and final presidential debate took place October 21; Richard Nixon, formerly a congressman and a senator for California and currently Eisenhower's vice president, was deemed to have lost to John Kennedy, then senator for Massachusetts.

[2] English playwright and screenwriter (1907–2005), often on Christian subjects; best known for *The Lady's Not for Burning* (1948) and *Venus Observed* (1950), and his translations of Anouilh, Giraudoux, and Ibsen.

novels[1]—two underlings, little butterflies out on a binge, happen on a lair of the great monsters; and, in course of time, they destroy the great monsters.

Some queen who is a high-school teacher at a very tough downtown school told the following stories about his pupils, whose ages are around twelve, or maybe younger. A boy says to him: "You're not queer, are you, Mr. A.? It's your husband who is." Another time, the teacher rebukes a girl who is chewing gum. "I don't mind your chewing," he tells her, "but stop blowing bubbles." At once a boy in the class yells out, "I'm Bubbles!" A ten-year-old boy says to the teacher, "Is that a big knife you have in your pocket or are you in love with me?"

October 28. Just got back from Santa Barbara to find a letter from Don—he has some more work and must stay at least until the beginning of next week. Well, I am glad of course; but now I enter a new phase of missing him. It becomes more wretched. I do not want to go to the [Albert] Hacketts' tonight, to the party for Angus Wilson but of course I must.

Last night I had supper with Douwe Stuurman at the little wooden house he built for himself at Isla Vista.[2] It is really one of the most glamorous places I've been in for a long while; standing on top of the short steep cliffs with the sea right below, and the island dimly in the background. The sun going down golden in a blue Monet haze; the waves breaking against the clay banks at the foot of the cliff; the boy (one of the students who live around) running and splashing through the foam. Douwe keeps a rope in his house to throw to people who get into difficulties with the rising tide. Many do. Two cars have been abandoned and lost.

Douwe's souvenirs: a Russian banner captured by the Nazis and then taken by the American troops, a shrunken head from Peru and a bust of [Albert] Schweitzer, a pair of wooden clogs from the days when they were still worn in the Dutch community in America where Douwe grew up. Douwe sleeps on a hard bunk bed right by the ocean window. He made everything, cuts down his own trees for firewood.

He feels that, after his second wife left him, he learnt to live alone, and I can see that he thinks of himself as a sort of father confessor and spiritual focus for the whole campus. He tells how,

---

[1] I.e., Faulkner's *Sanctuary* and its sequel *Requiem for a Nun*, which Richardson amalgamated for his film.
[2] Adjacent to the UCSB campus.

one night, he was sitting in a rocking chair before the fire and became aware that someone else was sitting in the chair beside him and he knew it was Death, his Death. So now he is quite easy with the idea of death and it doesn't bother him. A millionaire gave him a lot of very expensive hi-fi equipment and a T.V. set he didn't want. He knows several millionaires, and they come to him, he infers, seeking the peace they cannot find, because *they* live in big pretentious houses and *he* lives in his simple cabin.

Of course, I am exaggerating the shit element in all this. Douwe does have something; there's no question about that. One just sees a great danger in him of giving way to spiritual humility-pride. Perhaps he should admit his resentments more frankly to himself. His hate of his second wife. [...]

Later Howard Warshaw and a group of students came in, and we had a self-conscious sort of seminar. Howard was excellent, however. His attack on nonobjective expressionism (I may not have the name right, but it means abstract art) seems to me very sound. He points out that these people want to break with the past completely and start something new; and they don't care what associations you get from looking at their pictures. Howard says this is nothing new, because any painter who merely assembles objects and hopes that they will mean something—this Howard calls "naturalism"—is doing the same thing. And of course this is true of literature, too. They are trying to abolish the necessary triangle: the artist, the objective datum on which the art is based, the viewer. They want, as artists, to communicate directly with the viewers. But this—on the level of *maya*—is impossible. (Only on the level of the *Atman* is communication possible—i.e. *yoga*.) On the level of maya, you have got to have the object. The viewer has got to recognize the object in order to be able to appreciate the artist's rendering of it. (When I instanced a painting by Picasso, "A Man Leaning on a Table," 1915, and asked, "If I can't find the slightest trace of the man or the table, does that mean that Picasso has failed to communicate with me?" Howard had to say yes; but he qualified this by talking about artistic allusions in a way I couldn't follow.)

Anyhow, what I personally care about is that Howard is bitterly opposed to the cult of abstract art in art schools and the sneer with which representational talent is so often greeted nowadays—that the possessor of such talent will do well in advertising. I value this attitude of Howard's because it puts him on Don's side.

November 2 [Wednesday]. Had a telegram from Don this morning;

he's arriving back here on Friday afternoon. That means he will have been away three whole weeks, which must be the longest stretch of time we've ever been separated. Oh, I'm so deeply glad that he's coming back. But I'm certainly not deeply pleased by the way I've been handling my life while he's been gone. Drinking, idling, wasting time with people I didn't really want to see; and getting nearly nothing done on the novel.

Today I've been feeling sick in my stomach; I do hope I'm not going to get ill. That would be too tiresome. Probably I am simply run down from drinking and eating too much. I am not charmed with myself at all. Swami, with Krishna and Mrs. DePry, was in to have tea here this afternoon—the first time he has ever been in this house—and I had a guilty feeling that somehow he saw the state I am in. Well, never mind, I just have to snap out of it.

Of the people I've seen lately, the most interesting was John Rechy. We had supper and a long talk the other evening, and then he came again to discuss the latest episode in his novel, which I'd read. (It needs an awful lot doing to it.) One of the characteristic things about John is his fear of inventing; he wants to record everything exactly as it happened. So I spent a lot of time trying to convince him that this would be undesirable and anyhow impossible. But I do respect and like him; he quite fascinates me. He says quite frankly that he's an exhibitionist, and this makes it possible for him to hustle, etc. He is fascinated by mirrors; spends hours looking at himself in them. At the same time, his relationships are compartmentalized. He never told his engineer friend that he was a writer until quite lately. And, with his "mental" friends, he is exaggeratedly nonphysical; he hates to be touched, even in the most casual way. (I remember how Edward [Upward] used to laugh at me for this, at Cambridge.) I think he thinks of himself as being always in disguise.

I introduced him to Evelyn Hooker. They both took to each other immediately. Evelyn's motives were of course more interested than John's, because she at once saw him as an ideal expert informant to help her in her researches.

Two days ago, I definitely decided not to go to [L.A.] State College next semester. The two thousand they offer just is not good enough, and besides, I ought to get on with my novel; Laughton will probably be in my hair anyhow. And Byron Guyer[1] says that he can arrange a much better offer for me for next fall.

Huge excitement is stewing up over the elections. My Kennedy

---

[1] Professor of English at L.A. State, 1955–1978.

stickers have been scratched off the car twice, but I keep putting on new ones. There are Nixon stickers everywhere, it seems, and I am worried. So is Jim Charlton, despite the reassuring forecasts. Jim, says Tom Wright, believes that his own personal problems and anxieties will somehow all be miraculously solved if Kennedy wins.

November 9. Don's New York visit was just as much of a success as the first one. It now seems that his drawings will be on display at *three* different theaters—*Taste of Honey, Period of Adjustment* and *Little Moon of Alban.* He got back on the 4th. But he isn't at all well. He seems to be having constant attacks of my age-old complaint, spasm of the vagus nerve—at least, I *hope* that's all it is. He refuses to see a doctor. He is touchy and nervous and hostile, and then utterly sweet. And I just have to practise caring-not caring.

Worried because my ankle, which seemed all right, has suddenly swelled up again and hurts, after a walk on the beach yesterday.

Well, the toad Nixon is driven back into his hole, and rejected by his own home state, which is a special satisfaction.[1] I feel I want to triumph over that bitch at Santa Barbara, at Wright Ludington's party, who called Howard Warshaw "stupid." Those arrogant rich-bitches!

Have had a most gruelling weekend reading all kinds of manuscripts—including the huge novel by poor Alfred Weisenburger, to whom I wrote an unkind letter yesterday. I am ashamed of it. If I don't want to read these things, I shouldn't consent to do so. No justice in getting mad at their authors.

Ah, I'm so full of resentments, these days. Sick with them. I must get my calm back somehow. Tonight I have to take the Mishimas out to supper. They are going to Disneyland today. Mishima told me, "We also see the home of Mr. Nixon, and a ghost town—" he paused, "same thing!"

November 12. A day of heavy showers and strong winds. I have prepared my talk for the Santa Barbara temple[2] tomorrow. I only hope the rain lets up before I have to drive there. As usual, I feel a resentment against Prema, whom I always suspect of being in the background, whenever I'm burdened with one of these weary Vedanta chores.

---

[1] Kennedy's victory, November 8, was slim. In fact, when absentee ballots were counted a week later, Nixon was to prove the winner in California. See Glossary under Presidential Election 1960.
[2] At the Sarada Convent, Montecito; see Glossary.

At Santa Barbara, I see a lot of a student named Frank Wiley, because he is writing a novel—a queer novel about the UCSB campus in which he, the "I" of the story, falls in love with another boy. Quite aside from the natural sympathy I feel for him, he is certainly one of the brightest students in my seminar. A few weeks ago, I noticed that I'd been referring to him, in conversation and in my pocket diary, as "Grimm." I thought carefully about this Jungian error and decided that it was simply because of Grimms' *Fairy Tales!* I told this to Wiley. I don't know if I should have, or not. He didn't seem very surprised. But not very amused either.

I have to face it—my seminar is actually the least successful—perhaps the only unsuccessful—of my Santa Barbara activities. It's no good blaming the students for sitting around like muffins. It is up to me to toast them. I mean to do something about this, next time. I am going to ask each one of them a number of direct questions and see if we can't find out between us what is wrong. This is urgent, because we have already had eight seminars and only six or at the most seven more remain.

November 15. Yesterday, I again saw the green flash. Only this time it wasn't so much a flash; the sun disappeared and then a tiny nodule, like the very last bit of another sun, appeared, and it was quite sharply green.

This weekend has been difficult. Don has been suffering almost continuously from his stomach trouble. At last, yesterday, he made up his mind to go to Jack Lewis, who told him it may be an ulcer, but was so reassuring about ulcers in general and their relation to longevity that Don wasn't worried. Anyhow, he gave Don some medicine and, thank God, [Don] immediately felt better and his spirits greatly revived. I think the stomach had a lot to do with his behavior on Sunday—also the weather. Saturday, it rained heavily. Tom Wright provided a farce interlude by insisting on bringing us some firewood he didn't need. The firewood was quite waterlogged, and Tom, like a cheerful wet badger, arrived during a particularly heavy downpour, so we had to change our clothes and unload it into the garage. We had our best clothes on because we were about to go out to dinner with the King Vidors. This wasn't a success—although a very interesting man named Hendricks was there, the chief representative of *The Christian Science Monitor,*[1] who reassured us about the danger of a Kennedy upset due to

[1] Probably Kimmis Hendrick, Pacific News bureau chief from 1947 through the 1960s.

absentee balloting and uncommitted electoral votes[1]—because, as
so often, they practically ignored Don. Don stayed in town for
the night, but it didn't improve his mood; and when he got back
next day and we were having supper together at the Red Snapper,
he blazed up. Told me he wanted to be independent. Wanted to
go to New York for several months. That all I ever did was to
find ways of making him dependent on me. That he didn't see
why he should be grateful to me, because after all he had given
me so much of his life, and it was time that counted. I daresay
I could have taken all of this and realized how much and how
little it meant, if I hadn't been tired from a rat-race drive to Santa
Barbara and back. I had to talk at the Vedanta Temple about "The
Writer and Vedanta"—but, as it was, I got mad too, and asked,
what about *my* time? And so it went on. Don said he never feels
the house belongs to him. It isn't his home. Etc., etc. With much
hatred of me in his voice. And then, to top it all, we had tickets
for *The Threepenny Opera*, and it was ugly and crude and dirty
beyond belief—a parody of Brecht, even at his worst. We left at
the intermission, so as not to have to witness the spectacle of poor
old [Lotte] Lenya involved in this dreary horror.

Then Don began to say he was sorry and terribly humiliated.
And that he felt there was nothing inside of him fundamentally
but a "selfish little faggot." And of course I did my best to reassure
him.

He says he is desperate to get rich—earn money—and this is, of
course, partly because he wants to pay off the debt to me and be
free of this tiresome guilt-obligation. Whether he would decide,
when that was done, that he could and would leave me, I don't
know, but I think not. I believe that he still loves me, just as
much as I do him; and I still believe that this love will last. I know
that I am possessive and fussy. But I also firmly believe that I am
overcoming this fault, and that it would be absolutely possible for
us to ease into another sort of life together.

As always, I can help him by being more satisfactorily what I am
and therefore more independent of him or of anyone. Oh God,
when I look around this room! At the very least, fifty, maybe a
hundred books I haven't read! And then I should read all through
my boxes of letters, diaries, etc. There is so endlessly much material for hours of recollection and meditation. And then *japam*!

I believe more firmly than ever in Don. I believe in his talent
and his character, and I believe he will evolve into the kind of

---

[1] See Glossary under Presidential Election 1960.

person we both want him to be. I believe, furthermore, that he has taken giant steps in this direction already; and that therefore these outbursts mean much less than they meant three or four years ago. He is becoming more and more independent in the only way that matters—inside himself.

During the night, my nose bled quite heavily, all over the sheet. At the time, I thought it was just mucus from the sinus. There was a pleasant feeling of easement, afterwards.

November 19. Yesterday morning, at Santa Barbara, I got up early, not drunk—for almost the first Thursday of this semester—after a ghastly restaurant dinner and evening at the home of Dr. Girvetz, as guest of the philosophy department. (Dr. Girvetz is the victim of Douwe Stuurman's bitterest scorn, because, after establishing himself as one of the liberal hopes of the campus, he married a rich wife and sold out. And, as a symbol of his depreciated spiritual condition, he has gotten fat, says Douwe, and become a drunken glutton.)[1] The get-together at Dr. Girvetz's home was, like all such functions, far too large. Everybody's wife had been brought along, and also, said Douwe scornfully, the secretaries. (This was an obvious slap at poor Peg Armstrong,[2] of whom Douwe seems curiously jealous; chiefly, I can't help feeling, because he feels she is muscling in on our relationship!)

Well, anyhow, I got up early and drove along the shore, because I had time to spare before the classes—at which I had to speak on (1) E.M. Forster (2) children's literature. I stopped the car and got out and took out my beads and made japam. And I made a resolution—the words came into my mind—"From now on, I'll make japam every day until I die."

At Mrs. Haight's class on children's literature,[3] I rather surprised myself by holding forth in the most authoritative manner, as though I had been considering the subject for years. In effect I said: "The books I liked best as a child were mostly fantasies which

---

[1] Harry Girvetz (1910–1974), first Chairman of the Philosophy Department at UCSB, from 1957 to 1964, wrote books about liberalism and the modern welfare state, served on the California State Democratic Central Committee, and wrote speeches for Governor Edmund Brown in 1959 and 1960.
[2] Administrator of the Arts and Lectures program at UCSB; from 1959 to 1981, she planned theater, lectures, and arts events to complement the academic program.
[3] Genevieve Watson Haight (1904–1964) joined the English department in 1941 and mostly trained teachers of English. Her children's literature course had legendary popularity.

I could relate to my actual surroundings. I liked [Beatrix] Potter because I lived in an old house where there *were* rats; and where I could easily imagine little doorways and tunnels leading into 'the universe next door.' This, in a different way, was the appeal of H.G. Wells; science fiction based on very prosaic everyday settings. Then, in the case of Ainsworth, there was the real, twentieth-century Tower of London, where it was easy to find a doorway to the sixteenth-century Tower of Lady Jane and Bloody Mary....[1] H.C. Andersen is The Artist as Child. His account of life is written in terms of the fairy tale, but it is absolutely valid; the stark truth is told about suffering and love and death. The Little Mermaid must be considered in the same category as Anna Karenina.... This idea that children are pure and uncorrupted and that grown-ups have somehow lost their spiritual vision is just sentimentality. There are lots of children who are just as corrupt and insensitive and spiritually blind as the worst grown-ups, and children's literature is simply literature which speaks to the child's condition; relates to his known environment. Auden's fantasies were connected not with houses but with landscapes ... We cannot get back to the child's innocence. But the sophisticated adult can achieve another kind of simplicity which is maybe better ... etc. etc."

Because the Dangerfields wanted me to, I went to a Dr. Walter Graham in Santa Barbara, who is a famous bone specialist. He told me to take off the elastic bandage and never mind if my ankle *does* swell up. He predicts that it will be all right in another six weeks.

A beautiful brilliant day today. Don and I had hangovers after an evening at Doris Dowling's. She infuriated Don by treating him as so often—as my appendage. So we had the old fuss once more when we got home; I was blamed, for not protesting and for secretly liking Doris's flattery. But today all was well, and we walked on the beach—and my ankle has swollen very little, if at all. Tonight there are strong gusts of wind, and the bushes scratch squeakily against the panes of my workroom window; sometimes they seem to be furiously struggling to get in.

Now that I have this Thanksgiving vacation, I hope to do a lot on "Waldemar." Made a start this evening.

November 23. Have made quite good progress with "Waldemar." I ought certainly to be able to finish it during the holiday, except

---

[1] William Harrison Ainsworth's *The Tower of London* describes the tower in its account of the imprisonment and beheading of Lady Jane Grey, the nine-day child queen, by her cousin Mary I. Isherwood tells more about the book in *Lost Years: A Memoir, 1945–1951.*

that it is getting longer and longer. Well, I won't force it. I think something quite good is emerging.

Laughton and Terry are back and spend a lot of time in the house next door. But Terry is to go back to England because Charles refuses to "ruin his life"; i.e. he won't keep Terry unless Terry can get work here; and that'll only be possible in T.V. or movies—the modelling jobs are all in New York. Terry bows his head when this is said and looks sad but acquiescent. Charles plays it very big, enjoying the drama. Elsa, with her tour over and Ray Henderson about to get married, sits up at the Curson house and dares Charles to desert her. This, he agrees, he can't possibly do. "She's too old."

Iris Tree is taking off for Italy tomorrow or the next day. We saw her last night at Oliver and Betty Andrews's. She looked wonderful but seemed drunk and foolish, holding forth against the provincialism of Los Angeles. What she obviously meant was that she has found the atmosphere of Ivan's house provincial, as it certainly is. I think both he and she are relieved that she is going. And yet I feel sad. Iris has represented a bright flash of gaiety out here, none the dimmer for being ridiculous. Her ridiculousness is the most lovable thing about her. But then one thinks of her whole life, with all its flashy flutterings, and feels sad. Why? Not because she's what's called a failure—if she is and whatever that means. No, I guess just because this is one of those lives which put such an emphasis on youth. Still, I would hate it if Iris stopped dyeing her hair. When we said goodbye, she looked at me and said, "I do love you," and I know she does. I love her too. I am always glad to see her again after a separation; but I don't find I miss her very much.

News in the papers of poor Norman Mailer's breakdown. One headline said, "Author Mailer Stabs Wife." Don misread this as, "Arthur Miller Stabs Wife" and said to himself, "Why don't they mention Marilyn Monroe?" We agreed that, if it had really been the Millers, the headline would have been, "Marilyn Stabbed by Mate."

November 24. Have been working all day on the novel; still in my bathrobe at six p.m. Don is eating Thanksgiving lunch with his parents. He'll return for Thanksgiving supper with Jo and Ben. I haven't eaten all day—for the third day in a row. I have breakfast and then get along on coffee and a Dexamyl. My weight is down to 150 again, but I'm resolved to get much lower. And every day when possible I want to exercise.

No problem what to give thanks for, this year. Don. His success—even though that makes new problems. Having a novel to work on. Being in good health. Having this house. Having a job and the prospect of future jobs. (I think UCLA will work out).[1] So I do give thanks. And I will earnestly try to keep at my japam.

December 4. A gap, due partly to having had the typewriter serviced; partly to mere laziness.

On the 29th, I finished revising "Waldemar" and sent it off right away to Edward [Upward]. It isn't perfectly all right yet, but it's as good as I can get it until I have the whole book and can go through it relating all the parts to each other.

On the 1st, Don and I drove up to Santa Barbara with Jo and Ben, in their car. It rained heavily that day, which cut down the lecture audience and generally depressed us. However, the next morning, there was a marvellous after-rain clarity and all of the islands appeared, and the view from Douwe Stuurman's was at its best.

The evening before, we invited Douwe, Fran and Howard Warshaw to have dinner with us at a restaurant, with Jo and Ben. I got very very drunk, so did Don and so did Fran. Don complained next day, after we had gotten back home, that it was terribly inconsiderate of me to have brought Jo and Ben up with us, especially to be present at his first meeting with the Warshaws and Douwe. (Douwe had had a glimpse of him down here at Adelaide Drive, but not a proper introduction.) Jo is such a frump, Don said, and she must have shocked Fran, who'd been expecting me to have much more stylish friends. And then he accused me of aggressive, masochistic indifference. I had known perfectly well, he said, that Jo would make a poor impression on the Warshaws and Douwe—and yet I had brought her, largely for my own convenience.... This is at least half true. I *did* have misgivings about Jo and Ben and how they would fit in, and it *is* true that I dismissed them and thought, in effect, oh what does it matter? And of course there is my usual aggression of thinking: if they're *my* friends then they're good enough for anybody else in creation. It is also true that I knew the Don-Warshaw meeting would have gone off better without Jo and Ben. He would have had more opportunity to make an impression on them.... And yet I swear I didn't mean any harm!

Anyhow, after Jo and Ben had gone off to their motel—poor

[1] He was being asked to lecture there during the coming year.

things, they got gypped into paying *eighteen* dollars for the night, *and* damaged a muffler on the car, backing down the narrow El Bosque Road because I hadn't stopped them in time to make the sharp turnoff into the Warshaws' driveway—Howard asked to see Don's drawings and praised the portraits very highly. He did not like the nudes, and this was extra impressive and indicative of his perception because we had chosen to take them along to show him believing that they would especially appeal to him, being more like his own kind of work! Howard advised Don to practise working from old masters, in order to find out how they approached their subjects. But Don says that this idea is meaningless to him and that he can't follow it.

Don is going through a deep depression, with all his masochism in full play. After the success in New York—nothing. And the fear that there won't be anything. And the anxious feeling that he ought to be back in New York, angling for another job. Maybe he ought. It's impossible for me to make the decision for him. I only know that I don't want to go there, and indeed can't go there, at least not for more than a short visit. I *must* get on with the novel.

There's nothing to be done about all of this but wait and see and meanwhile sweat it out.

December 7. Don seems a bit better, though he still says he is "sad." I can't help recording this with a certain resentment, because *the effect* of it is a reproach aimed at me. I have somehow made him sad or allowed circumstances to make him sad or, at best, failed to prevent circumstances from making him sad. He, on his side, would cry out—has often in the past cried out—against my egotism in relating this to myself. It has nothing to do with me, he would tell me. At least—not very much. Well, we are both to blame. He *does* use his sadness—however much he may protest that he doesn't—to make my life as well as his own just that little bit more difficult. I say that we should try to make each other's lives more bearable, even if we have to pretend a cheerfulness we don't feel. But do I act up to this belief? Very often not. And my resentment against Don's sadness is of course selfish: it interferes with my comfort and forces me to stop being preoccupied with my own affairs and start being anxious about *him*. It *is* true that I want a smiling contented purring kittycat.

Yet—though I like to let off steam by writing a paragraph such as the one above—I also know that I do not really want the contented pussycat as a permanent companion; he would bore me to death. I want Don just as he is—*but* I also want him to be happy all the time. And that's impossible.

Tomorrow I go to Santa Barbara. On Friday with Douwe I drive up to San Francisco, where I'll read from *Down There on a Visit.* (I have just today decided to call the novel that, after all.) Saturday, I'll spend in San Francisco. Sunday we'll drive back and I'll come on back here—a long long haul.

Edward writes saying that he's satisfied with "Waldemar." Not really enthusiastic, I feel; but satisfied that its style is right for its contents. He can't be expected to say much more, until he has read "Paul."

Charles is in a great state about Elsa; she continues to be hysterical and obstructive. She wants to run Terry off the range, and yet she must know that, if the two of them were left alone together, she and he would only make each other more miserable than ever.

The last two days very windy, which makes me nervous and on edge. We still have no Christmas plans. Gavin may or may not return here before then; now that the Italians won't allow *Mrs. Stone* to be filmed there, he is obliged to stay on in England for rewrites. Tony Richardson is also a possible arriver. He may show up at any moment.

December 13. The San Francisco trip was really a great success, all except for the Writers' Conference itself.[1] That was a fiasco. To begin with, they had planned a banquet in honor of *Sir* Charles and *Lady* Snow, and the bastards went off to New York and didn't return for it or even write or wire excuses. When the British noblesse oblige, which is quite nauseating enough in itself, breaks down, then that's truly squalid. And all the worse in the case of the Snows, who are *posing* as aristocracy, waving his knighthood in the faces of the naive Americans, and glorying in having dragged themselves up out of the lower middle class.[2] (Why so heated, Dobbin? Do *you* want a knighthood? No, it's not as bad as that. But I suppose I even now resent these inflated reputations. The truth is, I want the English snoothood to break down just once and admit that, all kidding aside, I am the—greatest? best? no—just *most interesting*—writer alive today.)

Then they had failed to distribute the material which was to

---

[1] At Berkeley.
[2] C.P. Snow (1905–1980), novelist and Cambridge scientist, knighted in 1957 and made a life peer (Baron Snow of Leicester) in 1964, was a grammar school boy and the son of a shoe-factory clerk. His wife was the English writer Pamela Hansford Johnson (1912–1981).

be discussed among the student delegates; so the discussion broke
down. And they had failed to announce my reading, so the hall
was only three-quarters full. Well, that didn't matter. I chewed the
scenery just the same, nearly cracked the mike during the fishing
scene from "Mr. Lancaster,"[1] and made Mark and Ruth Schorer
really roar with laughter.

I like them both. Douwe says they're always fighting; but they
were very friendly to us. We drank a lot. They have a kind of
miniature funicular railway up the hillside from the garden gate to
their house. Mark is that very lean type with thin glossily brushed
black hair and a liquor-reddened face. And the half-amused, half-
challenging gleam in the eye which recognizes you as One of the
Gang.

A delightful drunken rainy Saturday, far from the sodden dreary
groves of Berkeley's academe, with Ben Underhill.[2] I felt enormously reassured to find that I can still enjoy this sort of thing so
much. Douwe was impressed, when he picked me up at Ben's
apartment on Sunday morning, to find me so cheerful and wide-
awake. He is terribly intrigued and thrilled by my goings-on, and
longs to hear even more than I'll tell him.

Both our drives were in beautiful weather. On the way home,
we again passed the young peace-marchers, still headed south and
apparently still in the best of spirits. We resolved to find out about
them, by calling the Santa Barbara newspaper office. I rather excited Douwe by saying, "Those kids must have a leader; maybe it's
someone whose name will be famous one day all over the world.
Maybe biographers will look back to this march as his first notable
teenage exploit."

I must say, I really do like Douwe. I'd been worried a bit at
the prospect of this long drive with him, but there was no strain
at all. We communicated very pleasantly. His bitchery gave just
the right spice to the conversation. He was full of the debate they
held sometime last week, "Is there a rational proof of the existence
of God?" He had spoken against the motion and caused quite a
scandal.

In fact, one of the letters I got on my arrival last week referred
to this:

I like your lectures very much, but you should not swear. Do
not use the word God at all, if possible. Dr. Stuurman said in
front of an overflowing audience: "God is nothing but a good

[1] First section of *Down There on a Visit*.
[2] Not his real name. An occasional bed partner. He was a schoolteacher.

word to swear by." He said it is embarrassing to talk about God in any other way in the twentieth century. You, of course, know that Stuurman is divine (He only identifies Atman and Brahman a little negatively). You are the man who could help him a lot.

This letter was written by Eleanor Pagenstecher, a well-known local crank.[1] It is related of her that she once went into one of the college offices and happened to hear someone refer to a self-addressed envelope. "What nonsense!" she exclaimed. "How can an envelope address itself?"

After leaving Douwe, I drove right on home and went with Don to a party at Walter Plunkett's,[2] for Hope Lange, her sister and brother-in-law,[3] and Glenn Ford. This was a terrible mistake. Glenn's welcome was more than embarrassing. And I insulted the sister and brother-in-law, who are psychiatrists, by refusing to play a silly game in which you have to say what certain things make you think of—a wood, a building, a stream, etc. "I just don't know you well enough," I said. "The wood, to start off with, reminds me of female pubic hair." Don was very angry with me for this, later.

But this morning he said, "I simply don't know how you stand me sometimes." And he told me that now, at last, he is beginning to feel real confidence in himself. It began quite suddenly last week.... Well—let's hope and pray.

December 19. Today, I read through "Paul" again. I think it's all right basically, but there are long boring stretches in the middle.

In the midst of my reading, Cindy Degener called from Curtis Brown to say that Lewenstein really wants to do the musical of *Goodbye to Berlin*, and that she was seeing Auden, getting in touch with Carter Lodge, etc.[4] Don says that I was extremely snooty and rude to her on the phone, and I fear this is true. Partly because it *was* the phone; partly because I hate being bounced into any project; chiefly because I just plain do not want to work on anything but my novel. I want to get money, yes—lots and lots of it; but not work.

---

[1] A Vedanta devotee at the nearby Santa Barbara Vedanta Society.

[2] Academy Award winning costume designer (1902–1982), then working at MGM.

[3] Minelda and Bob Jiras.

[4] Auden and Isherwood had worked on the musical before (see *D.1*); now the London impresario Oscar Lewenstein was interested. See Glossary under Curtis Brown, Auden and Isherwood's New York agency, where Cindy Degener handled drama, and under Carter Lodge who had rights in John van Druten's play *I Am a Camera*.

Don said later that he hates to see me being resentful and aggressive like this; he wants me to keep my aggression for him, just as he keeps his for me. "I'm jealous of it," he said. Don has actually been very sweet since last Saturday evening, when he blew his top—I forget why—and cut himself really quite badly on the ice tray from the icebox. We have had a talk. He says he wants an independent life but he doesn't want to leave me; I urge him to go to New York and spend three months there, till the spring, and see what gives. At present, he recoils from this idea, but partly because it scares him, I'm sure. As for me, of course I don't want him to go if he can be happy here; but better a thousand times he goes and comes back somehow reassured. Meanwhile, Don says that of course he's quite well aware how much progress he has made toward independence and how much he has learnt, though he is always bewailing his ignorance.

He has been reading Middleton's *The Changeling*, because Tony Richardson remarked he'd like to direct it.

Now, suddenly, there are two projects on foot: the *Goodbye to Berlin* musical and a possible film of *The Vacant Room*, written by Gavin and me and directed by Gavin.[1] As for Laughton, he's always a menace. Not one word comes to us from behind that avoirdupois curtain.

Yesterday evening, Don said, "Kitty's sick of his struggle—now he wants a triumph."

With one possible exception—I was so drunk that next morning I couldn't be certain—I have told my beads every day since I made that resolve. True, I usually remember them at the last moment and race through them, thinking of something else. Never mind—the habit is being reestablished, and that is all that matters for a start.

Compulsive Christmas letter and card writing. There is much aggression behind this; I feel I am slamming back the ball across the net and I think, "Take that—fuck you!" Don is so right; aggression is terribly bad for me. I may even kill myself prematurely with it if I don't watch out. And watch out means watch out *now*.

December 23. A new phase seems to have started. This man Russell McKinnon has offered to put up the money for Don to go to Europe, and this seems a kind of God's-will intervention; something which has to be accepted. So now Don is definitely

---

[1] Lambert had already helped Isherwood (in 1959) to revise this film script, originally written by Isherwood with Lesser Samuels in 1950.

thinking of going to the Slade School as soon as possible and spending as much as six months in England during 1961.

This, in its turn, has enormously improved our relations. Already we are living in the sad-sweetness of departure. I feel hideously sad whenever I think about it—especially when I wake in the mornings. I dread it and yet I know it may be the best, the only possible way for us to go on together. Don himself is afraid and anxious and terrifically excited. We only hope there won't be red-tape difficulties.

On the 20th, I started revising "Paul," five pages a day. So far have kept up my schedule. It's a kind of madness, but I must bust through, however compulsively, and get a big chunk finished this vacation. I want it to go off to Edward as soon as it can.

Besides, Laughton will be breathing down my neck soon. Terry went back to England today; returning in February. Things look better there; Elsa has a New York booking and will be out of his hair for a long while.

Beautiful winter sunshine—going to waste every day as far as I'm concerned.

December 26. A frightful hangover, caused chiefly by my unwillingness to tell Hope Lange, Glenn Ford and the others what I really thought of *Cimarron*, last night; so I got drunk at Hope's house after the premiere. Tried to cure myself this morning by going on the beach and into the water, which was deathly cold; also by taking one of the full-strength all-black capsules Carter Lodge gave me. This merely made me feel like death.

Swami's birthday lunch. Swami so angelic and radiant, all in white. "You don't have to tell me that you love me," he said to us, when the girls sang the gooey second verse of the Happy Birthday song.

Gerald [Heard] on the phone yesterday was in the highest spirits. He's just back from Hawaii. He says Aldous [Huxley] had cancer of the tongue and a surgeon told him half of it would have to be cut out and he refused; and then he went to Cutler[1] who cured him completely with an X-ray needle. Gerald said, "We have lost the art of dying." He is back on his favorite theory that "isophils"[2] represent a new mutation.

I have slipped badly on my revision of "Paul." That was only to be expected, I suppose, on account of the holidays. I must relax and get on with it quietly and not be frantic.

[1] Dr. Max Cutler, trained at Johns Hopkins and the Curie Institute in Paris; head of the Tumor Institute in Chicago before opening his practice in Los Angeles. He had previously removed a papilloma from Michael Barrie.
[2] Homosexuals.

December 29. A tearing wind, yesterday and today, which I hate; it makes me tense.

Yesterday morning, we called Stephen Spender long distance in London and asked him to ask Bill Coldstream if Don can be gotten into the Slade School without red tape, provided he isn't trying for a diploma or any academic grades. Stephen thought this might be possible, said he'd ask Bill, but that he was going away for a week. If we didn't get a cable from him within twenty-four hours that'd mean he hadn't been able to contact Bill and that we should have to wait a week. No cable has arrived, so we have to wait. And now we are both getting nervous, feeling we want a move to be made at once, as long as a move is going to be made.

But I'm still sick at heart, at the thought of his being away so long. My own life isn't going to be that much longer that I can afford to spend six months of it without him. Never mind, what must be must be, and I will try to fill in the time with work and self-discipline.

A new obstacle arises to my getting ahead with the novel. Laughton has suddenly decided to go on a reading tour and he wants me to help him prepare the material—including doing a specimen bit of Plato, the charioteer passage probably, from *Phaedrus.* He is an old man of [the] sea and would dearly love to spend every moment sitting on my back; and yet I can't refuse him and anyhow I'm fond of him. He had me weeping with laughter the other day, imitating God the Father as a Prussian bully and Moses as a clothing-store Jew, having an argument about the manna.

December 31. Just a word to round off the year. Tonight we are to go to three parties—Jerry Lawrence, Ivan Moffat and the big Charlie Lederer shindig, for which we've once again rented tuxedos.

It's a perfect sunshiny day. My chief resolution for next year, aside from the usual ones about work, time-wasting, needless anxiety etc., is to take more exercise, and especially to get on the beach and into the water more.

Barring nuclear war or personal illness, I suppose I shall certainly finish my novel before June 1. The revision is going ahead slowly but well and I don't think there are any great obstacles ahead. Indeed, the chief problem is how to begin the book—in other words, how to rewrite parts of "Mr. Lancaster."

This morning, Don—who has hurt his back at the gym—has gone to see Russell McKinnon and ask him if he could have the money to go to England right away, provided that Bill Coldstream says he may work at the Slade School. So all that will be decided one way or the other quite soon.

So, goodbye, 1960. On the whole, an excellent year; unspectacular, but I got such a lot of work done and was happy a lot of the time. Certainly one of the best years of my later life.

Later: 6:00 p.m. Since writing the above, I've been out and shopped; also written two more pages of "Paul." And Don has come home after talking to Russell McKinnon, who's apparently quite disposed to produce the money for him to leave for England in January, if this Slade School thing goes through. Also, a serious international crisis seems to be cooking up—Red troops invading Laos.[1] I suppose there'll be a lot of that, this coming year; even if this one fizzles out. One will just have to go ahead as if nothing but one's own job matters. "We are in God's hand, brother, not in theirs."[2]

Very good relations with Don, all this time. Yesterday evening—I forgot to record—we did something we haven't done in ages; danced together to records on the record player. A Beatrix Potter scene—the Animals' Ball.

1961

January 2. Yesterday was an almost total loss. Don and I didn't get home till five-thirty in the morning. We slept till nearly one and then Laughton arrived to talk about his reading tour and, specifically discuss what parts of the *Phaedrus* he shall use. Elsa came by too, briefly. Their relations are tense, and Elsa had called me the night before, begging me to talk to Charles, who was becoming mentally sick and dangerous again, she said. Charles showed no sign of this, and of course I continue to suspect that it's Elsa who's the sick one—but you can never be *quite* sure. When Charles had gone, Don and I ate at Ted's[3] and then came home and read. Don is reading Willa Cather's *One of Ours.* I read right through *A Lost Lady,*[4] finished *Confessions of an English Opium Eater*[5] and also read

---

[1] On December 30, Laos requested U.N. help, claiming North Vietnamese troops had crossed its northern border; see Glossary under Laos Crisis.
[2] *Henry V,* III.vi.
[3] Ted's Grill, a local restaurant.
[4] Also by Cather.
[5] By Thomas de Quincey.

'Tis Pity She's a Whore,[1] which I finished this morning.

Talked to Don about the doctrine of the apostolic succession as it applies to the status of the Pope. Don said he thought this was "rather marvellous." He keeps repeating that I know so much and that he fears he's stupid and plodding and will never learn anything. Which is ridiculous, because he has already learned an immense amount in a random, right-across-the-board kind of way. But I think he just likes to be reassured.

We're in a provisional phase. Don, I feel, is counting, rather desperately, on the Slade School project coming through. If it doesn't, he won't know what to do next. He writes a great deal in his journal and I would love to see what he's written. But I would never dream of looking. Not from a sense of honor but out of a sort of superstition—a fear of opening Pandora's Box. And yet, I must repeat, our relations are very good and I feel real love between us. It's just that he's desperately rattled.

On New Year's Eve, we went first to Jerry Lawrence's, then to Ivan Moffat's, then to the Lederers', then back to Ivan's. As far as I was concerned, the visit to Jerry was the most memorable, because I met there a very young blond boy named Mike Maffei, who is an acrobat and a champion archer, and has been a paratrooper. We started talking about the crisis in Laos, about which no one else seemed even to have heard, and Mike said how scared he was that he'd be sent out there. There was something about him, a kind of poignancy, which very few people have and which always moves me deeply. Don has it, and perhaps it's what draws me to him more than anything else. When it's related to youth and good looks it creates the aura of beauty. The merely weak are never poignant. Courage has to be part of it—the kind of courage which makes you feel its utter vulnerability. You feel how alone the person is, and how heartbreaking his courage is in the face of hopeless odds. I don't mean that I felt all this about Mike. I'm really speaking more about Don, and others I've known. But I got a hint of it, and I feel I must see him again.

At Ivan's, David Selznick appeared to be the only other person who was worrying about Laos. There seemed something almost symbolic in this pair of worriers, Mike and David. Or there did while I was drunk. I also had a long talk to the Lederers' young son, who was upstairs in his room with comic books all over the floor, and who showed me the rib of a whale, and a scale model of some prehistoric animal. Hope Lange and Glenn Ford were there;

[1] By John Ford.

Glenn very kissy, as usual. I'm sure I talked a lot of crap. But the evening was a success, both Don and I agree. At least—and this is saying quite a lot in regard to such evenings—it didn't leave a nasty taste in my mouth.

Thinking over it from a Proustian point of view, I remember that Lance Reventlow, who seemed, at the Lederers' party two years ago, almost the only attractive young male, now looked pouchy and sodden. Henrietta, Boon [Ledebur]'s ex-wife, still looked well, but her face had spread out and changed. Charlie Lederer was barely recognizable. There would be a Proustian sketch in my conversation with his young son—the polite but wised-up, maybe quite cynical little boy talking to the drunk middle-aged novelist. Both of them putting on an act—making conversation about natural history. The little boy of course getting a certain bang out of talking to *any* guest at this grown-up party from which he's been excluded—and *why*, since he's wide-awake and dressed. Because he hasn't got a tuxedo? (I forgot to mention that we finally decided to rent ours. Don much disliked this and said several times, "I feel like you feel when you're wearing rented clothes," when people asked him, "How are you?") As for my talk with Mike Maffei, Jerry tells me this morning on the phone that he was "quite intrigued" by me and said, "I don't care if he's a famous author or not, I like him."

I couldn't resist calling Mike on the phone after writing the above; and he really chewed my ear off—telling me all his opinions about the futility of war, etc. He isn't a bore, though, because all this has been seriously thought about by him, not just picked up out of a book. He is terribly concerned because all U.S. forces are now in a state of "red alert," and he has a brother, eighteen months younger than he is, who's a paratrooper and due to be sent off at short notice. This brother is his only living relative. So we ended up agreeing to meet sometime in the near future. I don't suppose it will lead to anything particular. That's not so important.

January 3. Hope Lange and Glenn Ford came to supper last night. It was a fiasco. For the first time in my life, almost, I let the barbecue fire get too low. And then I tried to pep it up again with more charcoal and lighter fluid, leaving the steaks on the grill. They were absolutely covered with ashes and nearly raw.

Hope was in an hysterical mood. She had cried in the car, coming down, and Glenn made things much worse by telling us this—adding that it was because of Don Murray and some decision he'd made about their children. Glenn wanted us to cheer Hope up. His idea of doing this was to clown around in the most embarrassing way. Then he got serious and talked about the crisis and how he had a bomb shelter. And how he was reserved, because he'd once shown a little bit of his real self to someone and had it smacked down, or sliced off, I forget which. (At Ivan's party, it seems, I had told Jennifer Jones and Glenn that they were mysterious, and Hope that she wasn't. This, for some reason, seemed to have pleased her.)

Glenn has a lot of aggression in him. His description of the rude Texan who had *dared* to grab him by the arm while he was eating his breakfast and had tried to drag him out of the hotel "to shake hands with my little girl." Glenn had told him to go to hell. The Texan had said, "We pay to see you, don't we?"

This morning I polished my "last lecture" (really, it might well be, if this crisis gets any worse) and worked on "Paul." I now see that a very important passage still remains to be written: the change in Augustus after his visit to India.

January 15. This afternoon, I'm to take part in the taping of the Oscar Levant program, then drive up to the Warshaws' because it's Douwe Stuurman's birthday, and really the only tribute I can pay him is just to take the trouble to do this.

Yesterday, Don got a frigid reply from the Slade School, signed by Coldstream's secretary, saying that his work would be considered next month for admission next October and that meanwhile he would have to fill out the enclosed forms, etc. etc. So this morning we again called Stephen Spender long-distance, and he thinks the whole thing is a mistake and that Coldstream will certainly let Don into the Slade, and that if he won't, then Don can most probably get into the Royal College of Art which is nearly if not quite as good. So Don's hopes, which were at zero, have shot up again. And now he's off to lunch with Russell McKinnon who seems about ready to produce the money for the trip.

Despite Santa Barbara and Laughton, who's been after me to work on two Plato bits for his reading tour, I have managed to write in the bit about Augustus Parr in India. It isn't quite right yet, but it certainly was necessary; in fact, it is one of the pivotal episodes in the story. So now I really do hope to get faster ahead. Especially as Charles will have to go to New York very soon to be with Elsa at her opening there.

January 17. The past four days it has been amazingly warm for the time of year; the temperature up in the eighties. Lots of smog in town; you could even feel it in your eyes down here. Tonight the surf is up, and you hear long menacing artillery-rolls from the beach. When I drove back from the Warshaws' yesterday the high tide had flooded the highway, along by Solimar Beach.

I got back to find that Don had been just as drunk as I'd been. He had been out to dinner with Paul Millard and Gavin Lambert and had been so pissed that Paul hadn't wanted to let him drive home. And when he *did* get home he left the lights of the Sunbeam on all night and the battery went way down to nothing; and today it stalled again and had to be fixed by Mr. Mead.

As for me, I still don't know *what* I did up at the Warshaws'; I only know that when I woke in the morning it.was nearly twelve; something that almost never happens to me. I felt like death all the way home.

Don was unimaginably sweet. It has suddenly hit him, what it will mean for us to be separated for six months. And now Russell McKinnon has definitely said that he will give Don the money just as soon as he can make arrangements about a school in London to go to. (We haven't heard anything more from Stephen yet.) I'm utterly sick whenever I think about his going; but yet I know it's the right, and even maybe the only[,] thing to do.

Abbot Kaplan and a colleague named Haas(?)[1] came around and talked about UCLA. They asked me to give three lectures there and they tried to get me down to two hundred dollars for each. I said definitely no, and they at once showed that they hadn't really expected me to agree, and I'm sure they'll go up to three hundred dollars. In fact, they'd probably have gone to four or five. Whatever anyone says, this kind of thing nauseates me; it is Jewy and vile and utterly shameful, coming from the representative of a serious institution of learning instead of an old clothes dealer.

Laughton is leaving for New York on the 23rd, so I won't have to go on with the Plato for a while. Yesterday and today both wasted and this is inexcusable. (Maybe I'll at least recopy the two last revised pages right now—yes by God I will!)

The only funny memory I have of the Warshaws' party (which was dull on the whole and, worst of all, dull for Douwe himself because he wasn't drinking) is of [a woman guest] whom I don't much like asking Fran Warshaw about her behavior at an earlier party at which she had passed out. She asked, "I hope I didn't do anything that wasn't *dainty*?"; meaning, like getting her skirts up above her knees, while she was lying unconscious on the floor.

---

[1] Robert B. Haas, on the staff of the UCLA University Extension.

Fran took me into the kitchen and showed me in horror that the cake had Douwe's name spelled wrong; Fran herself actually wasn't sure how it *should* be spelled! So I scribbled out the other letters of "Dowe," leaving only the D, and then we stuck candles over the spot to hide it.

January 20. Well now, suddenly, everything is fixed and Don is to leave for New York next Monday the 23rd. He'll stay there long enough to arrange with McKinnon's agent about the way his money shall be paid to him in London; then he'll go on there. Stephen cabled two days ago to say that the Slade will take Don any time he cares to start work.

Oh, I'm sick—sick with foreboding and anticipated loneliness. And Don is wretched about it too. We have never been closer to each other than during these past few days. There are moments when I think, *can* I bear it? But I must—not only that, but make something out of the experience; discipline and train myself. Not run around to parties getting drunk and looking for "consolation."

Japam, work—of which, God knows, there's plenty—and also physical training. I must try to get back into better shape. Today I'm well under 150 lbs., but so flabby. And I must be prepared for an attempted psychosomatic coup; getting sick in order to be able to call Don home.

Laughton leaves for New York on Monday, also; so I'll have a real opportunity to get down to hard work on the novel. So far, I've written fifty-four pages of the revised version of "Paul"; this is only forty-five pages of the first draft, but I'll be making very big cuts farther on, I expect.

After all the glorious weather, yesterday and today were smoggy and sorrowful. This morning, Dr. Haas and two of his colleagues from UCLA came back to see me. They have capitulated; I get my money. And we planned three readings for March, under the heading, "The Voices of the Novel."[1] My general line will be that a writer can be judged to quite a large extent by his tone of voice—just as you already form a judgement of someone, rightly or wrongly, by merely listening to his voice on the phone. The old *stichprobe*[2]—reading the first, last and middle pages of a novel—isn't so unfair as it sounds.

[1] The lectures were announced as "The Forefathers: Dickens, Conrad, Brontë, and Others," "Our Group and Its Older Brothers: Joyce, Hemingway, and Others," "The Young Novelists: Williams, Capote, Kerouac, Mailer, Bradbury, and Others," for three Sunday evenings in March.
[2] Random sample.

January 23. Don went off this morning. We both agreed we didn't want to go through the tension of an airport parting, so I just drove him up to the Miramar Hotel, where he caught a bus to the airport at 6:55 a.m. He should be in New York by now, staying with Julie [Harris] and Manning [Gurian].

I don't know how I feel. I've kept going all today on nervous energy and doing one thing right after another. Ideally, you could carry on that way for six months or until you dropped. I went to interview two gyms—Vic Tanny's in Santa Monica and Lyle Fox's in Pacific Palisades. I think I shall go to the latter, starting tomorrow. At Tanny's, I was high pressured. And the atmosphere, though sexy, is squalid.

Then I plan to get one of the bicycles fixed and ride it a lot. And of course there is my work, etc. I have even started preparing the material for another chapter of the Ramakrishna book. And I want to do a little meditating every day as well as the japam. And make an extra round of japam for Don.

Yesterday was so hectic. Don went into one of his last-minute whirls. At about 3:50 we tore off downtown to the County Museum to see a show of art nouveau which closed at 5:00! We made it, too, and saw nearly all we wanted to, because the show was pretty small anyway. Then up to Hollywood where we were just in time to see Buñuel's *The Young One.* Then for drinks with Paul Millard. Then, terribly late, to UCLA to see [Pirandello's] *Six Characters in Search of an Author.* And Don damaged the Sunbeam while parking and we had to be towed away afterwards. I was a bit cross or rather madly rattled about all of this, but then he was so sweet. We stayed up all night. He tried to draw me for the brochure of my UCLA lectures, but he couldn't do anything good. Then he packed and wrote letters. Our whole parting—all these last days—couldn't possibly have been more loving. For once, I haven't *one moment* of unpleasantness to reproach myself with. Don said, "One thing I'm sure of now—we didn't meet each other by accident." We both cried as the bus went off.

January 24. I can hardly believe it's still not yet forty-eight hours since he left.

Of course I'm still very schedule conscious; that's the first stage. Lots of japam—for Don too. And I got to the gym and had a good workout; and I worked on the Ramakrishna material and also (not enough) on "Paul." And tomorrow I'm going to talk to Don in New York; I sent him a telegram today asking him to call me.

Yesterday evening, I had supper with Jerry Lawrence. I told him that he is the only person I could think of to spend the first evening after Don's departure with; and the funny thing is, it's perfectly true. Nobody else would have been right—except Jo and Ben.

Tonight it started drizzling. Had supper with dull but rather sweet Bill Inge, and John Connolly, whom Don and I met years ago at our first New York Christmas; it was at George Platt Lynes's apartment and we all helped paint John's body for some masquerade party he was off to. Tonight we went to a fearfully dull and bad Mexican restaurant called the Caracol, which is however the same building as the fish restaurant, Marino's, where Jim Charlton and Ted Bachardy and Don and I had supper, that historic evening....[1] And, up at the Holiday House, where Inge and John Connolly are staying, there were three little kitties.... Well, I must get used to this sort of nostalgia.... What makes me feel bad and almost superstitious is that it's all somehow so reminiscent of "Afterwards"[2]—but then why not, since I wrote it?

The last two days, I've been wearing Don's sneakers. I like to have on something of his. Am now sleeping in the back room. I plan to take the sheets off the big bed and not use it any more till he comes back.

January 31. I haven't wanted to write in this book as much as I'd thought I should. This morning, I got a cable from London to say that Don has already enrolled at the Slade and is looking for a flat. There is a very faint chance that I might be able to get the assignment to write the screenplay of Graham Greene's *England Made Me*; in which case I could go there at the end of March or the beginning of April. Ivan Moffat is helping me try for this, because he knows [John] Sutro, the producer of it.

Sunday was bad—the day Don left for England. I let it get too late before phoning New York and then made desperate efforts to reach him through Pan-American at the Idlewild Airport[3]. There would have been lots of time to do this, but the people at the Pan-Am office were casual and careless beyond all belief. God help anyone who really needs to get in touch with a passenger! I never did get Don—only someone named Machardy, who wasn't even on the same flight!

---

[1] February 14 (Valentine's Day), 1953, when Don Bachardy first spent the night with Isherwood. The building is on the Pacific Coast Highway in Malibu.

[2] His sexually explicit 1959 short story (never published); see *D.1*.

[3] Now J.F.K.; Isherwood typed Idyllwild, like the town in the San Jacinto mountains where he stayed in John van Druten's cabin in the 1940s. See *D.1*.

After being stiff all over from exercising last week, I was better today and went back to the gym. I felt wonderful right after it; now I begin to ache again.

There's still an awful lot of "Paul" to do. I have revised fifty-eight pages of the original manuscript and this has turned into sixty-nine pages of the new version. If I could cover four pages of the old version every day throughout February, I could finish by the end, but this isn't nearly as easy as it sounds; because I write in so much new stuff. I do think it's pretty good, though.

February 6. At present it's a real effort and a bore, writing in this book. If I want to write anything about my life, it's letters to Don. I had a letter from him from London; full of homesickness. But now comes the difficult part—I must manage not to mind when he gets over this, as he will. I must want him to be happy in England. And I must get over the most basic part of my possessiveness—wanting him to experience everything through myself as an intermediary. What he is about to experience now isn't going to have anything to do with me—and this, I'd better not kid myself, will be painful.

Amiya and Prema down here to supper. Amiya yakked and yakked, retelling all her oldest stories. Prema sat rather sour, until near the end, when he'd had enough to drink and we told him he must become a swami and run a center on his own—not rush off to India when Swami dies.

Can't be bothered to tell about Monroe Wheeler's visit, or Bill Inge or Marguerite's party—though I must say I laughed a lot when Gavin told me that, after we'd left, Larry Harvey announced that he was secretly married to a British actor who'd come out to play *Mutiny on the Bounty*, and that John Ireland[1] was his mistress. He really is sympathetic.

Considerable excitement because John Zeigel is to come and stay with me. Well—

One thing I won't forget—driving up to the top of the hills with Monroe and Bill Inge and John Connolly yesterday, after Marguerite's party, and looking down over the city. All the platforms cut out of the hillsides, ready for pretentious French chateau-style houses "worth" eighty thousand maybe, but no more than slum dwellings because so crowded and viewless and altogether wretched. And I had such a sense of something spawning itself to destruction, spreading and spreading out until its strength is

[1] Canadian-born American stage and film star (1914–1992).

exhausted and then shrivelling up and dying, and then the rockets, or the new ice age, or the whole slab of coast cracking off along the earthquake fault and sliding into the sea; lost in any case. And the quickie promoters and real estate agents hustling to make their dollars before it happens. Such a sick sad knowledge that this is "Babylon the great city"[1] and it can't end well—and was never and could never be great, anyway.

In the middle of Saturday night, after the party at Jerry Lawrence's, Jim Charlton came blundering into the back bedroom drunk, just as he used to in the Rustic Road days[2], mumbling about how he'd realized he was a fascist and had got scared. In the morning, he seemed to take all this behavior for granted; and I begin to wonder if he isn't becoming a bit crazy like his mother. He certainly is full of the most dreary self-pity. Now he's all for leaving Hilde and ducking out from under the mess he's made—not that I have one particle of sympathy for Hilde herself [...]. I only know that I don't want Jim around as he is nowadays. He is the most ghastly bore and nuisance. While Inge, Monroe and John Connolly were at the La Mer restaurant last night, Jim showed up there—I'd idiotically mentioned we were going—and sat around and created gloom and boredom and some sort of dull-dog reproach for his sadness, aimed at all of us.

February 13. Bad. I feel sad, bored, impotent. I guess it is the dead of the year. The sun is shining here, but it's somehow chilly and dull. I'm not exactly stuck in my novel, but am making very poor time. Got to page 78 of the rough draft manuscript, which is page 89 of the revised version. Let's face it, I still have a good half of the thing to write. And now I have to devise this different setting for the seminar, in place of La Verne, the cabins on the Salton Sea. And it all takes so much *time!*

Johnny Zeigel is a very sweet boy—intelligent, though on the prissy, academic side, and capable of serious love. He stayed the weekend here—chiefly talking about Ed Halsey (whose intentions I suspect) and what their life is to be together from now on: are they to live in the Caribbean, and if so doing what? Yesterday we saw Charles, just back from New York and Elsa's show. He wanted me to tell him his scruples about calling Terry back were silly, so I did, so he called Terry, who was away for the weekend,

---

[1] Revelation, 18.21.
[2] September 1948–December 1950 when Isherwood lived with Bill Caskey at 333 East Rustic Road. Caskey was often away; see *D.1.*

and only succeeded in infuriating Terry's London landlady, because it was one o'clock in the morning, their time. Then we went to Laughton's house and drank and swam in the pool, and Charles sulked like a great baby and Johnny left feeling frustrated, and I went to see Paul Kennedy, which was a huge mistake, he is hopelessly sloppy and tacky and passive.

I have kept up the japam so far, and the exercising, BUT I MUST GET ON WITH MY WORK. Nothing else matters. Until I have done that, how can I go to England?

Don is supposed to be calling me tonight or sometime tomorrow, for our eighth anniversary.

Courage.

February 17. It's quite late already, but I want to get this book written in before I go to bed.

Today was the Ramakrishna puja and I went to vespers, which was an absurd mess, because Swami had decided that it would take too long, each one of us coming up to the shrine and being touched by the relic tray and given a flower and offering it. So he came around with the tray among the audience and touched us where he found us. Only the usual women wouldn't budge after they'd been touched (Prema claims one of them was dead drunk) and so a traffic block was created, and some got touched twice and others not at all.

Ritajananda is *not* going to Paris, and this rejoices Prema's heart. Now he hopes Vandanananda will be sent off to run another center. Ritajananda came out after supper and asked me, in the garden, if there was a feud between the center and Gerald Heard![1] I think he is being wised up, fast; and he is very anxious to be loyal to Swami.

I must say, I do love this house. I am really sad that I must go away and leave it all summer. If only Don could come back and we could simply stay here! I think it is the only place—except the garden house at Saltair[2]—which I have really liked for itself and rejoiced to live in.

Jitters about the novel. *Can* I finish it? Of course I can, but I must get busy. Criminal laziness today—I lay around finishing off

[1] Isherwood tells in *D.1* and in *My Guru and His Disciple* about differences between Heard and Swami over asceticism, women, the role of the guru, drugs, and mysticism. And see Glossary under Heard.
[2] At Evelyn Hooker's, 400 South Saltair Avenue, Brentwood, where Isherwood lived in 1952–1953.

the Oppenheim[1] thriller Chris Wood lent me. And now Dana Woodbury has lent me that de Sade book—*The 120 Days of Sodom*. But that looks like a bore—an ugly humorless French bore.

Laughton was here again today and we worked on the "wings"[2] passage from the *Phaedrus*, which Charles wanted to read aloud tonight to Taft Schreiber. We got it finished and he was enthusiastic. Now he talks about paying me extra for my work on the material for his reading tour. We shall see.

February 23. Stayed home and ate alone tonight. The first time in such a long while. If you stay home at night, you get all sorts of offbeat calls of which there's otherwise no record—like an offer of two free lessons with Arthur Murray's dancing school.

I'm sort of doubly lonely. Lonely for Don, as always. Lonely also a little bit for Johnny Zeigel. This is silly, but harmless and nice. I do feel he's a wonderful person—anyone who can love, properly, is wonderful. People like that always get me romantic over them.

When I went up there on Monday, I found him in a state of tension because Ed Halsey—who must surely be a self-centered ass—had sent him this cold, incredibly formal postcard, signed "your friend." In fact, the coldness was such that it seemed almost like the act of a mental case. So Johnny was desperate, and finally he was able, the morning of the next day—after we'd sat up nearly all night waiting until the circuits were open and gotten pie-eyed—to contact Ed on, I think, Grenada in the Windward Islands. And by this time the Mexican cleaning woman had arrived and could hear the conversation, and the line was so bad John had to shout, but he didn't care, he yelled, "Ed—*I love you* more than anything else in the *world*!" Ed seems to have pacified him, but somehow I didn't buy it and I don't think John did either, after he'd thought it over. He said, "I love him, and I don't care how old he gets, or if his belly gets bigger, but I'm not going to spend the rest of my life with a bitter old man."

Today, Bart Lord called up and told me that Ted Bachardy has gone to the mental hospital again. He allowed himself to be hospitalized voluntarily; but now he's in the violent ward, and the doctors want to give him shock treatment but his father refuses. The psychiatrist also says that he regards the outlook for Ted as being very black. Because the last recurrence came so soon.

---

[1] E. Phillips Oppenheim (1866–1946), Edwardian spy writer.
[2] I.e.,wings which carry the immortal soul upwards, toward Truth; see Henri Estienne (Stephanus) edition (1578), pp. 246–257.

What John's relation to Ed—or rather, Ed's contemptuous under-valuation of John—makes me feel is "like the base Indian threw [a] pearl away"[1] and I simply shudder to see this done, for Ed's sake, not Johnny's, because Ed is really the weaker and older and less likely to snap out of it. And that makes me resolve more than ever to look after *my* pearl. I had another long talk to Don on the phone yesterday. All seems more or less well so far. But I must not take anything for granted.

Am still rattled about the novel. *Can* I finish it in time? I must not mind if I can't, and yet it will be a terrible pity. I do want to get this revision done and sent off to Edward. I can rework it in London later, if I'm there.

· A very eager-beaver period. I decided to quit drinking at least until March 2, when I go up to Santa Barbara with Johnny to stay a night at the Warshaws'. And now I'm eating celery like crazy, because someone said Kinsey[2] discovered it was the only thing for potency. And then there's the gym.

February 24. Good work on the novel. I have now got to page 109 of revised version which is page 104 of the rough draft. The whole Salton Sea episode is written and it seems pretty solid.

Poor Olga Fabian![3] I had her to supper tonight. She is so garrulous and such a bore. How awful to be old in that sense! Took her to the Serbian restaurant, where the English waitress had returned and made it quite impossible to eat, with her continued interruptions.

No further news of Ted.

Tonight I felt so lonely that I called John Zeigel. Some of my loneliness has spilled off onto him, that's all; don't let's call it anything more. It only *feels* like love. And I must beware of annoying him with it. I woke him up tonight with my call, and he naturally wasn't any too charmed. The lonely are a public nuisance. He has had a good letter from Ed, and all is well, until the next time.

Now go to bed, foolish old Dobbin. I'll let you read for a while, but you're to be woken at seven. And given *lots* of celery.

February 28. A good day of work. I drove right through the whole

---

[1] *Othello*, V. ii. Isherwood typed, "the pearl."
[2] I.e., Alfred Kinsey, the sex researcher.
[3] Viennese actress (1885–1983); she played Fräulein Schneider in the original stage production of *I Am a Camera*, appeared in other Broadway shows, worked in T.V., and later had small film roles.

CPS camp[1] section in rough, making quite a lot of inventions. And it seems that, owing to big cuts in the original rough draft, I'm going to end up with page 129 of the rough corresponding to page 129 of the revised version or pretty nearly. So I'll have caught up with myself. And certainly the hashish section will be drastically shortened. So maybe about 160 pages will see me through.

This is a rat-race period. Even with brisk progress and sustained inventiveness, I shall barely get "Paul" done on time. And then there are the three UCLA readings. And the Vedanta Society eyes me sadly, waiting for another Ramakrishna chapter.

As for Sutro and his film, God knows. It looks as if there will be so little money in it and maybe it would be smarter to go to England and risk getting something else. I have written to ask Mr. Sidebotham if there is any money waiting for me, under M.'s will.[2]

No more news of Ted. He's still at the hospital, still under heavy sedation, apparently.

Don is to call tomorrow morning, I guess.

Good workout again at the gym today. I am losing weight, chiefly because I haven't drunk or smoked since the Claremont visit.[3] I suppose I shall on Thursday, when Johnny and I go to Santa Barbara and stay with the Warshaws. Wish I wasn't going in a way. I don't want to leave here till everything is finished.

Despite celery eating, I am, as far as I can judge, absolutely impotent. Is it old age, or just that all the gism has gone into novel writing, as it's supposed to? I certainly feel terrific, otherwise. Absolutely transformed since I restarted exercise; and I don't get a bit stiff any more. I haven't been in such good shape in a long long while.

I keep up the japam, despite entire "dryness"—a counterpart to impotence? I always make one round of japam for Don and this is the only one that I take trouble over. I feel I only want to pray for Don, not for myself. This sounds very noble; it doesn't feel so. And maybe it has some quite other psychological motivation. But I do think of him an awful lot. Always, when I do, the image of *openness, aliveness, tenderness* (in the early Quaker sense[4]) recurs.

[1] Civilian Public Service camp, for conscientious objectors during W.W.II.
[2] Isherwood's mother, Kathleen Isherwood, died in June 1960. Sidebotham was the family solicitor.
[3] I.e., to Johnny Zeigel.
[4] Softened and receptive—as to the light and power of God or inspiration of the Spirit—the sense which Isherwood learned from the Quakers during his early years in the U.S.

Despite all his egotism, he *is* more alive than most people. I love that look in his eye when he is drawing, so cool and yet hungry; he watches the sitter like an animal crouched to spring—well yes, of course a kittycat.

Am reading Shakespeare's *Richard II* with the pleasure one has in eating wholesome food after a trashy diet.

March 2. 1:00 a.m. The madness of art.

Just to record that today I got to page 124 of the revised version, which means that, today, I revised *ten pages*. I took three Dexamyls in the course of the day, drank God knows how much coffee, ate no supper and almost no lunch. Never mind.

A good talk with Don in London this morning. I felt so much love and assurance of love.

What happiness to have these two things. And what does it matter, honestly, what becomes of the work after it's finished?

March 2. 7:40 a.m. When I wrote the above last night—or rather in the early hours of this morning, I forgot to mention one thing which moved me very much and made me extra happy. In order to get on with my work, I had excused myself from going to Vedanta Place for supper, as I usually do on Wednesday evenings. So, around 7, Swami called and said, "I'm lonely for you, Chris." It wasn't that he was nagging at me to come after all; just an expression of love—a love that is without strings and therefore quite fearless; it doesn't hesitate to make gestures of this kind for fear of being a nuisance, or making the other person tired of it—fears which would cause the ordinary "lover" to hesitate. Yes, sometimes I hesitate even with Don.

Another thing I'd like to record; I've thought it often, these last days. How symbolic one's work is! What with the H-bomb, the population explosion, the menace of the lowering of living standards all over the world during the next fifty years, etc., etc., how less than probable that anything one writes now will, as one says, *last*! And how little that matters at the moment you write it! I write this book for Don, for Edward, for as many people as there are in Pacific Palisades, perhaps. All right! No complaints later about the critics, if you please!

And yet, at the same time, I can't help saying to myself: Boy, it's *good*!

March 7. Poor progress. I am now only on page 131 of the revised version, which is 130 of the rough draft; the gap has been nearly closed, as I anticipated. But there is still a formidable lot of work ahead; and I suppose, just about a month from today, I shall be leaving for London.

I talked to Don this morning. He said, "It isn't just that I love you; I *need* you so much." And this afternoon John Zeigel called and said, "I've been thinking about you all day."

What a strange period this has been! I went to Santa Barbara with John last Thursday. We came back Friday and he has been staying with me over the weekend. It was very delightful, having him around; but really it didn't add up to very much in terms of emotion. He is a good sweet boy, book-bright but maybe without much natural taste. He would fit quite happily into an academic job. Now he is definitely going down to Mexico at the end of the month to be with Ed, and then they will see what's to happen next.

Up at Santa Barbara, a great feeling of warmth and family snugness with the Warshaws, both of them. Also, in a different way, with Douwe. They were all, maybe, just a particle shocked at my bringing John with me. I had to do a great deal of talking about Ed.

10:40 p.m. Just back from having supper with Phil Frandson, Bart Johnson and [his friend]. An oddly depressing evening. Phil was really very interesting about modern technology. The evolution of hybrid creatures. The IBM machines which work all night and talk to each other by radio. But it all adds up to the horror of production for production's sake and the consequent stultification of our culture.

And, all the while, underneath this, was the stupidest nagging little ache of wishing John was around. Why? What for? Am I nuts? It doesn't really mean anything. Just nervousness.

March 15. Yesterday—which was the anniversary of my first visit to Berlin in 1929!—I finished "Paul." I haven't the least idea what it's really like, yet; but I know that I don't want to do anything more to it, right now. So the whole novel is now complete, except for the work of relating the parts to each other. That I'll probably leave until I get to England, and can discuss the whole thing at length with Edward.

Today I feel "restless and uneasy." In the night there was a violent rainstorm, the first in months, and now it's blowing frantically. And I have to go to Gerald's, and then come back here and pick up Kent Chapman and take him up to Vedanta Place, so he can talk to Swami about Vivekananda, with reference to this story he's writing, and then I shall spend the night at Paul Kennedy's and then drive out bright and early to Claremont to see Johnny, and go with him to a college production of *Hamlet* tomorrow night, and stay the night at his place and then, on my way back into town, go over to the passport office and apply for my passport, and then come back here. And then Ben Underhill will be coming for the evening and to stay the night.... Most of this program is quite fun, in a way, but as a whole it makes me nervous to think about.

And Don hasn't called yet; and that worries me. In fact, I'm going to try to call him, a bit later in the evening.

March 19. Don did call, but after I'd left, so I didn't get to talk to him until yesterday. Kent Chapman came up with me to Vedanta Place and that was quite a success; but he left a note on Friday saying that the people he has been living with can't or won't have him any longer so he is "homeless." I don't know where he is now, but I have an uneasy feeling I very soon shall. *Hamlet* was awful; we left after the first act. As for being with John, it is all very fine, but teasing is still teasing, however tastefully conducted; and I left him feeling frustrated and at the same time just the least little bit uncharmed. Especially since Ben Underhill showed up on Friday and was entirely sweet and non-teasing and fun.

Gavin is wildly enthusiastic about "Paul." He thinks it's the best thing I've done and altogether extraordinary. Gerald, I fear, is displeased. At least, that's the impression I get from Michael [Barrie], who put on a very grand, not to say snooty manner when I called him for news. He talked about himself and Gerald like a supreme court about to hand down a verdict. (I had never even given him permission to read it.) And, what was more, made difficulties about the date for doing this. It never seemed to occur to Gerald to call me on the phone.

March 23. No—Gerald did *not* like "Paul." I went to see him yesterday. He kept off the subject all through tea and forced me to bring it up. I suspect that his feelings are somehow hurt; but of course he wasn't about to admit this. He only said that—I can't remember his actual words, but the sense of them was—"Paul" was an anticlimax after "Waldemar"; it seemed a narrow, limited, trivial story and one lost all sense of the world crisis in it ... Well, I just don't think he's right, that's all. No use in getting steamed up. I must reread it and I must wait to see how it strikes Don and Edward.

Gerald went on to talk about the future. Although things look so awful and although, if several more countries get the H–bomb, war seems nearly inevitable, Gerald still feels we have a chance. He believes that the Demiurge who has been governing life on this planet may not wish to see the extermination of his experimental farm. The Demiurge very seldom intervenes in human history, but he may now do so; perhaps by direct telepathic action upon the various world leaders.... Gerald rather takes pride in his concern for the human race—because, after all, why should he bother, he says; he must expect to die within eighteen months or two years. I suppose he'll go on talking like this for the next ten or fifteen.

Am in an absurd tailspin about something truly trivial; a party I arranged to give on Saturday. The party was really to please Jo, because I know she resents having been left out of so many of our social events. So I started asking people, and at present the guest list reads: Aldous Huxley, Prema (to drive him), Glenn Ford, Hope and David Lange, John Zeigel, Evelyn Hooker, Terry Jenkins (Charles Laughton can't come; he's going to New York to see Elsa), Michael Barrie and Gerald Heard, Tom Wright. Alec Guinness and his wife may possibly come. So may the Stravinskys. And Ivan Moffat will come in later with some girl.... Having invited everybody, the question arose, how to feed them. And Jo said that, with such people coming, I ought to get in a caterer. This made me mad. Because I loathe caterers, and Jo's view of Hollywood High Society is so much a part of her fatal lack of style. And I'm doing the whole damn thing for *her,* anyway. So then Hope and Glenn offered to do all the cooking, etc.; and Glenn said, why not have the party up at my place, so I will know where everything is in the kitchen. And then we got to talking; and now finally it looks as if I'll simply buy food from a delicatessen and that will be that. I somehow hate going up to Glenn's, and it will add to all the trouble, not lessen it. But it seems as if it's going to be that way.

Tailspin anyhow, because of all that has to be done. This wretched Ramakrishna chapter, and the whole unsolved problem of the journey, and packing and arrangements ... Oh dear, how I hate it! How I wish that in some providential way, Don would come back—but without failure, regret, loss of face or any other disadvantage to himself—and that I wouldn't have to stir out of this house! But that's too big a miracle to ask. And I suppose I'll like it when I'm there. At least I'll be with Don; and that's not least but most.

March 27. The party cost $175 for the food—some dull ham and too few hors d'oeuvres—plus $54.35 for the drinks: $229.35 in all. And it'd have been much more than that if I'd let those stinking caterers get the drinks. Glenn was so strange about the whole thing; sulky and grand. And yet I knew he was thrilled to death having the Stravinskys and Huxley in his house. Well, at least Terry enjoyed it, and Jo and Ben, and John Zeigel—maybe a few more. To me it was sheer torture and I only thank God it is over; and now I just want to rest and relax, financially and otherwise, until I leave. I hope I will have the sense never never never to commit such a folly again.

After the party, John came back to the house to spend the night, and we reached a sort of climax in our relationship, and got by it rather successfully, I think. I don't really quite know how I feel about this yet, however. So I'll go into it later. Or maybe I won't. Anyhow, it's very good that he's going to Mexico in two days. I suspect some playacting on both sides. And yet he is a sweet boy and fun to be with.

My last reading at UCLA went off well, on the whole. I really was quite good, reading the end of *Ulysses*. And I gave Jo and Ben a big thrill by reading from *The Crackerjack Marines*.[1] This morning, a couple who were present sent me a box of red roses, as if I were an opera singer!

This evening, Vernon Old brought in his new wife, Doreen.[2] She is a pasty blonde, quite nice I guess but slatternly and sloppy. And Vernon, whose head is too big anyway, has ruined his appearance with a huge Victorian beard. I talked too much about myself and showed off and ended up by boring myself pissless.

Very tired now. Must sleep. Much work to be done on the Ramakrishna, if I'm to finish it by Wednesday and get it out of my hair.

April 2. Easter Sunday. I did finish the Ramakrishna chapter, and so that's that. I shall have to write two more, probably, while I'm in England.

This last week has had some quite nice things in it. To begin with I enjoyed John Zeigel's final visit. He met me at Musso Frank's for supper on the 28th, stayed the night and I drove him out to the airport at midday next day to get the plane for Mexico and Ed. I now feel I know him really very well. His mind isn't

[1] Masselink's novel.
[2] Not their real names.

very interesting, and I much doubt if he could write a novel, as he wants to; but maybe I'm wrong about this. He is also a complication maker and a bit of a masochist. He claimed—the night we got drunk and had the big showdown, and he said he left it to me if we should or not; as far as *he* was concerned, he wanted to; and I said no [...]. Yes, I suppose that's possible. Of course, fundamentally, we were on our party manners throughout. I guess, if there had never been a Don, I would have had a try at living with him; and I think it could have worked out. But then, so it could have with lots of people. That doesn't make him a Don, or even a Don-substitute, by a million miles. As for Ed, I still doubt if that will work out. He sounds like the most self-centered, boringly narcissistic kind of queen.

Then on the 30th, Howard and Fran Warshaw and Pepa[1] came and we had supper and got drunk. I was really quite enchanted by Pepa. She somehow reminded me of a very young Katherine Mansfield. Howard is very resentful because they aren't taking up his option, or whatever you call it, at UCSB. Shall he get another job, which he can—or devote himself to painting? Advised painting.

Then Ben Underhill came to stay the night of the 31st and we had supper with Alec Guinness, his wife, and the Goddard Liebersons.[2] Alec was very nervous—probably because he starts work tomorrow on this grotesque film *A Majority of One* in which he plays a Jap and Roz Russell[3] a Jewess. Anyhow, he made an absurd fuss, insisting that his wife (Marilyn?)[4] should change her dress. As a matter of fact, he was right, but still it sounded tyrannical and embarrassing in front of strangers. And then at the restaurant, Perino's, he insisted that he'd heard the waiter say, "They're only English," and "It doesn't matter about them." He was furious because he thought they hadn't given us enough caviar; so he dug more out of the pot. They ended up giving him a check for ONE HUNDRED SIXTY-FOUR DOLLARS.

Ben Underhill is a very strange creature. He utters grunts and little laughs and uses some rather shaming slang words, like "tubby" for bathtub. Under his good-humored grin and sleepy ways and sweetly simple sensuality, he really is odd. You feel strange depths or shallows. I could never possibly live with *him*. He would bore

[1] Fran's daughter.
[2] Pianist and composer head of Columbia Records and his wife, ballerina Vera Zorina.
[3] Irish-Catholic American star Rosalind Russell (1907–1976).
[4] Merula; see Glossary.

me terribly and make me nervous.

Last night, I went to a party at Phil Frandson's. It was really a bore, which I masked by getting drunk. That was the third drunk night in a row and I feel much the worse for it. Must lay off from now on in, until departure night, anyhow.

Mr. Mead's social club is called The Camelot Club, and its motto is Sociability with Distinction.

Just called Glenn Ford to find out what Lady Guinness's name is. He doesn't know. Significant?

April 6. 5:20. Don, I'm late, I haven't shaved or dressed and I have to call Eleanor Breese before Jo and Ben arrive; but I want to say one word to you, just in case it is the last. I love you. Never doubt that. Never doubt that you are everything to me. And never doubt—since, in any case, you'll hardly be reading this unless I am dead, earlier or later, that I am with you in whatever way one can be. I want you to know that I made japam for you every day while you were away. Maybe if I get to London I shall tell you this myself; maybe not. Goodbye my darling. I love you so. If anything happens to the plane tonight, I shall be thinking of you until the last moment. And beyond it. Yes, I believe that. Don't forget old Dobbin, who loves you so.

A Stay in England: April to October 1961

April 13. I got here nearly a week ago, on April 7. (*Here* is 11 Squire's Mount, Hampstead, NW3.[1]) And now I have rented this typewriter, which is going to be a bitch to work with, because I'm so used to the electric. But, bitch or no, I am determined to keep up a diary while I'm here, because I feel that this is going to be quite a memorable period, not necessarily a pleasant one, either.

I find Don desperately tense and full of his usual fears, plus a new one—that he won't ever be able to paint in oils. He loathes what he has done so far, though it seems good to me, because he has been told that it isn't painting at all but drawing colored. Like every American who comes here, he has been subject to British snoot. And of course even I can sense his utter failure to "make like a painter"; that is, do the sort of thing which corresponds to the French approach to writing, and which I detest. Well, we shall see how all that turns out. My only contribution can be to keep

[1] Richard and Sybil Burton's house.

my own wig on tight, and sit for him when needed. I'm doing that now; and today I made him mad because I have a tearing cold (caught from him) and my right eye dripped so much that I couldn't read—I've been reading the Shakespeare histories, starting with *King John*—and then I closed it and kept falling asleep. There's also an argument about the bed. It's soft and hurts my spine, but when I insisted on our putting the mattress on the floor, Don said that was too hard; and it certainly is, rather.

Underneath all this, great love, however.

The climate in this place changes so often it seems positively neurotic. Rain and sunshine every day. The pink blossom out everywhere and the vivid green leaves of spring. I rather hate the city at present, because of my cold. But this house is nice and we are right by the Heath; and up here there is little dirt and the air is clean.

Enough for a start. I just wanted to get the complaints off my chest. Already there's plenty else to report.

April 16. Mood much more cheerful—no doubt I'll be as variable as the weather here. My cold is over. I am getting a board to put under the bed tomorrow, so that both Don and I can be comfortable. Don has done some quite good work on the two portraits of me and is now doing a couple of self-portraits. We don't get out as much as I should like, but our life here seems very snug. I must start going to the gym again soon, or I'll lose what little I gained in California. (This typewriter continues to be a bastard.)

After a couple of talks with Sutro the *England Made Me* project is still up in the air. And of course I don't really want to do it. I don't really want to do the Berlin musical either. But that is not something to be admitted to yet; one most important consideration is, how much money is coming from M.'s estate. I shall see about this in a few days. Also, maybe on Tuesday, I'll get to hear what Edward thinks of my novel. That has number one priority. I must try to have it ready as soon as I can. Right now, I feel no interest in it whatsoever.

I had supper with Stephen the other night. The chief impression I got from him was weariness. He is only enthusiastic about Matthew;[1] hates England because it is so dark; wants to get a job in Athens; is sick of the eternal need to make money. But as always he had a couple of bitchy jokes which made me laugh. When he was on tour with Angus Wilson, Angus used to get him to speak

---

[1] Spender, his son.

first, then rose and said, "Of course, *we young writers* don't agree with Mr. Spender." Stephen thinks Chester [Kallman] is a bad influence on Wystan's work, has diverted him into the area of cleverness and private camp jokes.

Don seems more vulnerable, more nervous but also more alive here. He is like a burning fuse, almost; a fuse that is burning eagerly toward the point of explosion.

April 19. The weather is enchanting and quite as warm as Santa Monica on a mild day. Blossom blowing about in the street. I would really be about as happy as possible, if it weren't for the Cuban crisis.[1] I am also getting shooting pains down at the bottom of my spine. I wonder if this is because I have found a gym and it doesn't want to go to it? The gym is on Oxford Street, and one of its managers is a nice boy who worked for five years on Long Island as a tree surgeon. The place is too small and crowded—at any rate during the lunch hour, which was when I visited it—and there is a stinky red carpet on the floor which you feel has absorbed all the sweat and athlete's foot available; but it is probably the best to be found around this city.

Yesterday, I went down to see a Mr. Smith at a firm called Glyn Mills and Company on Lombard Street. He is in charge of the money from M.'s estate. It seems that Mr. Sidebotham wildly overestimated the amount that [my brother] Richard and I will get. He said the total capital was £20,000; Mr. Smith says it is £14,000! Still and all, I still think I may get about $15,000 after deductions, which is not to be sneezed at. And Mr. Smith gave me £75 right there on the spot; it was the interest which had accrued on my half-share since M.'s death.

Nothing more from Sutro. And now Tony Richardson is making frantically like he wants to do a film with me. This I'll believe when the contract has been signed and the money paid down. Tony's friend George Goetschius is said to be about to have a nervous breakdown and to need religion (from me, of course!); but Don and I suspect that this is largely Tony's mischief. He is fuller of it than ever, here. Right now he is filming *A Taste of Honey* in a yard in Chelsea and really in his element. I can see that this must be a dream life for him; and this kind of sloppy informal cheerful friendly bohemian movie work is certainly very appealing.

---

[1] Fifteen hundred U.S.-trained Cuban exiles landed at the Bay of Pigs on April 17, counting on a sympathetic uprising to overthrow Fidel Castro's revolutionary socialist regime; they were killed or captured by his troops.

The only trouble is, Tony is absolutely bent on upsetting the life pattern of everyone he comes in contact with, just for the sake of upsetting it. Right now, he is planning to take us off to the South of France or Italy at the end of May, so we can write the film there, in a week!

Richard is apt to come up to London the weekend after next, with his friend Alan [Bradley]. Well, that's all right.

Edward came up yesterday to see me. Sunburned from being in the Isle of Wight for his holiday, but very fat and I think not at all well. He complained of a razor cut on his face which has festered and he says that all his cuts and scratches do that. He was rather dazed by some antibiotics he had taken; and maybe this was partly why he seemed lukewarm and almost unwilling to discuss my novel. He has only reread "Mr. Lancaster" and now reverses the criticisms he made of it; criticisms which I agree with. Otherwise, he seemed concerned only with the tiniest details.

Still and all, these details are of value. For example, at the end of "Mr. Lancaster," he hadn't liked "most of the time, thank God, we suffer quite stupidly and unreflectingly, like the animals." (I'm quoting this from memory.) He found it sententious. And when I suggested "thank goodness" instead of "thank God" he said that changed the whole tone of the passage. And it does. And I feel convinced it's better that way. Again, he is bothered by the treatment of Hell and the doctrine of reincarnation in "Paul." As far as Hell is concerned, I'm sure I have only to make it clear that I don't mean the Christian Hell—and again he is right, because I don't. Well, we are to meet again soon, preferably after he has reread the rest of the book.

Next year, he plans to retire and start writing. But he does not dare admit this to himself, lest the writer's block shuts down on him again. He has to pretend he is just playing at writing. Really, I ought not to be so selfish and expect him to bother much about my novel. Because of course getting on with his own work must seem a desperately urgent matter, at his age, with so much ahead to do. I do sincerely feel this. And, in a way, I am actually more concerned to please Don than Edward. But Don, too, is desperate about *his* work and can't be expected to spend all his free hours concentrating on the novel. So the moral is, I'd better get on with it myself.

Miscellaneous: The youths of the city wore, and still wear, very long-pointed shoes called "winkle-pickers." But these are now going out of fashion in favor of blunt-toed shoes. But the fashion for long long hair seems constant; great fuzzy heads of curls. Sometimes from behind you can't tell a boy from a girl....

Overheard (or did I imagine it?) on one of the streets near here. A boy of maybe ten talking to his mother: "He was joking when he was six".... The utter fatalistic patience of everyone when a line has to be formed or a train or bus waited for. You feel the wartime mentality still very strongly here.... The smallness of Hampstead. The steep little brick streets. The tiny murderous purring cars and motorcycles, always ridden by learners with a big L,[1] it seems. They ride straight at you and no shit. You have to jump.... In general, life here seems tacky and lively and the people radiate a friendliness and willingness; all except for a shopkeeper or official type, which makes a face at you as if you had asked for the impossible and unspeakable.

From *Oscar Wilde and the Black Douglas* by the (present) Marquess of Queensberry: an extract from a letter written by Wilde to Lord Alfred Douglas in the summer of 1897 when he was at Berneval in Normandy after his release from prison—

"André Gide's book fails to fascinate me. The egoistic note is, of course, and always has been to me, the prime ultimate of note of modern art, but *to be an Egoist one must have an Ego*. It is not everyone who says 'I, I' who can enter the Kingdom of Art."[2]

April 28. It is just three weeks, practically to the hour, since I arrived in London. And, of course, it feels like three months. In many ways I am already utterly habituated—to the maddeningly faithless weather, to the endless taxi-riding (with drivers who quite often try to gyp you), to the starchy food, to the claustrophobic tube, to the general cheerful tackiness. There is much that is lovable here but thank God it is not my home. Never do I cease to give thanks I left it.

Don started again at the Slade, the day before yesterday. We have had several more or less frantic outbursts about his painting since my arrival, and there will be many more. I try to reassure him and above all to point out how masochistic most of his scruples are, and he sees this and blames himself and begs me to keep reminding him of this. I will, of course. I can even do it without much strain when I am feeling well. When I am not—when this place and the weather and the situation of not-being-in-California get me

[1] Referring to the red "L" on a white plate that British learner drivers must display.
[2] The 1949 book was by the 11th Marquess, Francis Archibald Kelhead Douglas (1896–1954), with Percy Colson, but he was no longer the present Marquess. Lord Alfred Douglas's father was the 9th Marquess. Gide's book was *Les Nourritures terrestres* (*Fruits of the Earth*, 1897).

down, then I feel it is too much and I am apt to be impatient and sulk. Oh, but we'll get through all right. I cling to japam. (One day, since I began, I have missed—and God knows why, just sheer *tamasik*[1] forgetfulness.) It is very significant that, the other day, Don asked me if I am still making japam for him, and I was able to answer yes. So it means something to him, too.

A great reassurance is that, following talks with Edward, I have made a really substantial beginning with the rewrite of "Mr. Lancaster"—and in many ways this was the biggest of the problems to be solved. I think I have seen how to open the series so that the whole book's purpose is announced and justified—at least, justified sufficiently; to justify too much would be fatal.

Nothing from Sutro about his film. Nothing from Tony Richardson about *his*. No other work news of any kind. Well, good.

Meanwhile, Cuba simmers, and de Gaulle is getting ready to shoot the rebel generals.[2] And poor dear little Gagarin seems already almost forgotten.[3] A lot certainly has been happening, these last three weeks!

Have seen lots of friends and lots of plays—but somehow I am not in the mood to write about any of that. Another day, perhaps—

May 3. Today I finished more or less the revisions on "Mr. Lancaster," and that's quite an achievement because it will probably have been the most difficult one to do. Otherwise I'm dull. It's this weather—how I loathe the greyness! I'll be glad to get out of this country and stop drinking so much and staying up so late. Perhaps I am just getting old, but I feel very little joy in any of these meetings. I go through the motions of being glad to see old friends, like Rupert Doone, Robert Medley, Freddy Ashton, and I am—in a way. But still, they are motions. I only really enjoy myself with someone like Jonathan Preston. Oh shit—why am I writing this? I'm just sulking because I've forced myself to write in this diary and I didn't want to. All right then, I won't.

May 15. I'm ashamed of the petulance of the last entry. Not that it's so important. So I am not having an utter ball—is that so terrible?

---

[1] Lazy. See Glossary under *guna*.
[2] Earlier in April, four retired French generals led an uprising against independence for Algeria. Of the military men captured, including other generals, some were imprisoned. Others were sentenced to death *in absentia* and later captured and imprisoned. None was executed.
[3] Soviet cosmonaut Yuri Gagarin orbited the earth on April 12, the first person in space.

I came here to be with Don and here I shall stay, as long as I'm useful and needed. In any case, I know I would be desperately unhappy inside a week if I were to go back to California without him. And I *am* happy, just being with him, most of the time, and when I'm not, it's really because he isn't around. Neither Jonathan nor anybody else is possible as a substitute.

As long as we're on this rather boring subject, let me just say this. I realize now, on this trip, that my longing to be away from England had really nothing to do with a mother complex or any other facile psychoanalytical explanation. No, here is something that stifles and confines me. I wish I could define it. Maybe the island is just too damned small. I feel unfree, cramped. I long for California. All right, you stupid old horse, so you long for California. Be thankful you can get back there sooner or later, and meanwhile, busy yourself, look pleasant, be pleasant, and make japam. (I have at least kept this up, including the japam for Don. Mostly, it seems utterly compulsive, and my memory plays me the bitchiest tricks, only reminding me to do it late in the afternoon or early evening. But I do do it—even sometimes going into the downstair toilet before going to bed, and making it while Don, having finished brushing his teeth, calls down "Dub-Dub?")

When I remember that "Dub-Dub," I suddenly realize what an idiot I am, to neglect, even for an instant, to value my luck and happiness.... Just imagine if anything happened to him! I suppose that's my trouble. I can't. It's as unthinkable, and as possible, as the H-bomb.

But I really opened this book today because there's something I have to discuss with myself—a problem connected with my novel. All this time I have been revising it, and yesterday I actually gave the rewritten version of "Mr. Lancaster" to the typist. Edward hasn't seen it, but Don read it and as Don disliked the first version—the only part of the book he *did* dislike—I was delighted when he liked the revision so much and felt I had met all his objections. I really believe it is one hundred percent better.

However, now we come to "Ambrose," and "Ambrose" opens up a basic question: what does Christopher learn about himself on the island, and why does he leave it?

Let's approach this from the beginning.

In "Mr. Lancaster," Christopher falls in love with Germany and resolves to go and live in Berlin.

He goes to Berlin, works the nightlife out of his system and becomes a political puritan.

At the beginning of "Ambrose," he has had a reaction from

this. He realizes, he says, that he has never been really involved in German political life, only an excited spectator. He is going to Greece for kicks—because it promises more excitement than going back to England.

All right, he goes to the island. The island isn't the right place for him. Why?

Because he's restless? Yes. He doesn't belong anywhere; that's what he finds out. He's on the same side as Ambrose and Geoffrey but he can't live their life.

Wystan objected to Christopher's reasons for leaving the island; the urge to go back and compete with Timmy North.[1] I certainly see that some of his lines are wrong—when, for example, he says he used to care about getting on and being someone.... But is this a wrong thought or merely a crude bit of phrasing?

May 18. Don's birthday. We are to meet later and go to see this review, *On the Fringe*—no, *Beyond the Fringe*[2]—which everyone says is so marvellous. And then later, maybe, go to see Garbo in *Marie Walewska*.[3]

I have been working all day revising "Waldemar." (I took "Ambrose" to Rashid Karapiet yesterday to be typed.) I alternate between excitement—feeling this is good—and feeling it's so sloppily written. There is something wrong with the ending of "Waldemar." I have to explain *why* Waldemar's leaving makes me feel I am free to go to America. Or, if I don't feel that, I mustn't say so.

A grey cold day. Grey feelings. So often, here, I get sad as a pig. Two days ago, Don had one of his outbursts; because he had been badly treated by John Gielgud. It was miserable. My feet ached. And we walked around and around; he was raging and bawling me out. Am I really a masochist, to put up with all this shit? Or what? I hardly know. I hardly know anything. I feel only partly alive; dull, old. I'm getting a kind of senile indifference. No—that's untrue. It's nothing new. Just a phase.

What do I want? Nothing. Just to be quiet and snug and to lie in the sun. But not to be fussed and made to go to Sicily to do it. Tony Richardson still hasn't gotten in touch with us.

---

[1] An imaginary Cambridge contemporary who is the author of a successful light novel.
[2] Peter Cook, Jonathan Miller, Alan Bennett, and Dudley Moore in the satirical revue they wrote for the 1960 Edinburgh Festival and which made them famous in London's West End and on Broadway.
[3] Released as *Conquest* (1937) in the U.S.

June 6. A nearly three weeks' lapse. And tomorrow it'll be two months since I arrived here.

Yesterday, I sent off one of the typescripts of *Down There on a Visit* to Curtis Brown in New York, and another to Dodie and Alec Beesley. The third has been with Alan White of Methuen over the weekend, and already I feel concerned because I haven't heard from him. Obviously he must have doubts and is showing it to someone else.

Well, I can't help that. It certainly has its faults. Parts of it—particularly "Paul"—are still sloppily written and I must tighten them up before they go to press. But I feel confident that the whole thing does add up to something, and that it has an authenticity of direct experience and is altogether superior to the slickness and know-how and inner falsity of *World in the Evening*. If people don't like it, I am sincerely sorry; but already I feel in my bones that I shall never repudiate it or have to apologize for it. So we'll see.

Today, we'll probably hear a very important decision: is Don to get a show of his drawings at the Redfern Galleries? If he is, that would of course be a quite considerable triumph—a justification in the world's eyes (and after all, let's face it, it is only the world that demands a justification) of his whole stay in England. But, if he has this show, it will probably be in October or at any rate sometime before Christmas, and Don will want me to be there with him; so the question will come up, shall we go back to California for a couple of months and then return, or shall we stick around here. And in either case I shall have to cancel my half promise to teach at Los Angeles State College next semester.

And now, having finished with the novel, at least for the time being, what am I going to do next?

Well, there's Ramakrishna. They'll be needing another chapter for the magazine before long and it would be good for me to get back to that. Good discipline. And a good irritant. Toiling at the Ramakrishna is just the way to irritate the Muse into providing me with a new idea for a novel or play. Of course, I do have a ready-made idea and a great deal of material for a short story contained in the hashish-taking episode which I cut out of *Down There on a Visit*. But I must wait until I'm absolutely certain that this *should* be cut out.[1] Cutting it out was entirely Don's idea, and, as of now, I do believe that he was absolutely right. The hashish scene deflects the interest from Paul to me at a time when we should be

---

[1] It was eventually published as "A Visit to Anselm Oakes" in *Exhumations* (1966).

centering down on Paul for the final scene. In this sense, it is a coyness, a flirtation with the reader, and therefore wrong.

June 7. Dr. Schwartz, the collector,[1] bought the manuscript of my novel from me at the Big-Boy cafeteria on Charing Cross Road yesterday afternoon. He appears to carry all his money in a bag, along with first editions. He produced fifty pounds and paid me, and then gave me an extra pound for autographing books and an extra ten shillings for taxi fare. When he talks to you on the phone, he always ends up by saying "chin-chin."

The taxi took me to a party given by Iris Tree and Diana Duff-Cooper[2] at the latter's studio. D.-C. (she should be called B.-C. . . .) is a rather rude woman. She spent the first part of the party making up for it and most of the last part making up and redressing herself for the next party. There was a sculptress, said by Iris to be an hermaphrodite, named Furi(?)[3] who appeared dressed like a male peasant in *Cavalleria Rusticana*[4] and announced that she had a hunger for life. Then we went on to an exhibition of paintings by an artistically underprivileged Australian and gradually got ourselves mixed up with a bunch of sleazy people who landed us in another of those eternal Italian restaurants. How I loathe this London regime of Italian food and sleepy red wine! By a great effort and thanks to Don's warnings, I didn't get drunk. But I always feel bad when I wake in the mornings here; and today I have a cough.

Wystan arrived this morning, and he and I went over to the Spenders' for lunch; he's staying there. It was good to see him. I felt a great stimulation. Wystan always brings you into the very midst of his life—so near, indeed, that it is out of focus, so to speak. He mutters about everything that's on his mind; feuds, unpaid bills, alterations to be made in proofs. Most of the time, you barely know what he's talking about.

He asked to buy one of Don's drawings of Stravinsky. But

---

[1] Jacob (Jake) Schwartz, a dentist living mostly in Paris, dealt in modern literary manuscripts from, roughly, the 1940s to the 1960s.

[2] Lady Diana Cooper (1892–1986), socialite and beauty, daughter of the 8th Duke and Duchess of Rutland, widow of politician and diplomat Alfred Duff Cooper, 1st Viscount Norwich (1890–1954); she was a longtime friend of Iris Tree.

[3] Maria Fiore de Henriquez (1921–2004), known as Fiore, born in Trieste, raised and educated in Italy, settled in London from 1949; she sculpted hundreds of portrait busts, including the Queen Mother, Stravinsky, Olivier, and John Kennedy.

[4] The opera by Pietro Mascagni.

Don was not charmed. He is sulking today. Is he a bit jealous of Wystan's friendship with me? Or cross because I went to lunch with Stephen, who is now private enemy number one?

June 8. Cross because I went to lunch with the Spenders. We had a long talk this morning, and I think it cleared the air. Maybe I never sufficiently realize how much I have intimidated him into feeling guilty, in the past. Also how terribly insecure he feels. Just because I in fact won't leave him, I have taken it for granted that he somehow knows this. He doesn't. And it's true—I have, subconsciously and even consciously, tried to make him feel guilty, again and again; I have a method of doing this which is positively hypnotic.

Of course, Don is also nervous at the moment because of our anxiety about the Redfern show. He called this morning, to find that Harry Miller hadn't taken the drawings in until yesterday and that the answer won't come till this evening. Suppose there's to be no show? We shall simply have to try to arrange one somewhere else.

To supper with Hester Chapman last night. A curiously old-world atmosphere of dusty brocade curtains and furniture. John Lehmann's sister, Helen, was there. She told a bitchy story about Jackie Kennedy's unsuccessful attempt to steal the Indonesian cook of the French ambassador in London. The whole point of this was that Jackie showed *American* brashness, crudity, stupidity, vulgarity—quite ignoring the fact that the Frenchwoman, who was much older and more experienced, showed infinitely greater vulgarity by calling in the *Daily Worker* and the gutter journalists to interview the cook. Also, she raised the cook's wages, thus tacitly admitting that he had been underpaid.

Walked on the Heath this afternoon and made japam and sat under a birch tree looking out over London. Such an "English" scene: far touches of pale gold light amidst the buildings (Turner), a steeple rising above swelling oaks (Constable), and a bright cumulus cloud against a dark thundery sky (Samuel Palmer). Some rain fell as I was walking home and I sheltered under a tree and was talked to by a weird little Filipino, accompanied by a big Dutch student. The Filipino asked me my profession, nationality, name, etc. and had heard of my books. We sort of flirted with each other; he was incredibly provocative. He is to be found, he told me, at the King William IV pub and an expresso bar called The Geisha. It was a very strange meeting, a sort of "recognition." I laughed wildly, as if with a familiar friend, and even found myself pretending to be

about to hit him when he made some joke about *Diane*.[1] Home
feeling unreasoningly elated.

Today I'm reading Alain-Fournier's *Le Grand Meaulnes*, the
libretto of Auden's opera *Elegy for Young Lovers*, and Jeremy
Kingston's novel, *The Prisoner I Keep*.[2] More about all of these.

June 9. Triumph! This morning, Loudon Sainthill phoned and
told us that Harry and his colleagues at the Redfern wanted Don
to bring around the photostats of his earlier drawings, as they had
almost decided to let him have the show. So he took them around,
and they did definitely decide. He thinks it was largely the fact
that he had gotten Stravinsky and Beaton to pose for him that did
the trick! He will only have one little narrow room—but it's the
Redfern, and the Redfern is THE TOP. And nothing that anyone
can say will alter that.

So Don is wild with joy. And I am spinning plans, how best to
exploit the victory and turn it into a rejoicing for our friends and
a rebuke to our enemies.

The show will be either in September or in February; we shall
be told in a few days. If September, I guess we shall stay on here
or go to Austria and stay with Wystan and Chester. As for Tony
Richardson, I fear we aren't going to work together; we just don't
seem to be able to come up with an idea we both like. *Le Grand
Meaulnes* is one of Tony's suggestions. That's why I'm reading it.
Too early to say anything yet.

Not a word from Methuen. And I haven't even heard that
Curtis Brown in New York has received the manuscript.

June 10. Grey, chill and windy today. I would like to go for a walk
on the Heath but can't; I must stay home because nice Mr. Daws
is coming to fix our front door, which keeps opening. The lock
doesn't catch.

A letter from a secretary at Curtis Brown's, New York, to say
the novel has reached them. *Still* nothing from Methuen. They
must certainly be doubtful about it and waiting until more of them
have read it.

Last night, I had supper with Eric Falk and his friend Bob
Jackson. It's strange how Eric radiates this feeling of goodness.
No—what I mean is that goodness is always strange; not that

---

[1] The 1956 Lana Turner flop for which Isherwood wrote the screenplay, as
he tells in *D.1*.
[2] Never published.

it's strange that Eric should be good. Rather to my surprise, he made a long speech about the persecution of the Jews and how he personally couldn't bear the thought of living in Germany or Austria now. He couldn't understand how Chester Kallman could live there. Bob Jackson told him not to be silly. They have a half-humorous bullying relationship. I got very drunk and came home later than I'd meant to. Don was out for the night. This morning he shows up and sings the blues because *I'm* going out tonight! I laugh at him and tell him he has a double standard, and then he laughs. And yet, absurdly enough, he really does. He doesn't see why I can't stay home by the fire with a good book. But I don't mind. I even find this rather endearing.

Interruption while Mr. Daws arrived and fixed the door. He only wanted to charge me half a crown! Made him take three shillings.[1]

Have finished Jeremy Kingston's novel. It is nicely written and has some good characters in it. And the love affair with the German boy is really touching. (I like it when they throw away his toothbrush so that they can both use the same one.) But at present it isn't a novel at all, merely a huge pile of building materials. One of Jeremy's faults is a kind of complacency about the mere fact that he is observing. He finds it fascinating just to think of himself observing things. But this fascination isn't, probably couldn't be, communicated to the reader.

Something I forgot to record. A sequel to the party on the 6th. After it, Iris Tree went out to dinner with the sculptress [Fiore], got drunk and told her, "You're in danger of becoming a type!"

Signs increasing of a big crisis over Berlin.[2] All I think is, I hope it won't interfere with Don's show.

Have finished the libretto of Wystan's opera. I think on the stage it should be marvellous.

As for *Le Grand Meaulnes*, it is all very nice but so far I'm not captivated; it is somehow too sweet.

Don said yesterday that he doesn't want to paint portraits. He feels drawing is the only way he can do that. If he goes on painting at all, he says, he might even become abstract!

I remember describing him to Johnnie [van Druten] and Carter,

[1] Before decimal money was introduced in the U.K. in 1971, there were twenty shillings in a pound and twelve pennies in a shilling. A half-crown was two and a half shillings or two shillings and sixpence.
[2] Premier Khrushchev and President Kennedy held talks in Vienna on June 3 and 4, reaching limited agreement on Laos but disagreeing on the future of Berlin. See Glossary under Berlin Crisis.

back in 1953, as "a sort of magic boy." I still feel that about him
now and then. Yesterday evening, for example, at Eric's, he absolutely sparkled like a diamond. He seemed a creature of another
kind, altogether.

June 14. Woke up this morning really black-depressed. I hate
this town. I feel caged in this country—even when we rode out
through the charming fields and woods on the train yesterday on
the way to see Forster at Coventry. I long to get back to California.
I hate the life we are leading here. Overeating and overdrinking.
Feeling stuffed and liverish all the time. It is all so desperate and
compulsive. Almost every night these parties. And Don usually
furious because of some insult and because he's rattled with exhaustion and feels he daren't let up for fear he decides that his life
is "useless."

Well, good. Now I've written all that, and I must consider
practical measures. At present Don needs me and I honestly don't
think he wants me to go back. So I shan't. That much is that.
Therefore it follows that I must change my life here as far as possible. I must not let Don's morning dawdlings in the bathroom and
at the phone hold me up. I must get on with my work right after
breakfast. Work means immediately the Ramakrishna book and
whatever comes up later. Also, I mustn't drink so much. Also, I
must exercise. Right here in this house, as far as it's possible.

We still don't know for absolutely sure about the Redfern date.
Harry Miller only *thinks* it will be in September. Otherwise, of
course, we should go back to California.

Morgan was angelically sweet and he really looked very well.
Only his movements have slowed down and his gait is shuffling.
Both Bob and May [Buckingham] seem elderly, too. (Bob's my
age.) Morgan seems chiefly interested in the censorship battle;
talked a lot about *Lady Chatterley* and the trial.[1] At lunch he said,
"I'll drink anything." Bob and May bored us nearly pissless talking
about the wonders of Coventry and all the new building which
is being done, and the concerts and plays which are being given.
It all represented an effort at self-reassurance and was therefore
touching; but they talked so much that Morgan couldn't get a
word in. And Don didn't get a chance to draw him.

Don furious because Mary Ure was rude to him last night, he

---

[1] Penguin published the first unexpurgated English edition of D.H.
Lawrence's *Lady Chatterley's Lover* in 1960, was prosecuted under the
Obscene Publications Act, and acquitted; Forster testified for the defense.

thought. (There was a rather ridiculous Negro from Venezuela there named Mitto Samson who has been a Satanist and seen live babies' throats cut and has written a book about it for Cape.[1] Maybe all his stories are true; but there's still something comic about him.) This morning, Don was penitent because he had sulked. And we always come back to the same thought: it's because I'm around and it bores him when people lick my ass. They do it nearly every evening, and I can honestly say it *even* bores *me*!

June 15. Don cheered up again yesterday evening, so so did I. We went to see a silly but not so bad play called *The Bird of Time*,[2] about Anglo-Indians, etc. in Kashmir; and then we went backstage and talked to old Gladys Cooper, who is really rather sweet. Don is going to draw her. She seemed genuinely pleased about this.

Don says he is miserable unless he can draw for three hours at the very least, every day. Any day on which he doesn't draw is a waste, and sheer hell. This morning he drew me, seven portraits in all. A couple of them among the best he's ever done. That meant I didn't accomplish anything much of my own, but that doesn't matter. According to my rules here, if I can help Don, let alone myself, that's a good day's work. Tonight he is going to stay in town. I'm having dinner with Jonathan.

It now seems definite that the Redfern show will be held sometime in September. So we must make plans accordingly. Maybe go with Tony Richardson and John Osborne to the South of France. But I still don't see any possible idea for a movie with Tony. Meanwhile, Jerry Wald will be in town before long; and Jim Geller thinks he is still interested in getting me to do *Ulysses*.

Then Chester Kallman will be coming and we finally approach the talks about the Berlin musical. I have a feeling these will end badly. Especially as I don't really like Chester, and as I feel the terms they are proposing are not fair to me: they want us to split three ways, while I feel that I should have something extra as the original author.

I haven't done one bit of work on the Ramakrishna book. I *have* done a few exercises, once. And I *have* kept up japam, almost completely—except that, the other day, I missed doing it for Don as well because of the lack of time.

[1] Possibly Arnim (Mitto) Sampson, a poet and specialist in Caribbean folklore, from Trinidad; evidently Jonathan Cape never published the book.
[2] By Peter Mayne (1908–1979).

Nothing from Methuen about the novel.

Heinz and Gerda and Christian [Neddermeyer] are coming here in a week's time. This is a big bore and nothing but a bore. I simply dread it. And, at the end of the month, I have got to go up and see Richard. Fate more or less ordains that, because I am to do this television show in Manchester, about the thirties, with Stephen and Cyril Connolly and, I think, Wystan.

A Frenchman(?) Swiss(?) named Miron Grindea has just been by. He edits a magazine called *Adam* and he wanted to know if I knew anybody who has £250 to help him publish Upward's novel in a double number.[1] As so often happens, I was furious when he insisted on seeing me and then liked him and was in no hurry for him to go. He arrived in a terrific flap because the street numbers here are not consecutive. This he called "criminal" and "tragic."

June 16. Hot and sweaty again. The papers full of the approaching Berlin crisis. My only feeling is, I don't want to leave Don while it's on. The only really unbearable thing would be to be separated from him.

This morning we went to a rehearsal of John Osborne's *Luther*. Albert Finney looked as if he is going to be really good. It's surprising how right he seems for the part. And cutting off his hair has vastly improved his appearance. He is still in some danger of his mannerisms, however. There were moments when he played the No-Neck Monster and the Hunchback of Notre Dame. But he is able to convey self-torture, constipation and fury.

June 17. At Dodie and Alec's cottage for the weekend.

The cottage looks as if it were right out of Disneyland, with clematis and Albertine roses climbing over the thatched roof and black and white fantail pigeons pecking around on the lawn. Lots of people who are driving by stop and frankly stare. But then the jets come screaming low over the fields from the nearby airfield, with their ridiculous bustling air of defending us, heel over just below the garden and head for their lunch in Africa.

Dodie looked like an old woman when I first set eyes on her on the station platform at Audley End; but then I stopped noticing and just remembered her average appearance. Alec seemed quite

---

[1] Grindea (1909–1995) was a Romanian critic educated in Bucharest where *Adam International Review* (for Arts, Drama, Architecture, Music) began publication in 1929. He took over in 1938 and brought the magazine to London in 1939. (Chapter 1 of Upward's novel, *In the Thirties*, appeared in *The London Magazine* in 1961.)

unchanged. He rattles on about country matters: the method of building cottages which is called wattle and daub, the way of plaiting hedgerows which is called cut and laid, the recent manufacture of metal scarecrows in Colchester[—]they are known by the old English name of maukins. He drove us by leaf-tunnel lanes so narrow that no one could have passed, to Lavenham, where there is a house built soon after Chaucer died. A German lady keeps a teashop and a bookshop there. We had tea and I found and bought two secondhand books by Maurice Sachs.

June 18. Violent indigestion-nightmares caused by all we had eaten—including the traditional fish cakes, of course. In one of the nightmares, I was attacked by thugs with switchblades, no doubt a memory of Jonathan's adventure in Glasgow.[1]

As we drove to Gosfield Hall, to see Phyllis Morris, Alec told how one of the thatched cottages caught fire, and the American firemen from the air base came to help, and one American asked him, "Why do you have grass on your roofs?" More twisting tunnel-lanes. One of them seems to dive right into the depths of the ground. This whole landscape heaves like the sea. It is so thickly wooded and twisty and secret that it seems far bigger than it actually is. You feel you are in the midst of an ocean of land. Partridges whirr up before the car. Ophelia-type brooks are everywhere, full of water lilies.

We suspect that Dodie maneuvered Phyllis into settling at Gosfield—because she, Dodie, thought how romantic it would be to live there and wanted to do so by proxy. Actually, it is rather a horrible old place, despite its "beauty" and "romance"; and all these crippled senile people accentuate its horror. Phyllis works hard at whistling in the dark to cheer herself, pointing out how "reasonable" the charges are and how nicely everything is arranged. Away in the distance on the lake, people were swimming and waterskiing. But what will it be like in winter?

In the evening, Don drew both Dodie and Alec. Dodie was quite imperiously concerned with her appearance. She knew what she *wanted* to look like, and that was that. Alec, on the contrary, said, "I don't care what you make me look like. You're the artist; I'm dirt," a remark which Don finds significant of their whole relationship.

We also discussed my novel. I'm still not sure how much either of them really liked it; but they raised very few points of criticism.

---

[1] Where, evidently, he was mugged, though this is not confirmed.

(All of these related to "Paul." "Ambrose" they seemed to like. "Waldemar" and "Mr. Lancaster" they ignored.) About "Paul," Dodie criticized the phrase "rosebud mouth." This I agree is not really a good descriptive touch and I think I'll change it to something like "well-formed" mouth. Then Dodie said that she thought ten thousand dollars was far too much for me to give Paul, because it suggested, *to the people here*, that I had been earning huge sums during the war. Now she may be right about the amount, but the idea that I should bow to British public opinion at this late date simply makes me angry. It can fuck itself.... Finally, Dodie thinks the ending of "Paul" is too abrupt. This may well be true.

June 19. We got back to town this morning. Don feels that the visit wasn't a success from his point of view. He was particularly shocked by a remark Dodie made while we were looking through old photographs yesterday evening, of the days at Malibu and Tower Road.[1] Don asked, "Did Bill Caskey take those?" and Dodie answered, "Yes, but Chris grouped them." Don went on talking about Dodie and Alec while we were eating lunch, and got annoyed with me because I was worried about a shopping list I had to make. He said, "You don't care how I'm feeling, as long as I show you affection." Now he has gone off to draw Vivien Leigh.

June 20. Right now, Don is drawing Wystan, who keeps talking to me as I write: Falstaff and Don Quixote are the only satisfactory saints in literature, etc. etc. Relations with Don are a shade better, but I think he would like me to go away for quite a bit of the time between now and his show, when he needs my moral support. It's the old story: he can't have any friends of his own as long as I'm around, because, even if he finds them, they take more interest in me as soon as we meet.

I am not going to comment on any of this, for the present. I shall try to write this diary like one of those French swine (Robbe-Grillet) who write a–literary novels, without psychology. I shall try to abstain from philosophizing and analysis, and stick to phenomena, things done and said, symptoms.

I forgot to record that, yesterday, a letter was waiting from Alan Collins of Curtis Brown, saying that he likes the novel and doesn't doubt that Simon and Schuster will publish it. So far, so good. Nothing from Methuen, however.

[1] Where the Beesleys lived in the 1940s; see Glossary.

Yesterday night, we had supper with Joe Ackerley. His flat, on top of the Star and Garter Hotel building in Putney, has a view right up the river. One night, a gigantic table was blown clear over the balustrade, crashed down into the street and made a hole in the sidewalk. It happened at night, otherwise several people might have been killed.

Joe grey and thin and so sad; his beloved dog Queenie has a cancer in her mouth and is very old and must die soon. Joe would like to go back to Japan. He has a low opinion of Bob Buckingham, whom he finds dull and stupid, but thinks May is a "great woman" who might under other circumstances have had a salon. Again, Don was offended. Joe didn't pay him the right kind of attention.

June 21. Reading Joe's *We Think the World of You* again, so as to write a blurb for his American publishers. It is a truly extraordinary book, not willfully fantastic but out of the ordinary simply because of the way it's felt and observed. It is also unshockingly frank, because Joe doesn't know how not to tell the truth.

Terry Jenkins, back from the States, called me. We had tea together and walked all the way from Gloucester Road to Sloane Square. I felt very much at ease with him. Terry enjoyed sitting in the square watching the people; he said he'd never done this before. He seemed unwilling to leave me, and I felt he was feeling terribly lonely for California.

Supper with Don and Tony Richardson, with whom we are making all kinds of complicated arrangements, to go to Nottingham with him to see *Luther* and to go with him to the South of France in August, where we shall share a villa with John Osborne who is writing a play about homosexuality which he thinks is his best.[1] Meanwhile, Heinz arrives, there are Wystan and Chester to be talked with about the musical, and today I'm having a drink with Truman Capote, and just this moment a note from Aldous, to say he's in London!

June 25. Yesterday we got back from spending a night with Cecil Beaton in the country. Very hot humid weather. It was nice when Cecil and Don painted and drew together in his studio, a neighbor's little girl. And I enjoyed seeing Desmond Shawe-Taylor and Raymond Mortimer again. We also visited mad Stephen Tennant

---

[1] *A Patriot for Me*, refused a license in England because of the subject and staged privately in 1965; it ran in New York in 1969.

in his house, which he has painted in various shades of pink and decorated by scattering books, clothes, bracelets and rings all over the furniture and the floor, partly like arrangements for still-life painting, partly like drunken unpacking.

Also, during our visit, Truman Capote, with whom I left my novel [on] Friday, delighted me by calling to say that it is the best thing I have ever written!

The Berlin crisis announces itself, far far ahead, in the late autumn. Things are not good with Wystan and Chester and myself. It's no use, I don't like Chester, and he infuriates me by doing crossword puzzles while we discuss the musical. He is no use. And I don't think we have a story. And I don't really want to do it, anyhow. Maybe I don't want to do anything for a while. I want to rest and daydream and think.

I do *not* want to go to Glyndebourne to hear their opera,[1] nor do I want to go to Weidenfeld's party for Cecil Beaton.[2] For both these functions we have to wear tuxedos. So we are going to buy them.

Don very sweet, these last three days. How perfect if it were always like this. And, whatever he might think, I should *never* get tired of it. I have had enough fusses for this lifetime *and* the next.

July 4. Once more, a lot of catching up to do.

On the 29th, in the midst of a ghastly heat wave which has just let up, I went up to Manchester to take part in a T.V. show on Granada, with Stephen, Wystan, Cyril Connolly, Arnold Wesker, and three young critics, Alvarez,[3] Hugh Thomas,[4] John Mander.[5] It was a mistake having critics, because we were separated from them doubly, by age and avocation. The general impression was that we old things were too frivolous and they were too serious. However, later in the evening, Malcolm Muggeridge[6] kept buying

---

[1] I.e., *Elegy for Young Lovers.*

[2] Weidenfeld & Nicolson was publishing Beaton's *The Wandering Years, Diaries: 1922–1939.*

[3] Al Alvarez (b. 1929), English poet, critic, novelist, educated at Oxford; he was then (1956–1966) poetry editor of the *Observer*, and he was about to publish *The New Poetry* (1962), an anthology.

[4] Historian and, later, Conservative peer (b. 1931); his many books include *The Spanish Civil War* (1961).

[5] Among his books on German and English culture and politics were translations of Lukács and Zuckmayer and *The Writer and Commitment* (1961).

[6] English newspaper journalist, author, radio and television broadcaster (1903–1990); he contributed to "Panorama," 1953–1960, and then hosted "Appointments with Malcolm Muggeridge" in 1960 and 1961.

us drinks in the Midland Hotel lounge and Alvarez made some girl and Thomas remained obstinately prune faced and I got plastered. Next day I joined Don in Nottingham and we saw Osborne's *Luther*, which begins well, with a weak ending, and a really good performance by Finney.

I'm not happy. I'm depressed, deeply. I hate this town and its climate. Relations with Don are bad much of the time; he resents my presence here and really would like me to leave, except that he knows I'll be useful when he has this show. He is only happy when he is painting or drawing—he is now trying a most interesting technique of painting in black and white. I'm not exactly sorry for myself; weary of myself, rather. Hemingway is dead; he probably did it deliberately, suddenly sick of it all, including his legend.[1] No wonder. I understand senile melancholia now. But I shan't give way to it, I think. I shan't abandon Don, though I may go back to California and wait for him. I know my path, whenever I think of it: I ought to get on with the Ramakrishna book and do my humble daily tasks. I'm really middle-aged now, and slow. I hate being rushed. I would rather be alone in the house, as I am right now at this moment, just slowly doing chores, check-signing, writing letters. The BBC called to know if I could say anything about Hemingway. No, I said.

Have seen Heinz and Gerda. Heinz impresses me quite a lot. I guess he told me all these war stories before, but I didn't take them in properly. He and Gerda went without food for eight days! And he stole beets from the fields. He made windows for the flat out of bits of glass stuck together. It is quite a saga. Heinz has a very deep voice, and he seems to have grown enormous—not merely fatter, taller and broader. He bosses Gerda around, but they seem happy together in a grumpy way. They talk mostly of saving money; how one can economize. Heinz says Gerda taught him to be thrifty. And apparently Christian is even thriftier than they are. They are much more like peasants than middle-class people. But Christian already has a car and they have an icebox and other modern contraptions. They describe Forster's flat in Chiswick, which he has lent them, as "primitive"!

One result of talking to them is that I feel I must rewrite the final meeting with Waldemar in postwar Berlin. I can make it much richer. Also—and this is really a consideration—I feel the present version would hurt their feelings unnecessarily.

---

[1] Ernest Hemingway (1899–1961) had depression and underwent electric shock treatment before shooting himself on July 2.

Berlin, Berlin, Berlin—the papers go on about that every day, bringing the crisis slowly and lovingly to the boil. They are so glad that it has to be done slowly. They can sniff at it and let it really simmer.

After my resolutions, I've been belly-aching again. I am a mess, a querulous old man. Well, I've fucking well got to stop. *No* excuses are valid for this kind of nonsense.

It is one whole month since I sent off the copies of my novel to Methuen and Simon and Schuster. No word from either of them. But, much more importantly, I have failed to do any work during this whole time. Have merely made a few notes for the next Ramakrishna chapter. Cecil Beaton is reading the novel now.

I hear from Iris that Ivan Moffat's engagement to Kate Smith still isn't official. And now Donna [O'Neill] has gone back to California and will presumably try to foul it up!

Delays over the payment of M.'s legacy, it appears. Something to do with my father's estate.

At present, it seems as if our Berlin musical project is certainly going to be dropped. The last suggestions made by Wystan and Chester simply aren't workable—that is, if I understood him correctly: they want it to be chiefly about an Englishman who comes to Berlin to find his boyfriend! The truth is, I suspect, that Wystan doesn't really want to work on the project at all. Well, I am better out of it. Tony Richardson is disappointed. He dislikes Chester, because Chester was too casual in his criticisms of *Luther*. Just dismissed it, according to Tony, with a few campy jokes.

Aldous, whom I saw on the 28th, was bitter about the misrepresentations in the Los Angeles press, reprinted by *Time Magazine*. Far from weeping or trying to rush into the flames, he and Laura, when they saw that the house was definitely [on] fire, simply turned their car around and drove away![1] The fire trucks didn't arrive until much too late. When they did come, the T.V. trucks had already been on the scene thirty-five minutes!

While I was in Manchester, I went into a chemist's with Stephen. He wanted a certain brand of mouthwash. And the young lady behind the counter asked, "Is that for cleaning false teeth?" Never, never would you hear that in London, much less Los Angeles.

Every day I manage to tell my beads. Usually at the last moment, in the toilet before going to bed, and once, in Nottingham, actually under the bedclothes in bed. I do it without feeling, compulsively,

[1] Their house in the Hollywood Hills was destroyed by a brush fire on May 12; see Glossary.

and yet I know that this represents the only thing I have that stands between me and despair. Trust to nothing else—*ever*.

Evasive behavior by Richard. He doesn't really want me to visit him at Wyberslegh, but he can't and won't admit this. His letters start off about how welcome I am. But then he says that he himself may not be present. And then that, anyhow, I can always come up if I want to take any of M.'s things—this is a bit of masochistic bitchery which recurs. Then finally there is a generalized outburst of resentment against the people who have made him play the "servant-host." He doesn't have to do this any more, he says, and he won't. He refuses to. He doesn't want to receive any more people at Wyberslegh; memories of the past are too painful. But he *does* want to see me—etc. etc.

July 9. Just back from a walk on the Heath. I should do this far more often, in fact every day. And now it is high summer, with the trees all shades of green, and their leaves ruffling and fluttering and waving and dipping with that extraordinarily complex movement which makes up the effect of a leaf-ocean which heaves and subsides. How good it is to be alone and relatively calm and still! I made japam, and then I sat down on a seat inscribed "Elsie, 1890–1958" and prayed for us both. People all around, dogs, fishing rods, bicycles, clouds.

I had a nice calm day two days ago, when I took Heinz and Gerda Neddermeyer on the river steamer up to Richmond. I wouldn't mind doing this quite often, but not with the Neddermeyers because they talk so much. I do like them, though; her too, which is surprising, I suppose. Last night I had dinner with them and Wystan and Chester. That was an effort too, until we got drunk and sang a little. (Incidentally, we seem to have abandoned the Berlin stories altogether as a possible plot for our musical and our now toying with a story about a woman who keeps a bar!)

I'm glad Heinz and Gerda came over, because seeing them again has made me take a much warmer view of the last episode in my novel; as it stands, it is cold and bitchy. The day before yesterday, I heard from Simon and Schuster that they are "really wild" about it but that they find "Paul" the weakest of the episodes. Still, they will publish it anyhow. And then yesterday John Barber[1] called and left a message that Methuen's are also very enthusiastic about it and

---

[1] British drama critic (1912–2005), for the *Daily Express* in the 1950s and for *The Daily Telegraph* from 1969 to 1986; in between, he was a literary agent at Curtis Brown.

will be sending me a letter. But why didn't they write before?

I keep catching myself humming; a senile habit.

Today I worked through a huge batch of letter answering.

How I love George Moore! It's the calm he projects. Am now reading *The Untilled Field*. Here's the last sentence of the story called "Home Sickness":

> The bar-room was forgotten and all that concerned it, and the things he saw most clearly were the green hillside, and the bog lake and the rushes about it, and the greater lake in the distance, and behind it the blue line of wandering hills.

How does Moore do it? Chiefly, as far as I can see, by his repetition of the word "and."

At supper last night, Wystan reminded me how, at the Manchester T.V. broadcast, the three critic-boys, Mander, Thomas and Alvarez, had all agreed that they never read any books just for fun!

July 10. Down to Liverpool Street this morning to see Heinz and Gerda off on the boat train. But they weren't there. I still don't know what happened to them.

Then to see Alan White at Methuen about my novel. They don't want any cuts or alterations at all. But Alan confessed that, much as he liked it, he liked *World in the Evening* better!

Last night we took Ken and Elaine Tynan to supper, and as Arthur Kopit[1] was there we had to take him too. This cost nearly ten pounds at a rude bad Hindu restaurant—or was it Persian?—called El Cubano. Didn't like Kopit, who bitched Andrew Ray and his marriage, etc. etc.

We got quite drunk in order not to be embarrassed while we discussed Elaine's terrible play, which she gave us to read.[2] Elaine herself wasn't embarrassed in the least. Then Ken told us that Jonathan Miller of *Beyond the Fringe* had told him about a café on the north circular road called The Ace, which is a motorcyclists' hangout. Miller claimed that there is a group there which gets orgasms from riding at top speed. You put on a black mask which leaves only slits for your eyes, tear off as fast as you can, have your orgasm and then throw yourself off your bike and get killed. You don't *have* to kill yourself, so long as you have the orgasm. But most people do, sooner or later.

---

[1] Harvard-educated American playwright (b. 1938); his black comedy *Oh, Dad, Poor Dad, Mamma's Hung You in the Closet and I'm Feelin' So Sad* was staged in London in 1961.
[2] *My Place.*

So Ken took us all up there in a taxi, costing fortunes, and of course it was deadly respectable and the only crazy people around were us.

July 11. This morning I have been putting in the corrections in the typescript of the novel which Methuen's had and which I brought back with me after seeing Alan White yesterday. It is loosely written in parts, but I feel that that's almost a virtue in this kind of autobiographical fiction. I still think it's *fun*, and much less fakey than anything else I've written.

Yesterday afternoon, Don drew Chester, bags under the eyes and all. He is curiously vain. Wystan fussed because they were getting late for a party at Stephen's—to which Don refused to go. He kept walking around, reading aloud passages from an account of a murder trial.

Later, Don and I had supper and went to see *One-Eyed Jacks*. Brando was mediocre and altogether it was a nothing picture. But some shots of the great Pacific surf smashing in over the beach made us both terribly homesick.

July 12. Chester and Wystan have been here and we've had another of these futile talks about the musical. I simply do not see one. As soon as we begin to probe, out comes the sawdust. It is a sheer waste of time talking about it. And it will be a sheer waste of time going to Glyndebourne tomorrow to see this tiresome opera. Not to mention the money spent on the tuxedos!

But what concerns me much more is that, reading through the typescript of "Paul," I'm disgusted by the passage describing how Chris gets involved with Augustus Parr. It is so sloppy. I think I must rewrite it altogether, taking out the smugness and sick weepiness. I shall have to get after this right away.

Last night a rather horrid little man named Bill Bridger, who is in the agency which employs Terry Jenkins, took us and Terry out to dinner. I don't quite know what he wanted—was it just to show off his expense account?

I was saying to Wystan how boring I found Sade's *120 Days of Sodom*, and he told me a most significant thing about it—how, when they first move into the house where the orgies will be held, and draw up their rules, one is that you are absolutely forbidden to laugh!

July 14. Yesterday morning, we went to Moss Bros. to get ourselves rented dinner jackets for Glyndebourne. Mine was a lot too large and gave me a sort of Empire waist, it was braced up so high. Walking through a department store later, on the way to the barber's, I suddenly came upon a schoolboy memory of how it *felt* to be going to get your hair cut, the connection with holidays and the excitements of shopping and the theater. London is full of memories like this, for me. They hang around in odd corners, just as smells do.

Don had some of his fashion drawing to do in the morning. This is the season of the "collections." We met Morgan for lunch at the Royal [College] of Art.[1] Morgan seemed tired. He only brightens when talking about love-lust; it is characteristic of him that the two are inseparably connected in his mind. This is really what differentiates him from any dirty old man. We talked of John Minton's picture of the death of Nelson, after Maclise, which hangs in the dining room.[2] The dead sailors are indeed sexy, and there is a beautiful glimpse of blue sky between tattered sails and keeling masts in the background. But none of the bodies seemed to me to be really properly lying on the deck. Morgan told how Gerda Neddermeyer had run out of the room at Cambridge, because she had been left alone there and had thought someone was coming. Morgan took this to be timidity due to her war experiences; but I think she was simply afraid she would have to talk English to them ... Morgan recalled how Heinz said at Ostend, "I have no objection to a plain breakfast, provided it is plentiful."[3] We talked about Heinz's relation to Gerda. The good-humored bossing on both sides. Gerda has cured him of smoking, drinking and gambling. But he counters by being the old-fashioned husband in the home. He won't help to prepare meals. "Let her do it; she's a woman." This is what I want to get into my revised version of the scene in "Paul."... Curiously enough, I felt embarrassed with Morgan after Don had left. We seemed to have little to say. He had disliked Tony Richardson's chauffeur who, he said, had made a sneering face when Morgan and I had kissed goodbye outside King's, the day we visited him on the way back from Nottingham. I also gathered that he doesn't like Tony much and resents Tony

[1] Isherwood wrote "Royal School of Art."
[2] Minton (1917–1957), who appears in *D.1* and *Lost Years*, taught painting at the RCA; "The Death of Nelson" (1952), after the 1864 wall painting of the same title by Daniel Maclise (1806–1870) in the Houses of Parliament, belongs to the RCA, but no longer hangs in the dining room.
[3] Probably in 1936, when Isherwood and Heinz settled in Ostend for a month; Forster visited them in September, crossing the Channel from Dover where he was staying with his mother.

having been offered the job of directing *A Passage to India*. It seems Morgan has some other candidate.

Even if *Elegy for Young Lovers* had been a masterpiece, it couldn't have made much impression on us last night. The getting to Glyndebourne, the getting a table to eat at, the bad seats (because we had to give up our good ones to Emlyn and Molly Williams whom we'd invited), the saying the wrong things to the right people and vice versa, and the getting home again; all this took up all my energy and attention. It seemed positively Mortmereish to be driving through summer afternoon countryside in dinner jackets. Don was furious because Natasha Spender kissed him and Stephen patted his back and called him Donny, without realizing how he feels about them. I think I offended Wystan and Chester by not gushing. Emlyn got a crick in his neck and drank most of the scotch out of a flask I had luckily brought with me. The set was the best part of the show; a nineteenth-century German Alpine engraving beautifully rendered, with spooky white Alps in the background and a carved wooden interior. As far as I was concerned, Hans Henze's music reminded me of pangs of arthritis, sudden and sharp and unpredictable.

Today, I've started working on the revision of the two parts in "Paul."

The war news is getting very bad. I'm only glad that I shall have the manuscript revised and with the printers before the real crisis starts. Because, then, I know I shan't feel like working at all.

Later: Stephen just rang up to ask me and Don to supper tomorrow night or lunch on Sunday. Got out of both invitations. No questions about Don's attitude. In fact, Stephen had actually called to say how extraordinary it was that neither he nor I nor Cyril Connolly (with whom Wystan and Chester had lunch yesterday) had been invited to the big party after the opera.

Stephen went on to say that he hadn't really liked the opera. "This kind of writing is the place where Wystan and Chester meet. I feel it's part of their contract with each other." He feels that Wystan isn't really interested in any of this material. "The young lovers" love is just what we call here in England "a tax-posture." Then there was a bit of flattery. "I was saying to Natasha at midnight" (why midnight?) "at least, whatever Christopher writes, it's always something he cares about."

July 15. Wystan rang up this morning from the airport to say goodbye. Luckily I'd called him yesterday at the Euston Hotel, so I wasn't in the wrong. But there was a feeling of haste and constraint and I don't think this was at all a satisfactory ending. Now he will be in Austria, and the obligation is on us to come there or not come there, as we choose. Nothing was said about the musical project; I guess it has been tacitly abandoned. Nothing was said about Glyndebourne either. I think I'd better write him a letter.

A dubious day with probably some rain later; but Don has gone off to Brighton to get some sun and sea air. He wanted me to come but I said I must stay here and get on with the revision of the novel. Also, I have to see Olive Mangeot later.

Worried about a cold sore on my lip which won't heal. There's no one here I know of or trust that I can show it to.

Maurice Richardson[1] called to ask if he might get me together with Gerald Hamilton for a talk on "Mr. Norris." He runs the "Londoner's Diary" in the *New Statesman*; and if poor old Gerald is mentioned in it, he will get five pounds.

July 17. I keep forgetting to write how really dismal and depressing this period is. The Berlin crisis never ceases for one moment to advance, and yet it is so slow; as one newspaper said, its pace is "stately." Neither Don nor I are happy. So we drink too much every night and eat over-rich food at restaurants and are strained and nervous. And yet I suppose we seem lively enough to most of the people who meet us.

At least, for the moment, I have some work to keep me absorbed—the final revision of my novel. But that should be finished very soon. I have only the last scene with Waldemar in Berlin left to write. When it's done, I shall be at a loose end.

Yesterday, we went with the Tynans to see Mort Sahl on BBC television. He didn't import very well. You felt he was firing in too many directions at once; anxious to hit all the targets everywhere. Then we went to dinner with them and Joan Littlewood,[2] who is a pretty bogus down-to-earther. She likes writers who whore and fuck and bugger, and aren't intellectuals. She thinks Shakespeare should be shelved for a hundred years, in favor of Ben Jonson. She

[1] British journalist, critic, surrealist novelist, and mystery and humor anthologist (1907–1978).
[2] Leftist English theater director (1914–2002); a founder of the Theatre Workshop at the Theatre Royal in London's East End, where she was the first to produce Shelagh Delaney's *A Taste of Honey* in 1958. Her productions of Brendan Behan, Frank Norman, and the musical *Oh, What a Lovely War!* (1963) transferred to the West End in the 1950s and 1960s.

urged Elaine Tynan to get out and see something of England and talk to real people. I objected that all people are real, *if* you can talk to them. She claimed me for an ally because I was born near Derbyshire, her favorite county, where they practise necrophilia more than anywhere else in the world.

The day before yesterday, we had supper at 21 Cresswell Place,[1] which is amazingly unchanged in its outward appearance. André was away. Olive, down from Cheltenham, was staying there. Sylvain [Mangeot] still lives there—they now have two bathrooms and are very comfortable. Edward [Upward] and Jean Cockburn[2] were invited too.

Olive is quite the old lady, in black; but she still exhibits her legs, which are still good. Sylvain is absolutely middle-aged, slow-spoken and sententious but well-informed and intelligent and interesting. Jean really looks very good. She has the appearance of a Hollywood star who has been made up for the twenty-years-after sequence; her oldness isn't very convincing. Olive and Jean are both family-obsessed. Olive talks chiefly about Hilda [Hauser], Phylly, Amber, Amber's husband, a "smashing" blond English policeman, and their daughter, who shows no trace whatever of Negro blood.[3] Jean talks of her daughter Sarah, who has one degree in classics and is about to get another in law, and who is really quite beautiful and at the moment plans to marry a rich feckless Jewish boy of whom Jean violently disapproves. Edward and Hilda [Upward] have just heard that she will be able to get this disablement pension,[4] and so they can both retire at the end of the year and Edward can get on with his next novel.

July 20. Today, while giving our bedroom a cleanout, I found the gold tie clip I bought years ago in New Orleans. I lost it quite soon after I arrived here and I had sadly given up all hope of seeing it again. This strikes me as a good omen. I like to think it is.

Don has gone off to Manchester by plane today, to draw Albert Finney. Then, I hope, he'll find out whether Tony really means us to stay with him in the South of France. Also, we have to get definite word about Don's exhibition at the Redfern, because we must renew our permits to stay in England. So a lot may be decided this weekend.

---

[1] André and Olive Mangeot's London house, in Chelsea.
[2] I.e., Jean Ross; see Glossary.
[3] Hilda Hauser was Olive's cook. Hilda's daughter, Phyllis, was raped by a black G.I. and gave birth to Amber. See Glossary under Olive Mangeot.
[4] She suffered continuing pain following a hysterectomy, but later recovered.

The day before yesterday, I finished the revisions on my novel, and yesterday I took the typescript back to Methuen's and sent a copy of the corrections to Simon and Schuster. I'm really quite pleased with them; I know I have cut a whole lot of sloppy stuff out.

Now, forget the whole thing, and get on with Ramakrishna.

Have seen Alan Bennett of *Beyond the Fringe*. We went to the show again and took him out to dinner afterwards. He is quite shy, with a non-U accent, and he doesn't want to be an actor. His chief preoccupation is historical research; he's preparing a book on the economic life at the time of Richard II. He was much hurt by Mort Sahl's ungenerous remarks about them. Sahl told an interviewer that they weren't funny, didn't have any positive point of view, didn't attack anything important, etc.

Also saw Gerald Hamilton, who is as spry as ever; always referring to himself as "poor Gerald." Also Cecil Day-Lewis, who is writing a Nicholas Blake play. Also Amiya, who described George's increasingly gaga behavior, his passes at little girls, his lapses of memory, his exhibitionistic masturbation.

Iris Tree now tells us that Ivan is definitely going to marry Kate Smith; but that they are waiting a while, to let Donna get used to the idea. Told Iris what I think of Donna. But neither she nor Don agree with me about this.

July 22. A day of stolid misery. Don is furious with me. He won't admit why, but it's because of my not coming back home the night he returned from Manchester. He won't discuss it, so I can't tell him I simply fell asleep. And anyhow that probably wouldn't make things any better.... It's no use saying all this is babyish nonsense, that I ought to rise above it and think of serious things, like Ramakrishna or at least the Berlin crisis. The truth is that when we quarrel like this, and I think that Don might leave me, I feel the most utter forlorn misery. My life wouldn't make any sense without him.

Reading the last two stories in *The Untilled Field* has depressed me, too; they are so bitterly sad.

Don said this morning that he feels he is losing all interest in other people; and he said that he now understands his mother and Ted much better. I asked him if he meant he thought he might go crazy. Oh no, he said, his danger would be too much self-control, not too little. When he is in this mood, he is absolutely *black*; there is no other way of describing it. You see in him such a terrible will to despair. It infects me. I sit watching him, all twisted up, feeling deathly sick.

But now I'm going to pull myself together and later I'll do some work on Ramakrishna, and eat at home tonight and try to get some sanity and balance back into me. When I ask myself, who would I like to talk to about all this, I find there is no one. That's good, because it means I have never discussed or shared my problems with Don with anyone.

July 23. After writing the above, I went to have tea with Paul Taylor, the boy with T.B. I once visited in hospital after he sent me a fan letter.[1] Was really shocked by his appearance when he opened the door. Sunken eyes and a blackness around them. He looked much worse than when I saw him in hospital.

He and his friend, who's an architect named Ian Grant, have taken a house on Ladbroke Square and redecorated it.[2] The staircas[e] wallpaper is purple. The dining room is black and orange. The drawing room is gold. They have pasted the walls with pictures of all kinds, portraits mostly. They are making a carpet, each panel different, with Victorian rose patterns. Ian's bedroom is Biedermeier, with a fur rug on the bed. Paul said, "It's so wonderful that we've got this place while we're still young. We'll have so many years to enjoy it." I felt like I wanted to cross myself, superstitiously.

Went back home, fixed fish cakes and frozen beans and then read all of Allan Monkhouse's [play] *First Blood*, the rest of Arthur Calder-Marshall's book for boys about Jack London, *Lone Wolf*, and quite a lot of Ainsworth's *Old St. Paul's*, which I got in a bookshop on Flask Walk. No work on Ramakrishna.

This morning, Don returned early, after staying out, and was quite happy again and we lay on the roof and sunbathed and he read me Elaine Tynan's short story in *The Queen*.[3] Tonight we are to hear her play read aloud. And tomorrow I go up to Wyberslegh till Wednesday to see Richard. . . . No time to comment on any of this as I have to rush out and join Don.

July 27. Just back from lunch with Stephen, William Plomer and Matthew Spender. Stephen incredibly bitchy as usual, claiming that the Redfern Gallery people are so dishonest that they tear prints out of books and sell them, and that he bought a Graham

---

[1] In February 1956; Isherwood tells about the visit in *D.1*.
[2] 41 Ladbroke Square; Grant (d. 1998), was an interior designer and collector and edited *Great Interiors* (1967).
[3] As Elaine Dundy, she published "City Manners" in *Queen*, Oct. 12, 1960, and "*Dr. Strangelove* and Mr. Kubrick," Dec. 4, 1960. No short story has been traced.

Sutherland from them incredibly cheap—their only stipulation being that he should take it away at once. He says he realized it must be stolen.... It's really a miracle he doesn't get sued much more often for libel. (In fact, I'm not sure he ever has been.)

Yesterday, I got back from Wyberslegh. I went up there on the 24th, as planned.

On the whole, the visit was a great success, I think. I found Richard and the house itself much more cheerful. The married couple, the Vinces,[1] keep the house cleaner, of course; though not as clean as I'd have expected. There are still cobwebs everywhere, and the books are grimy with the dirt of twenty years. But, aside from all this, there is a sense of release from tensions and the dead weight of M.'s age and illness. I slept in Nanny's old room, a little room overlooking the farmyard, and it seemed extraordinarily cheerful, joyful, almost, with a sort of childhood joy when I woke in the morning and heard the wind seething in the beeches, and looked out and saw the line of the moors, which is so different in different lights but has a sense of always-thereness like the sea.

Altogether, I felt a very strong nostalgic thing about the country. Jack Smith, the little farmhand of the Cooper period and hero of the saving of the farm from Nazi incendiary bombs,[2] drove us all around the Peak,[3] the day before yesterday; and it was so wild and wonderful, despite the tourists. I feel I want to buy an ordnance map[4] and pore over its place-names, as I used to when I was in my teens. (I found my old guide to the Peak, in a shelf, all white with mildew.)

Richard has lately become quite thick with Jack again, and goes down to his bungalow in Poynton, by way of a very "rats" road that passes over a level crossing of a railway in the middle of a wood.[5] Jack associates almost entirely with teenage boys—because, says Mr. Vince sneeringly, he's immature. One of them is a sixteen-year-old boy named Roger who has a huge mass of curly red hair and is very tall and the illegitimate son of an American G.I. in

---

[1] Hired by Kathleen Isherwood after her first stroke to keep house for her and Richard.
[2] See Glossary under Wyberslegh Hall.
[3] The Peak, Derbyshire (roughly between Buxton, Castleton, Matlock, and Ashbourne), now generally called the Peak National Park (since 1951) or the Peak District.
[4] British government survey map, highly detailed.
[5] Before they named it Mortmere, Isherwood and Upward called their fantasy world "the rats' hostel"; "rats," used as an adjective, meant spooky, weird, surreal.

the war. Jack calls him "matey" and they do gardening jobs for Richard together. Jack pronounces Derbyshire in the American fashion. He'll say, "Will you have another drink ..." of tea, meaning *a cup of.* He uses "Master" instead of "Mister"; and this not only in speaking of what he considers "gentry"—for example, he'll say "Master Vince."

It's curious to see how this environment affects Richard himself. He'll use some of these expressions, but nevertheless he is strongly class-conscious. He still speaks of people being "familiar." He feels that the Vinces are presuming; though often this merely means that they are bossing him—straightening his tie, urging him not to take so much beer or laxative—the former makes him drunk, the latter makes him leak shit into his pants and through them onto chairs. Mr. Vince *is* a bit smart-alecky and superior, born for better things, he intimates, than Mrs. Vince and his dull job in a Stockport store; he would "like to write." (How coyly and graciously people say this!) Mrs. Vince is nicer; big bottomed, outspoken, Yorkshire. He's from the South.

One day, Richard had two silver candlesticks sold, on the lawyer's advice. Later, he came into the house and overheard Mrs. Vince discussing the sale on the telephone, and saying how much he'd got for the candlesticks and how he was going to share the money with me. Richard was furious. Whom was she talking to? And how dared she tell *anyone* such a thing? He wrote in his diary (which he gave me to read): "Someone who is in my employ as it were; and someone in what is my house." (The *as it were* and the *what is* are so symptomatic of Richard's attitude, with all its humility-aggression!) He never did ask Mrs. Vince who she was talking to.

He has almost no front teeth now; just three or four yellow stumps. He is lean, despite the beer, because he eats so little. He has this strange kind of scalded red complexion; partly windburn perhaps and partly ruptured veins. His eyes are a very clear innocent blue. He laughs his screaming laugh more wildly than ever. He twists his head about, mutters to himself. Often he mutters, "*So* sorry, darling," as he used to, to M. He is full of the bitterest resentments. He has almost no interest in anything outside of himself and his world. He is not in the very least crazy.

Most of his talk is about Alan Bradley, "my pal and best friend," as he calls him in his diary. Richard goes to sleep at the Bradleys' place on the Old Buxton Road at least a couple of nights a week, walking home in the dawn over Jackson's Edge. Unfortunately the Bradleys were away in Guernsey while I was there. I would like to talk to them. Richard writes in his diary: "How I do love to see the constant demonstrative affection of the Alans for each other!" And he says, "Alan very amusing and splendidly light-hearted." When Richard stays there, Alan comes and tucks him up in bed, after tucking up his own daughter. Richard gets along with the wife and daughter much better since M. died.

(Don just got in and, rather than feel rushed—especially as I still have to make japam before we leave for the Court Theatre[1] for the opening of *Luther*—I'll stop here and go on tomorrow.)

July 28. Luther, last night, had its anticipated success. I don't think the play adds up to much. Don says he doesn't really like it. (And he has the, so far, almost unique authority of having seen it four times!) He doesn't like Finney's acting either. It certainly didn't seem nearly as straightforward as at Nottingham; the mannerisms are creeping back. He is a very *sly* actor. Sometimes the slyness in his face is right for the occasion. It was right, oddly enough, during the early scenes in the monastery; giving him an air of being partially possessed—the Devil seemed to peep out of him. At other times he just wasn't right, and his coldness became uninteresting. Now and then I caught such a startling glimpse in him of Laughton—that sly treacherous turnip-headed look which Laughton uses; a Halloween mask which might bite.

A big showy party on stage later, with lots to eat and drink, and a band and dancing. Loudon Sainthill and Harry Miller were there, and they told Don that his show is definitely on early in October; Harry couldn't remember the date. (This raises all manner of problems for me—and for Don—about British income tax; if you stay here over six months of any fiscal year, you may be liable. I won't fuss about that now. I'm going to Somerset House[2] to find out about our particular case, sometime next week.) John Osborne drunkenly friendly; calling me, "my dear friend!" Tony Richardson now talks about our coming to the South of France again, but probably not till later. (I phoned him this morning; he is quite pleased by the notices, but says that the real raves are only for Finney.)

Tonight, we have dinner with Angus Wilson and Tony Garrett and then go off with them in their car to stay at their cottage in the country, near Bury St. Edmunds.

Very good relations with Don since I got home. Only, this

---

[1] I.e., The Royal Court.
[2] Where the Inland Revenue has offices, in the east and west wings.

morning, I had to refuse to let him look in my little diary lest he should see Jonathan's name. It is too idiotic for words, and yet it was better than letting him. Don was hurt about this. If he could only realize how utterly unimportant all that kind of thing is. But no one can be expected to, of course—no one.

Well—back to my Wyberslegh visit....

It's hard to make out exactly how Richard spends his time. "I know," he said, in his mock-worried tone, "that I lead a terribly selfish and useless life." But he doesn't really feel guilty about this, I think; and why should he? (Yes, he should, I guess—he is a nuisance, a liability, a bit of driftwood which occasionally gets in the path of oceangoing vessels. But it's a very small bit of driftwood. And there seem to be plenty of people quite glad to devote time to him. And he *is* employing labor. And he isn't in the least stingy.)

He plays the piano more than he used to and even learns new tunes. He was laboriously and slowly playing "I Wonder Who's Kissing Her Now?"[1] over and over. He reads the books of the German Garden "Elizabeth";[2] most others he condemns as "depressing."

I wanted to bring back the first editions I gave M. of *Sally Bowles* and *All the Conspirators*; also the letters I wrote her. So we looked through the drawers for them. We found many of the letters—though certain batches were missing and may turn up later. Also a Victorian black-edged mourning envelope, inscribed in Granny Emmy's handwriting: "My dear Father's hair." Richard thought this as funny as I did and he screamed out like a macaw with laughter.

On the morning of the 25th, Richard and I walked down to Disley. When I exclaimed at the beauty of the woods at Jackson's Edge, Richard said that all this landscape depresses him. For him, it expresses the contrast between the past and the present. He says of himself that he feels "left behind."

All this sounds very poignant, but my overall impression is that he is making a good recovery from the shock of M.'s death. ("You must try not to mind too much," she told him, right at the end, when she was having slight heart attacks which left her breathless.)

---

[1] New only to Richard's fingers; it was a 1939 hit for Perry Como and later for Frank Sinatra and Bing Crosby, and it was the title song for the 1947 movie *I Wonder Who's Kissing Her Now* about Joseph E. Howard, who composed it in 1909.

[2] Elizabeth von Arnim (1866–1941), a cousin of Katherine Mansfield, published *Elizabeth and Her German Garden* anonymously in 1898 and twenty subsequent novels as "by the author of *Elizabeth and Her German Garden*" or just "by Elizabeth."

I think he dreaded my visit because he thought it would precipitate a big emotional scene which would upset him all over again. But I was much more prepared for this than he supposed and played it very offhand when we met; no brothers-in-bereavement stuff. I only embraced him when I left, and by that time we had established a very satisfactory post-M. relationship. I think, in fact, that he was really quite sorry to see me go. Quite sorry—not violently.

M. is buried, as he had told me, in the mound which is to your right as you come out the front door. There is no sign of the hole; the grass has grown over it again, and there is no marker. Though there *are* two—for two of the several cats which are buried all around her. It's admirably Buddhist in feeling. Only it's a pity that her ashes are in an urn. It will be dug up one day by strangers; and whatever they do with it will be inappropriate. I went out in the windy morning of my last day there and tried to dedicate the spot with *mantrams* to Ramakrishna.

Down in Disley, we drank beer at the pub. Richard drinks a lot of beer in the course of the day, but he was nowhere near getting drunk while I was there. He sometimes drinks ten bottles a day, he says.

As I mentioned already, he gave me his diary to read. Here are some items which pleased me:

Jack Smith's friend Roger is very untidy and has a way of "sprawling" in the chairs and putting his feet on a footstool. This offended Richard (and Mrs. Vince). So, when Roger came next, they hid the footstool!

During the night, it was discovered, the taps in two of the washbasins had mysteriously turned on and water was running. Richard half-seriously wondered if the Pussy could have done it. True, the Pussy seemed to have an alibi—that he had been sleeping all night on Richard's bed. *But had he, really?* And this bit of social moralizing:

> According to many people, Dr. B. is rather snobby and favors his wealthy patients—of course he and Jack are probably the same class really, but it is for the *most* part those who are recruited from that class who *are* the snobs of today.

Riding down on the train from Stockport back to London, I finished Angus Wilson's *The Middle Age of Mrs. Eliot*—with great difficulty. I had liked the first part, and Wystan, the so-easily-bored novel-reader, had recommended it strongly. But it seemed to me Angus does nothing with the character of the brother. It all pales away to nothing. And to compare, as one critic did, Mrs. Eliot to Madame Bovary! He must have been *sick*.

One detail about last night which I'd forgotten to put down: When they played "God Save the Queen" before the curtain went up, Alan Sillitoe,[1] who was sitting right in front of us in the first row, remained seated; but Ruth Sillitoe[2] stood up. There was something very pleasing about this. I felt I should refer to it later, but I didn't. Don and I agree that Alan is one mass of aggression. The way to get along with him is simply to be aggressive; that's all he asks of other human beings. It's Ruth who goes in for culture and intellectualism.

(Don just got home. He saw Harry Miller today, and his show at the Redfern is definitely fixed for October 3—till the 27th. The main exhibitor is Nicholas Georgiadis.[3] Don only knows his theatrical designs and says they're awful. But we hear that Georgiadis is "delighted" that Don is exhibiting with him. We met him, it seems, at a party.)

July 31. I really quite enjoyed the weekend with Angus and Tony. Angus is amazingly the country lady—perhaps the widow of a knight, or better still a dame. He enters into all the social life, belongs to the tennis club committees and the committee to save the old theater in Bury St. Edmunds, etc. And he devotes a great deal of time to social visiting. He feels he is *learning* provincial life and will soon be able to write about it.

His rather high voice causes telephone operators to mistake him for a woman. When they call him Ma'am and he corrects them, "*Mister* Wilson," they obediently repeat after him, "*Miss de* Wilson."

He tells how, when he was working at the British Museum, a mad Frenchman came to see him and announced that he had discovered how to understand the language of animals. Angus asked him what they talked about. Oh, said the Frenchman, nothing at all interesting—*rien que des bêtises.*[4]

A neighbor had been annoyed by the noise made by U.S. planes from a local airfield, and his little son had been scared. So he

---

[1] British poet and novelist (b. 1928), then known for *Saturday Night and Sunday Morning* (1958) and a short story collection *The Loneliness of the Long Distance Runner* (1959), featuring rebellious working-class heroes.
[2] American poet Ruth Fainlight (b. 1931).
[3] Greek-born painter (1923–2001); he trained as an architect, designed ballets, operas, and plays, and taught stage design at the Slade.
[4] "Nothing but nonsense," playing on *bêtises*, from *bête* for beast, fool.

phoned the airfield to complain. "And do you know, they had the cheek to answer me in American!?"

We got home last night. This morning Dick Gain arrived; the American dancer from Jerry Robbins's ballet whom the Burtons invited to share this house with us while the ballet is here. He is very young, simple, sweet, gossipy and domesticated, and I feel much reassured. I think we'll get along.

Don took more of his drawings down to Harry Miller at the Redfern, this morning. Huge enthusiasm. A gallery owner from New York happened to be visiting and promptly said Don must have a show in *his* gallery! Don, the eternal pessimist, puts this down to the number of celebrated sitters in the collection, not to the quality of the drawings.

August 4. On the eve of the Bank Holiday weekend, with rain threatening. It seems cruel to wish for it, and yet the Heath will be hell if the weather is fine.

The Berlin crisis stews on and on. Now an article in *The Spectator* says that Kennedy and his advisers don't expect "a great confrontation" on Berlin this year or even next.[1] On the other hand, Bertrand Russell, in a half-page advertisement in the *New Statesman*, says:

> Most people in this country as well as in other countries, appear to be unaware that the Governments of East and West are solemnly preparing by mutual vituperation to create a general state of mind in which the public will acquiesce in a large scale nuclear war.
>
> A large scale nuclear war, as almost all experts are agreed, means not only the extermination of nine-tenths of the population of Russia and the United States, but also what for us in Britain is peculiarly important, the total and complete extermination of the whole population of Western Europe and Britain.
>
> Perhaps, to be scrupulously exact, one should make one small exception. If it should happen that throughout the few days of the war the wind blew continuously from the West there might be a few dozen survivors in the Outer Hebrides....[2]

A nice man named Kenneth Allsop[3] interviewed me this

---

[1] "The U.S. and Berlin," by Richard H. Rovere, Aug. 4, 1961, p. 4.
[2] Aug. 4, 1961, p. 157.
[3] British poet, novelist, journalist and broadcaster (1920–1973); he reported for "Tonight," the BBC's first evening-news magazine, and was later book critic for the London *Evening News*. His works include *The Bootleggers* (1961).

morning for *The Daily Mail*. I have had quite a bit of notice taken of me lately. A BBC television interview offer, a BBC radio offer, an interview impe[n]ding from *The Yorkshire Post*, an offer to write about lust for *The Sunday Times* in their Seven Deadly Sins series—this last I declined, for obvious reasons.

Have got started reading for the Ramakrishna book. But it is very awkward working so far away from Swami; I need to keep asking him questions. However I can at least do a rough draft of the next chapter.

Don has been making a whole lot of drawings of me, for Methuen. Alan White wants to put one of them on the book jacket. He is also very much sold on the idea of an illustrated front. Don and I are dubious, rather. I'm sure the mirror idea is good, but making the faces look like me makes the whole thing somehow indiscreet.[1]

Great happiness with Don at the moment, and general harmony. As for Dick Gain, he couldn't bother us less. He stays in bed most of the day, resting up for the performances; and is most good-natured and friendly whenever we do meet.

Books I'm in the middle of reading now: George Moore's *Avowals*, Humbert Wolfe's *George Moore*, Cecil Beaton's *Diaries*.

August 5. So far, I love the Beaton *Diaries*:

> Papa observed about a horrifying residence surrounded by a hedge of repugnant shrubs. "That doesn't look a bad sort of place." I nearly went mad.
>
> How I abominate English seaside towns! When I'm on my own I shall never subject myself to such squalor.
>
> ... The smell of Daddy's filthy pipe insulted my nose.
> ... an old woman who arrived here yesterday. I thought her pert and perfect, an inspired little bird in smart London clothes.... She's almost bald but, clever little darling, doesn't wear a wig, simply parting her thin moth-eaten hair very slickly.

Heavy rain this morning. We always get up so late. Anyhow, I had a bad hangover-headache after drinking too much with Al Kaplan and his friend Christopher [Lawrence], and Don. Nicholas Eden was there at the beginning of the evening, such a great frozen monument to his Father and the Suez Crisis and the Old School; it

---

[1] The English dust jacket drawing shows Isherwood looking at a younger self in a mirror.

is truly shocking to see one of these young British fossils. They are older than all their ancestors, and they will breed sons even older than themselves.[1]

Dick Gain is the psychophysical polar opposite of Don and me. He is goony-vague and utterly relaxed at all times. (He claims he has stage fright but I can't believe it.) He sleeps about twelve hours a night, with a photograph of his friend Dick [Kuch] beside the bed. When we get home at night, he's sometimes writing to Dick—telling him, he says, what a "wonderful relationship" Don and I have. We're so "warm and human." Around eleven in the morning, he takes a long long bath, then goes off to the theater. He won't eat any of our food unless urged to do so. Richard Burton told them, "You make a very handsome pair."

August 6. Heavy showers have completely doused the holiday weekend so far. I was glad of them this morning, but now we have to go along to the [Donald Ogden] Stewarts' and it's a bore, and on top of that I left my raincoat at the Café Royal last night. Had supper there with Tony Richardson. (Don was out for the night.) I began telling him about my idea for an interview play; the basic idea of the Mexican *Down There on a Visit.*[2] I got quite excited and so did he, but I know that I don't really have any sound basic idea. I'll work on this.

Later we were joined by Al Kaplan and his friend Christopher. And we all got to watching a very brash large plump but rather cute young man at another table, who was entertaining a dark younger handsomer boy who was obviously very ill at ease and very drunk. Tony dared us to invite them over, and we finally did; and then they came back to Al's house. The brash young man [...] told us he was secretary to Lord Boothby.[3] He cross-examined each of us

[1] The 2nd Earl of Avon (1930–1985), educated at Eton, served in the Queen's Royal Rifles, later an ADC to the Queen and a Tory politician; he bred no sons and died of AIDS. His father, Anthony Eden—educated at Eton and Oxford, a W.W.I hero and Conservative M.P.—resigned as Prime Minister in 1957 after his use of military force to secure the Suez Canal against Egyptian nationalization provoked international condemnation.

[2] When Isherwood first conceived of the novel in Mexico in December 1954, south of the border was to be the Inferno, where a visitor would interview the main character.

[3] Robert Boothby (1900–1986), former Conservative M.P. and star political commentator on radio and T.V., was a life peer and a longtime advocate of homosexual law reform. In the early 1960s, he was implicated in a homosexual scandal with one of the gangster Kray twins, but won a libel action clearing his name. He professed himself heterosexual, married twice, had many affairs with women, and, reportedly, also with men.

in turn with drunken rudeness. He had never heard of either Tony or me; and his questions were chiefly designed to make us admit we'd "sell out" under pressure. Tony was rude to him right back. Al and Christopher were chiefly concerned lest the other boy, a rather charming cockney, should throw up all over their beautiful carpet. He didn't, but he slept all through our visit. It was a stupid evening, really; and once again I drank far too much.

August 8. A tearing wind today, with branches breaking off the trees. The air is very clear after all this rain. Shopping in Hampstead this afternoon, you could not only see St. Paul's but the hills on the other side of the city. A little girl said to her friend, "Look, you can see England!"

This morning, Don and I went down to Methuen's, where Alan White said they'll not only use one of Don's drawings of me but also one of his designs for the jacket—and pay him twenty guineas! The drawings were chosen and the whole matter settled in no time at all.

Very good relations with Don all this while; full of joy. He is staying in town tonight, and this is really a good thing; only I get annoyed when he makes dates at the last minute and I can't do the same. Jonathan is going to the theater tonight.

Lunch with poor old Edith Sitwell[1] yesterday. Most of the time she is perfectly logical; then she rambles off into an amazing tale of how her publishers, Longmans Green, (since abandoned) put workmen to boring holes in the wall next her bedroom, and how this caused an infection of her middle ear, and of how she called Scotland Yard and was with difficulty restrained from going around and slapping the faces of all the twenty-three workmen in turn.... With us at lunch was a trained nurse, an Irish girl; and a young writer(?) from Portugal, a bit of an ass-licker. Don joined us later. He hopes to get to draw her.

Tonight I shall stay home and eat fish cakes and string beans and read several books—*Quatermass and the Pit*,[2] *Avowals* and maybe start John Gunther's *Inside Europe Today*. The news is kind of fair and foul. Mr. Khrushchev talks of calling up reservists but pledges no blockade of West Berlin. And of course the papers are full of

---

[1] English poet, biographer, editor, critic (1887–1964); eccentric, upper-class hero of the avant-garde in Isherwood's youth.
[2] Nigel Kneale's six-part science-fiction T.V. play, first broadcast in 1955, and published by Penguin in 1960.

Major Titov and his seventeen turns around the earth.[1]

I have often intended to write something about this house, and I never have. I don't know if Ivor Jenkins was responsible, but whether he was or not it's really a disgracefully shoddy job. The woodwork is rotten and shabby already and the plaster is cracked in many places. The light fixtures couldn't be worse. Pairs of sham candles with crimson shades projecting from a fancy gilt fixture made to look like ribbons with tassels and bows. And there are wretched little miniature chandeliers. No coziness at all. But one must say one thing—the hot water's always hot!

We see very little of Dick Gain. His only offense is that he never cleans the ashtrays and so I have to keep doing it. But he's a good-natured sweet boy.

Architectural note which I failed to put down in its proper place. The walls of Angus Wilson's cottage are done with what's called "knapped flint."

Yesterday, I started playing at interviewing Don.[2] "What would you tell a young man who wanted to be an artist?" "Don't go to school." "What would you tell him to do, then?" "Never do any kind of work you don't want to do."

I'm so proud of Don sometimes that I could burst. And so, on an occasion like this morning at Methuen's, I put on a rather disparaging expression, like a parent who fears to show his pride. Of course, I know it's the most monstrous egotism on my part to be proud, to claim any part in what he has made of himself. Just the same, I do.

August 10. Another feature of this house; the absurd smallness of the upstair rooms. Two of them are barely more than closets. The whole place, with the ugly trashy rather theatrical unconvincingness of the downstair "living" rooms and the snug down-to-earth womblikeness of the upstairs, designed to hold nothing but double beds, makes you think of a very small whorehouse.

Am writing this waiting to go off and do a short interview on BBC television.[3] Don has gone to draw Margaret Leighton.

A party last night at John Lehmann's, ostensibly "for" me, but actually just a catchall. Oh, these seedy sophisticated youths! Of everyone there, I really only liked Joe Ackerley. We had him to dinner afterwards. But unfortunately an old friend of his, a French

---

[1] Soviet cosmonaut German Titov completed a twenty-five-hour flight, August 6 and 7, aboard *Vostok 2*.
[2] Preparing for press interest in his show at the Redfern.
[3] For "Tonight."

art critic who is a son-in-law of Matisse,[1] was sitting at the next table and had to be asked over. He was a surly old bore. When he talked about the painting of the young, he made a face like a spoiled gourmet in a restaurant who knows in advance he won't like anything they offer him.

Another triumph for Don. He went yesterday to see the art editor of *The Queen*, and they are going to publish three of his drawings in September, just before his show.

Have finished Cecil Beaton's *Diaries*. What a sad book! Not that it isn't amusing and entertaining. But it's sad because you feel—at least I felt—that this whole safari of Cecil's in search of The Real Right Set is in itself a frustration; and throughout it, he seems so agonizingly lonely. He is an extraordinarily heroic figure. In the last resort, he has *nothing* except his work. No friends. No alleviating vices. No real faith. Nothing. And he knows that.

And this brings back to me, once again, my own marvellous luck and underlying happiness. This is a wonderful time with Don; the best in months.

August 12. Une Vie.[2] I got a sudden presentiment yesterday and cancelled the date with Jonathan. The whole thing is getting too sticky, especially since Dick Gain told how he'd been to their dressing room and talked about me on T.V. It is too silly to describe—all this.... Anyhow, here I am, alone; because of course Don had meanwhile made himself a night in town. So I am bored and restless. Why? Because I bore myself. I don't really want to be alone. I shall make a terrible old man, I fear! So now I've called Oswell Blakeston,[3] because he sent me a note the other day suggesting we should get together for a drink, after all these years; and I'm going to see him after cooking myself a supper of sausages and peas.

Don told me the other evening I am sly. I said, "I suppose the truth is, I'm exactly like Dracula, and I want to suck your blood and make you one of the Undead, too." "If only I could believe that," said Don, "everything would be perfectly all right!"

Reading *Avowals*. I couldn't be less interested in the opinions of

---

[1] Georges Duthuit (1891–1973), a Byzantinist married to Matisse's daughter Marguerite; he helped Ackerley find pieces for *The Listener* and was a former lover of Ackerley's sister, Nancy West.

[2] "A Life," referring to Maupassant's novel *Une Vie* (1883) about the trivial sorrows of an unhappily married woman.

[3] English film and photography critic, travel writer, novelist, poet (1907–1985); his real name was Henry Joseph Hasslacher. He was once a camera boy at Gaumont British, where he possibly first met Isherwood.

Moore and Gosse about the various English novelists. What I like about the book is its atmosphere: the atmosphere of Moore's sitting room on Ebury Street; the fire, the slippers, the books, the maid and the tea. And the atmosphere of the book itself—it is the privately printed, signed edition with grey and white binding and uncut leaves. The book is numinous and makes me feel very close to the presence of the author. I don't think I could read it in a cheap edition.

August 15. Yesterday the East Germans closed the border in Berlin. Maybe by this time something worse has developed. I haven't been out to look. Rather than mope in crisis-jitters, I took a Dexamyl and wrote a draft of my catalog note for Don's exhibition, a draft of my description of *Down There on a Visit* and of my biography, both for the jacket.

Also walked out to Parliament Hill on the Heath. Marvellous view with sky threatening rain. Made a lot of japam.

We saw the Youth Theatre do *Richard II* last night, and soon left. They aren't even young.

We now have seats reserved on a plane to take us to Nice next Wednesday; God knows if we shall go. Tony Richardson admits that they have a water shortage at his villa because of the drought. And we also hear that Willie Maugham is dying. So presumably we can't go there.

Well, anyhow—Don is still being an utter angel. And that's *something!*

August 20. This morning I have been going over the jacket description of my novel, my autobiographical note and my catalog note for Don. The first two are all right now; the note for Don stinks. It's arch. I simply do not see how to do this. I can't hit the right tone.

The crisis roars on, with the dear Germans trying their hardest to pull the entire house down over their heads and ours. John Osborne has added his scream of defiance to the uproar: "There is murder in my brain, and I carry a knife in my heart for every one of you...." Shelagh Delaney and John Braine agree with him. Trevor-Roper and Priestley are snooty. Wesker slyly straddles the fence.[1] As for me, I warm to John for writing it; but why confine himself to the English politicians? The Russians are chiefly to blame,

[1] Osborne's "A Letter to My Fellow Countrymen" appeared in the *Tribune*, August 18; on August 20, *The Sunday Times* ran a page of commentary on it. See Glossary under Osborne.

anyhow. And that swine de Gaulle. And old gaga Eisenhower. Osborne should say to them, too, "I [would] willingly watch you all die for the West."

Meanwhile Jerry Robbins *still* plans to take his ballet into Berlin this coming week. (Dick Gain now gets so on my nerves, with his thick-skinned sweetness and light and failure to empty his own dirty ashtrays, that I can hardly bear to be in the room with him.) We are still supposed to leave for the Richardson-Osborne villa at Valbonne next Wednesday. Don is still at his very sweetest.

I'm beginning already to love Ralph Hodgson. His *Collected Poems* have just arrived. I love him while I'm reading his bad poems; that's the test.

At the beginning of last week, a journalist from the *Daily Express* called, wanting an interview. I said[,] No—your paper breaks confidence, and journalism's impossible if you do that.[1] He said, You're very hard; I've read everything you've written. (When they say that, I long to have them arrested and dragged into my presence for a viva voce examination; failure to score eighty-five percent minimum would be punished by a sound flogging.)

We had lunch with Chessy-cat[2] Cyril Connolly the other day. His bland talk of literary stocks and shares, relating to the sale price of one's manuscripts and first editions at the University of Texas. "If the libretto of *Elegy for Young Lovers* gets to America, Wystan won't be able to ask such high prices." Cyril says he's making a complete collection of everything written in the thirties. "One day, it'll be like having a collection of the Impressionists." He feels that his book reviewing is a sort of alternative oeuvre; something as good, in its own way, as the novels he might have written. But he doesn't believe this.

August 21. Don told me at breakfast that he had had a dream: he looked in the mirror and saw that his hair was long, down to his shoulders, and he thought that he wished he could dress it and fluff it out properly and then go to a masquerade ball. He said he was sure this dream had nothing to do with any desire to turn into a woman and thereby get a male partner. He said his fantasies are never connected with other people. I said that mine aren't either; they all have to do with travel. Having a yacht of my own, for instance.

[1] The *Daily Express*, without permission, quoted a letter from Isherwood to Gerald Hamilton at the start of W.W.II; this sparked public criticism of Isherwood and Auden for remaining in the U.S. See Glossary under Hamilton.
[2] I.e., grinning fixedly, like the proverbial Cheshire cat.

This morning, Tony Richardson phoned from the South of France to say that the villa is much smaller than he had supposed and that it's full of women guests. He suggested we should stay at a village eleven kilometers away, and come to the villa every day for food. Don indignantly rejected this idea, as I was hoping he would. So France is off. Of course, Tony says that we can come in two weeks, when all of these people have left. But one can no longer rely on him, I fear. He just doesn't care, and he treats everyone like an employee.

Supper with Amiya last night. She had been having a terrible time with George, so we took her to Kettner's to cheer her up. I do wish, though, that she wouldn't keep ascribing everything she achieves to Holy Mother's help. It makes us squirm. Actually she seems most at her ease with people like the man who owns the garage where she keeps her car. (He used to be an amateur actor and says he was in a production of *The Ascent of F6*, with Jimmy Stern's brother.[1]) He is brash and non-U, and though he calls her Lady Sandwich, you feel, as Don says, as if they might at any moment quite suddenly have sex, on the carpet.

August 25. Don has gone out for the evening and I've been eating at home alone. Tomorrow we go down to Essex to see the Beesleys; Monday we are supposed to go on to Cambridge to spend the day with Morgan Forster. But he's so vague that one just cannot be sure if he'll really be there.

The crisis is worse again; because the Russians are threatening to prevent the West from bringing in or supplying West German civilians in Berlin. It all seems too crazy even to record, but one must record it.

I am very bad. I do no work at all. I really must start, as soon as we get home from this weekend. Don is marvellously active and he has persuaded the Redfern to give him a bigger and better room for his show. Now we have sent off his second set of designs and portrait of me for the jacket of the Simon and Schuster edition of my novel.

Lunch with John Lehmann today. We discussed how much one should get from Texas for manuscripts, and similar matters. John says that Henry Yorke is now quite gaga. But John himself is becoming vague in a rather senile way. He tells me that Beatrix [Lehmann] won't see Don because she overheard him saying something awful about John at a party! I can't tell Don this; I

[1] Peter Stern, one of two younger brothers, acted with the St. Pancras People's Theatre in Camden Town, north London, during the 1930s.

promised I wouldn't, and it would only send him into a tailspin. I'm not sure I believe it, anyhow. I must find some subtle method of sounding out Beatrix—except that the whole problem couldn't interest me less. Is this my particular way of going senile—becoming so indifferent to so many people?

Do I have any birthday resolves? Work and pray. That's all there ever is. I am at least glad that I've kept up my japam pretty nearly completely since I made that decision on the beach at Santa Barbara.

Dick Gain left on the 23rd with the ballet for Berlin. I broke a tooth out of my bridge and Mr. Mackenzie-Young efficiently repaired it. Don profited by not having gone to France: he got to draw Lotte Lenya, and a Count Maurice D'Arquian who owns a gallery in Paris[1] and who paid him for a drawing. And on Monday he'll draw Forster, we hope.

August 30. I made an entry yesterday which I decided to scrap, because, after all, let's face it, Don, you are most probably going to read this one day, and it simply wasn't fair. I have no objection to making you feel ashamed of or embarrassed about certain bits of behavior, if that's going to help you in the future. But this was just a squirt of senile venom from me, which might as well have been aimed at de Gaulle or Tony Richardson or anyone else I don't or do know. It would only have pained you without reason. And so that's why the bottom of the preceding page has been cut off.

The venom was really all about my birthday. I suppose it is symptomatic of something, the way I always want to have a marvellous one and almost never do. This weekend was one of the worst in years. We should never have gone to Dodie and Alec's. It was sad unfun, and we overate, and I got crisis-blues and felt just utterly miserable. All Don did was to be tiresome, as when isn't he, about getting to the station; whenever we travel together, it is always a rush and this gradually rubs and rubs my nerves until I'm ready to scream. So I was hateful to him and then when I wanted to make it up, he wouldn't, at least he had that residual sneer-smile on his face. And then next day there was all the fuss of his drawing Dodie. He did a frank drawing of her, and she put on such an act—talking about not wanting to let her public see her like that. It was really *insane*. Don behaved very well and did a bad flattering one of her, which she loved.

Then at Cambridge, which we went on to on the Monday for

[1] International Galerie D'Art Contemporain in the Rue Saint-Honoré.

the day, Morgan got drunk and dozy at lunch and refused to let Don do more than one drawing, and so there were sulks about that, which I could hardly blame Don for. And Don kept saying that he would go stir-crazy there, and of course I agree, but we only had the afternoon to get through, and it would have been better not to fuss. We came home in a train with a prehistorian named McBurney who lectured us quite interestingly about caves in Africa.[1]

One of Dodie's neighbors said to her—after they'd talked about *Autumn Crocus* and her other early successes—"And you still plod along, eh?"

Must dash out now. More tomorrow.

August 31. This morning, we hear that Russia is to resume bomb testing.[2] Altogether, things couldn't look more ominous. Yet no one seems in a hurry to negotiate.

Don has a bad stomach upset. He couldn't be sweeter. He is working frantically, and now he is doubtful about taking any trip to the South of France. Theoretically, we go next Wednesday, with Tony Richardson, who's returning to London sometime this weekend. The possibility of seeing Truman Capote in Spain is off. He wrote today saying that he and Jack [Dunphy] are taking off at once for Switzerland. This may or may not be an excuse.

It's heartrending, how well everything would be going for us, if these were ordinary times. Why shouldn't Don have his little London success under an unclouded sky? He is doing so well. Everybody admires his work. He would so enjoy himself. So the old War Uncles have to step in and grimly spoil everything. If we are all blown up in the next few weeks, well, at least we shall never know the difference. But suppose we aren't? As I said in my diary at the time of Munich,[3] it's impossible, ultimately, not to loathe the Disturbers—*all* of them.

My tentative plans are, to leave England, about the second week in October, after the opening of Don's show. Otherwise, I shall have to pay British income tax. But if things look very bad then, or if Don wants me to stay on, then I shall.

[1] Dr. C.B.M. McBurney of Cambridge University, author of *The Stone Age of Northern Africa* (1960).
[2] The USSR, U.S., and U.K. had observed a moratorium on nuclear tests since late 1958; following their August 31 announcement, the Soviets carried out the first of ten explosions on September 1.
[3] I.e., the Munich crisis, August–September 1938; see August 31 in "Waldemar," *Down There on a Visit.*

Meanwhile, I have done at least one constructive thing: yesterday I restarted the Ramakrishna chapter. This is the ideal work for a crisis period.

September 2. Very true, but yesterday and the day before yesterday I failed to do any, and I doubt if I shall do any today. I'm in the grip of crisis sloth; psychologically, it's all *exactly* like 1938. Except perhaps that this time it's even more passive; the idea of an H-bomb makes one almost relax, in the miserable slaughter-house-passive way you try to relax before an operation. The chief sensations of the past three days have been: Russia's restarting of tests, the note from the stealers of the Goya ("Query not, that I have the Goya"),[1] and Mary Ure's baby. (Elaine Tynan tells me that Osborne is now trying to ditch Jocelyn Rickards in favor of a new girl in Venice.)[2]

Don went up to Stratford today, to do some more drawings of John Gielgud. Our relations are very good again, since we got back from the weekend. Yesterday evening, we went to see *The Best Years of our Lives*, and I think the scene between Myrna Loy, Fredric March and Teresa Wright about marriage "spoke to his condition." He said to me, "Am I fun to be with?"

Dick Gain sent us a postcard from Berlin. Not one word about the "situation"—only that the weather was good and Jerry Robbins had let him do [*Afternoon of a*] *Faun* and *The Cage*, which made him very happy. And, honestly, isn't his a much more psychologically healthy attitude than mine?

Thick fog today which cleared around noon. Don drew Ivor Jenkins and Ivor drove him down to the station, to go to Stratford.

I am mulling over a scheme to publish a book of my bits and pieces—short stories, articles, book reviews. I think I could link them together with some short autobiographical notes. It might be quite amusing.

September 5. What an absurdly jittery passage in my life this is! The crisis is quite bad enough, with Russia blandly popping off bombs and everybody else protesting, and East Germany about to

---

[1] Goya's portrait of the Duke of Wellington was sold to an American million-aire by the Duke of Leeds in mid-June, but a public outcry elicited a £100,000 donation from Isaac Wolfson to buy it back for the nation; when it went on display at the National Gallery in August, it was stolen.

[2] Ure's son was born August 31; she was still married to Osborne, but he was living with Rickards and beginning an affair with Penelope Gilliatt at the Venice Film Festival. See Glossary.

be recognized at any moment. And the weather is depressing and cold and grey. And, on top of this, we are creating extra tension for ourselves by messing around with Tony Richardson, who is Mr. Unreliable of 1961, or any other year. We are still *supposed* to go to France tomorrow. But shall we? Tony tells me this morning that he has something he must warn me about—he can't say it over the phone—and I think it's probably that George [Goetschius] is coming. George has just returned from America and seems to be rarin' to go. And then Don himself may be delayed because of a new job which has cropped up—another garment magazine in New York; this time they want men's fashions drawn. And so we go round and round.... And really, do I want to go to France? Willie Maugham is sick and can't see us—in any case, he has his daughter and her family coming. We get back to the ridiculous proposition that I am merely going to France in order to be able to stay longer in England. But, if this crisis blows up, I shall stay on anyway—income tax or no income tax—because I am not going to be separated from Don at such a time.

September 6. What Tony had to tell me was merely that there is tension down at the Valbonne villa because John Osborne had been cheating on Jocelyn Rickards with another girl, a model. I replied that I couldn't care less; Don and I were coming to get some sun. So here we are, committed to go this afternoon. At least, I am. Don may still decide to stop here because of the man from New York. And of course Tony is quite capable of backing out at the last moment. Well, we'll see.

Elaine Tynan, who is leaving for New York today, says she thinks Mary Ure's son really *is* Osborne's! According to her calculations Mary hasn't known Robert Shaw long enough.

These last days have been a specially totemistic period. Don said, "Do you think the Animals are withdrawing too much into their basket?"

Last night, a mysteriously happy dream. Just wandering about in a landscape. How seldom my dreams are absolutely free, as this one was, from anxiety!

America to resume underground atom tests, after third Russian explosion.

Bye now.

September 18. We got home two days ago, on Saturday. It's characteristic of modern jet travel that we spent two hours in the air, between Nice and London, and three hours being transported to and from the airfields and just plain farting around with officials and waiting. I must say, it is good to be back in London, with rain and cool winds and autumn setting in. The whole visit was a strain, especially as long as Tony was around, and what with the detensioning and some pills Pat Woodcock has given me my pyloric flap has ceased.

Today, I called the travel bureau and got a reservation on a BOAC jet out of here on Sunday October 15. That will just get me under the gate, as far as British income tax is concerned.

Here is the substance of some notes I took while I was at Valbonne:

(September 6.) Don drew H.E. Bates[1] in the morning. This was a commission arranged by the Redfern. We left for the airport at the last possible moment, after several drinks, with Tony, driven by his Polish driver. But all the speeding was of no avail; we were told we'd missed our flight. Tony called for the most important official available and proceeded, quite cold-bloodedly and with theatrical relish, to make a scene. "The BEA is the worst airline in the world." "In that case, Sir," said the official, who was big and uniformed as if he'd just stepped off a battleship, "perhaps you'd rather book on Air France." So we did, and got a jet which, leaving later, made so much better time that it arrived only twenty minutes after the British one. We had narrow seats, a kicking child in our rear, and no refreshment but tea and sugar cakes, but Tony was determined to be pleased with everything. And he was right. Because they could easily have gotten us on to the BEA plane.

La Baumette is a tall stone farmhouse with blue shutters, surrounded by vines and olive trees. It stands in a hollow with woods of pine above at the back of it; but from the highest point of the property, where the swimming pool is, you can see the sea.

Jocelyn Rickards came out to meet us, slaphappy with sufferings. John Osborne still away, with the girl. She told Tony and us, rather complacently, how she had cried on the shoulder of the old man, the grandfather of the French family which lives here and runs the farm.

Supper on the round stone table on the terrace under a mulberry tree. The table has a surface like cirrus clouds seen from a plane. We drank heavy red wine, and I snored and Don was furious and there were mosquitoes. A very tall cactus has grown right up and looks in at our bedroom window, like a phallus.

---

[1] English novelist and short story writer (1905–1974), author of *The Darling Buds of May* (1958).

The French family consists of the old man, who is nearly always drunk, his daughter and her two charming early-teenage sons. They make the most uninhibited noise at all hours, clattering up and down the stairs—they live on the top floor; and, as Don says, it is they who are living here, not us.

(September 7.) We drove down to the beach at Cannes. The huge ornate white hotels. The very expensive private beaches with their bars. The shark-faced hustlers. This is one of the eternal places; the gold coast which knew Cocteau and the twenties. Awful depression, caused by/causing pyloric spasm and stomach sickness. Misery, sulks. Cuthbert Worsley and John Luscombe are here.

When we got back to La Baumette, John Osborne had arrived. He seemed very cheerful, and so did Jocelyn. Later we heard that they had both, on meeting at the airport, wept for an hour.

Back into Cannes for supper at a restaurant up the hill above the harbor, the d'Arbutau. We had grilled sardines and a fish called dorade. Christopher Lawrence, Al Kaplan's friend, had a nationalistic outburst; he is from South Africa. Tony Richardson said, "I am of the People" and talked about the two sides of the barricades. Felt more miserable than I can remember for a long time.

(September 8.) Jocelyn is plump and pop-eyed and not terribly bright but there is something sweet about her. She doesn't seem bitchy at all. Her face in repose is curiously young and innocent. During breakfast she went back and forth between the terrace where Don and I were eating and the bedroom she shares with John. She took him coffee, etc., and reported, "He can't be left alone for long. He has to be distracted from lying with his face to the wall." But, again, John when he finally appeared seemed in the best of spirits. He is feminine and quite pretty, and he is obviously impossible in his dealings with women. With men he doesn't exactly flirt but you feel that he is somehow in the feminine relation to you. He kissed both Don and me when we met. Tony is always teasing him about his appearance, complaining of his spots, etc. They are like sister and brother.

Breakfast: long long rolls of bread brought from the village by one of the boys on the Lambretta,[1] coffee, jam and honey with wasps in it.

We spent the day lying beside the pool. Below us, the red rock-dry fertile landscape with the silver olives. The wind pouring softly through faint-scented pines. Clouds swelling menacingly snow-breasted above the dark thunderland of the mountains. Perfume

---

[1] An Italian motor scooter.

of pines and softly pouring air. Tony read *Passage to India*. I talked to him about my *Rashomon*[1] ideas for a Christ film, and he was interested, but not thrilled.

Our fearful mistake was to come here and stay with him. When you are at his mercy, he can drive you absolutely nuts. You have to do exactly what he says, every moment of the day. If you refuse, he asks "Are you all right?" as much as to suggest that your refusal is the first sign of an oncoming mental breakdown. He has a clacking sexless birdlike energy; he is a demon toucan. And yet, what do I mean when I write this? Merely that he knows exactly what *he* wants to do at every moment, and we don't; so we have to wait around for him. Nevertheless, he has reduced Don and me to nerve-raw neurotics. My pyloric spasm makes me feel as if I'd been kicked in the guts. I long long long to leave this place.

At supper, Tony's dirty-sadistic chatter, in which the others eagerly joined—John because he's childish enough to find it funny, Jocelyn because she will adopt absolutely any tone of voice necessary to keep in with the men. Much talk about "Lord Blowright" as they call Olivier now that he's married to Joan Plowright. Tony astonishingly venomous about him; told how Joan once threw up all over the table during dinner. He also teased John, by talking about castration, of which John has a horror. I asked what they thought about sterilization. Apparently this, too, was sinister and quite disgusting. So I refrained from telling them about my operation.[2]

Yesterday, Cuthbert gave me a typescript of a play by Terry Rattigan, who's about to arrive here. It's called *Man and Boy*, all about a financier who goes bust, like Kruger.[3] Tony read extracts from it aloud, with savage sneers. Cuthbert wants to get Tony to direct it.

(September 12.) Yesterday, and the day before, and the day before that, we spent on the beach at Cannes. We are now dark brown, and have drunk several gallons of gin and Campari. How I hate the thirsty sun and the salty sea and my hysterical gutache! We both long to get back somewhere where we can work. And

---

[1] Akira Kurosawa's 1950 film which tells about a rape and murder from four different points of view.

[2] In 1946, during an operation to remove a median bar from his urethra, Isherwood's sperm tubes were tied, making him sterile. See *D.1*.

[3] Ivar Krueger (1880–1932), Swedish engineer and match tycoon; he was involved in European financial reconstruction after W.W.I—borrowing in New York and relending to Germany and France; he shot himself in 1932, and huge irregularities were exposed in his companies.

yet, now Tony has gone, we feel we should make a serious effort to enjoy ourselves and get some value for our money.

Tony left in the middle of the night, with Al Kaplan, in order to be in London in time for a press preview of [his film of] *A Taste of Honey* today.

Yesterday evening, he was to meet Terry Rattigan and talk about the play. Because he thought he'd been trapped into this, he insisted on bringing with him Al Kaplan and an American decorator named Peter Gregson. "Fuck Rattigan!" he kept exclaiming. But the meeting was, after all, quite amiable. We had drinks on the terrace of the Hotel Martinez. A moppet sang "Because He Needs Me" and the song about liking being a girl.[1] Cuthbert said, "I was sitting on this terrace when the Second World War broke out, and I'll probably be sitting here when the third one does, in a few weeks' time."

(September 13.) Now at least we are free agents, and we have Tony's car to drive. But I'm still weary of the beach, and "relaxation." The *Herald Tribune* describes the great hurricane in Texas;[2] a crowd of rattlesnakes invaded one small town, fleeing before a tidal wave, and tried to swarm up the walls of the buildings.

John Osborne and Jocelyn left for England this afternoon. He has decided to take part in the antibomb demonstration next Sunday, and takes it for granted that he'll go to prison. Since Wesker and the others got a month—Russell's sentence was reduced to a week for medical reasons—John assumes the new batch will be treated even more severely.[3] He may get six months, he thinks. Their departure was solemn, with tears. Jocelyn embraced all the members of the French family, and wept. John looked smugly pleased with himself. He carried a large Bible in his hand. No doubt this was merely because they had forgotten to pack it, but, under the circumstances, the effect was that of the execution of a sixteenth-century martyr who goes to death with a proud smile certain of glory. "Madame a pleuré,"[4] said one of the Frenchwomen approvingly, after they'd gone.

(September 15.) We drove over to have lunch with Willie and

---

[1] I.e., "As Long as He Needs Me" from *Oliver* (1960) and "I Enjoy Being a Girl" from *Flower Drum Song* (1959).

[2] "Carla," September 11, 1961, one of the worst to hit Texas in modern times.

[3] Bertrand Russell's Committee of 100 was planning a mass sit-down in Trafalgar square on September 17. Some had been summonsed and sentenced in advance. See Glossary under Russell.

[4] "Madame cried."

Alan [Searle] at the Villa Mauresque. How I love that drive! The
ancientness of that narrow subtropical highway under the towering mountain wall, along which thousands of generations have
squeezed around the corner, between France and Italy. Willie
seemed deafer and perhaps a bit more shaky on his legs, but still
very much himself, every inch a Maugham. He embraced Don
and me with enthusiasm and remembered about Don's forthcoming show, and then we heard him asking Alan what our names
were! The martinis before lunch were staggering; but they didn't
seem to faze Willie, or Alan either, plump-thighed in tight white
shorts and a cricket-anyone? knotted scarf. Willie took a rather
strange bullying-affectionate attitude toward me during the beginning of our visit. Someone mentioned Berlin and he said, "L-lets
face it, Christopher, if it hadn't been for Berlin, where would you
be now?" "Certainly not sitting here," I said. And then Willie
started saying that he had believed I'd be one of the most successful
novelists of my generation—"and then you threw it all away."
The others were a bit embarrassed, and I pointed out that the
game wasn't over yet, and he should wait till he read my new
novel; but really we were on the very best of terms and I have
never felt fonder of him. Apparently, according to Willie's view, I
had "thrown it all away" for personal happiness and for Vedanta.
"And," said Willie, "I envy you." He went on to tell an oddly
self-pitying tale of how he had been humiliated in college by his
stammer; the class had laughed at him. The whole thing sounded
like a literary oversimplification. He was really composing a short
story. The Great Old Novelist who has sacrificed happiness and
love for his art confesses that he envies the younger brilliantly
promising writer who sacrificed his art for happiness and love. The
only thing that embarrassed *me* was that Willie implied that *he* was
without love, and this reflected on Alan.... But, really, we were
all too drunk to be taken seriously.

Willie also went on a good deal about Gerald Heard, whom he
seemed to think of as still a young man. Gradually Gerald emerged
as the villain who had seduced me from Art into Vedanta. Willie
assured us that *he* didn't believe in reincarnation. "Oh Willie dear,"
I told him, "how I wish I had your wonderful optimism!" No, said
Willie, he *knew* that after this life there is nothing; and he broke off
a bit of his bread to express the falling away of himself from life.
The idea of having to live his life again, he said, filled him with
horror. But Churchill, on the other hand, said that he would live
his life again gladly. ("Churchill's a vegetable," Alan said; and told
how gaga he has become, although he is actually some months

younger than Willie.) "Churchill always makes me go into a room ahead of him," Willie said, "he always insists that I'm his senior." And then back we went to the tragedy of Willie's life—which seemed increasingly farcical, as one looked around the marvellous garden of the villa and digested the excellent lunch.

Tony and Al returned from England in the afternoon. *Taste of Honey* a huge success, and Don's pictures are framed and exhibited at the theater.

We leave tomorrow, early.

September 20. John Osborne never did get sent to prison; only fined one pound. But he has stayed in the news by announcing to Jocelyn that he's going to marry this girl, the movie critic Penelope Gilliatt. There's a little snag; as of now, they are both married. We saw Jocelyn yesterday. Today, having heard the news, she is incommunicado again.

Don has twelve drawings on exhibition at the Leicester Square Theatre,[1] including all of the cast and Tony and Shelagh Delaney. The only thing is, they've left a small reproduction of Annigoni's picture of the Queen still hanging on the wall between two drawings of Paul Danquah!

He has spent most of today addressing cards announcing his show.

We are worried by the arrival of Mr. Burton, the man who adopted Richard.[2] For the moment he is staying with the Jenkins[es], opposite—but Ivor Jenkins has an uneasy embarrassed look on his face when he talks to us, and we fear we are going to be told that we have to leave because Mr. Burton needs our house.

Another memory of Cannes: Jocelyn saying, "I've got to go into the sea, so I can pee."

She tells us that Mary Ure's baby is definitely Robert Shaw's. What Tony Richardson told me, before we left England for Valbonne, that John was having an affair with a model, was false. Probably I just misunderstood it. The model had been invited down to keep Jocelyn company. But she sulked and fussed and had to be sent home again.

September 24. Only three weeks until my scheduled departure, and much to do. Don's show with all that that entails, and a new

[1] I.e., The Odeon, Leicester Square, the movie theater showing *A Taste of Honey*.
[2] Philip Burton.

cap to be put on my front tooth where the yellow one is, and a visit if possible to Richard, and to the Sterns, and the proofs of the British edition of my novel to be corrected, and Stratford to be visited to see Gielgud in *Othello*.

Today, Don has gone to draw Bryan Forbes again—the director who made the awful film about children finding a tramp and mistaking him for Christ.[1] I have fussed about, reading *The Sunday Times*—Gielgud on the theater—Dag's death and its aftermath[2]—gradually appearing signs that the U.S. is going to sell West Germany down the river, thank God.

My latest health worry: increasing stiffness in the joints of my jaws. (Marion Davies died of cancer of the jaw yesterday.)

I feel that Don is quite eager for me to leave, but I don't take this as anything bad; we both need a holiday from each other. He is under great strain now, naturally, with the show coming on.

September 27. A grey chilly wet day today—the kind of weather which is like an admission of failure: it's no good, we can't be bothered to go on trying to fool you, after all this *is* England and what else can you expect. One gets a horrid feeling that the six-month-long winter has started.

But here we are in high spirits. Don is so happy, because *The Queen* has just come out with two pages of his drawings; Stravinsky, Alan Sillitoe, Gerald Heard, Robert Stephens,[3] Angela Lansbury—not perfectly reproduced, it's true, and not very well arranged on the page, but still—Also, yesterday, we saw Peter Schwed from Simon and Schuster, a grey little Jew who looks very like Nehru, if Nehru were nobody in particular, and Mr. Schwed showed us a proof of the jacket for my novel with Don's drawings, and it really looks very good indeed.

On Monday, we had supper with Jocelyn Rickards. (Mary Ure had put us off, because she said she had a nervous peeling of the skin of her fingers. John Osborne, it seems, has a psychosomatic skin rash all over his face and crotch, which only goes away when

---

[1] *Whistle Down the Wind* (1961).

[2] Dag Hammarskjöld, Secretary General of the U.N. since 1953 and a friend through Lincoln Kirstein, died in a plane crash in Northern Rhodesia on the night of September 17–18; rumors suggested he was assassinated by Soviet or international mining interests for trying to restore order to the wealthy mining province of Katanga which had seceded from the new Republic of Congo.

[3] British actor and, later, stage director (1931–1994); he starred in Osborne's *Epitaph for George Dillon* at the Royal Court in 1958 and appeared in the film *A Taste of Honey*. First husband (1967–1975) of actress Maggie Smith.

he sees Jocelyn!) John is now down in Sussex, living in open whoredom with Penelope Gilliatt, and being moated around by the gutter press. A picture of Gilliatt, big-assed and plump but sassy, was being displayed on their front pages. John as usual roared to be left alone and allowed to get on with his work; he is just a great big huge girl. Meanwhile, Jocelyn acts shattered but brave. These people's capacity for taking themselves seriously would stagger even a Goethe. One wonders how they would behave if something *really* serious happened to them; maybe heroically.

The Rickards party—which included some dullish others, Alec Murray,[1] Richard Wollheim,[2] Diana Moynihan[3]—left us both with terrible hangovers yesterday. Nevertheless, Don had to draw and I had to go to the dentist. This was very unpleasant, as Dr. Peschelt[4] had fastened in that old yellow fang he made for me so tight that Mackenzie-Young had to give me Novocaine and hack it out with the drill.

On Sunday night we had supper with Jeremy Kingston and Rashid Karapiet. Jeremy's strange exhibitionistic indiscretions about his love life, in the presence of two other boys who came in after dinner and who, it seemed, had been involved. They and Rashid seemed embarrassed but they didn't protest. Yes, the more I think of that scene, the stranger it seems. I don't think Jeremy was being bitchy or taking a revenge on anyone. What makes him so shameless?

I told Jeremy my ideas for a Ganymede story. In telling them, I suddenly saw something—a technique of narration—which one could use on many kinds of material, but which would be particularly suitable for retelling classic myths. It is, in a way, a technique of describing one's effects, telling *how* one would tell a story rather than directly telling it. Hard to describe, and yet I know more or less what I mean. For instance, here's the way one might begin—

Very important not to overburden this with classical props. Let's start with two gigantic pillars and a pediment, simple, brutal, despotic, against a furiously blue oriental sky with black

---

[1] Fashion photographer, once a boyfriend of Rickards.
[2] Left-wing, Oxford-educated British philosopher (1923–2003), then teaching at University College, London. His published work reflects his interest in psychoanalysis and aesthetics.
[3] Née Dawson, model, waitress, and bohemian (b. 193[7]); her second, brief marriage was to Johnnie Moynihan and her third, in 1963, to jazz musician and screenwriter George Melly.
[4] Isherwood's regular Los Angeles dentist.

in its depths. That's all. And then the appearance of a figure. Naked, molten-gold, as if coming forth from a furnace. The young prince. The young arrogant animal, unconscious of the future, incapable of cruelty or love.

That's deliberately trashy, but it will remind me of what I want. The narrative has, of course, to be all in the present tense. It's a little like a film treatment, but cornier, artier, more visual.

Yesterday, after Don and I had seen Peter Schwed, we were walking around Piccadilly Circus and there was Bill Harris! Although his figure is good and he is perfectly healthy, he seems curiously old for his age, and shrivelled. Nervous tics and smiles, and self-conscious eye-rollings keep his face constantly without repose. He told us that he had been right through a treatment for T.B. and then discovered that he had never had it at all. It was nothing but an old scar which had healed itself. He is coming out to California soon so I shall probably see him there.

Eric Falk has offered very sweetly to put Don up at his place in the Temple, if Don can't live here any more after I leave. So that worry is removed. So far we haven't met Mr. Burton Sr., and Ivor and Gwen have had no direct news of [Richard][1] and Sybil in Rome.

I have been rereading some of the letters I brought back from Wyberslegh; the ones I wrote to M. during our stay in the Canary Islands and Copenhagen,[2] and the ones to her from Saltair Avenue. They are terribly dull, because I almost never tell her anything but mere happenings, never what I am feeling. Out of them comes such an odor of depressing minor anxieties and even more depressing minor hopes. Une vie.

September 29. This morning, a pre-exhibition flap. The Italian who is framing Don's drawings called to say that nineteen of them are the wrong size, because Don trimmed them, and that they will therefore have to have non-standard-size frames made for them. And Harry Miller of the Redfern has said that the gallery cannot pay for such frames, because they can't be used later for other pictures. So Don will be forty-five pounds out of pocket, unless he can absorb the framing cost by selling all of the nineteen drawings, which seems wildly unlikely.

I had to disturb Don in the midst of a fashion-drawing job to tell

[1] Isherwood miswrote "Robert."
[2] With Heinz Neddermeyer, 1934–1935.

him this. I also had to give Harry Miller the number and tell *him* to call Don. Don will probably be mad at me for the disturbance; it's the kind of thing he's utterly unreasonable about. But I can't help that.

Trying hard to read at least some of Angus Wilson's *The Old Men at the Zoo*, because he will be at a supper party this evening which we're going to. Christ—it is *dry*! In a sense, there is too *much* observation, that's what's the matter with it. Angus stands beside you with a pointing-stick, like a lecturer showing lantern slides, and he proudly calls your attention to every single god-damned nuance. He stage-directs the drama out of existence. I simply cannot imagine what got him interested in all these boring characters in the first place. No vitality, no fun, no joy.

We had a nice evening with Cecil Beaton last night. He took us to supper at the Mermaid Theatre and then to see *'Tis Pity She's a Whore*. Don had been drawing Morgan at Cambridge all day and returned late and exhausted and frustrated, because he didn't think any of the drawings were good; and because Morgan, although himself exhausted, had insisted on entertaining Don and in fact preventing him from returning on the train he had planned to take! One of the drawings *is* good, however. (And that reminds me, the drawings Don did of Jocelyn Rickards the day before yesterday are not merely good but fiendishly inspired; you want to roar out laughing at this dangerously demure puss-person, with her huge eyes flashing masochistic warnings and her wrist burdened with bangles. Like the V.D. army poster in Denny Fouts's apartment in Santa Monica, it could be captioned, "She may be a bag of trouble.")

Don and I agreed that Cecil seemed more than usually friendly; for almost the first time, I felt affection. We tried to reward him for this by telling him all about the South of France and the Osborne-Gilliatt-Rickards-Ure-Richardson pentagon. He chuckled with his curiously unbitchy kind of malice.

The play was a bore.[1] Partly because it *is* a bore; the incest seems merely arbitrary. It isn't made to mean anything more than a sex partnership which just so happens to be socially taboo. Also, it was very badly acted. Whenever anyone wanted to be more than usually Italian, he or she *yelled*. I strongly suspect that this technique is copied from the films of Visconti. And it's usually silly enough when he uses it.

---

[1] Directed by David Thomson with Edward de Souza, John Woodvine, and Zena Walker.

October 6. A gap, not so big but important because of what's been happening.

First, to add to my last entry, I have another memory of the evening at the Mermaid Theatre with Cecil Beaton.

The theater has two performances a night, at six and eight-forty, and so the theater restaurant serves supper in two shifts: before-the-later-show and after-the-earlier-show. When they decide that it's time for the before-the-later-showgoers to eat up and pay and get out, they ring a somewhat irritatingly loud cowbell, one of those church-like Alpine clangers. This annoyed Cecil, and he even told the waitress that he was in absolutely no hurry and didn't care if he missed the curtain. (As a matter of fact, there *is* no curtain at the Mermaid.) And even after we had left the restaurant he was still ruffled. "Imagine," he said, "her not realizing that *we* couldn't be herded in with the rest of the bell-followers!"

Well, now I'll put the rest of this down in order:

On the evening of the 29th, we did indeed see Angus Wilson and I was cowardly and didn't give the least hint of how I felt about *The Old Men at the Zoo.* Since then I have read the whole of it, and can only give thanks to God that I hadn't done so when we met that evening. Because it really does stink.... Angus made me feel a heel by offering to review my new novel in *The Observer*!

On Sunday, October 1, Don drew Francis Bacon. He has been trying to do this for a long time, and had asked Paul Danquah to ask Francis. Paul reported that Francis wasn't very keen on the idea, but he advised Don to go around to Francis's flat with his drawing board and act as though he had misunderstood and that they had an appointment. This Don did and it worked—Francis was very sweet and friendly and Don did three drawings, one of them really first-rate, and got it framed next morning, in time for the opening of the show.

On Sunday evening, we went around to the Tynans' to watch Ken's new T.V. show, "Tempo." It was a dead march; even the boys from *Beyond the Fringe* weren't up to their usual standard. We dashed off at the end of it, making a mystery of where we were going; actually it was to see *The File on Thelma Jordon* at the National Film Theatre. We both adore Barbara Stanwyck. Don and I achieve some of our moments of greatest closeness while watching films of this kind. I glance at him during some huge emotional scene and see his face absolutely lit up with delight and gleeful amusement. It's then that I feel this delicious subhuman animal snugness.

Afterwards we had supper with Maria St. Just. She seemed much nicer than in the old days—maybe because then she was on the defensive. Very funny about her two daughters, who, according to her, are pure Slavs and display incredible savagery. They have already gotten through fifty nannies. Maria described how she was with them at a railway station, to see a friend off on a train. They arrived late and had to run. One of the little girls fell down on the platform. Maria took no notice of her but ran right on, and this shocked the passersby. "But I knew," said Maria, "that in England there'll always be someone to pick up a child. And anyhow it served her right, after the way she used to turn that miserable pet tortoise of theirs over on its back."

On October 2, we had the party at the Redfern Gallery. It was a truly marvellous event. As I said to Don later, it almost if not quite made up for the bitter disappointment of the twenty-first birthday party Marguerite gave him, to which so many stars were invited and almost none showed up. Although there were bars both upstairs and down, the upstairs was practically deserted; Nicholas Georgiadis and Henry Cliffe[1] were left with their stupid old abstractions, and our downstair room, which was quite obviously the worst of the three positions, and which we had apprehensively nicknamed Room at the Bottom, This Way to the Tomb and Down There on a Visit, was crammed to halfway up the stairs.

Don was interviewed and photographed by the press, while I kept away in a corner, nearly splitting with pride. He was photographed with Amiya, who got into the act with both feet and nearly wrecked it. They stood beside her portrait and she leaned over toward him with the smile of a drunken amorous pig. I saw one of the photographers shrug his shoulders, as if to say that this was really more than he could take. Willie Maugham created a major sensation by appearing for a short while with Alan Searle. Forster and Joe Ackerley stayed quite a long time. Jimmy and Tania Stern said they would buy one of the portraits of Wystan, the one of me and maybe one of the ones of Stravinsky. There were a lot of commissions, including one by a well-known doctor who wanted his wife drawn naked to the waist. After it was all over, we took Francis Bacon out to supper at Gale's. He had stayed all through the show and somehow given it his blessing, although obviously he couldn't have thought very highly of this kind of work.

Next day, we spent four hours at the gallery. Quite a steady trickle of visitors. But Don was depressed by a feeling of anti-climax. Nothing in any of the papers about the show, except

[1] English painter and lithographer (1919–1983).

for *The Evening Standard*, which had some inaccurate stuff about Maugham and Amiya. Jimmy Stern showed up to announce sheepishly that they were only going to buy Wystan's picture, after all, because Tania had said they needed the rest of the money for a new bed! And some of the champagne had dripped down the stairs and made a wavy tide-line along the bottom of one of the already-sold portraits of Bryan Forbes's wife. We drank what was left of the last bottle of scotch with Harry Miller, who is actually very pleased with the way the show is going.

Later, because of Don's depression and our need to somehow make an evening, we called on Walter Baxter,[1] who has become a rather tragic self-pitying drunken figure with a philosophy of failure. What use was success, he asked. Oh yes, he *could* write again if he wanted to, but it would mean giving up drink, smoking and sex—and was it worth it? The only thing that interested him, anyway, was to record some of his very early sex experiences; and those couldn't possibly be published.... I felt we depressed him even more than he depressed us. He was eager to get us to leave him alone after dinner.

Last night, we saw Robert Moody, who has managed to escape from a couple of unhappy marriages and is now living happily with a not-so-much-younger-than-him woman named Louise Diamond. Robert has that same sly low-voiced confidential way of talking, and he seemed really pleased to see me. But how old he looks! A bald, slow-moving, wrinkled old man. Olive [Mangeot] was there, too. Incredible how young *she* looks. And her memory! But it was embarrassing when she talked so much about the past, because it left both Louise and Don out in the cold.

October 8. This time a week from today, I suppose I'll be in the air, somewhere between here and Los Angeles. I'm filled with dread and sadness, whenever I think of this.

Disappointment this morning, that there's no mention of Don in the Sunday papers. So far, Don has been called "accomplished" by *The Times*, which printed his full-face drawing of Wystan, and *The Arts Review* prints the one of Albert Finney, and says, "Bachardy is adept at varying the angle of head and pose, including shoulders and hands, and making a large, balanced work; his line is elegant and crisp rather than enquiring, impression rather than facial landscape drawing. On this account, his ladies, where their shape of

---

[1] Novelist (1915–1994), friend of E.M. Forster; he wrote *Look Down in Mercy* (1951) and *The Image and the Search* (1953).

face is not unusual, reveal more makeup tha[n] personality. The portraits of Auden are the best; they show the most investigation into the lines that life makes. No doubt this is not always what his subjects, whose work involves so many different projected images of themselves, require." (Michael Shepherd.)[1]

On Friday evening, Jimmy Wolfe drove us and a woman named Raemonde Rahvis[2] out to supper with Bryan and Nanette Forbes at their home in Virginia Water. I must say, I really don't like Jimmy very much, although he goes through all the motions of being friendly. It's all very well to say he's shy or scared; that's really what I don't like about him. And of course he has power and uses it, and is pretty thick-skinned and potentially ugly. He probably hates nearly everyone, underneath. He keeps referring to Tony Richardson's opinions, but he bitches him too. He said how both Tony and he had hated *Victim*. And "anyhow, I can't imagine anything less interesting than a story about homosexuality—"[3] Ah, how ugly! Disowning his nature like that, in the presence of outsiders. That's what causes persecutions. And then Jimmy went on to talk about Tony's sadistic mania for truth games. He once asked a man, "How long was it after your marriage before you started sleeping with boys again?" And the man hesitated and then replied, "Four months," and his wife cried out and got up and left the room, and soon afterwards they were divorced. Both Nanette Forbes and Raemonde Rahvis agreed that, in such circumstances, a woman wouldn't dream of telling the truth.

The Forbes[es] play it big, being happily married. Not that one doubts this, but oh, they do advertise! Bryan loves gadgets, and one reason they bought this house is that it is full of them. It used to belong to a millionaire who had orgies. A whole wall of the dining room sinks into the floor when you press a button under the table. I don't quite know why.

On the drive home, Jimmy and Raemonde talked gambling; they were going off together to play chemmy[4] at a club. Don and I went to a small loud smoky party Norman Mailer was giving at the Ritz. By this time, I was drunk.

Yesterday, we took Maugham and Alan Searle to lunch at the White Tower, and it was quite a success. Willie remembered the

[1] "Georgiadis, Cliffe, Bachardy," Oct. 7-21, p. 16.
[2] London fashion designer; from the mid-1940s she also created movie costumes.
[3] Dirk Bogarde plays a lawyer pursuing blackmailers who caused the death of a young man with whom he is in love.
[4] I.e., *chemin de fer*.

restaurant before the first war, when it was the Eiffel Tower and [George Bernard] Shaw and [Arnold] Bennett and [Augustus] John went there. The manager came up with a guestbook and said, "Mr. Maugham, we've waited years for your signature!"

They have decided to leave the Villa Mauresque and maybe settle in England; the only problem is taxes. As for his servants, they are all wealthy. Willie was alone one time at the villa and sick; he sent for one of them and said, "You know, I may not recover from this illness. I hope you're properly provided for?" The manservant merely smiled, as if this were an absurd question.

Willie was very lively and looked around at the other people in the restaurant with eager interest. "Do you think those two enjoy going to bed together?" He was also most interested in Don and his career and wanted to know exactly what proportion of his earnings Don had to pay the Redfern. Alan told a story of how, at Angkor Wat, he had leaned up against something massive in the dark and found it was a baby elephant. He is terrified of flying. On the plane from Nice to London, he had sworn to have no sex in London if the plane landed safely. Now he was sorry.

Joe Ackerley, whom we took to supper yesterday evening, tells us Stephen is hurt because Don didn't ask him to sit for him. Which was what Don intended.

Yesterday, the first batch of my proofs arrived. Today I've been correcting them. So far, they don't seem so bad.

(I didn't actually leave England until October 15, and, in the meanwhile, several things happened which I want to record. But I'll do that when I switch back to the volume of my diary I was using before this one, because I don't have any more paper this size.)

October 19. (This resumes after the interval of my stay in England, from April 7 to October 15. The diary of that is in a separate folder.)

Got back here on the night of Sunday the 18th. There was a heat wave on. At 8:30, at the airport, it was still over eighty degrees and I sweated in my thick suit which I'd put on because it was so cold in London, the first thick fog of the year, right down to the roots of the grass. We took off about two and a half hours late. Jo and Ben met me at the airport. They were as thoughtful as usual and had put flowers in the house and seen to it that everything was in apple-pie order. But I was dazed and sad. The misery of leaving Don in London was acute. We said very little about it, though, and there were no tears. I just felt utterly utterly wretched. And I still don't really know how Don is to live here; all his future as an artist would seem to be in Europe or New York. Well, we shall see.

When I unpacked my suitcase, I found amongst the clothes this note: "Kitty loves his dear more than anything in the world."

The last entry in my London diary is of October 8. So before I go on I'll put down a few things about the end of the London visit.

On October 9, I had lunch with Henry and Dig Yorke, and Don and I saw them again, for drinks, the evening before I left. I can't imagine what kept me from seeing them, all those months in England, because, as soon as I did see them, I realized that I like them both enormously—much better than most of the people we did see a lot of. Henry is terribly deaf and keeps lamenting that he has stopped writing. He is willing himself into being an old old man before he's sixty, and I suppose he'll succeed. Dig is very bright but she must mind, terribly. Also, we are told that the family business, which is now being run by the son, is failing. The son, Sebastian, has just broken up with his wife. What makes me like them? I suppose it's the thing which makes me like Chris Wood—an absence of shit. You feel truth in them.

On October 11, I went up to Wyberslegh to see Richard again, for one night. This visit was a success too, though a strain for me, because Richard was determined to expose the Bradleys (or the Alans, as he calls them) and me to each other. I also had to meet Alan's father and mother. We even spent the night at Alan and Edna's house, a council house on a side street off the Old Buxton Road hill above Disley; and that evening we drank with them at the nearest pub, the Ploughboy. I do see the snugness and consolation which Richard gets from them. They are on the level, I feel sure; but I got rather tired of the solemnity with which they kept telling me I needn't worry about Richard, because they had solemnly promised my mother on her deathbed, etc. etc.

So we stayed drinking and drinking. Richard drank pints of bitter—it's simply amazing how much he can hold—Alan drank Guinness, Edna and I drank gins and tonics—I sensed that "spirits" are considered somewhat ladylike, as they are in Australia. The pub stayed open nearly an hour after closing time. Toward the end, a "character" named Reg appeared. He is evidently a hero of Alan's and suspected by Edna as a bad influence. He said, "The Americans are living in a world of fantasy," and quoted W.H. Davies and Newbolt's "Play up! And play the game!"[1] Richard got depressed temporarily and exclaimed, "I'm so lonely!" But Alan efficiently shut him up. I gather he has to do this quite often; and he certainly seemed genuinely kind as well as firm. Earlier in the day, Richard had told me about the death of Jack Smith, who drove us around the Peak District last July. He died of a heart attack on September 1. Richard was with him until they took him away to the hospital. He was quite cheerful and not at all scared.

Waking up in the morning at the Alans', to the anxious voice of the radio. It would drive me absolutely nuts, and they don't even listen to it.

From Wyberslegh, I went down to Stratford and joined Don, and we saw Gielgud in *Othello* and stayed the night with him and drove back to London with him on the 13th. John was really no good at all as Othello; not even as "noble" as I'd have expected. The sets by Zeffirelli[2] weighed literally and visually tons; they crushed the play.

I do respect John, though. I even respect his vanity, because it is the humble vanity of a real honest to goodness professional.

Well, and now here I am, with lots and lots to do—restart the Ramakrishna book, get the Virginia Woolf article ready for *Encounter*, see if I can make a short story out of the hashish episode from my novel. I am stoking myself full of Tiger's Milk and vitamins. Am going to the gym again, and very stiff from the first visit. Worried about the muscular ache and stickiness in my jaws. If anything, it's getting worse.

[1] From Newbolt's poem "Vitaï Lampada" in *Admirals All* (1897).
[2] Franco Zeffirelli (b. 1923) designed and directed the production.

October 24. Today a colored man named Mr. Bayless, recommended by the Mr. Gardner who painted the house, recommended by Dorothy, came with a truck and three others and started clearing the hillside. I had to drive him to a place whe[re] he could rent a saw for the cypresses. The saw was gasoline driven, instead of electric to which Mr. Bayless is accustomed. This bothered him greatly. Just as we got home, he exclaimed that we had forgotten to rent an extension cord for the saw. I reminded him that the saw wasn't electric. He cried, "Ah'm slippin'." But he didn't seem to be. He is nearly seventy-one but he rushed about all over the hillside and reproved one of his assistants, a young Hercules, who was a lazy nigger and kept drinking water from the hose. Within three quarters of an hour, all the trees were down. It is sad to see the old Casa so bare; but it really made no sense, keeping the couple which were alive. Especially as Don still wants us to have a balcony. I heard from him yesterday, a very sweet letter of loneliness and not much news. But he seems to have lots of commissions.

Finished correcting the American proofs of *Down There on a Visit* and sent them off this afternoon. I keep worrying that maybe Tony Bower will object to the character of Ronny; but really he is quite sympathetic and merely a necessary *advocatus diaboli.*[1]

October 25. This morning, the saw Mr. Bayless had rented broke down. We had to take it back. The people were quite nice about this and indeed admitted that the saws were subject to defects. But still, I said to him, that didn't alter the fact that they were going to charge him for the time he had lost, bringing the saw back to be exchanged. Said Mr. Bayless, "That's what I'm harmonizing about." He will make some perfectly simple statement, like that it's a fine day, and then ask me, as if I had been puzzling for hours, "*Now* d'you get it?" He has a sort of wolf whistle which he repeats constantly. Poor old thing, today he couldn't even stand up straight, he had tired his back so much by working.

The removal of the scrub on the hillside reveals a number of empty pint bottles which have been tossed up there from the roadway over the years.

Worried about the Berlin situation, which looks nasty. I shan't have any rest until Don is back here from there. And then I shall worry about something else.

Saw Ted Bachardy and Vince [Davis] last night. They live in the

[1] I.e., devil's advocate.

midst of a sort of junk shop, largely composed of things Vince has made or painted. It is snug. But there was still a lot of strain and politeness between us. We aren't nearly at ease with each other. The conversation could only subsist on movie talk.

Missed the gym, on Tuesday. That's something I must never do. And now the field is clear. I have finished the proofs and have no other really pressing work; nothing I can't do in small stages. My next job is to rewrite the Virginia Woolf article.

October 28. A high wind, searching everywhere and causing unease and dryness. I'm depressed. My jaw still feels uncomfortable and worries me. (Russell McKinnon, whom I talked to the day before yesterday, tells me [his wife] Edna is dying of cancer of the jaw.) I miss Don more each day. Russell thinks he ought to stay on in Europe and study art in France and Italy. I don't want to stand in his way but I don't want to live without him, either. And now I feel disinclined to write the Virginia Woolf article. I simply do not have enough to say about her. But it's hard to make up my mind to tell Stephen this—when I am only too aware that part of this decision comes from laziness and part from a desire to get back at him for accepting and then rejecting "Ambrose"!

I love this house, though. And the calm of being here, among my books and belongings. Mr. Bayless finished the clearing of the hillside yesterday. It looks terribly bare and will probably erode badly in the first rains.

Yesterday I saw the Picasso exhibition at UCLA[1] with Jo and Ben and Dana Woodbury. It is truly marvellous, and I don't think there is any artist who has so fully explored High Camp. In the evening, I had supper with Jo and Ben[,] and Anne Baxter and Ranny Galt her husband were there. For them, Australia is obviously a purely subjective love nest. But we all three felt that, unconsciously, they gave one the most horrible picture of the Australians. For example, Anne was sick and couldn't attend her own birthday party. So Ranny presided and she stayed upstairs. All the women guests knew what had happened and no one went up to see Anne or tried to help her, because that wouldn't have been the thing to do. Anne does all the cooking for the two of them, but they never have the rest of the people on the "station" in to meals with them, because that isn't done.

---

[1] "'Bonne Fête' Monsieur Picasso from Southern California Collectors"; two hundred works loaned by fifty Californians, opened on his eightieth birthday, October 25.

The day before yesterday, I saw Gerald Heard, Michael Barrie and Chris Wood for supper. I forgot to mention that, when I met him alone, before this, Gerald told me that he feels an important change has come over him. I couldn't quite make out what this was, but it was apparently some kind of liberation. He feels he is becoming more and more prepared for death. He also repeated the story he told me before I left for England, that he saw a committee of drab schoolmasterish-looking men and they had just finished reviewing his case and had decided that he wouldn't have to be reborn; though he had only just scraped through the test. Gerald says he had this vision right after the automobile accident on Oahu.[1] He added, "I've never had an hallucination before in my life."

Chris was Chris, as always. He has grown another "benign" cancer on his shoulder. Gerald says, with a certain satisfaction, that this is the price you pay for years of sunbathing.

October 29. Yesterday was devoted to Colin Wilson, who came over yesterday noon with a colleague from Long Beach, where he has been lecturing. (He leaves California today.) He at once started drinking beer and talking about his own books. Few enough people can be honestly egoistic in this way, and most of those few are bores, because they have nothing to say and because they are vain in the wrong way. Vain about their mere success. Colin isn't a bore, because he is really intelligent and because, though he certainly *is* interested in success, you feel his interest is objective; it's in success as a phenomenon. (He spent a long time discussing the question: is there such a thing as bad publicity? He had once posed for photographs with Huxley on the steps of the Athenaeum Club.) Also, he has some mad, half-serious ambition to become "the literary dictator." While in the States, he has spent $1,200 already on phonograph records of operas. He doesn't understand why one should travel, believes in roots, dislikes America because there is too much space, etc. etc. An opinion every ten seconds.

He wanted to meet Henry Miller and Huxley. Miller came around to the house, with his thirteen-year-old son Tony, a very cute tough little blond boy with blue eyes. I liked Miller at once. (Jo and Ben had met him a few weeks ago, and again only a few days ago. They complained—or rather, Jo complained—or rather Jo and Alice Gowland complained—that Miller used such dreadful language and demanded so much to drink and also (said Jo) he

[1] A four-car pileup in July 1960 with Michael Barrie; in *D.1*, Isherwood says Heard was "quite badly hurt."

looked at Ben "in a funny way." This last I simply cannot believe, but never mind—) Yesterday, either because his son was present or he didn't know us, he behaved beautifully and didn't drink a drop. But what mattered was that he was so naturally sweet and really *wise*. Of course, there *are* traces of a pose: the homespun crackerbarrel philosopher who doesn't understand intellectuals. But he is genuinely intuitive and plainspoken and he has digested his experience. I like his narrow squinty eyes and his bald head.

We drove up to see Huxley, who is living in a house Laura's friend Mrs. Pfeiffer bought after her own was burnt in the great fire.[1] It is even higher up the same hillside; in fact right at the top of the lower ridge of hills, with a splendid view of the Hollywood lake-reservoir a short walk away. It's one of those old Hollywood houses, with a garden rising from ramparts out of the surrounding scrub; something secret about it. Only you can't help wondering why there shouldn't be another fire right there; there is a very considerable wild area between it and the lake. At this time of drought, the hills are a somber brown-black or green-black, like the landscape of a much colder climate; Colin was reminded of Cornwall.

Aldous looked rather withered. Colin was very brash and rather embarrassing. And Aldous's reticence made his blunt analysis sound like an attack. Colin kept saying, "I dealt with your novels at some length in my book," and asking, "Why couldn't you and Mr. Hemingway have understood each other?" And Aldous was just pained. It simply isn't in him to defend himself. What Colin was saying was that Aldous and Hemingway put together would have made one really great novelist, and this, of course, sounded rude although I knew that Colin, in his own ungracious way, was paying a sincere compliment—a greater compliment than I could honestly pay Aldous myself, because I don't think that either Hemingway or Aldous could have produced the wherewithal to create one really great character, even if they had pooled all their resources.

Laura was nicer than I have seen her before; good-humored with Colin and playful but not patronizing.

As we drove back down the hill in the car, Tony Miller said, "When all of you intellectuals get talking, you never listen to each other and you never stop to think what you're going to say next. When we kids talk, sometimes we won't say anything for maybe

---

[1] The same fire that destroyed the Huxleys' house in May; see Glossary under Pfeiffer.

five minutes." But then he added, "I guess that's maybe because we don't have much to say." Henry Miller was delighted at this. He told us that Tony hates books and writing and only cares for football and surfing and that he wants to be an engineer. He had made Tony write a hundred words on "Why I Hate Books."

Later we went to see Charles and Elsa, who were at the next-door house. Charles rather drunk, recited a scene from *Advise and Consent*; his accent wasn't quite good enough.[1] Then we had supper with Gavin and Tom Wright, just back from their trip to Arizona and New Mexico, and we went on to see Gavin's new house on Sumac Lane, which is really quite something. Colin was taken back to Long Beach by a not-quite-sober female friend, attractive.

Today I had lunch with Howard Warshaw, who was in town clearing up various matters after his mother's death. He had drawn a wonderful picture of her on her deathbed, in a notebook of colored sketches he is keeping; mostly studies from Rembrandt. I felt more than ever that he is a remarkable artist.

He told me that his mother kept, from his earliest years, a book written *to* him, addressing him as "You," even when he was still a baby and speaking always of his future and then of the impact of his various doings on her life. The whole thing was written on the assumption that he wouldn't get to read it until she was dead, and indeed he didn't. Wouldn't this be an excellent framework for a novel? Howard's mother kept a similar book for his sister.

Something I'd forgotten to record. The satisfaction with which Aldous told us how Corbusier built a large glass structure for the Indian government which looks marvellous but heats up to 140 degrees in the hot weather and is continually busting its air-conditioning.[2] This was to illustrate the architect's contempt for the people who have to live in his houses. (Aldous leaves for India shortly.)

Colin Wilson said one shouldn't learn to speak any other language but one's own. He had been thrilled by the photograph of Henry Miller's daughter Valentine, in the paperback called *The Intimate Henry Miller*. Although we were late for the Laughtons, he insisted on stopping off at the Millers' house in Pacific Palisades on the way back from the Huxleys' and seeing her. She was a pale little girl in curlers. Rather embarrassed by the admiration of the "elderly" Colin.

[1] Laughton was playing the southern senator Seab Cooley in Otto Preminger's film.
[2] Evidently one of the buildings Le Corbusier (1887–1965) designed for the new capital of Punjab, Chandigarh.

Today I have written to Stephen and told him I can't do the Virginia Woolf article. I think I am being quite honest about this.

Worry about my jaw. Depression after reading the *Los Angeles Times*, which is full of fallout shelters. I *must* lay off the newspapers. The newspaper reader dies many times before his death, the nonreader not nearly so often.

A story told me by Michael Barrie: Jesus and the Blessed Virgin go out to play golf. The Blessed Virgin is at the top of her form, drives and lands on the green. Jesus slices and lands in the bushes. A squirrel picks up the ball and runs off with it. A dog grabs the squirrel, which still holds the ball in its mouth. An eagle swoops down, picks up dog, squirrel and ball, and soars into the air. Out of a clear sky, lightning strikes the eagle, which drops the dog which drops the squirrel which drops the ball, right into the hole. The Blessed Virgin throws down her driver and exclaims indignantly, "Look, are you going to play golf or just fuck around?"

October 31. Very sad. It's a grey day and cold. And no word from Don. Of course I know he's busy but I can't help feeling anxious just the same. This is one of the days when you feel all of the six thousand miles between here and London.

In Russia they've exploded the biggest bomb ever and taken Stalin out of his tomb.[1] There's a brush fire near Pasadena. And it's Halloween, which means moppets.

Yesterday, Prema and his friend Ram from British Guiana and Swami Ritajananda came to tea. A purely symbolic act; but Ritajananda wanted to come and he is so sweet and going away soon to preach to the ghastly French. Then I had supper with Gavin at La Mer, who told me that Speed [Lamkin] once said to him that his Jimmy[2] was the only person he'd ever lived with "of my own class"!

Sunday night, I had supper with the Bracketts. Xan[3] wore black. It was like *Mourning Becomes Electra*. And then Ilka Chase[4] and Dorothy Parker and the boy who plays Dr. Kildare[5] and rather

[1] Stalin's body was removed from the Lenin Mausoleum on October 31 and buried under the Kremlin wall after thousands of murders during his purges were exposed October 28 by the publication in the USSR of Khrushchev's 1956 secret speech to the Communist Party's Twentieth Congress.
[2] Davison, tall, good-looking, and wealthy.
[3] Larmore, his daughter.
[4] Columnist and actress (1905–1978).
[5] Richard Chamberlain; see Glossary.

nasty Jack Grate[1] and others showed up, and turned it into a very old and creaky drawing-room comedy. Watched *The Power and the Glory* on T.V. Olivier was awful and I could barely recognize Julie Harris. Then I was alone with Charlie and he began to tell me that [his friend]'s brain wasn't really affected—only one lobe—and he could be cured, and the woman was going to recover, and the whole thing had been greatly exaggerated, etc. etc.[2]

November 2. Still nothing from Don. I had so hoped there'd be a letter today. I'm worried, although I know how hard it must be for him to find time to write. And this morning Russell McKinnon called to say that he and his wife are going over to Europe almost at once and that he'll be seeing Don. I sort of fear that they may persuade Don to stay on indefinitely, which is what Russell wants.

I miss Don more and more every day; without him, life seems really quite meaningless. And I'm increasingly anxious about my jaw. The dull ache continues and there is a sore right up in the back, apparently caused by my bridgework. Yet I don't want to go to either Dr. Lewis or Dr. Sellers; they are so gloomy and hospital minded. Patrick Woodcock would cure me if anybody can.

And now I'm feeling a kind of horror of California—though I remind myself that I was horrified by London too. The California horror has to do with the advertisement life which is lived here. The smiles, for instance, of the women in a bank this morning in Pacific Palisades; smiles that advertised Courtesy and Customer Handling and Financial Integrity and Friendliness. Oh, I am sick at heart.

Saw Gore Vidal last night. He is in town for a few days only. He keeps visiting at the White House. He thinks Kennedy is the most normal president we have had this century. That he is calm and undismayed and that he still enjoys his job, despite all the headaches. Kennedy wants him to be his Minister of Culture, if Gore can figure out what a Ministry of Culture should do. Kennedy is well aware that Gore's private life might be brought up—but, as Gore says, really there is nothing concrete to bring up except his novels. So maybe it will go through.

Gore doesn't think there will be a war over Berlin, but he greatly fears the growth of fascism in this country and its victory

[1] Not his real name.

[2] Charlie Brackett's friend had spied on and then attacked a woman playwright who was a tenant in Paul Millard's building; the friend was arrested and eventually sent to a mental institution, the Menninger Clinic, where he recovered and married. Later he became a painter.

in a few years. Meanwhile, he has written a play about the last Roman emperor of the West, Romulus Augustulus. And he was busy writing a scathing review of the military novels of Evelyn Waugh.

November 5. A glorious day. I feel absolutely sick with misery. No word from Don. It's not so much that I really think anything awful has happened to him as that I long for a word. Without him my life is pointless.

Jaw still bothering me. And I have the shits. I am at the lowest ebb, despite the vitamin pills which a man at the gym sold me for ten dollars a hundred.

This afternoon I have to go off to Trabuco for two days. I hate leaving this house, simply because a letter might come from Don tomorrow. I am a mess.

November 8. Got back from Trabuco yesterday. Came down here to see the extent of the fire—L.A.'s biggest ever—which has been burning all through the surrounding hills from Bel Air to behind the Pacific Palisades. (There is ash everywhere, and if you open any of the windows it begins to drift in. This morning the World War II bombers kept swooping over the ridge which backs the view from my workroom window, dumping borate on the flames. They fly daringly low. Once I saw a great tongue of fire shoot up and lick at one of them. It is moving to see them being used for such a sane purpose.) Then, yesterday evening, I went back up to Hollywood and attended Kali puja, for no reason in the world but to please Swami. I hate the puja itself as much as ever—no, not hate, but it is quite meaningless to me, with all these posturing women fixed up in the saris. Even Sarada, with her hair loose on her shoulders but oh so elegantly arranged, seemed theatrical. Sat next to Jimmy [Barnett], now down at Trabuco, and gossiped cozily in whispers, waiting for Swami to asperse us with Ganges water. This he did vigorously, looking as if he were ridding a room of flies with DDT.

When I got back yesterday evening, there was a packet of my lecture transcripts from Santa Barbara, plus some photos a boy in London took of me and Don, plus a long letter, full of love and work and so all is well. The packet took so long to arrive because it was opened by the customs, which is rather embarrassing, considering the tone of the letter. But probably all this Kitty and Dobbin stuff bewildered them.

Aside from the worrying about Don, the visit to Trabuco was

a great success. It is very agreeable down there now. Franklin [Knight] (Web Milam's cousin) the only *brahmachari*,[1] Jimmy, Eddie [Acebo] the Mexican boy, Len [Worton] the British ex-sailor,[2] Richard Thom and a Richard (Epstein?) the son of one of the Epstein brothers who used to work in movies.[3] Franklin sort of holds the place together, as solid as a barn. Jimmy says he is much happier down there, because the work is more varied. Eddie wishes he could study more; and he complains that at the meetings of the Vedanta Society[,] Trabuco is always slighted and given the smallest allotment of funds. Len just gets on with the job, navy style. Neither of the Richards will stay, probably. Richard Thom is as sly and enigmatic as ever. Richard Epstein is frankly only there on a short visit, "to acquire merit"; he is a good-looking, [...], compulsively talkative boy with a sloppy figure, a chain-smoker.

Ritajananda, who came down with us, is leaving this week to take over the French center in Gretz. He is marvellously placid. Vandanananda is still sulking whenever Swami is present. But he made a point of saying, "Welcome to Trabuco," as soon as we arrived. A possessive act. And it obviously gave him great satisfaction that I went out walking with him both mornings. The Trabuco ranch is baked hard and grassless by the sun. There are big patches of cactus, full of rabbits and snakes. A fire probably couldn't even take hold here. It's going back to desert. But, in a few years, when the water is piped in, the whole surrounding country will begin to be built over.

As we sat in the cloister, with that marvellous, still empty prospect of lion-golden hills opening way into blue distances and the line of the sea, I said to Swami: "You're really *certain* that God exists?"

He laughed. "*Of course*! If He doesn't exist, then *I* don't exist."

"And do you feel He gives you strength to bear misfortunes?"

"I don't think of it like that. I just know He will take care of me.... It's rather hard to explain.... Whatever happens, it will be all right."

I asked him when he began to feel certain that God existed. "When I met Maharaj. Then I knew that one could know God.

---

[1] Evidently, Knight was still preparing to take his *brahmacharya* vows; see Glossary.

[2] Others recall he was an ex-soldier; see Glossary.

[3] Julius (1909–2000) and Philip (1909–1952) Epstein wrote screenplays together, including *Casablanca* (1943), with a third writer, Howard Koch. Philip died of cancer, and Julius subsequently worked alone. Richard was Philip's son, raised thereafter by his uncle.

He even made it seem easy.... And now I feel His presence, nearly every day. But it's only very seldom that I see Him."

Later, when Ritajananda had joined us, he said, "Stay here, Chris, and I'll give you *sannyas*. You shall have a special dispensation from the Pope." He said this laughingly, but I have a feeling that he really meant it. I said, "Swami, that would be a mistake worthy of Vivekananda himself." Just the same, it staggered me.

Swami says that the Hindu astrologers predict that the world will come to an end next February 2. (Rosamond Lehmann told us the same thing in London, only according to her it was merely the destruction of California and it was to be on February 5, I think.) However, even the astrologers are praying that it shan't happen. I remarked to Swami that Ramakrishna had predicted another incarnation for himself on earth and that this, in itself, contradicted any such prophecy. He agreed.

November 12. I now hear from Don that the New York show of his drawings at the Sagittarius Gallery has almost certainly fallen through, and that he'll probably be back for Thanksgiving. I can't be sorry that he's coming back, but it is bad about the show, because he ought to follow on from the London one, right away. Maybe we'll go to New York for Christmas and arrange something. In any case, there's always the prospect of going to the White House, unless Gore is being wishful.

This jaw thing of mine has gone on and off, and for the time being I've decided to do nothing about it, at least unless it gets worse. I can never forget Kolisch's advice (though I certainly haven't taken it), "Never go to a doctor again. It'll only be necessary once, and then it'll be too late."

Then there was a blow from Sidebotham in Stockport. He says the Estate Duty Office has discovered some new grounds for taxing M.'s estate and that he may have to ask me for part of the $14,000 back. This makes me angry as well as dismayed. Wrote him a chilly letter hinting that I had practically committed all of the money and that he might not get paid back for a long time.

With Dana Woodbury's help, the yard is gradually being cleared. Yesterday I broke my glasses. I keep dipping into Mark Schorer's life of Sinclair Lewis; I'm drawn to it by horror-fascination. He calls it, "an American life," and it is, in the most horrifying and depressing sense. The boyish drunkenness, practical jokes, lack of self-assurance—even the awful shaming skin trouble.

Coupled with what Howard Warshaw told me about his mother's journal (see October 29) I keep thinking of a possible father-son novel, about Don and me, more or less. What puts me off, at present, is fear of being sentimental, and also the mistrust of presenting one relationship in terms of another. But the answer to this latter objection is: why do you have to think in categories of relationships at all? Why not simply describe *a* relationship?

November 14. Yesterday morning, Nehru's press secretary, or whatever he was, called me and invited me to lunch. "The Prime Minister's suite," he said. For a moment, I thought he'd said, "The Prime Minister's sweet," and was somewhat at a loss for a reply until he added, "at the Ambassador." I went off rather in a spirit of martyrdom, expecting a mob including Robert Hardy Andrews.[1] But no—there were only fourteen of us, and that included everybody, Nehru, his daughter,[2] a whole bunch of Indians, Will Durant and his wife,[3] Irving Stone and his wife,[4] Marlon Brando and Danny Kaye![5]

Nehru looked old and skinny. He didn't wear his little hat. At first he was very quiet and attentive, watching. His manners were much better than those of the other men present; he never sat down while any lady was standing. He seemed tired and slightly irritable but not viciously so; it was just that he didn't think in our terms. He is like an old schoolmistress, and all the other world leaders are boys in his class. It's almost incredible that such a person could be one of the most influential men on earth; it is also extraordinarily reassuring.

Mrs. Durant, who is like a very very bad imitation—so bad that one is embarrassed for her—how *dare* she try to get away with that accent[?]—asked him, "Mr. Prime Minister, do you still enjoy doing your job?" Nehru winced ever so slightly—it was a Krishnamurti kind of wince—and said, "Well—one would have thought that the word *enjoy* was hardly the most suitable one,

[1] American reporter, newspaper editor, novelist, radio, T.V. and movie writer (1908–1976) of war and adventure films, including *Tarzan Goes to India* (1962).

[2] Indira Gandhi (1917–1984), who first became India's Prime Minister in 1966, two years after Nehru's death.

[3] Ariel Durant (1898–1981), co-author with historian and philosopher Will Durant (1885–1981) of the last four volumes of his eleven-volume *The Story of Civilization*. (They won a 1968 Pulitzer Prize for volume ten.)

[4] Stone (1903–1989) wrote biographical novels; *Lust for Life* (1934) about Van Gogh and *The Agony and the Ecstasy* (1961) about Michaelangelo were made into films. His wife, Jean, was an editor.

[5] American singer, comedian, Broadway and Hollywood star (1913–1987); he hosted a T.V. variety show from 1963 to 1967.

under all the circumstances—" When Durant talked about "The West," Nehru said, "Will you tell me, please, why do you use that expression? What *is* the West?" Brando (who was wearing an absurd little ducktail curl at the back of his neck; his sailor's pigtail for *Mutiny on the Bounty*) said that the nature of man is evil. When I objected, later, that the nature of man has produced the Buddha etc., he answered that such people were really produced by a repression which was due to fear.

Nehru quoted with approval what Vinoba Bhave[1] had said to him, that politics and religion are both out of date, they must be replaced by science and spirituality. "In ten to fifteen years' time," he said, "We'll be either different or dead." He found Russia much more stable than ten years ago, with relatively greater freedom of speech, and rather conservative. Most of their slogans were just lip service to the Marxist dogmas. All their admiration was for American technology. Russians and Americans were very alike; by nature they should be friends. The Chinese, on the other hand, were in their first revolutionary fervor. Also, they consider every one else on earth barbarians. When they are on top, it is natural for them to expand. The USA was more socialist than India.

Khrushchev had said to Nehru, "You're such friends with the Americans and English, can't you bring us together?" When Khrushchev came to visit India, he was delighted by his reception. He expanded and became a different person. But then a British journalist was rude to him and he was furious and aggressive again. When Nehru told him he shouldn't get upset so easily, he said that Russia had been isolated so long, fearing attack, that Russians are suspicious of everyone. Nehru says that Indian nationalist propaganda always discriminated between the British imperial system and the British as individuals[—]so effectively that an Englishman could walk through an Indian crowd in the midst of the rioting and not be harmed.

Durant talked about the warmongering of the *Los Angeles Times*. Stone, who had been touring the country, said that Americans don't want war. He agreed with Durant that there is an evil competition in the business world to scare the public into buying fallout shelters, even calling it *a patriotic duty* to do so. I quoted the radio announcer, the night before last, who told us as though it

[1] Indian spiritual leader and reformer (1896–1982), a disciple of Gandhi; he walked 40,000 miles during the 1950s asking landlords to give their land to poor farmers and villages.

were grave news that the fallout from the latest Russian bomb test hasn't been nearly as big as was anticipated.

We had a miserable lunch of tepid overfried food. Durant was the only total vegetarian present. No alcohol except sherry. Someone managed to spill ice cream all down the back of my jacket. Nehru parted from us very agreeably. I haven't the least idea why we were invited. The only serious question Nehru asked me was what I was writing now. I told him about the novel and then mentioned the Ramakrishna book. He just barely flicked an eyelid, though this would have seemed a good conversational opening for us. And the questions he asked the others were equally unsearching ... Ah well, no doubt he got something from just watching us.

Supper yesterday evening with Evelyn Hooker. The fire the other day got so close that she moved out "The Project"[1] into cars, and it took days to put back again. The people next door have just had a fallout shelter installed. Evelyn says that the treatment of homosexuals in Oslo is actually worse than in London. True, it is legal but there is very strong feeling against it socially and in business, and the homosexuals have to have their clubs more or less in secret. There are no queer bars. Evelyn wrote a paper about "The Homosexual Community" to read to a psychological convention in Copenhagen last summer. I must say, she makes it sound madly glamorous and thrilling, a mixture of the Mafia, the Foreign Legion and Alice Through the Looking-Glass. She refers to a heterosexual as "a representative of the dominant culture"!

Glorious weather today. I toted wood up from the walk below, where the men had left it after cutting it up for firewood, and stacked it against the retaining wall behind the house. Then I lay in the sun. Then I went to the gym, where Richard Egan[2] works out in a hooded sweater with a mackintosh pair of pants over it, presumably to make him sweat that much extra. Dogged work at the Ramakrishna book; right now, it's boring me nearly to the point of paralysis. But one can always *just* struggle on.

November 16. Talked to Pat O'Neal[3] in the steam room at the gym. Chiefly about Kevin.[4] How aggressive he is, and how he hates being so small and how his I.Q. is just below that of genius,

---

[1] I.e., her notes and questionnaires on homosexuals.
[2] American movie actor (1921–1977), briefly a leading man in the 1950s.
[3] Patrick Ryan O'Neal (b. 1941), stuntman and amateur boxer, later a star as Ryan O'Neal in "Peyton Place" on T.V. (1964–1969) and in *Love Story* (1970) and other movies.
[4] Kevin O'Neal (b. 1945), his brother, also a T.V. and movie actor.

and how he loves to box. Kevin is working out like a maniac, these days, because he is building himself up after an operation he had two months ago, on his bladder. It was taken out and manipulated to get it into the normal shape, and now he is permanently 4-F.[1] He has three scars branching out from his penis—two old ones at the sides which are because of hernia, and the new, vertical one in the middle which goes straight up from the root of his penis to his navel. Pat called Kevin in to show me these. Kevin said, "Three scars, all pointing to the same thing—that's symbolism." Pat said that sometimes he wants to kill Kevin; the rest of the time he loves him so much that Kevin kids him about it.

Working out is already having its effect on me, though I haven't lost one ounce of weight. But I do feel very good, all except for my jaw, which is just the same as ever. I get scared about it, and then I relax and decide to postpone having it looked at until Don gets back.

Yesterday, I went to see Judith Anderson and Bill Roerick doing their snippets from *Macbeth*, *Medea* and *The Tower Beyond Tragedy*.[2] I was chiefly impressed by Bill, whom I've never seen act before; he held his end up very well, has a nice resonant voice, and humor. But afterwards, when we talked in the dressing room, he told me about some money he had just inherited or otherwise acquired, and said that he was so happy because now he could help struggling young actors, etc.—and oh, it was false, false.

This from Aronowitz and Hamill's *Ernest Hemingway: The Life and Death of a Man*. I think it's one of the craziest sentences I ever read: "Cuba, of course, was then, as it remains, only ninety miles away from Key West...."

Have just remembered how Bill Roerick also told me how he had prayed for Morgan, when Morgan was sick in Cambridge, and how he feels that Morgan is somewhat changing his agnostic attitude in his old age. He told a wonderful story of a dream Morgan had. Morgan dreamed he had died, and went upstairs to a room full of corpses of hideous old men. The corpses all got up and welcomed him, and he forced himself to embrace and kiss them. And then they told him to go downstairs again and out into a garden, where his mother was waiting.

November 24. Last Sunday, I think it was, Glenn Ford came around and talked about Hope. How they'd had an affair and she

[1] I.e., classed by the draft board as deferred from U.S. military service.
[2] By Robinson Jeffers, who also made the adaptation of Euripides's *Medea*.

said he was the best sex she'd ever had and she loved him, and how now she said she was out of love with him and didn't want to see him again and how nevertheless she did keep seeing him and they'd have a marvellous time together and then she'd say it was no good, he frightened her, she didn't want to get involved. Glenn said he couldn't sleep and couldn't eat. He loved her so, he demanded nothing from her, but she ought to marry him because otherwise she'll go to pieces, she'll destroy herself. She's running around with young kids; mind you—he's not saying she's necessarily going to bed with them. But just the same, it's all wrong, she's nearly thirty, she's got two children, she ought to get wise to herself before it's too late. He told me all this low voiced and self-righteous and weepy and presently he actually burst into tears, moved to the heart by self-pity. I couldn't feel sorry for him, but I was helpful and encouraging, and now I'm stuck with interviewing Hopey tonight, buying her supper and pumping her to find out if she really loves Glenn or not. Ivan Moffat and Kate feel sure that she is really tired of him; he's too solemn and noble. No—I *am* sorry for Glenn. How can I not feel for him, having so often been in the same boat? But I do see how dreadful to be dull, and how hopeless it is, as soon as someone decides you are.

Jaw the same; neither better nor worse.

Don may be coming home next week. I cabled him asking him to do that and then we can plan about going to New York.

Laughton and I have reopened the Plato project. It looks very promising. I haven't shown him what I've written, yet. He and Elsa are at dagger points; poor Bruce[1] looks like a mortician's assistant. However, Terry is coming in January, to go on another reading tour with Charles.

Saw Gerald Heard yesterday. He was in the highest spirits. Told me that the Luces don't expect war now. But in September they were very seriously alarmed. Clare Luce had told him how you know when a crisis is really bad. First, the volume of short-wave radio communication increases enormously between the capitals of the countries involved; then suddenly it stops almost altogether. This is the fatal moment when a decision has been taken. There is to be no more compromise; so embassies and other agents stop asking for any more secret instructions.

Yesterday was Thanksgiving. As always, there are only two basic reasons for me to give thanks: Prabhavananda and Don. Though, of course, I'm glad that I have written such a good novel. I do

[1] Zortman, Laughton's secretary and a disciple.

believe it *is* good, this time. Had supper with Jo and Ben. Gavin was there, also Bill Harris, who has been staying down at La Jolla with his mother. We talked about the "great" days of the Canyon, which was a bore for Gavin and anyhow sad; especially because I got the feeling that Bill had only really lived then. What an amazing animator of other people Denny [Fouts] was, and Bill Caskey!

Have attacked the three Henry VI plays of Shakespeare again and am on the second. They are interesting as long as you don't stop reading for long. Have also just read right through the poems of Cavafy. Have decided that, for the Enjoyment of Literature course at LASC,[1] I'm going to give them *Macbeth*, *Wuthering Heights*, *A Farewell to Arms*. My twentieth-century British books will be [Forster's] *Where Angels Fear to Tread*, [Greene's] *The End of the Affair*, [Lawrence's] *The Virgin and the Gipsy*.

November 26. When I had supper with Hope, the night before last, I got a pretty strong impression that she won't go back with Glenn whatever happens. But of course I couldn't tell him that. Why in hell should I? I compromised by telling him to play it very cool when they were together, always be gay, never make any demands, etc. This is utterly impossible for Glenn, anyhow. His heartbroken attempts at gaiety are the weepiest thing imaginable. Even while he still did have Hope, they were nearly as depressing.

Last night, I took Stanley Miron out, the nice and attractive Jewish doctor we met in London.[2] Well, he was even nicer and more attractive and he assured me we'd had a great time and that I was one of "the most beautiful people" he'd ever met. Funny, I can remember so clearly, the last person who said that to me was Eddie From, back in the forties. Maybe it's a Jewish thing to say. I drank far far too much and saddened myself hideously today. There's nothing that saddens me more than putting on my act, as I did last night. Yesterday it rained quite hard, today it's clear and beautiful; but Don is absent from the sunshine. I am needing him so much, and somehow uneasy that he will delay his coming.

Enough for now. Tiresome selfish old Charles is coming in to talk Plato. Plato bores me pissless in my present mood. I am idle and jittery, and, despite all my sweating at the gym, I can't even get down to 150 lbs.! But at least I didn't fail to do a page of Ramakrishna today. I must go ahead and at least finish this chapter.

[1] Los Angeles State College.
[2] He settled in San Francisco, where he ran his practice and where, for a few years, Isherwood and Bachardy continued to see him whenever they were in town. Stanley Miron is not his real name.

November 29. Grey today. Yesterday afternoon, such a strange ghost-berg of fog came stealing in over the surface of the smooth sunlit bay. It looked as if it were actually floating on the water. Dorothy, who was here cleaning, was quite scared by it. And last night it was thick all along the coast and the airport was closed down.

Maybe that's why I haven't heard from Don. I was hoping for a letter today. On Monday morning, two days ago, I got a cable: "Definite offer for show January 2 to 13 will go to New York mid-December writing Julie asking if we can stay with them letter on way love Don."

I'm delighted and thrilled, of course; though disappointed momentarily because I'd been hoping to see him here by the end of this week. Also, we won't really be together if we stay at Julie and Manning's, and a hotel would be so expensive.

Now, more than ever, I must crack the whip and get something accomplished. I am very worried about my throat and jaw. At least, I go through fits of worrying every day. Last night, I had supper with Morrie Blumberg, the Jewish writer who lives down on Mabery Road. His wife is an ex-nurse and she scared me by talking about the nursing of patients with a "terminal" illness; it seemed like an omen. This morning I woke feeling terrible and quit[e] sure this thing is steadily growing worse. Suppose I do have to have some ghastly operation, and then die six months later, being jollied along and told it's rheumatism or sinus trouble? Well, unless this thing gets really much worse, I am going to leave it until I get to New York. I cannot bear the prospect of being shipped off to the Cedars of Lebanon without Don. And then of course he would have to come out here and it would wreck his show. At least, if I were in New York, he could go ahead with that and still see me. Also, oddly enough, I feel that Wystan would be a great support, at such a time. And I do not want to see either Lewis or Sellers.

At seven this morning I was yanked out of bed to receive a special delivery letter—from Dr. Abraham Kaplan of UCLA, asking me to sign a telegram to the *Los Angeles Times*:

THE RECENT TRIALS OF LEADERS OF THE MOSCOW AND LENINGRAD JEWISH COMMUNITIES AND THE LONG PRISON TERMS TO WHICH THEY HAVE BEEN SENTENCED, ARE CAUSE OF GRAVE CONCERN OVER RENEWED ANTI-SEMITISM IN THE SOVIET UNION. IT IS ESPECIALLY DISTURBING THAT ON THE

VERY DAYS WHEN STALIN'S UNSPEAKABLE ACTIONS WERE
AGAIN DENOUNCED BY KHRUSHCHEV, THE SOVIET GOVERN-
MENT SAW FIT TO TAKE STEPS APPARENTLY DESIGNED TO
STIFLE WHATEVER LITTLE SPECIFICALLY JEWISH LIFE HAS
SURVIVED THERE. WE APPEAL TO THE SOVIET AUTHORITIES
TO RECONSIDER THEIR POLICIES TOWARDS SOVIET JEWS, SO
AS TO RESTORE JUSTICE FOR THOSE SENTENCED AND FOR THE
WEAK AND DEFENSELESS MINORITY THEY REPRESENT. MEN OF
GOODWILL EVERYWHERE ARE SHOCKED BY OFFICIAL PERSE-
CUTIONS OF ANY PEOPLE. IT IS NOT ENOUGH FOR THE SOVIET
UNION TO PROCLAIM HIGH PRINCIPLES OF THE EQUALITY OF
PEOPLES. THE WORLD HAS A RIGHT TO EXPECT THAT IT WILL
ALSO ACT ON THEM.[1]

I called Dr. Kaplan and told him I wouldn't sign this because
it was "too general" and I added that I had made up my mind to
sign only petitions on local and particular issues. Kaplan was very
nice and said he quite understood this. But of course there was
much more to it, as far as I'm concerned. I cannot help feeling
that this is simply not my affair. The occasion is badly chosen, too,
from a tactical point of view, because the Jews are accused of other
crimes and the evidence on either side is extremely vague. I can
honestly say that I wouldn't sign a similar document relating to the
persecution of the Soviet queers, either.

Morrie Blumberg was terribly upset last night because one of
these right-wing fascist groups has actually been using a swastika
on its propaganda, with the slogan "We're back!" I said I thought
it was a good thing, because by doing this they have overreached
themselves and will surely alarm most people who might otherwise
be impressed by them.

Jo and Ben have Jo's daughter and her husband with them this
week. Jo and I had a talk about this on Monday, during a walk
on the beach. It seems that she has never told her other friends
here that she has a daughter (and a son); not even Peter and Alice
Gowland. And now she is terrified that they will find out. "Not
that I mind their knowing," she says, "only to go back and open
up all that old stuff again—Chris, I just *can't!*"

There seems much more in this than meets the eye. *Why* does Jo
have to conceal this? Perhaps because of Ben; she says that he hates
to have it talked about, and didn't like seeing the daughter Betty
[Arizu], though now he says he doesn't mind. Is it because he has

---

[1] See Glossary under Abraham Kaplan.

hated to admit to himself that he has married an older woman? Is he, or was he, afraid of being kidded about it? And therefore unwilling to have to admit to Jo's advancing age by seeing how even Betty is getting older? [...] a delightful time must be being had by all.

December 4. A time of uncertainty. Yesterday, Julie Harris called me from New York to ask, "Where's Don?" Well, it turned out that in her vague actressy way she had never written him to say we could stay with her; so he had cancelled his reservation! Now we'll see what happens. Meanwhile, John Zeigel is arriving the day after tomorrow, expecting to stay a couple of nights here before going on to San Francisco.

I am uneasy and restless, as I always am when I am about to be uprooted. And yet of course I *want* to go to New York; or rather, I want to see Don and his show.

Tomorrow, I should finish chapter 12 of the Ramakrishna book. The Laughton project hangs over my head, and it may well end with my backing out of it, one way or another. I think I would, if it wasn't for the money, which would have to be paid back, or partly.

I have a sort of Shakespeare craze; a real appetite for him, such as I've never had before. The more I read of him, the more I seem to want to. Have run through *Richard III*, *Timon of Athens*, *Henry VIII*, and now I think I'd like to reread *Cymbeline*, *Pericles*, *A Winter's Tale*, maybe *Titus Andronicus*.

Cannot lose weight. I'm at 153! And yet I exercise like crazy and am really putting on muscle and getting much stronger, even in this short while. As for my jaw, I don't know. I don't think it is any worse; but not really better, either. The basic difficulty is still there, and the only difference is in my attitude to it, on different days, in different moods.

Yesterday evening—or was it the evening before?—I saw the green flash again. Vivid and sharp and localized.

December 9. A week of uncertainty: where was Don? But this morning I got a letter from him. He is still in England and may not be coming over until toward the end of next week. He will cable me when it's definite. Meanwhile, I can't quite make up my mind—should I go to New York before he does, so as to be on hand to meet him?

I have been very bad and not accomplished anything much. I did however get chapter 12 of the Ramakrishna book done. Now

I am haunted by the Plato project. Can I really do it at all? I find it so hard to settle down to, when I'm apt to be leaving any time.

John Zeigel was here, for one night, en route from Mexico to San Francisco, where Ed Halsey has bought an apartment house. This time, he seemed much less interesting, almost dull. Partly, perhaps, this was because I was worried about Don and so couldn't give him my full attention. But I think I did detect the stultifying effects of Mexico and of Ed. John seemed much more political, and more argumentative, which bored me.

Ben Masselink has been coming with me to the gym. Jo went too, yesterday, but today she has a bad cold. They won't be leaving here until after Christmas, because of Ben's T.V. story. I had supper with them last night, and they fixed a bouillabaisse, maybe the best anyone has ever eaten anywhere. Marvellous, anyhow. Jo kept moaning how she wished Don was there, went on so long about it I got quite depressed.

Oh God, John Zeigel did make Ed sound boring! Whenever John steps out of line, Ed proceeds to get psychosomatically sick and lies in bed refusing to speak!

Tentative decisions about my teaching next February: I will tell my writing class that I don't want completed short stories, just passages of description, dialogue, etc. Also, they must all agree to having their work read out, *with their names mentioned*, in class. If they don't want that, they can't come. For myself, I'll promise not to let anyone audit the class. *Anyone* who tries to sit in without having written something must leave the room.

Gerald Heard praised Pepys and Sterne warmly, last time we were together. So I have started reading Pepys and I like him very much. This is only an abridged edition; maybe I'll get the entire diary.

On Tuesday (5th), Jo and Ben and I drove up to Chalon Road in Bel Air to see the fire damage. The hills burned grey looked like wrinkled grey elephant's hide, and the burned bushes were like ugly black hairs on it. God, the ruins were ugly! A long winding street of houses burnt to their foundations, with only the brick chimney standing. Twisted water pipes and blackened cars and iceboxes and a huge litter of rubble and wire and smashed glass. Many of the lots have been cleared already for rebuilding, and their owners have planted notices, PRIVATE PROPERTY KEEP OUT, though there is still nothing to keep out of. After this, we drove up the Hollywood Hills because I wanted to show Jo and Ben the walk I took the other day with Aldous and Colin Wilson and Henry Miller, looking down on Lake Hollywood. We started out and suddenly, around the corner of the hill, came Aldous himself, just back from India, full of scorn for the [Rabindranath] Tagore centenary and of admiration for the non-neurotic way children are raised in the Orient; you never hear one cry, he says. He looked very thin, very grey and old, but he has the stiff-limbed energy of a stork. He walked up and down the hills, never stopping talking for a moment and never seeming out of breath.

December 16. Here I am in New York. Without warning. For once I took off on a journey without leaving any "farewell" note in my diary. Chiefly because the departure was so sudden. On Monday morning last, the 11th—having still heard nothing from Don—I decided it would be best to get to New York anyhow that week, lest I should be prevented by the holiday rush. After all, I knew I could stay with Julie and Manning, and why shouldn't I be there to welcome Don's plane? So I went up to the travel bureau in Santa Monica and bought a ticket for Thursday 14th.

But, no sooner had I got back home, than a Western Union boy arrived, grinning all over his face—maybe because the telegram he brought said that a stray kitten was wandering around mewing for a horse, and the horse should be fetched at once. It was signed "N.Y.S.P.C.A."

So I got my ticket transferred to that night, ate supper with Jo and Ben and took off a quarter past midnight, in a huge three-quarters empty jet as dull as a Statler hotel. Arrived here early on Tuesday morning, spent hours stalled in the morning traffic jam, in a grey drizzle, and arrived to find Don looking radiant and about seventeen years old, despite all his worries and infuriations. He fixed me breakfast in the presence of Peter [Gurian], grown larger and more spoilt, and his nurse [...], a false-sweet woman who bitchily pesters you to let her do things for you and then complains that she's overworked—Manning had already warned Don of this. Manning is still Manning; I can't see him ever changing, and that certainly *is* rather a relief in itself. As for Julie, when I caught sight of her for the first time later in the day, I saw her transformed, at least to some extent, into the French maid she is now playing;[1] just as she transformed herself into Sally Bowles and Joan of Arc. She still looks wonderful. Her voice has assumed a slightly conscious mock-solemnity, somehow reminiscent of Garbo's.

Well, Wednesday it turned cold and clear. We bustled about, enjoying our being together, and Don bought a tuxedo, because

---

[1] In Marcel Achard's *A Shot in the Dark*.

the man who owns the Sagittarius Gallery where he is going to exhibit, Count Lanfranco Rasponi, had organized a ball for the night of Thursday 14th, and had told Don he should come to it and meet people who it would be advantageous for him to draw. That evening, we went to a party given by Gore Vidal and Howard Austen. Gore was leaving next day to go to Hartford for the out-of-town opening of his play about Romulus Augustulus. The party was a crush, and the only good that came of it for us was that Don arranged to draw Myrna Loy. Then we went on and met Marguerite Lamkin. She looked older [...]. She took us off to a claustrophobically dull party and then arranged that we should be taken out to dinner with her by one of her admirers, a blond unpleasant man named Norman Hickman,[1] to the Colony, where the food was nearly uneatable. [...] She went on and on about Freddie Ayer, the logical positivist, who, according to her, is in love with her and wants to divorce his wife and marry her. And she succeeded in maneuvering me into saying that I should refuse to meet him, because I knew she wanted to provoke an argument between us and, frankly, I thought his kind of philosophizing was a stupid and childish waste of time. Now, Marguerite had absolutely no motive for stirring up all this mud. It wasn't even in any way in her own interests [...].

Still, I wouldn't have hated that evening with such positively metaphysical violence if I hadn't been in a thoroughly toxic condition. That night, and next morning, I was in a high fever and everything in my throat was swollen. I went to see David Protetch, the nice youngish doctor who is a friend of Wystan and the Stravinskys and is said to be slowly going blind.[2] Aside from feeling ghastly and shaking with chills—I would have sent for him to come to me here except that I didn't want to scare Julie and Manning at the thought that I might infect and incapacitate her, not to mention their precious Peter—I was deadly scared because I thought this is the showdown; he'll find that I have cancer of the jaw. When he said that he'd better take some blood and have it sent in to the lab for tests, I was so weak that I nearly fainted. But it was great to get back and into bed and let the shivers warm into fever. And the sore throat yielded at once to penicillin and antibiotics. I dozed through the day, feeling the most voluptuous appetite for

---

[1] Manhattan stockbroker (19[20]–1989), educated at Yale, decorated twice during W.W.II. Later, he was associate producer for two Hollywood films, wrote quiz books, and married a Palm Beach divorcée.
[2] Protetch was diabetic, careless about treating himself, and a hard drug user; he died of pituitary cancer, evidently a consequence, in 1969 aged forty-six.

sleep. Don came back at midnight disgusted with himself, saying he had behaved in such a feminine way, waiting to be shown what to do. Rasponi had said to him, "You should be dancing," Don had answered, "I don't dance" and Rasponi (who'll be sorry for this later, if I have my chance, the old vicious snobqueen) said, "What else do you go to a ball for?" Don, quite rightly, feels it was feminine or at least naive or at worst masochistic of him not to have been prepared for all this to happen. But it was not for Rasponi to talk to him like that.

The next day, yesterday, was better; because I felt better and because Wystan came and because Don discovered that you can buy boys' suits, of a size that is exactly his, at less than half what men's suits cost; so he went out and bought three and asked them to hold a fourth! Wystan told me that Lincoln Kirstein was offered and refused the Ministry of Culture (does this uncover another of Gore's lies?), syphilis among New York boys has reached epidemic proportions (one boy infected 200 people!), he believes that Shakespeare's sonnets were written to several quite different boys. His conversation always has this quality; information spiced with gossip and vice versa. Either one alone gets boring.

In the afternoon, I called Dr. Protetch, who told me that not all of the tests had come in but that so far they showed nothing significant. Obviously, he wasn't alarmed. Today I have been up and around the streets. It is bracingly cold. My throat is almost perfectly well. My jaw feels much the same as ever. I had very bad stomach flutters in the night and again today. And I notice how poor my eyesight is getting. I think I am in a deep dip physically. In California I staved off the crisis for a while by going regularly to the gym—we can't seem to find one here that doesn't cost a fortune—and being able to choose exactly what I ate and drank. But what really matters is, can I be of use to Don here? Otherwise I'd be far better off in California, getting on with my work. I very much doubt if I can work here, though I must try my hardest.

December 22. This week has been spent rushing around, buying Christmas presents and addressing envelopes to send the catalogs in, for Don's show. I have been in curiously good health. It's as if my two days in bed merely provided a needed rest, and the penicillin cleaned out a lot of lurking toxins. Anyhow, I have been on the go every day without ill effects, and it even seems to me that my jaw is better. Don, however, is suffering from stomach cramps.

Have met a lot of the people at Simon and Schuster's, and won a battle to keep the jacket of my novel just as Don designed it.

Julie is marvellous in her part, but what a crappy play! It is tragic to think of her confined in it for the next year. A disgraceful waste of her talent. If she were on the stage in England, she would have played at least half a dozen parts this year, and still done the same amount of television.

Lincoln took us to the ballet. He was very funny about a command performance of *Macbeth* they did at the White House. The guest of honor was some ruler or other from Africa (or was it Arabia?); in any case, Lincoln was suddenly told that it wouldn't be tactful to do *Macbeth*, because this man had murdered his predecessor in office! Lincoln told them he was sorry, that was the only Shakespeare available; so they did it, and the guest sat there quite unmoved and greatly enjoyed it. There was one crisis, however, when Lincoln found that the White House security police had confiscated the daggers, because no weapons are allowed in the neighborhood of the president.

Needless to say, I have done nothing on the Plato project. But at present I'm excused by myself from that, until I have finished addressing all the envelopes. I hope to finish them off right now, before going to meet Don at the theater and see *Daughter of Silence*.[1]

Tomorrow, we go down to the country to spend a night with Gore Vidal and Howard Austen; then back and to Wystan and Chester's Christmas Eve party.

December 27. It was quite beautiful at Gore's, because, when we woke, the snow was falling and the river was full of grey waves and the bare trees had marvellous tints of pale pink and yellow, against the snow. Both Don and I felt a great warmth from Gore and Howard, though on very different wavelengths. Gore is always cool, Howard is the eager loyal spaniel who irritates him with his fussing.

We made it back to New York in time for Wystan and Chester's party, which was embarrassing for various reasons. The food was very badly cooked, sweet tepid venison and mushrooms that smelt of piss. And Chester had asked a lot of people Wystan disapproves of, so he wouldn't talk to them; and this subtly *pleased* Chester. Then Wystan talked embarrassingly about theology. "We're not told the size of Christ's cock, but we know that it would have been the average size for that particular time and place, neither

---

[1] Adapted by William Morris from his best-selling 1961 novel; Emlyn Williams was in the cast.

smaller nor larger." Also, he refused to let Billy Vinson[1] have a third cocktail, simply because he had decided there wouldn't be time before dinner. I think Billy Vinson (who feeds lice off his bare arm for some scientific institute at Boston) is a spiteful bitch, who really loathes Wystan in his heart. But justice is justice.

Christmas Day was pleasant. We saw a quite good film called *From a Roman Balcony*, with a sexy Italian boy in it[2] who looked very like Ted Bachardy, and Graham Greene's play *The Complaisant Lover*, in which Googie Withers[3] was marvellous, and we went to two parties. Yesterday, we were to drive down to Philadelphia with the Paul Newmans and Howard to see Gore's play *Romulus*. Don couldn't come but I went. The play was charming and intelligent Shavianism, maybe the best thing Gore has written, but also very poorly acted and directed. I came back on the dreary late train, got home at 2:00 a.m. to find that Don, in one of his manic states, had been working for eight hours without a break, finishing off the last of the envelope addressing, writing-in, sealing and stamping. I helped him for about an hour; then we went out to an all-night post-office on Lexington and mailed them all and then had vodka martinis and Don ate. Today he went off early to a pair of sitters, Mrs. Douglas Fairbanks[4] and Se[r]ena Stewart,[5] and we are to meet later at Sardi's with Tennessee, Frank [Merlo], and Tennessee's folks, and all go to see *The Night of the Iguana*!

This New York visit is even more madly rat-racy than usual, but I think I shall look back at it as a happy time. Don and I are getting along marvellously, despite all the strains.

December 28. Rain, turning to wet snow. The town is as gloomy as a fjord. The lights from the great towers shine palely through clouds. Down on ground-level it is raw and dirty, and all the delights and promises of the metropolitan bazaar can't conceal the

[1] John William Vinson (191[6]–1979), professor of microbiology at Harvard, was an authority on tropical and venereal diseases and also a poet. He was a friend of Kallman.

[2] Jean Sorel (b. 1934), star of the film, is French-Canadian, but the film was Italian and dubbed into English.

[3] English actress (b. 1917, in Karachi), mostly on the London stage and in British films; from the 1950s she worked in Australia where she appeared in *Shine* (1996).

[4] Mary Lee Epling Hartford (191[1]–1988), former wife of A&P heir Huntington Hartford, and second wife of actor Douglas Fairbanks, Jr. (1909–2000).

[5] Daughter of wealthy New York socialite and beauty, Mrs. William Rhinelander Stewart, Jr; Isherwood wrote "Selena."

fact that we're on a wretched wintry island in a flat ugly land, too far north.

Don's rat race seems far more desperate here than it did in London. We staggered up from our beds dazed with sleep and already far behind schedule. There is just barely time for breakfast at the place around the corner on First Avenue, and then Don is off, darting through the traffic—the lights are *always* against him, it seems—with his awkward drawing board and his kit bag full of brushes, inks and pencils. He admits to the feeling that, if he were to stop rushing, he wouldn't be able to work at all. This is probably true—at least, as long as he believes it.

As always, however, breakfast is one of our best times together; the time when we really lose ourselves in conversation. At other times, Don is so apt not to be listening, because he is in a daze of thinking of three or four other things simultaneously.

Last night, we saw the preview to Tennessee's *Night of the Iguana*. Neither of us could make much out of it; it seemed just to wander along. Bette Davis was unexpectedly vulgar in blue jeans and a red hair-dye, but she added a bonus of disgustingness by having her shirt open and displaying the inner curves of her deflated breasts. Margaret Leighton was most distinguished, but without much character. Poor Tennessee was frantic as always; at openings he is like the character in Stevenson's "The Suicide Club."[1] There was a strange scene as we left the theater: a drunk man who had been an assistant stage manager or something and was now no longer working there, stood at the entrance yelling at everybody. No doubt he had a grudge because he'd been fired. But Tennessee took this personally and shouted threateningly at him. And later, on the street, Tennessee bumped into a blind man, who got very mad, "Do you want the whole sidewalk to yourself?" Now Tennessee plans to leave for Jamaica, or Tahiti, or Europe, right after the opening, with an old friend who has just been released from an insane asylum.

December 30. *The Iguana* got quite good notices, nevertheless; so Tennessee has postponed his departure until after the New Year. It's cruelly cold again. How I loathe the cold and the heat, and more than either, the whole silly cult of "seasons"! Don says he hates New York. But he loves London, and I think we shall have trouble when we go back to California. Never mind, we'll cross

---

[1] Presumably the young man giving away the cream tarts, who must eat any refused by the strangers he accosts.

that bridge when we come to it. At present, all is harmony. I am very happy to be here with him, and very miserable to be here. I have a nasty sniffly cold, and I feel so lazy and tired. I don't want to do any work. I sprawl on the bed, as soon as Don has gone out drawing, and skim through books that don't really interest me, Hesketh Pearson's book on Tree, Duggan's novel about the emperor Elagabalus.[1] Outside, behind this house, they have made a huge excavation for an apartment building. The men on the project start working with steam drills around eight o'clock and continue all day. At least they have been doing that—standing on a horribly narrow ledge about thirty feet above the ground and destroying their own foothold with the drills. Now I think they are about through and will start the building, which'll probably be even noisier. This frantic building and tearing down is going on all over central Manhattan. The streets have great gashes in them, covered over with timbers or iron plates, so they can be used. And, as usual in winter here, the gratings in the streets keep fuming away volcanically. I am getting terribly slipshod again with my japam. One day, shortly after I got here, I missed doing it altogether; entirely forgot. This forgetfulness, which is really a sick resistance to japam itself and all it stands for, makes me so mad with hate that I could actually flog myself, the way those crazy old monks used to. But, of course, that's exactly what "it" wants. The only way to deal with it is to remain calm and peacefully determined to continue. One night, I had to make japam after we'd turned the lamp out, under the bedclothes.

Dashed down to the framer, Mr. Ohms,[2] this morning, to wheedle him into framing Don's stop-press drawing of Tammy Grimes.[3] He was sulky, but quite easy to handle when we found we had both lived in Berlin. Then up to the Sagittarius Gallery to deliver a batch of envelopes to Mr. Nuti, who is supposed to be looking after it. He arrived to do this just as I drove up in a taxi, at 11:15! I have a feeling that this whole operation is unspeakably sloppy. And I'm already preparing the row I propose to have with Count Rasponi, just as soon as he can be of no more use to us. I plan to tell him, "You're just another of these wop closet-counts!"

December 31. Well, here we are at the end of this year. As far

---

[1] *Beerbohm Tree: His Life and Laughter* (1956) and Alfred Duggan's *Family Favorites* (1960).
[2] Carl Ohm.
[3] American actress (b. 1934), she became a Broadway star in *The Unsinkable Molly Brown*, which opened in November 1960 and was still running.

as Don and I were concerned, it was certainly a good year; one of the best. It was also a year in which the world came quite near to blowing itself up. One of the most experienced Western diplomats in Moscow, when asked what he thought was the most significant development of 1961, answered (according to *The New York Times*) "we survived." And 1962 is just the same in prospect. Don and I have every reason to be optimistic. We have more money than ever before, or shall have: maybe as much as thirty thousand dollars when all of M.'s legacy has been paid off—not to mention our three-quarters-paid-for house. Don's show won't be town-shaking, but it is sure to add to his reputation, and I hope my novel may do well, too. Our relationship has never been happier. (Don even said, yesterday evening, that he had "a wonderful life," and that's something he very very seldom admits to, since the earliest days of our being together.) On the other hand, all this tiny private world, like billions of others, may quite easily be destroyed before the year is over; and it's probable that total war will, in any case, not be avoided in the course of the next ten years. Indeed, the chief hope for survival is in the development of apter weapons which will localize destruction and keep it down to the measure of purely military requirements!

Having said all of which, the answer, as always, is Forster's answer: Get on with your own work, behave as if you were immortal.[1]

1962

January 6. Rain today. Dark sad weather. I wasn't able to make any '62 entries before this one because of preparations for Don's show and the breakdown of his typewriter, which has only just come back from being fixed.

On January 1, we saw *The Play of Daniel* which is beautifully done and I guess enjoyable if you like twelfth-century religious drama with quaintness and chanting in Latin. It was in a church. Wystan, who had written the narration, luckily didn't appear and embarrass us.[2] Don ate candy—his early movie upbringing makes

---

[1] "The people I respect must behave as if they were immortal and as if society were eternal." In "What I believe," *The Nation*, July 16, 1938.

[2] The thirteenth-century musical drama was edited by Noah Greenburg for the New York Pro Musica, of which he was director. Auden's verse narration was spoken between episodes. This version was first performed in 1958.

him quite unable to see that eating candy in a church, or a theater for that matter, is different from eating candy in a movie. He left early in desperate boredom. I felt I had to stay and was rewarded by getting a ride home with Pavitrananda (of all people!). Later, Julie and Manning showed us the kinescope of her T.V. performance in some of Housman's Victoria plays.[1] I thought Julie was marvellous—until the makeup in the old-age scenes got so thick that she simply could not act through it—but Julie put on a strange almost hysterical show of self-disgust when it was over. She said she was "vapid" and "shallow."

On the 2nd, Don's show opened. We stayed on duty at the Sagittarius Gallery from two-thirty till seven, lunchless. Count Rasponi showed up, but he obviously didn't give a damn. He is surprisingly undistinguished, prissy and languid and clerklike, like some unpleasant official at a passport office. Wystan, Chester and Lincoln showed up right at the start, followed by Julie and Manning. But in the middle of the afternoon there was a horribly dead period and my heart began to sink. Then, an hour and a half before closing time, Marguerite appeared, bringing with her socialites and journalists; and cameras flashed. Betsy von Furstenberg[2] was cross because her portrait wasn't hung. But the day was saved. One way or another, Don has now sold sixteen drawings! There are about thirty-five altogether.

I met Freddy Ayer at a party given by George Weidenfeld. (I didn't want to go to it at first but decided I should in case it was somehow helpful to Don.) I was so intrigued to know what he had told the Cabinet in Washington about philosophy that he promised to tell me, and he did, yesterday, at the apartment of a girl named Jean Hannon,[3] who fixed lunch for us and also listened. Freddy is very lively and argumentative and pleased with himself, and he can be surprisingly lucid about his ideas. There was really nothing he said that I'd disagree with, and I do see that his kind of semantic analysis of keywords like Honor and Justice and their implications would be enormously useful to a group of politicians who were seriously trying to do their best. Freddy attacked the favorite Washington cliché that, "We have to find a valid and

---

[1] Based on Laurence Housman's *Victoria Regina* (1934), for which Harris won an Emmy; kinescope was a pre-videotape method of recording from television.

[2] German-born actress, aristocrat, beauty (b. 1935).

[3] Artist and illustrator, born in Boston, then in her mid-twenties; she trained in Boston and Paris, worked for *Show Magazine*, and became Mrs. Gordon Douglas when she married in the late 1960s.

inspiring philosophy which will be our answer to the philosophy of communism.... etc. etc." "Nonsense," he says he told them, "You don't need any new philosophy, you've got a perfectly sound one already, good old-fashioned liberalism!" Jean Hannon put in her oar and said didn't Freddy believe that everything was really for the best, and this gave Freddy his cue to denounce Christianity, which he did with huge gusto. But, really, all that stuff is so utterly beside the point, or, at any rate, just first base. Either you can know God or you can't. What in the world matters beside the question of the validity of the mystical experience?

As for Don and me, there is a cloud on the horizon: the return part of his airline ticket back to Los Angeles, will expire on January 19. So Don says I should take it, pretending to be him, and he should take mine when he wants to come home, pretending to be me. When I resisted this, because I don't want to leave so soon and because I rather hate the embarrassment of possibly being found out, he became furiously angry—that was the night before yesterday, and admittedly we were both very drunk—and said he didn't want to see me any more if I wouldn't do what he said. Now he is sort of sorry and yet he can't bear to think of sacrificing the whole cost of this part of the ticket, as we might possibly have to, according to the girl at the Pan-American office I talked to yesterday. As for me, I feel more and more that I must just sacrifice self-will, even when it is "justified," in my dealings with him. If he says I should go on the 18th, I guess I'll go.

Certainly, being here depresses the shit out of me. Because I have nothing worthwhile to do. I fill up my time by fetching and carrying for Don, and I don't begrudge that, but it's nevertheless a sort of laziness. I *cannot* see how to handle the Socrates material. The idea of having a theatrical group like Laughton's doesn't work, because there is no script for them to keep referring back to. Taken as a whole, the evening wouldn't mean anything. They seem to be setting out to discover how Socrates should be acted. But do they ever in fact discover this? *Could* they? Isn't the question meaningless?

January 12. My face is healing up nicely now, from the marks of where Don slammed the taxi door on (in?) it, the evening after I wrote the above entry. This has been one of our very worst weeks ever, and I am still quite uncertain what he really wants—as indeed he is himself. Well, never mind all of that. I have every intention of surviving somehow. *I shall return.*

Meanwhile, I'm definitely planning to leave for California right after the meeting of the Institute of Arts and Letters,[1] on the 24th, and I'm seriously considering going by train. "Nostalgia enters the train somewhere about Albuquerque," Gore warns me, "disguised as an Indian." But I would prefer those kind of blues to the stark despair of riding out to the airport and climbing into a jet. Besides there is the rather tempting prospect of being met late Saturday night by Prema, driven to Vedanta Place, put up for the night and reading the Katha Upanishad at Swamiji's breakfast puja. At least that'll be an auspicious start.

January 20. We seem to be sailing into calmer waters. Partly because my day of leaving approaches. I *am* going by train. I hope this isn't a mistake. At any rate it will be an experience.

The day before yesterday, I got the first copies of my novel. It looks very attractive, except for the ugly little colophons which American publishers love so dearly.

That evening we had an extraordinary supper at a restaurant, as the guests of Alan Pryce-Jones.[2] To start with, he hadn't figured out that there would be thirteen of us, so he had one of the tables separated by a couple of inches from the rest, and put Don at it, face to face with a very dull young man whom he had to try to amuse all through supper. This made Don furious. When a French Rothschild baroness tried to talk to him, he told her, "We're at separate tables." Then, the French, who included that Belgian politician Henri Spaack (Spaak?),[3] were lumped together in a mob, then came Wystan and the wife of Ian Fleming,[4] and the [...] girl who entertained Freddy Ayer and me to lunch, Alan Pryce-Jones and myself. Wystan, who was in great spirits, held forth against the French. (The ones at the table certainly all understood English.) Their language is hideous, he said, and should be forbidden. They have no critics—except Valéry and Cocteau. Their whole outlook on life is wrong. This rather scandalized our neighbors—the French pretended not to be listening; but the women were excited

---

[1] Isherwood had been a member since 1949; see Glossary under National Institute of Arts and Letters.
[2] Eton and Oxford-educated Welsh novelist and literary journalist (1908–1987), editor of *The Times Literary Supplement* 1948–1958.
[3] Paul Henri Spaak (1899–1972), then Minister of Foreign Affairs for Belgium; previously Prime Minister (1938–1939, 1947–1949), top U.N. and NATO officer, and chair of signatories of the 1957 Treaty of Rome, establishing the European Economic Community.
[4] Anne Charteris (1913–1981), also widow of 3rd Lord O'Neill and former wife of Esmond, 2nd Viscount Rothermere.

and pleased when Wystan said that all administrative posts should be handed over to them, because they were "on the side of life," and that men should devote themselves to the sciences and the arts. Wystan also very much interested me by saying that he woke up every morning delighted to be alive!

I think this was the most ill-assorted, idiotically arranged supper-party I ever in my life was at.

January 23 [Tuesday]. I don't imagine I'll have time to make another entry after this one, before I leave. In any case, I have to return the typewriter. So goodbye to one of the nastiest, most miserable phases of my life. I hate this city anyhow, and I've hated being here this time because of the way Don has acted. Right now, he is nerve-strung almost to screaming point and it is misery to be with him. I'm sure he hates me and I rather hate him, I mean on the surface. Underneath, things are more or less as they've been for years. Whether we shall go on living together, and whether we ought to if we do, remains to be discovered. There is absolutely nothing to be done about this, as of now. We must both sweat out the time till I leave on Thursday. The crisis is acute and yet I really do not know how serious it is. For that reason, aside from any other, I could never go to anyone else and say, "Don and I are splitting up." If we do, I think almost everyone who knows us will be sincerely amazed. Which is good.

The Stravinskys arrive here this afternoon. I doubt if we shall see them, as Igor is said to be exhausted and maybe seriously ill. David Protetch is even going to meet him at the station.

Yesterday, Marguerite interviewed me, because she may get a job if she sends in some specimen interviews to some editor. Meanwhile, a boy named Andy Warhol drew my feet[1]—he is preparing a book of foot drawings of well-known people! The drawing was too chi-chi stylized (rather Cocteauish) to have any kind of scientific interest, I should think. I liked him, though.

Lunch yesterday with Max Schuster, Peter Schwed and Sean O'Crairdain the Simon and Schuster publicity man, in a private dining room in a club at the top of the RCA building. Talk about the man who wrote a book claiming he'd been to Tibet and had his psychic third eye surgically opened by lamas. One New York publisher had said, "We need that book like a hole in the head."

[1] Warhol (1929–1987), worked as a commercial artist during the 1950s and drew shoe advertisements for fashion magazines; his early portraits often show only hands or feet. His pop art career took off in 1964 when he exhibited his *Brillo Boxes* at the Stable Gallery in New York.

January 28. Just to record that I'm back at last here in my beloved Adelaide Drive. It has been a perfect day, except for the hot desert wind and even that I have enjoyed. It has given me the beginnings of a tan.

I got to Los Angeles at 11:15 p.m. last night on the Chief and Prema met me and drove me up to Vedanta Place, where I spent the night in one of the apartments of their apartment house. This morning, just before six, I saw Swami, and then we went into the shrine for Swamiji's breakfast puja and I read the Katha Upanishad; vain, I have to admit, of my rendition. Then I had breakfast with Swami—he has been pestered by another of these madwomen; she broke into his room in the middle of the night and later wrote letters accusing him of forcing her into *samadhi* against her will, and of teaching her masturbation by remote control. He gave me *mahaprasad*,[1] a grain of rice from Jagannath Temple,[2] which, said Ramakrishna, is, like Ganges water, Brahman made visible. Swami has a whole store of these grains and takes one first thing in the morning, every day.

I left New York on the Manhattan, at 1:35 p.m. on Thursday last, the 25th. After we got out of the city, there was snow all the way, right across the Plains and the mountains, until we began to descend toward Needles.[3] Chicago was like some awful raw-black stark snow-slushy city on the tundras of Soviet Siberia. But how snug it is to shit in your roomette, watching the land roll by! I tried very hard on this trip to enjoy everything. I sat for long spells in the vista dome, just watching. But alas, I feel that I have dulled the organ of my delight. Only occasionally nowadays will it respond.

Read O'Hara's *Sermons and Soda Water*, which is readable but somehow grubby, and Ian Fleming's *Dr. No*, which I really enjoyed until the boring tortures at the end, and finished off *The Scarlet Letter*, with effort but a certain respect.

Yesterday evening, at sunset, in the midst of my first scotch, as we rolled through the California desert, I wrote this draft of a letter on the envelope of my ticket:

Won't write a Dobbin and Kitty letter. That's sentimental. Though it's a beautiful poetic sentimentality. But I think we can

---

[1] Especially sacred consecrated food; see prasad in Glossary.
[2] In Puri, Orissa, eastern India, a traditional place of pilgrimage. It houses an image identified with Krishna and his brother and sister.
[3] California town on the Colorado River at the borders with Nevada and Arizona.

still talk to each other by our own names—I mean, even when we're not mad.

I realize I am deeply selfish. You admit that you are. But that doesn't stop me loving you. And perhaps we would get along better on the basis of being admittedly selfish. I said a true thing when I said I didn't like being "good" any more than you like being "bad."

You are so much the reason for my life—my writing, the house, my teaching. You say, that's just accident. Anyone could have been the reason. No. You know that's not true.

My selfishness is that I want you to stay with me. Your selfishness is that you ask yourself, couldn't you do better; considering you are young. So my selfishness is really much more sinister than yours.

Am writing this halfway through my first scotch of the evening, in the vista dome, going through the desert beyond Needles. But I'll copy it out if I mean it tomorrow....

I did mean it but I didn't copy it, because this afternoon Don called from New York. He now plans to come home in two weeks instead of a month. (I find that, as of now, I am ever such a tiny bit disappointed, but that will pass, I know.) Also, he is considering buying six chairs for the dining room at the cost of six hundred dollars. So he certainly seems to mean to stay!

A Jew on the train [...], in advertising. He sells advertising spaces. And studies biology on the side; cell structure. And writes poems. Is married, for the second time, to a two-time-loser wife, and has six children (I think). He judges, Jew-like, every man, including all the poets of the past, who didn't settle down into a well-adjusted marriage. Doing so he calls "finding youself." This is the great heresy of the psychiatrists, and it has brought us to the population explosion. He bored and bored me. I hid from him in my roomette. But he caught me at meals.

February 12. I have been bad, drinking too much and staying out too late, and thereby knocking myself out next day and failing to work. Then there has been this massive rain storm: three days of solid downpour, plus the beginnings of it on Wednesday and the considerable parting showers yesterday. Now the sun's shining—at least, down here at the beach. But we're threatened with another storm in a couple of days.

Don wrote again, very sweetly, but puzzling and disturbing me a little because he utterly ignores all the things he said to me in

New York. It's just Kitty and Dobbin as it was in the beginning. He is supposed to be out here before or on Wednesday, our anniversary—he referred to this, unprompted, in his letter. And then what? My guess is that we shall postpone the next crisis by building him some sort of studio. He won't have to face the problem of his work until it is finished. Well, at least we have some money to do this with. Just heard from Sidebotham today that he is sending another $5,000 (approximately) as the balance of M.'s legacy. And then there is still the money from the trust fund. So the money position is exceptionally good. About $21,000 in the bank here. Not a penny of Simon and Schuster's advance touched. Earnings soon due to me from L.A. State College. Not to mention the possibilities of doing *The Vacant Room* with Gavin, doing "The Beach of Falesá" for Richard Burton, getting money if Carter negotiates a deal for making *I Am a Camera* into a musical. And then there are Don's earnings to help out.

What do I feel about his coming home? Gladness. Yes, always, in spite of the strain and hostility. Much as I have come to love this house and to enjoy, in small doses, living alone in it, the whole affair would still have no reason to exist without him. He is the ultimate reason why it's worthwhile bothering at all.

And what's left, if Don goes out of my life? Swami and Ramakrishna: yes. As much—more so—than ever. My japam has been getting more and more mechanical. But, when I told Swami this, he didn't seem worried. He assured me that I will get the fruits of it sometime or other; and I really believe this. The only thing that sometimes disturbs me a little about his teaching is the idea that we—all of us who have "come to" Ramakrishna—are anyhow "saved," i.e. assured of not being reborn. This disturbs me because the idea seems too easily optimistic. But then—who am I to talk? Swami says it, and I do honestly believe that he somehow *knows*.

Last week, I started at L.A. State College. I feel dissatisfied with my Twentieth-Century British Literature course. I hadn't prepared it properly and was sloppy. Understanding Literature went better. And the Creative Writing, which I was looking forward to with a sinking feeling, seems off to a good start, because nearly all of the students are cooperating with great energy.

(Incidentally, the drive to college last Thursday was a nightmare. The cars on the freeway threw up a mist of spray and exhaust fumes so that you couldn't see which lane you were in. I remembered with fury how Gerald, the day before, sitting safe at home in his garden house, had acted gently superior because I complained of the weather and had said how beautiful the rain looked through the oleanders. Fuck aestheticism!)

A good word-of-mouth beginning for my novel, which is now out in a number of bookstores, although not to be officially published till the end of the month. Jennifer and David Selznick, Ivan Moffat[,] Alan Campbell and Dorothy Parker, and Chris Wood are particularly enthusiastic. Also, *Time* sent a man to interview me, and apparently they don't do this if they are going to give a book a bad, or even a cursory notice.

Yesterday, the tooth I had recapped in London started to disintegrate—a big bit of it crumbled and broke off and now the ugly metal peg is showing. And today's Lincoln's birthday[1] so I have to wait until tomorrow before I can get it fixed, and tomorrow's a school day.

Some raw material for a novelette:

X., a middle-aged writer, comes to a college campus as guest professor. Part of his duty is to read and criticize students' manuscripts. The first student who brings him a manuscript is a tall blond somewhat prim and owlishly bespectacled boy of twenty, named Y. The manuscript is a partly finished novel and when X. reads it he finds that it is a story of frustrated homosexual love, set on what is obviously the campus of that very same college, with characters who must almost certainly be based on real students. X. is hugely intrigued. He even wonders if Y. is playing an elaborate practical joke on him. X. has never made much secret of his own homosexuality and it is probable that the students may know about it. Is Y.—maybe in collusion with friends—using this manuscript as a bait, hoping X. will jump at it and commit some amusing indiscretion? X. doesn't underestimate the cruelty of the young, but he is inclined to doubt this. He senses that Y. is a bona fide homosexual. Nevertheless, he resolves to play it cool. When Y. comes around to discuss the manuscript, X. is the detached, objective unshockable, disinterested professor. He tells Y. calmly that the story needs more action; the two boys in it ought actually to go to bed together. His tone, as he makes this suggestion, is positively clinical. Y. agrees—but X. can't figure him out: is he surprised, shocked, amused? He is deadpan and demure, and thanks X. most politely for his advice.

After this X. and Y. see quite a lot of each other. They go for long walks and talk in a relaxed and almost intimate manner—but always only about books, poems, plays. X. is very well aware that

---

[1] I.e., a national holiday, honoring Abraham Lincoln.

the eyes of many people are upon him and he is determined not to do anything which might give rise to the least breath of scandal. Just the same, Y. *is* more intelligent than the majority of the students, and the tacit knowledge of each other's homosexuality is a bond. X.'s conscience is eased by the fact that he doesn't find Y. in the least attractive.

By degrees, he finds out quite a lot about Y.'s background. He is the son of wealthy parents from the East and has always lived in an atmosphere of comfort, big cars, parties, travel. He knows stage and movie stars, plays bridge with them, and goes dancing, but obviously without any particular thrill. X. becomes aware of another side to Y.'s persona: the prim student must become a different person, worldly and sophisticated, when he's back at home. Y. also admits casually to drinking a lot at parties, and X. gets the impression that somewhere in him there's a surprising amount of wildness. Altogether, X. no longer feels quite so sure of anything about Y. Maybe Y. isn't even homosexual. He always comes to X.'s seminars in company with a girl named Z., and X. keeps meeting the two of them together around the campus. Z. is no beauty, but she's intelligent and has a kind of sweet eagerness. You can see that she is violently stuck on Y. He doesn't seem particularly responsive to this, but maybe that's just male superiority or a fastidiousness about showing affection in public. The relationship between them rather irritates X. Because Y. seems to be a "deserter" from the homosexual ranks? Maybe a little. But much more so because Y. and Z. whisper together during the seminar, exchange notes and altogether fail to help make things go. This is all the more irritating because the seminar group is ill-assorted and X. has to work hard to keep them even halfway interested. Y.'s failure to help seems a betrayal, considering his relative intimacy with X. X. can't but resent this and he stops going for walks with Y. And this isn't the worst. Toward the end of the semester, a skit appears in the students' magazine, making fun of X.'s seminar and the atmosphere of boredom and embarrassment which surrounds it. The skit is poorly written and malicious rather than funny. From various clues, X. is quite certain that it is by Y., or, more probably, by Y. and Z. in collaboration. X. is a bit hurt but also puzzled; he hasn't been prepared for this hostility, it even rather intrigues him. Is Y. anxious to disown him, for Z.'s sake, he wonders. But of course it is beneath his dignity to show Y. and Z. that he has even noticed the skit. And, when term ends, X. even casually tells Y. that he is welcome to call him and bring him the novel, when it is finished. He does this not because he particularly wants to see Y.

again but because he has the impulse to show Y. and Z. that he's not a poor sport or a bad loser. Y. agrees to come, but he doesn't seem much pleased or impressed by the invitation.

Many months later, however, Y. calls X. and asks if he may send him the novel, which is now completed. X. says yes and the novel arrives. It is clumsily written and dull. But X. can't help being interested in it, just because of its subject matter, and he feels a revival of interest in Y. He invites Y. to dinner. When they have both had a good deal to drink, X. says that, now he is no longer on campus, he wants to ask some personal questions about the book. Y. is ready to answer them. He tells X. that he was indeed in love with another boy student, just like the one described in his novel. But the other student, though fond of him, was not interested sexually and nothing ever happened between them. This leads X. to ask Y. if he has had *any* sexual experience at all. Yes, says Y., but nothing satisfactory. Once, after a beach party at night, he has managed, without much enjoyment, to screw Z. on the sand. And several times he has been down to a queer bar on the other side of town. There he has met a hustler who has sex with him in a parked car, always asking to be paid his ten bucks in advance. Obviously, these furtive visits to the sex underworld thrill Y., purely as adventures. He wears his oldest clothes and shudders with fear that he'll be recognized and involve his prominent family in a newspaper scandal. But he doesn't much enjoy the hasty commercial act itself. Y. tells X. all of this with the utmost frankness, and this evening is far more intimate than any of their previous meetings. When X. has finished his questions, Y. asks him, "Why did you single *me* out, on campus?" X. is rather surprised by this question. He reminds Y. that it was he who started their relationship by bringing this manuscript. X.'s answer seems to disappoint Y. a little. Later, just before leaving, when Y. is very drunk, he says something incoherent about being much more daring when he's sober, and he leaves X. with the impression that he is somehow frustrated. It does of course go through X.'s mind that Y. may have been trying to make a pass at him; but he dismisses this as another example of his middle-aged vanity. And, anyhow, Y. doesn't interest him sexually. Which is not to say that he wouldn't have been flattered and pleased if Y. *had* propositioned him. The Young are still the Young.

Another considerable time lapse, and Y. calls X. and says he has written half of another novel. May he bring it around? When Y. arrives, X. is at once struck by the change in him. His appearance has enormously improved. He has filled out. He looks positively

handsome. He has a new sexiness. Also, he seems full of vitality
and joy. He announces right away that he is going to take X. to
the same restaurant to which X. took him at their last meeting.
He has left college now and is just about to go to Europe for a
long stay. He tells X. that Z. has gotten married to someone else
and implies, with smirking self-satisfaction, that this was on the
rebound; she was trying to forget him. Y. also implies that he now
fully accepts his homosexuality and is determined to live in his own
way, not according to other people's rules. He is wearing a sexy
but blatantly faggoty sport shirt, and this seems like a declaration of
his independence. When they get to the restaurant, Y. wants to sit
with X. for a while at the bar, and X. learns that this is another act
of defiance, because, only a short while before this evening, Y. has
visited this same bar and someone has made a remark about him
which he has overheard; Y. was referred to as "that queen." X. is
amused and touched by all of this. For the first time, he really likes
Y. For the first time, also, he finds him attractive; quite powerfully
so. At dinner, he pays Y. a couple of physical compliments, on
his shirt and his appearance. Y. is drunk already and talking rather
wildly. His declarations of queerness are mixed in with curiously
puritanical statements. For example, he says that he hates going to
men's rooms and he can't pee there if anybody else is present. This,
he later confesses is because some kids once laughed at him—or he
fancied they did—because his cock was so small. As dinner goes
on, Y. gets more and more disturbed. There is something, he tells
X., which he wants to say, but can't. X., of course, encourages
him. Finally Y. blurts out, "I want to go to bed with you." By the
time they've reached this point, X. has pretty well guessed what Y.
is about to say. He responds with the tact of long practice, telling
Y. that he's most flattered, that he'd love to, but asking Y. if he
is really sure that's what he wants. Y. says, "It must be, because
I'm shaking all over. I always shake like this when I go down to
that bar." So X. tells him he can relax now, because it's all settled;
they'll do it as soon as they get home.

On the way back in the car, X. tries to remove any fears Y.
may be suffering from. He holds Y.'s hand firmly like a doctor or
places his own hand on Y.'s thigh. He suspects that Y. may fear
impotence, so he says that one never knows how these things will
work out—but what does that matter? They'll just play around
for a while and have fun. Nothing drastic has to happen. Actually,
X. is also guarding himself. He's not quite sure if Y. really attracts
him enough to make him potent. Suppose *he* can't do anything?
That's sure to hurt Y.'s feelings. Well, it's too late to worry about

that now, X. reflects. They will just have to go through with it and hope for the best.

When they get back to the house, X. insists on going to bed at once. Y. is a little nervous, but he undresses as he's told and throws himself down on his belly on the bed, naked except for his jockey shorts. Obviously, he wants X. to tear them off. "I defy you to excite me!" he giggles. He now seems ridiculously feminine. But more attractive than ever to X., who has no trouble at all in responding. As they take hold of each other, Y. becomes absolutely transformed. He seems to go almost mad with joy. It is quite obvious that this is the first time he has ever experienced a compatible sex act. He screams until X. fears the neighbors will hear him, then breaks out into wild laughter. Once or twice, he brings out some half-sincere remark about his own ugliness, but X., with the earnestness of lust, reassures him: he is a gorgeous boy, perfect all over. X., by this time, is nearly as wild as Y. Y.'s terrific innocent crazy lust has excited him far beyond his normal powers, and his lovemaking is twice as convincing as his words. At last, when they are both thoroughly satisfied, X. asks Y. if he enjoyed himself. Y. says, "This is the first time I ever realized I had a body."

At breakfast next morning, X. watches Y.'s face for any sign of regret or distaste for what they did. But there is none. In fact, Y. is fairly beaming with satisfaction. "I wish I could stop myself smiling," he says. They part, agreeing to meet again, on one of the few nights left before Y. leaves for Europe. It is understood that there'll be more sex. In fact, when X. says he wishes they had started having sex together at the college, Y. seems to agree with him. Y. leaves the manuscript of his new novel for X. to read.

X. can't stop dwelling on this amazing scene. He keeps seeing Y.'s face as it looked at the crisis of their enjoyment, and then picturing it overmasked by the prim face of the bespectacled student. He feels that something truly beautiful has happened to him. He has been privileged to assist at a rite of spring, as it were. He tells himself that Y. will never be able to forget him. He will have been the first. And Y. will never quite be able to recapture that thrill again. Meanwhile, he starts, in an indulgent mood, to read Y.'s manuscript. Not at once, but gradually, he is overcome by another kind of amazement. For this novel is *good*. Indeed it is amazingly good. How could the author of the first novel have produced this? It is crude and awkward in places, yes; but it is the work of a writer of serious talent. X. is enormously excited. And relieved. As a matter of fact, he now realizes, he has been rather dreading that novel. Suppose it had been bad—as bad as he'd been expecting it would be? How could he have told Y.? As it is, the excellence of the novel presents another kind of problem. Will Y. believe him when he praises it? Thinking things over, X. sees that he and Y. have gotten themselves into an awkward tangle. They are in two relations to each other at the same time: teacher-student and lover-lover. X. thinks it highly probable that Y. will, by the time they see each other, have fallen in love with him. Why not? It would be the most natural thing in the world. Maybe Y. will even decide not to go to Europe but to stay with X. X. doesn't want that. He feels pleasantly romantic about Y., but nothing more. Also he desires him, but that will soon wear off. If he can get another good hot night with Y., he will have had enough. But he does want that night. He wants to get Y. to bed as soon as possible. He even feels that the novel is a nuisance. Why did Y. have to produce it just at this stage in their relations?

Nevertheless, X.'s enthusiasm for the novel is genuine and he takes the first opportunity to call Y. Y. is out and he gets through to Y.'s mother. Obviously she doesn't know anything about him, but X. indulges in one of his characteristic fantasies. He imagines himself defending his conduct with Y. to Y.'s mother; telling her that he has done more for Y. in a single night than she and her husband have accomplished in twenty years. Then X. reflects that Y., if his mother tells him of the call, will fear that X. is canceling their date. Y.—in love for the first time—that college student doesn't count—it was mostly imagination—will be terribly upset. He will give way to all his old inferiority feelings, will decide that X. doesn't, could never have, loved him, that his cock is too small (actually it is enormous) etc. etc. X. gets quite sentimental as he paints this tragic picture for himself. Y. is so sweet and unprotected and helpless. He mustn't be made to suffer, even for an instant. So X. calls again. This time, Y. has returned. He seems dazed rather than pleased when X. tells him how wonderful his novel is. Then X. explains that he is calling to reassure Y. There is nothing to interfere with their date that evening. Y. answers without much enthusiasm, "I shan't be able to stay very long. My parents have a party." X. gets a slight shock on hearing this. But he decides that Y. is probably talking in this casual offhand tone because his mother is listening.

However, when Y. arrives that evening, X. sees at once that another change has taken place. Y. is certainly pleased—deeply pleased—that X. likes his novel. But his manner is ever so slightly abstracted. He is impatient to get back to the party. X.'s fantasy picture of him, a sensitive withdrawn boy, awakened to first love and trembling to meet his love again, in grateful lust, now seems absolutely idiotic: the senile invention of an old conceited fool. There is obviously no question of sex. X. is wise enough not to hint at it. But, when Y. says he must leave, X. can't help asking, "You aren't feeling badly, are you, about the other night?" "No, why should I?" Y. asks, with a certain indifference. "I just thought you might be." "No, I've been thinking about it. You made me feel very peaceful." And that's all. X. gets the impression that, whatever it was that seemed to Y. to be happening, it was happening only to him. X. has played scarcely any part in it. Maybe it's the same with Y.'s writing. Maybe that's Y.'s whole attitude. People exist to serve him, with literary advice or a stiff cock, according to his mood, and that's that. As Y. is getting into his expensive sports car, X. says, with slightly acid intonations, "Well you certainly surprised me—twice!" Y. smiles. "Perhaps I'll surprise you again. I'll have to think of something." And then he says, "You make me feel peaceful. You're my tranquilizer." And with that, off he drives—to the queer bars of Cannes and the Spanish Steps and clap and crabs and lots of fun, no doubt. And X. goes back down to his house, reflecting that the Young are the Young are the Young, but that the Old, thank goodness, are tough—as they need to be.

February 14. Talked to Don last night; he called from New York. He isn't coming back today, which makes me sad. But he does seem eager to return and he said it would probably be on Friday. He has lots of things to attend to first, and I honestly don't believe that this postponement is because of anybody else—though I never quite discount this possibility, suspicious (and guilty) old thing that I am.

Rain in the night and more expected. I hate the prospect of it tomorrow, on my way to school. The classes are going fairly well—particularly the creative writing, which continues to be fun. Henri Coulette disappointed me by speaking with great reserve about my novel, though maybe that was partly shyness. However, Wystan wrote today saying that it is my "finest to date" and that Paul is "quite magnificent." (Coulette said "Dostoevsky in a beach towel.") And Mark Schorer called from San Francisco after I had gone to bed last night. He was very enthusiastic though admittedly drunk.

It seems that I won't have to appear at the trial of the bookseller who sold *Tropic of Cancer*, after all.[1] I had warned the lawyer that

[1] Bradley Smith, who ran a bookstore on Hollywood Boulevard, was the first

I might turn out to be more harmful than helpful to the defense, because I should be bound to admit under cross-examination that it's not just Miller's book which I don't find pornographic; I don't believe that there is any such thing as pornography. And I don't believe, for the matter of that, that the young can be corrupted by talk about sex.

Yesterday, at college, Selznick called me from New York. He wanted permission to show my book to "several top-flight dramatists" and the real reason for this is that Jennifer said she would love to play Maria in the "Ambrose" episode! David was very impressed because the switchboard operator at the college referred to me as Dr. Isherwood!

A marvellous passage in a letter from Richard. The first I've had since I left England. "Am sitting at the kitchen table. It feels very peaceful and forgotten and far away here, like it usually does—an atmosphere at once sad and calming and cheering—then a distant rumble of a plane over at the aerodrome about twenty miles from here, being what they call 'tried out.' It sounds like a distant heavy express train rushing across a viaduct. There has just been a violent downpour of rain, a cloudburst. Now it's bright sunshine again."

February 17. Just after six. It's getting dark. The weather looks pretty good. I have just finished making up the bed in the front room. Don is arriving tonight.

What do I feel? Great joy. Oh yes, I know this is the beginning of a whole new maze of problems. I know that before another week is out we shall have tied ourselves into new knots, and the palpitations of my vagus nerve will have started up again. (All this while, since I've been back here, they have stopped.) Well, never mind. What must be, will be.

Stephen and Natasha also got in today from Mexico. They are staying with Evelyn Hooker. I should be glad of this, if it didn't pose a problem with Don. He doesn't want to see them, or even to have them come to the house.

Actually, Stephen rather put me to shame by buying two copies of my novel and giving one to Evelyn. He is just exactly as usual, and his accounts of his tour in Latin America ditto. Natasha seemed a bit sad and grumpy.

Richard writes again today to tell me Uncle Jack [Isherwood] is

to be prosecuted in Los Angeles for selling the 1961 Grove Press edition; academic colleagues and acquaintances of Isherwood testified for the defense along with the literary editor of the *Los Angeles Times*, Robert Kirsch. See Glossary.

dead. He died on February 6. He had left a wish that there should be no flowers and no "communications." But I felt I had to write and did, some crap or other. There's no duty so horrible as family duty.

Yesterday evening—chiefly because Gerald had urged me to do it—I invited Will Forthman to have dinner with me. He is a strange withdrawn creature nowadays. Not withered looking as yet, indeed he hasn't changed a bit; but at the same time he seems ingrowing. He spends weeks and weeks up at Margaret Gage's, toying with a dissertation on William James which obviously bores him pissless. He doesn't want to go back to teaching and plans to spend another year—doing what? Gerald says he is burningly ambitious. If this is true, he must be very unhappy, because he hasn't really found anything to succeed at. Because I was so ill at ease with him, I took him to La Mer and spent a lot of money. When I gave him Pouilly Fuissé, he said. "Wine is a good sedative!"

February 23. Don has been home nearly a week now, and we have been very harmonious. He even agreed to meet Stephen and Natasha. There was no great reconciliation, but things passed off smoothly.

I see in him a certain change. It's odd, but this very short separation since New York seems to reveal it—maybe just because here we are at home and relaxed and can study each other. What I see is a reserve. He doesn't seem so childishly open as before. I really don't know if this is "good" or "bad." It is simply that he seems more in control of the situation.

However, the problem which we both expected does seem to be arising: can he work, all by himself and without being under pressure? We are considering doing some kind of reconstruction on this house, so as to provide him with a studio downstairs. But that in itself is no solution.

Three sayings:

Vera Stravinsky: "All I want is a cat, a fire and silence."

Don: "Stephen has one of the great virtues of a conversationalist—he never thinks before he speaks."

Me (quoted to me—I'd altogether forgotten saying it—by Evelyn Hooker, as a comment I made on her researches into the lives of homosexuals): "Evelyn, you think you're going to discover a secret. All you'll discover is a nature." (Evelyn says that this remark has greatly inspired her!)

February 28. And to these I must add another saying, also by Don and also about Stephen. Don was remarking how extraordinary it was that they were able to fill Royce Hall with people anxious to hear Stephen's lecture the other day. I said, "Well, after all, he had the attraction of being a visiting fireman."[1] "A fireman's the right word," said Don, "because he certainly dampens the evening."

At breakfast this morning, we had a talk about frankness, because I had belatedly told him something about Gavin which displeased him and had later tried to soften the effect of this by making excuses for what Gavin did. I replied that I simply didn't want to have him falling out with Gavin. Why not, he wanted to know. Because, I said, I wanted us all three to get along together. Don said this was a demonstration of my possessiveness. No, I said, nothing so sinister—it's simply a matter of convenience.

Nevertheless, by and large, Don and I are on the best of terms since his return. As for what I told him about Gavin,[2] I most certainly didn't do it to make mischief; mischief between them is against my deepest interests. Why *did* I do it, then? I guess, because I find it terribly difficult to withhold anything from him.

A searchingly cold wind after the rain. I have never known it so bitter. We had supper up at the Selznicks'. David is still in New York. Kate Moffat kept telling me how happy she is with Ivan. I hope this isn't dark-whistling. I nearly lost my voice, reading *Macbeth* with my students. But college is going quite well now.

Only one thing's wrong: I still am not getting on with my work.

March 5. I still am not getting on with my work. I have to repeat that sentence. Nagging at myself hasn't seemed to help. And yet, it is terribly hard to see just why I don't do more. Because my immediate objective—getting the Ramakrishna book finished—is a question of willpower rather than of inspiration. It would still be possible to finish at least the rough version for the magazine before my next birthday, and I must make up my mind to do that at least.

What prevents me? True, we've been getting up late in the mornings, but that's far more my fault than Don's. He is usually the one who makes the move. And now he is determined to restart working, including going to school. If he takes night classes, then I ought seriously to try working nights, which I easily could on the mechanical first drafts of the chapters and the material gathering.

The first review of my novel, or rather the galley proof of it,

---

[1] Spender served in the Auxiliary Fire Service in London during the war; "visiting fireman" in American idiom is any important visitor requiring attention.

[2] Bachardy no longer recalls what this was.

arrived from England this morning; Julian Jebb for the *London Magazine*. Very favorable, calling "Ambrose" my finest work yet. May it be a good omen.

Just spent the afternoon being tape-recorder interviewed by a trio from USC. That's a typical waste of my time.

March 18. Rain. The Canyon wet and sad. Don, who was sick in bed yesterday with a hangover stomach, is out drawing someone. Yesterday I cooked a pork supper for Bill Inge, Mike Steen, Jerry Lawrence and a friend of his from San Francisco named Lou Bennett. It was dull. Jerry dutifully read my reviews—the good ones from Cyril Connolly and Gerald Sykes, the bad one from the *New York Herald Tribune*. Actually, the *Herald Trib.* one was just a conditioned reflex puritan attack on the subject matter; the only sinister aspect of it was that the editor should have given the book to a man like that, who could do no other than react according to his nature. Later on in the evening Stanley Miron arrived, flashing enthusiasm, but it was late and everyone else wanted to get to bed. I was quite unable to be what Stanley expected, the all-wise, all-knowing uncle to advise him about his love life.

I remain bogged down, or nearly, in the Ramakrishna book. Oh, it bores me so! And there is so little time to do anything on it. I have to keep going on my college work. This weekend for instance I must reread *Wuthering Heights* and then tell the students how Heathcliff evolved from Byron and Poe.

Don: "Gavin has a vague way about being rich. He lets it pile up behind him but he never turns around."

March 25. Yesterday, the fine weather really began again, and today is just as beautiful. Don has gone down to the beach but I have strong-mindedly stayed here in the house and have just finished sorting out all the letters, etc. Now I want to do some work.

Don is going through a bad period because he can't make up his mind to start painting. Or rather, he *has* started but without the least appetite and he doesn't know whether to force himself to go on, or what. The temptation, in a way, is to have another show here, of drawings. And yet, why shouldn't he? Why should he force himself to paint? And so the devil of tamas plays with him. And there's so little I can do to help, although I know this process inside out and have been going through it for the past forty years.

The one way I *can* help him is to work, myself. And God knows I have every cue to do that. Not only have I the Ramakrishna book and the homework for my college classes, but I also have a new idea for a novel. It really came to me yesterday. I won't write it down here, though, because it would be hard to find later. I'll put it in the big book.

The score on *Down There on a Visit* is still unsettled. *New York Times, Saturday Review* are good; *Time* will sell copies. Nothing yet from *Newsweek, New Yorker*. I know Dorothy Parker will be favorable in *Esquire*, suppose Stephen will be in *The New Republic*. In other words, the only serious slap was in the *New York Herald Trib*. And, in England, Cyril Connolly was very good in *The Sunday Times*, the *Manchester Guardian* was very good, *The New Statesman* sort of reproving but grudgingly favorable. Angus Wilson delivered a schoolmasterish stab in *The Observer*. Haven't seen the rest of them.

Now the only thing to do is forget the whole affair and get on with the next project. I still feel very good about *Down There* and am glad I wrote it. It is good and I know it is good and so do nearly all of the people whose opinion I care about. So what more do I want?

Dorothy Parker and Alan Campbell came to dinner with us last night. It was a flop. The two of them, alone, seemed lost and dull, like two people who are having dinner together dully in a dull restaurant. It seems as if they only become animated when there are lots of others around. Dorothy sits so sadly, with a kind of aristocratic sweetness, a Lost Lady, the bags hanging down under her eyes. Somehow, we just could not cheer her up. The only time they showed animation was right at the end of the evening, when we got on to the subject of famous murders.

I shall call this new novel project provisionally *The Englishwoman*.

March 28. Yesterday and the two previous days I have made notes for *The Englishwoman*, and I really do feel there is something there, a sound basis for a novel or novelette. Not only that—I also have the feeling, which I always have when I'm confronted by a theme which is really right for me, "I'll *never* be able to do justice to *that!*"

So I feel much less guilty now about work. Especially as I have at last ground out a rough draft of chapter 13 of the Ramakrishna biography. And school is going well.

Morgan doesn't like my novel. He wrote a sweet letter about this, and I don't mind. I really don't. It's funny, although I am such a dedicated disciple of Forster and have learnt so much from him, I don't particularly respect his taste.

March 30. Joe Ackerley came to supper with us last night, brought by Gerald Heard and Michael. Noticed how deaf he is. But otherwise very bright and sharp. And resentful—like most of us old things, alas. Apparently he is seeking wisdom from Gerald, that unreliable oracle. Joe wants to know what to do with the remainder of his life, or so one infers. Yet he didn't seem at all eager to meet Swami, when I suggested it.

Lots of provincial reviews are coming in now. The bad unfair ones depress me far more than they should. I realize that I was counting on a resounding success and vindication. Silly old Dub!

Don is being absolutely angelic and this has been one of the very most harmonious bits of our life I can remember in a long long while.

Today he's taking his drawings to show to Rex Evans, to see if Rex'll offer him a show at his gallery. Don partly wants this, partly doesn't. He dreads all the work and arrangements but welcomes it (I suspect) as an escape from this compulsion to paint. At the same time, he at last admits to rather liking a painting of his mother which he is doing now. (Glade very nearly blotted her copybook for keeps, by refusing to sit for him. But Don spoke to Ted who spoke to her, and now they're quite friendly again.)

Today, as I was sitting in my bathrobe writing a letter to Jim Charlton in Japan—I have just read the diary he sent me—the bell rang and a young man named Fred Watkins was outside, wanting me to autograph his copy of *Down There on a Visit* for him. How surprised he'd have been if he had known how something like this pleases and reassures me! Now I feel all rubbed the right way. No doubt Watkins thought of me as slightly annoyed at being interrupted and reminded of a celebrity which has grown tiresome to me and offered an admiration I am sick of!

Jim's diary is self-conscious and often pompous; yet it moves me, because I know Kyoto and him. I can picture him there so vividly. And I understand so well his necessity to act a part for himself, in order to make his life there a little more amusing. But again and again he has to admit that it is so damned *cold*!

I have also just called New York to tell a Mr. Stephan Wilkinson that I cannot review Iris Murdoch's new novel, *An Unofficial Rose*, for *Cosmopolitan*. By the first page of chapter 2, I was already licked—bored senseless. How *dare* these British family novelists assume that one wants to hear about their dreary characters!

April 1. Already! I am slipping behind, I feel; way behind schedule on all my projects. I haven't even kept up to my minimum of entries in this diary—January, February and March have all been short, and now I should make nine extra entries for April—seventeen in all—in order to catch up. And of course I shan't.

Thick fog in the Canyon this evening and it's getting quite dark, although it's only a little after four. Don has gone out to deliver a couple of drawings—marvellous ones—of Angela Lansbury's brother.[1] Rex Evans will give him a show next September, if he wants one. He can't quite make up his mind. Meanwhile, there is still the possibility of his going east in the summer or late spring. I shall probably stay here. I want to work uninterruptedly, on the Ramakrishna book and, I hope, on *The Englishwoman*—no new inspirations about that, so far.

The funny thing is, I'm happy. Largely because of the harmony with Don, which continues, and just because I'm here in this house I love. Last night was a bugger, though. I had to lecture on "The Writers of the Thirties" over in Monterey Park, at the public library. The library is run by an Indonesian named George Anang,[2] who is plump and aggressively race-conscious. "You Caucasians," he keeps saying. He has a wife and family back in Indonesia—according to Dorothy Parker, who came with Alan to the lecture—and here he has a "Caucasian" girlfriend, named Paule, pronounced Powly. She is young and quite pretty and chic, and a terrible cook; they fixed tasteless curry after the lecture.

I'd thought I had organized the material very well, but I hadn't; I found myself running overtime, panicked, and ended up saying next to nothing. Anang was nervous because the American Legion[3] had picketed Dorothy's lecture (the previous one) and he had feared violence and sent for the fire truck so the mob, if any, could be squirted with water. This time there was no picketing. But a John Birch Society member asked me—after I'd said I was a pacifist: "Is there nothing *worse* than war?" Dorothy assured me that I handled this question very well, by going off on a sidetrack and explaining that I didn't believe in the political efficacy of pacifism. But actually I was so tired and stupid by that time that I simply did not realize what he was trying to do—provoke me into saying that it is better to be Red than dead.

Fred Shroyer was busy licking Dorothy's ass, trying to get her to collaborate with him on an anthology of short stories. He has quite

---

[1] Bruce Lansbury; see Glossary.
[2] A board member of the San Pedro Public Library and, later, librarian of the Urban School in San Francisco (1968–1976); he collected and published moon legends from Malaysia, Indonesia, and New Guinea (d. 1999).
[3] A U.S. military veterans' organization.

abandoned me, I think. I suppose I haven't come across as far as his own work is concerned.

When I got home, around 1:30 a.m., there was poor Don sitting up working on a statement of his 1961 expenses for the income tax declaration.

This morning, I paid off the Russian gardener and his wife [...] who the Stravinskys recommended. I just coldly said I didn't want them any more. Vera—who ended by admitting that she wasn't satisfied with them, either—had told them we were going away to Europe. But I didn't want any lying. The wife is very snoopy and would certainly have found out the truth. It's just that I had taken a violent dislike to them. They were charging us fifteen dollars more a month than the Stravinskys. The wife seems always to go along on the job and stick around, snooping. She has a dishonest face. A bad pudding-face. And when they were here, they killed a lizard, a poor little thing, for no reason. My stupidity in hiring them in the first place cost us nearly forty dollars. And the stuff they have planted looks as if it were dying already.

April 5. Thick coast-fog tonight; you can barely see the lights in the bottom of the Canyon. It is snug being down here in the fog, cut off by it from the rest of the city. Los Angeles seems very far away. And, to add to the snugness, we are going out to supper with Jo and Ben, so won't be leaving the fog-pocket.

This morning, on waking, Don told me he had had a dream. He was trying to draw Lotte Lenya, and all sorts of people had come in, and also some dogs, and Lenya had started saying that she was tired, and Don became more and more frustrated, and then I arrived, and he "screamed at" me.

Yesterday, Michael Barrie drove Gerald, Joe Ackerley, Don and me to Laguna Beach and San Juan Capistrano. We were supposed to look at the flowers. Certainly, the hills around Laguna Canyon were a beautiful bluish green, with thick patches of lupines, and mustard, and white flower-clumps, like streaks of late snow, which Gerald and Michael call elyssium—I can't find that name in the dictionary, however; only alyssum. But the rest of the drive, down through Long Beach, was a wretched cavalcade of billboards, telegraph wires and poles, fluttering pennants on used car lots, gas stations, hot-dog stands, etc. etc., and I could feel Joe thinking, "This is their America, is it—well they can keep it." And the Capistrano Mission was drab and sordid in another sort of way: you felt the sordidness of the cynical reverence of all the millions of tourists. As the result of this drive, Don is inclined to feel he doesn't want to go away at Easter. I don't care—indeed, I'm quite glad.

As we were driving back, a woman in another car shouted at us and we stopped and she told us that we had dropped an envelope, back on the outskirts of Laguna. This was an envelope addressed to Dr. Gerald Heard, and it came from an institution called The Spiritual Frontier Fellowship. It contained several copies of *One* and the *Mattachine Review*.[1] Needless to say, Gerald was quite eager to get it back; he had brought the magazine to lend to Joe. Well, we drove back into Laguna and there the envelope lay, having been run over by dozens of cars already. The moral of this story is, don't say horrid things about women, as Gerald had been doing throughout the drive, with our approval.

In the evening, I went to the gym. It was empty, except for an elderly (my age) man with thin red hair. When he was through in the steam room he went over to a mirror and rubbed his face carefully with some kind of cream. "I can't afford to get old," he told me.

Then I joined Don and we went downtown to see an indifferent French film about vampires. Downtown Los Angeles seems to be getting more and more squalid. The hash joint where we ate was full of hustlers and boys dressed as gangsters. The movie theater was chiefly a flophouse for snoring bums; it stank sour. You felt this was really a potentially quite dangerous part of town, and yet it was one block away from the Biltmore.

April 7. Hot yesterday and even hotter today. Don and I went on the beach. Today I feel lousy—a big pyloric flap. I'm exhausted, want to lie down all the time. Sick to my stomach and yet can't throw up.

Yesterday, we went to the movies with Ted and Vince. Ted has just confessed to Don that he has been sentenced to a three-week prison term, for shoplifting. This was the third time he'd been caught. He went to the market, got a lot of groceries, then saw there were long lines waiting at the cashiers' desks so tried to leave without paying. Unfortunately, the market had just installed shoplifter-watchers. Poor Ted is trying to find some excuse for asking for a leave of absence which will fool his employers.

Vince told me that he has a friend in my twentieth-century British literature class named Thomas Pierson who says *he* likes

---

[1] Published respectively by the homosexual advocacy groups, One, Inc. and the Mattachine Society.

my way of teaching but that most of the students are puzzled and bored. I find that this hurts my feelings very much. In fact, I find myself far more sensitive to criticism, lately, than I'd supposed I was. I am really depressed by all the niggling sneering notices of my book, although I know only too well that this is just what I should have expected. I hit The Others with everything I'd got, and now I expect them to love me! As for the hostile critics on "my" side, well—let cowards flinch and traitors sneer.

Don, as we were having supper at the Casa Mia last night, looking sadly at Tom Calhoun,[1] "A thing of beauty is not a boy for ever."

Don is still being an angel. As far as our relations are concerned, this is a time of great happiness. But he still hasn't solved his problem about painting. And now, over the phone, he has agreed to do *another* show this year—at Larry Paxton's gallery in New York,[2] next December!

April 8. Another beautiful day, though a bit hazy, and since it is a Sunday the beach is crammed. Don has gone down, to meet Ted and Vince; I'm staying up here in the house. I must work—at the Brontës, at Ramakrishna, and at my letters. Also, I'm sick. This morning I woke up to find I have a discharge. This is probably some trifling infection and I won't do anything about it for a day or two, in the hopes it'll go away, but it is a nuisance. Also my jaw is very sticky again and aches dully, especially when I wake in the morning.

Ivan and Kate Moffat came last night to supper. We barbecued lamb chops with kidneys on the barbecue, the first time we've used it in more than a year. It took much longer than I'd expected—about thirty-five minutes.

The party wasn't a success. It's strange, a meeting of this sort; four people who could become quite intimate, and in fact know far more about each other than they'll ever admit, and yet they meet as if they were going to play bridge. They play as partners, hiding their cards. Ivan and Don—or rather, the Ivan team and the Don team—got into the most boring argument about *Judgment at Nuremberg*, and I made a needlessly sweeping statement that Stanley Kramer[3] was "worse than Louis B. Mayer," and altogether the

---

[1] A neighborhood acquaintance who did some remodelling work at 145 Adelaide Drive in 1959; he appears in *D.1*.

[2] Paxton worked at the gallery, but didn't own it; he sometimes lived in Los Angeles.

[3] Director of *Judgment at Nuremberg*.

atmosphere got disagreeable. It improved later, however, when Don showed his portraits of Kate to Ivan, and a photograph of the drawing of Iris, over which Ivan raved. And then Don offered the drawing of Kate which Ivan had liked best, as a wedding present.

How compulsively Ivan talks! He feels that Hollywood is done for. And indeed America, too. The Common Market[1] will lower the standard of living here, labor will try to combat automation without success, unemployment will mount, and people will get increasingly disgusted and turn to the reactionary Right. Ivan and Kate seem to be considering a move to Europe.

At breakfast this morning Don said that he thinks his mother is already adjusting herself to the prospect of his father's death. His father has begun acting old, and he has had (apparently) this skin cancer and now believes he has had a heart attack. He is a year younger than I am.

At last we have found a gardener who seems satisfactory. His name is Mr. Edwards and he works for Gavin Lambert. Yesterday he cleared the flower beds on the terrace and planted junipers and cypresses. It looks very tidy and nice, but the tidiness accentuates the cracks in the cement and makes the place look alarmingly dilapidated.

April 11. The day before yesterday, I went to see a doctor who Gavin recommends, Dr. Alan Allen. Simply because I cannot face the fuss and long-drawn-out ritual of a checkup by Dr. Lewis. Dr. Allen's office is on a very tacky part of Pico, and although it is new, in a small apartment-house building, it has the kind of newness which turns almost instantly into slum-shabbiness. I was interested to find how strongly I react to this. However much I realize that all this front in Beverly Hills is *only* front, it's nevertheless hard to believe that any doctor who doesn't have it can be good. And yet Dr. Allen did seem most efficient. He says I have a prostate infection, which is, in a way, much more tiresome than clap; harder to get rid of. I have had to lay off drink, which depresses Don, because he doesn't like to drink unless I do.

Tomorrow will be the last day at L.A. State before the Easter vacation. I hope to make a start on a rough draft of *The Englishwoman.* Also, I *must* get on with the Ramakrishna. More shameful and needless delays.

[1] I.e., the European Economic Community (EEC), functioning from January 1958.

April 16. The day before yesterday, Don made another of his declarations of independence. He has got to have a studio of his own, here at the house, and his own telephone, and his own money and his own friends. I'm making him sound tiresome, but actually this outburst wasn't hostile; was even full of love. And he quite realizes that he has to do nearly all of the getting himself. He only asks of me that I shall understand. Well, I do—and I sincerely believe that things would be much better if he could achieve all these objectives. The trouble is, some of them are really opposed to other deeper wishes, or perhaps one should rather say fears, in his nature. For example, he would do much better to have a studio away from the house altogether, though conveniently close. Why not in Santa Monica? When I ask this, he says jokingly that he wants to keep an eye on me. And I suspect that this isn't entirely a joke. He is afraid of leaving me *too* much alone. He doesn't want *my* independence.

Oh, but what do I care? He was out last night, for instance, and really I am just as glad as long as I have plenty to do, as I have just now. I have been getting ahead with the Ramakrishna book, and I hope to make a start on a rough draft of *The Englishwoman* this week. Maybe on Easter Sunday—an auspicious day for such starts.

For days and days there has been thick sea-fog in the Canyon, with only a little hazy sunshine in the late afternoon. This suits me for working. The rest of the city swelters.

Saw Dr. Allen again last Saturday. The infection still hasn't cleared up, which is really tiresome. He has given me more pills for it.

April 19. Another foggy morning. I can hardly remember the sunshiny ones any more. Don was out last night; haven't heard from him yet. Yesterday, I finished chapter 13 of the Ramakrishna book, the one about Keshab Sen and the Brahmo Samaj.[1] I feel a desire to get on and finish it now. Maybe when the young disciples start to arrive it will bore me less.

Dorothy Miller has just called to say that she "blacked out" in the market, and the doctor says it's high blood pressure so she's going east to Atlantic City to stay with her sister for several months and have a good rest. Poor darling old Dorothy, she sounded scared, and yet you had the feeling that she was making up her mind to die. Luckily she fell on a basket of toilet paper so she didn't hurt herself.

---

[1] Nineteenth-century Hindu reformer and the non-sectarian religious and social movement which he transformed.

The Stravinskys left yesterday—for Seattle Fair,[1] Toronto, New York, Paris, South Africa, Germany—all in ninety days! Igor seems very gentle and tiny and smiling and old; but Vera says he has been frantically irritable about everything, these last weeks. The latest book of his conversations with Bob Craft has infuriated the critics, whom it attacks. "We have *so* many enemies now," Vera told me on the phone. "We are in such a revolt—like young people." Igor, when Don and I went around to say goodbye to them, talked about Form and Content—which is more important? No one can tell. I got a picture of him very carefully looking up words in the dictionary and meditating on their meaning. T.S. Eliot had asked him to set "The dove descending breaks the air...." to music. Igor found this very flattering; I was surprised how pleased he was.[2]

I have already bought the pink paper—to differentiate the manuscript from the blue on which I roughed out my novel and the yellow for the Ramakrishna book—so I can start *The Englishwoman* on Easter Sunday. I haven't an idea in my head; but that's all right.

Reviews of my novel continue to come in. I don't really like any of them. None are intelligent. And I am shocked to find how vindictive I feel toward Stephen, Angus, Kingsley Amis, etc. I am a mass of resentment, nowadays, and getting steadily worse. *And it is resentment for the sake of resentment.* I like to chew on it. For example, last night I dreamt that I had a violent outburst against Edward Teller, the physicist. Now I certainly disapprove of Teller, who is, or seems, pro-bomb, but I don't disapprove of him that much. And even as, in my dream, I was ranting against him I was also aware of my ignorance; I was aware that, if challenged, I wouldn't be able to say *exactly* why I disapprove of him![3]

What can I do against this? Japam. There is no other remedy.

Don is reading *How to Know God*.[4] I don't know what significance, if any, this has. No doubt he'll tell me later.

A decision: before I go on with the Ramakrishna book, I'll read the rest of the material over again. I must do this in order to be able to organize it better.

Aren't you lucky, you stupid complaining old thing? You have a *dharma*. How would you like to be a retired bank employee, condemned to die slowly of "security"?

---

[1] The 1962 Seattle World's Fair.
[2] He set "Little Gidding" from *Four Quartets* as *Anthem*.
[3] Isherwood evidently identified with Teller's colleague, Robert Oppenheimer, whose professional judgement was questioned by Teller at public security hearings in 1954. See Glossary.
[4] *The Yoga Aphorisms of Patanjali*, translated by Prabhavananda and Isherwood.

April 22. Well, happy Easter. At least I have done something to make it a productive one. Today I started a draft of *The Englishwoman and* another chapter, the fourteenth, of the Ramakrishna book. Nothing to be said about *The Englishwoman* yet—and probably not for a long time. The opening is nothing but a barrage to "soften up" the main enemy positions. Then comes the attack itself; and only then shall I know if the positions can be captured.

Marvellous weather and consequently the Canyon is overrun. How I hate public holidays! Don has gone to lunch at the Bracketts'—we were both asked. His motive is to get an introduction to Alice Faye.

Am worried about my jaw. I have kept forgetting to mention it, but it really seems to have become quite chronic. It's all around the back of my neck too and I feel a pressure on the sides of my temples. As for the prostate infection, Dr. Allen seems to think that's cleared up, but he still doesn't want me to drink. Privately I am just as glad that I can't, but I sulk about it and tell Don that if I can't drink I can't endure to see the people who bore me—and that means nearly everybody. How truly boring *I* am getting! I feel that I bore Don and I really want to see him go out and enjoy himself, so he won't become sick of living here. I am unfit for company most of the day. I should just work and work.

Yesterday we got up early and drove Jo and Ben to the plane for their week in New York. Don remarked that whenever Jo fixes herself up she only makes herself look that much older. She had streaked her hair light and dark blonde. It is amazing what a peevish tone she allows herself when she is crossed over trifles. Because they hadn't got seats by the window she really behaved as though it were Ben's fault.

Yesterday I read Graham Greene's *The End of the Affair* for my class. It depressed me deeply. Such a sense of varicose veins! The British sex-gloom, and then the cruel Catholic puritanism. Don got home late and noticed my sadness; and then after supper we were trapped by Mike Steen into going to the house where he is now staying with friends, by the steps at the bottom of the Canyon, and seeing two films they had shot: outings at Yosemite and Morro Beach of a San Francisco motorcycle club. Oh the joylessness of the camping! It was the other half of the Graham Greene joylessness. Between them, they unmade my day. Don says we offended Mike Steen and his friends. He says that, when we are not being amused, we are very grand and feminine and distant. Probably true, and if true how tiresome of us!

Perhaps I have been very unfair to Graham Greene. I reread the end of the last chapter just now and, after all, it is beautiful, and the whole book is, at worst, marvellously ingenious and at best, I guess, very touching. It's ridiculous to claim that Sarah is a great character. She is only a mold into which a greater writer could have poured greatness. But to have made the mold, even—that's quite something.

April 25. Grey again with only faint gleams of sun. My jaw bad the last two days. I'm worried and yet I don't want to go into a medical fuss about it. Yesterday I felt really toxic and miserable but I went to L.A. State nonetheless and did quite a good day's work—particularly a good rendering of my version of Tolstoy's *Father Sergius* and a description of how the "reassuring" type of writer takes you by the hand and leads you step by step from a familiar into an unfamiliar situation. (Cf. Hemingway, leading you into a game hunt or a battle; and, if he's in a place you don't know, he tries to persuade you that you *do* know it—"You know how it is there early in the morning in Havana, etc.")

Mr. Kimball Haslam the builder came this morning and I went with him to apply for a permit to build our balcony, but we were told we must have an engineer's drawing of it first. The officials at the building department employ a technique of extreme distaste. You hold out your form to them. They glance at it with aversion; they don't want to touch it. Then they take it with a prefabricated frown of discouragement—"Let's see what's wrong with *this* one." Their movements are very very slow. But Mr. Haslam's Mormon patience is not to be ruffled. He merely winks at me. When he is describing what he'll do, he always puts it impersonally, "A person might cut a door in this wall...."

April 27. Woke this morning with a miserable head and back-of-neck ache. I feel really sick—jaw, gums under a kind of pressure, and also the pylorus flap. The misery of getting old and worn-out. Weariness and longing for dead sleep. Have I got cancer of the jaw, or a tumor, like Jerry Wald, pinching nerves in my spine? Or is it just a new acting up of the old broken-down machine?

The creative writing class yesterday. Nick Barod very militant about the Un-American Activities Committee, and triumphant because CBS had favored the pickets in its reporting and shown most unflattering pictures of the counter-picketers, who looked, said Nick, like a Little Rock, Arkansas antisegregation crowd.[1] One of the women wrote a piece about an intrepid lady racing-driver who was feminine and petite and this set the boys off talking about the typical lady racing-driver who is masculine and dikey. Don Vucetich talked about "fag hags." I think he's a trifle crazy. He seemed aware of the impression he'd made, because he stopped me later and explained that all this semester he has been suffering from amoebic dysentery.

Then I had a long talk with Michael Rubin, with sweet blue eyes which are out of alignment and a Jewish profile. He has a quite exceptional air of innocence, likes e.e. cummings and body-surfing, writes poems, wants to be a marine biologist. His father is a dentist. He admires Kennedy but disapproves of the crackdown on U.S. Steel, because he believes in free enterprise.[1]

On my way back, I stopped into the Beverly Hills Martindale's and was told that the novel is doing well, picking up fast. A very young boy, not more than thirteen, was asking for Dante; recommended the Ciardi translation—just as if I had studied the whole thing dozens of times.

Party given by Frances and Albert Hackett, because they are going away to live permanently in New York. They have sold their spacious non-home to Ray Milland.[2] One of the female guests enthused to me about England. Their next-door neighbor there was a market gardener who was imprisoned for the theft of scaffolding. He had been doing it for years. To all this and much more I listened in an almost unbearable non-daze of utter sobriety. The misery of having a clear head!

Don has just heard that the Cassini job[3] is on again and he may very well have to go to New York. He is sad about this, doesn't want to leave here. We are very happy again now.

April 28 [Saturday]. Very depressed yesterday, because I felt terrible and because I had been so negative: I said no to three different

---

[1] Suspected communists were questioned in Los Angeles, April 24–28. Possibly, Isherwood meant to type, "anti*de*segregation crowd." See Glossary under House Un-American Activities Committee.

[1] U.S. Steel announced a six-dollar-a-ton price increase after pushing through a non-inflationary wage agreement with the union. On April 11, Kennedy publicly attacked the executives for their greed; privately, he threatened an IRS audit.

[2] Welsh-born actor and, later, director (1907–1986); he won an Academy Award for *The Lost Weekend* (1945) and had his own T.V. show in the 1950s.

[3] Pen and ink drawings of American fashion designer Oleg Cassini (b. 1913 in Paris, known for dressing Mrs. Kennedy as First Lady) and of Merle Oberon in a Cassini evening gown, for full-page ads in *Vogue* and *Harper's Bazaar.*

people—a *Newsweek* journalist wanting "anecdotes" about Stravinsky, a woman who wanted me to appear on a panel with Gavin Lambert and Dorothy Parker, and a girl from L.A. City College who wanted "quotes" from the lecture I am to give there next Wednesday. In each case, I had definite reasons for refusing—I have decided never to supply journalists with anecdotes about my friends, especially over the phone; I was irritated by the woman's approach and her endeavor to make me feel a heel for refusing; I don't think of myself as giving lectures with quotes in them and anyhow the girl was just being lazy, she wanted to save herself the trouble of actually going to the lecture and listening to what I was saying. Still and all, I felt guilty; a mean grouchy old thing. And I disliked myself.

At six-thirty in the afternoon we picked up Jo and Ben from the airport and gave them a fish supper, barbecued red snapper. We both felt the trip to New York hadn't been a success, but there was very little to be found out from either of them. Jo rattled on about Julie and Manning and the plays they'd seen and the heat and the shops, etc. etc. Ben said very little. One had the impression of him lurking around the bars during the daytime while Jo was working.

After they'd gone, I unwisely said how worried I am about my health, and this merely alarmed Don; he fears becoming nurse to a permanent invalid. He told me angrily to go to the doctor. This morning he was sorry, but I did go anyhow. The prostate infection has cleared up. When I told him about my throat and jaw and the back of my neck he arranged at once for some X rays to be taken; they were done this morning. So on Monday he will be able to tell me something.

April 30 [Monday]. Not long after I wrote the above, I glanced at my appointment memo from Dr. Allen's office and saw that he isn't seeing me again until a week from Monday. I took this to mean that he doesn't regard my case as serious, and was correspondingly relieved. And now today I have just been talking to him on the phone. He has looked at the X rays and says the whole thing is arthritic. Not much to be done, except for heat treatment and massage. No surgery is indicated.

Today has been glorious. I got the car serviced, then came back here and lay in the sun for an hour, reading James Hogg's *Memoirs of a Justified Sinner*[1]—the first on my first list to read. Also included are [Stendhal's] *The Charterhouse of Parma, The Walnut Trees*

---

[1] *The Private Memoirs and Confessions of a Justified Sinner* (1824).

of Altenburg,[1] *Restless House (Pot-Bouille)*,[2] [Gertrude Stein's] *Three Lives*, [Faulkner's] *The Wild Palms*, [Hawthorne's] *The Marble Faun*, [Virginia Woolf's] *Mrs. Dalloway, The Stranger*,[3] [Djuna Barnes's] *Nightwood*. Only two of these—*Parma* and *The Stranger*—have I already read right through, and that was ages ago. I am liking Hogg very much.

And now I have finished another page of the fourteenth Ramakrishna chapter, two more pages (to 9) of my novel, and some notes for my lecture at L.A. City College on Wednesday. And when this is written I'll take off for the gym. Then home to fix myself fish cakes and beans. And then Don will be back from a long day's drawing—Dorothy Parker, John Dall(?)[4] and his group of colleagues.

Felicidad.[5]

The night before last, we went to a party at Alan Pa[k]ula's because Don hoped to get Myrna Loy to let him draw her again. This didn't seem to work; she was evasive. Kate Moffat alone in the middle of the room, embarrassed yet not wanting to go back to Ivan's side—an admission of failure. So I talked to her and we *mimed* having marvellous intimate fun. Of such is the kingdom of earth. The guests included four famous wrecks, Eddie Fisher,[6] Tony Curtis,[7] Troy Donahue,[8] John Stride (the Old Vic Romeo).[9] But Rock Hudson looked good.[10] And there was a strikingly

[1] *Les Noyers de l'Altenburg* (1948) by André Malraux, translated by A.W. Fielding in 1952.
[2] Percy Pinkerton's 1953 translation of Zola's novel, which includes an introduction by Angus Wilson.
[3] Stuart Gilbert's 1946 translation of Camus' *L'Étranger* (1942).
[4] American actor (1918–1971), educated at Columbia, on Broadway from the early 1940s. His films include *The Corn Is Green* (1946), *Rope* (1948), *Gun Crazy* (1950), and *Spartacus* (1960). He lived across the street from Parker and Campbell with his lover Clem Brace.
[5] Happiness.
[6] American singer and actor (b. 1928); he starred on his own radio and T.V. show until 1959 when he divorced Debbie Reynolds to marry Elizabeth Taylor.
[7] American actor (b. 1925); he had recently starred in *Sweet Smell of Success* (1957), *Some Like It Hot* (1959), and *Spartacus* (1960), but his marriage to actress Janet Leigh was ending.
[8] Blond, blue-eyed star (1936–2001) of teen T.V. series "Surfside Six" (1960–1962) and "Hawaiian Eye" (1962–1963) and of similar films with Sandra Dee and Suzanne Pleshette.
[9] British actor (b. 1936), Romeo opposite Judi Dench in Franco Zeffirelli's 1960 stage production. In Hollywood, he had character parts. He sometimes returned to the London stage and later developed a T.V. career.
[10] Hudson (1925–1985) was nominated for an Academy Award for *Giant*

handsome boy who was Preston Sturges's son.[1] A little too nobly independent and boyish-manly to be quite quite true, however.... And, speaking of that, Arlene Drummond, who has come to clean for us in Dorothy's place, is too spiritual to be quite true either. One night, she heard Jesus speak to her.

May 4 [Friday]. On Monday morning (probably) some people (probably boys) came into the house and took—three pairs of slacks, one of Don's, two of mine; an opened bottle of scotch, ditto of vodka; a silver coin from Thessalia, Thessalonic League, 196-146 B.C. (Double Victoriatus. Head of Zeus with oak wreath to right. Reverse: Pallas Itonia to right.) The coin was given Don several years ago by Harry Brown, and neither of us ever liked it. Also missing, a pair of nail scissors and an old and dirty jockstrap. Our doors were all open of course, as usual. Maybe if they'd been shut the thieves would have busted in anyway and smeared the walls with shit and wrecked the place as they so often do, apparently.

Today I finally reported it to the police. Had to lecture on Wednesday, so no time before.

Don gloomy and cross last night. This morning he said all the things he'd been feeling seemed quite unreal. He has just finished Vivekananda's "The Real Nature of Man."[2]

Have been piddling around today getting nothing accomplished but tiresome chores. So now I can't go to the gym. Must work instead on the Ramakrishna book and if possible the novel. We were to have had supper with Gavin and he has just called it off. I'm afraid Don won't like that and be cross again. Patience.

Wilbur Flam came to my lecture on Wednesday—it was at L.A. City College—so did Mary Herbold. I saw both of them later. Wilbur first. I drove up with him into Griffith Park and we sat under the trees. Such meetings consist largely of speeches for the defense. One justifies one's life since last heard from. Wilbur and his wife Bertha are going to live [abroad], because they can live with Bertha's mother there and it will be cheaper. Wilbur hasn't been a success so far at his artwork—he does collages. But now he hopes to do better with children's books. He complains of lack of

(1956) and had more recently appeared in *Written on the Wind* (1956), *The Tarnished Angels* (1957), and *Pillow Talk* (1959).
[1] Probably Solomon Sturges IV (b. 1941), known as "Mon," son of the playwright, screenwriter, director (1898–1959) and his third wife, Louise Tevis.
[2] A lecture included in *Jnana Yoga* and later in *Vedanta: Voice of Freedom* (1986) with a foreword by Isherwood.

sexual interest, but fears [living abroad]'ll stimulate him again, and in the wrong direction. There was little comment I could make. Naturally I feel that he has fouled up his life. But maybe there wasn't much to foul up.

Mary Herbold, curiously enough, doesn't seem nearly so fouled up. Of course she is a screaming bore, but she's a lively old thing and has kept in touch with the world and other people, and I respect her. She dwells a great deal on her sex adventures, which must have been mostly in the mind.

Yesterday, at L.[A]. State, they had chariot races; the chariots pulled by boys. One of the drivers was in drag; a girl's bikini. A fat student was overturned at high speed from a bathtub mounted on three wheels but wasn't hurt.

May 9. The day Daddy was killed in France—or was it? I have forgotten. Is this awful? No—not particularly. But at least I'm remembering him today.

Have finished a rough draft of the fourteenth chapter of the Ramakrishna book today, and am on page 16 of the new novel. I still don't know what to think of that. But today a Japanese girl character whom I'd introduced for no reason apparent to myself, just as a friend of Colin's, suddenly acquired symbolic status as another Foreigner, stuck midway between being a nisei in an America she despises and the alternative of going west and being a Japanese in a Japan she doesn't know. This is promising.

We heard from Ted this morning; they let him out of the Honor Farm[1] in the small hours. He was rather sweet about it. He said, "I even got a kick out of riding home on the bus."

The night before yesterday, we had supper at Michael Barrie's, and Gerald was there in his greatest form. He went on at great length about Aldous's tongue cancer and the genius of Cutler the surgeon. He is sure Cutler could have saved Maria.[2] He says that Maria, just before she died, told him that she had no idea if Aldous really loved her or not. Then Gerald got on to the subject of Europe and the Common Market. He owned to have been quite wrong about Europe. Now he believes Europe is going to be great again. The Russians missed their chance by being too clever and sly and suspicious. Only Europe could take it—*it* being "the religious, the political, the economic and the psychological revolutions"—

[1] A minimum-security correctional facility where inmates worked in the laundry, kitchen, and garden; probably the Los Angeles County Honor Farm.
[2] According to Huxley's biographer, Sybille Bedford, Max Cutler pronounced Maria Huxley's cancer hopeless in 1954.

"without going mad." There was much much more which alas I have forgotten because I got drunk. I got drunk because this was the first time I was allowed to drink (grudgingly) by Dr. Allen after exactly one month. Having x-rayed me he has found a lot of arthritis in my spine and neck, but there's nothing to be done about that, except suspend me in traction like Swami.

Don agreed that the evening with Gerald was a real success. "It's so seldom," he said, "that you feel you're in the best of all places you could be."

May 11. I'll try to remember to note down here the various pairs of linked events which keep happening to me. Partly because I want to see if they do, in fact, *keep* happening. One's memory plays tricks and the only way to be sure is to list them. Here are three recent pairs:

Following the burglary at our house, in which three articles of clothing were stolen (i.e. the three pairs of pants) there was a burglary at our laundry, in which a lot of clothes, including a box of ours, were stolen.

In my last entry, I mention that I have had an idea for my novel about a Japanese girl character who might play quite an important part. That same evening, I picked up a hitchhiker—a thing I very seldom do nowadays—and he was a Russian-American boy who had lived most of his life in Japan and even spoke English with a Japanese accent; in other words he, like my projected character, was a sort of nisei. What made this coincidence odder was that the boy didn't look in the very slightest Asian, so I can't have picked him up because of any subconscious association of ideas.

This morning, at the Department of Building (where I finally got a provisional permit to build our balcony, subject to the approval of an inspector on Monday), I was accosted by a man, a builder presumably, who knew my face from seeing me on television. I left the building and had just reached my parked car when I was accosted by a young man who knew me from television and had been listening to my Santa Barbara lectures on the radio.

Gerald calls these pairs of events *synchronicities*, because one of their characteristics is that they must happen right after each other. He thinks that they are parts of a cosmic pattern which we thus glimpse but can never hope to understand.[1]

---

[1] Carl Jung coined the term and lectured and wrote about it several times, including a long essay, "Synchronicity: An Acausal Connecting Principle" (1952).

May 13. Cold windy weather, although the sun shines. I wish I could get out in the sun, but it's too chilly and there is so much to do. Yesterday I slipped badly; I should have pushed ahead with the revision of chapter 14 of the Ramakrishna book, and I also failed to get on with my novel just at a time when I ought to be doing something every available day. During this part of the "climb" there's a great danger of freezing onto the cliff face if I don't keep moving.

Ted and Vince are here right now, looking at Don's drawings. Ted remarking on the recent dearth of new movies, said, "I couldn't have picked a better month to go to prison." I complimented him on his pretty shirt, and Vince looked slightly quizzical, which made me wonder if Ted had stolen it!

Last night, Don and I went to supper at Henri Coulette's apartment in Pasadena. A disaster. We started late, because Don got back late from drawing the Bracketts, having spilled a bottle of ink on the floor, and then we ran into a huge traffic jam on the Sepulveda Pass, and then Don wanted to see "In a Lonely Place" on T.V., so we left early. There weren't enough martinis, there wasn't enough food and there were too many guests. I don't think heterosexual parties are workable, anyhow, just as conversation groups. If women and men mix, they should dance and flirt; they have very little to say to each other. An unmixed group of men or women would have been far livelier. And, oh dear, the academic atmosphere with its prissy caution! One man said that very few people in America "had the background" to be able to appreciate *The Brothers Karamazov*. WHY? If this were really true, it wouldn't be a criticism of America but of Dostoevsky.

Sure, I am prejudiced, but I feel always more strongly how ignoble marriage usually is. How it drags down and shackles and degrades a young man like Henri, who is really sweet and bright and full of quiet but powerful passion. The squalid little shop, the little business premises, you have to open, and the deadly social pattern which is then imposed on you—of dragging some dowdy little frump of a woman all around with you, wherever you go, for the next forty years. Not to mention the kids. It is a miserable compromise for the man, and he is apt to punish the woman for having blackmailed him into it.

Must reread Strindberg on this subject, and read his *Confessions of a Fool*. Have just finished Camus' *The Stranger*, which irritates me merely; Camus is such a dreary mind. Am enjoying *The Charterhouse of Parma*.

May 18. The building inspector came this morning and met Mr. Haslam our contractor here, and it really looks as if the balcony and even the studio may be started before too long. Don was inclined to be sad about his birthday—he was mad at me last night because I hadn't told Jo and Ben, with whom we were having supper. But he cheered up a bit when I gave him a shirt, a tie and the kind of English oatmeal biscuit he likes. Now he has gone off to draw Dorothy Parker again and to have supper with his folks. We are supposed to meet for a movie later.

I feel rather wretched about him, and yet I know there's almost nothing I can do. This is a period without glamor. He blames me because his birthday isn't marvellous, and I would blame him under the same circumstances.

Two more pairs of linked events, except that one of them was a triplet—Don spilt his drawing ink on three different occasions during the last week; and two unlikely people have given me off-beat religious works to read—Arlene Drummond our new maid gave me *Deep Meditation; A Very Simple System from the Ancient Vedic Culture of India Ideally Suited to the Tempo of Modern Times; Transcript of a Talk and Questions and Answers by the Maharishi Mahesh Yogi* and Mr. Haslam gave me The Book of Mormon!

Yesterday, at our creative writing seminar, I tried reading bits of the Mexican version of *Down There on a Visit*. It seemed miserably poor and thin. I don't know how I am going to make it sound interesting enough for the writers' conference.

May 24. Another pair of events: yesterday, I got two film offers—from Zachary Scott[1] to do a film about the life of Lafcadio Hearn or based on one of his Japanese ghost stories, and from a man named Julian Lesser,[2] to write and speak a narration for a documentary on the life of Gauguin. The latter tempts me more than the former.

Yesterday, I went to see Gerald in a small hospital on La Brea. He has had an operation for hernia, not serious. Of course he knew all about Hearn, how he had lived in a sewer (where?) and how he had become disgusted with Japan after the Japanese cut his professor's salary in half because he had become a Japanese citizen.[3]

[1] American actor (1914–1965), on stage in England and on Broadway; he became a movie star with *The Mask of Dimitrios* (1944) and *The Southerner* (1945).
[2] American movie producer, son of Sol Lesser (1890–1980) who produced Tarzan and Dick Tracy movies and Westerns.
[3] Hearn (1850–1904), born in Greece and educated in England, was a

Gerald has a low opinion of Hearn's ghost stories because they have no cosmology. He was in the highest spirits and only winced in pain when he coughed and it hurt something "down in the plumbing." He expected to get out today or tomorrow.

Up at Vedanta Place, Prema remarked that he wished things could go on just as they are now for years, because everything was very harmonious: Swami well and Prema is enjoying what he is doing now. Maybe this was an important confession psychologically, because I fancy Prema has been very restless until recently— yearning to get away or become a swami or something. Anyhow, "harmonious" isn't quite the word I would have chosen, for mad Marlene[1] has been sending more anonymous letters with torn-up photos of Ramakrishna or Swami in them, and the question "Is the farce still going on?" And Tito [Renaldo] lowers and glooms, and the other day at breakfast jumped up exclaiming that if somebody didn't stop hexing him he would pack and leave!

May 25. Am writing this after an evening of reading through the test papers and trying to grade my students—what dull work! Only Janvier and [Glenn] Porter stand out so far—Janvier because he is sincerely interested, Porter because he is simpatico-crazy. Now I'm going down for a late supper at Ted's [Grill] with Don, who has been drawing with Jack Jones.

Don had a very good synchronicity this morning—two actresses, both in films, Jane Fonda and Eileen Heckart, want to be drawn.

Yesterday afternoon, at L.A. State, I had to have tea with Dr. Vida Marković,[2] who is head of the department of English at the University of Belgrade. She had especially asked to see me, because she so much admires *The World in the Evening*! And certainly she seemed to know it much better than I do. What she particularly admired was the character of Elizabeth Rydal and her psychology when dying. She seemed very bright about other British writers, which was reassuring. A very handsome, strong but not mannish woman, with a wide humane forehead. You felt she could have run the Yugoslav underground single-handed. She wore a filmy summer dress which seemed too insubstantial for her. She was accompanied by a young man from the State Department with

journalist in America, then, in 1890, settled in Japan with a Japanese wife and taught English literature at the Imperial University, Tokyo.
[1] Marlene Laurence (d. *circa* 1980), an eccentric Hollywood devotee with an unpredictable temper; she worked at Kodak.
[2] Novelist and literary critic; she published work in English and in Serbo-Croat.

a ruddy face and a moustache. I don't know if he was a cop.
Elizabeth Sewell[1]—as well as Shroyer, Coulette, Jean Maloney[2]
and a Dr. Collins—also took part in the meeting. I do like and
really admire Sewell. She has a bright dry British intelligence
which goes with her non-nonsense face and humped back. We
chattered away, disposing of Snow and [Lawrence] Durrell, and
paying tribute to Compton-Burnett. Once I caught the eye of the
State Department man and he winked at me.

Yesterday evening, Don and I had supper with Bill Roerick.
Tom Coley[3] is recovering from a nervous breakdown, during
which he would wake up in the night to see a man in the room—
a man in an old yellow cloak or wrapper, who had no face, or
rather a skull with skin stretched over it. Don was sure that all
this was a mechanism to get rid of Bill and stop living with him.
Tom is now down near La Jolla. Bill was a bit saintly about it all.
However, he indulged in a little bitchery of Bob Buckingham,
whom he accused of putting on airs and acting artistic—in other
words, forgetting his place. Bob's son Robin is dying of a kidney
disease. There is nothing to be done for it.

May 27. A detail I forgot to add to the last entry—this specimen
of academic market-jargon: "Last February I wrote inviting you
to speak at the University of California at Berkeley in August
and you indicated an inability to accept at the fee we then could
offer.... etc. etc." In brief, having tried to jew me down to some
ridiculous and impudent offer of theirs, they now, with anal
squeals, agree to pay my fee. So I shall go up there on August 30.

More grading all yesterday. It's the dullest work. Not even many
amusing mistakes.

I also forgot to record a talk I had on Thursday last with Richard
Naylor, who is perhaps the brightest of my creative writing group.
He was much worried about the feeling of not belonging to The
Others. If he became a writer wouldn't that cut him off from
them? On the other hand, he has this uneasy feeling that writing is
his dharma—he didn't use that word—and that he ought to force
himself to work. I developed a theory—I don't know if I really
believe in it—that one should abandon oneself to the will of Art

---

[1] American poet, novelist, critic (b. 1919); her books include *Paul Valéry: The
Mind in the Mirror* (1952) and *Signs and Cities* (1968) with a Bachardy drawing
of her on the cover.
[2] A professor of English at L.A. State from 1960 to 1984.
[3] American stage actor (191[4]–1989), Roerick's longtime companion; he
appears in *D.1*.

in the same sense in which one speaks of abandonment to the will of God. No doubt this is a very dangerous doctrine for the merely lazy. And yet perhaps it would reduce the would-be writers if they would wait for "the call" instead of rushing in to mess up a lot of paper.

Yesterday evening, a party at Jack Larson-Jim Bridges' house. The usual complaint, too many people eating awkwardly on laps, and Roma wine served, a headache in every glass. John Kerr's[1] drunken smile, respectful in the presence of Culture (me); Romney Tree's anxious smile which seems to advertise, but without much confidence, her way of life.[2] Betty Andrews insisting that we all come and see her in some gruesome play about Henry VIII. A plump dyke brusquely objecting because Don said the fight between the cats during the credit titles of *Walk on the Wild Side* was cruel, and Don disgusted with himself later because he had made a silly crack about dykes in general with particular reference to Barbara Stanwyck....[3] Oh, why record any of this? The thing to remember is that Jack and Jim are really very sweet and likable, and if you don't like catchall parties, well, why go?

May 29. Yesterday I had supper with Tom Wright. Gavin was there and told me, to my horror, that Paul Kennedy may be dying of cancer of the lung. However, I called Barbara Morrow today and she says the tests are negative so far, though they are still not certain it may not be bone cancer.

Today I sort of wound up my duties at L.A. State, including telling Byron Guyer that I am definitely not coming back there in the fall. It was rather sad to say goodbye to some of the students, although I know them so little—sassy David Smith, handsome melodramatic Nick Barod, silly mystic Glenn Porter, grinning fattycat Charles Rossman. A long talk today to Richard Pietrowicz about his surprisingly good fragment of a Roman novel. He says his parents are illiterate and cannot imagine why anyone should want to write. He has no money and has got to go into the army shortly. Should he risk everything on writing this summer and

---

[1] Harvard-educated American actor (b. 1931); he starred in *Tea and Sympathy* on Broadway (1953) and in Hollywood (1956), played a T.V. lawyer in "Arrest and Trial" (1963) and "Peyton Place" (1965–1966), then became a real-life lawyer.

[2] Tree was married to Brad Fuller, and they had a child, but Fuller was companion to American poet and translator Hugh Chisolm; Fuller and Chisolm appear in *D.1.*

[3] Stanwyck played a lesbian bordello mistress in the film.

trying to get the novel finished? Yes, I said—which was what he hoped I'd say.

Bill Roerick came by yesterday at his own suggestion for a drink, with a young man named Carson who had been somehow connected with the Judith Anderson show. Roerick full of vanity, talking about how he had never thought himself good-looking even when young, etc. etc. However, he told a very funny story about the British government being bankrupt after the last war, so a Mr. Cohen, a troubleshooter from America, is called in. Mr. Cohen asks to see King George all alone. He says, "You've still got Australia and Canada, haven't you? Vell—here's vat you do—put them in your vife's name."

Tom is preparing to leave on another trip to Mexico and Guatemala. If this is a success, he thinks he may settle down there for a year or so. He is absolutely caught up in the mystique of Mayan ruins. But, at the same time, he knows all about the stock market, which took a huge dip yesterday but recovered strongly today. Gavin, also, was worried about the market. And to think that never, in my whole life, have I had anything to do with it!

May 31. A beautiful day, after the wretched one yesterday for the Memorial Day holiday. Antisocial as I am, I can't help being rather glad about this. Anyhow I am in a bad mood because I have a foul cold. It won't be better for several days I fear.

Charming David Rubin appeared at the door this morning, having hitched a ride all the way from Covina or wherever he lives, to bring his term paper—and now I find half of it is missing! This afternoon I plan to get the car serviced, read the rest of Gavin's novel, visit Paul in hospital and decide whether or not to buy Simone de Beauvoir's autobiography (volume 2) which sounds interesting. I am not going to the college this week again but I shall go next week and pick up the rest of the papers and grade them and turn them in. And that will be that.

The Englishwoman has been crawling along. Now I must fetch a whip to her. And I must get chapter 14 of the Ramakrishna book done before next Wednesday.

Talked to Bruce Zortman yesterday. Charles [Laughton] remains in New York, mysteriously involved. Bruce thinks it isn't really his health but some dreary blackmailish business, but he was so very vague that I couldn't quite understand what he meant.

June 2. The day before yesterday I went to see Paul and found him very depressed, with charley-horse pains in his legs. He said, "I'm twenty-eight and all I've got to show for it is a shadow on my lung and an enlarged liver." It was so sad to see him, all alone in the hospital—the Temple, which, I must say, is one of the nicest, perched on the top of a hill—and hiding a dreadful anxiety: what *is* the matter with him? The doctor admits that he doesn't know. Yesterday, on the phone, Paul told me that they have discovered something in his bone marrow which "ought not to be there."

As for me, Dr. Allen discovered a cyst in my ear. He didn't seem alarmed about it though, just said he'd look at it again in a month's time.

Another pair of events: two checks Don received within a couple of days, from Count Rasponi and from Dorothy McGuire,[1] had a discrepancy between the sum expressed in words and the sum expressed in figures.

Read Gavin's novel the day before yesterday. I don't like it. It's false. Just a piece of gingerbread, which wouldn't matter, except that it takes itself seriously. Of course I couldn't possibly tell him this. Luckily, I hit on some criticisms of the ending with which he entirely agreed.

Don and I went to a party at Gavin's last night. Jane Fonda was there with her phoney Greek boyfriend,[2] and old Barbette, whom I liked, and Dorothy Parker with Alan. I'm afraid Dorothy really is a dead loss to us all; something has been permanently smashed and there's no use hoping it will ever work again.

After the party, drunk, Don told me he wants me to go away to San Francisco and leave him alone all summer. He still wants it this morning, though less aggressively. I wish *he* would go away. If I do, I know I won't be able to work and I shall just waste several months. Still, I must seriously consider it, because I realize that his reasons for wanting to be alone are serious, however selfish he may be in seeking this solution.

Don also said this morning that he would like to have a mantram and he wished Swami would give him one. Swami, of course, would like nothing better than to initiate him—and get him into Trabuco for that matter—but only after he has attended several months of lectures and classes. It's the first time Don ever said this.

June 5. In the late afternoon of the 2nd, after I made my last entry, Elsa called from New York to tell me that Charles was going into .

---

[1] Bachardy drew her half a dozen times on two different occasions and later drew her son and daughter, but she never bought a portrait from him, so the check was for something else.

[2] Andreas Voutsinas; see Glossary under Fonda.

hospital next day to have one of his kidneys removed. Today, Michael Barrie told me that the radio news says quite definitely it's cancer; Elsa wasn't specific about this. But she did say that Charles is terrified and that he is quite unprepared to die. He doesn't want to come back here; he talks of going to Europe. He wants to discuss things with a Zen Buddhist. He dismisses Vedanta, saying that it's "Indian." Elsa told me all this with perhaps just a faint tone of "you see, he doesn't need you, he needs me" and she said that she felt she had become very strong. She is still slightly resentful because Charles loves seeing Terry, who has settled in New York now, it seems.

As for Paul, he has been vomiting, and now the doctor is taking another sample of marrow and they have also given him a T.B. test.

Oh, the horror of modern medical death! The hospital has become a sort of earthly purgatory which you have to pass through.

Don terribly upset, desperate to have a life of his own but not sure how to set about it. I feel I should go away for a while and yet I do not want to, with all this work to be done. Don seems to pin his faith on the building of the studio in the garage, but I am doubtful. Surely it won't be that simple?

In the presence of all this suffering, what do I do? Get drunk at the Selznicks' on Sunday night and incapacitate myself for all of yesterday. So now I have decided to give up drinking altogether for a long while. I am sloppy and fat, despite the gym; and yet everything demands that I should be disciplined and alert. I ought to become more and more of a workhorse; it is the only happiness for me. There is not all that much time left, and I have such an infinity of things to do. Right now, I have fallen way behind and must postpone the finishing of chapter 14 for another week—not to mention getting on with *The Englishwoman*. Still lots of test papers to grade.

June 7. The test papers are all graded now, except for four or five which are waiting for me at the department office. I shall go to the college tomorrow, grade them and hand in the rest of the cards, and that will be the end of it.

Yesterday evening, Don came up with me to Vedanta Place, talked to Swami and told him he wanted a mantram. Swami seems to have been pleased and surprised—as well he might be, after nearly ten years! He told Don how to meditate and said he'd initiate him next December.

Woke this morning feeling really toxic. My whole gut is sick, and my tongue burns. There are times in the day when I feel so awful that I would gladly go to bed, and yet I perk up, work or go to the gym. I guess this is just being middle-aged.

Forgot to say that I was disappointed in *The Charterhouse of Parma*. It is so smart-alecky, and Stendahl has a vulgar show-off mind. No wonder Balzac liked it. Toward the end—as in the *Splendeurs et misères des courtisanes*—you feel that the author is getting bored.

Have also finished the two first Claudine books.[1] There are really first-rate things in them. And however one may be put off by the frequent whiffs of cunt and dirty drawers one must remember that this is artistically right for the subject matter. Colette never gets bored with *her* story.

Swami is being threatened by mad Marlene, who is now in San Francisco. So the boys have fixed up a buzzer system, between his room and the monastery. If Marlene arrives in the middle of the night and starts to smash in Swami's door, he merely has to flip the switch and the entire monastery is alerted like a fire station. The boys can be over in eighteen seconds!

June 10. I don't know if this will pass for a synchronicity: I have been twice "honored." Just heard that the Mid-Century Book Society has chosen *Down There on a Visit* as its July selection. Or rather, half of it. The other half being—irony of ironies—Iris Murdoch's *An Unofficial Rose*, which I refused to review because I found it absolutely unreadable! Also, tonight, I am to be given an honorary degree or some kind of award, by USC at what they describe as a "banquet," at a rather tacky French restaurant on La Brea. Kirk Douglas and Laurence Olivier are the other guests of honor, but they are pretty certain not to show up.

Paul Kennedy says he feels better; but he had some mysterious treatment which knocked him out a couple of days ago. And, when I phoned, the nurse I talked to said it was a shame because "he's so young"—which sounded bad, to put it mildly. The Laughton mystery deepens; Charles and Elsa are said, by their Eddie,[2] to be coming home this week, and Charles hasn't had the operation. Again, this may be good, may mean that it's too late to operate.

My grades are all in, and nothing remains but to start "vacation" work—i.e. get the hell on with Ramakrishna and the novel. I have been very bad, not doing any of this today, but Florence Homolka came in and photographed us both. She is a huge blundering but

---

[1] *Claudine à l'école* (1900) and *Claudine à Paris* (1901).
[2] Chauffeur and helper.

not unsympathetic cow—a poltergeist, Don says. She knocked down a picture, trampled the flower beds, nearly wrecked the table. And she takes pictures so slowly that you can't hold the pose.

A new symptom: my tongue burns nearly all the time—as if I'd been smoking, though I haven't smoked at all.

June 14. The "banquet" turned out to be nothing more than an end-of-term dinner for the USC chapter of the National Collegiate Players. We had to pay for all of our own drinks except the first one, and then sit through a regular prize-giving—best actress, best actor, etc. Then an old fuddy-duddy got up and introduced me, informing the audience that I'd spent two years in a monastery in India and written a one-act play called *Mr. Norris*. Also, for some inscrutable reason, he quoted my translation of the Chaitanya hymn, "Oh mind, be humbler than a blade of *glass*"! One of our hosts addressed Jo as "Mrs. Masculine."

Paul Kennedy *has* got cancer; Vic Morrow told me. He doesn't know this yet. I went to see him on the 11th. Strange and terrible how he has already lost so much weight and turned dull orange yellow and looks shrunken. But he is in high spirits because the pain in his legs has gone. They are giving him radiation treatments every day. He throws up a lot and doesn't care for food. His nose bleeds. But he plans eagerly to get out and go to parties. His legs are horribly thin.

Gerald is also an invalid still, but a recovering one. He has a marvellous tailored robe which Michael has had made for him: tight-waisted, with full skirts and hanging sleeves—a kind of number which Ivan the Terrible might have worn. He talked of the absolute necessity for legalizing euthanasia. I cheered him by telling him that homosexuality became legal last year in Czechoslovakia. (Read this in *Encounter*—along with the less cheering news that *An Unofficial Rose* has been chosen by the Book Society.)

Yesterday, Mr. Haslam and two assistants started work on the garage and the balcony. There is something shocking about the way workmen attack a building; it seems cruel. Don agreed. He had seen a bulldozer *deliberately* gather its strength together, dash across a lawn and smash into an old house—which made a horrible creaking screaming noise, Don says; adding, "It was like a fight."

My tongue still burns. Worried about this.

June 18. Another double: George Sandwich died last Friday and Vishuddhananda, the head of the Ramakrishna Order, died yesterday. Swami got through to Amiya on the phone. She was drunk, said George had died in her arms and she had tried to get him to chant the name of Holy Mother.

On Saturday we had the usual ghastly Father's Day lunch for Swami.

Elsa called yesterday. She and Charles are back here. Now it seems that the cancer is doubtful; there may be something different the matter with him. But really it is impossible to get anything definite and reliable out of Elsa, so determined is she to melodramatize every instant of her life.

Dull foggy weather. My tongue burns; no better, no worse. Don is in a flap. Have I ruined his life? Or not? Or what? Last night, he saw a man sneaking around the house. This must have been about one a.m. We turned on all the outside lights. No further sign of him. This morning, a letter from the woman who wants him to do the drawings for Cassini. Shall he refuse to draw Princess Radziwill[1] from photographs? It is against his whole method of work; but he doesn't want to lose this connection because it offers independence financially, which means independence of me.

But another double came up; a good one for him—they are holding an exhibition of portraits of contemporary writers at Cheltenham[2] and want some of his to exhibit; and Rex Evans wants two of his drawings for an assorted summer show. This would help advertise Don's one-man show in the fall.

My mood is bad, as you see. Sour and worried. I am not getting on with my work. Of course the builders are a disturbance. But one can't help rather loving the amazing Danish boy Evan, who sings and jokes and skips around, shovelling concrete and banging at plaster and staggering under loads of lumber as if he really knew that work is Mother's Play. In contrast, the fattish degenerate face and figure of the non-Mormon assistant, sucking at a cigarette and loafing.

June 21. An absolute chain reaction of synchronicities at the gym today: A short while ago, I told Lyle Fox the story about Jesus and the Blessed Virgin playing golf (see October 29 of last year; page 130) and Lyle has been retelling it ever since. This afternoon he told it to one of the men who work out there, and this man remarked, "Now I must go see *King of Kings*," (which happens to be playing now in Pacific Palisades) and, just as he said it, in

---

[1] Caroline Bouvier (b. 1933), socialite sister of Jacqueline Bouvier Kennedy; then married to her second husband, Prince Stanislaus Radziwill (1914–1976).
[2] I.e., Cheltenham literary festival.

walked Jeffrey Hunter![1] Neither Lyle nor the other man told him the story, feeling that he had probably become allergic to Jesus jokes.... In due course, Hunter went into the steam room with a copy of *Esquire*. After he had left, I went into the steam room and the magazine was lying there. I picked it up, opened it at random, and there was an ad with the caption "The Inferior Golfer"!

A minor synchronicity this morning: I met both Frank Wiley and Jim Charlton in the magazine shop near the library. (I was there because I had time to kill waiting for the eye drops to take effect so Dr. Bierman could examine my eyes and prescribe some new glasses for me.) Frank Wiley has become a naval officer and is about to go off to the Orient and serve on an aircraft carrier. Jim is very vague and grand and treats everybody and everything here as so much non-Japanese trash.

Last Monday, George Huene and Barbette and Gavin came to supper. It was tragic to see Barbette, who is now crippled as the result of some form of paralysis, tottering as he made his way over the debris of our future carport, and saying to George, "I can't balance as well as you can."[2]

Talked to Paul Kennedy on the phone. He is so delighted because he is getting out of hospital, probably, at the end of the week. And yet he is losing weight still and can't sleep at night and still can't eat properly. His hope for life is heartbreaking.

Saw Doctor Allen yesterday. He couldn't explain the soreness of my tongue, but took blood for tests. It *might* be due to anemia, he said, or syphilis, which is very prevalent just now. I asked him why, and he answered, "Because society is so disorganized." When he was pricking the tip of my finger to get the blood, I apologized for wincing, and he said that lots of people are sensitive about their fingers. A colored woman had come into his surgery with her hand wrapped in a cloth. The cop who came with her—there had been some accident—handed Dr. Allen a package, remarking, "These are her fingers." When the woman saw them she fainted dead away. Dr. Allen told me this as if it were an amusing example of hypersensitivity. And yet he isn't at all the brutal type of doctor.

June 23. Talked to Dr. Allen this morning. He says the blood tests show nothing alarming, so that is that. The burning in my tongue continues, but I just won't think about it. I really wasn't in the

---

[1] "Jesus" in Nicholas Ray's 1961 remake of the Cecil B. DeMille silent movie about Christ.

[2] He had been a tightrope walker; see Glossary.

least worried about the syphilis, although I know, theoretically, that almost anyone can get it and it is a nuisance to cure. I don't even worry about heart attacks, sclerosis, T.B. or strokes, although strokes run in our family and I shall very probably end up with some, and they can be unspeakably terrible. No—all my fears are centered on cancer. I thought I might have cancer of the tongue; just as I thought I might have cancer of the jaw. That is my obsession. And yet cancer i[n] our family is very rare, almost unknown. Only Aunt Esther [Isherwood] had it, very late in life.

Don was miserable yesterday because he tried to do this job for Cassini—draw a fashion model in a copy of a dress belonging to the Princess Radziwill and fake in her face from photographs out of *Life Magazine*—and he couldn't. He has sent them a telegram saying so, and offering to pay for the model. Personally, I believe that this refusal will raise his prestige in the long run; but naturally he can't see this.

I felt great satisfaction yesterday because I went in a truck with Mr. Haslam looking for filler dirt for the floor of our new carport, and when we found some, on the slope of Chautauqua, I was able to keep working right along with him, shovelling it into the truck. I couldn't have done this if I hadn't been going to the gym. It impressed him, I think. He says work is part of the Mormon religion, which is why the boys who work for him are so lively and active.

The Seven Arts[1] have offered me this job adapting *The Night of the Iguana* for the screen. I don't want to do it. So I have asked for more money—fifteen thousand instead of ten. Geller doesn't think they will accept. Yet we do need the money, because we have the most ambitious schemes for fixing up the house, with the help of this nice but dull tall girl, Margot Smith. Last night we took her to *Lolita*, which has been made into a fascinating film by Kubrick, brilliantly acted and, like the book, all about nothing.

Work on the novel again yesterday, after a long interruption. I got a good gimmick-idea, about voyeurism. But I *must* keep at it every day. Also I started another chapter, the fifteenth, of the Ramakrishna book.

June 25. Yesterday, after what seemed like a whole age of fog, the sun shone. Don and I lay on the deck, which still has no railing and seems as insecure as a flying carpet, with the wind blowing up between the floorboards and the whole Canyon floating in the air around you. Today, the building is going full blast. A plumber

[1] Independent production company.

is putting in Don's studio john and cement is being laid down in the carport and the man has come to arrange the railing for the deck. And the sun is glittering on the ocean.

Bruce Zortman told me on the phone that Charles definitely does have cancer and it can be only a matter of time. I talked to Paul, who is out of the hospital and staying with the Morrows. He can't sleep well and he only weighs 118 pounds.

Don just got a call from New York and now they want him to go there and draw Cassini himself. So his rejection of the Radziwill fake-portrait seems to have worked for him, as I suspected it would. He may leave at the end of the week.

Nothing from the Seven Arts people about *Night of the Iguana*. Not that I care. I just want to go right on doing what I'm doing. Got in a very good spell of work yesterday on the novel and the new chapter on Vivekananda.

Don is being an angel. My tongue burns and my jaw aches, and I would be perfectly happy if—if I wasn't me.

June 27. This glorious sparkling weather continues. The workmen have now put up the trellis over the deck, casting a barred shadow. Don is in raptures. The framing of the view gives him exquisite pleasure and now he keeps saying how happy he is here and how happy he is with me. And so, of course, I am happy too.

We are still waiting for a word from New York. Now it is possible that Cassini will come out here and Don can draw him here. Don is praying that this will happen; he dreads going to New York, and indeed this would be a most awkward time for him to go away, when so many details about his studio have to be decided.

Talked to Al Spar[1] yesterday about the *Camera* musical deal and some kind of rather mysterious merger of Don's and my finances which Al is planning, to save us income tax. Al is so funny, the way he talks. After explaining something to me, he wanted to say, "Let me recapitulate." What he actually said was, "Let me recapture myself."

Yesterday afternoon, for the first time since his illness, I saw Laughton. I had prepared myself for a ghastly shock, but actually he didn't look so bad; thinner, and very serious, with a walrus moustache, but his color was quite fresh and rosy. He walked round the pool several times with his male nurse. (Who, I hear this morning over the phone from Bruce Zortman, got drunk after I

[1] Lawyer-accountant for Carter Lodge, who recommended him to Isherwood.

left yesterday on vodka, fell out of bed, peed all over the floor, and was packed off by plane to New York.) Charles said at once, in his typical abrupt tragic voice, "You know what's the matter with me, don't you?" So we talked quite frankly about death, God, etc. Charles told me, however, that cancer hasn't been definitely diagnosed; this may or not be so. He said he was appalled how unprepared he was to die. The only thing that had helped him was thinking about some of the Japanese temple gardens he had seen on his trip. I got the impression very strongly that Elsa is fatal for him, and so is all this medical solicitude. What Charles needs is to play "his last great role." He should actually appear in a film, even if he had to be wheeled on to the set and given shots of adrenalin. At best, it would cure him; at worst, it would be a mercy killing. To lie in bed, moping, not even being able to make up his mind to read, is simply crawling toward death. I see all this vividly, and yet, would I act any differently? I think I did help him a bit, however, even if only just by being outspoken.

Charles said, "It's a ghastly irony, isn't it, that I watched Albert Brush[1] die, and now I have to go through the same thing? The last two weeks, Albert was like an animal."

June 29. Sloth. It is twelve-fifteen and I still haven't done anything this morning except read. Nearly all my activities are compulsive—making japam, going to the gym, keeping this diary, working on the Ramakrishna book and the novel (which has crawled as far as page 33). Not compulsive is lying flat on my back on the couch in my workroom and reading Ian Fleming; have just finished *Moonraker*. And why do I like Fleming? He's not all that good but he has atmosphere. It's a world. You can enter it and have fun.

Why do I do all these other things, then? Because, if I don't, I start feeling awful. Life is such a *drag*. Charles and Paul dying. Don out for the night, and soon, it seems, going to New York on the Cassini job. Supper to be marketed for and prepared for tonight, because Doris Dowling and Gavin are coming and we are all going to a film I don't particularly want to see, *Pandora and the Flying Dutchman*. Mr. Haslam has now stopped work on the house and this may drag on, I suppose, for weeks. And I don't really care about any of that, either. Only if it pleases Don. What do I want? To idle, rest, lie in the sun—I never do—have a couple of

---

[1] A friend of the Laughtons and of Jay de Laval; Isherwood met him in 1945. He is mentioned in *D.1* and *Lost Years*.

sex adventures maybe—I never do—sleep, sleep, sleep forever....
Sloth. It means nothing.

Talk with Gerald, the day before yesterday. He says he feels the
Buddhistic position is unsatisfactory. It's not enough just to want
to end sorrow. And now, he says, we know that the universe is
going somewhere; it's not just meaningless. I urged him to write
his memoirs, maybe in the dialogue form of the Stravinsky-Craft
books. ("I am the vessel through which *Le sacre* passed.")

Today I have to pick up my new spectacles. And that reminds
me of something which oddly enough I never recorded here.
Sometime in May I got a fan letter from a man named Ronald
Flora,[1] from Boston, who said he was coming out to the Coast
and would I see him. He did come and I met him at the Georgian
Hotel where he was staying. He knew Maugham and seemed a
fairly intelligent rather pissy-assed languid queen. That was on
June 3. Then, out of the blue, on June 14, I got a phone call to
come over at once to an apartment on a street very near here,
Palisades Avenue. So I went and there was a detective and a man
from the Georgian Hotel, and they were bullying Flora, whom
they called Smith (that turned out to be his full name) and they
forced him to confess to me how he had run out of the hotel in
the middle of the night without paying. He was shaking all over
and the detective was really sadistic. Well, it ended up with my
having to pay forty dollars, rather than see him taken off [to] prison
on the spot, and this I did with a very bad grace. He promised to
call me, but he never did. And then I found I'd left my spectacles
at the apartment and when I went around there I was told that he
has left for the East. Temporarily losing these spectacles goosed me
into getting some new ones prescribed, and now I shall have these
as spares, which is all to the good. I remember that the detective
acted very matey to me, and remarked that the H-bomb tests had
undoubtedly ruined the weather! Why did I never record any of
this? Because it is the sort of thing which bores me. Such a dreary
little tale.

July 2. Just talked to Barbara Morrow on the phone. Paul has had
a relapse. He is in bad pain and terribly upset and keeps crying.
The doctor wants to have him back in hospital and give him
drugs. He says he won't go. Barbara still doesn't know what will
be decided.

Last night, Elsa came down to 147 [Adelaide Drive], with Ray

[1] Not his real name.

Henderson. She is very bitter against Terry, who is here now. Says he won't sit with Charles and thinks only of gadding about. Says Charles called all the doctors together, when he was in hospital in New York, and made a great confession about his sex life. She thinks he is obsessed with guilt. At the same time, he tells her he loves her, and kisses her on the mouth. She says he was treated by Ernest Jones,[1] many years ago, because of a scandal in the park; and then left Jones. And a short while ago, when there was a notice in the paper that Jones had died, Elsa told Charles and Charles said, "Who was he? Never heard of him."

Oh, it was so ugly and horrible to listen to her. It hardly mattered if she was lying or not. It was horrible in either case.

Beautiful weather. Margot Smith came yesterday and we planned all kinds of things for the house. As usual, my puritanical nature made me feel an almost superstitious dread of spending all this money for such a purpose. I can only do it by firmly reminding myself that it is entirely for Don's pleasure, reassurance and morale. He has got to feel that this is his home, created by himself. I would really feel much better if I put the house in his name.

It now seems fairly definite that he'll leave on the 11th for New York. I am keeping on with my tasks, the Ramakrishna book and the novel. I feel as if the darkness has crept in very close, and yet I am well, functioning and my life with Don is happy. So cling to japam and watch and pray, especially for Paul and Charles.

July 6. This morning I was practically woken up by some journalist on the phone telling me that William Faulkner died today and asking what were my reactions. "He was a very great writer," I said, in a voice like a tape-recorded answering service. "I am proud to have known him but I only knew him very slightly."

On the evening of July 4, Don and I went to supper with Gladys Cooper. Robert Morley[2] was there, and Cathleen Nesbitt.[3] Morley is amusing but pompous. At first, the pomposity seems part of an act; then you begin to suspect that it's a double bluff. He is a pompous man pretending to be pompous in a slightly different way in order to cover his pomposity. He made some good

[1] Welsh-born psychoanalyst (1878–1958), biographer of Freud, author and broadcaster. President for many years of the British and the International Psychoanalytical Societies.
[2] English character actor and playwright (1908–1992); he was married to Cooper's daughter, Joan Buckmaster.
[3] English stage star (1888–1982), also in character roles; she appeared in a few movies and in "Upstairs Downstairs" on T.V.

remarks, though. One of them was, "It's never difficult to play a part, it's only difficult to get it." After supper and drinks, I surprised him and myself by a violent outburst against Russia. The outburst was no good, because I didn't explain what I really loathe about communist morality; its betrayal of Marx in its interference with the private life. But I do like Gladys—she is such a wonderfully good-humored woman and so full of energy. It seems strange to see her now, running around cooking and joking and fixing drinks, and then look at those marvellous theatrical photographs of the twenties which show an infinitely languid creature in an elegant sack, being fervently kissed on the cheek by her leading man while she gazes at the camera with an expression of frigid, rather absentminded purity.

This afternoon, Jerry Lawrence and I drove to the hospital to see Paul Kennedy. He is terribly thin and his stomach is distended, because o[f] the tumor on the liver. But he was quite lively and got out of bed and came with us into the visitors' anteroom. When Jerry Lawrence offered to help him pay the hospital bills, he began to cry.

July 10. Don left late last night for New York. We went to the airport in lots of time, for once, to catch his eleven o'clock plane, and then got so engrossed talking in the bar (where there is no public address system) that we missed it! So then we had to wait an hour and forty minutes for another one, which stops at Chicago, and Don proposed we should go up into the new Sky Room (or whatever it's called) for drinks. They wouldn't let us in because I hadn't got a tie on. So we bought one at the gift shop; it was expensive but really very pretty. And then we talked about the mantram Swami gave Don. Don says that it has made a tremendous difference to him already, repeating it, and that the words of popular songs, which used to run in his head all the time, now fill him with a kind of horror. He also said that he is afraid that this whole thing may take over, that he may find himself getting in too deep and thinking about nothing but finding God. I said well that was something which would happen if it was going to happen.

What with God and everything, I really don't know exactly where I stand with Don right now, but it seems that, in some way or other, we are happy together as almost never before; never, at any rate, since the earliest time, and that was in such a different way. He made such a strange declaration of love, that night we went up to Margot Smith's romantic little house, perched up on the hill above the Golden State Freeway intersection and confronting the mountains. And the next day, when I asked him if he had meant it, all he would say was, "I'm a man of my word." Last night, he said he was very happy with the life we are leading now, and asked if I was, too. At the same time, he said he wished we could speak frankly about *everything* that we did. I said this wasn't desirable. He said, "But I get to know almost everything you do, anyway."

The other evening—Sunday—we had Evelyn Hooker to supper. She was very optimistic about the liberalization of the sex laws. She made us laugh describing a doctor (in St. Louis?) who is researching the orgasm. He makes people masturbate and even have sex with a partner right in front of him and his female secretary, and he wants Evelyn to sit in on this. She is terribly embarrassed at the mere thought of it! He has also constructed a dummy penis with a camera inside it![1]

July 19. Much to be told. To begin with, Don is still in New York but may return tomorrow unless there are more people he has to draw. On the 15th, Gene, Ben Masselink's brother, died of a heart attack at Taliesin East.[2] So Jo and Ben have gone over there for the funeral. The remodelling of the house continues steadily. Today they have started plastering and the new kitchen floor is being put in. We also now have the lath house[3] up outside the front door, complete with gate. It is certainly private—in fact, it seems doubtful if anyone will ever again find his way into the house. This morning, at 7:00 a.m., I was roused out of bed by strange sounds and discovered that the plasterer, unable to get in any other way, had climbed the ladder onto the deck! He seemed to believe quite seriously that this was the way we go in and out of the house! On the 16th, I finished the fifteenth chapter of the Ramakrishna book but have failed to do any work at all on the novel. Hope to restart this today. On the 13th, Jerry Wald died of a heart attack at forty-nine, the same age as Ben Masselink; another pair of events.

[1] Evidently William Masters and his research assistant, later wife, Virginia Johnson. They worked together from 1957 and founded the Reproductive Biology Research Foundation in St. Louis in 1964. Their bestsellers include *Human Sexual Response* (1966) and *Human Sexual Inadequacy* (1970).
[2] Ben's older brother, Eugene, an artist and architect, helped Frank Lloyd Wright run Taliesin in Spring Green, Wisconsin (which Isherwood calls Taliesin East) and Taliesin West in Scottsdale, Arizona. He appears in *D.1.*
[3] Made of thin wood strips spaced to let in light and air.

Last Sunday, Gavin and I had lunch together out at Malibu, with Hugh French. James Mason was there and it seems that he is involved in an affair with a married woman.[1] He arrived with her and her daughter. They went out on the beach together. Suddenly, a youngish, rather unattractive man appeared and, without even introducing himself or saying hello, walked right past us through the yard and out to the beach to find them. This was the husband. It seems there was quite a fuss. A good scene for a film; the man who is so preoccupied with his jealousy that he strides right through a big formal reception, say, wearing a bathrobe.

Gavin and I looked in on Doris and Len Kaufman on our way home. Their horrible Johnno had a sick-joke syringe with which you can make-believe stick yourself and draw make-believe blood. While we were there, two women came in to look at the house with a view to renting it. (One thousand dollars a month! When they first heard this they left at once; but then they thought it over and came back again.) When the women went into Johnno's bedroom, there were rubber horror-heads lying on the bed, and a big horror-hand trying to get out of the chest of drawers. For the first time, I rather liked Johnno.

Have bought *Calories Don't Count* and had a short spell of trying to eat three meals a day with hamburgers, etc., cooked in safflower oil. Now I am back on my old bad habits, though being a little more careful than I was, and eating gluten bread.

July 22. Don got in from New York at 8:30 last night looking marvellous. I mean, lit up—not like a cocaine addict but like an electric light; positively brilliant with nervous energy. The moment he is back here I find myself being kept waiting, wasting time, not being able to do things my way. The only difference is, this house and the alterations and all the fuss and noise and dirt of the workmen, immediately begin to have a meaning and be worth the trouble. Yes, he is a marvel.

A ghastly contrast to him was Charles Laughton yesterday. He had come down to the house next door for the first time since his illness began, with Elsa and the Australian nurse. And immediately on arrival he had started abdominal pains and had to be given lots of painkiller; by the time I got over there he was lapsing into open-mouthed snoring sleep. But later he woke up and we sat together for a while. I held his hand. He said, "Do you think I shall ever get well?" and I asked him, "Do you want to?" "Yes," he said, "I

---

[1] He was married himself at the time. See Glossary.

don't want to go." I urged him to get Terry to come back here and stay. The tears ran down his cheeks. And then, in the next room, Elsa was telling me that she won't sell this house whatever happens, only the Hollywood one and the one at Palos Verdes. And she discussed who should live with her. "I don't want to run around looking for lovers." She seems to think only of the future, after his death; and she is irritated by his "false hopes." Charles kept repeating that he would keep coming down to this house and doing a little more each time and "spreading himself." His fight for life may or may not be hopeless but it is perfectly natural, and I will never forget the obscenity of Elsa's determination to see him buried. It would be amusing to see her change of attitude if she suddenly found herself in the position of the wives of the Peruvian Inca, who were always strangled when [the husband] died.

Gerald, whom I saw on the 19th, says Max Cutler says we nearly all of us get cancer, only the system usually throws it off. He leaves for Europe on the 24th with Michael. He wants to do a tape with me, discussing his chief interests and so indirectly describing his whole life. An autobiography along the Stravinsky-Craft lines. He hinted slyly at Michael's bossy possessive attitudes. But I'm sure he's fond of Michael. As a matter of fact, this may well be one of the typical relationships of old age. There is a person, if you're lucky, who fusses you and bosses you and hustles you around and thereby keeps you alive. You know this and are grateful to this person and at the same time you bitterly resent being pushed around. I am sure that Gerald would be really relieved if he didn't have to go to Europe at all.

July 23. I forgot to note one of Jim Charlton's theories about Japan, namely that one of the chief differences between Japanese and Americans arises from their attitude to masturbation. The Japanese have no shame whatever about this, and Jim believes this is because they are capable of loving their own bodies and don't think it's shameful to do so. He theorizes that all this masturbation makes the Japanese incapable of aggression in their sexual attitudes. An English (or American) guest professor noticed that his students went to the bathroom regularly, almost every hour. He asked, did they have stomach upsets? No, they explained quite calmly, they had to masturbate, in order to relieve their nervous tension.

I also should have written something about Ricky Grigg and his Chinese wife whom he calls Sandy. I met them at Jo and Ben's last Friday. They go back to Honolulu in a couple of days. Although

Ricky is such a great star among teenage surfers, he doesn't give you the impression of a star personality. And of course he is much more than a star; he is a genuine hero. That rescue he did in Hawaii, of his fellow surfer, when no one else dared go into the surf and he himself had just escaped from it utterly exhausted and half drowned, would certainly have got him the Congressional Medal of Honor if it had been done in a war. He is quite disappointingly skinny; not at all an imposing physique with his clothes on. And though he is obviously intelligent, he is quiet and without projection. Oddly enough, he reminded me a little of another (lesser) hero, Dominguín.[1]

[...] but, so far, she and Ricky have always made up again.

This morning a James Simpson called me from New York because he has written a biography of a boy named Ramsey who was at the Mitre when I was at the Hall, at Repton. I couldn't remember him and asked what he had done to get a biography written about him. Simpson said, "He's Archbishop of Canterbury!" (Ramsey is slightly younger than I am.)[2]

Still struggling on with *The Englishwoman*. Utter vagueness as to what it's all about. Doesn't matter.

July 30. Sunshine at last, after so many cloudy days in the Canyon. Walked to Santa Monica and back along Ocean Avenue park with some guy (Jackson?) who is studying the history of the Ramakrishna Order.[3] So no work will be done today. None was done yesterday, because we taped the recording of *F6* at KPFK.[4] The others seemed to like my performance as the Abbot, but I think I was pretty hammy.

On the 24th, a woman ran into the back of the car as I was standing at a stoplight on Hollywood Boulevard. I got a slight

---

[1] Luis Miguel Dominguín (1926–1995), Spanish bullfighter; in *D.1* Isherwood mentions his deadly rivalry with Antonio Ordóñes.

[2] Arthur Michael Ramsey (1904–1988), Bishop of Durham (1952–1956), Archbishop of York (1956–1961), Archbishop of Canterbury (1961–1974), and afterwards life peer. The Rev. James B. Simpson's book was *The Hundredth Archbishop of Canterbury* (1962). The Mitre and the Hall are boarding houses at Isherwoods' public school, Repton.

[3] Carl T. Jackson, later a specialist in U.S. intellectual history and a professor at the University of Texas at El Paso, was writing his UCLA Ph.D. thesis, "The Swami in America: A History of the Ramakrishna Movement in the United States, 1893–1960," published as *Vedanta for the West: The Ramakrishna Movement in the United States* (1994).

[4] *The Ascent of F6*, Auden and Isherwood's 1936 play, for broadcast on local radio on November 7.

back-jerk. I was absolutely furious, in a senile unhumorous way, which disgusts me to think of. I made her get out of her car and come over to mine to give me her insurance company's address. (I don't regret that part of it.) A synchronicity: on the same day, Jo hurt her knee ice-skating with Ricky Grigg. Also, on the 27th, another accident due to crass stupidity: an elderly man, who was shovelling dirt out of the vacant lot next door proceeded to shovel it all over my car and the seat. Again I was furious, despite the good resolutions I had made after the accident. *This must stop.* This kind of rage is the ersatz vitality of the elderly.

We nearly but didn't buy a bead curtain like a Monet painting, really beautiful, but costing five hundred dollars. We did buy a dining-room table, a chandelier and perhaps some chairs.

The deck now has a railing around it.

The Stravinskys went off on another of their immense voyages—Santa Fe, New York, Israel, Venice, Russia. Vera told me over the phone that she dreaded Russia. "It will be vodka, kisses, embraces, lies."

Drove down to Laguna for the "monks' picnic." Swami described how Maharaj once put him into a state in which he "talked and talked" without remembering anything he said. Only, right at the end of it all, he found himself addressing Maharaj in the familiar form instead of the polite form—the second person singular, as it were. Then he heard Maharaj ask him, "*What* did you say?" and he hastily corrected himself. He thinks that Maharaj was asking him about his former lives and had put him into a higher state of consciousness so that he could recollect them.

July 31. Such a brilliant windy day, with everything sparkling. Elsa called to say that Charles went into hospital last night in acute pain because the cancer has got in his spine. He had an operation on a disk which may or may not save him from paralysis. Afterwards, they could not get him sedated, even with morphia, for a long time. Elsa asked me, did I think Terry really cares for him. According to her, the doctor is unwilling for Terry to be brought out, because the situation shocks him. It is the wife's place, etc. etc. I can hardly swallow this.

Oh God, shall I have to die like that—and in the Cedars of Lebanon? The Cedars is the last horror but one. The last is Forest Lawn. "A dark hospital and a detested wife...."[1]

---

[1] Cf., Shakespeare, *All's Well That Ends Well* (II.iii), "A dark house and a detested wife."

Meanwhile, life being as heartless as it is, I have had a splendid day so far—got five letters written, said my beads, did a page of Ramakrishna and a page of the novel, went to the gym, watered the kentia palms, talked to Mr. Haslam, decided with Don and Al Spar to call our business merger Bee-Eye Enterprises (so we can have a drawing of a bee and an eye as a trademark).[1] Now John Schenkel is coming to talk about his writing.

Reading *Mrs. Dalloway*—that wonderful passage about how Septimus Smith goes mad. And *The Wild Palms*. Faulkner's sentences are beyond belief. You think it *must* be a parody. Finally, on page 76, I said to myself, really *this* time the Master really has flipped his wig—*this makes no sense whatsoever*; and then I found that pages 77 to 92 in my copy were missing!

August 4. Another brilliant day. But I am blackly depressed. And as usual my depression demonstrates the insecurity of my toehold on sanity and happiness. All that has really happened is as follows:

A cable from *Time and Tide* asks me to do a review of Edward's novel,[2] quick quick quick. I went to see Paul Kennedy in hospital yesterday. I had a talk with Elsa in which she expressed strong subconscious alarm that Charles might not be going to die yet. I took part in a reading of *F6* in front of a small audience;[3] Don thought I was bad. Don announced that he was going to sleep in the front bedroom. This morning I opened the door and he screamed at me. Kent Chapman is coming to take up the whole afternoon; then Florence Homolka is coming with photographs; then we have Bill Inge, Ned Rorem, and some young actor to dinner. Otherwise, everything is JUST FINE.

It seems that Paul may really recover. The cancer has cleared up in his lung (I don't really understand what this means) and in his liver. But he is having these agonizing cramps and his weight is down to 108, and he has been put on the critical list and got the last sacrament, and then more last sacraments from the hospital priest, who gave him what Paul describes as "a cheap rosary" and a little colored booklet of the stations of the cross and the various indulgences. At the same time, the doctor urges him to walk around! He is very weak, a jaundice-yellow, with dark blotches all over his body and the marks of hypos, and his eyes are brilliant blue in yellow eyeballs and when he looks at you his gaze freezes on you; this

---

[1] For Bachardy-Isherwood Enterprises.
[2] Upward's *In the Thirties*.
[3] August 3, at the radio station KPFK.

I suppose is the morphia. He was perfectly logical and didn't seem scared, only quietly bitter that this should have happened to him.

August 7. Am down at Laguna Beach, staying with Swamis Prabhavananda, Vandanananda, Satprakashananda, Pavitranananda and Krishna and Arup at the Camel Point house.[1] Came down on Sunday night. It was a sudden decision taken Sunday morning, the day before yesterday. Don was still in his bad mood and refusing even to tell me what it was all about—I mean, specifically, this time; of course it was about our relationship and his freedom, as usual. So I said it would be much better if I got right out and left him alone to think things over, and he said yes it would. So I called Vedanta Place and suggested coming down here. Now, this morning, Don called me and said he wanted me back and suggested coming down and picking me up this evening. So that's how it's going to be, though Prabhavananda protests violently.

Well, I have had two nice days in the sun and the ocean water and got some tan and wasted a lot of work time. I have however skimmed through Edward's novel again and admire it far more than ever; I think it is a masterpiece, truly massive.

Yesterday, three of Satprakashananda's devotees from St. Louis came to visit us; mother and father and a twenty-six-year-old son, named Bergfeldt. The son was skinny and grinning-eager with pink-rimmed bespectacled rabbit-eyes. Swami Prabhavananda has decided that he shall be Satprakashananda's personal attendant. He really appeared as a sort of demon matchmaker, and though I laughed like hell I must say it was a fiendish idea; the poor boy wanted to join Trabuco or the monastery in San Francisco.[2]

Much excitement about Marilyn Monroe's death.[3] Everybody talks about it; from Ted Bachardy to the Swamis.

August 11. Paul Kennedy died on the evening of the day I wrote the entry above. It wasn't directly anything to do with the cancer. That was clearing up. But he was utterly exhausted and then he developed pneumonia. He died without pain, in his sleep. Vic

---

[1] On Camel Point Drive, the oceanfront house of a devotee, Ruth Conrad; Swami often stayed there in the summer.
[2] Bill Bergfeldt became a monk, Atma, at the Vedanta Society of Southern California and later settled at the Hollywood Vedanta Society. Previously he was a devotee at the Kansas City Center. He does not recall serving Satprakashananda.
[3] On August 5. Isherwood describes his several meetings with her, at parties, in *D.1*.

Morrow, whom I talked to yesterday, says that when they saw him last Sunday he was more lucid and in a better mood than they had seen for a long while. The awful thing was, I never called the hospital on the day after I got back here from Laguna, as I certainly should have. This was partly because I felt like staying away from cancer a bit longer; I *still* haven't called Elsa. But also because I genuinely believed Paul was on the mend, at least temporarily. So did everyone else. When I did call the hospital yesterday, the switchboard put me through to a nurse, who said, "I'm afraid the news isn't too good; he was buried today." So I even missed the funeral. I can't pretend I'm sorry about this, especially as Vic Morrow told me they had wanted me to say something about Paul.

Now we hear that Harvey Easton has cancer too.

Jo and Ben have a weird story about Peter Gowland. Some time ago, he was given a plaster statue by Mae West. It was supposed to be of her, but it was actually a beautiful and far younger naked woman. Mae West wanted Peter to photograph it for her. He kept putting this off, until last Sunday the 5th, in the morning he decided to do it. He set up the statue in the garden with a suitable backcloth. Just as he was about to start photographing, out rushed his daughter from the house exclaiming, "We just heard on T.V.—Marilyn Monroe is dead!" And, as she spoke, there was a terrific gust of wind and the statue fell and broke into several pieces.... Peter mended it later; so well that he didn't have to confess the breakage to Mae West![1]

On the 8th, Don and Aldous and I went down by helicopter from the National Aeronautics and Space Administration office in Santa Monica to the North American plant in Downey, to be told about the moon rocket which they are building there.[2] Don and I chiefly accepted the invitation for the sake of the helicopter ride, and it was even more exciting than I had expected. The ease and abruptness of the ascent is like flying in dreams. It is as if you merely make an extra effort of the will—symbolized by the roar of the engine—and suddenly the ground tilts away from you and you are soaring. I was also reminded of Rembrandt's drawing of the angel leaving Manoah.[3] Several people who happened to be passing watched our ascent into heaven with expressions of pleased amazement.

[1] Mae West (1892–1980)—film star, playwright, sex bomb, wit—was about to have her seventieth birthday, on August 17; Monroe was thirty-six.
[2] North American Aviation, Inc. built the Apollo spacecraft for NASA and the Saturn V rockets which launched them.
[3] See Glossary under Rembrandt.

The city was shocking in its uniformity; all those roofs and little yards and bug-autos and occasional glittering green pools, so much of it, stretching away and away, you never saw the end of it. Only, in the background, the big mountains appearing behind smears of yellow smog.

The eager-beaver executives at North American were intimidated by Huxley's ghost-pale introspective intensity. He was like a ghost they had raised to speak to them of the future. And they didn't much like what they heard. Aldous held forth with his usual relish on the probability that the astronauts would bring back some disease which would wipe out the human race. And then he described the coming overpopulation of the earth. You felt that these people had bad consciences. They were making a fortune for their firm because the government will aid and abet them in playing gadgetry. So they keep reminding all who will listen that *if* they can go faster than the speed of light, and *if* they can reach an inhabited planet, and *if* the inhabitants of that planet are ahead of us in technics, and *if* they are willing and able to communicate to us what they know—why, then we shall be able to make great strides ahead.

The expert from Texas who stumbled over his own technical vocabulary as though he were what he sounded like, an illiterate Bronx truckdriver. (He was much the nicest of them.) We were told how you "abort" the flight if it isn't going right. How the nose cone will descend on the moon while the other section remains in orbit around it and how then the two will rendezvous. We walked through the plant; very hot. Security posters: a naked baby with his arm raised, saying, "I swear I locked my file!" A framed motto: "Remember: *American* ends with *I Can*." We had to wear name badges in our outside pockets. The office of the director had stars in the ceiling; but on the wall was a collection of antique spurs and bits.

The day before yesterday, I finished the review of Edward's novel and sent it off. Not good enough but not bad.

The house is nearly finished. The carpet laid, the front bedroom all painted, the studio ready except for the toilet and heater. I won't say anything about the problems with Don. They are still acute. And my tongue still burns.

I never seem to record anything nowadays about my "spiritual life." That's hardly surprising because it has never been less evident. I keep up japam regularly, with very few omissions. But I have no sense of being in contact with whatever is there. Even when I was down at Laguna and got to meditate in the shrine with Prabhavananda and Pavitrananda, which ought to produce a "field" if anything would, I still got no sense of contact. Ought I to worry? I don't know. On the one hand I'm conscious of my own sloth and slackness, on the other I know that one mustn't demand sensation—do your duties and have faith that they are producing unfelt results. Just the same—

August 15. Jo, on the phone this morning, told me how depressed Ben is, these days. Everything he writes gets turned down; and even when he has a relative success, like the novel or getting that television job, it never really pays off. He feels he ought to give up writing and get a job. Meanwhile, Jo's knee, which she injured skating with the Ricky Griggs[es], has swollen, and so she has to go and get sonar treatment and sit with her legs in a bath of whirling water. The girl who had been operating this for her, asked her about her shark's tooth ornament and this led to descriptions of Tahiti and the revelation that Jo is a designer. And the girl was thrilled. "Those poor girls," said Jo, "don't have any glamor in their lives."

Meanwhile, Don succeeds and succeeds. He even made Merle Oberon clap her hands and call in the servants to admire one of the drawings he did of her today. And this week he has made—I think it's around five hundred dollars! And he is about to draw Charles Boyer for his play.[1] On the other hand, deep deep gloom. But I have resolved not to dwell on this, unless it's something out of the ordinary. This is a bad month, for both of us.

Don's studio is nearly finished. We are considering going to New Mexico, in order that Don shall draw Igor before they leave. That would mean we'd have to go there this weekend.

I have failed to get on with my work. Stomach flap and depression. Uneasy sleep, in which my dreams consist of rigid objects without any apparent significance, like sticks. This is very disagreeable; it verges on the frightening discomfort of actual delirium, when your thoughts *hurt*.

A sweet letter and a very warm and understanding review of *Down There* by Dachine Rainer[2] for a British magazine called *Anarchy*.

[1] French stage actor (1897–1978), a Hollywood leading man in the 1930s and, during the 1950s and 1960s, a regular on a T.V. drama series he co-produced. The play was *Lord Pengo* by S.N. Behrman; Bachardy's drawing was for a poster to promote the Broadway opening.

[2] Poet, author, anarchist (1921–2000), imprisoned as a C.O. during W.W.II; he co-edited with Holley Cantine a collection by imprisoned C.O.s, *Prison Etiquette, The Convict's Compendium of Useful Information* (1950), with a preface by Isherwood.

August 18. Vera Stravinsky put us off going to Santa Fe by telling us how hot it was and how crowded and how she and Igor and Bob were beset by journalists, agents, etc. Evidently she thought we would be a nuisance, but I didn't resent this, particularly as I didn't really want to go.

It is warm and beautiful. Today we went on the beach and in the ocean; both are dirty, but we are getting good tans. Don has been to draw Boyer, who was pompous and uncharming and limited him strictly to one hour. So Don didn't do well. When Boyer was asked to sit again, he made difficulties. However, Don has just reported the whole thing to Janet Adrian, who will report back to Paul Gregory,[1] who will maybe put the screws on Boyer.

Temporarily, at least, I am down a little below 150 lbs., the first time in ages. I have passed page 50 in *The Englishwoman* and did a big swatch of the sixteenth chapter of the Ramakrishna book. Also an analysis of all the slips and mistakes on the first page of Bart Johnson's novel. Nearly three pages of them! I'm afraid he will be shattered. But he shatters rather too easily, anyhow; and sloppiness must be treated drastically if it is ever to be cured. Now my next distasteful task is to tell William Hoopes what I think of *his* novel.

August 22. This morning the man has come to put in the drapes in the front bedroom and in Don's studio; and other men are putting mirror into the wall of the dining area. And tonight, Jerry Lawrence, Larry Paxton, Jack Larson, Jim Bridges and Max Scheler[2] are coming to dinner, and Gavin is coming in later. Gavin leaves in a couple of days for England because he is going to write *The Night of the Iguana* screenplay. (They have decided to make the picture in England and so they want a writer who has a British passport; this is the reason given for letting me out, and I daresay it is really the true one.) This afternoon, Wyatt Cooper is bringing Gloria Vanderbilt for Don to draw. Wyatt is thought by Don to be scheming to marry her. This means that I may have to keep him entertained while Don is drawing her.

I have said I won't complain until things become really unbearable. So I won't complain.

A pair of linked events: after silence from my ex-students all summer, I hear today from Glenn Porter, Frances Yampolsky,

---

[1] Gregory was producing *Lord Pengo*; Janet Gaynor, widow of Gilbert Adrian, recommended Bachardy to him. See Glossary.
[2] German photographer, introduced by Herbert List in Munich during Isherwood and Bachardy's 1955 European trip. He appears in *D.1.*

Nick Barod. Also, I discover that Glenn and Frank Wiley are both on the *Princeton!*[1]

Reading *Mrs. Dalloway*, which is one of the most truly beautiful novels or prose poems or whatever that I have ever read. It is prose written with absolute pitch, a perfect ear. You could perform it with instruments. Could I write a book like that and keep within the nature of my own style? I'd love to try.

A fan in Oregon named Jack Rosen has sent me a hand-drawn birthday card with greetings from the characters in my Berlin novels—"to our 'Poppa' ..." Of such is the kingdom of heaven.

Harvey Easton definitely has cancer. It is in his liver. He won't see anyone. His wife June is pregnant.

To live every day as if this were going to happen to oneself. And yet not to let it happen.

Great satisfaction from working out at the gym; but I'm barely losing weight. I still can't get down below 150 and stay down. But my hips are much smaller.

August 24. What was wrong with the film John Huston and Wolfgang Reinhardt made about Freud?[2] (We saw it yesterday evening.) Just exactly what I had expected would be wrong. Freud's lifework is presented in the pat terms of a neat logical discovery. "If that is that and this is this, then that-this must be—No, that's wrong." (Something happens, and the discoverer gets a flash of insight.) "Ah, *now* I see. The answer, ladies and gentlemen, is—" And the discovery is discovered and remains discovered, forever and forever.... This would be perfectly all right if you were describing the invention of a certain kind of corkscrew. But this method (which has been the method of all Hollywood biographical films dealing with scientists) utterly belittles the greatness of men like Freud, who were not inventors or necessarily discoverers, even, but explorers. They dared—that's what matters. Scott dared to approach the South Pole. Freud dared to descend into the Inner Night. If they really discovered anything, if they succeeded or if they failed—that's secondary.

Afterwards, Jack Larson and Jim Bridges took me to supper at Frascati's. (Don had to go to a party at Romanoff's, given by Gloria Vanderbilt.) As soon as we sat down, Jack went and telephoned Monty Clift in New York, because he was waiting to hear

[1] Aircraft carrier *USS Princeton*, then training with marines in Okinawan waters.

[2] Reinhardt and Charles Kaufman wrote the screenplay *Freud*; Huston directed.

how he had looked in the film. Jack has to call him after every one of his films. As a matter of fact, Clift didn't disgrace himself. He had very little to do, except to be seen watching the patient, or asking him, "And then?" or, "Why?" And talking of why, *why* must non-Jews play Jews nowadays? It is the most bogus kind of liberal nonsense—that we're all the same really. We fucking well aren't.

Then Jack Larson talked about Paul Kennedy, whom he'd known quite well, for many years. How Paul was AWOL from the marines for months and how he finally gave himself up and was thrown in the brig. And then his life of poverty in Hollywood, sleeping in all-night movie theaters. I should have asked more, if I hadn't been sleepy and sluggish. I do like Jack and Jim more and more, the more I get to know them. But chiefly Jack. Jack seems to me a genuine man of goodwill. And he is a great authority on old Hollywood. I wish I could see more of him but there's never any time, it seems.

Was horrid to Don this morning, blaming him for getting me into the *Freud* showing and the consequent obligation to write to or telephone John Huston. I told Don that he is the real tyrant, not me. This is the kind of talk which is neither true nor false. We are both tyrants whenever either of us gets the upper hand. No need to get nasty about this.

August 26. My expectation of life is now seventeen years. But how well aware I am that it may end any time now. That aggressive Negro girl I met at the party in New York at Christmas said I would die after writing one more (very successful) book; a year or two after writing it, I think she said.

Am I afraid of death? Yes and no. I am terrified of going over the edge, knowing the illness is terminal, knowing the doctors and nurses have got me in their cruel kingdom of insufficient pain-killing. But I have faith, too. I believe that I shall be somehow sustained. If there's an afterlife at all. And that's still a little bit of an *if* for me. Not a big one.

I am not pleased by the way I'm growing old. My face is ugly with tension and resentment and lust. It is not beautiful at all. Old and ugly, and I am plump around the middle, despite all my exercising. I have got to curb my resentment somehow; it is wearing me out.

Do I hate Don? Only the selfish part of me hates him, for rocking the boat. When I can go beyond that, I feel real compassion, because he is suffering terribly. I still don't know if he really wants to leave me, or what. And I don't think he knows. Last night, he had a drunken fit of crying, over in the studio, so loud that I could hear him in the house. I went over and he said to leave him alone, he liked to cry. I really felt he was on the edge of a breakdown. But then, this morning, he appeared with birthday presents inscribed in the Dub and Kitty idiom—two shirts, white socks, a watchstrap and a beautiful Japanese model horse, white with trappings of orange, green and gold. Now he has gone off to draw Selznick.

Last night, Hope Lange and Glenn Ford came to supper. I burned the steaks (which were anyway far too small) on the barbecue. It is so symptomatic of the gloom we live in that this accident, which would seem merely funny to happy people, became a disaster like something in an Ibsen play. Hope and Glenn were as constrained as ever, despite their alleged get-together. You felt Hope was his prisoner. So that was depressing, too.

I had looked forward to a day on the beach, to freshen up my tan before I leave for San Francisco on Wednesday. But it's one of those maddening days of beach fog with blue sky a mile inland.

Have just finished *Mrs. Dalloway*. It is a marvellous book[.] Woolf's use of the reverie is quite different from Joyce's stream of consciousness. Beside her, Joyce seems tricky and vulgar and cheap, as she herself thought. Woolf's kind of reverie is less "realistic" but far more convincing and moving. It can convey tremendous and varied emotion. Joyce's emotional range is very small.

Am on page 56 of *The Englishwoman*, and yesterday I finished the rough draft of chapter 16 of the Ramakrishna book. Now I must write to Olive Mangeot, one of the very few who remembered my birthday. The others: Glade Bachardy, Jerry Lawrence.

This from a Chinese fortune cookie I was given at the House of Lee in Pacific Palisades a couple of nights ago: "Don't write curt notes to people who have failed you"!

September 10. I had meant to begin a new volume after that last entry and after my trip to San Francisco. But somehow this isn't the moment to do it. I would like to open on a happier note, or at least on the sensation or illusion of a new start.

Things are bad with Don, but at least we had an air-clearing talk the other day. I ought to go away, of course, for several months and leave him here to find his bearings. Not to do this is to force *him* to go away, and this is wrong because he is the one who didn't feel really at home in this house, and now that he has his own studio he should be free to enjoy it.

Then why don't I go away? Because it is such a lot of fuss and I don't want to leave *my* home and above all my books. I want to stay here and get on with my work, in my own tempo. I can leave him alone now, of course, much more easily than before; we have much more privacy. But not enough. Our life together is all off on the wrong foot and I am not at all certain it can get back off it.

Aren't I bad for him, now, under any circumstances? Probably. He only needs me in his weakness, not his strength; and he hates me for supporting his weakness. But meanwhile I am useful, and the show is beginning at Rex Evans's gallery a week from today; in fact we are in the midst of an emergency, which makes an excuse for doing nothing till it's over.

Today he is getting a new car, a Corvair. I am trying to get a Volkswagen as soon as possible.

All this time, my tongue has been burning. Now I'm told by Dr. Allen to see a specialist about the little tumor in my ear. It looks slightly larger to him. So I'm going tomorrow, maybe to have it cut out. Otherwise my health seems good. At the gym I feel very strong, but I still can't lose weight.

Only yesterday did I make a very faltering restart on the Ramakrishna chapter.

I was in San Francisco from August 30[1] to September 4. Had meant to write about this today, but Anthony Brown[2] just called to tell me about his book on the British wartime propaganda newspaper, and now I have to go out.

September 12. About San Francisco first. I really don't have much that I particularly want to note down. This strange guy [...] gave me a sweater which had belonged to John Cowan.[3] I had a little bit the feeling that, in giving it to me, he was passing on an unlucky object—the way you are supposed to have to get someone to accept the runes which have been cast on you. Anyhow, I left it, unintentionally, in Stanley Miron's car.

Then the trip I took with Ben Underhill to the White Horse

[1] The day he delivered "A Personal Statement" as part of a lecture series "The Writer at Mid-Century: The Moral Crisis," sponsored by the University of California at Berkeley Extension.
[2] Anthony Cave Brown (1929–2006), British journalist for *The Daily Mail*, *The Times*, and the *Manchester Guardian*; he later published popular books on military intelligence and espionage.
[3] A blond beauty celebrated in Santa Monica Canyon during the 1940s, briefly a boyfriend of Jay de Laval; Isherwood tells of his own relationship with Cowan in *Lost Years*.

Ranch, up in the hills near the Napa Valley. It belongs to Wakefield Baker and Ted Sheridan. (A sort of Maugham story there: how these two most eligible bachelors in San Francisco disappointed so many mothers by setting up housekeeping together. And then the misery which crept up on them.) In the night, a doe fell in the swimming pool and one of the ranch dogs dragged it out by the throat and killed it and partly ate it. The doe's screams were horrible, like a human's.

Thom Gunn and his leather jackets and his bar life. He hasn't written poetry for nearly a year, but he doesn't seem rattled. There is something very tough in him. There is a bar called the Why Not which has a boot party on Thursdays.

Maybe I'll think of other things. Not now....

A beautiful day, and I am so relieved because the specialist—a Minnesotan Swede, I think, with a brutal face and a certain effeminacy of pose—told me the thing in my ear isn't a tumor at all but an exostosis, and that it won't have to be operated on unless it interferes with my hearing. The hearing test; sitting with headphones on in a soundproof room. Gradually you hear the delicate tiny fairy jingles. So pretty. You could write music for them.

But Laughton is probably dying. The kidney is too far gone to operate on. I talked to Elsa today and shall most likely go to see him on Friday. And Harvey Easton is certainly dying.

The pursuit of publicity for Don's show. Don was saying this morning how revolting this is. He says he's getting sick of doing portraits. Or rather that he wants to find another way of doing them. "So they'll look like me instead of looking like the sitters."

The Cuban crisis is cooking up big.[1] Don has turned down Anthony Cave Brown's idea that he should be commissioned by NASA to cover the next American earth orbit at the recovery point in the Pacific. What they needed was a sketch artist, of course.

September 14. Yesterday I reread my novel, the fifty-six pages I've written so far. I am discouraged; very little seems to be emerging. Maybe I really have to sit down and plot a bit before I go on. I do not have a plot and I don't even know what I want to write a novel about.... No, that's not quite true. I want to write about middle age, and being an alien. And about the Young. And about this woman. The trouble is, I really cannot write entirely by ear; I must do some thinking.

Supper with Tom Wright yesterday. Gavin was there. Tom

[1] See Glossary under Cuban Missile Crisis.

leaves soon for a two-year stay in Mexico, Guatemala and else-where in Central America. There is something very Beatrix Potter about him. He is like a self-contained eccentric animal, very much alone and quite satisfied to be so. You can imagine him puttering around the rain forest and the Mayan ruins, with his stammer and his stoop and his mild southern amiability, a creature altogether alien and yet quite able to take care of himself, avoid snakes and wild beasts, escape infection by the judicious use of pills, even cope with the most ferocious Indians—the kind who have killed many a trained anthropologist who spoke their language perfectly, knew all about their customs and religion, but didn't have Tom's saving, smiling adaptability.

September 16. On the beach today; it was beautiful, though the beach itself is foul with the trash of the untidy summer. Tomorrow is Don's show, and he is working frantically now trying to do a self-portrait of himself because Rex Evans wants to put one up outside the gallery. Meanwhile I have been writing letters, to Frank Wiley, to consumptive Paul Taylor in London, to Amiya. What a weary labor! And really it is nothing but slamming a ball back across the net. Frank is maybe the only one who deserves a letter because he is stuck out there in the navy; and yet it's precisely to him that I can't write the things he'd want to hear, because of this prissy censorship and spying. So I sent a long liter-ary chat about Faulkner and Woolf which I could barely finish for boredom.

Last night, Don went out to dinner with the Claxtons and got drunk and fell on the way back to his car and raised a huge bruise on his thigh which is paining him dreadfully. So that's one more cross to bear tomorrow. But he has been so sweet, the last few days. All that is such a mix-up. Perhaps I should just offer the whole thing up to God's will and stop worrying. But I can't help feeling that refusing to worry is somehow a betrayal of Don. (I'm tired and writing nonsense, I think.)

Elsa still makes excuses for me not to see Charles. She may not even realize what she is doing. And of course I am not madly eager to see him. When I do it will be sad and painful. Japam, japam; there is nothing else to be done about anything.

September 18. Don feels that the opening of his show last night was a great disappointment, because only one portrait was defin-itely bought—Lotte Lenya, by Gavin Lambert. But Glenn Ford publicly declared that he intended to buy the portrait of me (a strange, rather leering one, which Don finished only about an hour before the show opened!) And today we hear there is a buyer for the Huxley, and a nibbler after the Tennessee Williams.

The party was certainly well-attended, though there were a lot of freeloaders and queen-bums. Gladys Cooper came, and Dorothy Parker, and Shelley Winters, and Glenn and Hope. But that cunt Oberon did not come, after tricking Don into hanging the picture of her that she likes, and Connie Wald didn't come, because of some mysterious disaster, and Lee Gershwin[1] and Doris Dowling didn't come. And that bastard Mike Connolly[2] didn't come, although Jerry Lawrence even asked him to dinner to get him. The greatest gaffe of the evening, on my part, was to inscribe a copy of my novel to "Tyler," because I had gotten Keats Tyler's name turned around. Don followed my example. So we have decided to send him another copy, properly inscribed. (I had thought of doing this and dismissed the idea, and then Don suggested it, on his own.)

This morning we went on the beach and discussed *The Englishwoman*, and Don, after hearing all my difficulties with it, made a really brilliant simple suggestion, namely that it ought to be *The Englishman*—that is, me. This is very far-reaching, but I shan't go into it here, I'll write an analysis of the idea within a day or two, in my big flat planning-book.

September 22. The day before yesterday, I got a black Volkswagen sedan, and the day before that Don got a Corvair, very handsome, a kind of wine red, black upholstery with silver buttons. They fill our little carport and Don has to park diagonally because I have the jitters about backing out and have already scraped my whitewall tires and made a tiny dent in the wing. But I am very pleased. The Volkswagen and I understand each other. It reminds me of the Consul, which I used to describe as a very loyal little car. The V.W. has the bouncy loyal eagerness of a small dog. It doesn't really care how you treat it. It isn't very bright but it is cheerful and that's such a relief after the gloomy neurotic moods of the Simca. You never knew how it would be feeling from day to day. It hated me, and I rather hated it for its French sulks. But it was sad to part from the Sunbeam Talbot which had shared so much with us. Those first spins with Don up the coast highway

[1] Leonora Gershwin (d. 1991), wife of Ira Gershwin (1895–1983), the lyricist brother of composer George Gershwin.
[2] Columnist from 1951 to 1966 for *The Hollywood Reporter*.

to the beach where we used to be able to swim naked, and the Monument Valley trip, where it figured in several early photographs. And how beautiful it was when it was young and horizon blue!

A few more of Don's drawings have been sold and Henry J. Seldis wrote a good notice of the show in the *Los Angeles Times* of yesterday. He said: "It is an impressive appearance by a highly skilled and perceptive draftsman who captures the personality as well as the appearance of his sitters with elegant lines." So Don feels better. But he is still wrestling with post-exhibition lassitude. He resolves to start going nights to art school at the beginning of next week; also to paint during the day.

And I resolve to get the hell on with the Ramakrishna book and restart my novel. I should never have stopped it; but now that I have this different approach I can't go on from where I left off.

The Cuba-Berlin crisis continues. And there will be a crisis like this every year until we blow each other up, or the deadlock is broken—by England's declaration of unilateral disarmament; an act of political genius which is nearly but perhaps not utterly unthinkable, since it would satisfy at least one national appetite, the desire to be nobler than thou.

September 23. This morning we went over to Gavin's to look at John Hart's[1] interview with me on T.V., and instead, although it had been announced in the papers, there was an interview with a Negro comedian named Dick Gregory.[2] Not one word said about the interviewee for next Sunday.[3]

Last night we went to the theater at the Uplifters—it's actually a gym with folding metal chairs—to see two plays by John Mortimer, *The Lunch Hour* and *I Spy*. Ghastly, and made ghastlier, as usual on such occasions, by Don's jitters and fury. He can't help taking such things personally. It is he who has been insulted and injured.[4] Moyna Macgill[5] was very good however. And an actor

---

[1] American radio and T.V. journalist (b. 1932), for CBS and, later, NBC, where he became John Chancellor's understudy on "NBC Nightly News" in the 1970s and 1980s.

[2] Gregory (b. 1932) became famous in 1961 for his Chicago Playboy Club skits satirizing racism; later he was a civil rights activist, anti-war protestor, anti-drug campaigner, and author.

[3] It was Isherwood, September 30 on KNXT.

[4] Bachardy thought the plays banal and could never bear watching professionally accomplished friends humiliated by amateur productions.

[5] Irish stage actress (1895–1975), mother of Angela, Edgar and Bruce Lansbury. She played small character parts in Hollywood films and appeared on T.V.

named Ben Wright[1] gave the evening an unusual distinction by falling right off the stage during the blackout between the last two scenes. In the darkness, we heard this awful crash and then a great groan. But he had only sprained his wrist. He rather touchingly begged the audience's pardon as he was helped out.

The weather has much improved. Today, and four times during the past week, we have been on the beach and in the ocean. Blazing hot sun, dirty water. Don has a stomach upset and sore throat. During lunch at Ted's, we discussed what he should do. Russell McKinnon has hinted to me that he would produce more money for Don to go back to England. But Don says he doesn't really want to go. He doesn't want to go to New York either. And he most decidedly doesn't want to go to Rome or Paris. So then I had to point out to him that the logical alternative remaining is for me to go off somewhere while he stays here and works. I could tell at once that this was what he wanted. But where shall I go? I probably could get enough work in New York; maybe even in San Francisco. But that would mean compulsive drinking and running around. Suppose I went to Trabuco for two months? Hideous boredom and loneliness, but maybe the meditation and lack of liquor would work wonders, and I should get on, *faute de mieux*, with the Ramakrishna book and maybe also with my novel. (Remember how I had that extraordinary breakthrough with *The World in the Evening* when I was at Trabuco in January 1953, shortly before my life with Don started.) Well, I'll give this idea serious consideration.

September 29. I certainly won't be able to go away in October, because so many things have come up for me to do. Lectures at Riverside, Garden Grove, and UCLA (a discussion with Jerry Lawrence) and another public reading of *F6*. As far as Don is concerned, I can't make out how much he minds. He sulks, rushes off and spends the night out, and then is quite nice. He says he is in despair, but this only means what it means for most artists of any kind; he doesn't know what to do next. I'm sure he must go out much more alone. He says he wants to go out dancing with girls. I say, well how marvellous. And then he says, but I don't know any girls and I can't dance any more—seeming to imply that I am stopping him. When, actually, one of the earliest things

Bachardy drew her portrait several times.
[1] Wright (1915–1989) was British; he guest-starred on American T.V. from the late 1950s through the 1970s.

I urged him to do was take dancing lessons, and he did, but then lost interest.

One thing I realize, I must stop being even in the least bit altruistic; because that is false anyhow. I must not make big gestures; just good-humoredly nudge him into greater freedom.

Harvey Easton is dead. The guy who runs the gymnasium in his place is so sloppy that he still uses the sign Harvey used to use: Harvey will be back at.... And then a clock face, so you can indicate the time. The day after the funeral, the sign said: Harvey will be back at 3:30! This is one of the most striking examples I have ever heard of the startling horror which can be achieved through quite simple normal "harmless" insensitivity.

Akhilananda of Providence and Boston also died a few days ago.[1] His assistant, Swami Sarvagatananda, is out here now, taking refuge from the madwoman who has dominated the centers for years, and of whom they are all terrified. When I was up at Vedanta Place last Wednesday, a woman named Dorothy Louis (Shraddha) was raving against this woman and also goading Swami Prabhavananda, saying that he was afraid of her, rather in the manner that women in the Icelandic sagas goaded their men on to undertake a killing by jeers at their cowardice. It was viciously ugly and one of the very few times I have seen Prabhavananda really rattled. You realize how a woman like Shraddha could degenerate into just such another domineering madwoman as the Providence one. There are women like that wherever a religious group gathers. The only person who can hold them in check is the priest or minister or swami in charge.[2]

We have had all our planting done now. It was done by Mr. Graef of California Flowerland and two assistants (one called Angel) at the beginning of the week. On the days when all the shrubs have to be watered, it takes more than an hour. But the place is really beautiful now. The hanging boxes of plants give such style to the balcony.

Yesterday I went to the Cedars for the second time to see Laughton. He seemed weaker and he is starting to get hallucinations again. He told me that he believes the doctor is practising witchcraft and is trying to get control of his (Charles's) mind. "Of course," said Charles, "he could only have it for very short

---

[1] He ran the Vedanta centers in both cities for many years.
[2] Shraddha, a former ballerina, and her husband Bob Louis, had lived in India, where they became close to Swami Sarvagatananda. Sarvagatananda became head of the Boston and Providence centers for the next forty years. The Louises remained devotees in Santa Barbara until their deaths in the 1990s.

periods"; and I realized that Charles was mixing up his witchcraft fantasy with show business. This became more evident when he asked me, "How much do you think I'd be paid as a witch?" I told him, a great deal. This seemed to please him.

On September 25, I restarted my novel. I have absolutely no idea how this will go yet. But I feel I am nearer the mark, calling him William[1] and writing in the third person.

October 3. Last night we went to the Ringling Brothers circus, with Barbette. It was in the huge new sports arena, way down-town, and the place was not one quarter filled. It had never struck me before how the Circus is a symbolic play about Life. That sounds heavy and Germanic; what I mean is that the Circus is exactly *like* Life. The Circus audience is much less attentive, gen-erally speaking, than other audiences. It crunches and munches and slurps soft drinks and talks to itself, and its attention—like the attention of The Others in Life—is only momentarily captured. Indeed, it is made almost impossible for the audience to attend properly, because different things are happening most of the time in the three rings; you cannot concentrate. A sexy girl with long blonde hair is balancing outstretched on something, in a not very difficult pose; but she is watched. In the meanwhile, a dear little Japanese has a billiard cue on his chin, and a chair on top of that, and his wife sitting on the chair (no longer young), and he is twirling colored rings around one arm and juggling flaming torches with the other hand, and keeping a rubber ball balanced on the toe of one foot. It has taken him his whole life to learn to do this, no doubt; and maybe he can do it better than anybody else in the world; and who gives a damn? There was one act in which a young man rushed around setting plates spinning on pointed rods; and when they were all spinning he rushed around catching them before they slowed down and fell off. (One of them did fall, and shattered.) And this was a perfect symbol of the Rat Race, the Age of Anxiety. And then the disorganization and irrelevance, the sheer chaos of Life, expressed by the sudden invasion of the clowns, the frantic hurry in which many of the acts are performed, the meaninglessness of most of the animal acts—*why* should bears ride bicycles?—and the abrupt exit of even the star performers, walking quietly away, unfollowed by the spotlights, but perfectly visible to the onlookers, who nevertheless don't applaud, and indeed pretend not to see them. The clowns

[1] Later changed to George.

are a curious mixture; half of them wholesome nursery types, like Popeye the Sailor Man, the other half abominations from the world of nightmare—things with snake-necks and tennis-ball heads, heads which are cut right off, creatures which split into halves and walk off separately. (They would be even more abominable if their designers had had the genius as well as the intention of a Francis Bacon.) And then there are great engines, absurdly imposing when you consider the idiotic tasks for which they are made; the cannon, for example, which shoots two people into a net—*why*? The animals which seemed best adapted to the mood of the Circus were the elephants—and yet their monumental poses are a kind of parody of all classical sculpture.

As for the trapeze artists, their art is something else again: high camp about Death.

The wire walker who makes fake slips (and some real ones); that's one approach. His wife was watching, and wincing each time he seemed about to fall. Attendants held a miserably small net underneath him.

The other approach is the classic style and grace of Gerard, the aerialist. He swings by his heels. He wears a magnificent cape, more feminine than masculine in style, which he takes off before going aloft, in tights with a diamond belt, naked to the waist. Barbette introduced us to him. An utter lack of vanity. No noticeable nervousness, although we met him first before the act. A blond, fairly good-looking, unremarkable, muscular boy in his middle twenties, I guess, who had put on a certain amount of fat. His friend Cesar, a Filipino. Cesar was in college when he met Gerard and joined the circus. Gerard taught him to do a low-wire act and to juggle. He is so good that he was featured in Madison Square Garden; there is no room for him on the program here. Cesar is Gerard's assistant in his act. He has declared that, if Gerard falls, he will throw himself underneath to break the fall as much as possible. They have been together two or three years. (Once, Gerard slipped and only caught the trapeze with one heel and had to swing right back in that position. He thought he would fall, but he recovered himself. He earns $550 dollars a week. Barbette says he is going to teach Gerard some new tricks when they are in Sarasota, Florida, for the winter. He says that the circus is so big that it requires very big showy movements.)

We took Barbette out to supper afterwards at a Mexican restaurant he recommended, the Taxco, on Sunset.

When we were talking the evening over this morning, while having breakfast on the deck, Don said that, if the Circus symbolizes the meaninglessness of Life, then it follows that to have a job in the Circus is the most meaningless work of all.

We watched the unattractive wife of one of the neighbors on Mabery Road go out and look in the mailbox. Don said, "What can she be expecting, except bills?"

October 9. We had a picnic with Gerard (Soule)[1] and his friend Cesar, and Jack Larson and Jim Bridges and Gavin on the beach yesterday. It wasn't rewarding. Gerard was remote, and only responded to Jim, who has the unfortunate mannerism of creating "secret" conversations. Speed used to do it; and there's no doubt, it always indicates a certain amount of bitchery—in this case, bitchery of Jack. You manage to suggest that your talk with the other person is a sort of conspiracy, even if you are discussing the weather. You talk to him in a low voice, so you can't be heard by anyone else in the group, and if possible you draw him aside, take him for a short walk or retreat into the middle distance, but always (this is most important) remain in full view of the others. (My writing this suggests that I'm beginning not to like Jim. I think this is true.)

As for Gerard, he was passive; he merely let it happen. I found it quite impossible to talk to him without awareness of his predicament—i.e. that he could easily be killed that very day. In other words, I thought of him as someone incurably sick. They were moving the circus to San Bernadino and giving a show that evening.

Cesar has a kind of Asian wriggly femininity, with a cruel little giggling laugh. He giggled scornfully over Barbette's scheme to found a permanent circus here, with real style and red velvet seats. The circus, he said, is for children, who eat popcorn. Was his bitchery partly jealousy of Barbette, because of Barbette's influence over Gerard? Probably. Cesar would be jealous like a real Asian wife, who puts poison in the rival's food.

Don said afterwards that he had disliked the picnic because he feels "One should take the beach seriously." We were ignoring it and all the beauty of the fall day—although Don and I did go in the water. We turned our backs on it and ate greasy precooked chicken and drank wine.

The chore of watering all our plants is coming to an end. We get the gardener tomorrow. But the boxes of geraniums and the

[1] Possibly Soules (1935–1991), born in Canada, raised in Detroit; he was a circus star for nearly four decades. After a bad fall in 1964, he developed a dog act, "Poodles de Paree." He was murdered in a Las Vegas hotel.

hanging wooden baskets have to be watered every day, and they drip all over the deck and wash away the blue paintwork.

October 11. Spoke too soon. Mr. Shikiya showed up but had misunderstood what we wanted and merely rooted up a few weeds without watering any of the shrubs. And now his next visit is postponed till next week because his truck has broken down. Maybe he will be no good to us.

John Zeigel called this morning from the airport on his way to Mexico to tell me Ed Halsey was killed last weekend near Yuma in a head-on car collision. He died instantly. John is already studying at Cal. Tech.[1] and all I could say in the shock of the moment was that we would meet when he got back. He is living at Claremont and commuting. Also today I get a strange rambling tactless letter from Bill Robinson, telling me this news too and saying that he hopes I will help John, who was in a bad state even before this happened because his relations with Ed had become "grim." I don't quite see Bill's motives in writing this.

Two gloomy future threats, one short-term, one long—: even the *New York Times* (which we now get) admits that there will very probably be a major Berlin crisis within the next few weeks; Edmondson[2] has told Jo and Ben that there is a plan to make an overhead freeway above the existing highway, with a cloverleaf at the Canyon entrance. This would make the Canyon "uninhabitable" from Jo's viewpoint. But anyhow everyone agrees that this can't happen, even if the scheme is passed on, within the next ten years.... And meanwhile—?

Meanwhile, get on with your work. I must say, though, Ed's was an excellent end, under the circumstances. He probably didn't even see it coming; and was spared all the dreariness of the break with John.

At the Mission Inn, Riverside, where I went yesterday to lecture, there was a convention of air force officers. It was sad, how ugly they were; big pie-faced men with misshapen bodies. Visitors to a phoney-antique establishment like this one have a curious air of reverence, they are so anxious to admire, but always from a distance; the past is utterly remote from their consciousness. I suppose everything pre-twentieth century seems to them a little mad. I saw one couple who had achieved such a state of alienation from their surroundings that they were watching the pool attendant as though he was a South Sea island savage.

[1] The California Institute of Technology, in Pasadena.
[2] A Santa Monica realtor.

Just for the record—my tongue still burns quite a lot of the time, and my jaw is still sticky.

October 16. In my last entry, I forgot to mention something I heard while spending the evening at Vedanta Place on the 10th. It seems that there is an Englishwoman who has just written a book about her search for a guru in India and her failure to find one. In this book, which has apparently been accepted already by Gollancz, she mentions Ramakrishna, saying that she was much drawn to him and to the Ramakrishna Order but that, of course, he was "a homosexual pervert" who had to struggle violently before he overcame his lust for Narendra. (All this, I repeat, I heard at second hand, and it is probably quite [a] bit distorted.) Anyhow, this Englishwoman was staying up at the house across the road from the Santa Barbara convent which is used as a retreat for female visitors, and somehow Swami got to hear about and then read the manuscript and of course he hit the ceiling and the Englishwoman was terrified and withdrew the offending passage and implored *him* to be her guru....[1] All this would be merely another funnyish story in the Vedanta Society tradition and hardly worth recording; but I do so because false or true, it made me think and see something which I had never seen so clearly before. Which is—

When Swami used to teach me that purity is telling the truth I used to think that this was, if anything, a rather convenient belief for me to have, because it meant that I didn't have to be pure but only to refrain from lying about my impurity. Well, that's the minimum or negative interpretation. But, thinking about it in relation to Ramakrishna, I saw this: that the greatness of Ramakrishna is not expressed by the fact that he was under all circumstances "pure." No. And even if he was pure, that didn't mean that he wasn't capable of anything. You always feel that about him—there was nothing that he might not have done—except one thing—tell a lie. So, when I hear this story about him and Naren, I do not say to myself, "He was incapable of it." I say to myself, "I know it isn't true, because, if he *had* felt any lust for Naren, he would have been incapable of not telling everyone about it." This seems to me basically important.

(Of course, there is always the possibility that he did confess to having had lust for Naren and that this confession has been suppressed by his biographers. But that is utterly unimportant. It has

[1] Anne Marshall's *Hunting the Guru in India* was published by Gollancz in 1963; it contains no passages about Ramakrishna and homosexuality.

nothing to do with *him*. Anyhow, I don't for a moment believe it was so, because of all the other statements he did make about Naren, and his attitude to him.)

Much might be written on this subject—I should like to bring it into the biography but I doubt whether Swami will want me to. And yet, if I don't write something, and leave in all these references to Ramakrishna's getting into drag, etc., there will be the usual reviewers' sneers. It's funny that I, who am steeped in sex up to the eyebrows, can see quite clearly what Ramakrishna's kind of purity is capable of, and that most people just can't. I suppose it's having been around Swami so much *and* understanding camp. I am privileged; far more than I realize, most of the time.

The day before yesterday, a beautiful soft watering-can rainfall greatly helped our planting. Someone—probably the girls from opposite—pulled the Private Property sign off the tree at the top of our steps and threw it down the slope. I promptly got out the spare sign we had and nailed it up there. Let them get a glimpse of British obstinacy.

That night we had supper with our next-door neighbor but one, James Covington. Unfortunately I got too drunk too soon, so I don't know quite what to think of him. Don likes him. His stories of law and movie business, always implying considerable wealth; and of his persecution by thugs from the Mafia, because the producer of this movie borrowed money from them and wouldn't pay interest. Hence his "bodyguard," Frank, from the marine corps. He refers casually to two marriages, both broken up. He has a collection of third-rate seventeenth- and eighteenth-century drawings, which Don thinks have been sold him for twice their value by a crooked dealer. He *is* a strange creature.

Yesterday, I finished chapter 16 of the Ramakrishna book. I think there are six, maybe seven more to write. The novel creeps along; I am merely treading water. And now I have to prepare for this reading at Garden Grove next Sunday.

October 22. I don't know why I'm starting a new volume just at this particular point.[1] The day isn't auspicious—nothing memorable but the death of Cézanne. And today is foggy. Rather snug, as foggy days always are in the Canyon, inviting to household chores. Don has movie clippings all over the bed in the back room—which has just returned there from the studio, because Michael Barrie said we could have the double bed in the front part of his house, so Don is taking that—I have just tidied out the two top drawers of my desk, anything to avoid serious work but it did need doing. The only thing is, Don is going into town for the evening, maybe the night, and I haven't made a date and can't make up my mind if I should or not. I weigh the possibilities of Jo and Ben, Gavin, even maybe Jim Charlton (whom I haven't seen in ages) against the snugness of staying home and reading Anaïs Nin's *A Spy in the House of Love* and Laura Huxley's *Recipes for Living and Loving,* etc., etc.

Yesterday on the beach—which was warm and beautiful—John Zeigel suddenly appeared back again from Mexico. Ed was killed because he and their friend were in a Peugeot, the lady who ran into them wasn't killed because she was in a Cadillac. John says he still can't realize that it has happened. I chiefly sensed in him an anxiety not to be excluded by the rest of us—as we tend to exclude the bereaved. We are to meet soon.

Someone—the daughters of the man who lives opposite?—has/ have torn down our Private Property sign for the second time; this time it has disappeared altogether. Question—what is the anthropological approach? Not to put up another sign and just wait to catch a trespasser, I guess. But the temptation to keep on putting up signs until you win the obstinacy match is very strong.

A whole week of no work, except preparations for the reading I gave last night at the Garden Grove High School auditorium. (This was looked on by the organizers as a historic event, because it was the first lecture given under the auspices of the future Irvine branch of U.C., UCI,[2] which has at present no buildings, no students, almost no faculty except the chancellor and some other administrators, nothing but one thousand acres of land.) I have got to get on with Ramakrishna. And I must keep at the novel, just

---

[1] He continued the new volume for four years and labelled it "October 1962–September 1966."

[2] I.e., the University of California at Irvine.

for the sake of provoking a breakthrough. Don, meanwhile, is getting desperate because he cannot make himself paint. Jack Jones is being very sweet and helpful, encouraging him to do this. I think Jack regards it as a kind of therapy and feels good, because he has been the patient so often himself and wants to reverse roles.

The nice red-headed carpenter, Bill Flinders, who is a member of Kim Haslam's team, has just fitted in the new door. He is quietly pleased with himself because he made the lock work, after hours and hours.

October 23. Despite what I wrote above, I couldn't have picked a more momentous day to begin the new volume. Indeed, I had barely finished yesterday's entry when Gavin told me on the phone about the decision to blockade Cuba.[1] So today has had the all-too-familiar crisis atmosphere, listening to the news. There has even been the familiar kind of crumb of comfort—Khrushchev in Moscow seemed calm, went to the opera and applauded an American star in *Boris*![2] At the gym, it was admitted that the Cuban speech at the U.N. was clever and that there are two sides to every question.[3] If we are to be fried alive, it seems funny to be working out; and yet that's precisely what one must do in a crisis, as I learned long ago, in 1938. I have also been prodded into getting on with both my novel and the Ramakrishna book today, and I have watered all the indoor plants. Now I must write to Frank Wiley and Glenn Porter, before I go to have supper with Gavin.

Last night, I went with Jo and Ben to see a show on La Cienega of paintings by Jack Dominguez.[4] Dull little primitives, with the paint laid on very thick. A champagne party was in progress, and the room was crammed with boys in dark suits taking an immense interest in each other and no interest at all in the pictures—alas, their stupid sneering good-looking faces! Jo was appalled by the hypocrisy of it all and kept exclaiming that never *never* could she face the ordeal of exhibiting her work. She'll face it, though—and why not? I'm sure Jack Dominguez would rather be bitched than unhung.

---

[1] In order to turn back Soviet ships carrying more weapons; see Glossary under Cuban Missile Crisis.
[2] American bass Jerome Hines (1921–2003) sang the lead in *Boris Godunov*, October 23, and Khrushchev even visited his dressing room.
[3] The Cuban representative, Mario García Incháustegui, characterized the Soviet weapons as defensive, asserted Cuba's right to defend itself, and denounced the blockade as an "act of war."
[4] Longtime resident of Santa Monica Canyon.

Anaïs Nin is the most egocentric writer alive. She sees herself as such a dear darling tiny little thing, before whose charms everyone and every moral standard must give way. Because she has genuine power, this attitude seems fun and sympathetic; not in the least repulsive.

October 28. According to the news today, Russia is going to remove the missiles from Cuba. This seems rather too good to be true. But of course it is just one move in the long wrestling match. I feel such a curiously strong loathing of Castro—something to do with his beard, his sincere, liquid-eyed beard. I should like to see him forcibly shaved in the U.N. I wish Groddeck were here to explain this to me. After all these years, I am reading [his] *The Book of the It*; Wystan used to rave about it in the twenties, but it was somehow "his" kind of book and I seldom if ever read that kind. Now I love it. He is just the right kind of psychological writer for me. Oddly enough, the statement which has made the most impression on me so far is that there can never be great love (on the human level—though he doesn't make this qualification himself) without great hate. Obvious, but how often forgotten, or rather, how often desperately denied by me!

This fog has choked the Canyon all week, and it has begun to get us down. Yet we are very harmonious again. (In relation to this, I must say I think Don instinctively understands Groddeck's proposition about love-hate much better than I do. I mean, I think he would be prepared to accept and live with it, if I would, too. But I am sentimental—in the worst possible way—the way my mother was ... I have a great deal of that attitude which makes women say "not before the servants," "not before the neighbors." Only the "neighbors" in my case are some kind of an internal audience. It is all part of my playacting. I must try to keep thinking about this, thinking it out; maybe I shall discover something of immense value.)

Don has been trying to paint—flowers from Margaret Gage's garden—with Jack Jones. He is in despair. He says that now he doesn't take Dexamyl he only has downs, no ups. Yet a lot of the time he is as lively as he ever was. Sometimes I feel so strongly that I really do not even begin to understand him, and of course that is what has made living with him so fascinating, all these years. (Ted came in while I was writing this—and he said to me, in the tone of one "sane" person speaking to another, "You must keep him from getting depressed again, like he was a while ago." Maybe we all see ourselves as healers nowadays, no matter how nuts we are.)

Today I'm in quite a good mood, because (a) the crisis seems to be easing off (b) the fog is thinner, with gleams of sunshine (c) I have practically gotten over a short vicious cold with acute sinus-aches, which hit me on Thursday night. I felt miserable and did no work whatsoever. Tonight I have to appear with Jerry Lawrence at UCLA in a "public dialogue" and answer such earthshaking questions as Who am I as an individual / Who are we as a nation / How do I know who I am? I fear Jerry is going to gag it up in best radio style and make asses out of us both. He is terribly flattered to be in on this sort of thing—much more flattered than he'll admit.

November 1. Jerry did gag it up, as anticipated—oh he was in-decent. But I didn't really care, and a lot of the audience liked it. Jerry's old Jewish momma was there, and told me afterwards, "When he was born, I was in heaven." She also said she wanted to kiss me, so I kissed her. Jerry is provoked by her malapropisms: "They've given John Steinberg the Nobel Prize. I loved that novel of his, *The Wrath of Grapes.*"

Fog again, and crisis again, because Castro won't let in the U.N. observation team. Russia said to be going ahead with dismantling the rockets, however.

Tea with Gerald. Talk chiefly about psychosomatic medicine, and the id.

Supper at Vedanta Place. Swami told me he had had a dream in which Premananda had said to him, "If I had had you beside me, I could have conquered the world." Prema is not going to Boston after all, because the swami there feels he can deal with this mad-woman who has been terrorizing them all.

I feel that my novel is developing in a new way, as a much simpler structure, a day in the life of this Englishman. More about this in due course.

I have asked Abbot Kaplan of UCLA to find out for me about the possibility of a lecture tour in Australia. This is one of those Lord's-will things. If it develops, I guess I'll go.

Don this morning at breakfast on the deck, in the fog, wearing his dark glasses. I told him he looked like a photograph in one of the Stravinsky conversation books: "Bachardy at Yalta, 1901." This pleased him, but he is in another negative phase, can find nothing to admire or like anywhere. Ted had depressed him by saying to him yesterday on the beach, "I'd have left Vince long ago, but I don't know what he'd do without me."

Yesterday came the news that Nehru has removed Krishna Menon as Defense Minister, because of India's unpreparedness for the Chinese invasion. Swami said, "He should have been lynched!" Swami is very patriotic about the crisis; thinks India should take a stand with the West.[1]

November 4. The day before yesterday, terribly hung over after a night of drinking following the art show in the Canyon at which Jo showed her watercolors, along with Renate Druks, the girl-friend of Ronnie Knox the football star—she looks like Isadore From in drag with a long wet black witch-wig—I drove out to Trancas and went in swimming early. Haven't done that in ages and ages. The same day, we went on the beach at noon and swam again. It was sunny and warm. But now the bad old fog is back and it's Gloomsville. Don has several boys and girls drawing in his studio. Alas, the water isn't working; the pipe burst, that same day, and won't be repaired till tomorrow. Don in despair again. I have just written to Mark Schorer to find out if there's any possibility of my teaching in the San Francisco area next semester. I wish *he* would go away, but actually, once I get myself off, I shall profit by it, I know that.

Charles has been given some new drug, called something like leukocristine.[2] So far it hasn't had any effect on him, good or bad. But there is another cancer patient at the Cedars of Lebanon, a girl, who is said to have been in a far worse state than he is and who has now almost no pain at all, as the result of taking it.

November 9. Well, all sorts of things have been happening. Chief and best, Dirty Dick Nixon has been flung out of politics. In defeat he showed his yellow poison-fangs.[3] The Cuban crisis isn't really over. But the government is keeping quiet at the moment— till the rockets are removed.

Yesterday afternoon, I went to see Laughton. He looked at me with open blue eyes, didn't know me. His brother Frank(?)[4] and Elsa were there. And a Dr. Wilson came by to see him. This Dr. Wilson is approved of by Elsa because he doesn't belong to the

---

[1] China invaded on October 20; see Glossary under Sino-Indian War.

[2] Leucocristine, or vincristine, extracted from periwinkles (vinca), used in chemotherapy.

[3] On November 7, Nixon lost the California governor's race. In what he called his "last press conference," he announced, "You won't have Dick Nixon to kick around any more."

[4] Frank Laughton (1907–1964), youngest of the three Laughton brothers; he ran his parents' Pavilion Hotel in Scarborough, North Yorkshire, where the boys had spent their childhood.

"Cedars of Lebanon Gang" and isn't a Jew. He is a large pale man who seems to move under the shadow of death, but without being exactly sad or solemn. He obviously thinks Charles is dying. He says he is very close to coma. Wilson has, on his own responsibility, countermanded both the cobalt radiation and the leu[c]ocristine. I think this pleased Elsa very much. I kept watching her face. It was smug and sly. Wilson kept referring to Charles as "poor man."

Since it was no good my sitting with Charles, I had time on my hands and so I drove up to the Griffith Park Observatory to watch the sun set. Astonishing, how empty and wild the hills still seem. As I stood there I felt, as I have felt so often, why don't I spend more time in *awareness*, instead of stewing in this daze? How precious these last years ought to be to me, and how I ought to spend them alone—alone inside myself, no matter who is around.

Supper with John Zeigel. He still seems terribly shaken. He described how the highway patrolmen came in the middle of the night to bring him the news of Ed's death. Before they told him, they said, "Don't you want to sit down?" "Then," said John, "I knew." He is disgusted by [Ed's friend] who came from the East to see him and expressed nothing but concern about his share of Ed's money. John says that Ed was about to change his will when he was killed. He wanted to cut [the friend] out of it again.

November 11. Mark Schorer has written back and says that I can get a Regents' Lectureship (most probably) at Berkeley from April 15 to May 15. This might not be a bad arrangement, because it looks as if Don will be away quite a bit, anyhow, in the earlier part of the year. Certainly he'll be at Santa Barbara for his show during part of January-February,[1] and the Phoenix show may follow right on after that. He would have to stay at both places during the shows and maybe for a while after, to draw people on commission.

At the moment all is peace and affection because I am leaving in an hour or two to go with Swami to Trabuco until next Wednesday. Now suddenly Don says he doesn't know what he will do while I'm away!

Last night, we went up to supper again with the Huxleys and Mrs. Pfeiffer and her charming adopted children. Laura is all excited about the publication of her book and I have had to write a blurb for it. Aldous told about his visit to Memphis and the

[1] Twenty-two portrait drawings at the Santa Barbara Museum of Art, January 29 to February 17, 1963.

southern aristocracy there. They still talk about darkies but claim that integration has been achieved without any fuss—except for public swimming pools.

Gerald just got on the phone and told me that, during his six days at Long Beach, he gave fifteen lectures! The news that Laughton is dying sent him off into the usual philosophical meanderings.

I have been disgraceful about work. Nothing done yesterday or today.

November 16. Yesterday was a lost day. It was so freezing cold and I had such a shocking hangover after supper at the Larmores'—who are on the wagon!—that I wasted it feeling miserable and reading a very poor novel by David Stacton called *Old Acquaintance*. Only in the evening did the clouds lift and I had supper with Jo and Ben and persuaded them, I don't know why much less how, to go to Ceylon next winter. I shall do no work today because I have chores and then we have to go to supper at Carter and Dick [Foote]'s and see their film about Bali, with, no doubt, Dick cavorting in the monkey dance.

Laughton's brother Frank smuggled in a Catholic priest who gave Charles the last rites. Elsa is outraged—she suspects that he has signed away some money to the church. He mumbled something about having signed—but didn't say what, and anyhow it surely wouldn't stand up in court. He also mumbled, "I feel I want to join the mob," and "Catholics are all alcoholics." Elsa wants me to see him and try to find out what *did* happen. She is busy shopping for cemeteries.

Don, bitching John Zeigel (as usual): "I suppose he gave you tart blanche?"

And, talking of bitchery, someone described Connie Wald's new marriage as, "A funny thing happened to me on the way back from the funeral."[1]

The visit to Trabuco was a great success—went there on the 11th, came back on the 14th. I found I could, if not meditate, at least sit through the meditation periods without getting the jumps.

I took a lot of notes while down there. Most of them I can't be bothered to transcribe. But—

Vandanananda is accused of being much too interested in girls. Also, he shocked Santa Barbara by giving a lecture in which he

[1] Jerry Wald had been dead for only four months. Stephen Sondheim's musical, *A Funny Thing Happened on the Way to the Forum*, opened May 8, 1962 on Broadway.

said, "This is the meaning of *tat twam asi*[1]—if you're a ballet dancer, then tat twam asi, that's what you are—" This upset Sarada so
much she went to bed!

Swami is concerned about [one of the monks], who is always
getting sick. [The boy] says he sees "little people." Swami says
these are psychic phenomena on a very low plane; they often appear just before death.

Adrian Wolheim[2] objects to work. Does one have to work to
be spiritual, he asks. Swami tells him, all right, walk around all day
thinking of God. But don't expect to get anything to eat.

Swami said to me, "Just think, you might have been a swami by
this time." But then he added, as he has never done before, "But
perhaps you are more useful like this."

Franklin [Knight] says the doctor is impressed by the way
Vedantists die.

The boys seemed really quite pleased to have had me there. I
must go again soon. There are seven of them at the moment—
Franklin (the only brahmachari), Eddie [Acebo] the Mexican boy,
Len [Worton] the British ex-sailor, Tom Battle (shy and quiet),
Adrian Wolheim (rather crazy and unlikely to stay long, he hitch-
hiked all around India), Bill Bergfeldt the sickly boy and Tony
Eckstein, a little Jewish ex-marine whom everyone likes, he is
hardworking, friendly and quite bright.

November 19 [Monday]. The last few days, the weather has switched
to dry and bright, with strong gusts of wind drying up the hanging plants on the deck. At least it's much more cheerful than the
fog. This morning, at breakfast on the deck, I kept thinking that
gloomy old we didn't deserve this view. Except that we can be
very amusing about our gloom. Don made me roar by saying,
"The view from the brig."

I dread Mexico, though. There will be ghastly scenes and the
most tiresome confrontations. On the beach yesterday John Zeigel
showed up and was eager to arrange to meet us down there during the holidays. This would be just what the undertaker ordered.

On Saturday night, I was having supper with Bart Johnson and a
friend of his who teaches at the same school, [...]. We were talking about [Katherine Anne Porter's] *Ship of Fools*, and [the friend]
asked what did I think of her writing? I said, "With many writers,

---

[1] "Thou art that," one of the Hindu "great sayings" stating the oneness of
the soul with Brahman.
[2] Not his real name. He was a trained engineer, considered brilliant.

one can instantly say how you imagine them dressed, when they're at work—I don't mean, literally, but ideally, symbolically, judging from their style. I see Miss Porter taking a perfumed bath and then sitting in front of the mirror for an hour, fixing her hair and making up her face, and then putting on an exquisite, very low-cut evening gown without sleeves, and then elbow gloves, and then earrings and necklaces, and rings over her gloves—and then sitting down at her desk to write."

I did a big swatch of work on the novel, Saturday. I am still excited about it; in fact, I sat up till nearly four, Saturday night, with Gavin, talking about it—which meant that I was too hungover to work yesterday. But it does seem to me almost infinitely promising; that is to say, it is a possible *form* for a masterpiece, if only I could write it like a master!

Don has started using color in his drawings. Not coloring them, but working with brush and pencil alternately.

November 20. A good day—perhaps the beginning of a new epoch. Castro has given way about the bombers,[1] Kennedy has called off the blockade, the Chinese have suggested a cease-fire with withdrawal. True, this last is thought to be a trick, but still and all it is sort of good.

Last night, I drove down to Long Beach to see Glenn Porter. He is staying with three other sailors off the *Princeton* in a flea-trap apartment house on Daisy Avenue, just back from the oceanfront. When I got there the scene was crazy and rather wonderful. The cheerful Jewish sailor was making love to a girl on the couch, the married (or anyhow involved) sailor was watching his wife/girl combing her/his/their little daughter's hair. The reckless Swedish sailor was getting drinks from the kitchen. And Glenn and I talked about Rilke! They had been terribly drunk the night before, had smashed the mirror in the living room and poured beer over people descending the stairs. No—that was the previous night, I guess—because Glenn told me that, on the Sunday morning, still terribly drunk, they had driven into town—the Swede driving at ninety all the way—and made it to the Vedanta Society where Glenn (not the others) had heard Vandanananda's lecture. As we talked about this and other things, a cockroach ran across the floor and was killed by the Swede. When we left to go to supper, the

---

[1] The missiles were already being shipped back to Russia, but Soviet IL-28 bombers remained in Cuba; Castro agreed, on November 19, to their removal.

married sailor gave me a carton of cigarettes. His wife/girlfriend is Hungarian and speaks German; when we returned, she came out and spoke to me in German, saying she hoped we could talk German to each other another time.

All this was so old-fashioned! The girls I saw around the place were real floozies from any play or film called The Fleet's In. And Long Beach didn't seem to have changed since the war. Blazing lights and big buildings along the beachfront and then miles and miles of dark tacky shut-up streets, until you get to the lighted artery of Pacific Ocean Highway.[1]

Glenn looked marvellously well and healthier, although he spent whole weeks inside the carrier not even seeing the ocean. He said at first you got claustrophobia; now he rather loves it. But he said, without the least sarcasm, "I find it a bit difficult to write, in that house," and he also has to endure constant prodding from his buddies to get drunk and get girls. He is a strange boy. He says he hasn't had sex for the past year. When he used to study at the Pasadena library, he often got so tense that he would go outside and climb a nearby building, but he could never get quite to the top, because there was an overhang! He assured me that he *hadn't* talked much about me to his buddies, but the Jewish boy said he had, and added, "I was expecting a little guy with a notebook."

Another perfect morning. I went on the beach and in the water.

November 22. Here's my twenty-fourth Thanksgiving in this country. Very very much to be thankful for. Things couldn't be going better at this particular moment. We are saved from atomic war in Cuba. The Chinese and Indians have ceased fighting, as of this morning, at least temporarily. Swami is Swami. Don is Don; and our life for the past three days has been most happy. We are both well, though both weighing more than we like—Don 142, me 151. Don did some really stunning work yesterday evening, though admittedly not in color, as he would have wished. I creep on with my novel, or novelette as I now suspect it to be, and I know there is something there. Also, we have this beautiful house—*and* about $34,000 in the bank—before taxes!

About Christmas in Mexico, I feel: let His will be done. If there is no good reason not to go, I will go, and just pray that it isn't a disaster—or rather, pray that I will be able to take it if it is a disaster.

As for San Francisco, that still isn't certain but I would look

[1] I.e., Pacific Coast Highway.

forward to it. And I would quite look forward to Australia, if that were to materialize.

Today is hazy with pale sunshine. Yesterday night was the worst coastal fog I can ever remember.

Saw Gerald yesterday. I asked him about his great phrase, "the novel written in protoplasm," but he was vague, said it was somewhere in his handwritten material which Michael hasn't yet revised or typed out. This led to further disclosures about Michael's possessiveness. Because of it, Gerald isn't sure if we can have our tape-recorded conversations together, opening the way for a memoir. If Michael is asked to do anything which isn't entirely his project, he just puts it off *sine die* or "forgets" it.

We talked about morality. How nowadays people tend to think of religion as meaning only a set of ethical standards. I said I don't go to Swami for ethics, but for spiritual reassurance. "Does God really exist? Can you promise me he does?" Not, "Ought I, or ought I not to act in the following way?" I feel this so strongly that I can quite imagine doing something of which I know Swami disapproves—but which I believe to be right, for me—and then going and telling him about it. That simply isn't very important. Advice on how to act—my goodness, if you want that, you can get it from a best friend, a doctor, a bank manager.

What does matter is to make japam and pray.

Up at Vedanta Place, Swami has become a true Kshatriya.[1] Not only are the Chinese to be run out of the whole area; he demands Tibet. And an all-out military alliance with America and England. I think he'll be really disappointed if this truce leads to peace ... Well, there you are: that's the other side of the coin. I disagree with Swami's attitude, ethically; but what does it matter? Not the least bit. That's not what our relationship is all about.

And shall I confess? Deep down—no, not deep, about half-way down—I do feel a certain satisfaction at Kennedy's stand on Cuba, the temporary disadvantage to Soviet Russia, the folly of the Chinese, the involvement of India with "us." Yes, I feel it—but oh, what infantile nonsense it all is, really! Early yesterday morning, the phone rang, and it was a cable from the London *Sunday Times*: would I write five hundred words on what is best and what is worst in the United States? I replied no (thank you); because, the moment you try to think this out, you find that it's easy to say what is worst but when you get around to what is best, the United States doesn't own, isn't responsible for, any of it. And this is true of all countries.

[1] Warrior; Swami belonged to the second Hindu caste, the warrior caste.

Our plants seem to have survived the windstorm of a few days ago. The hanging redwood baskets on the deck are trailing profusely with that pale green delicate wispy plant (what *is* its name?) and the geraniums in their boxes are thriving, although one is broken, and the bottlebrush trees below the house seem all right.

November 27. Hazy sunshine, after two more days of sea-fog. Am in a winter mood, waiting for some little nudge of spring to get me moving again. Mexico is a problem. I feel sure it would be a mistake for both Don and me to go there together; there would certainly be friction. Don can't stand the least discomfort of travel, any delay, any boredom—and all of it would get blamed [on] me. I wish he would go alone, but he says he doesn't want to. I would be prepared to go, but Don says he can't possibly stay here. I am making him sound tiresome, and of course he is; but this isn't a fair statement of his problem. He wants out—not permanently, but for at least several months. I, on my side, have resolved not to be noble, because that's the most annihilating kind of aggression. I will not "nobly" leave this house just for his convenience. If he wants out then he must be the one to get out. On the other hand, I am ready to go when there is something interesting to go to, like San Francisco or Australia.

None of this is as tragic as it sounds. We are still deeply fond of each other and I quite expect we shall go on living together, after a period of adjustment. Most of the freedom Don is looking for could actually be achieved right here, living with me. He doesn't realize that yet. Okay, he can find it somewhere outside and then come back.

The day before yesterday, Frank Wiley came by with a shipmate [...]. They seem to be having a lot of fun together. You got a sense of the intensely provincial atmosphere of the carrier; they might have been two spinsters living in a nineteenth-century English village—a monosexual village, however. [The shipmate] was pleased, after several gins and bitters, because he still remembered to say "wall" instead of bulkhead. Two expressions, "out of phase" (out of whack), and "scuzzy" (spelling?) meaning horrible, tacky, a mess; used for example of girls.

Charles is being moved back home today. He protested violently against this; I suppose he thinks Elsa is planning to murder him. Scott Schubach told us this last night, we had dinner with him and his friend [...], a sweet little boy, whose face twitches. Scott is an incredible bore; he talked all evening about his lysergic acid experiences. He lives in this huge rambling house which is stuffed with antiques. He sleeps on a bed with goat feet (Venetian). Lots of classical columns, busts, inlaid cabinets, lattice work from casbahs, marble tables, Moorish cushions, drawings by [Hans] Erni, Khmer buddhas, etc., etc. He does most of the housework himself, and cooks. He must be very very rich.

On Saturday, Stanley Miron was down here to see his folks and brought with him the sweater of John Cowan's which [someone] gave me and I left with Stanley when I was up in San Francisco. It is very tacky and moth-eaten. Don didn't want me to wear it, feeling that there might be some kind of a curse on it, but I thought, after all, what harm did Johnny Cowan ever do or wish me? So I wore it yesterday.

Peter Quigley has written an article called "A Glimpse of Isherwood" for the *Irish Times* in Dublin. Here are two of the "glimpses."

> Medium short and still boyishly well proportioned, he cuts a workmanlike figure. With his trim haircut, California suntan and much laundered "fatigue" shirt and trousers Isherwood in his fifties looks more like a retired Rommel than a widely read author....
>
> At the foot of the steep driveway we stop again in the cold wind. He looks very much alone, standing with shoulders hunched and eyes peering out at the wintry light of a dying day from beneath the sun-bleached and bushy eyebrows which are characteristic of his late fifties.[1]

November 29. After these last dull days of fog, the wind got up last night and blew in terrific gusts. One of our geraniums has been broken off and I don't think the T.V. antenna will stand up much longer.

Today Don is drawing Arthur Laurents who is here for a few days about movie work. He loathes this town and keeps saying so, sometimes amusingly, sometimes merely bitchily. He described a party at a producer's house at which he got into a discussion with a girl about Dufy. "What other painters do you like," she asked him, "I mean, in the same price bracket?" No one else who was listening seemed to find this at all funny, Arthur says.

Yesterday evening, when I went up to Vedanta Place, Swami brought up the subject of Don's initiation. He is ready to do this quite soon, on December 18. I'm not sure how Don feels about

---

[1] Not traced in the *Irish Times*; possibly the article was printed elsewhere.

it; maybe a little uneasy and scared of getting himself in too deep. All he tells me is that he doesn't want to have to meditate on Ramakrishna. I assure him that Swami won't insist on this.

We are still undecided about Mexico. I think Don wants to go and he can't really understand why I don't want to go with him. (That's natural, because he obviously can't be expected to understand how *inevitable* it is that he will make scenes, as soon as something on the journey doesn't suit him.) And of course, as always, I feel cruel and selfish and start saying to myself why don't I go and risk it?

Yesterday, I showed Don the first twenty-eight pages of this second draft of my new novel. He was far more impressed, even, than I had hoped. He made me feel that I have found a new approach altogether; that, as he put it, the writing itself is so interesting from page to page that you don't even care what is going to happen. That's marvellous and a great incentive to go on with the work, because I always feel that Don has a better *nose* than almost anyone I know. He sniffs out the least artifice or fudging. He was on his way out after reading it, and then he came back and embraced me and said, "I'm so proud of old Dub."

What still bothers me very much, however, and makes me hesitate to go ahead, is the problem of plot. How much should there be? How entangled should William be with Charlotte, and with her son Colin? The point is, there are two strands of styles interwoven in this sort of writing—the lyric, sub specie aeternitatis thing which observes William like a wild creature, an antelope, with his daily habits and his whole symbolic meaning as a type, and then there is the mere plot approach, which ties this particular individual William up with this particular individual Charlotte and Colin. Too much of the second is death to the first.

The day before yesterday, I think it was, Florence Homolka died. I feel really sad about this. She was a bumbling tiresome clumsy creature, but sweet and kind and quite talented, and I had known her oh so long, right back into the Caskey era, when we used to go to her house for evenings with Chaplin, etc. She seems to have died quite suddenly, of what the radio said was a "respiratory ailment."

November 30. To see Charles at the Cedars yesterday afternoon. Elsa wanted me to go, to influence him if possible to agree to come back to the house, which he has violently refused to do. When he was told I was coming, he said that he had something he wanted to talk to me about. He was sleepy and in pain but quite lucid. He said, "The preoccupation is with death, isn't it?" What he really wanted to ask, though he didn't put it directly, was whether or not I approved of his having seen the priest. I told him I certainly did. He said he would like to see another priest, a better one, but he didn't make it clear in what way better. I tried to tell him that it didn't really make all that much difference if he got to see another priest or not. He should speak to God, ask for help. Because God is there. "I know," Charles said. And then, either before or after this, he said that having seen the priest had already helped "quite considerably." He kept dozing off and I was holding his hands and praying to Ramakrishna to help Charles through his suffering and dying. I even said, which I have never said before, "Do it for Brahmananda's sake, for Vivekananda's sake, for Prabhavananda's sake," and somehow this was "put into my mouth," it seemed. All mixed up with the praying—which moved me and caused me to shed tears—were the caperings of the ego, whispering, "Look, look, look at me, I'm praying for Charles Laughton!" and then the ego said, "How wonderful if he would die, quite peacefully right now at this moment!" It is most important not to make these confessions about the ego as though they were horrifying. They are not—it is mere vanity to pretend that the ego doesn't come along every step of the way; it is there with you like your sinus and its instructions are no more shocking than sneezing.

The really important question is, *why* should one pray to Ramakrishna for Charles? Does it do any good? Granted that Ramakrishna is "there," available, only waiting to be asked, shouldn't one simply tell Charles to ask him, or ask Christ, or whatever avatar he believes in? I must ask Swami about this when I see him next.

Coming out into the world of the healthy, on Hollywood Boulevard that evening, with the chilly wind making the Christmas decorations swing from their moorings, I must say it did seem most horribly important *not* to have cancer. Even hustlers without scores, shivering at corners and maybe needing even the price of supper—ah, how lucky they seemed!

December 4. I have decided to go right ahead with the novel and finish this draft; it's the only way I shall find out more about the inwardness of the story. Am also trying to make up for lost time on the Ramakrishna book.

The last three days have been beautiful and peaceful, outdoors and in. The last two evenings we have spent at home, reading; a thing we haven't done in a long while. Mexico is still on; and I feel better about it now, because I believe that Don means to make it a success.

There isn't anything else of interest to report. We met two nice boys, both painters, who are friends of Jo and Ben: Paul Wonner and Bill Brown. They have studios on the vast empty top floor of an old building in Ocean Park, and their work is influenced by Francis Bacon and (a little bit) Keith Vaughan. More about them later, I hope. Frank Wiley came by to be drawn by Don and left the first part of his new novel and his journal, on the title page of which is written in rather beautiful script: "Herein is contained the journal of the sentimental education of Franklin Evelyn Wiley Jr., Ensign in the United States Naval Reserve; appended by extracts from a forthcoming work of purest fiction." (I suppose that last ungrammatical bit is to guard against snoopers; but just the same the thing is much better lying around here than on that aircraft carrier. So fictional it isn't.)

December 8. Swami, when asked about prayer, said that it is good both for you *and* for the person you pray for; and he added, "You see, when you are speaking to God like that, there are not two people, it's all the same." He also said that all that was needed was faith that the prayer would be answered. You didn't have to be a saint. If you had faith, then it would be answered. He said this with that absolute compelling confidence of his. He made you feel he was quite quite sure of what he was saying.

Don is definitely to be initiated on the 18th ([Holy] Mother's birthday) before we leave for Mexico.

Last night we went to supper with Michael and Gerald, and got drunk. Don says he watched me and was aware how determined I am to get drunk. He said, "I got the feeling that the alcohol wasn't even really necessary."

Gerald has insisted that if the magazine takes his essay on death it must be anonymous. I can't help feeling this is some kind of malice against the Vedanta Society—testing them to see if all they want is his name on the cover. Meanwhile, he doesn't at all insist that I shall take his name out of my manuscript which I think will be published by the society as *An Approach to Vedanta.*[1]

Talking about drinking with Don gave me an idea for my novel. William should be an elderly man in the morning, a mature man at noon, a youth in the late afternoon, a baby at night. You could say that about me. I wake feeling definitely my age. I get working,

[1] It appeared in 1963 from the Vedanta Press.

drink coffee, take Dexamyl and feel much more alert and creative. I go to the gym and often develop quite surprising energy; I am almost youthful. And then, at night, there's this urge to get drunk, to let go altogether and let the others look after me, like a baby.

The novel is actually going quite well. I do wish I could rattle off some kind of a rough draft of the whole thing before we leave, but I fear that's quite impossible. There's amazing richness—or rather, amazing opportunities for discovering richness—in the material.

The Stravinskys are back. Talked to Vera this evening. Haven't called Elsa for several days, and feel guilty about this. Charles is back home.

December 14. The American Express, which was supposed to get the Mexican train tickets, has goofed (Don hates that word) and so now we shall have to fly direct from here, on the 23rd, which is hateful but at least gives more time. I might even really be able to get the draft of my novel finished. I have already reached the scene at the gym, which leaves only the supermarket, the supper with Charlotte, the meeting with Colin in the bar, and whatever else follows that. If the finished novelette is to be as long as *Prater Violet*, that would be approximately 120 pages. This draft wouldn't be more than seventy, I should think.

Charles is said by Elsa to be right on the brink. He nearly died last night, of some infection in his lungs. She is anxious that I shall be present and read at the funeral. Am going to see them this afternoon.

The 11th and 12th we spent up at Santa Barbara. Don drew Thomas Storke, the editor who got the Pulitzer Prize for exposing the John Birch Society,[1] and Judith Anderson (I mean, Don drew her, not that Storke exposed her!); we also saw Douwe Stuurman, Geo Dangerfield, and stayed with the Warshaws and got even drunker than usual. Next day we had lunch with Wright Ludington and saw around his mausoleum of a house. It seems almost incredible that he should deliberately have designed a gallery for his paintings which has to be lighted at all times by electricity. Indeed, it seems incredible that he could live in that building at all. Yet the stillness of the hillside is magical. And the view of the mountains and the ocean. And the safety of the sun-trap wall by the pool. And the secrecy of the little stone garden among the olive trees. Oh God, he is so dull. But very well-disposed toward Don.

Judith Anderson is rattling around in an even less habitable

[1] In 1962, for his editorials in the *Santa Barbara News-Press*.

house, in a valley back from the sea along the Ojai road. The country is marvellous; a pocket of the old California. But she lives there in such an uncomfortable grim British way. And she hardly seems capable of fixing even a cup of tea. Don did two wonderful drawings of her. We drove back to town and saw Gavin's film, *Another Sky*. It rather haunts me. It is far too long, but photographically beautiful and it has a kind of unemphatic relentlessness, like [Antonioni's] *L'Avventura* or *La Notte*.

The Stravinskys came to supper on Monday evening, along with the Huxleys. Bob Craft told us that Igor and Vera were quite transformed while in Russia.[1] They were so happy to be speaking the language in which they were really fluent. All their pride in Russia emerged—especially, of course, Igor's. Igor, like Picasso, is still really a tolerated exception in the arts; the authorities still don't approve of what either of them stands for. Igor was chiefly pursued by young people, to whom he is an avant-garde champion. But more of all this, I hope, tomorrow night, when we have supper with them, at their house.

December 17 [Monday]. On the 14th, I saw Elsa and Frank [Laughton], but Charles was unconscious. Frank says he said, during a brief lucid interval, "I've fucked my whole life away." That evening, I was to have met Don at Musso Frank's. He didn't show up until hours later. I got very drunk. We finally had a late supper down in the Canyon and I told him that I wasn't going to Mexico if I had to fly. I also told him that I was psychic and that I could see he had a nun as his familiar. He was rather impressed by this. I simply cannot remember or imagine what made me say it.

On the 15th, Don said he was going away out of town for the night, with a friend. Vera had called and asked if we would bring Gavin to supper. So I went with [Gavin] alone, after telling him that I wasn't coming to Mexico. He was sad, but very nice about this. I got terribly drunk again. After we'd returned from the Stravinskys' (I can't remember *anything* they told us!) Elsa called to say that Charles had just died, about half past ten.

Yesterday I saw Elsa and Frank and it is all fixed that I'm to be the speaker at the funeral on Wednesday. I had supper with Frank Wiley and got drunk again. I really must cut this out; my hands have started to shake.

Today, Don has decided that he doesn't want to go to Mexico

---

[1] Their first visit in fifty years, to mark Stravinsky's eightieth birthday, with performances and public celebrations, including a meeting with Khrushchev.

without me. So here we shall stay. I look forward with appetite to getting a lot of work done. Tomorrow is Don's initiation.

December 20. Don's initiation duly took place. It imposed the usual states of aversion and boredom on him that nearly everyone goes through under the circumstances: the long boring puja first, the devout women, the reek of Sunday religion. He didn't stop for the end of the *homa* fire, or lunch. And now, like I did, he has forgotten his mantram and must go up to Vedanta Place to check it with Swami! Never mind. The deed is done, and of his own free will. And that's all that matters for the time being—maybe for years to come. It will catch up with him.

Yesterday was Laughton's funeral. Ray Bradbury wrote Elsa a letter which could be nominated for the all-time slime-and-honey prize. I hope to get a copy of it later. But these are my impressions:

> Dear Elsa—I am a writer, but today I have no words. This morning, my second daughter came into my room crying, and told me that Charles was dead. And now all I have to offer you is my daughter's tears—

I'm sure Ray thought this was exquisitely beautiful. And, indeed, you have to be a very very good writer to produce such horror.

Elsa said that she wanted the Mitchell Boys Choir[1] to sing in Latin, "Because English is so full of repetitious words, like God." As a matter of fact, by barring the Catholics, she merely let the Protestants in through the back door. She asked for a "non-denominational" service and got the usual Episcopalian thing.

I really don't care to dwell on the streamlined horror of the ceremony itself; at the Hollywood Hills branch of Forest Lawn. They have constructed an Early American church with a tall steeple (copied from the one at Portland, Maine, where Longfellow went as a boy, the brochure tells us) right in the midst of this San Fernando Valley scenery: pylons, Warner Brothers, the T.V. station on top of Mount Hollywood, and a fine dim view of the mountains through smog. A whole crew of attendants, with white-topped caps, looking rather like the crew of a yacht; they take off their caps to a funeral procession as though the owner were

---

[1] Led by Bob Mitchell and featured on the soundtracks of such films as Bing Crosby's *Going My Way* (1944), *The Bells of St Mary's* (1945), *White Christmas* (1954), and the Disney cartoon *Peter Pan* (1953).

coming aboard. Miles of electric cables. Flowers arranged as if in a florist's shop. The truly obscene contrast between the nicey-nice church behavior and all the cameras and newsmen outside, sticking their lenses practically down Elsa's throat, even while the service was still going on. The coffin was surprisingly heavy, though there were experts to aid us official pallbearers: Raymond Massey,[1] Taft Schreiber, Lloyd Wright,[2] Jean Renoir,[3] Bill Phipps and me. I think Bill Phipps was the most genuinely upset one present. But when I got a little weepy over the "I am the Resurrection and the Life" speech, there was a movie camera whirring away at me instantly, like a rattlesnake. Elsa said later, "I wish it had been a grey day, it softens the face in the newsreel shots." She was really very nice, though, and extremely professional and brave. Frank said she had broken down violently the night before.

I said some "words of appreciation" and read three bits out of *The Tempest* ("rounded with a sleep," "I'll drown my book,"[4] and the last half of the Epilogue). The acoustics were excellent, and I know I was good. At my very best, Don said; and Elsa was genuinely delighted. So I feel I didn't let Charles down.

Do I miss him, Don asked me. Yes, I do indeed—or rather, I will, now that all this evil fuss is over. Funerals are deadly for flattering your vanity. I am actually still preening myself over my theatrical success, and a little disappointed that I haven't been called by anyone and complimented!

An argument with Don yesterday because of his mania for making up his mind at the very last moment: he wouldn't say whether or not he wanted to go to the funeral, and of course it was I who had to do the arranging about this. So Don said—more to punish me than anything, I think—that he still might decide at the last moment to go to Mexico with Gavin on Sunday. Well, if he does, he does. I shall make out all right.... And, despite this, I must say that the last three weeks or so have been quite unusually harmonious. I think we are gradually discovering a new way of living together which might work almost indefinitely.

Ted is showing signs of going mad again. He is full of hysterical

[1] Canadian-born, Oxford-educated actor (1896–1983), on the London stage from 1922 and in Hollywood from the 1930s; he played Kildare's mentor Dr. Gillespie in the television series "Dr. Kildare" (1961–1966).
[2] Frank Lloyd Wright, Jr. (1890-1978), architect and landscape architect settled in Santa Monica; son of architect Frank Lloyd Wright (1869-1959).
[3] French film director (1894–1979), son of the impressionist painter; he directed Laughton in *This Land Is Mine* (1943), about the French Resistance.
[4] From Prospero's parting speeches, IV.i and V.i.

enthusiasm, even about the weather, which is still wretched; and will not go to bed at nights. Don says he looks terrible.

December 26. My encounter with the Bill Bopp situation, and the subsequent quarrel with Don on the way to the party next evening, the 22nd, are not things I want to dwell on yet. Maybe all will work out for the best—but I don't know that, and I don't even want to think it. When I suffer, I suffer as stupidly as an animal. It altogether stops me working. I am ashamed of such weakness.... Well, that's enough of that. The only thing worth recording is the (not at the time, though) farce of our losing our way, simply because I was so rattled, on the new freeway over Sepulveda and having to go right back to Sunset and do it all over again.

Christmas (which I seem to hate more every year) was placid and almost joyous by comparison. The last two days have been cold but very beautiful. Don and I lay on the beach and talked affectionately. I think he would really love it if he could discuss *everything* with me. But, alas, I am neither the Buddha nor completely senile. I have my limits. I *cannot* help minding. When I finally stop minding I also stop caring. I don't give a shit.

Don forgot his mantram. But today we went to Swami's birthday party and Swami wrote it down for him. Don hates to destroy the paper it was written on, but Swami told him to.

A Dr. Jim Lester called me. "Mr. Isherwood, I heard *The Ascent of F6* on the radio. It interested me very much—perhaps not quite for the usual reasons. You see, I am a psychologist who is going along on the Everest expedition to observe the effect of hardship and tension on the climbers at high altitudes"!![1]

I am slowly getting started up again. Both on the novel and Ramakrishna.

Swami looked absolutely radiant. He told us that his best birthday present had been "a visit from Maharaj." He had woken at five this morning, gone to the bathroom, gone back to bed and had an (apparently) long visitation dream of Maharaj, between then and seven o'clock. He couldn't say where the encounter had taken place, here or in India. He had been dressing Maharaj. The wearing cloth was crumpled. He was impressed by the beauty of Maharaj's skin; it was golden and shining.

I never knew before today that Swami suffers from feelings of

---

[1] Lester took part in the National Geographic Society ascent of Everest in 1963.

sickness quite often after initiating people. "But," he told us, referring [to] the last initiation (Don's) on the 18th, "I didn't feel anything bad that day; they must have been all good people."

On the 21st, Don's car was stolen. He had left it outside a restaurant in Hollywood with the key in it. The police say there's an eighty-five percent chance of getting it back. But no news yet.

Don (last night): "I love Dobbin's muzzle when it isn't a crossword muzzle."

December 29. Don's car was found yesterday. Last night, after the Stravinskys had had supper with us and left, we drove downtown to pick it up (as we hoped) and found it all smashed up in a garage. Someone must have had his head banged forward right through the windshield. The lid of the trunk is sprung so Don couldn't get the framed drawing out of it which he has already sold, but he could feel that the glass was broken. The garage people said maybe it will cost 250 to 300 dollars to repair.

The Stravinskys and Bob arrived full of flu shots and in the mood to get drunk. They did, especially Igor. During supper, he kept rubbing his hand first on the white coral on the table, then on Mirandi [Levy]'s skin, as if for contrast. After supper, he fell down but did not seem to have hurt himself.

It was the Stravinskys who asked if they could bring Mirandi with them. She is terribly worried, poor thing, because she has a tumor, her second, which can't be removed till next week. The first one was nonmalignant, however. Oh dear, she is so vulgar though and really such a bore.

Don drew King Vidor yesterday and Dr. Myron Prinzmetal, the heart specialist.[1] King wanted to be made handsome, because he's a vain old Hungarian lady-killer. Prinzmetal wanted to be made ugly because he's a Jew. Prinzmetal said to Don, "I'm the most important person you ever drew."

Arlene [Drummond], when she came in to clean yesterday and heard about the recovery of the car, said to me, "You worked that through your Masters, didn't you?" She is wild for spiritualism, theosophy, and, no doubt though she doesn't admit it, voodoo.

Saw Gerald, yesterday afternoon, just back from Christmas with the Luces in Arizona. He talked exclusively about cancer, which is really a comedown after flying saucers and lysergic acid.[2] The

[1] Prinzmetal (1908–1987) was Connie Wald's new husband. In 1959 he helped identify a form of angina pectoris, "Prinzmetal's angina."
[2] The Luces had been experimenting with LSD.

day before yesterday, Frank Wiley brought his much-described stepmother Alexandra[1] to visit us; they stayed on and on, drinking throughout the afternoon. I was terribly disappointed in Alexandra, so was Don. The usual pretty-pug spoiled American girl face. She is awkwardly tall and the legs are too thin and she has no style whatever. Oh, *los ricos*![2] I felt about her as I feel about so many Americans with money who are neither Jews nor Negroes nor any kind of a minority, that they are the real White Trash. Trash in the sense of worthless; they hardly exist. The minority members are so much more alive, however hateful or charming they may be is beside the point, than they are. Alexandra had, as she put it, "stolen" one of the drawings Don did of Frank. It never occurred to her to ask to pay for it. So I had to call Frank yesterday and be rather icy, reminding him that Don is a professional. (Alexandra seemed to think, anyway, that this drawing was just a rough worthless sketch for a finished portrait!) Frank went off with our book on Miró the other day, and he hasn't returned that, either. And now I have to tell him that his new novel is a frost, at least the way it is now.

December 31. Goodbye to this frightening and tragic year. Not that I ought to complain personally. My health has been good. I have published a successful novel. And inherited all this money from M. Also, I have got this idea for a new novel which will keep me busy a long while.

But oh, the cancer creeping all around! And the rockets rattling. And always the feeling: there's worse ahead.

A bad year with Don. And yet, despite all omens, I still believe we may get through this phase to some new kind of happiness together.

At least one marvel has been achieved. His initiation. Yesterday he said: "Today I made so much japam I'm slitty-eyed."

I wish my reader a *significant* 1963.

1963

January 3. Don left this morning by jet for Phoenix, where he's to draw Mrs. Luce, Mrs. Wright[3] (whom Gerald goes on calling *Mrs. Wrong* so persistently that the joke seems ugly and senile) and in

---

[1] Not her real name.
[2] The rich!
[3] Olgivanna Wright (1898–1985), widow of Frank Lloyd Wright.

general prepare Phoenix to be aware of his show. He hated flying, poor pet, and went off very subdued. I suspect some complication with Bill [Bopp]. Don's sudden revelation about the Bowles experience in Tangier is also a bit mystifying to me.[1] He was so very anxious I shouldn't tell anyone and said he even regretted having told me. But this I find impressive, on the whole, and a good attitude to take. "Hide it as you would hide the news of your mother's unchastity."

An Indian astrologer told Swami a few days ago, "You saw your guru in a dream; soon you'll be seeing him in person." He added that this does not mean Swami is going to die. Of course, Swami *has* seen Maharaj in person already, in a vision in the shrine.

Threatening phone calls to the girls in Santa Barbara: "You bitch—tell your Swami to get out of this country in twenty-four hours or we'll burn the temple to cinders!" They have called in the police, who take the matter quite seriously and have even been patrolling the area by plane. Swami, telling this story, said, "You beech—"!

On New Year's Day, we saw *Lawrence of Arabia*, which is one of the most marvellous films I have ever seen. Both of us kept bursting into tears at the sheer visual daring of it. Gerald and Michael came to supper in the evening. Gerald, despite the senile "Mrs. Wrong" joke, was anything but senile. Have seldom seen him with such vitality. He says he has lived under three cultures: the Christian anthropomorphism, with sin and hell; the Freudian revolution, which "put psychology back sixty years"; the new ecological culture, which understands that knowledge does not depend merely on taking things apart, because the whole is greater than the sum of the parts—this culture, he says, goes beyond tragedy to meta-comedy.

New Year's Eve at Glenn Ford's: a $500,000 house in which there is absolutely no privacy. The master bedroom is so big that you would feel you were on a raft in the midst of the Pacific. The pool room is so small that the players can't take proper shots. The imaginary holiday warmth of people you will see only once in your life. But we got to the Stravinskys' in time for the midnight toast, so I hope the New Year started propitiously.

Don, before leaving this morning: "I ought to say to Henry Luce, *Time* must have a stop!"

A fascinating question (which only occurs to me since he has

---

[1] When they took hashish with Paul Bowles in October 1955; see Introduction p. xx and *D.1.*

left): Is he conscious of any relation between the Tangier experience and his decision, seven years later, to get initiated by Swami?

January 8. Don got back yesterday disgusted by the small-town pretensions of Phoenix. Mrs. Wright had behaved like an empress, Mrs. Luce like a duchess. Still and all, he brought back some very good drawings, plus the feeling that, by and large, Dub is necessary. Our reunion was very happy.

It is nothing against him to say that, while he was away, I had a sort of Indian summer of fun. Unfortunately, this also involved far too much drinking. The weather became suddenly perfect—tonight we are told to prepare for small showers; on Sunday, which I spent mostly at Jerry Lawrence's, we all stood watching the sunset with a kind of awe—it was like the beginning of a new and golden age. As for that midnight swimming escapade on State Beach,[1] it was only curious because it paralleled the scene in my new novel. Otherwise it belongs to the how-silly-can-you-get department and we were certainly lucky the cops didn't come.

Two letters. One in the would-be grand manner, from Herbert Samuel Crocker, 282 Camino al Lago, Atherton, California:

> You may recall that early last year you were kind enough to inscribe a book I sent down to you and later send me a card. In reviewing my correspondence sent and received in the last year (which I find, in several ways, a telling guide to my progress and attitudes), I again find your card. As I may have said at the time, this was the first time I had written an author personally unknown to me to ask this favor. I shall keep your card and remember the considerate attitude it showed.

Also, a near-hysterical letter from Phil Griggs, taking me to task for writing that the relation between Rakhal[2] and Ramakrishna was "even closer" than that between Ramakrishna and Naren.[3] He tells me this is going to be "misleading to millions of future readers," and adds, "Now I think you surely know that I have no wish to dim in the least the glorious status of Maharaj—I am myself his grandchild, and my head is *forever at his feet*. But I have it from many swamis of the Ramakrishna Order (some of the oldest) that

[1] A separate episode; Lawrence lived on Malibu Beach.
[2] Brahmananda, also called Maharaj.
[3] Vivekananda, also called Swamiji.

Swamiji, far from being just 'a chief disciple of Sri Ramakrishna' as I have seen it put, was the darling of his heart."

Ashokananda undoubtedly put him up to this. Phil asked me not to tell Swami and I won't. But I wrote him a come-off-it-Mary letter I dearly hope he will show to Ashokananda.

Ted and Vince are definitely splitting up. Ted is crazy again. He suddenly showed up this afternoon, carrying a kitten he has named Angel. Don wasn't there. Ted said Don is afraid to see him at present—he can tell. We were watching the sunset and he suddenly said, "Is this the twilight zone?" and pretended to throttle me. But I just laughed. Ted says he left his job today. He is going to be a dancer and/or an artist like Don. I got rid of him as soon as I could.

January 12. Terrific wind. The plants on the deck taking the usual beating. All the paper in the house curling up at the ends. The cattails by the front door starting to seed and blowing about the living room. Gavin home from Mexico, coming to see us this evening; which is very good. And very good that I have done another fairly extensive day's work, on Ramakrishna and the novelette (which *really* begins to excite me now).

Evelyn Hooker's trial began today, but only with legal formalities. Her attorney appealed to the judge to dismiss her case because of lack of evidence, but she won't know anything until February 1.[1]

Ronnie Knox came in to see me the day before yesterday about this story he wrote. But his own story is much more interesting. His real name is Raoul Landry Junior; his father is French, an atomic chemist (I'm sure that's not the right term) who puts some kind of coating on rockets. His mother left Landry for Knox who is a snappy dresser and who convinced her, quite wrongly, that he was going to make them all rich. Knox is very possessive, and insisted on Raoul calling himself Ronnie and taking his name. (It is significant that Ronnie hasn't switched back to his real name, and that he still keeps slipping up and calling Knox "my father" instead of "stepfather.") Knox decided, when Ronnie and his sister (a year and a half older) were in their teens, that the sister should become a movie star and that Ronnie should be a star athlete. Ronnie succeeded, and ended up in pro football. The sister didn't succeed, though she got a contract with Howard Hughes. Even after high school, Ronnie told his stepfather that he didn't want

[1] She was accused with four others of conspiracy to obtain a criminal abortion; eventually, the case against her was dropped. See Glossary.

to play football in college, but this was ignored. The other day, he went to see the doctor who attended him when he got a concussion playing football in high school. The doctor told him, "If you had been my boy, I'd never have let you play football again, with the injury you had; but it wasn't my responsibility—I told your stepfather all about it and he made the decision. And I guess I was wrong, because here you are. You survived it." This is one of the things Ronnie can't forgive. In due course, he threw up his football career and broke with his stepfather and mother (who said, "Ronnie, you're so *cold*"). The sister broke with them later. Then Ronnie took up with Renate [Druks], and the sister married a cripple. (*Just to show them*, I can't help feeling!) Now Ronnie has very little money, except that he sometimes gets jobs in T.V. He is going back to school, UCLA, to get his degree and be able to teach.

Phil Griggs has replied to my letter; much more calmly.

January 15. Cold but beautiful weather, and every day seems to be getting a little warmer. Jo and Ben are back from Florida—so Ben can start work on a T.V. story. We are going to have supper with them tonight; so snugness is reestablished in the Canyon.

Today I finished that wearisome chapter on the direct disciples of Ramakrishna; twenty-four pages! And I have snapped into my novel again. I should get the first draft finished sometime in February.

Things pretty good with Don. Bill Bopp is somewhere in the background. I don't ask about this. He will tell me when he's ready to. Last night I dreamed he read this diary. I hope he won't, because it would upset him.

Gerald and Michael came to supper last night. Afterwards, Michael showed his photos of Italy and France. God, what a bore! He is the most conventional kind of photographer. View over Florence from Fiesole; view of St. Peter's from our hotel, etc. etc. The only amusing shots were of Gerald in a gondola in Venice, wearing dark glasses and looking madly incognito. The worst of the evening was that Gerald got no opportunity to talk. It is so tiresome that you always have to invite Michael too. And he simply enrages Don.

Al Spar has sent us a bill for legal and accounting services—$720.17. He really *is* insane. And now I have all the hateful embarrassment of confronting him and making a fuss.

The day after tomorrow, I shall be reading the Katha Upanishad once more at Swami's prebreakfast puja. That makes it nearly a year (except that last year it fell on the 28th) since I returned to Los Angeles from New York. I haven't been away properly since then. And it hardly seems to have been a moment. During this time, I have written *only five chapters* of the Ramakrishna biography! What excuse have I to demand that my life shall be prolonged—while Paul Kennedy dies at twenty-eight—if that's the most use I can make of it? Still, objects the Defending Angel, he *did* write 126 pages of draft on his new novelette.... True—but by this time I ought to have finished a *complete* first draft of it, *and* finished the biography.

January 21 [Monday]. A slight cold has transformed me into a senile sniffling invalid. Hope I shall snap out of it as quickly as I started it.

Swami has been sick—nothing serious, it seems—but it meant that I had to give a talk on Vivekananda yesterday at the temple. I think it was one of the best I've ever given anywhere on anything; at least, that was my illusion. I shocked them in the right way. My main thesis: "A saint must always be judged guilty until he is proved innocent." Hence the immense importance of Vivekananda: he was by far the severest test to which Ramakrishna was ever subjected. In order to believe in Ramakrishna's faith, we must first believe in Vivekananda's doubt. Etc., etc.

Two worlds department: While I write this, a very attractive young guy of the Will Rogers type[1] is pruning the eucalyptus tree outside my window—for Elsa. He keeps seeming about to fall; though only to the length of his safety rope.

Much as I often don't like it, the Bill [Bopp] situation certainly does seem to make Don behave better around the house. At the same time, Don was quite displeased by my evening with Bill Brown on Thursday last! I am waiting to see if he'll make a fuss when I suggest framing and hanging the drawing Bill gave me.

Later. I feel strongly moved to add something to what I wrote about Don; it sounds so cold and unkind. My unkindness is a sort of senility, really; a lack of juice. Because I do love him dearly, which means that I sometimes hate him also, but almost never, in all my memory of our ups and downs, have I seriously planned to get rid of him. He is terribly complicated, nervous, talented, affectionate, frank yet quite capable of telling the most drastic lies—and, let me face it, I like him that way. I think I want someone to look after

---

[1] Rogers—homespun comedian and movie cowboy (1879-1935)—was half-Cherokee Indian, rangy, thatch-haired, easy-going, and clean-cut.

me and humor me and wait on me hand and foot—but I don't;
or I should get one—it isn't difficult. So, I really should make it a
rule never to complain in that cold elderly way because he won't
do just exactly what I want. I'm allowed to hate him, yes—that's
human.

There are only two prayers that I keep wanting to pray, and do
pray when I think of them. I pray that Ramakrishna may come
into Don's life, more and more; so that, when he starts losing his
looks and getting older, he will have something that really supports
him. And then I pray—and this I ought to do increasingly, because
it is horribly important—that I may be helped to leave the body
when the time comes, to let go, and none of that Laughton horror.

It was a beautiful sunset, but now it's cold as hell. Although I
feel bunged up, I have worked on the Ramakrishna book and my
novel; so let us rejoice. Now I am going down to Ted's [Grill]
to eat alone. Don is with his folks. When he gets back here, we
are supposed to spend the evening addressing envelopes for the
announcements of his Santa Barbara show.

Oh, and speaking of Ted's reminds me that Ted [Bachardy] has
been passing bad checks. They caught up with him and threatened
to prosecute, so Glade had to make the money good.

January 29. Got back yesterday afternoon after a night spent at
Wright Ludington's at Santa Barbara. A horror dinner party, with
the Warshaws and the Austrian woman called Ala, and the fat
woman she lives with.[1] But it is not good enough to moan about
how boring they were, or how I saw that Fran Warshaw is really
no different from Peggy Kiskadden. No—the truth is, I must not
go on drinking like this: I fell down in the garden on our way
to the absurd guest bedroom which is about a hundred yards long
and two yards wide. I had already strained my side somehow, I
guess by a fall I don't even remember. No excessive self-castigation
about this, just a resolve: I will stop smoking when I drink. That
alone, I know, cuts down the toxic effects amazingly. Also, it will
stop this cough.

Don just phoned that he will be back tonight but not until eight
o'clock, and as we are eating with Jo and Ben there'll be some
moaning at the bar about this, I fear. He stayed on in Santa Barbara
for the opening of his show today.

Gavin feels hopeful about this man Oderberg[2] he went to today

[1] Alice Story and Margaret Mallory; see Glossary under Mallory.
[2] Dr. Phillip Oderberg, a psychotherapist practising in Santa Monica.

about his attacks of panic and tears. Meanwhile he relies on a new drug which takes them away whenever they start to come on. Last night, we had supper together and he told me that he heard from some New York girl that Gore has started feuding with Bobby Kennedy.[1] He wrote an article about Bobby which *Esquire*(?) was supposed to publish, and then along came Salinger[2] and told Gore that he must warn him, if the article was published, his income tax returns would be drastically re-probed for the past fifteen years! I don't know how much of this I believe. In any case, I'm sure you can't go back *fifteen* years, because of the statute of limitations. The girl also said that Gore, in disgust, is retiring to write novels in Rome.

Robert Frost died today.

Apparently, the Brahmananda puja the day before yesterday was a rare occasion. Prema says that Swami seemed to be filled with power. "He kept blessing people," Prema told me, "and you felt he could really do it!"

February 1. Light rain, yesterday and today. Put the plants out on the deck. Everything got a good drink. My novel is racing along, although this last part is in some ways the most difficult. I have written nearly ninety-five pages. Will probably finish this draft with 110. Oh, the joy of having a project! However much you may say it's not really important—and even believe that it obviously isn't, sub specie aeternitatis—who cares? It's marvellous—just the joy of invention. It's the joy of finding yourself not yet impotent.

Last night, the Stravinskys took us, with Bob Craft, to Jerry Lewis's restaurant.[3] To find a good restaurant on the Sunset Strip is as much of an achievement as to find a good hamburger joint. This place is furnished most ornately with hangings of blackish-plum color and dangling baroque cherubs. The manager actually kissed Igor's hand, and of course every other word was Maestro. We drank champagne. Not smoking turns drinking for me into a real pleasure. I must never do it any more. Igor talked about having *schwarze Gedanken*,[4] but admitted that they could be taken away by Librium(?). (Question: Should one do this?) When he was composing in the twenties, he drank wine from southeast Spain.

[1] Robert Kennedy (1925–1968), the president's younger brother, was then U.S. attorney general.
[2] Pierre Salinger (1925–2004), then President Kennedy's press secretary.
[3] Owned by and named after the comedian (b. 1926).
[4] Black thoughts.

Now he says two double scotches are his limit. There was some undercurrent disagreement between Vera and Bob about the next volume of Stravinsky-Craft conversations, because they are being held up to include memoirs of the Russian visit last year, and Vera claims that Bob is misrepresenting what Igor says and feels about it.... Oh yes, I know what the Jerry Lewis restaurant reminds me of, the paintings of Francis Bacon, both have approximately the same background color. It gives an atmosphere of elegant horror; almost unthinkably sophisticated for this town. Alas, we hear the restaurant is already folding.

When I was driving to Vedanta Place the day before yesterday, I thought I'd turn off Sepulveda on Mulholland and drive through the hills down on to the Cahuenga Pass. This would be fun by daylight but it was getting dark, and I ended up turned all around and having to go down Laurel Canyon into the valley and come laboriously through to the pass along Ventura Boulevard!

Swami has a new project which excites him: to get some young swamis from India, train them at Trabuco and then send them as assistants to the various American centers. I see definitely that he does not want to produce U.S. swamis to head U.S. centers. He seems to feel that Americans wouldn't take them seriously.

Don is going through another desperate struggle to paint. But, although he is under such strain, he couldn't be sweeter.

February 5. After the rain, we've had glorious weather, beginning with a baby heatwave the day before yesterday; the temperature going up to nearly eighty by nine in the morning.

On the 1st, we went to a party at Glenn Ford's, at which the clairvoyant, Peter Hurkos, gave a demonstration. He wasn't specially good but we felt he was absolutely on the level. An awkward bulky Dutchman who sweated profusely. Glenn mismanaged the party by inviting 150 people and making an asinine speech describing himself as Hurkos's "disciple." There were speculations about Linda Christian and Hope—would they fight? Neither Linda nor Hope was about to, of course—they couldn't care less: Glenn should be so lucky. (A day or two later, Don drew Linda, who told him Glenn had asked her to marry him. Don thinks she will, if he's serious.)[1]

On the 2nd, I had supper with Frank Wiley, who has now

---

[1] Linda Christian (b. 1924), half-Dutch Mexican-born actress, known for her turbulent love life, had been married to actors Tyrone Power and (briefly) Edmund Purdom; she did not marry Ford.

departed for the Orient on his carrier. My God, he is stingy! Again he let me pay. And I had to go around to the apartment next day to collect the Miró book he borrowed.

On the 3rd, I finished the first complete draft of the novelette. I have made notes about this elsewhere. But I do think I've got something there. Don, as always, was very helpful. He finished it today, and talking to him showed me a lot of things that are wrong. I said I would like to get started on the rewrite immediately, whereupon he said then why didn't we give up our trip to New York. Of course I can't help thinking to myself that this is at least partly because he doesn't want to leave Bill. Don *said* that he doesn't want to interrupt his efforts to paint, and I'm sure that's true too. Well, I suppose it means we won't go. I was actually rather hating the idea of going, stay-at-home that I am—and then it will be so cold there. But I have been looking forward to seeing Wystan—on whom *Time* is doing a cover article. Their Bob Jennings interviewed me for it yesterday, and I told him I thought *Time*'s corporate image is that of a neurotic woman so full of venom that she's incapable of praising anyone or anything even when she sincerely wants to; the bitchery just slips out. However, I did manage to promote the idea that they should use one of Don's drawings of Wystan. It can't be the cover unfortunately; Bouché is doing that.[1]

February 6. I forgot to mention that, a couple of days ago, Dr. Allen taped up my right side, saying that I may have a cracked rib. He was very relaxed about this, didn't seem to want it x-rayed or to think that it mattered much if it's cracked or not. Maybe today it is a bit less painful; but the tape makes me itch and feel dirty and unappetizing.

This morning, Don made it quite clear that he *does* want to go to New York, so I suppose we'll go. I feel unwilling because of the flying and because this means I can't start the new draft of my novel until nearly the beginning of March. However, I must anyhow first read through my diaries for possible bits I can use. I have started doing this. Oh my God, it is so depressing! The sheer squalor of my unhappiness.

Am getting into a flap about the Don-Bill situation. Last night I had two, if not three, dreams about them. This is so utterly idiotic.

---

[1] Bouché painted the portrait, but *Time* never ran the article because, as Edward Mendelson records, "the managing editor objected to honoring a homosexual" (*Later Auden*, p. 452).

And meanwhile Don—no doubt largely because of this—remains quite unusually sweet and affectionate. I ought to be grateful, really. Oh—idiocy!

February 9 [Saturday]. Every night this week, except Tuesday, Don has been out with Bill. Today it is pouring down rain and they have gone to Santa Barbara to see about his exhibition. Bill is living alone now in his own apartment, and Don took him some of our plates; admittedly, not ones we use any more. I am wildly miserable, but only in spurts. What I am miserable about is the feeling that Don is gradually slipping away from me. To go to New York with him at this time, especially in order to "celebrate" our anniversary, seems grimly farcical. I don't feel I have the heart for it. Also, to make matters worse, I have been reading through all these diaries and feel absolutely toxic with their unhappiness.

I have written this down, but with misgivings. Maybe I should stop doing this. Wystan in *The Dyer's Hand* says, "Most of us have known shameful moments when we blubbered, beat the wall with our fists, cursed the power which made us and the world, and wished that we were dead or that someone else was. But at such times, the *I* of the sufferer should have the tact and decency to look the other way."[1]

Actually, under the misery, I feel almost glad, that the screws are being put on me like this. It is the only way I can ever hope to get through to "the ending of sorrow."[2] One thing is vividly clear to me: there is no question, here, of finding any kind of a solution to the situation on the personal level. I can only find a solution through prayer and japam. What will actually *happen*, as between Don, Bill and me—that's really quite beside the point.

So, courage, Dobbin.

Jim Charlton is being a help. Quite unconsciously, because he knows nothing of any of this. I saw him again last night.

February 16. Poor old Jo just called. She is terribly worried because she has had dizzy fits and now the doctor says her jaw is badly infected. She is sick. She is dropping behind. Oh—I do feel for her so.

Not that I am sick. My ribs seem better. I have stripped off the plaster and I'm just off to the gym for a mild workout. We are not

---

[1] "Hic et Ille" in "The Well of Narcissus."
[2] "The Yoga of Renunciation" (XVII) in the Bhagavad Gita; in their translation, Prabhavananda and Isherwood have "The end of sorrow."

going to New York, thank God. I would have hated the cold, not to mention the other rat-race aspects of the trip.

Since our tenth anniversary celebration the day before yesterday (Don cooked meatloaf and we showed several of our old home movies for the first time in years) I feel much much better about everything. Not only because Don says he would never under any circumstances live with Bill but because I realize I was quite wrong in thinking that he is becoming alienated from our life together. Indeed, Bill is quite probably the best thing that could possibly have happened. Much more about all this later.

On the 11th, I started the second draft of my novelette. This is going ahead quite briskly, but oh, the work ahead!

A nice placid evening at Dean Campbell's yesterday. He is a terribly gracious liver, but sweet and naturally generous.

Here's a joke I made which I want to record because it's topical and I wonder how much sense it will make in, say, five years. Someone told me that they are making a film called *The Fall of the Roman Empire* and added, "They've left out *The Decline and*." I said: *The Decline* is directed by Antonioni; *And* is directed by Buñuel.

February 24. Jo's jaw is better, but she has headaches every night. Don is getting over a cold. He still has a cough. Most of this last week, he has been around; so I asked him, "Aren't you seeing Bill nowadays?" He said, "I don't want to talk about it." "If you and Bill have split up," I said, "I'm sorry—because, after feeling all sorts of different ways about this, I now realize that it's probably the best thing that could have happened." This pleased him, I could tell. And now today they have gone off together to the beach. Sooner or later I suppose I shall find out just what the score is, was.

In a day or two, he will be driving up to Stanford to get his show opened. And then he'll be going to Phoenix,[1] and to New York. I'm just as glad. We are quite harmonious, by and large, but I need a little rest from him. Long enough, at least, so that I'll miss him.

To Jerry Lawrence's today. The last two days have been glorious, and I thought I'd have a beach day of agreeable youthful atmosphere. But tiresome Jerry had invited old Louis Untermeyer,[2] who's a sententious bore, and a boring couple (John Weaver and

[1] The Santa Barbara show moved on to Stanford University, opening March 1; the Phoenix Museum of Art opening was planned for April 16.
[2] American writer and anthologist (1885–1977), he published over a hundred books, including collections of Frost (a close friend), Eliot and Pound. In 1961–1962, he was poetry consultant to the Library of Congress.

his wife;[1] maybe not boring really but silenced by Untermeyer). And then Mrs. Untermeyer[2] had a raging toothache, which old Untermeyer blandly disregarded, leaving Weaver to organize an emergency dental visit.

In the can, Jerry has arranged books with "suitable" titles, such as *You Can't Take It With You*,[3] and two copies of Charles Lindbergh's book, to spell out *We We*.[4] How I hate that picture of the line of sailors peeing off the dock into the sea![5]

At Vedanta Place last Wednesday, Swami retold me the story of how he met Brahmananda, with the various stages of his involvement. There was one episode which, he told me, he has never told anyone else: one time he went to see Brahmananda at someone's house (Balaram's?[6]) in the days before he became a monk and was still a student in his late teens, he suddenly felt an overwhelming desire to go and sit on Brahmananda's lap. This made him ashamed, so he ran out of the room without speaking to Brahmananda.

Since the 11th, I have been working steadily on the second draft of the novelette. I'm afraid I may be overwriting it a bit, but it certainly has much more meat this time and is expanding without my having to pad it. The only snag is, I don't see how I can possibly finish it before I have to go up to Berkeley, and that will be a very serious interruption.

February 28. The day before yesterday, Don went up to Santa Barbara in his car. He was planning to stop the night there, with William Dole,[7] and then go on to Stanford yesterday. Haven't heard from him yet.

Meanwhile the weather is heaven and I am quite happy, especially as I am having "great openings" on the novelette. Probably for this same reason I feel a disinclination to write anything here.

No news about Berkeley yet. Have just identified a quotation Gerald wanted from Pope:

---

[1] Weaver (b. 1912), a novelist and journalist, contributed to *Harper's*, *Atlantic Monthly*, and *The Saturday Evening Post*; his story "Holiday Affair" became a 1946 movie. His wife, Harriet, assisted Untermeyer with an anthology *Stars to Steer By* (1938); they were lifelong friends.
[2] Bryna Ivans Untermeyer, Untermeyer's fifth wife, from 1948 until his death.
[3] Probably Moss Hart and George Kaufman's 1936 play, published in 1937.
[4] I.e., his 1927 bestseller, *We*, about his life and his transatlantic flight.
[5] A framed photograph.
[6] Balaram Bose, wealthy householder disciple of Ramakrishna; Ramakrishna and his followers often visited his house in Calcutta.
[7] American collage artist; he exhibited at the Rex Evans Gallery. Bachardy stopped only briefly, to see Dole's work, which he admired.

Tired of the scene Parterres and Fountains yield,
He finds at last he better likes a Field.

It's from Epistle IV of the Moral Essays, to the Earl of Burlington,
on The Use of Riches. Why Gerald wanted it, I don't know. I
neglected to ask him; and that is precisely one of my defects which
I can do something about: I'm not nearly curious enough. I didn't
ask John Zeigel (whom I saw the same night Don left, at Pasadena,
where he's living now) nearly enough about his present feelings
toward Ed Halsey. And this was inexcusable, because he told me
a fascinating thing, that he has willed Ed to appear; and that Ed
has appeared, two or three different times. Although he willed it,
John doesn't feel that this was any kind of autohypnosis. Because
Ed appeared within light, and John, going counter to his will, felt
afraid of the light and didn't try to penetrate it. If he had done so,
he feels he could have seen Ed more clearly; but he was afraid of
being swallowed up by it. (Or is this an interpolation of my own?)
His chief impression, however, was that Ed is very happy. (Gerald
says that there seems very little evidence, according to the best
belief of the psychical research people, for unhappiness after death.
No signs of the Catholic purgatory. But, says Gerald gleefully, no
doubt the Catholics themselves suffer in it. They don't realize it's
all made of cardboard and mirrors.)

An extraordinary tale told me by Prema. Two members of the
Vedanta congregation were both drunkards. The other night, the
wife called in hysterics that her husband had come home drunk
and she had strangled him in self-defense with a judo hold, after he
had attacked her. Prema went around and the police were there.
The wife wept and took Prema into the bedroom and confessed all
over again, even offering to demonstrate the judo hold. The police
certainly know all this; but they haven't even taken her down to
the station for questioning and it doesn't seem there will be any
prosecution! Of course, they may be convinced she is lying....
Prema was quite thrilled, and subtly pleased to have had such an
adventure. He talked to the police photographer who was taking
pictures of the corpse—it lay right there on the floor. "You must
see some ghastly sights as a police photographer," Prema said; and
the photographer answered, "Yes, and I bet you see a lot of things
as a church secretary!"

Last night, Swami warned us strongly against making japam
while you are feeling any kind of resentment toward anyone. He
even seems to think it might harm that person, after the manner
of black magic. I shall have to watch this—indeed, I have been

getting horrifyingly careless about my thoughts during japam. This morning, instead of trying to stick to Ramakrishna, I thought all the time about Swami—sitting up in his chair, meditating in the shrine, etc. This worked quite well.

March 6. Splendid weather. Mood ditto. This is one of the famous-last-words periods when it seems as if Don and I had it made for the rest of our mutual life. (If we really had, it would be two other guys, and a bore.) He said yesterday, "Dub used to be my jailer, now he's Kitty's convict." The Henry Kraft[1] situation, into which I never probe, seems to make him permanently happy and at the same time much fonder of me—in all ways. Well, good while it lasts!

Novelette progresses steadily, though not fast enough. Berkeley is fixed. My rib seems all right, but now I have a curious condition like varicose veins in my calf.

Forgot to mention that someone (a woman) gave Swami a book about the trials of Oscar Wilde. Pagli asked him what the book was. He said, "You see, in those days, people were sent to prison for homosexuality," and then he added, "Poor man!" And to Prema he said, "All lust is the same."

March 20. I am making another entry here, after this long long lapse, out of a feeling of duty. I don't really want to. Partly on account of Don. This is a strange period, and I feel I don't want to make any statement about it until it is over. Seriously, it is possible we might have parted by the summer. And yet our frankness with each other might equally well lead to a much better relationship. It is very good, in any case, that I am going away to San Francisco so soon, in about three weeks.

Wystan has been staying with us. He arrived on the 16th, left today to continue his lecture tour. Of course he was an awful nuisance and stank up the place with smoke and had us drinking pints. But he is marvellous and strong. I don't think I could possibly undertake a tour like his.

I keep on at the novel. Slowly but fairly surely. Only external accidents will prevent me from finishing this draft, at least. What I have written so far—thirty-four pages—I quite like.

March 23 [Saturday]. Ben Masselink has had a *second* book accepted! Something to do with Tahiti,[2] which I haven't read yet.

---

[1] Not his real name.
[2] *The Danger Islands.*

He much admires Jim's half-finished stories. Talking about them on the phone today, he said how Jim simply regards them as a means of making money, in order that he can go back to Japan. Deploring this, Ben said, "That's the only reason you're writing, for sort of a source of love."

We saw Dorothy Tutin yesterday. Jerry Lawrence brought her by to see the house; then we went to his house for lunch. Don really loathes Jerry and perhaps this colored his attitude to Tutin; he says he dislikes her. She is false, and looks "like a stale bun," and her accent is wrong, "She is no more U than they are." Oh yes, of course she is false, poor wretched little thing. I felt sorry for her, though, with her alcoholic father and her leaky barge on the Thames. She longed to stay here, but she has to go back to England tomorrow—she has been touring in *The Hollow Crown*—and get ready to play in *The Beggar's Opera*.

Since the 20th, we have both given up drinking; we plan to stay on the wagon until Don gets to Phoenix and I to San Francisco. Or approximately. It is really much better. You are bored more, but the pain stops when you leave the bores; you don't hate them next morning for causing your hangover. And, with me, it also automatically means quitting smoking—I still have this strange thing of only wanting to smoke when I drink. And that is even more valuable.

I want to get this (eighteenth) chapter of the Ramakrishna book done before I leave; and reach page 50 at least of the novelette. Not at all impossible.

Last Wednesday, Swami remarked quite casually that he is seventy. It came as a big shock to me; somehow, I'd been playing around with the idea of his being only 68–69. He looks marvellous, however. I can quite clearly remember a similar shock, at the end of the thirties when I realized that M. was seventy. And look how much longer *she* lived! May it be a good omen!

Yesterday, while Dr. Stevens the dentist was drilling, I experimented with the music you can listen to through headphones. There is also a kind of jarring noise which masks the noise of the drill and has been found to be partially anesthetic in its effect. You can control the volume of the music and switch on the noise yourself. I found that, by bringing in the music (which was "light classical") very loud at certain moments, I could create the atmosphere of a silent-movie love scene which was so absurd in relation to the drilling that it made the drilling itself absurd. It also turned the down-looking faces of Dr. Stevens and his cute nurse into a pair of medical lovers from a television serial. This game amused me so much that I was laughing all the time—with my eyes only, of course, because my mouth was full of instruments. I tried to begin to explain how I felt to Stevens and the nurse, but realized they just weren't going to understand. Stevens is backing two of my upper jaw incisors with gold; otherwise, he says, they will grind themselves down and crack up.

Have just been out on the deck to watch the sunset. This evening, the sun is already setting almost "offstage," at the landward end of the headland. All summer, it will go down behind the mountains.

April 1 [Monday]. We went off the wagon after a week. Never mind, it was valuable while it lasted, and I think has made us both a bit more aware of the messiness of alcohol.

Don has decided not to have the show at Phoenix. He is still in a terrible state about his work, but keeps right on after it. He'll never show me anything.

This morning he left for San Francisco, planning to stay with Stanley Miron and pick up his drawings from Stanford tomorrow. I'm planning to leave for San Francisco a week from next Friday, the 12th.

Meanwhile I plug on at Ramakrishna and hope to finish the eighteenth chapter before I leave. After this will be a chapter on M.'s Gospel[1] and Ramakrishna's teachings. A chapter called "The Last Year," which I hope will cover everything till the death. And a final chapter about the doings of Vivekananda, the founding of the Mission and Math, and what (very briefly) has happened since, down to the present day.

The novelette is at page 41. I know I am off on a digression about Huxley's *After Many a Summer*, but that doesn't matter. I'll just keep writing until I write myself out of it again. My target is to reach page 50 before I leave; but this isn't so important as I mean to take the manuscript with me and work on it up there, at least enough to keep the pot simmering.

April 7. Have finished the eighteenth chapter of Ramakrishna, *ma longue et lourde tâche.*[2] As for getting to page 50 of the novelette, I'm not going to sweat for that, just see if it happens or not. Right now, I have five more pages to go.

---

[1] Mahendranath Gupta's *The Gospel of Ramakrishna*.
[2] My long and heavy task, from Alfred de Vigny, "La mort d'un loup" (The Death of a Wolf," 1843).

And five more days to go, here. Don and I are sort of quietly waiting it out, until I can leave. I don't know all that's happening to him and maybe I shan't find out until much later, if at all. Don brought back the bed to his studio, yesterday. He doesn't appear to be seeing Bill. I am miserable about all of this, but not very. I am resting from being miserable. I wish I didn't have to go away and yet I know it will be good for me. I need a thorough change of scene and spirits. I have been in this house too long.

Yesterday afternoon, we went up the coast to see Renate Druks and Ronnie Knox. Anaïs Nin and Rupert Pole were there; a pair of November-May couples, to which Don and I made a third. Rupert lectured us interestingly about the different kinds of chaparral on the hillside above the house. Sumac grows back quickest after a fire; it grows from its roots, which seldom get burned. The forestry people plant mustard after a fire to hold the hillside together. Then there's sage, and pea vine and yucca and Indian paintbrush. These plants have to do with very little water, so they cover their shoots with wax to hold in what they get for as long as possible. But this wax is highly inflammable; when there is a fire, it makes it burn all the more easily.

We took Cecil Beaton to dinner, at Sinbad's. I do like the people there. They always seem to be celebrating each other's birthdays. Last night, they brought in a huge banner congratulating the bartender, "Happy birthday, dear Okie." Cecil told us how much he likes and respects George Cukor. He feels Ivan Moffat is making a fearful mistake by remaining in England and abandoning his Hollywood career. Princess Margaret is a little bitch, he says, amusing but utterly unreliable, and Jones[1] is just an operator. Ivan is putting too much trust in British Society, Cecil says, and it will let him down.

Am reading Calder Willingham's *Eternal Fire* with real joy. What a delightful book! Also *The Waves*. Ramming my way through this. It is just as nothing to me as it was before; the greatest wasted idea in the history of literature. But I would rather be bored by Woolf than by anyone else.

More from Rupert Pole: when the chaparral is virgin and has never been burned, it grows to a height of about twenty feet. This is known as "elfin forest." But this kind of growth is very rare, and only found in the canyon bottoms. Even before the area became

---

[1] Antony Armstrong-Jones (b. 1930), English photographer and designer, married Princess Margaret, younger sister of Queen Elizabeth II, in 1960 and was created 1st Earl of Snowden in 1961. They divorced in 1978.

populated, with the resulting fire hazard, there were fires caused by lightning. There is no elfin forest to be found anywhere in the Angeles National Forest.

April 14. Arrived here (2424 Jones Street, San Francisco) in the Volkswagen on the afternoon of the 12th. The drive was quite beautiful—through the Grapevine Pass, and via Bakersfield and Fresno and Oakland—a fine day, not too warm, with the Sierras ribbed with snow beyond the farmlands. It was 403 miles door to door, and it took me from 7:45 a.m. to 3:20 p.m. The Volkswagen went like a charm at a steady seventy-five when called on, and Bill Brown's directions were so exact that I didn't make one single mistake.

This house is intimidatingly *moderne* and grand, but still and all wonderful to be in by oneself, especially during the day. At night it is unhomely and creaky; it may well be mildly haunted. I shall perhaps describe it later. But at present I am concerned with my psychological convalescence. Oh, I did so need to be alone! Now I am resolved to get on with my work, I mean my own work; and to exercise—I am hatefully fat. (I just bought the Royal Canadian Air Force book, which will be good because it doesn't require any equipment. I'm really shocked to find how out of breath I get. These hills are really a workout, after car riding in Los Angeles.) Oh yes, I am happy to be here.... As for my duties at Berkeley, Schorer, whom I talked to on the phone this morning, was vague and uneasily breezy. He obviously hasn't arranged anything, and probably won't.[1]

April 26. I have refrained almost superstitiously from writing in this book. I didn't want to break the spell of contentment. I have so liked being here—getting up before seven, making japa on the roof, doing the Air Force exercises (without missing one day, so far), shaving, fixing my daily fish cakes and coffee, and then getting to work, with walks in the town and sunbathing a few times, otherwise it has been rainy. I am keeping along with my novel and the Ramakrishna book and reading—*Eternal Fire, A Clockwork Orange*,[2] *The Waves, Salt*,[3] and an anthology of contemporary American poetry. My only failing has been drinking much too much and smoking ditto. Otherwise I should have lost

[1] Isherwood gave at least two lectures during May, "Influences" and "The Autobiography of My Books."
[2] By Anthony Burgess.
[3] By Herbert Gold.

more weight and be in better shape than I am; but just the same I manage to get up and down those hills. I like this house, mainly because the carpets are so thick and there is radiant heat under the floor when you walk barefoot and the kitchen is so big and clean and modern. The big daubs by Mason and Frank[1] do not grow upon one; and by and large the house is imaginary, a decorator's exhibit in a Modern Living show. Never mind.

Am starting to think a lot about Don, miss him, wish he'd write. But I won't pester him. Why does he seem unique, irreplaceable? Because I've trained him to be, and myself to believe that he is? Yes, partly. But saying that proves nothing; the deed is done and the feelings I feel are perfectly genuine.

I shan't be going back for another twenty days or so. I won't make any resolutions yet about all of that. We'll see. At least I have proved to myself that I can still live alone and function. In some respects, I have never felt so truly on the beam. Only I should pray more. Pray for Don, for myself, for all of us, and for Ramakrishna's help now and in the hour of death.

May 3. Don wrote a couple of days ago, saying that he is "going through an awful time.... Something is terribly wrong and not only do I not know what to do about it, I don't even know what it is that is wrong, or why. Fits of doubt and gloom keep descending. I try to fight them off but I seem to have fewer and fewer weapons." He ends, "I don't want you to worry about me. I must do this alone. I must get through by myself. And I try hard to love you instead of just needing you."

Well, of course I am terribly worried. I am even losing my confidence that this will end all right—though I wrote him a reassuring letter.

I have suggested that I shall stay on here in the city for a while, so he can have more time alone. But I can't honestly say I hope he agrees to this, because Frank Hamilton now says he is coming back on the 20th, which means I should have to ask Ben Underhill or someone to let me stay with them.

Aside from this, and a sore on the roof of my mouth, all is joy.

A long, not unboring but nevertheless sweetly peaceful day with Bill Whitman. Only, when we walked in Golden Gate Park, I was suddenly blue, overcome by memories of Vernon [Old], of

---

[1] Mason Wells, who owned the Jones Street house, and Frank Hamilton (b. 1923) his younger companion, were both painters. Wells was the brother of Jo Masselink's friend, Kady Wells, also a painter, mentioned in *D.1.*

Caskey, of Don. The past came crowding in, as if this had been some certified breeding ground for nostalgia, like the Tuileries. It was only tiresome when he asked me questions about Flaubert.[1]

I am behind with my novelette, but making progress.

Gavin Arthur's[2] story about Wilde. He had boasted that there was no subject he couldn't be witty about. Some clubmen challenged him on this suggesting the Queen. Wilde picked up a glass of wine, said, "The Queen, God bless her, is never a subject," drank and broke the glass.

May 10 [Friday]. Well, I got a bad back. It's a bit better now. Stanley Miron says it's due to a virus, not an injury. The back was the beginning of a truly great Blue Period. On Monday I talked to Don long-distance and he said hesitantly that he *would* like me to stay on up here. But he arranged to come up for his birthday. I was sad as hell and drank far too much. Gave a very successful last lecture on Wednesday, feeling like death, but drank lots more at lunch with some friends of Ben Underhill, whom we had cultivated in order [that] they should buy Jim Charlton's Japanese prints. (They didn't.) And then the news that Larry Paxton died of diabetes on Tuesday morning because some Christian Science woman persuaded him he should drop the insulin.[3]

So today the funeral. The family asked me to speak, although I didn't really know him that well. It was ghastly. The corpse transformed into a well-dressed made-up wax doll, lying in an open casket. The ex-show-biz mother drunk I think and seeming not to know what was taking place. The very sweet teenage half-brother, Ken; a big football boy but innocent like a child. He put his arm around me and held me all through the service, and kept stroking my hand. I don't think he had the very faintest suspicion that anyone might think this strange. I responded of course; and I think he was actually getting some sort of support by treating me protectively. But very few American boys would dare behave like

---

[1] The Tuileries Palace was repeatedly sacked by the Paris mob, beginning in 1792 when Louis XVI and Marie Antoinette were imprisoned there, and repeatedly restored, until the Communards burned it down in 1871 leaving a colossal ruin and the extensive gardens. In *Sentimental Education* (1869), Flaubert fictionalized the 1848 attack on the Tuileries, at which he was present.

[2] Chester Alan Arthur III (1901–1972), merchant seaman, astrologer, psychic, early gay liberationist, and grandson of the U.S. President; he wrote *Circle of Sex* (1962), cataloging sexual preferences by star sign.

[3] Paxton's mother was a Christian Scientist and raised him as one.

this. What with a backlog of hangovers and the general feeling of upset, my voice came out raw and strange and of course I couldn't help noticing that this was theatrically effective. Well, why not? I spoke poorly, though, and Donne's "Death be not proud" wasn't really such a good piece to read; when old Mr. McKenzie suggested it over at the college I thought it would be marvellous. When I'd finished speaking I was trembling all over, and Ken squeezed me tighter than ever. There was a long draggy service by an old minister who read as badly as they always seem to; and then the awful procession past the corpse. Larry's Greek friend burst out into kind of gasps of despair and fury; he struck the coffin with his fist. Larry's aunt began a kind of protest and had to be dragged away by the others. When it was over, I pointed to the coffin and said to Ken, "That's not Larry, you know," and he said, "I know it isn't." These were practically the only words we exchanged, and it is very unlikely that we shall ever meet again.

The tomato juice which Larry and I bought together when he came to this house, ten days before he died, is still in the icebox. And I have the funny photograph Ken Wagner took of the two of us, wearing derby hats.

Today, I hear that Frank Hamilton is returning on the 15th, bother him. This means moving to Ben Underhill's, then going to the motel with Don, then moving back to Ben's.

May 18. If I am going to make entries during this period I have got to be extremely factual. Otherwise this will be a sickly tale of Self.

I drove down back home on the 15th, because Don said he didn't want to come up to San Francisco after all; since we didn't have the Wells-Hamilton house.

On the 16th, they put up a new and higher (by five feet) telephone pole outside the house and this cuts right into the ocean view. First reaction to this: a sick rage on the verge of tears and a sense that life in the Canyon is nearly finished. Called the telephone people; nothing to be done. But Don wants me to go on trying; find out, for example, if the poles could be moved and rearranged.

· Yesterday, I rushed downtown to Kazanjian's and bought him a ring with an Australian sapphire, dark blue. This morning at breakfast he shed tears, said he couldn't accept it. Our relationship is impossible for him. I am too possessive. He can't face the idea of having me around another ten years or more, using up his life.

I said I absolutely agree with him. If it won't work, it must stop.

Now he has gone out. (Tonight we have a birthday party, with Bill Brown, Paul Wonner, Cecil Beaton; and Dorothy Miller, just back from San Francisco, cooking.) I cried a bit. Then drank coffee, felt a lot better, and began figuring. Don should start by getting a studio away from this place, where he can stay whenever he wants to. Also, he should go to a psychiatrist. (This is his idea.) And we must start thinking about selling this house.

Today, I have done some work on the novelette; the first in a week. I must keep hard at it from now on in.

June 2. Diary keeping at this time seems definitely counterindicated. (Though maybe I'll want to make a couple of entries in the calm of Trabuco, where I am going with Swami and Prema tonight, until the 5th.) But I will just record that things keep on. Part of Don wants to run me right off the range and wreck our home beyond repair; part wants to keep on and see how things work out. There have been moments of warmth, especially following the appearance of Henry Kraft on the scene, because this time I behaved better. And there have been relapses—such as this morning, following Elsa Laughton's coming to supper last night. Don says she is absolutely evil. She and Cecil Beaton nearly got into a fight over Isadora Duncan. But all I see is a miserable stupid half-crazy uneducated old bag, who can't help bitching.

I have now reached page 77 of the second draft of the novelette, bringing it up to the beginning of the big scene with Charlotte. I do so want to get all this work squared away before I face a complete break with Don—if there has got to be a break. But of course this is the voice of my personal convenience speaking, so he hates it.

Oh shit, I am so weary of all this!

June 6. Back from Trabuco yesterday. When I returned, Henry Kraft was at supper with Don. The scene worked out as well as could be expected, so today all is fine again. Don says he can't make his case against me stick. He asks forgiveness. I forgive (no shit) and so we go on. At least I have done a lot of work today: Ramakrishna book, plus article on Brahmananda (the latest chore wished on me), plus the novelette, plus letters, plus gym.

Now I'll rattle off some notes I made at Trabuco.

(June 3.) Walk with Prema, under overcast skies; right down to the gate, then back and down the hill in front of the monastery, along the edge of the ravine; fallen trees (from some terrific wind?) and indignant but very timid young bulls. Prema horrified when he found that Usha is to get sannyas this summer (before *him!*). Swami had told no one of this until it was passed by the Belur Math. Prema thinks this was Swami's final test of him (which I don't believe). Also that Swami knows Usha is the strongest person there; in an emergency she would take over. (This sounds as grim as preparations for the functioning of the cabinet after the president is dead, in a rocket war.) Poor Prema is so sick with hate that he feels he'd like to stay on in India and maybe not come back here. At the same time, he said he was regretting our conversation. But he couldn't stop. At present, he says, Santa Barbara is run by Prabha, and she is southern and regards Ramakrishna as a nigger who ought to be kept in the background, not featured. That's why they don't have a statue in the shrine up there, as Swami had planned. Prabha feels that the community is against the Ramakrishna cult and she points in evidence to the acts of aggression committed from time to time. The latest, some kind of a sex doll; Swami was vague. It was left on the doorstep.

Light rain falling. The frogs in the lily pond in front of Swamiji's statue. The noisy blackbirds. The dog got skunk secretion all over him and lay outside the shrine and you could smell him.

Watching Swami, bald-headed at the back, huddled before the shrine, I thought, *He's been doing this all his life, he isn't kidding.* Such a tired old sense that life is routine; and yet I could pray to Ramakrishna for devotion, for help at the hour of death, and for Don, all with really quite considerable faith. I felt that I was sort of storing up something which could become apparent later, back here, at home.

Reading Coward's *Present Indicative*, Cocteau's *The Imposter* (recommended by Larry Paxton, and with the photo of the two of us inside it and some Ruskin. My quotation of the Buxton quotation was wrong and better. It's: "... in order that every fool in Buxton could be in Bakewell in half an hour."[1] Also the Tolstoy quotation: "Why do prostitutes and madmen *all* smoke?"[2])

Prema camera happy, and collecting shots of monastic life to show them in India, snaps Krishna without permission, as he brings the food offering from the kitchen to the shrine. Krishna furious, spins right round, Prema gets him nevertheless, Krishna growls, "What do you think you're doing?" Talking about this to Swami,

[1] Ruskin was criticizing the railroad for blasting track through a scenic rocky valley between the two English towns; see "Joanna's Cave" in *Praeterita* (1899), vol. 3.
[2] In "Why Do People Intoxicate Themselves?" Tolstoy's preface to *Drunkenness* (1890) by Dr. P.S. Alexeyev.

who told me, the first time Prema appeared, Krishna said, "Who's that detective?" And Swami added, "He *is* a detective, in some ways. He's very curious." Then he described the jealousy of Gerald Heard and Kolisch when Swami took them both to the Shanti Ashrama, before the war. "I felt like I had a pair of jealous wives."

(June 5.) This morning, Swami said that he had "the intense thought that *I am the Self in all beings*, so how can one harm anyone? It's a wonderful life, if you can feel that.... I say, *Oh, Lord, don't test me!*" (He explained later that he meant he didn't want to suffer, as some people do.) While we were there, the Pope died,[1] after great sufferings. Swami said this was perhaps because it was his last life. Sometimes karmas must be destroyed in this way. He told me, "Pray for devotion *and* knowledge. Say *Not I but Thou.*" But he added that it was no use praying that God's will be done, because it will be done, anyway.

His three experiences: a vision of the Impersonal God at Puri[2]—lost outer consciousness, saw nothing but light, no images, no people worshipping, heard a voice in English saying God, God, God. Sujji Maharaj[3] grabbed his arm and told a priest to hold the other one. Later, Swami asked him, "How did you know what was happening to me?" "Because I've lived with Maharaj." Then, back in his bedroom in the old house in Hollywood, before the temple was built, he saw [Holy] Mother, "very powerful." After this he was dazed for three days. Then once in the temple, he saw Swamiji, with the Others more dimly behind him.[4] This was in the nature of a reassurance, because Ashokananda had just attacked him (and me) for the alleged slurs on Swamiji in my introduction to *Vedanta for the Western World.*

Oh yes, and that remark of Len's that the monks have a joke amongst themselves. They look at a young boy with a young girl and say, "Are they *really* happy?" You are supposed to expect the conventional answer, *No*; and then the monks answer, "You're damn right they are!" Swami couldn't get the sense of this joke at all. Actually it is profoundly British.

June 18. I think Don has finally decided to go to a psychiatrist; Gavin's.

[1] John XXIII (1881–1963), Pope since 1958.
[2] I.e., the town of pilgrimage in Orissa.
[3] Swami Nirvanananda, a direct disciple of Brahmananda (Maharaj); Treasurer of the Ramakrishna Order and later Vice President.
[4] I.e., Vivekananda with, behind him, Brahmananda, Sarada Devi, and Ramakrishna.

Gloomsville morning with drip-fog; it will probably brighten too late. Paul Latouche appears to have gone through Don's drawings and taken away the ones of himself and Rob.[1] So today for the first time Don used the padlock on the door.

Alan Campbell and Jim Geller dead. We went to see Dorothy Parker on the 15th, with Gavin. She was very pale, thin and somehow tough. A lot of people there making bright conversation, to keep her as it were afloat. Right after Alan's death was discovered, and the coroner was still in the house, James Larmore arrived drunk and sang.... I guess Jim's funeral will be tomorrow. I shall miss him; he was that unusual thing, an honest man. Have already arranged to have Hugh and Robin French for my agents. Maybe they can get me a job in England or Italy this fall, after the draft of my novel and the draft of the Ramakrishna book are finished. That would relieve the situation here. Henry Kraft visits regularly, but doesn't seem to dispel the gloom for more than an hour or so.

Have reached page 84 of my draft, and the midst of the last chapter but two of Ramakrishna.

Paul Latouche, after a whole morning with me alone on the beach: "What's your last name, Chris?"

I tried to make Mr. Shikiya the gardener understand that Don has taken this studio room in the same house with Bill Brown and Paul Wonner at Ocean Park. But he merely took this to mean that Don is drawing portraits of people in a park, somewhere!

A sign of the times: I look out the window and see two young boys about to climb over our gate. One of them is a Negro. I think, better not interfere. I tell this to Dorothy Miller. She laughs. She walks with a stick now and has applied for old age pension; she's sixty-five. She says she has to lie down most of the rest of the day, after a day spent working for us. But I feel this work is good for her, psychologically, and anyhow it is a way of giving her money.

On the 25th, I'm to speak at one of the sessions of the Pacific Coast Writers' Conference, at L.A. State College. This is from the pep letter addressed to all speakers by Leon Surmelian, who's directing it: "I know you will do your share to create a relaxed, informal atmosphere, combined with high seriousness of purpose, with a bit of friendship and charity thrown in. You are a dreamer among dreamers."

---

[1] Bachardy drew Paul Latouche (not his real name), but there are no drawings of Rob; he met them both on the beach.

June 23. Last night I got drunk at Bruce Zortman's and sideswiped some car on the way home and bashed up the Volkswagen. I was too drunk to go out and look to see just what I did to the other car—it must have been one of those which were parked at the entrance to the lane leading to Adelaide out of San Vicente, or, horrors, maybe the one belonging to our neighbors, the Marion Hargroves! Some worry and guilt about this.

But such a heavenly day today. Don and I went in the water and I rode quite a big wave.

Prema was terribly upset by Rechy's *City of Night,* which I loaned him. He said, "I was sick for two days." Couldn't quite make out if this was disgust or lust.

The young Jewish boy who comes to the readings at Vedanta Place asked Swami earnestly had he done wrong—he gave a man at work ten bucks and the man spent it on liquor. Swami was amused.

Swami was impressed because Queen Elizabeth and Prince Philip had gone to Victoria Station to welcome Radhakrishnan,[1] which surprised me. But Swami said, "A few years ago, they'd have just said he was a native."

Don has started going to Gavin's psychiatrist, Dr. Oderberg. He likes him, but has great difficulty telling his problems. Don says, "When I try to write them down, they suddenly seem so ridiculous."

The other day, Don found that his drawings of Paul Latouche and his friend Rob had been extracted from the closet in his studio. Paul firmly denies having had anything to do with this. So now we don't see him around. And Don keeps the studio locked. (Note: one of Paul's favorite expressions is "Are you ready for it?" He says this after he has told you something surprising. It's equivalent to "Can you believe it?")

We have been trying to sell the Danish chairs, without success.

I forgot to record Dean Campbell's story of hearing a man yelling outside the window in the middle of the night. Two other men, who were with him and had presumably been threatening him or actually beating him up, then ran away. The man sat on the side of his car for a long while; then he got up and took off his pants and his undershorts and threw the shorts away. Then he dressed again and drove off. In the morning, Dean found the

[1] Sarvepalli Radhakrishnan (1888-1975), President of India, 1962–1967; he had been a professor of ethics and Eastern religions at Oxford, 1936–1952, and held other posts in the Indian government. He was knighted in 1931.

shorts were full of shit—presumably because the man had been so scared.

July 26. Yesterday I finished, or rather, came to the end of the novelette—128 pages in this second draft, plus one line on page 129. Now I have put it away, and I hope I shan't weaken and look at it again until both Don and Gavin have read it.

As I was getting toward the end, I had the idea of calling it *The Survivor*—but this title has been used in various forms at least three times recently. Don suggests *Making Do*, which is a sort of Henry Green approach. I quite like it but am not sure.

Beautiful weather, now, for this long while. Today I went in swimming, with Don and Henry Kraft.

Beautiful news too, as far as it goes: the test-ban treaty was initialled in Moscow yesterday.[1]

Talked to Aldous on the phone this morning and wished him a happy birthday. Largely because Chris Wood, who came to dinner last night, told us Aldous had said to Peggy Kiskadden that he has no friends. Also, I was concerned because Peggy (that's to say Bill [Kiskadden]—who, amazingly, isn't dead yet) thinks that they ought to have cut Aldous's tongue right out, when he had that cancer, and that now he has it in the throat. Aldous admitted to being slightly hoarse still. But he leaves tomorrow for a trip to Europe, starting with some conference in Stockholm. He seemed quite cheerful and pleased I'd called.... Alas, I can't help still hating Peggy and feeling that she longs for Aldous to have cancer, in order to prove that Laura has neglected him and in general to get one more recruit to her squad of dying men.

Dorothy Miller, who now cleans for us every week—not that we need it, but she needs the money—is plagued by a bluebird when she goes outdoors. It swoops down and pecks at her hair. So she wears a handkerchief tied over her head. I have a hunch that she regards the bird as some kind of an evil omen.

I have been bad about going to the gym but pretty good about doing my daily Canadian Air Force exercises. Have now reached A+ on chart one. I hope to reach C+, which is the ceiling for my age group, by my birthday. Of course I plan to go on beyond that, at least to the ceiling for the 45-49 year group.

Another birthday target, to have a rough draft of the two final

[1] The Limited Test Ban Treaty, prohibiting nuclear testing in the atmosphere, outer space, and under the sea, was signed by Britain, the U.S. and the USSR.

chapters of the Ramakrishna book. So far I have written six pages of the last chapter but one.

August 2. Dorothy refers to Don as "Mr. B." We were talking about Gladys Cooper. She asked, "Is that Mary Pickford?" (She was thinking of Pickford's real name being Gladys Smith and no doubt supposing that people as "in" as Don and me would refer to her thus!)[1]

Gavin has read the novelette and seems to like it a lot. But he is concerned about George's identity. He feels that George's way of speaking and his attitude to his college job are so absolutely me that one cannot accept him as an independent character. This may well be true. But I'm not sure that anything can be done about it. Perhaps it will be better to publish this as an admittedly flawed work than to try to create a fictitious George and end by losing all the madness and gaining only a completely convincing dull character study.

In bed, on Monday night, Don was silent for a long while. I thought he had fallen asleep. Then he suddenly asked, "How about *A Single Man* for a title?" I knew instantly and have had no doubts since that this is the absolutely ideal title for the novelette, and I shall use it, unless someone snitches it.

Don has been going steadily to Dr. Oderberg. He feels that progress is being made, though he gets bored talking about his dreams. Certainly his whole mood has changed. No more gloom. But this is no doubt largely due to Kraft, who is now quite an institution. Also to association with Paul Wonner and Bill Brown. Don paints away at their studio and says he is at grips with his great problem—*can* he paint and if so does he really *want* to? No solution appears in sight, but the point is that he is at grips. Before, he was only fearing to get to grips with the problem. Also, he keeps on with his japa and feels that this is producing results. (He broke his beads yesterday.) So, altogether this would seem to be a very crucial and on the whole productive period for him. As I tell him, he is one of the few people who are doing something about their problems on all four levels—physical (he goes to the gym), psychological (Oderberg), artistic (Wonner-Brown and his painting), spiritual (japa). Don's only doubts are about earning money. Should he go to New York, and become the new Bouché? I say, get to a point with the painting first.

[1] Isherwood and Bachardy knew both movie stars, but Pickford (1892–1979) only slightly.

And me? I'm melancholic. But more of that another time or never. What I have to do is finish the Ramakrishna and get started on my new project—a book of autobiography along the lines of the lectures I gave at Berkeley: the autobiography of my books.

August 9. Have heard from Edward, saying that the novelette is "absolutely wonderful and it has made me extremely happy" but adding that the Charley episode is an anticlimax and that the Ronny episode, though much better, isn't quite up to the first seventy-five pages and the ending. (Actually page 75 is in the supermarket scene, so maybe he doesn't like that, either.) Then he says, "The book as a whole cuts the reader to the heart, and dazzles him too. And I think your new manner comes off 100 percent. One gets the feeling from the start that you are totally at home in it."

So now I must reread the book—that is, as soon as I have finished the twentieth, last-but-one chapter of the Ramakrishna book. All the time that I've been writing about Ramakrishna's cancer of the throat my own throat has been sore and I've been hoarse. I hope ending it will cure me!

Yesterday I got the release arranged by Ben Alston from Mrs. Helen Burd, the lady whose car I sideswiped on June 22. Now it is over, I can admit that it has been a great worry to me. In fact, I superstitiously didn't want to write about it before it was settled. It does seem to be, now, since the police have dropped the case and Mrs. Burd doesn't even know now who did it. She has simply acknowledged receipt of the money to pay for the damage. She even told Alston to tell his client that he was "a fine upstanding boy" for coming forward and paying up! The fact remains that this little caper cost me nearly nine hundred dollars—damage to both cars and Alston's fee of five hundred. And all I need have done was to sleep on Bruce Zortman's couch, or for that matter take the most expensive hotel room in town—including taxi fare there and back to my car next morning, that would still have been at least eight hundred and some dollars cheaper!

Frank Wiley is having to resign from the navy [...]. Much more about this, no doubt, later. He's in San Francisco.

August 16. There's so much to say, but nowadays I am just too damn busy to feel like writing here. I am in one of those crisis states: I function but I'm far from well.

My throat is still sore and hoarse, although I finished the chapter containing Ramakrishna's death, this morning. Dr. Allen examined me, it's true, the other day, but then I am inclined to think he's much too casual. Gavin says he failed to detect Gavin's amoebic dysentery.

I went to see Allen because of the rib I broke in the car wreck on the 10th.[1] He didn't think it worth x-raying; didn't even strap it up. Ah, well. Enough money has been squandered already on my foolishness—around $1,500!

Edward wrote a second letter, however, saying that this novelette has "even outdone your best." So that makes up for much misfortune and I really feel eager to rewrite it, now.

Don followed up his invention of *A Single Man* by finding me an adjective for the cremation scene in the Ramakrishna book. I wanted to suggest that the waters of the Ganges kept flowing past and offering no security, as it were, to the mourners. So Don thought a little while and said, "How about—the *inconstant* waters?" When I asked him where in the world he got that from, he said *Romeo and Juliet!*[2]

Am off to the monks' picnic at Laguna Beach tomorrow. Sunday I have to speak up at Santa Barbara temple, repeating what I said about Vivekananda.[3]

Now I'll get ahead with the draft of the last Ramakrishna chapter, but not rush too much. Am getting too compulsive.

Don is very conscious of the existence of this "old black book," as he calls it. He's sure it's full of criticism of him. I tell him, well, when I die, all he has to do is burn it. Very hard to tell how he is getting along. His mood is a sort of cautious pessimism, regarding his painting. As for Oderberg, he is being a bore right now, because he takes too much interest in Don's parents.

Lines composed while walking around the block, supposedly making japa:

> But, when so sad thou canst not sadder,
> Counting thine every vice and crime,
> Cry: I'm sure bad but these are badder—
> Goldwater,[4] Teller, *Life* and *Time.*

August 20. Have now gotten started on the rough draft of the

---

[1] Another accident, on the night of August 9, with a Mrs. Ratner.
[2] II.ii, where Juliet warns Romeo not to swear by "th' inconstant moon."
[3] I.e., in his draft of chapters 15 and 16 in *Ramakrishna and His Disciples.*
[4] Barry Goldwater (1909–1998), arch-conservative, pro-nuclear Republican senator from Arizona; he made the cover of *Time* in June 1963 as a likely presidential candidate for 1964.

last Ramakrishna chapter. Like all the other chapters, it turns out to be rather more difficult to do than I'd expected. Also it will probably be quite long.

This morning I finished rereading Frank Wiley's book [...]. It really is *quite* good. I'm going to send it to Harper's,[1] to a man who wrote me the other day, named Roger Klein. He says he played Fritz in *I Am a Camera* at Harvard.

My throat continues to worry me. It won't clear up, although I'm now gargling with salt and water. I still feel it's connected with my writing about Ramakrishna's throat cancer. Not only did I happen to finish that chapter on the exact day of his death, but yesterday Lee Prosser sends me a batch of folders about the lakes and caverns around his hometown, Springfield, Missouri, and among them is one about the Meramec Caverns, where Jesse James and his gang used to hide out. On this folder it says that an old man claimed in 1948 to be Jesse James and that his claim could never be disproved, and that he finally died in Texas in 1951, aged 103—on August 16![2]

Over the weekend, I went to the men's picnic with the swamis at the Camel Point house in Laguna and also spoke at the Santa Barbara temple. Driving home, Prema told me he definitely plans to stay on in India, if he possibly can. Either he will settle down as a spiritual recluse in one of the monasteries, and "maybe become spiritual"; or he will find some worthwhile project connected with the Ramakrishna Mission and give himself up to that. Here he feels rejected. The business with Usha still hurts him terribly. He told Swami how he felt, but Swami didn't make the speech Prema doubtless hoped for—didn't tell him he is indispensable here, didn't beg him to come right on back here after taking sannyas.

Don has done some very interesting paintings of dolls. One or two of them are curiously poignant. You feel the tragedy of their not being human—just as one occasionally feels the tragedy of some human being's not being *more* human.

I got the Volkswagen back yesterday, all nicely painted up and straightened out, to give me another chance to be a grown-up driver.

Henry Kraft has left [his photographer boyfriend]'s at [the photographer]'s request, and gone to live with friends. What will

---

[1] Harper and Row, the publisher.
[2] The bank robber, Jesse James (1847–1882), was almost certainly shot in the back of the head by a member of his gang for a $10,000 reward; the body was identified by his Civil War wounds.

he do now? Don is inclined to be severe; doesn't think Kraft is serious about his photography.

August 22. In swimming yesterday with Bill Brown and Don, the other side of Pacific Ocean Park pier, near their studio. The water much cleaner there and the beach nicer. Very little pain from rib. Bill has been helpful about Don's doll paintings. Paul Wonner is gloomy and always tired. Fear some liver trouble. To Dr. Allen about my throat; I worked myself up into a state of alarm. Mustn't it be at least a growth on the vocal cords. Again, Allen seemed very casual, though I described my symptoms. He peeped down my throat for an instant, admitted it was inflamed, gave me a shot of penicillin. Today the hoarseness is as usual, though the inflammation has gone. I get another shot tomorrow.

Keeping on steadily with the rough draft of the last Ramakrishna chapter. Another letter from Edward today, again praising the novelette. He writes, "I dream that you are now beginning to tap an immense reservoir of experiences which for one reason or another have had to be dammed back until now"—which certainly sounds thrilling, if true! He also says that he wouldn't mind if George were further disguised, that is, made less like me, "for reasons of nonliterary expediency."

Both *Time* and *Newsweek* wrote short bitchy notices of Clifford Odets's death, and both quoted the old pun "Odets, where is thy sting?" So Gavin and I composed a telegram to the editors: "We protest against your remarks about the late Clifford Odets. An important American playwright deserves more than a per-functory dismissal with a tastelessly exhumed pun."[1] Signed, John Houseman, Gavin, Jerry Lawrence and Bob Lee, Dorothy Parker, Lenny Spigelgass, Gore and me. And just now, as I'm writing this, a woman from the *Time* office out here calls to know if the telegram was authentic!

September 3. Thank God the holiday is over, though the fine weather is too. Yesterday a delightful lunch with Joan Houseman, whom I now definitely like. Gavin Lambert's staying with her, recuperating from this Mexican bug.

Painfully slow advance on the last Ramakrishna chapter.

[1] Odets (1906–1963), a one-time member of the Communist party, wrote *Waiting for Lefty* (1935), *Awake and Sing* (1935), *Golden Boy* (1937), *Rocket to the Moon* (1938), *Night Music* (1940), *Clash by Night* (1941), and *The Country Girl* (1950) and worked in Hollywood, where his best-known screenplay was *Sweet Smell of Success* (1957). He appears in *D.1*.

Have been inscribing *The World in the Evening* for Henry Kraft. I am giving it to him for his birthday along with a copy of the French translation, so he can use the one as a crib for the other.

The photo taken in Hawaii of a lunch at a Japanese restaurant given to a lot of famous actresses. All of them are mugging at the camera except dear old Gladys Cooper who is eating ravenously with her face nearly down on the plate. That's so like her.

Remember how Paul Wonner winced at the restaurant when that rich silly ass stuck one of their little plastic swords (for spearing olives in cocktails) right into a *kachina* doll[1] he had just given Paul and Bill. Paul and I both experienced the horror of this outrage, and at the same instant. It was a terrific rapport.

One day, Don told me that Dr. Oderberg had completely taken my side against him; but he wouldn't explain how or why.

Gore Vidal gave me the *Reader's Encyclopaedia of American Literature* for my birthday. Its only value judgment on my work is to say that *The World in the Evening* "frankly disappointed reviewers, who found it 'commonplace.'" For this reason I had never bought the book before, which was childish of me.

[...], the man who gave me coffee on the night of the accident with the two other cars,[2] phoned me later to say that his friend [...] wanted to see me before deciding if he should testify that I was drunk at the time of the accident. I'm pretty sure this was his own idea. Because I gave him a big tip for driving me home, he thought he could do a bit of blackmail. Kind of saddening. I was very firm, however. Told him to talk to my lawyer. Haven't heard from him since.

September 4. I took Prema out last night to a sort of farewell dinner at the Malibu Sports Club place. How he dwells on the past! The party at Glenn Ford's, for instance. In Sydney, they are renting a hall for him to speak in. Usha promptly said, "I hope the Vedanta Society isn't paying for it." His suspicions that [one of the monks] is making the scene with one of the devotees, and that Mark is impotent.

Rain in the night. I feel utterly exhausted, these days, and my throat is bad again. The rib seems to be getting better, however. I *long* to return to the gym and do my Canadian Air Force exercises.

Today came the news of Louis MacNeice's death. I really hardly

---

[1] A carved and painted figure made by the Hopi Indians of northern New Mexico and Arizona to represent a supernatural being. Wonner and Brown collected the dolls and revered their religious and historical significance.
[2] The August 9 accident.

knew him, but he is the first of the Old Guard to fall. At fifty-five.[1]

Reached page 10 of the last Ramakrishna chapter today. A third, maybe, or more.

Don in a very bad mood about his work—after temporarily recovering from a very bad mood about me. And Paul and Bill will be leaving the studio because they can paint in their new house. So more trouble ahead!

September 19. Well, yesterday I finished the last chapter of the Ramakrishna—the longest and cruellest of all my Vedanta chores. It's marvellous that writing it didn't make me lose my faith altogether. Don winces at the very idea of reading it, and yet I must have a cold-eyed critic who isn't simply an atheistical idiot.

Both he and I hugged Prema last night, saying goodbye. This pleased him, touchingly. He left for Honolulu this morning.

Have told Gore I'll dedicate *A Single Man* to him. This pleased him, too. Don, that wizard at name finding, has picked Kenny as a substitute for Ronny and Doris as a substitute for Ruth, in the book.

Have been greatly worried over my throat, but I saw Allen again this morning and he maintains it's an infection, nothing more. In fact, he talked in quite a relaxed manner about people who have premalignant nodules on the vocal cords.

Very warm, after heavy rain. A good workout at the gym; the first since my rib was hurt. I can't feel it now. The Republican poster had been taken down, so I didn't have to protest to Lyle, as I had made up my mind to.

Have now definitely said I don't want to have to meet Henry Kraft any more. I should never have done so in the first place. That kind of thing is messy and was messy in the days of Lord Byron, and always will be messy. Unless one simply doesn't give a shit.

Which reminds me of a terrific squaresville thing Henry said ages ago about love. "For love you need four things—will, determination, integrity, responsibility."

David Smith (whom I like) describes how a colored queen talks: "Well, cakes, I'll just tip over parkside and find me a chico chunk (an attractive Puerto Rican)." "Chunk" means anyone attractive; it's full version is "chunk of life." Other Negroes are called "kabukis."

[1] MacNeice (b. 1907, Belfast) was at Oxford with Auden and Spender and collaborated with Auden on *Letters from Iceland* (1937). He was a university lecturer in classics, a BBC writer and producer, and published verse, verse translation, autobiography, and plays. He appears in *D.1*.

September 26. The heat is beyond belief. It rises through the floor-boards of our balcony; at breakfast this morning it was stifling.

Last night, Swami pressured me into saying I'd come with him to India for the Vivekananda centenary. I feel miserable about this. The idea fills me with blank horror. But it would only be for about three weeks. Starting December 19.

Anyhow, it makes a reason for going into a drive to get both books ready for publication before then. I am working grimly at *A Single Man.* The Ramakrishna has to be typed up for me first.

Don is doing oil portraits, rather more happily. His private life is obscure; but will be revealed no doubt before long.

My throat is still just the same, and I am worried about it, of course. Altogether I am low on energy and melancholy and toxic. But much of that could be cured by drinking less, cutting out all smoking, taking exercise regularly and praying. (I never miss japam, but it is utterly mechanical.)

In case *A Single Man* is later thought to be a masterpiece, may I state that it bores me unutterably to reread? Going through it is really a grind.

October 8. For the first time, this evening, the sun set to the west of Point Dume, in the ocean. The beginning of fall.

October 23. Last Saturday, the 19th, I finished the third and final (I trust) draft of *A Single Man.* It had taken me a little less than a month; I started it on September 21. The day before yesterday, I sent off typescripts of it to Simon and Schuster and to Methuen.

Jerry Lawrence, the first person to read this final version with a fresh eye, seems to like it enormously; but he still says, as all the others did about the second draft, that the scene with Charlotte is weaker than the rest. I know I have improved it a lot; and there is probably nothing more I can do to it which would make any significant difference. At least, it's weakness is in itself effective up to a point; because, if you've been let down, this is apt to make George's sudden decision to run off to The Starboard Side all the more amusing and exciting; and perhaps the scene with Kenny will benefit by contrast.

Vera, on the phone today, warns me we shall find Igor much slowed down and slower in the uptake. But she says the doctor says this isn't just old age; it's because he will take so many pills, particularly tranquilizers. We are to see them on Monday; the first time in ages.

Chris Wood has finally lost his little dog Penny. She died while they were in New York, the other day.

Today I began revising *Ramakrishna and His Disciples* (as I think it will have to be called). Tedious work. I have to keep taking out words expressing vehement overemphasis, such as absolutely, completely, utterly.

October 31. Still sweating it out. Not a word from Methuen, even that they've received the typescript. [Alan] Collins of Curtis Brown in New York has read it and likes it, but doesn't understand the scene with Kenny! Ah well, that aspect of the whole thing—whether people like it—anyhow, people like Collins and ninety-nine percent of the population—doesn't seem so important, right now. I am almost certain that it is my masterpiece; by which I mean my most effective, coherent statement, artwork, whatever you want to call it.

I still have this thing in my throat. And, psychosomatically, it gets worse every Wednesday when I have to read to the family up at Vedanta Place. A *passionate* psychosomatic revolt is brewing against the Indian trip. I am almost capable of dying at Belur Math, out of sheer spite. I will not surrender my will; be made to do anything I don't like. With the Don-Henry Kraft situation, this is not apt to produce a real explosion; because any concession I do make immediately puts Don in a defensive position, and I can get back at him for making me make it.

I seriously believe that I am, beyond all comparison, nastier and madder than ever before in my life; although still capable of occasional gentleness with Don and, of course most gracious and charming to strangers who rub me up the right way. My "religious life" consists in making japam without the faintest devotion, and indeed mostly while thinking of anything else in the world including my resentments. Now and then I get around to asking Ramakrishna to give me devotion "even against my will." And this is not shit. I still believe. I still know that this is all that matters. And yet—

Of course, I'm aware that part of this state of mind is due to the phase of intensive work I have been going through. When I work, I declare a "state of emergency" during which I'm allowed to behave much worse, and during which I always have the feeling, "How dare they upset me in any way, while I'm getting ahead with this sacred and important project?" Well, the work is over now, all except for revising the remaining sixteen chapters of *Ramakrishna and His Disciples*. This is most necessary and I think I can really improve the book a lot by cutting out all the preaching and nursery-school explaining of which I've been guilty; but, still and all, it isn't strenuous, and now I ought to be simmering down.

Igor, when we saw him, seemed lively and quite quick-witted, but Vera and Bob assured us that this was a "good" day, and that he was seldom up to this standard any more. Aldous appears to be really very sick. Laura, on the phone, sounded truly distressed. But she still spoke of his illness as an "infection." She asked me to come and see him.

Mexico is still on, but probably not until the end of November. Don may go to New York; or stay here, over Christmas.

Swami amused us by announcing that *The Leopard* was the worst film that ever existed.

Dodie Smith's new novel has arrived.[1] Have just begun it. It seems terribly harmless and readable. Wrongly readable. Like T.V.

I suppose I should go and work out at the gym. I kind of hate doing this until I actually do it; and it takes up so much time. I am fairly strong in some ways, but still can only barely do nine press-ups, though the Air Force manual says my age group should be able to do fourteen! You should be able to do nine at eight years old.

My Francophobia is more violent than ever. How I loathe Genet! I said to Vera and Bob, "Genet is someone who really got crucified—and then he comes back and lies about it!"[2]

Reading V.V. Rozanov (because I really must return the two books on Saturday to the Stravinskys, after keeping them several years.) I have never heard love-hate for the Jews better expressed.[3] It seems strange to hear him say that the Jews are going to get possession of Russia; and now here we are, saying the same thing about America.

November 1. Said yesterday:
"Kitty's month is nearly over."

---

[1] *The New Moon with the Old* (1963).

[2] Genet (1910–1986) was a foster child, teenage criminal, and convict. His 1943 autobiographical novel *Our Lady of the Flowers* and Sartre's 1953 study *Saint Genet: Actor and Martyr* were published in translation in the U.S. in September 1963. Stravinsky and Craft had met Genet in Paris in October 1962, and Stravinsky let him use, for free, two pieces of music ("Histoire du soldat" and "Octet") for the film, released January 1963, of his play *The Balcony*.

[3] Isherwood evidently had the volumes translated from Russian by S.S. Koteliansky and published in England in 1927 and 1929 respectively: *Solitaria* (1912) with excerpts from the *The Apocalypse of Our Times* (1918) and *Fallen Leaves* (1913 and 1916).

"What's so special about October? Why is it Kitty's month?"

"All months belong to Kitty."

Chris Wood likes the novel. He didn't know how he should pronounce Geo.[1] He was doubtful about, "Who are you trying to seduce?" Shouldn't one say "whom?" (This seems to me absolutely grotesque.) He apologized because the part about Jim's absence made him think of Penny.

Jealousy: Not what they do together sexually. But the thought of their waking in the morning, little pats and squeezes, jokes, talk through the open doorway of the bathroom. For that one could kill.

Jo, when I failed to watch Ben's T.V. show on "The Eleventh Hour,"[2] reminded me how I'd failed to see *her* show, too. And—to make me feel extra bad—she told what pains she'd had last night, after eating nothing but clam chowder.

Yesterday afternoon, on my way to the gym: In Pacific Palisades, the kids are encouraged to write Halloween inscriptions on all the shop windows. But it is all Disneyfied, rendered harmless; and the kids themselves are carted off to the park to have organized fun there, and not annoy anyone.

This morning a cable confirms that the novel has arrived at Methuen's.

November has begun. India is only one month and nineteen days away. *Must* I?

I did do fourteen press-ups yesterday at the gym, after all!

To meet Bruce Zortman yesterday evening on the UCLA campus, to see a performance of Pirandello's *Right You Are (If You Think You Are)*. I thought I'd put in a couple of hours doing research on some quotations in the library. The library was open all right, but there was no one there of sufficient authority to renew my library card. So I was stuck on campus for about an hour and a half, waiting for the show to start. Question: What do you do when you are alone? It is disconcerting to find what a bore one is to oneself. And prayer is so tiring. The whole brain begins to jingle.

Pirandello's urbanity. His delight in a kind of weepy maddening very Jewish kind of masochism (Signora Frola). The worst error in taste: Lamberto's laughter at the end of each act. But the ending

---

[1] Jee-oh.

[2] NBC drama series featuring psychiatrists who helped people in their eleventh hour of need. Masselink's episode, "Oh, You Shouldn't Have Done It," was aired October 30, 1963, starring James Coburn. Isherwood and Bachardy owned no T.V. set until 1964.

is fun; and this morning the play has quite a pleasant taste in my mouth.

November 5. Aldous is dying. I went to see him yesterday morning, at the Cedars of Lebanon. He looks like a withered old man, grey faced, with dead blank eyes, speaking in a hoarse voice, hard to understand. But his mind seems to be as good as ever—that marvellous instrument, about to be swallowed up in the ruins and shattered.

Laura looks haggard. She says that he does not realize how sick he is. And Aldous certainly gave me that impression; though of course it may be his way of softening the blow. He talked with a kind of petulance about being old. He was angry with the so-called arthritis in his back; this is an area where the cancer is spreading. He said to me that when one is old one is almost absolutely cut off from other people. But I think he enjoyed my visit. I told him the story about Jesus and the Blessed Virgin golfing,[1] I described my difficulties with the Ramakrishna book, I went on about my horror of India; in fact, I said everything I could possibly think of. Aldous on his side spoke very interestingly about Rozanov, remarked that all the African new nations would soon be governed by their armies, and was only unable to remember one name: Puri. He seemed most interested when I told him that my character has gotten worse as I get older. He was amused.

Gerald, who was to see him today, asked me anxiously what he should talk to Aldous about. Gerald can't help showing that he is faintly disapproving; Aldous, he feels, should be thinking about death and consciously preparing himself for it, not pretending to himself. (Laura told me that Aldous still speaks of things they are going to do together next year.)

In contrast, there was Igor, who came to dinner three nights ago and seemed nearly as good as new. He asked me so sweetly to observe what a good healthy pink skin he had for his age—and only one of those old-age freckles on his hands!

A novel is just being published called *A Singular Man*, written by J.P. Donleavy. On top of this, its chief character is called George, and the book begins with his getting up in the morning! This is a real misfortune; as I'm nearly sure the publishers will want me to change the title.

Jo doesn't like the book, I'm sure. Ben, rather surprisingly, does. In fact he takes an uncharacteristic, almost arty attitude towards it;

[1] See Oct. 29, 1961 and June 18, 1962.

for example, he says he's so impressed by its symbolism: the symbol of the Road, which appears successively as the freeway, the road of life narrowing to the width of Doris's bed, and Charlotte's description of her sister as being like a road. Ben also says, most revealingly, that, "The greatest thing in the book—in fact, in any book I ever read—" is Kenny's line, "I'd have liked living when you could call your father Sir." (A line which, incidentally, in my opinion, skates to the very brink of corn, Salinger-corn, and maybe falls in.)

No news from New York or London yet.

I have been very low; largely because of drinking too much. My vitality is sapped and I'm depressive and paranoid. Terribly violent resentment of Henry, whom I'm nevertheless now committed by Don to seeing quite often. I suffer in a way that is utterly grotesque. That must stop, of course. But how can it be stopped? Again and again, I come to the idea of having a friend, a real confidante, of something nearer my own age. But that only means Paul Wonner.

As the result of all this, I have hardly worked on the Ramakrishna revision. Plenty of chapters have been typed up, but only six are even roughly revised. Every day I dodge work. I have no heart for it.

November 6. After a quiet evening with Don, eating at Casa Mia, drinking nothing but a bottle of rosé and smoking nothing, I woke this morning refreshed and altogether more optimistic. Also we had a soothing heavy rainfall during the night. Today is windy and brilliant. I have restarted the Ramakrishna revision. Also done my Canadian Air Force workout. Terrible difficulty, still, with the fourteen press-ups; otherwise everything is fairly easy. (Have just realized I've been doing the press-ups wrong; maybe they'll be easier the proper way—starting on the floor!)

Aldous was so sick yesterday that Laura asked Gerald not to come. But he will be moved back home today, just the same, she says.

Last night, while we were reading—Don hating [James Baldwin's] *Another Country* and calling it dull and self-conscious—I found in Rozanov's *Solitaria* the passage Aldous quoted to me when I saw him. It's when Rozanov says "private life is above everything.... Just sitting at home, and even picking your nose, and looking at the sunset.... All religions will pass, but this will remain: simply sitting in a chair and looking in the distance."

The feeling in this seems to me kind of Zen.

A mad picture sent me by a fan named Dolores Giles. It's called "Pregnancy: The Fourth Month." Shall try to give it to Jim Cole.[1]

November 11. On the morning of the 7th, Peter Schwed called from New York to accept the book. He seems to like it very much indeed; said it is one of my best. He only suggested cutting one line, about wiping the belly dry, and wasn't positive about that, even. He isn't sure if the title need be changed. They will talk it over.

An exhibit from The Age of Innocence; this letter from Lee Prosser—

> I hope my letter finds you as happy as I am. I am the most happy individual alive! I finished my novel, and my only hope is that it's art. I hope it's a good one. Would you read it and advise me? I would like to know what you think.

Talking about alternative titles for my novel with Don in the car. I remembered *Paul Is Alone* and wondered if *George Is Alone* would do. Something wrong. How about *He is Alone*? Or maybe *It is Alone*? It couldn't be It, I said to myself; because It is the Atman. And the next moment, Don said, "It couldn't be It." He had thought exactly the same thing. At such moments, our rapport seems almost supernatural.

But, alas, these moments have been few, lately. Terrible mood-storms about Henry, since my last entry. One thing came to me very strongly. I must be more positive in my attitude; not just wait to be offended and take offense. So I called Ben Underhill in San Franciso and asked could I stay with him. He is coming down to L.A. on business at the end of this week. Maybe I'll go back with him for a few days. Meanwhile, I have tried to warn Don that he is actually starting to destroy the cohesive element between us, the "ultra-clay." I don't think he quite understood me or quite believed me, if he did understand. But temporarily—after staying out three nights in a row—he is all solicitude. I have said definitely that I do not want to see Henry any more, under any circumstances. I also asked Don to please take the photo of Henry into his studio. He had put it up on the desk in the back bedroom. . . . God, how I hate lowering the boom, like this! And yet it is, ultimately, the only decent and truthful and friendly way to act. The alternative is sulks and silent reproach.

---

[1] An attractive blond friend who served in the military and once sat for Don Bachardy.

Aldous nearly died, a couple of nights ago. Yet he still seems unaware of his condition. He said to Laura that he was worried how he would spend the rest of his life, if he couldn't write: and he implied that he expects to live at least five more years. Cutler doubts he will last through this month. Gerald says that Laura gave him lysergic acid a little while ago. According to Gerald, this ought to have made him realize his condition; but apparently it didn't.

November 30. Such a strong disinclination to write anything about Black Friday the 22nd. But I ought to. To remind myself.

Don and I were still in bed, around eleven, because we had had Cecil Beaton to a farewell supper the night before (he left for New York next day en route for England) and Paul Wonner and Bill [Brown] and Jack Larson and Jim [Bridges] had been there too, and we had stayed up late. Henry phoned (even now, I mind that it was Henry—he seems to take possession of everything—pushing in in his thick-skinned German way) to say that the president had been shot. And we plugged in Harry Brown's old radio, which we otherwise never use, and lay listening to the reports coming in and soon confirming the death itself.

When Roosevelt died, I was sad but thought, Goody, we'll get the day off from the studio.[1] This was quite different. Just disgusted horror. When I think of it, I keep being reminded of that time in Calcutta when, in the so-called first-class hotel, some kind of black slime began to ooze out from under the toilet. Also, there *was* the feeling—journalistic as it may sound to say this—that some sort of nationwide evil was functioning. It *was* something we had all done with our hate.

Aldous seemed an anticlimax; I suppose partly because it wasn't our fault. He died without pain. At the end he asked for lysergic acid, and was given it. His mind was quite clear. The day before, he had finished and revised an article on Shakespeare and religion which Laura says is very good.

On the following Sunday, Laura and Rose, Maria's sister,[2] and Maria's mother,[3] and Rose's son Siggy,[4] and Matthew Huxley and Peggy Kiskadden, and a few others including me, all went for a walk down toward the reservoir—instead of having a funeral. It was Matthew's idea. Peggy and I were polite but barely spoke.

I'm just writing down anything that occurs to me....

[1] April 12, 1945, as Roosevelt began his fourth term in office.
[2] Rose Nys de Hauleville Wessberg.
[3] Marguerite Baltus Nys, known as Mère.
[4] Sigfrid Wessberg, then about twenty, the son from Rose's second marriage.

Heard at last, this morning, that Methuen's *are* taking my novel; but this was only from the agent. Meanwhile, the day before yesterday, I finished revising *Ramakrishna and His Disciples*, and shall take it up to the center tomorrow, just to get it out of the house. I know it's not really right, yet, but I can't do any more.

Don feels the president's death passionately. He really burns with despair. This is partly because he is feeling miserable anyway. He can't paint. And (I suspect) Henry is either away or they have had a quarrel. But I refuse to ask him about this.

I keep thinking: Well, the books are done now. Maybe I shall die soon, as the colored girl in New York said I would. If not, let's wait anyhow till this Indian horror is over, and then see what gives. Life goes on, or stops. If it goes on, it will change for me.

December 11. Beginning of the countdown. A week from today we are off.

The usual sloth which follows finishing books. Now that I have got *Ramakrishna and His Disciples* out of the house, and there is nothing more to be done about *A Single Man*, I can hardly bring myself even to write a postcard, or to go to the gym or to walk on the beach, though the weather is heavenly, though cool. (Looking out the window, I feel, like the man in the Icelandic saga, "Beautiful is the hillside—I will not go!"[1])

This morning, a blow. Roger Angell of *The New Yorker* has refused *A Single Man*, either whole or in part. "While I can believe this novel, I don't find it particularly interesting."

Nothing yet from Methuen.

Yesterday evening, we went up to see Laura Huxley and Virginia Pfeiffer. They are really nice, both of them; and now I feel our friendship will survive Aldous's death and increase in strength. Also the children are enchanting. The only children, Don says, he has ever liked. Went over the proofs of Aldous's last article (for *Show Magazine*). If some frog like Romain Rolland were writing his life, I'll bet the book would conclude, "The day before he died, he wrote the last word of his last essay. It was— Shakespeare."[2] (Actually Aldous dictated the second half of the article, in a ghostly scratchy voice which we listened to on tape. Don said that, listening to it, you felt the great barrier between him and

---

[1] *Njal's Saga*, ch. 75.
[2] The penultimate chapter of Rolland's *The Life of Ramakrishna* (1929), which Isherwood knew well, concludes with this account of Ramakrishna's death: "According to his utterance of faith, 'He passed from one room to the other ...' And his disciples cried: 'Victory!'"

Laura; she sounded like a journalist.) But Laura spoke beautifully about Aldous, saying how ridiculous it was to suggest that his life was unhappy; he was always so full of enthusiasm. Virginia criticized Stephen's article about him.[1] (So did Gavin.) Stephen says that Aldous never got over Maria's death and the burning of his house. The publisher had suggested John Lehmann should write the biography. Laura asked me what I thought of the idea, so I had to tell her that John disbelieves in, and is aggressive toward[,] the metaphysical beliefs which Aldous held. All he would describe would be a clever young intellectual who later was corrupted by Hollywood and went astray after spooks.

December 16. Gore says that once he was at a horse show with Kennedy, sitting beside him. He remarked to Kennedy, "How easy it would be for someone to take a shot at you—and, of course, if they did, they'd be certain to hit *me.*" Kennedy grinned and said, "That'd be no loss." Gore thinks that Kennedy was beginning to lose his grip at the time when he was killed. After the death of his son,[2] he became subject to crying jags; and he seemed to be losing his confidence that he could succeed in getting people to see things his way. Gore approves of Johnson but feels sure he won't live through his full term if reelected; he will have another heart attack.[3]

Don is in New York. Talked to on the phone last night, he was very despondent. Why on earth had he come there, he wondered. And he finds Marguerite's apartment so depressing.

Woke up in a big flap this morning; travel-dread gripping me. So, following Gavin's advice, I have started taking Librium in advance. The idea is that I shall be riding high before the take-off. So far, it doesn't make me sleepy, like Miltown.

Dorothy (who moves in here on Friday to keep house) said, as we embraced, "For God's sake be careful!" I know she is having big death presentiments. "You do the travelling and I'll do the praying," she said.

This afternoon, I had a sudden desire for an ice-cream cone. This sucked the detachable plate out of position, and, before I

---

[1] "A Unique Person and Talent," printed underneath Huxley's obituary in *The Sunday Times*, London, Nov. 24, 1963, p. 8.
[2] Patrick Bouvier Kennedy, born prematurely on August 7, lived only two days.
[3] Lyndon Johnson (1908-1973), Kennedy's vice president, who automatically succeeded him, had survived a massive heart attack in 1955 and eventually succumbed to a second. But he was re-elected in November 1964.

could stop myself, I was munching it up. So now I have to fly to Dr. Stevens first thing in the morning. I strongly suspect the diencephalon of trying to give me appendicitis[,] because I swallowed the jagged corner of the plastic backing.

(What follows is transcribed from a pocket diary I kept during the trip. I should first say that we took off, as planned, on the morning of December 18 and arrived in Tokyo on the evening of the 19th, because you lose a day crossing the date line.)

It is no annihilating condemnation of the devotees—about fifty of whom had come to the airport to see us off—to say that they would have felt somehow fulfilled if our plane had burst into flames on take-off, before their eyes. They had built up such an emotional pressure that no other kind of orgasm could have quite relieved it. The parting was like a funeral which is so boring and hammy and dragged out that you are glad to be one of the corpses. Anything rather than have to go home with the other mourners afterwards!

Swami wouldn't leave until Franklin [Knight] arrived; he had had to park the car which brought the boys from Trabuco. The fact that it was he who arrived last seemed to dramatize his role as The Guilty One, and his farewell from Swami was a sort of public act of forgiveness. He was terribly embarrassed, with all of us watching—especially all those [women] who know what he did.[1]

So we got into the plane at last and it took off. Swami said, "To think that all this is Brahman, and nobody realizes it!" I sat squeezed between him and Krishna; the Japan Air Line seats are as close together as ever. Despite my holy environment, I couldn't help dwelling on the delicious doings on the couch, yesterday afternoon. I didn't even feel ashamed that I was doing so. It was beautiful.

A fierce hot breeze blowing through the Honolulu airport. People looking impatient of their heavy loveless leis. Then the long long flight northwestward, passing from Wednesday to Thursday through the almost infinitely extended afternoon, with the red sun dying so slowly over the ocean cloudfield. Tried to read Cather's *Song of the Lark*, but could concentrate only on *Esquire* magazine articles—Calder Willingham's reply to Mailer; Mailer's threats to write a novel; Gore on *Tarzan of the Apes*.[2] Very good food on the plane. Swami took a drink, but I refused; determined to

---

[1] He behaved inappropriately towards a woman outside the congregation; police were reportedly involved.
[2] Willingham, "Aftermath: The Way It Isn't Done: Notes on the Distress of Normal Mailer," pp. 306–308; Mailer, "The Big Bite," p. 26; Vidal, "Tarzan Revisited," pp. 193, 262–264, *Esquire*, Dec. 1963.

keep this trip dry. In the evening, there were steaks. When Swami and Krishna said they couldn't eat them, they were given stuffed chicken from the first class.

December 20. At the Hotel Nikkatsu. Last night I slept more than ten hours. Swami slept badly and has spent most of today in bed.

Today, for the first time, I felt a real intimacy with Krishna as we shopped and wandered around Tokyo. He bought a camera and a tape recorder. I made an excuse to stay browsing in the Jena bookstore because I wanted to look at the U.S. physique magazines. There was a swarm of Japanese teenage boys giggling around them.

There's a blizzard up north and the atmosphere is bracingly cold. I feel wonderful. Wore my new painful shoes to stretch them.

The city is being torn apart for the 1964 Olympics. Deep crevasses in the streets where a subway system is being put in. The workers still wear cloven socks and baggy knickerbockers, and you still see tiny old women in trousers toting huge loads of bricks. The traffic is as mad as ever. And the air around the palace is just as blue. And the dollhouse bars in the narrow lanes are just as inviting. I would like to live in this town. All the stores are sparkling with Christmas—more or less of a camp here, presumably, and for that very reason far more attractive than in the States.

December 21. Swami still feels unwell, fears kidney trouble, has fever and pains. He says he'd go right back to California if it wasn't for all the people who'd be disappointed.

I don't feel anything about anything, particularly; no doubt because of the Librium. If this is how "ordinary" people feel, well, good for them.

A brisk walk with Krishna, who slyly admitted he wanted to get a plug of some type for his recorder. His shopping has an air of juvenile naughtiness. We found the plug and then peeked into the Imperial Hotel. I felt a wave of sentiment for the old place. Wystan and I first saw it in 1938,[1] and my memory clings to an improbably symbolic tableau: under a chandelier (which certainly isn't there nowadays though that in itself proves nothing) stand two figures in uniform, a Japanese officer and a Nazi gauleiter; in fact, The Axis. As I regard them, the chandelier begins to sway— and this is my very first earthquake!

[1] On their return from China where they observed the Sino-Japanese conflict for *Journey to a War*, they sailed in mid-June from Shanghai to Japan, then on to Vancouver, New York (by train), and England.

The Japanese genius for life is expressed by the perfectly har-monized tone and texture of blue doors and off-white ceramic in the men's room at the Tokyo airport. But the Air India plane is crowded and shabby. There is also something squalid in the fact that it goes all the way from here to New York. Unchic.

A five-hour delay in awful Kowloon, with its bright trashy cleaned-up slums. Then the plane was filled so full that I thought we'd never get off the ground; most of them disembarked again at Bangkok. The Calcutta airport in the dead of night. Swamis, flower garlands, *pranams, namaskars.* Prema and Arup.

As we drove through the empty lanes and streets to the Math, I felt a magic begin to work. You both smell and feel the strange perfumed softness of India. Two men seated before a charcoal pot at the roadside. A booth, brilliantly bright and noisy, in which a *kirtan* was being held. People of the dust. Houses of the dust. Dust to dust.

December 22. Belur Math is far more delightful than I'd remem-bered it. The light is so soft.

At three-thirty this afternoon, the grounds are crowded. The people just sit on the grass or peer into the temples. The strollers don't embarrass the worshippers or vice versa. Some of the women are in very bright saris. All kinds of craft pass along the swiftly flowing river; small steamers, high-prowed barges, boats with huge square sails like junks, galleys rowed by standing oarsmen which look as if they were straight out of Cleopatra's Egypt.

Soft, soft brown, these people. Prema loves them. Says he never wants to go back to Hollywood. Arup isn't so charmed.

Swami, enthroned in Shankarananda's former room, seems like the head of the whole order[1]—more kingly, gracious and assured than any of the others. He showed us that exact spot—on the upper balcony outside his room—where he first met Maharaj. Swami had been looking into Vivekananda's room, next door, and Maharaj said to him, "Haven't I seen you before?" (Which, inci-dentally, was exactly what Holy Mother said to him when he met her as a young boy.) Swami told Maharaj, no, they hadn't met. Then Maharaj told him, "Take off my socks," and he asked, "Can you massage my feet?"

A small black cow walks by as I write this, sitting under a tree.

[1] Shankarananda (1880–1962) was President of the Ramakrishna Order from 1951 until his death. Isherwood met him in India in 1957 and writes about him in *D.1* where he spells his name "Sankarananda."

Various groups shoo it away. They are not respectful at all. Ah, the horizontal evening light; it makes the pink and yellow houses on the opposite shore look like gaily painted toys. Children play shouting around the porch of the temple of Maharaj. The factory chimneys are old-fashioned and not ugly. Sitting on the grass, under this tree, I am almost absurdly in the midst of India; yet quite quite isolated.

I am staying at the guesthouse, just outside the compound, along with Nikhilananda's party. Nikhilananda and Prabhavananda both eat with us. I foresee friction between them. A huge vegetarian lunch. Now that I've cut out liquor, my appetite is enormous.

Later—because at that point I was stung by several red ants and had to get up off the grass.... My room in the guesthouse is bare but clean, with its own bathroom. You are brought buckets of hot water to slosh over yourself and then shower with cold from a shower. It all floods over the floor, and in the morning you lock the door between bedroom and bathroom and open the outer door of the bathroom so the boy can come in and clean it. Many such precautions against theft; when the girls from Santa Barbara were over here, one of them had her clothes stolen through the window by means of some kind of a fishing pole.

The door of my bedroom is protected by a bolt like a Tower of London dungeon's, with a huge padlock. Also, you can bar it with a wooden bar from inside. I take a peculiar pleasure in doing this, not because I fear midnight intruders but because the bar gives me a sense of snug individuality in the midst of all these surrounding millions of people. It is very snug to be barred in, and then get inside the mosquito curtains into bed, and read.

Nikhilananda's party consists of: the Countess Colloredo, who is British-American, rich and known as Nishta (Nikhilananda, who is a fearful snob, always calls her "Countess"), she acts as his secretary-hostess; Mrs. Beckmann, a timid rather sweet woman, who is the widow of the painter Max Beckmann; Chester Carlson, who is the president of Nikhilananda's New York Vedanta Society and has invented some kind of process for duplicating manuscripts which is used everywhere;[1] Al Winslow, a young doctor, maybe queer, who blinks a lot and is utterly Nikhilananda's slave. They are rather like characters in Forster.

To vespers with Prema and one of his special buddies, the young

---

[1] Here Isherwood added a superscript "x" for a marginal note, "Xerox." Carlson (1906–1968) invented electrostatic copying, later called xerography, towards the end of the 1930s. The first photocopier came out in 1958.

and dramatically handsome Swami Aranyananda.[1] The singing was even more thrilling than I'd remembered. Prema is mad because someone here has prepared a list of all the swamis and brahmacharis of the order, omitting the American ones, as though they didn't count! (I mean, the American brahmacharis.)

I have to sit next to Nikhilananda at meals, which I hate. He embarrasses me by making conversation. His face is ravaged with nerves. Yet he is admirably opposed to the Indian weaknesses; fatalism, love of chatter, and indifference to social abuses. He bullies Al Winslow and the countess with an arrogance which I have noticed already in several of the swamis—racial aggression toward the Westerner?

This time I am playing it very broad with pranams. As an elderly man, I'm not expected to show such respect to the young swamis, but I do it anyway. (*My* kind of aggression.) And I have to jump backwards every now and then to prevent someone doing it to me!

Up on the roof after supper. The sooty smoky night air. Plenty of mosquitoes. Air raid sirens go off at irregular intervals from the nearby factories, for no special reason.

December 23. Woke feeling wonderful, and did some exercises. No more Librium. Unicap vitamins. Still eating too much.

Nikhilananda is an anxious egotist. His behavior jars on Swami, though they say very little directly to each other. Swami said rather wistfully to me, "I can't make conversation." I think he was afraid Nikhilananda would impress me. This kind of jealousy is his "last infirmity."

Visited Swami Madhavananda this morning. He was sitting up in bed, solemn with sickness, in his flapped cap. I cannot feel drawn to him or even respectful of his holiness. He just seems sulky.

Prema and I walked down to the little Howrah post office[2] which is just outside the Math gates. Prema was grimly determined to get his letters registered there—and, by God, he did too—despite all those brown pushing hands with *their* letters. No one had the least sense of obligation to stand in line and take turns. I think Prema was bent on proving that he can live in India on Indian terms. This is very much part of his attitude toward the immediate future.

Al Winslow is tall, slim, wide assed, boyishly pretty and eager

[1] Not his real name. He later left the order and married.
[2] Belur is in north Howrah, West Bengal's second largest city and its district headquarters.

beaverish, with curly hair and a pale face dark ringed under the
eyes. Mr. Carlson is middle-aged, but has the smile of a brainy boy,
a boy inventor in fact. He can fix anything, from the plumbing to
the electric light. He once took lysergic acid and saw a man's face
change into different faces belonging to different historic periods. I
walked down with them to the college which is right next to the
Math. We were welcomed by Swami Gokulananda, one of the
Pious Pig type, who made me promise to speak to the students.

We had to climb over the gate, because the gates into the Math
compound are locked during the afternoon and at night. Getting
the gates opened is going to be a constant problem, as the guesthouse is outside.

The wonderfulness of Prabhavananda is that he seems every bit
as much himself here as he does in Hollywood. He doesn't make
any concession to the environment.

His room, like most of the rooms around here, has the glamor
of Victorian Hindu; the funny old photographs and stuffed furniture. The Leggett House, in which both Prema and Arup are
staying, has charming old fanlights of colored glass, bottle-green
louvered shutters, doorhandles made in the shape of hands.

Great care must be taken to avoid getting your shoes stolen
from outside the temple.

Mosquitoes not so bad as last time I was here; but they bite.
Mylol is said to help.

Nishta is nice; a big woman, handsome and friendly toward me
with a mannish good nature. Maybe she's lesbian and has a thing
with Mrs. Beckmann, who is certainly the most feminine of creatures.

December 24. Tried to write to Don today, but could say nothing coherent. I feel dazed with "unreality"—which simply means
unrelatedness. They have pulled me up by the roots, flown me all
these thousands of miles and dumped me down here. I can't be
transplanted. But I may not die if I'm moved back promptly. Am
getting fatter at an alarming rate. There's nothing to do but eat.
Fat, lonely, bored.

This morning, on the river, two truly huge haystacks on rafts—
floating mountains of hay.

We drove into Calcutta, where I made a reservation for Rome
on BOAC for January 7. That seems centuries away. Swami
changed back into western clothes for the outing; he looked dapper
and ridiculously out of place. What crowds and crowds of people!
This is what most of the world is really like—overpopulation,

near starvation, utter squalor. Old trucks, bullock carts, rickshaws, little closed cabs such as Ramakrishna used to ride in. Holy men smeared all over with ashes. And the skinny wandering bulls.

Arup is sick, with the shakes. Nikhilananda has nosebleeds due to his high blood pressure; he has been told to stay in bed for a week. Wish he would!

Prema told me a strange tale about bands of transvestites who roam around the villages. (Though, as a matter of fact, he first saw them in Benares and took them at first glance for raddled old whores, then realized they were men.) The second time he saw them was at Kamarpukur. They are supposed, according to someone who informed him, to be "neuter." They have ways of knowing when a woman in the neighborhood is going to have a child, and then they come around and "do something" (unspecified) to the husband. They are terribly malicious. You must never cross them, or they'll revenge themselves on you. They sing and dance to amuse people.

Came out of my room around 5 p.m., after finishing (and enjoying) Waugh's *Ordeal of Gilbert Pinfold*. The river and the people in their bright colored clothes were just a river and people and clothes. I might as well have been in any foreign town anywhere. Hard to convey the strangeness of this unstrangeness. Let's put it that I felt as if I might easily have returned to my room, come out again and found myself, this time, in Cuzco. It would have made that little difference.

This evening, after *arati*,[1] we had a puja for Jesus. They built up an altar on a side aisle of the temple, with shelves of fruit and cakes, all surrounded by a picture of the Virgin and Child. I had been told to read the birth of Jesus from Luke and the Sermon on the Mount from Matthew. Was much disconcerted because they had given me a Roman Catholic bible, with different words, such as "our supersubstantial bread." So I had to keep transposing and improvising when I couldn't remember the King James. Then Prema spoke, in his somewhat now-my-dear-friends American way, but quite well. And then Prabhavananda spoke, on the Lord's Prayer. All the dark faces of the swamis, listening. I knew exactly how I ought to be feeling, so I didn't feel anything at all. But the ceremony wasn't in the least revolting. They sang a couple of songs to "Sri Isa"[2] in Bengali, which had the merit of taking Jesus right out of the Episcopalian church and putting him back in the

[1] Ritual in which lights are waved in front of a deity or a holy person.
[2] Revered Jesus.

middle of Asia, where he belongs. Krishna recorded it all on his new Jap recorder.

December 25. How nice to wake here, on my refreshingly hard bed, under the mosquito net! Waking up is helped by the rude crows, and a bird which emits liquid tropical whistles; and the factory sirens, and river-boat hooters and the noise of trains crossing the Vivekananda Bridge; and then finally a great smashing clash of buckets and pans as the help starts to get ready for breakfast, which we eat at 7:30 a.m. There is a rather sanctimonious head servant named John. He has been butler to many grand families and has worked in England, I believe. He is a Catholic and attended midnight mass yesterday.

Last night, in my dreams, my ordinary life suddenly caught up with me—I was surrounded by my playmates, Jack [Larson] and Jim [Bridges], Bill [Brown] and Paul [Wonner], Gavin, Jim Charlton, [Mark Cooper],[1] etc. And Don was there too. And *that* was so much reraller than this oriental backdrop. I felt very happy.

This morning, at breakfast, Prabhavananda tried one of the swamis' yellow flapcaps on Krishna. He looked very good in it. Probably he will henceforth wear it always.

Arup is still shaky. Swami is very concerned because, he says, if you aren't perfectly well, they won't give you sannyas. One of the hazards of the ceremony is that you have to bathe in the Ganges, and in the middle of the night, too. Telling me this, Swami drops his voice, as though he were describing some obscene rite of sexual initiation, instead of the most ordinary of holy acts.

We breakfast Britishly, on porridge, scrambled eggs, marmalade, strong tea and lots and lots of hard toast.

Nikhilananda is said to be recovering already.

This morning was astonishingly cold. (It has warmed up now, around noon.) Borrowed a sweater from Arup, rather than one of Prema's, because I felt this would make him feel a bit more included. Arup is isolated by his illness, which is really nothing but psychosomatic India-horror. Prema has almost forced him into this role by grabbing the role of India-lover. The greatest possible demonstration of India-love is not to get sick here.

A weirdly skinny oldish man who is a journalist and also connected with the Ramakrishna-Shivananda Ashrama at Baraset came to see me because he has always treasured my remarks about

---

[1] Not his real name. He was a boyfriend of Jim Charlton, and a model for the main character in Charlton's novel, *St. Mick.*

the guru in "What Vedanta Means to Me." He really did have the clippings with him and I couldn't help feeling flattered, although he was as embarrassing as hell about it. He brought me a book he has written, called *The Patter of Asude*, which is described by the blurb as "the funniest book ever." Also a Christmas cake, very small and heavy and hard. This he wanted me to eat right away. He offered me a rusty knife to cut it with. But Swami prevented this; promising him, however, that we'll meet again next week when we visit Brahmananda's birthplace.

Walked with Prema down by the ghats beyond the Math property. Old tumbledown dark crimson houses, with French statues in their gardens, and broken walls and vines climbing over everything. Pools full of water flowers, open stinking drains, white cows, lanes that wander to a sudden end, choked with rubble and garbage. Down at the ghats, brown-skinned youths scrubbed their faded paper-thin wearing-cloths and changed them without ever exposing their sex. They dunk their heads in the cloudy brown river water, swill it around in their mouths and spit it out. They have good wide shoulders but wretchedly thin legs. Prema says he is never troubled by lust in this country. We talked about the Franklin scandal. Prema believes he was guilty.

All over the Math grounds, they are putting up *pandals*, canvas tent halls with a bamboo framework, for speeches and mass meals on the day of the Vivekananda birthday puja, January 6. Also, in the field in front of the guesthouse, they have dug latrines and shit holes.

This afternoon, Swami, Krishna, Prema and I went to the opening of the Women's Congress. A stunning bore. All the speeches are in English and most of them you just cannot listen to; the Bengali accent and the droning delivery keep nudging you over to the brink of sleep. Thousands of people. On the platform, the Maharanee of Gwalior, a plump lady in widow's white.[1] They kept saying, "She has come all the way from Gwalior to attend this meeting," which was hardly tactful towards us world pilgrims. Also a *pra[v]rajika*,[2] with her shaven head looking like a small brown bespectacled nut. Also a lady named Maria Bürgi from Switzerland, wearing an improbable green toque and suffering from acute

[1] Vijaya Raje Scindia (1919–2001), widow of the last ruling Maharaja of Gwalior and Rajmata (1916–1960). She represented Gwalior in both houses of the Indian parliament and was committed to social service and mass education.

[2] Nun who has taken final vows; see Glossary under sannyas. Here, Isherwood spelled it "prabrajika".

enthusiasm. And that dreary old ass, Yatiswarananda, as chairman. The only gleam of joy was in the presence of Swami Aranyananda, who is really one of the handsomest boys I have seen in this part of the world. He comes from [...] the southern tip of India. His magnificent, nearly black eyes, very dark skin and fierce white teeth. His smile is fierce, tigerish, and challenging; but his eyes regard you with a languishing intimate sweetness. You can imagine him using phrases of classical oriental endearment like "soul of my soul" without the least embarrassment.

The front of the platform was lined with pots of pointsettias. On the wall back of the stage were three crude paintings, of Vivekananda, Ramakrishna and Sarada Devi, all of them decorated with garlands. Of the three, Vivekananda has been noticeably "favored"; his portrait is larger and more boldly executed than the others. On either side of the portraits were huge objects which looked like the fans used to fan some late Roman emperor. Whenever a speaker began to speak, a technician would promptly appear with a screwdriver and adjust the mike to his or her height, getting between speaker and audience and ruining the effect of the opening lines. I got a delegate's rosette to wear: orange with blue and purple ribbons enclosing a soulful picture of Swamiji inscribed, "Every soul is potentially divine."

In the middle of the street, a dead cow, killed by a car. This is said to be most unusual. As we drove home, Prema wanted to buy *Life* and *Time*, to divert poor Arup, whose fever is up again. Swami protested, because it meant delay in our getting home to supper. Why, he said, couldn't Arup show a little renunciation, one week before taking sannyas? But Prema quietly and firmly bought the magazines anyway.

December 26. This morning had a strange bright unhealthy chill; the sun burned one side of your body while the other shivered as if in a tomb. Arup is dreadfully hungry and worried about his health. He is confined to his room. Prema is primly healthy and pleased with himself.

Prema has taken hundreds of photographs for the Ramakrishna book. We spent the morning looking through them, sitting in front of the Leggett House on the marble bench by the entrance stairs. The benches are backed by sort of stone bolsters. White stone dust comes off on your pants. The charm of the little garden plot along the river embankment. A fountain full of green scum, supported by three swans below and above by two cupids, one of them headless. The gardeners are working on the chrysanthemums, dahlias and roses. At the foot of the embankment, discarded leaf plates and broken earthenware cups are agitated by the lapping river waves. A young swami who has just taken his bath wrings out his wet *gerua* cloth and hangs it up to dry in the breeze.

Prema, with his usual crushing frankness, remarked that he has been reading through the Ramakrishna book and doesn't think it's really "great." I agree with him, of course. But I added that I could probably draw a much better portrait of Ramakrishna to a sympathetic stranger one evening when I had had a few drinks. There is that in me which will never write its best to order. Deep down, something has always been resenting the censorship of the Math and Madhavananda's comments.

Meanwhile, thank God for the genius of Willa Cather. I am relishing every single page of *The Song of the Lark*.

This afternoon, I wanted to get into the Math grounds before the gates were opened. But I couldn't climb over the side gate because there were so many people standing there. And the main gate was much worse: a crowd of maybe a hundred, including two cows. So I had to wait my turn. A tiny child begged cross-legged on an outspread mat. An adult beggar exhibited his deformed hand. Merchants squatted behind white cupie-dollish figures of Ramakrishna and framed photos of Vivekananda and Rabindranath Tagore. Quite well-dressed middle-aged men crouched to piss in the ditch alongside the lane, exposing ugly naked shanks. The *dhoti* looks so proper from in front; then, from behind, you glimpse the bare legs; and the effect is indecent because it seems unintentional.

White cranes perch on the backs of the cows, apparently searching them for lice. Anyhow, the cows seem to like it.

Tension at supper this evening. We asked Mr. Carlson and Al Winslow to get us a birthday cake for Swami. But it was a great mistake, because we embarrassed him in the presence of Nikhilananda, who smiled in a superior manner, to remind us that sannyasins are not supposed to recognize birth and death. Nikhilananda told corny Jew stories. Prabhavananda looked small and sad, at the other end of the table. And the cake was hard as rock.

December 27. A tedious interview with an ass of a journalist named D.P. Tarafdar, of the *Amrita Bazar Patrika*. He wrote down everything very slowly in longhand. Then Prema and I dashed into town to pick up some film for taking pictures of the Parliament of Religions delegates. Prema rejoiced that he will soon be retiring from all these concerns into the seclusion of preparation for sannyas. It was hot and loud in Calcutta, with a hint of the weary heat of the coming months and the smell of sewers and septic dust. How easily one can lose one's vitality here!

This afternoon, I talked at the Ramakrishna Mission College. They had fixed up the gymnasium for the occasion; the stage was part shrine, part oriental parlor. On the back wall was a thing like a monster valentine, enclosing Vivekananda's portrait. Below this were a number of basketwork lotuses. Incense was burning before them. Downstage were a draped couch and a draped table, with a stick of incense burning on it, right under the noses of us speakers, as we sat on the couch, garlanded by the students, like gods. (Luckily we were allowed to take the garlands off again; they were terribly hot.)

I was fairly good, I guess. Not that I said anything much, but it came through without hesitation, and good and loud. Many of the students had their arms round each other as they listened. They were thin, pliant-waisted youths with dark mocking eyes and smiling teeth and, quite often, moustaches. Then Swami was asked to come up on to [the] platform and answer questions. He was all silver and gold—silver hair and gold skin with a silvery light on it, and the blending yellow of his robe—and again his greatness was revealed. He told the boys that their college would be a success only when it produced at least a dozen monks a year. He was adorable—so amused and teasing and yet quite quite serious. By an awful effort of piggy peg-toothed Gokulananda, Nikhilananda was not asked to speak. He sulked a bit about this, but I will admit that he had the grace not to sulk afterwards at supper.

After Prabhavananda, there was Justice P. B. Mukherjee, who spoke for nearly an hour, flipping through the pages of a manuscript which could probably have lasted [two to three] hours if read in its entirety. When it was over, Swami Gokulananda, without the faintest trace of irony, said, "All's well that ends well." We were then served grey sweetened milk tea and oranges and cookies. There were also glasses of water. Swami, who is determined that I shan't be poisoned on this visit, said quite loudly, "Don't drink it, Chris."

December 28. This morning, the *Amrita Bazar Patrika* carries a piece on me by Tarafdar without one word in it of what I actually dictated to him. It begins, "Why did distinguished writer Christopher Isherwood become such a strong admirer of Swami Vivekananda (to the extent of banging his fist on the table in raptures, as he did in Calcutta on Friday)?"

Then Gokulananda arrived, to announce that my talk had not been taped. Could he have a copy of it? Told him with sadistic relish that there was no copy; I never write my speeches. But he then produced a short version of the speech taken down in longhand by one of the students—quite inaccurate, but earning a big B for emotional blackmail. Now I'm obliged to go back to the college and redictate the whole thing to a tape recorder. Fuck them.

Today, I finished *The Song of the Lark*. It's certainly one of her greatest.

Why do I feel such an intense eagerness to leave this place, and this country? I count the days. It *is* an experience, being here. I *am* getting something out of it, I know. And yet I strain like a leashed animal to escape.

Today, Prema, Swami, Krishna and I ate lunch with the swamis, in the monastery dining room, sitting on the floor. You eat with your right hand; mustn't use your left. I finally had to sit on mine, because it kept flying up to my mouth to help the other. At intervals, one or other of the swamis would start a chant. Often these chants sound lively and aggressive, like political slogans. I noticed particularly one sturdy old monk, who walks around with a pilgrim's staff; he chanted with such an air of game toughness and sturdy enthusiasm. You saw him as a young boy and now as an old man. He hadn't changed. He had taken his vows and he would go through with this thing to the end with unquestioning loyalty and faith. Such a comical old man, his chin nearly meeting his nose. All he has done—all—is to take what Ramakrishna said quite literally; and so has no problems, and no money and no fame, and is maybe a saint. One out of dozens.

But I hate floor eating. It is messy and unsnug. Couldn't get my plate clean of the dull runny tepid food.

Along with the interview in the newspaper was a photo Prema just took of me, making me look lean faced, sly and shifty eyed, rather like Oppenheimer.[1]

To the college, where I taped my talk—it turned into something entirely different. The students sat around bright eyed, but they didn't really understand; when it was Swami's turn they asked him to speak Bengali. Later, Prema gave his lecture on the Vedanta Society of Southern California, illustrated with slides. The slides gave an extraordinarily strong impression of luxury, cleanliness and lack of clutter. In contrast to here, even the freeway looked tidy, and the shrines seemed so sleek with polish and well carpeted,

[1] Presumably Robert Oppenheimer, the physicist.

they were like comfortable hotel rooms, and the flowers and trees were so luxuriant. The audience was more or less the same as yesterday's, except for some small kids who lay bundled up, two to a wrapper, on the floor under the screen. Prema spoke excellently, but they didn't really understand either him or the pictures. (The scenes from Disneyland might as well have been visions obtained through mescalin.) But then there was a Bengali film showing the procession which inaugurated the Vivekananda celebrations, last winter—an endless straggling confusion of cars, banners, military cadets, cows, musicians, trucks, political speakers—and this they truly understood. It was *their* Vivekananda—he appeared again and again, as a photograph, as a cardboard cutout, as a plaster statue— imposing his presence through sheer campy absurdity—made into a god in order not to be taken seriously. (How he would have raged against the editors of the Delhi centenary volume; it is a jungle of misprints![1])

Just before supper, the lights in the guesthouse failed and stayed failed. Bed in the dark.

December 29. Swami, Krishna and I have just moved into Calcutta. During the Parliament of Religions, we are to stay at the International House of the Ramakrishna Mission Institute of Culture, because it will be easier for us to get to the meetings. It is very grand and well planned, but the floors are grimy with fallen dirt, as in New York. However, we have air conditioners.

The place is run by Swami Ranganathananda, a very handsome well-built middle-aged man, who is reputed to live entirely on milk. There is also a sort of hostess, Mrs. Bouman, who is Dutch and tall in her winding sari, and rather like Virginia Woolf. And there is Aranyananda, looking older today and unshaven, but still beautiful. I took the dust of his feet, and he tried to do the same to me. When I jumped backwards, protesting, he said, "We regard you as more than a swami."

We had lunch at the institute—rather delicious British-type fish rissoles—and got involved with Dr. Miroslav Novák, who is head of something called The Czechoslovakian Church. (He calls himself a Protestant, but other information seems to suggest that he is Russian Orthodox. At the meeting he wore a black Protestant pastor's robe on which was embroidered in red what appeared to

---

[1] Evidently *Swami Vivekananda Centenary Memorial Volume* (1963), edited by R.C. Majumdar and published in Calcutta.

be the Holy Grail.)[1] Didn't like him. Too much a *faux bonhomme*.[2]

Aranyananda introduced us to a Dr. Roy, who is a surgeon and lives here. He will get us a negative of the complete cremation picture for our book. It is far more impressive than the two-thirds which are usually printed. The gazes of the mourners have a focus, and the corpse isn't at all shocking; it is nearly buried in flowers. As Swami says, many crucifixion scenes are far more gruesome. But no doubt there will be fusses. The Math itself is opposed to printing the whole picture. We shall see. This is something I'm prepared to take a strong stand about.[3]

After lunch, Aranyananda was in a flap, because he had to produce biographies of the foreign delegates. So I helped him correct their awful English. An adorable brahmachari named S[h]ashi Kanto,[4] who rooms with Aranyananda, made this task seem even lighter.

But there was nothing light about the inaugural session of the parliament. It began at 3:30 p.m. and went on for three hours. Next to the hashish experience in Tangier, this was the least endurable time stretch I have ever known. Not one of the speakers bothered to project; they droned out their written speeches as if they were saying mass. There was an audience of about eight thousand people, and I doubt if eighty of them really understood English. They sat there with—no, one can't call it patience—with the inertia of cows. Nikhilananda, who was next to me on the platform, fidgeted openly and didn't bother to applaud. He really is a very second-class swami, but I find his disgust humanly sympathetic. When it was my turn, I spoke too loud and too urgently—rather like a communist speaker in the thirties.

After the meeting, Swami, Krishna and I were taken to the Calcutta Club by a Mr. and Mrs. Gupta. He's a cricket-anyone? English-type Bengali; he even wears a kind of blazer and knotted scarf. He has some important job managing the port of Calcutta. Her name is Mallika.[5] She's American, plump, pale, blonde,

[1] The Czechoslovak Church began as a reform movement among Roman Catholic clergy in Czechoslovakia and was officially established under a Czechoslovak Patriarch in 1920. Novák was the fourth Patriarch.

[2] Man of pretended joviality.

[3] The "Death Picture of Ramakrishna," taken August 16, 1886 and showing more than fifty devotees standing behind his body, is printed as Plate 19 in *Ramakrishna and His Disciples*.

[4] Isherwood typed Sashi Kanto, but hereafter he typed Shashi Kanto, showing the pronunciation.

[5] Mallika Clare Gupta, author of a privately published book about Swami Gnaneswarananda and co-author with Irene Ray of *Story of Vivekananda*, for children.

humorless. The Calcutta Club was founded in 1909[1] by an Indian who was refused admission to the regular British club—but it is more British than the British, charmingly old-world London club atmosphere. A weepy young rich drunk who was a friend of theirs was introduced to Swami and confessed embarrassingly. He was playing a scene.

Then we went on to the Star Theater, to see the old paintings of Ramakrishna and Girish Ghosh backstage.[2] The backstage part of the theater is probably very much as they knew it; the rest has been modernized. They were doing a modern play by Debnarayan Gupta (whom we met) called *Tapasi*. The plot was wildly complicated, with old-fashioned coincidences; recognition of a blind mother, first deserted by her husband and then reunited with him when he too goes blind. Sheer Dickens. But the acting was so lively and enjoyable. Without understanding a word, you could see how naturally theatrical these Bengalis are, and how unnatural it is for them to put on a ditch-dull show like the Parliament of Religions.

Afterwards, the Guptas took us to supper at the Sky Room, quite a grand restaurant, but the food was inferior to the guest-house. A mixed clientele—two bald-headed Britishers out on a spree with Hindu girls; two club-type Britishers obviously trying to pretend they were in London; a mixed-up family of Eurasians, some of them beautiful.

As night falls, a truly hideous smoky fog closes down on the city, from all the charcoal pots on the streets and the soft coal fires in houses. Back in my room I really wondered if I should be able to breathe. In these few hours my shirt has become filthy around the neckband.

December 30. Swami didn't sleep much. He complains of the smoke but won't go back to the Math. Showed him an astonishing folder I found waiting for me last night when I got back to the institute; a bunch of pages of lettering and pictures in ink and watercolor, headed "Jesus Christ Writes to Christoper Isherwood in Calcutta." There is a small snapshot of a skinny bearded young man, naked to the waist, grinning in a slightly mad self-conscious way. On the other pages, various facts are revealed—that Jesus

[1] 1907.
[2] Girish Ghosh, Bengali actor, playwright, songwriter and Ramakrishna devotee (1844–1912); Ramakrishna attended several performances when the theater opened in 1884 and thus became patron saint of Bengali drama, with his portrait backstage in most Calcutta theaters.

is opposed to the partition of India, that he has broken with St. Peter (who, it appears, is now in London), that he was helped by Cyril Frederick Golding to get a job in *The Times of India*'s layout department. Some quotes: "Peter came and established the Rock of Hatred in advertising. I joined him on the Rock and got what I wanted." "During the War and the riots of Calcutta, Peter loved God as his own child and raised his wages to a thousand silver pieces per month; but Christ, after looking at the cruelty of man, resigned from the services of Peter and took up the pen for judgment." However, on the next page, Christ says, "Oh, why bring tears in the eye of an old man! He suffers anyway." All this rings faintly homosexual, especially as the last page refers to "John." Christ approves of Sir Christopher Wren, Beethoven, Mozart. He disapproves of Nero.

I think he's quite likely to show up today. I rather wish he'd come and make a scandal at the parliament and brighten things a bit.

I forgot to mention yesterday that Swami is conspiring with Dr. Roy to bribe the custodians to hand over a coat belonging to Ramakrishna which is at Mathur's former house.[1] Roy says it isn't being kept properly, and Swami wants to bring it to the Hollywood Center.

Calcutta is a pale faded yellow city—all strong color has been burned, parched out of it by the sun. At night it is crowded but cheerless, under its pall of dirty smoke. A poor wretched place; the joyless street of six million people. Looking out the window at dawn, you see bent figures in wispy smoke-colored garments moving silently about like emanations of the smoke, as they light their fires to create more smoke.

Ranganathananda showed us around the institute. He is the Monsignor Sheen type,[2] very handsome, grey haired, youthful, fanatically energetic, fiercely ambitious, socially alert. He tells us he keeps his health by doing *asanas*. He no longer sticks to his all-milk diet but still limits himself to just a few vegetables.

The institute is really well equipped and admirably efficient. Poor students can get meals almost for nothing and study all day in

[1] Mathur Mohan (d. 1871), also called Mathuranath Biswas, a devotee who persuaded Ramakrishna to settle at Dakshineswar, and who provided for his material needs.
[2] Fulton J. Sheen (1895–1979), American Roman Catholic priest, a bishop from 1951 and, later, archbishop. He broadcast "The Catholic Hour" on NBC radio beginning in 1930, and from the start of the 1950s had a weekly T.V. show. He also published a number of books. Isherwood typed "Sheean."

the library. There is a meditation room with nothing in it but an electric light projecting from a kind of lingam. It is in the shape of a flame. Swami, who doesn't really like or trust Ranganathananda, complained that it was much too strong; you couldn't concentrate.

Ranganathananda is greatly excited about the growth of interest in Vedanta among the Japanese. He showed me a letter from a young Japanese who is coming here soon to join the order and work at the institute. His eyes gleam with fanatical delight as he tells about this. I feel he thinks he is running the order single-handed.

Aranyananda and Shashi Kanto were also around; they came with me to the parliament for my speech. Shashi Kanto says his name means Moon Beauty or Husband of the Moon. He is from near Bombay, a big boy of about eighteen, bulky and yet graceful in his cocoon of white muslin. The cropped hair and little topknot suit the charm of his long sensitive affectionate nose and dark soft velvet eyes. He seems utterly incapable of anything but love. He finds all manner of excuses to be around us. Every day he washes Swami's and Krishna's gerua clothes.

At the parliament, I found that two or three of the speakers were missing; there were only Prema, and a Captain Bhag Singh,[1] and the president of the day, Dr. Chatterjee. Prema talked on "Vivekananda Through the Eye of an American." He was good but much too brief. He seems quite a dour elderly figure on the platform; projecting grim austerity. He even reminds me a little of de Gaulle. As for myself, I was pretty good. I pretended to myself that the audience could understand me, and indeed they seemed to—probably because I talked a lot of political stuff about Vivekananda and the English, the oppressors in their bondage to the oppressed, etc.

When it was over, a tiresome peg-toothed swami (of the Gokulananda tribe—swamis fall into quite recognizable physical groups, I notice) tried to sick the journalists on to me. But Aranyananda charged them like a little tiger and made a way for me through to the car. As we drove back to the institute, he was very indignant because there had been a translator to render the gist of the talks into Bengali, and, said Aranyananda, he hadn't done so but had wandered off into remarks of his own. Aranyananda said that the audience *did* understand English and would certainly have booed the translator if they hadn't been intimidated by the pictures

[1] A founder and editor of the *Sikh Review*, published monthly in Calcutta from 1953.

of Ramakrishna etc., which made the pandal into a shrine where you had to behave yourself.

December 31. Just before going to bed, I started to get the gripes and shits. I shivered a lot and couldn't sleep all night. Lying awake in the dark, I was swept by gusts of furious resentment—against India, against being pushed around, even against Swami himself. I resolved to tell him that I refuse ever again to appear in the temple or anywhere else and talk about God. Part of this resolve is quite valid; I *do* think that when I give these God lectures it is Sunday religion in the worst sense. As long as I quite unashamedly get drunk, have sex and write books like *A Single Man*, I simply cannot appear before people as a sort of lay minister. The inevitable result must be that my ordinary life becomes divided and untruthful. Or rather, in the end, the only truth left is in my drunkenness, my sex and my art, not in my religion. For me, religion must be quite private as far as I'm publicly concerned. I can still write about it *informatively*, but I must not appear before people on a platform as a living witness and example.

Luckily, Swami's sister came to visit him in the morning with her son-in-law, who is a doctor; Dr. G.K. Biswas. He examined me and gave me pills. I felt achey and sleepy.

Three women got me into a corner at the end of the balcony and started to ask me about karma, reincarnation and so forth. Another little woman joined in. Another came and got my autograph, after giving me a New Year's card. And merciless Aranyananda, after shooing away a man who said he was the son of a millionaire, got me to correct a translation of a speech made by one of the Japanese delegates.

Then I was fetched to go out to Narendrapur, where the mission has a huge project; schools, clinics and a farm. This was more fun than I'd expected, because Winslow and Carlson and Prema's two friends who are staying at the Great Eastern, Bill Chapman and Jay Taylor, all came along. I very much liked Swami Lokeswarananda, who runs the place; he reminded me of Dore Schary. Also, there was a nice solid young bra[h]machari from Pavitrananda's center in New York, called Amul (his name is Clare Street) who has been here three years already. Also a handsome and sexy nineteen-year-old boy from Cheshire, named Mark Vallance,[1] who isn't a devotee

[1] Mountaineer (b. 1944), son of two Unitarian ministers; later a Base Commander for the British Antarctic Survey team and President of the British Mountaineering Council.

yet but has come here to teach English—or rather, his very no-shit Midlands accent. He plans to start reading Vivekananda's works as soon as he has finished [Hemingway's] *For Whom the Bell Tolls!*

Sick as I was and groggy from the hot sun, I was hugely impressed by the Narendrapur project. It makes you feel that India isn't in such a bad way after all. The government favors the mission because it is one of the very few social service agencies where there is no graft. Vivekananda was absolutely right; you simply cannot do this work without dedicated people. For others, it is too tiresome, so they turn into crooks. Lokeswarananda told us how he went to the ministry prepared to ask for 25,000 rupees for his project. But before he could speak, the minister told him, "Look, Swami, we admire your work, we respect what you are aiming for—but we simply cannot let you have more than half a million. It's no use arguing. That's our limit." So he got half a million.

It was lovely to be out in the clean country air. Afterwards I came home exhausted and lay down and napped on the bed. Aranyananda tried to wake me to perform some new chore, but I acted dazed-sick and he went away.

1964

January 1. Slept well and woke feeling much better. I still am resolved to tell Swami I won't give any more religious talks; but I'll do so only after my talk at Belur on the 6th; and I'll offer to give two talks about this trip, in Hollywood and at Santa Barbara, and also two readings on other Sundays while he is still away.

Like a marvellous omen of joy for 1964, the first person to appear at my door (while I was shaving) was brahmachari Shashi Kanto. He had come to wish me a happy New Year.

Dr. Biswas looked in to see me after breakfast. I have to admit it gave me a slight jolt when he told me that he's the senior medical officer in a leprosy clinic! He says the disease can now often be cured and always arrested, but that there is still a great deal of it around. There is still no law to compel lepers to report themselves and be treated. People from all classes get it—usually during childhood, from infected nurses. People with European blood are hard to cure; they have no immunity to it. Often they must be treated for the rest of their lives.

Swami presided at the parliament yesterday. He says Nikhilananda tried to persuade him to speak first (the president is supposed to speak last) and then leave. This was because Nikhilananda plans to do this when he presides, and he wanted to have a precedent created for him.

View from my bedroom window, on to the street outside: A large building, once a wealthy family mansion, now broken down and overcrowded. The plaster has fallen away from it in great pieces, exposing brickwork. Its green shutters are faded. Small trees grow from crevices in the upper balcony. A bamboo pole is fixed across the main entrance, to hang laundry from; a crow perches on it. There is more laundry in another part of the balcony which has carved balustrades and some faceless figures which may once have been lions. The trunks of the four tall palms in front of the house are stained with smoke. An old woman fans a charcoal brazier. A young man pees against the wall. Barefoot children wander back and forth. Along the street a white cow passes. Then an incredibly stringy calf. (The cows belong to people. The bulls were let free in the streets as sacred creatures, whenever a loved person died.) A bridge connects the main house with a garden house. It, too, is ornate, with broken Corinthian pillars, but now it is roofed with corrugated iron and bamboo matting. Two long saris, of different shades of green, are drying from the rail of the bridge.

After lunch today, Swami told Maria Bürgi to stop wearing hats altogether. (She had on a truly weird contraption, just like a slipped turban.) Bürgi explained that she wore hats out of respect for our sacred surroundings. Swami replied that that was merely an idea of St. Paul's; it doesn't apply in India. Swami also advised Bürgi to wear gerua on her head, if she must wear anything. The result will probably be appalling.

I have a new badge now. The same ribbons and general design, but this one has the motto, "Mother, make me a man."

I presided at the parliament this afternoon, after a flirty tea party with Aranyananda and Shashi Kanto. Such languishing looks, delicate hand-touches and flashing glances are perhaps only possible for the absolutely innocent. Though I'm not sure if Aranyananda is quite as innocent as all that. I feel he has been around.

Our session of the parliament was fucked up by the non-appearance of Mr. Humayun Kabir, the minister for petroleums and chemicals. There is a possibility that he may have heard that Swami Sambuddhananda (the organizer of the parliament, who has the tact of a hog and the voice of a bull) referred to him, by a slip of the tongue(?) as Mr. *Hanuman* Kabir—which could be construed as a deadly insult since Kabir is a Moslem. Making a monkey out of him, literally!"[1]

My speech was better than the other, though less well received. When I had reached the very last sentence of it, Sambuddhananda handed me a written message, "Continue for fifteen minutes." Because they had realized that Kabir wouldn't show. I ignored this, and stopped. This kind of behavior is enormously insulting, however unintentionally so; typical Hindu thick-skinned bossiness.

When I got back to the institute, I dropped and broke my glasses. So Prema had to be phoned at the Math to send my other pair by Sujji Maharaj, who is to meet us tomorrow on our way to Sikra Kulingram, Brahmananda's native village. Meanwhile, Krishna temporarily mended the broken frame with adhesive tape. He cut his finger doing so—and somehow this seemed touching, a kind of bloodshedding for me.

Maria Bürgi appeared at supper hatless, with a ribbon around her hair. She looked very good.

I was in the bathroom, brushing my teeth, when Aranyananda appeared. He obviously wanted to talk, so we did, until half past twelve—that is, for more than three hours.

He started with anecdotes about his relations with Shankarananda. (At least, I *think* it was Shankarananda. Anyhow, it was some very senior swami of the order—which one hardly matters much, since Aranyananda's attitude to the whole thing was so subjective.) Such studies in monastic psychology exceed by far the sensitivity of a Proust. Aranyananda—his eyes blazing with remembered passion and also with satisfaction at his own hypersensitivity—described how, after waiting on the swami faithfully and faultlessly for months, he made one little slip—forgot to get some medicine the swami had ordered. Next day, he was told that the swami had been very annoyed. So Aranyananda became furious and went into the swami's presence spoiling for a fight. But the swami somehow conveyed to him by a glance how much he loved him. So all was well. The motif of a loverlike need for reassurance kept recurring. You are equally ready to leave your guru and the monastery for ever, or to fall at his feet in tears. Such scenes could obviously become as necessary to one as playing Russian roulette. They would have to be repeated at least once a week.

I then asked Aranyananda how he came to join the order, and he gave me a description which was built up move by move and word by word. The curious thing about the story—since his must,

[1] Hanuman is the Hindu monkey god.

after all, be accepted as a genuine conversion, not a caprice—is that Aranyananda apparently wasn't influenced by any living human being. He was more or less of a freethinker, surrounded by very intellectual brothers and sisters, most of whom have subsequently made for themselves brilliant scientific careers. His father (a [literary] scholar) and his mother would have been horrified if he had told them he was planning to become a monk. So Aranyananda had to run away from home, which he successfully did. He didn't even have a friend of his own age who would have understood him.... And what made Aranyananda decide to become a monk? Simply reading the works of Vivekananda!

The night he left home was very carefully planned. And yet this boy, who was renouncing a loving family and financial security and all the good things of the world, very nearly missed his train because he couldn't, at that hour of the night, find anyone to carry his suitcase, and to carry it himself would have been a loss of face! Aranyananda quite saw how funny this was.

And now, says Aranyananda, he is blissfully happy at the institute, and he and Ranganathananda and Shashi Kanto quite often joke together like equals, making jokes "below the belt" so that onlookers are quite shocked. He says Ranganathananda can skim pebbles with terrific force, using his bent-back middle finger as a catapult, and he can take water in his closed hand and squirt it for astonishing distances.

Behind all Aranyananda's stories there is a certain suggestion of "see what a tiger I am—yet I'm as gentle as a dove if you treat me right." Also, there's a good deal of name-and-fame awareness. In what other station of life, he asks, would you find famous men and women actually taking the dust of your feet? He frankly delights in this. And he told me, encouragingly, that I should become far better known by my book on Ramakrishna than by any of my novels. (Incidentally, he thinks Romain Rolland's book is supreme; all I can hope to do is be the next best.)

I see Aranyananda and Ranganathananda as two of a kind. Aranyananda understands and thoroughly approves of Ranganathananda's ambitiousness. Shashi Kanto is different. He told Swami that he wants to give up work and spend his time in meditation. But Swami told him to stick to work for the present.

January 2. The kind of sweat that breaks out on you on a warm smoky morning here is like the unnatural sweat you sweat after taking aspirin.

This morning, Swami, Krishna and I left the institute by car and drove to Shivananda's birthplace.[1] There we met Sujji Maharaj and went on with him to Brahmananda's birthplace, Sikra Kulingram. It is a tiny village out on the flat paddy-fields of the Ganges delta, enclosed in an oasis of lush trees and very bright blossoms. There is a shrine there, and a guesthouse.

Throughout the drive, I felt awful[.] Partly upset stomach and headache, but chiefly rage against the Parliament of Religions, the Ramakrishna Math, India, everything. This is a very deep aversion which I have been aware of from time to time ever since I first got involved with Vedanta. It has—as far as I can figure out—nothing directly to do with Ramakrishna, Vivekananda or Swami. (Did Roman converts to Christianity loathe the Jews all that much the more?) Anyhow, it all expresses itself in the old cry of the ego, *I'm being pushed around!*

When we arrived, we were told that Krishna and I would have to share a room. Did I mind this? No—I honestly don't think so. But I immediately said I had a headache and wanted to lie down. Maybe I *did* have a headache, but what I really wanted was time to figure out what I was going to do next. I realized that I was going to make a scene and I needed time to rehearse it. Presently, I was through with the rehearsal so I got up and began walking around, feeling better already. It was quite warm, with a brilliant blue sky. The leaves were flashingly green, the flowers were vivid. Dark smiling children sat among them, half hidden in the shadows.

I found Swami sitting with Sujji Maharaj (whom, at least for the moment, I disliked, as being one of the pushers-around). I took Swami aside and asked him if I could have the car drive me back to Belur Math at once. Swami seemed bewildered, as well he might be. He said gently, yes, of course—but wouldn't I have lunch first? On such occasions, he seldom asks leading questions. If you want to make a scene you have to make it all by yourself, under your own steam. So now I did. I said, approximately, "Swami—it isn't just that I'm sick—I feel awful about everything. I've made up my mind: I can't ever talk about God and religion in public again. It's impossible. I've felt like this for a long time." (Already, I had withdrawn the concessions I had previously planned—to agree to talk in Hollywood after I get home. Some instinct told me that this ultimatum must be drastic or it would make no impression at all.) "I suppose I've wanted to spare your feelings, but that's not right, either. After all, you *are* my guru—you have to be responsible for me anyway—and you're probably a saint. Anyhow, you're the

[1] Barasat.

nearest thing to a saint I have ever met. So why shouldn't you be told how I really feel? It's the same thing, really, that I told you years ago when I was living at the center: the Ramakrishna Math is coming between me and God. I can't belong to any kind of institution. Because I'm not respectable—"

At this Swami laughed, more bewildered, than ever. "But, Chris, how can you say such things? You're almost *too* good. You are so frank, so good. You never tell any lie—"

"I can't stand up on Sundays in nice clothes and talk about God. I feel like a prostitute. I've felt like that after all of these meetings of the parliament, when I've spoken.... I knew this was going to happen. I should never have agreed to come to India. After I promised you I'd come, I used to wake up every morning, feeling awful—"

"Oh, Chris—I'm *sorry*. I shouldn't have asked you—"

"You know, the first time I prostrated before you, that was a great moment in my life. It really meant something tremendous to me, to want to bow down before another human being. And here I've been making pranams to everybody—even to people I've quite a low opinion of. And it's just taking all the significance out of doing it—"

"But, Chris, you don't *have* to do it. Nobody here expects it of you—"

All this time, we were walking up and down in the brilliant sunlight, along the path between the ranks of glossy dark leaves, with Krishna somewhere in the middle distance, and Sujji Maharaj and the others on the porch of the guesthouse, and the hidden children watching. I felt that everybody knew a scene was taking place. I also felt that I was acting hysterically. Indeed, I couldn't have looked Swami in the eye while I was saying all this. But I didn't have to, because I was wearing the dark glasses belonging to Jim Cole. (I brought them with me to the airport to give back to him—he left them at our house—but then he never showed up, so I had to take them along on this trip.)

Swami had barely understood a word. He was quite dismayed. "I don't want to lose you, Chris," he said. I told him there was absolutely no question of that. That I loved him just as much as ever. That this had nothing to do with him. But still he didn't understand. He looked at me with hurt brown eyes. I felt rather awful and cruel—but not very. However dishonest all this may have been in one sense (for, after all, by taking this stand, I am saving myself one hell of a lot of work and annoyance) at least its expression was honest and frank. It was far better to have spoken than not to have spoken. The boil was lanced and I felt better immediately. Sujji Maharaj received the news that I was going back to the Math with his usual slightly cynical impassivity. "Can't take it, huh?" he was thinking. He often used to look like this when I was sick during my last visit. We had a silent embarrassed lunch at which I ate only rice. Then the chauffeur drove me back into Calcutta in a cloud of red dust. Through the eyes of my relief, India suddenly seemed charming. The long fruit market alongside the street of De Ganga village, where we were stalled behind produce trucks. I almost loved the dark-skinned country people, so completely absorbed in the business of their world and shouting at each other in angry voices without anger and with campy fun. And I was so happy to get back to my quiet room at the guesthouse.

When I told Prema about all of this, he was most understanding. I hope he'll be able to explain things to Swami in due course.

January 3. At lunch Nikhilananda talked at length about John Moffitt and his defection from the order to become a Catholic.[1] He often refers to Moffitt; it's obvious that he feels guilty and responsible for what happened and is trying to forestall criticism. Nikhilananda is very sympathetic at such moments, because he really does seem to cover all Moffitt's reasons for leaving him, including Nikhilananda's own bossiness and constant belittling and humiliation of Moffitt by loading him with menial chores and failing to acknowledge the huge extent of his literary help in the books they published. But, as Prema pointed out, what Nikhilananda *doesn't* take into account is that Moffitt really *is* drawn to Christianity and prefers it to Vedanta and Ramakrishna. Rather than admit that, it seems, Nikhilananda will blame himself.

Not knowing Moffitt (I'm supposed to have met him once but I don't remember it) I picture him as a weaker brother of mine. I think that he, like me, is prone to do more and accept more responsibility than he really wants to, and then to have violent reactions in which he goes to the opposite extreme. (Once, after nursing Nikhilananda with the utmost care, while he was sick in the country, he suddenly walked out on him, leaving him all alone.) I say he is weaker than I, because I know enough about myself, usually, to let off steam before the pressure gets dangerous.

---

[1] Moffitt, a writer, was a monastic follower of Ramakrishna for twenty-five years. His nine books include several volumes of poetry and a memoir of his spiritual journey.

As for the bossiness of swamis, whole books could be written about this. They are nearly all arrogant—lacking in manners—by western standards. They push their disciples around. (Of course, I'm only speaking of the kind of swami who has disciples.) They push the younger swamis around, even. When you see Nikhilananda bullying Al Winslow or the countess, you feel he is compensating for what the British did to India. (Nikhilananda, to his credit, was an active anti-British terrorist in his teens and got sent to a concentration camp.)

Went to the Cultural Institute for a social tea arranged by Ranganathananda; just another concealed lecture. I found Swami there, in bed with a cough; very rumpled and sad. He had become sick again at Sikra Kulingram. The country dust is blamed; but I got a strong impression (later confirmed by Prema) that the sickness has a lot to do with me. This is perhaps the only respect in which Swami can be described as sly; he is absolutely capable of getting sick to make you feel guilty, though I doubt if he realizes this—it is purely instinctive. Since I couldn't possibly admit that I know this about him, all I could do was to be extra sweet, and at the same time absolutely firm about my decision. (As I said to Prema, I think the Hindu national technique of wheedling has even been developed in relation to God. Their motto is: All's fair in prayer.)

The atmosphere, even in Swami's bedroom with the air conditioner working, was thick with smoke. Never have I known it worse. Ranganathananda tried to get me to stay the night; but I had nothing with me, so could excuse myself. Ranganathananda was belittling Swami's illness and telling him he should do asanas. (He can lie on his stomach and bend so far backward that he looks straight up at the ceiling.) Swami was saying he's too old to travel. I suspect that he'll cut his trip short.

So I had to double for him and myself as guest of honor at the grim tea reception. It was held in a room containing an enormous circular daybed, on which a sultan surrounded by a dozen reclining wives could easily sit. Ranganathananda tried to make me sit there, with the fifty (or so) guests around me. This I wouldn't do. But after tea I had to let myself be publicly quizzed by him. He made everyone shush, and then asked me questions; I had to answer in a voice loud enough to be heard by all. Among other things, he tried to get me to say that I disapprove of the dirtiness of modern literature.

Later the conversation became more general. Dr. Roy, wriggling girlishly, told us how he had seen a yogini, a woman of about thirty-five, levitate in a small mud hut. She did a lot of deep breathing, then took one great breath and rose into the air from the chair on which she had been standing. Dr. Roy and his friend, a skeptical chemist, passed their hands under her feet.

Noticed, on the drive into Calcutta, the huge wheels of a bullock cart with a dwarfish driver sitting between them, pointing his stick at the bullock with the gesture of an enchanter pointing his wand.

January 4. Today, Prema and Arup got their heads shaved, in preparation for sannyas. They were very coy about this. On the one hand, they didn't want to expose their baldness; on the other, they wanted to tan the indecent whiteness of their skulls. The shaven Bengalis, with their brown skins, look perfectly natural. Prema told me he feels that this ceremony was a crossing of the rubicon. Now he is really committed. He and Arup have already been issued their gerua clothes, all neatly folded ready to be put on.

All over the compound, the preparations for the big celebration on the 6th are almost ready. There is an arch over the entrance gate, numerous big pandals with blue and white striped curtains, tented entrances to the shrines of Brahmananda and Holy Mother, hung with glass chandeliers, a pavilion containing all the books about Vivekananda in every language. In several places, the cobra and swan emblem has been put up in wickerwork.

A little golliwog-haired professor named Naresh Guha, whom I met at the tea yesterday, and who is writing a book on Yeats, came to take me to Javadpur University,[1] where I gave a question-and-answer talk to the students. It went quite well. They seemed to understand everything I said, and they reacted. Among other things, I talked a good bit about Huxley.

Later, I stopped in to see Swami briefly. He says he is better and is coming back to Belur tomorrow. Returned to Belur after visiting Guha's apartment and meeting his wife. They had a copy of [Burroughs's *The*] *Naked Lunch* and reproductions of a Van Gogh fruit tree and a Rouault.

At supper, Prema and Arup were eating as much as they could possibly manage, because their fast begins tomorrow morning.

Prema says he feels that no one should take sannyas until he

[1] Guha (b. 1923), Bengali poet and essayist educated partly in the U.S., taught comparative literature at Javadpur; among his publications is *W.B. Yeats: An Indian Approach* (1968), about Yeats's knowledge of Indian philosophy and religion.

has been "smashed" (I think that's the word he used; anyhow he meant, until the ego has been smashed). He feels that his conflicts with the women at the Hollywood Center—particularly with Usha—were a form of disciplining by Mother Kali.

January 5. Today I feel fairly well. The shits have continued until now, but this morning's stool was thicker. What distresses me is my dullness. I feel nothing, nothing but the dull senseless urge to get the hell out of here. Such is my longing to do this that I'm not even nervous about the flight; a sinister sign. I feel only partly alive. Jacked off this morning, and not because I really wanted to—just out of meanness. I'm mean and sullen.

Reading Tolstoy's *Resurrection*, after finishing Balzac's [*La*] *Peau de Chagrin*, a pretentious bore; Balzac is such an *ass* (no pun intended!)[1] Tolstoy's indignation is always fun, even when all else fails.

In the morning, Sadhan Kumar Ghosh, who wrote *My English Journey*, P. Lal, Jai Ratan and Kewlian Sio came to see me.[2] We talked in the guesthouse dining room. P. Lal writes poetry (a bit soppy), is tall and big, quite handsome though with a cast in one eye, married and a college professor. Ratan and Sio write short stories. Sio is a Catholic and some kind of a Chink. Neither of them talked. The other two were sort of teasing–flattering, in Indian style. They wanted to know how I had come to be able to write such beautiful prose. What was my secret? Ghosh's book is quite amusing and bright, but it has a vicious attack on queers.

Swami came back in the afternoon. He has a swelling on the side of his face which he is trying to reduce with hot compresses.

January 6. A terrific wailing and drumming burst forth at about 4:30 a.m., announcing the Big Day. (Prema calls this "snake-charming music.") After that, there was kirtan till breakfast. Jacked off as a protest and went back to sleep.

---

[1] Balzac's title, *The Wild-Ass's Skin*, is already a pun: the hero receives a magical piece of the skin, also called *shagreen*, which answers his every wish but shrinks as it does so, causing him extreme *chagrin*, for when it disappears, he must die.

[2] Lal (b. 1929) was founder in 1958 of the Calcutta Writers Workshop, which publishes contemporary literature and translations, including his own translation of the complete Mahabharata from Sanskrit into English. Ratan (b. 1917) and Sio were among the seven writers who launched the workshop with him. Ratan, also a college teacher and later a business executive, is a prolific translator of novels and stories from and into English, Urdu, Punjabi, and Hindi.

Found Swami worried about his face. He has told Prema, "I want to have my *mahasamadhi* in India," so Prema is worried too. Arup fell asleep—which is strictly against the rules during the sannyas fast—and dreamed of pork chops.

The Math grounds were crowded all day. They were patrolled by thin-legged police in shorts; a whole encampment of them have moved in. Thousands of devotees were fed on leaf plates. Loudspeakers shouted. Kirtan singers wailed. One of the Swami's sisters came to visit him, with a tribe of grandchildren. The little girl and the baby boy had their eyes made up—darkened with kohl to protect them from the glare of the sun. A line all day on the stairs to view Swamiji's room.

In the afternoon, there was a meeting. Despite his poor health, Swami presided and spoke. I spoke too—my last speech on religion, I do trust, anywhere.

When we got back to Swami's room, he held out his hand and asked me to massage it. I did my best, telling him that I'd never massaged a hand before. He answered, "Why can't you do something for Swami you never did before?" He was in his "baby" mood. He kept dozing off but wanted us to stay in the room with him. Of course one had the suspicion that maybe this was a kind of inspired playacting. Wasn't he perhaps in a high spiritual mood and giving us the privilege of serving the "It" which had taken him over? I hate this explanation because it sort of embarrasses me; but I don't discount it. In that building, with Vivekananda right next door, it made perfect sense. In that atmosphere, the edges of personality get blurred, and Swami becomes a little bit Brahmananda-Vivekananda-Ramakrishna.

Swami said to me this evening, "I can't believe you're going, Chris."

Nikhilananda (in a good mood at breakfast this morning), "This is the country of self-destruction."

The countess and Mrs. Beckmann are exultant—because they had been present at a special puja in Vivekananda's room at which he had been "fed." Also, a little, because I *hadn't* been there. Women attach extraordinary importance to such occasions. I can never quite believe in this kind of religious enthusiasm—but that's merely because I seem incapable of it.

The guesthouse gate is locked tonight, so I'm excused from going out to the Kali puja in the temple, as I'd unwillingly planned to do.

January 7. Woke with a sore throat to the noise of snake-charmer music over the loudspeakers. But this poor old snake couldn't rise. However, I did get up at 7:30 and went to look for the new swamis. I met Arup first, near the Leggett House, by the pandal in which the monks have been eating. He was embarrassed and delighted when I prostrated, and hugged me. He is now Swami Anamananda. (Anama means The Nameless One; this is a kind of side-reference to Arup, which means The Formless One.) Arup looks absolutely marvellous in his gerua. The gold flame-color brings out the blue of his eyes and the fairness of his wrinkled skin. He looks very tall and very old and spiritual; the abbot of a monastery, at the least. And it is with the benevolence of an abbot that he raises his hand in blessing, and murmurs, "Bless you," whenever anyone takes the dust of his feet.

I walked with him toward the office, and presently Prema came by, in a group of other new-made swamis on their way to beg alms. (You are supposed to do this barefoot, but the real point is not to wear leather on your feet, so Arup was allowed to compromise by wearing rubber sneakers.) I ran out to him and prostrated and he hugged me warmly; the onlookers were much edified, I felt, to see us Westerners playing the game according to their rules.

Prema is Swami Vidyatmananda. (Vidya is knowledge itself; vidya-atman is the soul of knowledge. To all intents and purposes, Prema could just as well have been called Vidyananda; but some other swami has that name already.)

Then it was time for Aranyananda, Ranganathananda and Shashi Kanto to leave. Nearly all of the swamis (including Prabhavananda) had been present at the sannyas ceremony during the dead hours of last night. Ranganathananda wanted me to come back to the institute with them and see a documentary film on Vivekananda, but Aranyananda whispered in my ear, "Not worth the candle." (How typically Indian to use this faded slang!) So I declined. I gave Shashi a great big hug, which surprised and delighted him. I prostrated before Aranyananda and then hugged him. But he was a couple of degrees cooler. He really is quite a cool-blooded creature.

Later I went in to see Swami. He was being massaged by his attendant, a tall athletic and attractive young swami they called Ramesh. Swami said to me, "You see—I massaged Maharaj, so now I get massage!" Then Prema came in from begging alms, with his cloth full of damp tepid food. Both Swami and I had to take some. I nearly gagged on mine, and I noticed Swami took very little, though he remarked that this food must be very pure!

Then Gokulananda came in with two of his college boys. Swami began telling the boys they should become monks. "I tried to think

lustful thoughts in Maharaj's presence, and I couldn't. I tried deliberately. But such an experience will not be possible again until the
Lord comes back." "Run away from home," he told one of the
boys; then, turning to Gokulananda, "Swami, get him a railway
ticket to Madras. Otherwise, he will get married to a little girl—"
Turning to the boy, with a kind of inspired affectionate teasing
tone, "Yes—you will get married, and then you will say, I got
married because my mother cried!" Then he added, "Write to me
when you join the monastery—not before!"

Then Gokulananda sent the boys away and started to ask Swami
some personal advice. (It was Swami who persuaded him to join
the order.) So I went out of the room. Vivekananda's room was
open; a swami was cleaning it. I went in and prostrated and prayed,
"Give me devotion to you, give me knowledge of you—even
against my will. And be with me in the hour of death." And I
prayed the same for Don. Then I touched my forehead to the
bed. I went out on to the balcony where Swami first met Maharaj,
and prayed the same prayer. A swami was bathing in the Ganges
below, pushing aside the floating water hyacinths before he im-
mersed. . . . Later that day, I brought my beads and touched them
to the spot on the floor of the balcony where I guessed Maharaj
and Swami must have stood.

Talked with the countess before lunch, about the bitterness of
the masses in this city. At the Great Eastern Hotel you are not
supposed to tip; but the management doesn't pass on the service
charge to the help, and they are so mad that they'll only bring you
one shoe, etc.

She also said that the Parliament of Religions was attended only
by rich bored people who had nothing else to do.

I felt lazy in the afternoon, so I stayed in my room instead of
going to Dakshineswar with Al Winslow and Carlson. (Winslow
actually put on his trunks and went swimming in the Ganges!)
Then I packed and sat with Swami, who was feeling much better.
But the doctor want[s] him to have his lungs x-rayed when he gets
to Madras.

Suddenly, it was time to go. I had said all my goodbyes—to
Madhavananda sitting listless in a steamer chair; to Yatiswarananda
in a half-lit room, too dim to read in, with earsplitting music coming from a nearby loudspeaker so that you had to shout at him; he
must have nerves of steel. I talked to Prema, who is very happy
about everything. He plans to stay in India for at least a year, as a
troubleshooter for the order, getting projects organized, etc. He
says Arup says he'll go back to Hollywood, eat his three meals a

day and lead a spiritual life. "Like Elder the pumpkin cutter in the Gospel,"[1] Prema commented. He is just as sour and bitchy as ever; it is strange to hear this bitchery proceeding from those austere-looking lips. When he complained that his dhoti keeps slipping and Arup remarked that his doesn't, Prema said, "Perhaps you have more so and sos to hold it up with."

Rather to Prema's dismay, Swami has ruled that henceforth he must be called Vidya, and Arup Anama. But I doubt if this will stick. Too many people are too used to the old names. (Incidentally, what a very real austerity this name changing is!)

Krishna volunteered to come with me to the airport. Also Gokulananda, maybe prompted by Swami. There was a big delay, because the kirtan was still going on, and Krishna had left his tape recorder running on the musicians' platform. We waited for them to stop and they didn't, so Krishna finally had to remove it in front of the whole audience. But we had gallons of time anyway.

Then, as we passed the office, Nikhilananda was standing there with a group of swamis. Nikhilananda ordered Gokulananda out of the car and thrust a swami from Singapore into it—all this in Bengali without a word of explanation to Krishna and me. Nikhilananda had also forbidden Al Winslow to come with me as he'd wanted to. I think this was sheer love of bullying, but this was no time to protest.

When we got out to dreary Dum Dum, I persuaded Krishna and the swami to leave me alone, fairly soon. For some stupid reason, I didn't hug Krishna on parting. I ought to have—I know he would have liked it. Krishna said, with a grin, "I suppose you're going to write all night?"

January 8. We took off from Dum Dum about twenty minutes after midnight. The plane, BOAC, had come from Sydney and there were a lot of Australians on board—large beefy men in white shirts with sleeves rolled to the elbows, as if for cricket; they had brick-red faces, and gave a collective impression of cockney Scottishness. It was deathly cold on board; and, though I had three seats to lie down on, I couldn't sleep. Because of my cold, the descent at Karachi was horribly painful. The mucus seemed to get into my ears, and I was, and still am[,] rather deaf. They didn't

---

[1] *The Gospel According to Sri Ramakrishna.* Elder is "neither a man of the world nor a devotee of God." Old, unemployed, and with nothing to do, he is available for household tasks such as cutting pumpkins for cooking. (By tradition women in Bengal do not cut them.) See Aug. 20, 1883; p. 281 in Swami Nikhilananda's translation.

make us get out of the plane, thank goodness.

Now we are airborne after another landing, around breakfast time, at Damascus—a city in a desert, and made out of desert. Brown mountains in the background, with some snow. Bracingly cold outside, even in the sunshine. At first, the officials didn't want me to get out and merely walk around the plane; they wanted us all to go to the transit lounge in a bus, and buy things, I suppose. But it was so wonderful inhaling deep breaths of the thrillingly clean air—the first air since Tokyo—and there was even something exciting about watching the cleaners at work; the modern counterpart of changing horses at an inn. Baggage being lifted down through the trapdoor; shit and dirty water and towels being carted away; fresh food arriving in containers. Two uniformed Britishers, maybe pilots, pulled some kind of a plug on a long stem down from the lower surface of the wing. A mechanic then brought them a jar of water which they examined very carefully, like doctors examining urine. It looked beautifully clean however.

We are scheduled to arrive in Rome at 11:00 a.m., their time. And Gore and Howard will be waiting for me, I hope, like the Two who come to conduct the dying man into his new life....[1]

(That's the end of the diary. I stayed in Rome two nights, with Gore and Howard, at their apartment. On January 10, I flew to New York and stayed with Don at the Hotel Chelsea. On January 23, I flew back to Los Angeles, and have been living at home since. Don stayed on in New York to draw various people, some of them for *Glamor* magazine.)

---

[1] In the opening scene of *The Dog Beneath the Skin*, the Two are opposing presences who mark the passage of time and witness the activities of men, making life possible and limiting it. In the sannyas ritual, the monk dies and enters a new life of stricter religious devotion; Isherwood's rebirth takes place in the opposite direction, abandoning the monastery for the world.

February 11. Don returned home in the evening of the day before yesterday. As Dorothy, who came yesterday, said, "The household is completed." And we were truly all delighted to see each other. When Don isn't here, my life simply isn't very interesting. He creates disturbance, anxiety, tension, and sometimes jealousy and rage; but never for one moment do I feel that our relationship is unimportant. Let me just recognize this fact, and not bother about making good resolutions. He will behave badly; I shall behave badly. That's par for the course.

I have had to omit all the things that happened in Rome, New York and since I got back here. Maybe some of them will come back to me. For instance . . .

There was a minor earthquake while we were away. According to Dorothy, it sounded "like as if the Chinese were coming."

Wystan telegraphed me to say *A Single Man* is "by far the best thing you have done." To Don, however, he added three criticisms. (1) That George stays far too long in the bathroom. (2) That there is too much made of the homosexuals' right to be regarded as a minority, in the same category as the Negroes and the Jews. (3) That Wystan was shocked when George thinks that he will "make a new Jim."

As far as I can make out, criticism (1) was based on the fact that Wystan never stays long in the bathroom; (2) arose out of Wystan's feeling that my upholding of the homosexuals was indirectly anti-Semitic; (3) meant Wystan refuses to believe that this is my own attitude toward human beings.

Wystan told Don on another occasion that he thinks I dislike Chester because I am anti-Semitic. Not a word of this to me, of course. His most startling dictum this time was that the only art form truly appropriate to the nineteenth century was opera, and that therefore Verdi and Wagner are greater than Dickens, Tolstoy, Degas, Tennyson, etc.!

I have frittered away eighteen days since my return (as if I had so many left!); and I still have lots of mail to answer. But now I will get down to work. My first job is to go through my diaries and find all references to Huxley, and then construct my article for the memorial volume.[1] This isn't a waste of time, because this is all research for my own autobiography. Then I want to consider

---

[1] *Aldous Huxley, 1894–1963: A Memorial Volume, Together with his last Essay, "Shakespeare and Religion"* (1965), edited by Julian Huxley.

the idea of a short novel based on Prema taking sannyas. More of this later.

What a wonderful life I have, really! How very seldom do I do any thing I don't want to do. My only afflictions at present are ill health. Right now, I'm troubled by what may be the remains of my Indian stomach upset. The muscles keep twitching and the gut aches, off and on. I'll go see Dr. Allen as soon as I have the time.

Don is busy designing the jacket for the English edition of my novel. The deadline is February 14.

February 18. They have definitely taken Don's driver's license away but Ben Alston thinks he can get it back, after a re-hearing of the case; he is chiefly being punished for not having attended the first hearing.[1] Meanwhile I shall have to drive him around, and this is bound to lead to friction. Yesterday, he told me I was behaving *too* well, because I didn't get frantic when he kept changing his mind about where he wanted to go.

Perfect weather, though cold at nights. We have had a very happy time since he got back; but now there are storm clouds. Was he right to have cancelled the Phoenix show? What is he going to do next? How about painting?

I am skimming through my journals looking for references to Aldous—there aren't nearly enough of them—before I start my article on him for the memorial volume. This idea of Methuen's that I shall do a book of bits and pieces is also very stimulating; and it's the kind of project I can easily work on in the midst of writing a movie script. (Let's hope I get one to write! Both Burton and *The Loved One* are still possibilities.)

March 8. All this time has passed, and yet there is little to report. I have been offered an appointment as a Regents' Professor on the UCLA campus, which makes me respectable, I suppose, and would bring in $10,000, and would be quite convenient, because I could do it next spring, from this house, with very little sweat. Shall probably accept.

Nothing from Burton or Tony Richardson about the movies. Have just finished revising the final typescript of *Ramakrishna and His Disciples*; so it should finally get off to Simon and Schuster and Methuen. Now there is the Huxley article.

---

[1] Bachardy had three traffic citations in one year and was therefore summoned to a hearing which he missed because he was in New York.

Don still without a license. Yesterday, for the first time, he took a chance and drove to the gym.

Still this icy wind and brilliant weather. I'm sick of the cold. Also, I have a worryingly prolonged attack of pyloric spasm. Am taking pills for this. If it doesn't get better, Allen wants to x-ray.

Last Monday, we bought a T.V. set. It is rather a joy. At least, one can toy with the idea of seeing this or that movie, and you know there is always entertainment if you're bored stiff.

March 13. This morning, I read Don some poems of de la Mare. He liked "All Hallowe'en" the best. A bright windy day after yesterday's rain. Yesterday I saw Dr. Allen again, and convinced him, almost against my own better judgment, that I'm really all right. I don't know if I really am, but I do know that I want to be. I want to work with Tony Richardson on *The Loved One.* Part of me at least is full of springtime vitality.

Letters, manuscripts to read, all sorts of chores. Relations with Don very good; the television helps a lot, in an odd sort of way. It gives us a new vice in common—watching the ends of old films in the middle of the night and thus getting up late, next morning.

Tony Richardson is scheduled to arrive on Sunday; so then I hope things will start to happen.

Don has joined the Lyle Fox gym. Still no word about his license.

March 15. Last night, Don spent the first night out since he has been back. He arrived home late this morning in good spirits, so I hope this is going to help.

As for me—well, there's Bart Johnson, sort of. I wish that would work out better, because it would be so damned convenient. Which is probably just why it won't.

The day before yesterday, I saw Gerald Heard. He thinks that we are all losing our memories because of the spastic shifts of the magnetic pole. He told me what Chris Wood already told us as a deadly secret—that Margaret Gage is selling her house and thus turning him out. According to Gerald this is a sort of revenge; she feels she has been treated badly and her friends assure her that this is so. They regard her as a great seer, a leader with a spiritual message, and they tell her that she has been too much under the domination of Gerald. She has a new friend who is a psychologist and who encourages her to act like a young girl, although she's near seventy. In the evenings she wears short ballet skirts. Gerald also believes that Michael did a lot to set Margaret against him—he is so bossy.

I asked Gerald what he is going to do. He was very vague. "After all,' he said, "I don't wish to say this melodramatically, but the fact remains, I am dying."

He described himself as being "floored" by *A Single Man*, which he and Michael have just read. He thinks it is by far my best book. "Now, obviously, you can write anything." So he advises me to deal with *awe*. Cites *Outward Bound*,[1] and Chesterton's play *Magic*.

Last Wednesday, when I was up at Vedanta Place, Usha remarked that someone had been into the bookshop and asked for a guidebook to the temples of India. So I said, why *didn't* they stock a guidebook to India? And Swami grinned and said, "No, Chris—I will not deliberately send anyone to his death." He is full of such cracks at present and behind them you feel a real resentment; he keeps declaring that he will never never return there. That he couldn't meditate at all while he was there, etc. And yet he also tells how he went to meditate in the shrine of the Holy Mother at Dakshineswar and was aware that the image was alive!

March 29 [Sunday]. Here we are at Easter. Well, at least I've worked all day; my traditional celebration. The outline for *The Loved One*. I've been on payroll since last Wednesday, March 25, and it certainly is fun. Tony Richardson won't be here much longer, however. He goes back to England soon and then returns to direct, in a couple of months.

Don still hasn't got his license back. They took it away from him and sent it to Sacramento, and of course he drives, nearly every day, so the worst may be expected. No use dwelling on that.

Nothing more I want to say now.

My next chore: the article on Huxley. I just finished reading through my diaries to find all the references to him. It's rather shocking, how seldom we met.

This morning I dreamt that Igor was dead. But the corpse could talk. This dream was somehow reassuring.

May 26. Don left yesterday at noon, by plane for London. He'll stay there four or five days, then join Lee Garlington and his friend in Egypt.[2] From there they'll go to Greece, Austria and elsewhere.

---

[1] The 1923 play by Sutton Vane.
[2] Bachardy had a sexual interest in Garlington, once close to Rock Hudson; Garlington's friend was a travelling companion only.

This is Don's "birthday present" for his thirtieth birthday. He said he wanted to do it "with my blessing."

When we got to the airport, the entrance to the plane was guarded by two cops. I said to myself "a bomb" but didn't say so to Don lest it should worry him. Now we hear that two of the Beatles were on board and the authorities were terribly afraid of a mob demonstration.

On the 22nd, I was laid off *The Loved One*, because my screenplay is finished and nothing more can be done until Tony Richardson returns and starts work on it. I don't think this is merely a brush-off. However, when Robin French went to John Calley and tried to up my price for the future, Calley turned him down flat.

Now I am beginning to think about the bits and pieces book. The night before Don left, we were talking about a possible title with Gavin, and Gavin suggested *Digging up the Past*—or rather, he said, "What a pity you can't use it!" And then that made me think of *Exhumations*; so I shall call it that, provisionally.

My latest symptom: shooting pains in the groin.

A very vivid dream which I had about a month ago.... I was standing with some others, including Don, on the terrace of a house high up on a steep hillside above the ocean. (I think it was Joseph Cotten's house, but this didn't have any significance in the dream.[1]) Attached to the side of this terrace—though quite unrelated to it in architectural style—was a wooden platform, a balcony without handrails.... Suddenly there was a tremendous blast of wind, and this platform was blown clear of its supports. Because of the updraft, it remained almost motionless, however, swaying slightly and hovering in empty air like a helicopter. There were four people on the platform: Arthur Loew, Natalie Wood, Sarada and another woman (unidentified). We all gasped, for they were obviously doomed; it could only be a matter of moments before the platform fell. What was so shocking was that they were quite near us, only a few feet away, and yet beyond all possible help.... The chief interest and vividness of the dream was in the behavior of the victims. Arthur (whom I don't actually like much) behaved with a kind of heroism. He obviously wanted to cheer Natalie up and keep her from thinking of her imminent fate, and so he grinningly crossed himself, thus alluding to it and yet taking the curse off it, as it seemed to me, by his deliberate sacrilege. Also,

[1] In *D.1*, Isherwood expressed sorrow that Cotten was drunk at a party after his first wife died and left him alone in their house; Bachardy had just left Isherwood alone, and their house was also on the side of a steep hill above the ocean.

there was the gallows humor of the self-conscious Jew making this Catholic gesture. And Natalie smiled bravely back at him. These two were playing parts, both for each other's benefit and for ours.[1] But Sarada, meanwhile, was obviously and frightfully scared; she had turned white with terror. (Did I think at the time, or was it later, that it was shocking to see that the thought of Ramakrishna gave her no support at all?) The third woman was neutral; I don't know what she was feeling.... Well, all this was quite appalling and yet at the same time exhilarating, as any ghastly accident is to the spectator. And then the wind swirled the platform away, and it fell, far below, on the ocean highway and, I think, caused a huge traffic pileup.

Well, I reopened this record, which is all I really wanted to do. I am in a brisk housekeeping-choredoing mood which always immediately follows a parting from Don. You might call it a mild form of shock.

June 7. Just about to take off for a trip to Big Sur with Bart Johnson. Why in hell did I agree to this? I couldn't want it less.

Nothing from Don yet.

The weather is clearing, after much greyness. Dorothy blames it on underground atomic testing. "They've shaken the veins of the earth." She said she was tired, last time she was here, so I made her drink some bourbon. After this, she laughed wildly because I told her how Don will mop the floor with the sponge meant for the dishes.

Still this pain in my groin, through the left nut and down my leg. Also pain in my little finger, which is serious because it interferes with typing. I'm afraid it is the arthritis spreading.

Chris Wood has a new dachshund named Beau.

Ted, still nutty, claims he has found an agent who wants to get him into a Las Vegas show.

Have finished typing "Gems of Belgian Architecture" for my *Exhumations* book. An advance copy of *A Single Man* arrived yesterday. It could be worse looking. The type is good.

Arup refused to do some domestic work for the girls up at Vedanta Place, saying, "How dare you ask a swami to do that!" Swami told him off, saying that a swami should be humble, helpful,

---

[1] Loew, movie writer and producer (1925–1995)—a grandson of the founders, through his mother, of Paramount and, through his father, of MGM and Loew's Theaters—was once linked romantically with Wood. He reportedly offered her the starring role in *Penelope* (1966) to pull her through emotional and professional troubles. He is mentioned in *D.1.*

gentle, etc. etc. Prema also is in the doghouse because of his intriguing to get sent to the Paris center.

Jo says that Ben still refuses to see Betty [Arizu]'s children. The very thought that they exist upsets him.

June 18. Starting to feel very low and sad, because I miss Don so. Also because the discomfort in the groin persists. Dr. Allen saw it and took it as calmly as usual.

Big Sur was magnificent but Big Johnson wasn't. I behaved badly, but made up for it later, I guess. Can't be bored to relate all this.

Working on *The Loved One* again since yesterday, at the nice pool house of the house Tony Richardson has rented. Jan [Niem], the Polish chauffeur, has a respectful-sassy relationship to Tony, Bud[d Cherry] and Neil [Hartley], throws them the pool ball he bought at the filling station.

September 7. Labor Day. A restart after a big lapse.

What's to report? They are shooting *The Loved One*, with dialogue about ninety-nine percent Terry Southern's; all that's left of my script is some of the skeleton. And now I have finished a first draft script of *Reflections in a Golden Eye*, and Tony says he's delighted with it. So now the decks are cleared for my own work. All I have to do is get the proofs of *Ramakrishna and His Disciples* corrected.

Then I can get on with *Exhumations* and think about my new novel.

Despite the sour reception of *A Single Man* in this country, I still feel very good about it. Not so much as a work of art but as a deed. I feel: I spoke the truth, and now let them swallow it or not as they see fit. That's a very good feeling, and this is the first time that I have really felt it.

The only other thing I feel like reporting right now is some table talk of Tony Richardson's. This was mostly said on August 17, while Don and I were having supper with him and Vanessa (she has gone back to England now). A few of the remarks seemed aimed at Vanessa. However—

He said that now he has lost interest in the theater. He wants to do movies. In the theater, you have to keep carefully to the interpretation of the author's text. In the movies, you are much freer. The script is something you can depart from. You are free to improvise. Also, you are not so much at the mercy of the actors.

But actors are wonderful, because they accept life. When stars get old and are no longer stars, they accept this and take little jobs and don't complain, as other people in other professions would.

Richard and Liz Burton are completely corrupt; they think only of money.

Samuel Beckett is a great writer. He has real compassion.

Chekhov is as great as Shakespeare.

Brando has a Japanese girlfriend. She appears at meals but leaves at once when the men are talking business. That's the way women should be.

September 18. The day before yesterday, I went up to Vedanta Place and Swami and Vidya and Vandanananda and Usha and another girl who has been proofreading went right through the proofs of *Ramakrishna and His Disciples* and incorporated all our corrections and my changes, and so now, aside from checking the captions under the photographs, the whole work is done and the rest is up to Methuen.

India had a last straw to throw on my back—after all this while, they wrote to say that they hadn't got my talk straightened out, because it was never properly recorded. So now they want me to rewrite it. No, I told Swami. Whereupon he said he would do it. Oh, the blackmail! So, of course, I had to say that, if he did it, I would revise it later. Didn't even bother to look and see *what* talk it is.

Five days ago, I woke up with my back hurting. It has hurt ever since, not really getting much better. Dr. Allen gives it heat treatment and I take pills to relax the muscles. Oh dear, it is so tiresome being sick! I seem to go from one ailment to another, without a pause, and of course that means I'm toxic, physically and mentally. I do wish I could snap out of this. I am such a mess. And for no reason. I have money, fame, a happy home. Don is being marvellous. Tony wants me to do more work for him—either Marguerite Duras' *Le Marin de Gibraltar* or Colette's *Chéri*.

I must try to get back into some sort of regular meditation, however brief. I must try to prepare myself for death. I must try to be less of a cantankerous nuisance and more of a public convenience.

September 26. I finished the screenplay of *Golden Eye* on the 4th and since then all I've done has been the final correcting of the Ramakrishna material. Well, and why shouldn't I take a holiday? I have certainly earned one. But the truth is, I am bad at holidays. Instead of relaxing—whatever that means—I just idle without joy and (consequently, I do sincerely believe) get sick. My back still hurts, but it is better than it was. The X rays showed a disc which has worn thin. Sooner or later, barring accidents, it will fuse and then the pain will stop.

The fire up at Santa Barbara came right to the eucalyptus grove at the edge of the convent land. The girls were evacuated, taking the relics from the shrine with them. Then the wind changed. Vidya and the other monks from Vedanta Place went up there to help. Vidya is under some sort of a cloud. Apparently he lied to Swami, but I haven't heard the whole story of this yet, because Don was with me the last time I went to Vedanta Place, so Swami didn't talk about it.

Am still waiting to get the English translation of the Duras novel, *Le Marin de Gibraltar*. Then I must make up my mind, do I want to do a screenplay on it. If I do, I shan't go to New York with Don on October 8 or 9. I don't really want to go, because anyhow I should be quite inactive there, and I ought to get on with my book of bits and pieces. I wrote to Alan White, asking if they would approve the title *Exhumations*, but no answer yet.

Have been making a tape for Don of various poems. He wants to play them to himself while he is painting in the studio.

Don is now getting quite enthusiastic about his painting—partly because Paul Wonner and Bill Brown have at last told him that they like it and think he ought to exhibit some of it. But the New York gallery (the Banfer) only wants drawings. Dr. Oderburg tells him that he ought not to stop painting at this time; so there is the problem of trying to work at it while he is in New York.

October 1. Thick fog in the Canyon all day, and my back and ribs as bad as ever, but somehow I felt gay and full of love—not only for Don but also for Budd Cherry, Phil Anderson and most everyone else on earth who isn't old, hideous, pro-Goldwater or otherwise impossible.

Lyle Fox massaged my back, probably without any effect but never mind. Told him this story—I think I got it from Paul Wonner:

A young man boards a plane, sits down next to a lady, takes out a copy of *Playboy*, pins up the two-page photo of a nude cutie on the back of the seat in front of him and jacks off, looking at it. When he has finished, he wipes his cock with his handkerchief, puts it back in, takes out a pack of cigarettes, turns to the lady and asks very politely, "Do you mind if I smoke?"

October 28. Don left for New York on the 15th. His show opened there yesterday at the Banfer Gallery.[1] Stephen Spender called me last night to tell me that he had been there and there was quite a big crowd.

Strange weather. At lunchtime yesterday it was warm enough to go in swimming. Later it rained. Today it is grey and more rain is said to be coming. I am plugging away at *The Sailor from Gibraltar*. There is an awful lot of it and I still really do not know just what I am doing. The chief technical problem is the fragmentation of the flashback.

Neil Hartley told me today that Tony should be through shooting *The Loved One* by Thanksgiving.

I still feel a sick foreboding about the elections, despite all the pollsters who declare that Johnson has it in the bag. It is the mere smell of Goldwater that sickens me. Horrible to think that he got even this far toward being elected.

Supper with Cecil Beaton last night. He was very gleeful because George Cukor had made a poor showing at the big press conference which was held for the opening of *My Fair Lady*. The opening is tonight, and I have not been invited, despite the fact that I am such friends with Cecil and on good terms, even, with Cukor and Rex Harrison—not to mention Audrey Hepburn.[2] Well, I don't regret it; it would really bore me to go, without Don. Also Tony Richardson (who, like me, hasn't been invited) is showing the Jean Genet prison film tonight.[3] Even to miss seeing that again doesn't break my heart. I would far rather be doing what I am doing—going to Vedanta Place and then looking in on Bob Rosen. Isn't that typical of me!

Rib still hurts; lower back more or less all right. A nasty lump in the mouth cleared up as soon as Dr. Stevens filed a bit off my lower bridge. He is now preparing my $400 upper bridge, which is to combine the three separate bits.

Am fat and drinking too much, but feel a good deal more energy

[1] Pencil and ink portraits of Auden, James Baldwin, Leslie Caron, Glenn Ford, Forster, Gielgud, Paulette Goddard, Geoffrey Horne, Aldous Huxley, Isherwood, Vivien Leigh, Anita Loos, Simone Signoret (featured on the announcement), Spender, the Stravinskys, Barbra Streisand, Virgil Thomson, Gore Vidal, Collin Wilcox, Natalie Wood, and others.
[2] Hepburn (1929–1993) starred as Eliza Doolittle opposite Harrison; Isherwood first met her in New Haven in 1951 when *I Am a Camera* and *Gigi*—adapted for the stage by Anita Loos and with Hepburn in the lead—were playing at the same time before opening on Broadway. She is mentioned in *D.1.*
[3] *Un Chant d'Amour* (1950).

since I went back on the high potency vitamins. As usual, I am bored without Don—nothing bounces off anything; it just falls flat to the floor. He won't be returning till the 15th, at the earliest.

Reading wonderful Byron (his letters) and a drag-queen auto-biography Gavin lent me, called *Mr. Madam*.[1] And now Peter Viertel's novel, *Love Lies Bleeding*, has arrived—a wretched title and I do so hate bulls and their annoyers.

October 30. Budd Cherry took me to the Kirov Ballet last night, and to supper at Perino's, where we sat in splendor in the best banquette because, apparently, the waiter had mistaken Budd for a Dr. Cherry who is one of their best customers. Budd told how Tony resists all possessiveness and security, and how he uses people. All this with a despairing affection; for Budd still feels that Tony is the greatest artist he ever met and the most marvellous person, and he loves him. Actually, I think Budd's inviting me was in itself a gesture—not exactly of defiance but of self-assertion; he was determined to show Tony that he can still have his own relationships, even with Tony's friends. And, of course, I'm quite ready to play along with this, if it makes Budd feel better.

When we got back to their house, after the ballet, Tony was engaged in a typical piece of mischief—trying to persuade a boy who supports Goldwater to come to a party on Tuesday night, to share in the (presumed) Democratic triumph.

Neil made me a big speech about how they all love me and wish I would work right along with them on all of their undertakings, and why don't we see each other more, etc. Well, fine—but, while this strokes me up the right way, and would make me purr deeply if I were a pussy, I see the deep workings. No doubt Neil felt that Tony had been ungracious and might lose a valuable assistant in consequence. But the truth is, I prefer Tony's ungraciousness. This American business warmth makes me nervous, because I know so well how quickly it can cool. All I have to do is fail, once.

I was about to add, "Just the same, I like Neil"; but, God, what a meaningless word "like" is! You can say you love people or that you, temporarily, desire them. The rest is really mutual convenience—are you going my way? Do we both want to get drunk, or double date, or see that play or movie, or vote for Johnson, or visit San Simeon, or talk about Proust—all right, then, we'll do it together and call ourselves friends. And if we keep going each other's way over a long period, well, something else may start to

---

[1] *Mr. Madam, Confessions of a Male Madam* (1964) by Kenneth Marlowe.

happen—and that's love. Or is it, even then? Is there any love until there has been friction and a clash of wills and an understanding that one does *not* agree on everything? Until, in fact, the mutual convenience relationship has been broken.

The Kirov Ballet was the squarest theatrical performance I have ever seen. They danced with exquisite precision, very very slowly, with glum faces or pained smiles. (One girl kept making exits which always seemed just a shade too long—it was as if she kept dashing for the stage door in an effort to defect and being dragged back again.) Only one man, in red boots, had a big American grin on his face as he jumped. So the audience loved him, though some of the others jumped higher. I think it was chiefly the boots.

Tony says all ballet is dead except for the New York City Ballet, and that he refuses to see any other.

Yesterday was the last day of shooting at Greystone. Now the house will either be torn down or remodelled and made into an art gallery. Such a huge ugly expensive place; such a waste of perfectly good building materials. Really, its existence was only justified during the few weeks that it was used for this film.[1]

At the gym, Vince Eder, the Chinese boy—or is he Japanese or American Indian or all three?—was wearing a sweater with FUCK written across its chest, backwards. Vince and a couple of others had had these sweaters lettered for them at a sports shop. This is the kind of thing that just would not have been possible, five years ago.

Truman Capote has been staying in town. I saw him three times and read the first three parts of *In Cold Blood*. I don't know what I really think of it until I have read it all, but it is terrifically impressive. (Reading in *Show Magazine* that he was born with the name of Persons, I thought that a good title for his autobiography would be *Persons Known*.)

Truman is wonderful to be with. (I do *almost* love him.) We were comparing our fantasies of revenge on our enemies, and I told him how I always start my trial of the chief criminal by killing a couple of his relatives right there in front of him, to show him I'm not kidding. Truman said, "I know exactly what you mean—just to sort of establish the mood."

November 1. Got drunk yesterday because I spent the afternoon

---

[1] Greystone, said to be the largest house ever built in Beverly Hills, was purchased and preserved by the town and continues to be used as a film location.

drinking with Budd Cherry and listening to more of his woe. (The most interesting piece of new information is that Tony is absolutely passive sexually!) Then I went on into Hollywood and drank some more at dinner with Bob Rosen and Phil Anderson, with the result that I fell asleep during the Belmondo movie, [*That*] *Man from Rio*. Woke up this morning deeply depressed by this sloppy behavior. But I *have* managed to do some work on the *Sailor from Gibraltar* script this afternoon.

Don called this morning. It seems that his show hasn't amounted to much. But I could tell that he is having a good time. His mood was good. I am starting to miss him terribly. Just the mere blankness of his not being here. But, when I am in this state, it is really better that he isn't here.

To work!

November 15. Don called this morning and told me that he is definitely going to take on this assignment of drawing the twelve principal dancers of the New York City Ballet. So he will stay in New York for several weeks yet; maybe right through till Christmas, when he would join me at John Goodwin's in Santa Fe. Furthermore, Don says, if he comes home now he'll feel depressed and disappointed, because his show hasn't done well; only four drawings sold. He says he is determined never again to have a show of drawings. And that he hates Ferdinand, who runs the Banfer Gallery.[1]

Just at first, I felt a slight pleasurable excitement, because I shall have another five weeks alone, and being alone is always a challenge. But already there is the aching realization that I won't see Don for another five weeks; and that is a most awful long time. Without him, I'm not complete.

On November 1, I woke up with a hangover and was cross with myself and vowed to cut out drinking until Don returned. I have done so and already my face looks a whole lot thinner, though my weight hasn't dropped. But now that Don isn't returning, I fall back on a codicil to my vow, which was that, if Don didn't return, I'd stop drinking up to Thanksgiving.

Last night, a dream which is obviously related to the dream I describe in my entry for May 26, this year. I (maybe Don too) was in a theater, in an upper balcony, quite high. And Laurence Harvey was balancing on a ledge in front of us. After a while, he

[1] Tom Ferdinand, the "fer" in Ban*fer* Gallery, ran it with Richard Bennet, the "Ban."

tried to climb back into the balcony and he slipped and fell. I was so horrified that I looked away. When I looked again, I saw that Harvey had somehow been caught by a man (Jimmy Woolf?) who was sitting on some kind of protrusion from the wall; perhaps a lamp fixture. Anyhow, it was obvious that this person hadn't got a secure hold on Harvey and that Harvey was just about to fall. Also, it was evident that Harvey knew this. Knowing that he had only a moment to live, he kissed the man who was holding him, on the mouth. Then he fell. I saw him lie on the floor of the theater, presumably dead.

As in the former dream—there was a delayed moment, a death-moment, before the actual death; and there was heroism of the same kind—a gallant gesture in the presence of death.

I think I have about two more weeks' work to do on *The Sailor from Gibraltar*.

November 23. Swami, when I saw him last (on the 18th), told us: "Think about death—and you'll know what to pray for."

Saw the Stravinskys on the 19th. Igor has an abscess on top of his rupture, so he couldn't wear his rupture belt and had to come down to supper holding on to his side. Later, after he had gone up to bed, I went to say goodbye to him, as they are leaving for New York, Boston, etc. He lay there with his icons beside him, reading a book by Alexander Werth on Russia during World War II.[1] He said, "What is the use of the Germans and the Russians?" Meaning, as he then explained, that, as you read a book of this kind, the whole concept of nationalism seems to become meaningless.

Am now in my eighth week of work on *The Sailor from Gibraltar*. I still hope I can finish at the end of it, despite Thanksgiving. But Tony still has to tell me what he thinks of the first draft. I am to talk about this with him tomorrow at the airport, where he is shooting. *The Loved One* is also nearly finished.

Since November 1, I have stuck to my non-drinking resolution. I guess I can hold out, now, until Thanksgiving. But it has been tough. Everybody I meet on any evening—except the people at Vedanta Place—urges me to drink or very faintly suggests reproach because I let him or her drink alone.

I can sense that Don is likely to return at almost any minute. I talked to him again yesterday. I miss him, and yet I don't want him coming back in a defeated, frustrated state; that only leads to fights.

[1] *Russia at War 1941–1945* (1964).

If he does come back, he will be defeated, because that will mean the New York City Ballet project has failed to materialize.

Stephen Spender spent the night of the 20th here. He is coming back for Thanksgiving. He is in the midst of a lecture tour. His energy and complete uncomplaining acceptance of his chores are really admirable. Natasha has had an operation for breast cancer; it will be years before they are certain there won't be a recurrence. Stephen is taking her to North Africa after New Year's. I feel that he accepts all this as a punishment for his sex activities. Now he has got to stay close to her all of the time. He let drop something about their discussing "If this is all my fault or not." I couldn't help remembering Cyril Connolly's remark in *The Unquiet Grave*: "The true index of a man's character is the health of his wife."

Stephen is still very handsome, with his florid face and wool-white hair, but the middle of him is a suitcase coming unpacked. He hates Wystan, and says Wystan is one of the most famous people in the world. (This after I had remarked that Wystan's stock seemed to me to be very low. I[t] occurred to me that I think this because I am rooting for Wystan; and that Stephen thinks what he thinks because he is rooting against him.) Stephen says that Americans always seem to have a clock on their heads from which you can read your exact degree of success at that particular moment. Here again, I felt that Stephen was imputing something from his own character. Stephen's vice is ambition. Mine is vanity. Stephen would never bother about his personal appearance, I imagine. I bother very little about whether or not I have succeeded; maybe because I feel that I have, according to my rules. What *I* am concerned about is whether or not other people recognize the fact of my success. And this concern arises from vanity, not ambition.

I now feel that, *according to my rules*, *A Single Man* is a masterpiece; that is to say, it achieves exactly what I wanted it to achieve. I keep dipping into it and always I feel yes, that is exactly the effect I was trying for.

1965

January 1. Last night, I dreamt about an atomic war, a dream I almost never have. Was worried rather than terrified. Things kind of straightened themselves out at the end.

Last night, Don and I went to a party at the Lederers'. The guests seemed to be nearly all oldish. Many encounters with the long forgotten and long avoided. Everyone looked older, except mad Mrs. [Bronislau] Kaper. She looked just the same as always; like a powder puff which has been used so often it's getting bald.

Bill Inge drank a glass of champagne, without going into an alcoholic tailspin. He danced a lot and was merry. I drank lots of champagne; nothing else. I am quite off hard liquor, at present. Haven't been really drunk since Thanksgiving.

Today we walked on the beach and lay in the weak sun. Beautiful and clear for the first time after all the rain.

Have been trying to sort out the articles for *Exhumations*, getting them into order.

David Selznick told me he did not vote for Goldwater; thinks he's hopeless as head of the party. He thinks the Republicans are licked for years to come. He thinks Johnson will be one of the great presidents; but that he would not have been if Kennedy hadn't come first and been assassinated. He thinks there is a period of prosperity ahead; that "The Great Society" will be realized.

January 7. Don flew back to New York yesterday—presumably for at least two months, probably four or five. Before he left, I told him that this short time together has been the best I have ever had with him. He said, "Lately I've been thinking that the Animals haven't seen anything yet; they still haven't had their golden age." I said, "They'd better hurry."

Last Sunday, the 3rd, we had lunch at the Selznicks'. David believes in a new Republican hopeful, John Lindsay.[1] He told a story about Sam Goldwyn explaining why he had voted for Murphy as against Salinger for Californian senator. "This man Preminger, he doesn't have a platform. All he keeps saying is, how well he knew Kennedy. Hell, I knew Kennedy—a whole lot better than he did.... That was before the assassination, of course."[2]

This morning, I really got started on the bits of commentary for *Exhumations*.

A vile reactionary pamphlet called *What Kind of a Country Are*

---

[1] Lindsay (1921–2000) a New York City congressman since 1959, was elected mayor in November 1965; he left the Republican party during the race for his second term and then tried, unsuccessfully, for the 1972 Democratic presidential nomination.

[2] George Murphy (1902–1992) actor, dancer, and Republican, was elected senator from California in November 1964. He beat Kennedy's former press secretary Pierre Salinger, who was the Democratic incumbent as a result of being appointed to replace Senator Claire Engle after Engle died in office. Preminger directed and produced *Advise and Consent* (1962), based on the Pulitzer-Prize-winning novel set in the Senate.

You Leaving Me? was sent by Coast Federal Savings. A photo of a cute little boy is on the cover. He is supposed to be speaking: "It's no use pretending you don't see me.... I'm not very smart yet, but I'm smart enough to see what you are doing to the country in which I must grow up and support my family.... What makes you think it would be 'bad' for me to have to make my way in competition with others? Where did you get the cockeyed idea that the man who earns *twice as much* should be taxed *four times as much*? (I know where you got it: right out of Marx.) Where did you get this idea that government can take care of everybody? I don't quite know what you mean by the word 'conservative,' but if it means what I think it means, *that's me! Aren't you ashamed!!!*"

So I went into their office and told them that I was drawing out our deposit, because I didn't choose to be dictated to politically by people I was doing business with. The young clerk I spoke to was obviously pleased by my attitude.

These savings associations are having New Year's open-house parties. Lots of old folks are sitting around in their offices, enjoying free coffee and cookies and collecting free ballpoint pens, calendars and toy balloons for the kids.

February 7. Got back from New York yesterday. I was there seeing Don from January 26 onward.

Now, at least, I have plenty to occupy me. This afternoon I have to give a talk to the people at *One* magazine, "A Writer and a Minority." And on Tuesday I have to start at UCLA, and on Wednesday I have to give a public lecture there, etc., etc.

The *Exhumations* manuscript is with Methuen, and Don will deliver it to Curtis Brown for Simon and Schuster, as soon as he has read it.

New York was as dirty, cold and brutish as usual; but the visit was a success, as far as relations with Don were concerned. Not that we didn't quarrel a bit—if we hadn't, it would only have meant that we were on our party behavior. There is still the old problem of our seeing other people together. But I do feel that the whole thing between us has strengthened and changed out of all recognition, since this time two years ago.

Our chief grounds for dispute: the ballet, Tennessee Williams. Don is now an ardent ballet lover, and why the hell shouldn't he be? But I am going through a phase of being bored by it, and so I resented having to go there—the evening I arrived, because I was tired; the evening they did *Harlequinade* with Rouben

Ter-Arutunian's set,[1] because I honestly thought it dull and unin-spired. Don said, "Don't be so *pleased* that you don't like it."

The ballet is now more than ever Lincoln Kirstein's kingdom. After all these frantic intrigues, he has captured the golden State Theater for his very own,[2] and he has megalomaniac schemes to make it the most famous theater in the world. The two giant Nadelman groups in the foyer are his declaration of war.[3] In due course, he will have Watusi ushers, and the New York Park Department will supply him with orchids twice a week. A special cat house will be built at the zoo in which his collection of cat statuettes and paintings will be housed, as well as a live collection of domestic cats.

George Balanchine is about to retire, and Jacques d'Amboise[4] is in disgrace, because he wants to quit the ballet for several months to earn more money in some musical show in Florida. The con-ductor is in disgrace, too. But Lincoln will find replacements for all of them. Don is to have a special exhibition of his drawings at Stratford[5] this summer and at the ballet next winter.

Lincoln gets up at four every morning. Before lunch, his day is well on toward its end, so he starts to drink. The only trouble is, he stays up nearly as late as everybody else. He announces that this is the Age of Peter Rabbit—pornography is out. He is sick of the ballet and wants to become a great librettist and to write plays like Shaw. He now speaks of himself as a millionaire; once he said he has three million, another time it was fourteen million. He is lav-ish with his promises, and I dare say he often does pay off. Don is supposed to get twenty-five hundred for the Stratford drawings; so far he has been given five hundred. Lincoln says that Don will be the Sargent of our time, and that he shouldn't be bohemian but dress stylishly and live in extreme elegance. That was why he ad-vised Don to take this two hundred dollar apartment, and it must

---

[1] Ter-Arutunian (b. 1920), Soviet-émigré set and costume designer for ballet and opera, was also designing *The Loved One* for Tony Richardson.
[2] The New York City Ballet moved into the New York State Theater at Lincoln Center in 1964.
[3] "Circus Women" (1931) and "Two Female Nudes" (1931); Kirstein introduced Isherwood to Elie Nadelman's sculpture in 1947, when he was pushing for the 1948 retrospective at the Museum of Modern Art.
[4] American dancer (b. 1934), trained at the American School of Ballet, a versatile and long-established principal.
[5] Connecticut, at the American Shakespeare Festival Theater, where the ballet performed in the summer. Bachardy's portraits of the actors in the festival company (including Lillian Gish) appeared in the souvenir program, but there was no exhibition.

be said in Lincoln's favor that he did offer to pay the rent. But Don refused, and camps out there in squalor and depression.

Tennessee says he hasn't had sex, or indeed an ejaculation, since last May. He still talks incessantly of Frank [Merlo]. He seems to think of himself as a failure; he has had two flops, he says, but he really only means the two unsuccessful productions of *Milk Train*.[1] He says that the cancer changed Frank's whole personality; he became very aggressive. He said to Tennessee, "You bore me." He is worried because he has a theory that if two people live together and one of them gets cancer the other gets it too. I asked him if he ever dreams of Frank. He said, "I don't dare to."

Don felt that Tennessee was disregarding him and slighting him; he was very angry about this. He felt the same about Monty Clift. And of course Monty *was* rude, though the hostility may have been subconscious. When Don came in, he said, "You weren't invited," and when Don left, he said, "Goodbye shitface." Clift certainly is, even at his best, a dismal kind of degenerate, with a degenerate's ugly unfunny aggressive attempts at humor.

Running around New York fairly martyred my feet. Every day, I had to put foot powder on them. One day, after I had very carefully dusted on the powder and put on my socks and shoes, I realized that I had absentmindedly been using Babbo, which I had taken out of the closet to clean the bathroom floor with!

While I was in New York, Lincoln went over to London to see Churchill's funeral.[2] He found that most of the people he met didn't want to watch it, even. But Lincoln got drunk and wandered around with a bottle of bourbon, weeping. He was one of the few who stood on the pier when the coffin was carried on to the launch on the river. I told him I have composed a last sentence for a Churchill biography: "The great ceremony was over at last, the huge crowds were left behind, and the coffin was carried on to the launch in the presence of one single weeping drunk American millionaire." Lincoln loved this. Just before I left, I went through the extra poems he has written and made a list, advising him which ones should be included in the new edition of his *Rhymes of a PFC*.[3] He wants to arrange some of the rhymes for a performance by actors on the stage. We are to be partners in this.

[1] *The Milk Train Doesn't Stop Here Anymore* ran for two months in New York, from January 1963, to bad reviews; Williams revised it, but the new production, in January 1964, lasted only four days.

[2] He died January 24.

[3] I.e., private first class; the poems were written while Kirstein was serving in W.W. II. See Glossary.

February 14. Twelve years since that memorable February which started with the fire in the Hookers' garage; and then the reunion with Ted and hence the meetings with Don; and then on the 14th the party at Jerry Lawrence's. And then Ted going crazy, and my trip downtown to get him, and his being committed to Camarillo....[1] Don called this morning and I wanted to say something to him about these twelve years, but I couldn't—that sort of thing is for college presidents. I'll try to write him something tomorrow. I suppose the real point is that the twelve years in themselves don't matter. All that matters is what we have now.

However, just to celebrate the occasion, today I started the new novelette about the two brothers who meet in India just as one of them is about to become a monk. I know almost nothing about any of this but I may just as well make a stab at it. Something will emerge—something quite different, probably.

When I spoke at One Institute[2], on the 7th, Gerald introduced me by making a speech in which he insinuated that I was going to talk about the Triple Revolution; his latest toy. This was his usual semi-deliberate bitchery, but I got myself out of it. And it was good that he mentioned the whole thing because then Michael sent me the pamphlet and I read it and was able to use it in my lecture on the 10th at UCLA; which was standing room only—in fact, they said they turned two hundred people away![3]

Still this wretched chilly weather. We are now being told that the causeway across the bay is to be started in a year's time. This depresses me out of all measure. And how silly. It seems that I must dread something. Stop me from expecting atom war, and I turn to cancer. Turn my mind off cancer and I fasten on to something like this.

February 22. A heckling letter from [a man], who was in the audience at One Institute when I spoke. He more or less takes the "fire next time" Jimmy Baldwin tone of voice. And all because I cautiously said that I didn't think the Negro problem was quite the same as the homosexual problem, and that I didn't think the

---

[1] See the entry for March 6, 1953 in *D.1*.
[2] I.e., One, Incorporated; see Glossary.
[3] A committee of thirty-two academics and activists—including Todd Gitlin, Tom Hayden, Irving Howe, Dwight Macdonald, Gunnar Myrdal, Linus Pauling—published *The Triple Revolution*, about rapid advances in cybernation, weaponry, and human rights, and called for commensurate changes in attitudes and policies; they sent the pamphlet to President Johnson, leaders of the Senate and the House of Representatives, and the Secretary of Labor on March 22, 1964.

measures taken by the Negroes were necessarily those which we should take to solve it. So I am accused of fence-sitting. I can't say that I feel guilty of this, because I do stick my neck out quite far, in my own way. But of course a man like [that] (who, incidentally, is on the police force and claims that he has told them he is homosexual) sees "the struggle" in terms of group action. What is interesting in [his] letter is the statement that, "This is a war of minorities outside the society of the community, against a minority" (he means the police) "that has been given a certain responsibility as designated servants of the public, who have long ago forgotten that role, to think themselves the master, the public their rightful servant, and fair prey when it suits their purpose."

Bill Legg, at One Institute, is going to publish this letter in some sort of circular which the institute puts out for its members.

Caskey came and had lunch with me yesterday, at Ted's [Grill]. He had called to tell me that he had read my novel and that he is leaving the country because he can't stand it here. He will sell everything and go first to Australia and then Europe. I felt very unwilling to see him and yet it seemed somehow wrong not to, after all these years.

He seemed greatly changed; and not just physically. His jaw is jowly and his eyes are pouchy and terribly tired, and he has a pot. But what I chiefly got was the sense that he had become a much older man. He was so utterly self-obsessed; exclaiming peevishly against America and Palm Spring and the rich. Everybody drinks too much, he said, and he has to drink too, because they all bore him so. He hardly referred to anything in the past, and showed very little interest in what I was doing. He only mentioned Don once, with some conventional praise of his drawing. He spoke with a heavy Irish sententiousness; an Irishman laying down the law in a bar. What made him seem so unyoung was, in a way, his lack of bitchery. When we went into Ted's, I looked up at the window of our old apartment opposite and it seemed absolutely incredible to think that we ever lived together there, or anywhere else. It wasn't disagreeable, being with him. Indeed, we could agree, for instance, in loathing Goldwater. And we even laughed quite hard, especially when I told him that Richard has become a Rosicrucian.

February 24. Another sweet letter from Don today, suggesting that we go to Monument Valley at Easter, just as we did twelve years ago.[1] One of my daily meditations ought to be on just exactly

---

[1] May 4 to May 10, 1953; see *D.1.*

what it means or should mean, to me, at my age, to have Don in my life. It is nothing less than a blazing miracle. My situation is infinitely less usual than that of being a millionaire. It is less usual, even, than that of being a major artist. It is so extraordinary that perhaps, after all, I had better *not* dwell on it, or maybe I shall be reduced to grovelling terror of losing him.

Am very discontented with myself in relation to my class at UCLA. It is getting completely out of my control; partly because it is too big, partly because I have failed to make an appropriate program. I see now that I must do this. I must have something that I keep getting on with, whenever the interest starts to lapse.

Pagli (spelling?) at Vedanta Place had her room burgled the other day. Telling the story, she remarked, quite seriously, that she felt humiliated because her room was so untidy.

Lately we have been having some wonderful sunsets, with violet tints in them. Gerald says that these are the last of the "Krakatoa sunsets." He claims that the dust from the explosion of Krakatoa has been drifting around the world since 1883; sometimes it circles the southern hemisphere, sometimes the northern. It has moved up from the southern hemisphere now for the last time; soon it will disperse altogether.

March 15. Grey, cold. Am in bad shape. Got drunk two nights ago at the Masselinks' and fell and hit my head against the bathtub. It still hurts, and also I have a bad cold and feel very depressed, old and exhausted and generally impotent. Am way behind with my novel and with my letter-answering chores. I just long for Don to come back.

Jack Larson called today to say that one of their dogs ran out of the house last night and got killed by a car. It was probably scared by our freak thunderstorm, a single flash of lightning and a terrific crash. (The rest of the storm was miles away over the mountains.) Jack says that he and Jim were up all night, and he started to cry as he told me about it. "I haven't a clue," he kept saying, "what to do when things like this happen." All *I* felt, I am sorry to record, was pleasure that dog wouldn't bark at us again—yes, and a slight irritation at the fuss Jack was making. But I'd have understood and been most sympathetic if it had been a kitty.

Tony Richardson was here last week. He ran *The Loved One*, and the Filmways people didn't like it.[1] They even got Terry Southern

---

[1] Filmways, an independent production company where John Calley was an executive, was co-producing with MGM.

to call Tony and suggest alterations. But Tony's morale remained quite high. He says that he feels, for the first time, that he has done something good. Also that he has a great creative period ahead of him—to be followed, he implies, by his death.

Some interesting information gleaned from Sheldon Andelson.[1] Entrapment, legally speaking, is when you are persuaded to do something which you would normally be disinclined to do. Enticement is when you are persuaded to do something you are anyway inclined to do. According to this definition, a homosexual cannot be entrapped into committing an offense related to homosexuality; he can only be enticed and enticement is permitted to police officers.... Evidence collected by listening devices, tape recorders, etc., is only valid if it has been done while you are overtly talking to the victim. You can record a conversation without telling him you are doing so; but you can't bug the house or sneak up to the outside wall and listen in while he is talking to someone else.

Prabhavananda, when I saw him last, on the 11th, told me that he now feels quite indifferent whether he goes on living or dies. It is all according to Maharaj's will. He is aware of Maharaj all the time. He quoted Ramakrishna's simile of the magnet, saying that he now feels the attraction of the stronger magnet, drawing him away from everything else. "You know, Chris, He is everything. Nothing else matters."

And I, what do I feel? Stupid, dull, unfeeling, fat, old. Utterly unfit to associate with Prabhavananda, or with Don either. A stupid old toad—and not even ashamed of being one. Too dull for shame.

March 20. The weather is warmer, my toothache and other aches are better, so I'm pulling out of the depressive phase again.

Have just been looking through the draft of my novel, *A Meeting by the River*, of which I wrote twenty-five pages and then stuck. It is all wrong. Not just that every word of it is wrong, which is to be expected at this early stage. The method is wrong.

Why is it wrong? I think because the overt confrontation—what the brothers actually say to each other—isn't what deeply matters. By staging these meetings as dramatic scenes, with or without insight into what the two of them are thinking, I have put a wrong emphasis on the story.

---

[1] A Beverly Hills lawyer; in *D.1* Isherwood mentions that Andelson arranged bail for Ted Bachardy in 1957.

What seems to me (at the moment) better is to tell the story through letters and a diary or diaries. Maybe Martin, as the outgoing one, should write letters, while Leonard, the introspective one, should keep a diary. Can the whole story really be told in this way? That's what I have to find out.

Letters mean confidantes. I have to work out carefully, before starting, just what Martin's situation is, how many people he will be corresponding with, what their relations are to him, etc.

The problem of the story itself remains unchanged, however. This confrontation doesn't only take place in order to display the contrasted characters of the brothers. It should also produce some results in their lives. The confrontation must at least alter Martin's and Leonard's attitudes toward their lives to some degree. And, although nothing really dramatic can happen at the time of the meeting in India, it is quite possible that some very dramatic things happen later on—Martin enters into a new relationship, or doesn't, Leonard leaves the order, or maybe they meet again under quite other circumstances, or one of them kills himself. So perhaps there is an epilogue.

I must also consider the possibility that there are more than two central characters. Does Martin come to India with a companion, a wife or a mistress or a sister or some other man? I'm afraid of this because it seems to take away from the dramatic simplicity of the confrontation; but it must be considered as a possibility.

One thing is for sure. There must be an awful lot of background to this meeting. I have to know all about the lives of Martin and Leonard from the time they were born right up to this point, with all their involvements.

What *do* I know?

I think they are half brothers, with a mother as their only surviving parent; the stepfathers are dead or quite disappeared. They were both born in England, and I think the mother is there still. But Martin emigrated to the States, where he is doing well or rather, impressively and successfully, at something or other. Yes, what the devil *is* he? It is important not to sneer at him by making him something vile, like a Madison Avenue boy. I'm even afraid of making him connected with the movies, although that would be the easiest for me to describe. If he *is* in show business, it must be made clear that he isn't just a hack. He must have his serious side ... As for Leonard, I think he first went into some sort of social service, maybe with the Quakers and then met a swami of this order, not in India but somewhere in Europe and lived with him there for a while before coming out here. I feel he hasn't been in India long—if only because that would have had a kind of effect on him which I don't feel able to describe. (One great advantage of the letter and diary method of narration is that I don't have to be so precise about the nature of the order, the other monks, etc. Thus I can avoid too great similarities to the Belur Math.)

Provisionally, I'm inclined to think that Martin has three correspondents; all women—his wife, his mother, and some girl for whom he may be about to leave his wife. Owing to the shortness of Martin's stay in India, these women won't be answering Martin's letters, which is very good. It keeps them in the background. They will be subjective characters; projections of Martin's consciousness. And the physical background—the monastery and its surroundings—will also be subjective. Martin will describe it, from his point of view. Leonard will take it more or less for granted. That, too, is helpful. It is what I've wanted from the beginning—that the setting of this narrative shall be, as it were, "mental." I am emphatically *not* writing about the humors and quaintnesses of the Orient. This is not a realistic novel.

Yesterday I went to the Stravinskys', where a German named Liebersohn(?) was making a film documentary of Igor's daily life.[1] Gerald, Michael Barrie, Mirandi Levy and I were to play intimate friends who drop in at teatime. Gerald of course hogged the show; he could hardly be stopped talking. But Igor did have some fascinating and rather touching moments, when he told us how he loves the act of composing, he is so "content" while he is at it, it doesn't matter to him if the work will be performed. Sometimes he gets an idea and notes it down in the middle of the night. He doesn't so much care where he is, he can work anywhere, provided there is not too much noise. He seemed to be just what Aldous once called him, "A saint of art."

The day before yesterday, we had this awful memorial dinner to Aldous, at USC, in the presence of about four hundred "Friends of the Library," and then—Laura, Cukor, Robert Hutchins and me—got up on the platform under blinding lights and were filmed as we reminisced about him. It went quite well, thanks chiefly to Cukor, who really *is* a director.

This is from the brochure they put out beforehand, announcing the dinner:

The wondrous works and warm ways of the late, great Aldous

---

[1] Rolf Leibermann (1910–1999), a Swiss composer, then General Manager of the Hamburg State Opera; his film was *A Stravinsky Portrait* (1966).

Huxley will be the subject of an evening's "conversation" for the University of California's Friends of the Libraries and their guests Thursday night.... First there will be a dinner with a menu that the gaunt genius always favored—a couple of vegetables "not overly cooked," and good meat. The dessert will be one which Mr. Huxley himself created.

Needless to say, the food was foul and cold.

While I'm documenting, here are the titles of some lectures scheduled currently at UCLA: The Electrical Conductivity of a Partially Ionized Gas. Growth Mechanism in the Young Polar Front Cyclone. Do Fibroblasts Make Antibody? Play of Inhibition Within the Motor Cortex. Regulation of Extrarenal Electrolyte Excretion in Birds. Methodology for Social Technology. Advanced Structural Concepts.... It is stunning to think that Aldous could probably have talked intelligently on any one of these subjects!

April 2. Rain and thunderstorms, since the day before yesterday. The radio says the rain will continue through tomorrow, which is a shame, because Don is coming back and I would have loved for him to be welcomed by sunshine.

He is literally everything I have in this world. He is what keeps me alive. When he isn't with me, I am in a partial coma, most of the time. Which is okay. Because the alternative would be pangs of misery because of our separation. I eat abnormally and am fat, but not as fat as I would be if I drank. Haven't touched any kind of alcohol since March 15.

The new line on my novel is the right one, I believe. So far, it seems to be. Not that I am progressing very easily. I need to invent so much about the offstage characters. Am *very* tempted to make Martin a bisexual counterfeiter— I mean, to make him cheat on his wife with a boy. This further suggests that Martin seduced Leonard when they were young, and that Leonard, knowing himself to be homosexual, has decided that it's wrong and that he must cut out sex altogether. Hence his inclination toward celibacy.

A little while ago—this year, anyway—Laura Huxley was visited by a medium from Seattle(?) who told her that Aldous was about to give her positive proof of his presence. The medium then told her to take down the third book on a certain shelf and look at the 23rd line on the 17th page. Furthermore, he made her go into two other rooms (in Jinny Pfeiffer's house) in which there were books, take down the corresponding book and look at the same line on the same page. Here are the results:

The first book was a report by the PEN Club on a conference held in Brazil in 1962. Aldous had been speaking and this was the beginning of a speech made by someone else to thank him. *"Aldous Huxley no nos sorprende en esta admirable comunicacion...."*[1] The second book was about parapsychology. The indicated line read: "Parapsychology is still struggling in the first stage. These phenomena are not generally accepted by science." The third book was *My Life in Court* by Louis Nizer. The passage referred to the libel action brought by Quentin Reynolds against Westbrook Pegler:[2] "... it suggested that he was a slacker who, though six feet five...." (Aldous's height, says Laura, was six five. However, Gerald and Michael maintain that he was only six four. This may or may not be bitchery. They both had to admit that they were nevertheless greatly impressed.)

What is astounding is that *anything* referring to Aldous and the circumstances of the test could have appeared within such narrow limits of reference. Even if the medium had been a cheat and had been left alone in the house all day in order to fake the test, he would have needed an awful lot of luck to produce these results.

Had supper with Jim Bridges last night. His story of his first great love affair in school: the only place they could make love in safety was in the college theater, high up in the flies, under the roof. One night they risked screwing on a bed on the stage—it was the set for *Death of a Salesman*—and were nearly caught by a cop.

Jim says that when the dog was killed Jack actually screamed and beat on the wall and talked of suicide. I couldn't help showing that I found this absolutely beyond my understanding. I can understand a lonely bachelor like Chris Wood feeling the loss of a pet as much as the loss of a human being. But Jack.... I cannot help suspecting his behavior was a kind of aggression toward Jim.

(N.B. The medium's name was Keith Rheinart.)

April 14. Don has been back here since April 3. We have been very happy together—although, in a way, everything has been against it. Don isn't well, there's something wrong with his liver. He has to see Dr. Allen again about this on Friday and I'm worried, of course. Particularly because I suspect Don isn't telling me everything Allen said. He did let drop that Allen wanted him

---

[1] "Aldous Huxley does not surprise us in this admirable communication...."
[2] Reynolds, W.W.II correspondent and author, successfully sued Pegler, right-wing Hearst columnist and once a W.W.I correspondent, for calling him a coward, a liar, and a degenerate in a 1949 column. Louis Nizer was one of Reynolds's laywers.

to go into hospital, but that he refused; then Allen seems to have told him to stay in bed, but Don is running around and working as usual. The only point on which he has obeyed Allen is that he has given up drinking, which he doesn't do much of, anyway.

No doubt also partly because of this thing with his liver, Don has been terribly depressed about his painting. He says he doesn't know what to do with it. He wishes he had a lead, a gimmick, a line—something which would make him able to do a whole series of paintings in the same manner. This makes him envy many other painters he knows, even when he doesn't admire their work at all.

It is absolutely impossible for me, involved in this as I am, to have any opinion any more about his painting—even if my opinion about painting in general were worth anything, which it almost isn't. I am rooting for him so hard that the problem appears purely psychological, not artistic at all. I cannot help noticing that he refuses to show his work to people who *might* know. Yes, it's true, he did show it to Bill Brown and Paul Wonner and they were lukewarm. But that was all. It simply is no test.

Well—we just have to sweat this out.

Talking of Paul and Bill, we never see them any more and I guess we never shall. I have hardened my heart against Paul. I find his sulks thick-skinned and tiresome.

Other news. Mike Leopold was mad but better now; he tried to cut his wrists and throat. Budd Cherry was sad but better now; he is staying with Gavin.

I keep on with *A Meeting by the River*. It may well be only a short story, or nothing. But I will bulldoze a draft through to the end.

We have had the longest spell of rains I can remember. Today is warmish and springlike, but they say there will be clouds and showers again tomorrow. The two cypresses that Mr. Garcia transplanted have both died, so we have two more. Everything else is growing like mad.

To my astonishment, when I sat down and made a list, I find that I have talked to twenty different people already about their writing. One or two, I don't even remember what they had written. Here is a list of the remembered ones: some of them are students at UCLA, others come to my class from off campus.

Dinny Johnson. I think she's married to a professor at UCLA. Young. A friend of Alison Lurie. She wrote a worthless but somehow quite readable novel of musical-chair sex. Very sympathetic.[1]

---

[1] Diane Johnson (b. 1934), then married to a professor of medicine with

Ira Sohn. Very superior good-looking verbose intelligent Jewish boy. Is writing a novel without any dialogue. Admires Conrad. Okay.

Mike Oppenheim. Really talented playwright. One-acter about a maternity ward, called *The Population Explosion*, black comedy. Very sour and disgusted because they won't perform it at the drama department.

Lee Heflin. The kookiest of them all, bearded, works at the new research library. Very talented, I think, but apt to wander off on Gertrude Stein trails. Writing queer novel. Also paints. The only one who has had me to his home. My best supporter in the class with bright questions. Treats me with formal politeness, Sir and Mr. Isherwood. Paints abstract. I like him and suspect him of being remarkable.

David Arkin. Wrote a play, not so hot. Made dates with me and didn't keep them, so I write him off as rude casual Jewboy.

Hillary Russell. Has written two somewhat piss-elegant queer novels about a fatal love ending in murder and suicide. An older man with a southern accent, very pleasant to talk to. Interesting relation with a straight son who nevertheless isn't shocked by the novels and gives his father advice on them.

Cliff Osmond. Big fat actor. [. . .]. Talented.[1]

Deena Metzger. Swarthy Jewess. Wrote turgid love novel. But she is bright. I rather like her.[2]

Frederika White. Very attractive girl, partly Negro. Seems very bright. Hasn't shown me any work yet.

Ramaswamy. Very handsome small middle-aged Indian professor. Treats me like a guru. Interested in films. I taped a bibliography of my works for him. A bore but sympathetic.[3]

Edward Peters. A middle-aged union official who has spent his life working for labor groups. Like him very much. Has written very interesting studies of labor negotiators.

Margitt Michel. An old Jewish woman who has huge charm.

---

whom she had four children before divorcing, was writing a dissertation on George Meredith's poetry for a UCLA Ph.D. in English. She published her first novel *Fair Game* in 1965. Later novels include *Persian Nights* (1987) and *Le Divorce* (1997). She co-wrote the screenplay for *The Shining* (1980) with Stanley Kubrick.

[1] A Dartmouth graduate studying finance and theater history at UCLA. He had small roles in Billy Wilder's films in the 1960s, including *Irma La Douce* (1963), acted and wrote for T.V., and directed *The Penitent* (1988). He also taught acting.

[2] Novelist, poet, playwright, essayist; she later became a healer and popular public speaker.

[3] Professor Iyer Ramaswamy taught genetics; he later returned to India.

Lesbian? She roams around the beach and watches people and writes little sketches and fables.

Ursula Moore.[1] Crazy and really a menace. She has hallucinations about plots against her activities: she is trying to promote a universal method of conversing through gesture, without language, and she believes President Johnson will order it introduced all over the country, in all schools. By a fiendish coincidence, she is the manageress of the apartment house in the Canyon where Lee and Mary Prosser live.

Thomas Victor Siporin (accent on the *or*) is an ugly-attractive young student who plays baseball and basketball although one of his arms is deformed and the hand no more than a bent claw. He is very bright and writes funny-Jewish stories. He is a mathematician. He calls me Chris. I like him very much.[2]

Keith Gunderson is a professor in the philosophy department. He is young and cute with bristly short hair. Writes verse in curvy lines. I like him.[3]

Christine Dickson. A strikingly pretty girl who has money of her own (this came out as the result of my questions) and lives independently. She may have talent; can't tell. She intrigues me.

Robert Livermore. A strange big-headed young man who is experimenting with a friend in hypnosis, and experimenting in writing a drama in pseudo-Shakespearian dialogue. I am doubtful about his talent but he is certainly bright. He intrigues me a lot.

Miss Paz[4] (don't know her first name). Is from Argentina. She has talent. Writes stories. Takes the square atheist line—belief in the supernatural is a sellout from reason. The worst of these people is that they have to balance their lack of God with humanism, and this compels them to pretend that they love humanity a whole lot more than they actually do. I rather dislike her. She looks like she may grow a moustache later.

Violet Hamilton. Wrote a story about cockneys, while she was in England. Nothing special, but not bad either. Quite like her.

---

[1] Not her real name.

[2] Siporin (1942–2001) had polio in childhood. He soon started law school at UCLA, got his degree in 1968, and later worked for the American Indian Movement.

[3] Gunderson had a Princeton Ph.D. and was an Assistant Professor; his poems appeared from a small press in 1976, followed by a book set partly in Santa Monica, and he also published academic work. He later taught at the University of Minnesota.

[4] Not her real name.

April 19. I had intended to make an entry yesterday, because it was Easter, but there was no time. I did at least get on with my novel—I still have absolutely no notion if it adds up to anything or not—and with some revisions of the notes to *Exhumations*. These should be finished in a few days.

Beautiful weather at last. We were on the beach. The water full of surfboarders, about fifty of them. Which reminds me to mention that allied sport, skateboarding. It is almost incredible, but I believe I have never mentioned it here. *The Christian Science Monitor*, while recording that a son of John Houseman is among the American youths who have just introduced the skateboard into Paris, says that skateboards started in the United States as early as 1960. Ben Masselink thinks even earlier.... Anyhow, their long trainlike sound has become one of the basic Canyon noises. The little kids are sometimes incredibly graceful and adept; they'll take off at all hours of the day, even in their Sunday suits before leaving for church. Michael Sean[1] rides a skateboard right down our hill. His technique is pretty good, but he looks ridiculous because he is so big and a grown-up.

On the 15th, Don and I drove downtown on the newly opened Santa Monica Freeway. Downtown at night now seems mostly very clean and empty. Big shining new glass office buildings with no one in them; almost like models on show. We ate at the nostalgic old Clifton cafeteria—the one with the redwood decor: the other has been torn down. I must say, the food reminded me of that breakfast in jail when I was waiting to be bailed out on my drunk driving rap. And then we went to see *Youngblood Hawke* (perhaps the most ridiculous film ever made about writers and publishers) in a rather wonderful old theater called The Globe. It looks as if it had been legit. These downtown visits are especially moving with Don, because they recall the days when he used to come there with his mother and Ted to see films on Saturday mornings, from their home with the palm tree near the railroad station in Glendale. He and I went to look at it, once, when I was leaving by train for San Francisco. The palm tree had been removed.

June 13. This morning Don called from New York, as he has done every Sunday. It now seems that he won't be returning till near

---

[1] Professional name of Allan Carter, aspiring actor and a protégé of Bill Inge. He played a film scene opposite Joanne Woodward in *The Stripper* (1963), adapted from Inge's play *A Loss of Roses*. Isherwood typed, "Shawn."

the end of this month. (He went back to New York on May 20.) This seemed quite a blow, which is unreasonable of me, for he hasn't even been gone a month yet, and he must finish his drawings of the New York City Ballet for this book, even if Lincoln ends by never publishing it.[1] I have just finished making him some brownies, which I will mail off to him tomorrow or the next day. It was quite painful to do this because I kept thinking about him so intensely while I was doing it.

As soon as Don goes away, I get older. My legs are very thin, with white spots all over the shins—chalk deposit? They dry out and I have to keep them oiled.

Many thoughts of death. Chiefly the sadness of having to go. I remember Laughton saying, "I don't want to go."

Well, work is the great tranquilizer. Today I did six and a bit pages on *A Meeting by the River* in rough. I have eighty-six pages of the first draft already written and I'm pleased to think that I managed to do most of this while working at UCLA. I gave my last class there on June 1, but I still have a lot of manuscripts to read, and more that have been sent me by non-UCLA writers.

Jack Larson is playing Androcles in a local production of [Shaw's] *Androcles and the Lion.* I am very much at ease with him and Jim Bridges (who is working on a screenplay of the original Mary Shelley *Frankenstein* story) and of course with Gavin. We, and Budd Cherry (who has been staying with Gavin and not being a success—he will leave soon) all went last night to a newly opened French restaurant called Eve's in a cellar on San Vicente. Discovered too late it is a snob place, no menu and they charge you nine dollars a head, and the domestic wine costs five dollars! We got about eight courses—but *I* didn't, because I had to leave after the fifth to talk to a youngish man named Charles Sweeting who was in my class and wrote a chapter of a novel for me to read. He was leaving today for England, hence the rush. He is one of those excessively polite people that you just know are awful bitches. He described several people as "a Trinity man."

[An acquaintance], thanking me for letting him and his cute friend [...] sleep in the front room because it was late and [the acquaintance] was very drunk, describes this, over the telephone, as "a fraternal gesture."

Reading bits of Marchand's *Byron.* I want to read *Don Juan* right

---

[1] The drawings were published—on separate loose sheets assembled in a souvenir portfolio—but they were never sold at the ballet; see Glossary under Kirstein.

through. Also finish [Apuleius's] *The Golden Ass*. God, am I sick
of these manuscripts! They are a kind of punishment, I sometimes
think, because I refuse to read most of the best-loved authors of
our day, such as Wouk, Bellow, Malamud, etc. (Last night, Gavin
had an outburst against the Jews—their utter thick-skinned indifference to other minorities. Jack and Jim were a bit shocked but
rather thrilled. I sat there thinking smugly, well, *I* didn't say it, *this
time!*)

June 22. Yesterday I finished the first draft of *A Meeting by the
River*; it has taken me exactly three months, to the day! I have just
finished reading it through. There *is* something in it. But it seems
quite boring in parts. Perhaps it needs cutting down to a long
short story. It's now 110 pages—let's say a bit over 35,000 words.

The boringness is partly due, I think, to generalizations. I could
cure it by going into more detail. Also, the line of development is
unclear. The graph of Martin's feelings about Tom does not relate
very much to anything that happens at the monastery. One doesn't
know *really* why Martin makes such a play for Swami and asks
him about Hindu philosophy. And Martin's tone is all wrong. He
writes in such a fuddy-duddy elderly style, and his falseness is made
so apparent that it's corny. Leonard's style is a bit better but could
be improved. These, of course, are things you don't expect to get
right in the first draft. Especially with my method of thinking and
inventing while I write. This means that everything is spur of the
moment and therefore not my best.

And yet—something has been created. This confrontation problem has been solved, basically. You feel the importance of it.

I think I was wrong in saying, earlier in this diary, that something "dramatic" has to happen as the result of the confrontation. I
now very much doubt this. The obvious thing would be for Tom
to kill himself, but I don't buy it. It would prove nothing.

Leonard's professional background seems all right, but I question Martin's. He should be more important. A producer, if in the
movies at all. Or a director. (Tony Richardson would suspect it
was him!)

I had a dream early this morning that there was a vast earthquake here, a really tremendous one, but I wasn't really scared.
An earthquake *is* foretold by astrologers at the end of this month!

Well now, I have finished the novel for the time being. Don
must see it, and maybe even Gavin, before I attempt a rewrite. So
let me not lapse into laziness. I should get ahead with all my other
tasks. Thank goodness, I finished that huge bugger of a manuscript

by Edward Peters yesterday—548 pages about union politics, quite instructive and even interesting in a way, but suffocating in bulk!

July 3. I dreamed that I was in a bedroom that was haunted, talking to a rather attractive boy who had been sleeping there. The bedroom was quite light and unsinister, but very strange. The house it was in appeared to be on a very steep hillslope, almost a precipice, with the result that the window in the back wall was very high and the window in the front very low, and a strong breeze blew through the room, coming up from the depths below. The style of the architecture was like that of a fortress.

The boy was about to leave and I was going to take the room over. I felt a bit heroic and pleased with myself about doing this, also fairly confident in the power of the mantram to protect me from the ghost, but not entirely confident. Underneath I was somewhat scared.

I told the boy (expecting that this would shock or repel him) that he ought to pray every night before going to sleep in such a room. He answered that he did pray, but that it didn't help at all.

Then he went away and I was left alone to sleep in the room. (It did not seem to be night, however.) The ghost didn't appear, but it somehow spoke to me or put a thought into my head. It said, "I want you to sleep here, so I can get into your dreams."

This really frightened me and I tried to cry out. I woke myself by doing so, and immediately heard a soothing noise from Don, who was himself half asleep. This was a happy surprise, because Don has been home such a short while that I haven't yet got it deeply into my consciousness that he's here. As a rule, I'm aware he is here even while I'm asleep, or so it seems to me. That doesn't necessarily prevent me from having nightmares, but I had a feeling that I wouldn't have had this particular nightmare if I'd known he was here!

He returned from New York two days ago.

David Selznick died on [June] 22.[1] I wrote my last entry in this diary before I heard the news. In fact, Jennifer had been talking to me on the phone that morning about going there to supper on the evening of the 23rd, and it can only have been a short while after our talk that she got the news of David's attack at his attorney's office. An hour later he died in the hospital.

The funeral was on the 25th at Forest Lawn. I was one of the pallbearers. It was a sad grey day with drizzle. George Cukor gave

[1] Isherwood mistakenly wrote July.

us directions; he was having a great time, and talked so loud that someone came in to shush him. He told Sam Goldwyn, "Now Sam, you're to go in front of the coffin, with Bill Paley.... The rabbi will go first, of course." "Why should the rabbi go first?" Goldwyn asked. This may just possibly have been a deadpan joke, but he certainly seemed a little gaga, vaguely smiling and telling everyone, "You look good." He said to me, referring to David, "He was very fond of you," and then added, "We're all very fond of you." So I had to forgive him. If I'm not careful, I soon won't have any mortal enemies left—except Peggy Kiscadden—and I'm no longer even sure how to spell her name![1]

Katharine Hepburn[2] read Kipling's "If." When she got to the last line, she turned toward the coffin and said, "... You'll be a Man, my son!" I later heard that George Cukor thought this a supreme touch of artistry. I thought it farcical. One expected David to put his head out of the coffin and exclaim, "*Now* she tells me!"

July 4. Yesterday I forgot to record that, on the evening of June 30, I got a traffic ticket for going through a red light. It happened largely because the engine of the Volkswagen kept stalling and I darted forward to avoid letting it die on me. Anyhow, the cops followed and signalled me to stop, so I turned right off Santa Monica Boulevard on to Colby, and stopped behind a parked car. While one of the cops was writing out my ticket, the other strolled over to the parked car, flashed his light in and found a young man, with a beatnik beard, more or less passed out inside it. The young man was dragged out and apparently the cops decided he was high on some kind of dope. A car with two young plainclothesmen arrived, evidently summoned by radio, from the station nearby. The beat became semiconscious and searched for something among the litter of clothes in the car, maybe it was his license. While he was doing this, I saw how the uniformed cop half drew his gun from its holster and stood ready with his finger on the trigger. One saw how easily a suspect might get himself shot, especially not knowing that he was covered like this.... I felt a bit guilty about the whole affair. I had actually led the cops to the spot. Otherwise he might have sobered up and never been caught.

Thick white sea-fog, again this morning. We have been cursed

[1] Kiskadden.
[2] American actress(1907–2003). Selznick produced her first film, *A Bill of Divorcement* (1932), and her fourth, *Little Women* (1933). Both were directed by Cukor and were among her greatest successes.

with it nearly every day for at least the past three weeks. Laura, Tom Van Sant and others are coming down for the fireworks, but will there be any?

I'm worried about my ear, it is sore. Perhaps the beginning of an inflammation.

Don has gone to his studio for the first time. Let's pray he will be able to work, this visit. The fog disappoints him bitterly, of course. He has been so looking forward to beach life, while in New York. As for me, I'm simply happy to have him here. As I've realized so often before this, without him I am not altogether alive.

July 9. The sea-fog has continued like a curse, but yesterday and today it partly yielded to weak sunshine. This morning we even went on the beach.

Don can't paint and feels utterly frustrated but, as he says, for some strange reason not in despair.

As for me, I'm marking time. I want to rewrite *A Meeting by the River* but I don't feel a great urge to, and so I don't press Don to finish reading it. He has to do that first and we have to discuss it before I can start.

On the 7th, I had supper with Jennifer, alone. I had been afraid it would be sad and embarrassing, but it wasn't; in fact, I felt I was really communicating with her for the first time. She told me that David had had a dream—a nightmare, in fact—shortly before his death, in which he felt that he was shut in on all sides, with only one way out—toward India! Two clairvoyants foretold his death. One said that someone very close to Jennifer would die, the other named David. David himself, after his heart attack, said to her, "When you have looked around the corner, you know that there's nothing to be afraid of." But Jennifer says she still can't quite believe that it has happened. "I keep expecting the telephone will ring, and it will be David calling from New York." (When I told this to Don, he gave a kind of gasp and I suspected that he had had a sudden glimpse of how *he* might react if *I* died!)

Jennifer says that she thinks she will stay on in the house. She feels closer to David there. She would like to act, but doesn't quite know how to go about it.

About the funeral, she commented on the false impression which the rabbi gave by his address. He tried to make out that David had been a zealous and orthodox Jew, "a proud Jew" was how he expressed it. He told how David had been about to crown his career by undertaking a great project—a project which would be greater than *Gone with the Wind*—to film John Hersey's *The Wall*.

But Jennifer says that David gave up this project a long time ago, because he realized that the problems of the present-day world were far greater than the problem of any one minority.

July 24. Today the weather is beautiful but the sea stinks like bad cabbage. I went into it, without joy. I have lost my joy in this beach. It is always crowded. I am developing my crowd obsession alarmingly.

Don says he hasn't been able to paint since he returned here. He wants to live a more independent life, sleep out in the studio, eat breakfast alone, etc. He feels overpowered by my being around. Aside from this, our relations are good. Indeed, they cannot be otherwise for long, because we are really not each other's problems. Don has to come to terms with success-failure. I have to come to terms with death.

Julie Harris to supper yesterday evening. I drove her back to the motel near Warner Brothers Studios, where she is making a film called *The Moving Target*,[1] in which she is tied to a chair and tortured. She told me that she may leave Manning. She has found another guy. She says she has been attracted to someone in every play she has been in, and often they have slept together. Julie's attitude to adultery is so solemn and weepy that it seems almost like a gag. I couldn't help laughing. But there is cruelty in my amusement. I aggressively refuse to take the woes of heterosexuals seriously. So Peter grows up without a mother—am I to fall on my ass?

Am reading Paul Goodman's *Making Do* with delight. It is a real human marvellous modern serious fun novel. He redeems single-handed the drivel of the other Jews.

Don loves *A Meeting by the River*, Gavin seems impressed, though with reservations. They seem to agree that Leonard (all these characters should be renamed) is undeveloped. More about this. More also about my problems with regard to the two autobiographical books; for I think there must be two—*The Autobiography of My Books*, and a piece of straight narrative, based on my diaries of the first years of my life in America. (A good place to stop would be the production of *I Am a Camera* in New York at the end of 1951; a period of twelve years.)

Dorothy Miller came back and cleaned house for us today; the first time since she got sick last year. Needless to say, she took the line that the house was almost irreparably filthy. "I just want to get the hard dirt out of it." Her best saying so far: "The best religion

---

[1] U.K. title; in the U.S., the film was called *Harper* (1966).

of all is the Jewish religion, because you only have to make one payment." I don't know what she meant by this and don't want to know. It has the charm of utter mystery.

July 27. Am slipping into a do-nothing phase. I have to have my daily task or I start to ramble and putter and hum. Weather beautiful, sea-stink less, relations with Don much better. I heard him tell someone on the phone, "I've broken the curse," meaning that he had managed to paint something he didn't actively dislike—I think it was a picture of me.

The Stravinskys, Gerald and Michael came to supper yesterday. Michael brought a poodle puppy with him, although Don had told him over the phone that I can't stand having dogs around in a room, they break up all coherent conversation. I was furious. Don said this was unworthy of me. Michael is such a bitch, he said, and should be ignored; your being angry pleased him. Anyway, the puppy *did* distract everybody and the evening was a mess. Igor seemed feebler and quieter and more withdrawn than I have ever seen him before. But this may just have been one of his bad days. Gerald rattled away about unidentified aerial objects.

I suppose my disinclination to begin *The Autobiography of My Books* is simply that I know I ought to plan it in advance, and this is what I most hate doing. Vaguely, I see what it is that I want; I want an artfully rambling, seemingly casual, in-and-out book which wanders along with the inconsequence of conversation, jumping from subject to subject and yet follows a line of thought. Perhaps I could evolve a pseudo-conversational style, abrupt, slightly incoherent, yet not too obviously faked and not too badly naturalistic. A style which avoids "actually," "well, now," "I mean," "kind of" and other such locutions and yet has the abrupt *in medias res* quality of real living speech.

August 10. Still slipping. One of the many things I have failed to learn in this life is how to take a proper holiday. I either work or putter and waste time guiltily. Not that I deserve a holiday just now. I have lots to do and not so much time to do it in, either; because I may quite well have to take a movie job in the near future, that is, if one comes up. If I don't earn some more money, we shall have to dig into our savings.

(Note. Right now, we have more money saved up—forty-five thousand dollars—than I have ever had at any time in my life. Even allowing for inflation. But insecurity knows no limits. If it were forty-five hundred thousand, I'd still be fussing.)

There is no reason why I can't go quickly ahead and finish the second draft of *A Meeting by the River*. However poor it may be, I can still make it a great improvement on the first. I ought to do at least two and preferably three pages a day.

As for *The Autobiography of My Books*, that remains a problem. The tone is still wrong; I can't hit it. But I should go ahead, because collecting the material together is at least a valuable first step. (Dodie, in a letter this morning, advises me to write the book about my life in America first. She thinks that if I do this *Autobiography*, I'll never write the other; and she says a direct autobiography is more interesting, anyhow.)

The night before last, on the way back from having supper at Malibu with James Fox (who is rather a doll, in all senses of the word), Don began getting such violent pains in his chest that he couldn't shift gears, I had to do it for him, as we were in the midst of traffic and couldn't stop and change places. Toward two in the morning, the pain got so acute that he could hardly breathe. Finally, I called Dr. Allen, that public benefactor, who arrived quite quickly, in perfect good humor, and gave Don a shot. Apparently it wasn't heart, or lungs or any other sinister ailment, only a muscle strain—oddly enough, because the chest is one of the strongest parts of Don's body; he has been developing it at the gym all these years without ever hurting himself before. He seems almost entirely cured, now.

While he was drowsy and a little high from the shot, Don told me that he really does love Ramakrishna, but can't feel much for Swami. He has tried to ask Ramakrishna to make his presence felt. And he believes that Ramakrishna has shown him that the signal that he is present is pain—psychological pain just as much, if not more, than physical. This struck me very much, because I used to feel a certain joy in the sense of alienation I often had from Ramakrishna while I was living with Caskey. That was pain, too. I am so happy that Don is getting this kind of experience. It is better than anything else I could wish for him.

Also, a few days ago, he did two paintings of Budd Cherry which really pleased him.

A William Linville (did we ever meet him?) sent me the program of an amateur performance of *I Am a Camera* in Bangkok. With this heartrending comment: "A hot wet night. Lizards all over the walls. A miserable production."

After another long spell of cold sea-fog, it has been very hot here. Last night, for the first time this year, we ate supper out on the deck, barbecuing sausages on Old Smokey. Then I went

down to the Masselinks' to borrow Jo's electric hot pad to put on
Don's chest. Passed a boy, very young, naked to the waist, very
drunk. He shouted at me, "Do you know it?" How perfectly that
expresses the characteristic mystique of this period!

On August 3, Julie came to supper again, this time with her
guy—it's this actor Jim Murdock, who was on T.V. in "Rawhide"
(I think). He seemed an almost entirely imaginary character;
neither Don nor Paul nor the Masselinks could make anything of
him. Now Julie has gone back to New York, and we don't know
if she will leave Manning or not.

Finished *Don Juan* yesterday. Except for the thirteenth canto,
I like the earlier part much more than the later. But what a truly
modern work!

August 26. What is there to be said at sixty-one? Sixty is monumental (if anyone cares; with me they didn't) but sixty-one is just
the beginning of late middle age. It only becomes memorable if
you die. This is the thickest part of the cancer zone.

Otherwise there's just the usual command: there will be no retreat from this position.

Am struggling on with *A Meeting by the River*. But it may very
easily be that I'll have to rush off to join Tony Richardson in
Rome and work on *Sailor*. More of this shortly.

August 27. Bill Brown showed up unexpectedly yesterday afternoon with two bottles of champagne from Vera Stravinsky, who
had remembered my birthday. This was not only pleasing and
touching, but it presented a good opportunity to seal the peace
pact with Bill. I invited him to come with us to see *Midnight*[1] at
UCLA and have supper at home later.

Bill says that Bob and Igor have quarrelled terribly. That is to
say, Bob wants a split. So he spoke rudely to Igor at a rehearsal of
their concert for the Hollywood Bowl. Igor said to him in French,
"Please do not insult me in front of the musicians." Igor has been
trying to make it up ever since, but Bob won't. It's heartbreaking,
that they should quarrel like this, after so many years, so near the
end of the road. (Don says I'm being sentimental, and that such a
split up can't in itself be considered heartbreaking or the reverse.)

Midnight is one of those heartless-sentimental farces about rich
titled people and taxi drivers in Paris at the end of the thirties

[1] The 1939 film directed by Mitchell Leisen; Charlie Brackett and Billy
Wilder wrote the screenplay, from Edwin Justus Meyer's story.

which somehow reek of the oncoming war; just as the fashions of
1913 make you feel that the First World War was already inevitable. This is all part of the culture of the doomed. But it was very
funny in places.

We drank Vera's champagne, cooked on Old Smokey and ate
on the deck. Old Smokey seems to get hotter and hotter. Last
night, before I lit it, I found sparks in its ashes from the night
before!

Today I have been pushing ahead with a draft of the third section of *A Meeting*. I have to keep reminding myself that this is only
the second draft and that I don't have to bother about solving *all*
the problems this time around. But I want this draft to be good
enough to send to Edward for his opinion. It still seems awfully
thin to me, especially in the characterization.

October 2. The day before yesterday Don (who was planning to
leave for New York this weekend) had Rex Evans down to look
at his paintings. He'd been putting this off all summer, because,
whatever one may think of Rex's taste, it is a sort of verdict.
However, Rex liked some of the paintings very much and offered
Don a show in January, only suggesting that he should paint some
nudes, full-length figures and groups to give variety to the present collection of head-and-shoulder portraits. Rex also suggested
having one room full of drawings and the other of paintings, but
I think he could be talked out of this idea. Then, yesterday, Don
phoned Lincoln Kirstein in New York and found out from him
that the portfolio of reproductions of Don's ballet drawings, which
is to be sold at the State Theater, won't be fully manufactured
until January. So it seemed obvious that Don should stay right
here, get on with his painting and go to New York later.

Don quite agrees to this plan and I am only sorry that I spoke
too soon. Yesterday at breakfast, when we knew about the Rex
Evans show but not yet about the portfolio, I urged him not to go
to New York but stay here and work, adding that going to New
York was an unnecessary expense and that we're short of money—
i.e. we are going to have to draw on our savings (which are admittedly higher than ever before, just over $45,000!). It would have
been much better if I'd kept my mouth shut, and let him make the
decision all by himself.

Now we are definitely at the end of the holidays. Not because
we haven't been working all this time, quite hard, but because
this brings to an end our summer period together and we have
both been looking toward the detensioning of a separation, even

if a short one. This summer together has been, on the whole, one of the best periods we have ever spent together, and I hope we will be able to extend it without getting on each other's nerves. (I mean really getting on each other's nerves, friction is inevitable and even absolutely necessary, like roughage in food.)

Well, anyway—

I am nearly through with the second draft of *A Meeting by the River*. It ought to be ready by the middle of this month, to be sent off to Edward, maybe also shown to Gerald Heard. Then I'll think about one or other of the autobiographical books.

But I also very much want to get a movie-writing job, now that they definitely don't want me for *Sailor from Gibraltar*. (Am I disappointed? Not terribly. It would have been exciting and fun, but life on that yacht[1] would have been a real psychological boot camp I know.)

Very hot here, with marvellous beach days, the water bracingly cold, the sun scorching. Smog in town.

This morning I counted (very roughly and liberally) the number of words in *A Single Man*, *Prater Violet* and (when the present draft is completed) *A Meeting by the River*. Estimates: *Single Man*, 53,504. *Prater Violet*, 37,422. *A Meeting*, 37,620.

October 22. I finished the second draft of *A Meeting by the River* quite a long while ago, on the 10th. Since then, I've been in the usual state of drifing and idling and yet feel guilty about it. Why can I never learn to take an ordinary holiday? I suppose I just do not believe in holidays.

I sent a carbon of *A Meeting* right off to Edward, and today I have an answer from him already. He feels that there should be much more drama, by which he means tension between Patrick and Oliver caused by Patrick's efforts to get Oliver to leave the monastery.

Now Edward may be right. I mustn't dismiss this suggestion as being merely square, which is what it at present seems to me to be. As I have been seeing it, Patrick's opposition to Oliver's vocation is quite largely a kind of teasing, and not fundamentally serious. My feeling is that Patrick is really incapable of being serious enough and passionate enough to take any drastic steps to get Oliver out of the monastery—*and that that is his tragedy*. I am quite ready to agree that many of the moves and countermoves in the present draft of the novel are wrong, and perhaps Patrick's fundamental

---

[1] Which Tony Richardson hired for filming.

indifference ought in itself to be dramatized more strongly. Perhaps Oliver should even reproach him for it. I think Edward is on much firmer ground when he says that Oliver ought to agonize more. Yes, I can see Oliver in agony because of the struggle Patrick has started up inside him; and Patrick not really caring—maybe even a bit incredulous, first amused, then apologetic, when he dimly sees how much suffering he has caused Oliver.

Meanwhile, it's very very hot, with smog right down to the beach two mornings ago. Gavin just returned from Europe. Yesterday morning we went with the Masselinks to see the Ikeya-Seki comet—drove up into the Brentwood hills before dawn—but didn't see it.[1] I still haven't had a definite offer from Pa[k]ula on that movie job, but have been asked more or less officially to be a Regent's Lecturer on the Riverside campus next semester.

Will try to keep this journal better for a while.

October 25. Old Dobbin, after reading in the *Los Angeles Times* about the year 1985 and the population explosion and all the mass horror in store, stuffs himself with corn nuts, takes an afternoon nap and wakes to find himself too late for the sunset. It's as though he had missed an appointment. In his dirty red silk wrapper he sits down at the table on the deck, in a hot wind, gulping low-calorie coke and staring in a daze at the afterglow. After sunset already, and he doesn't *really* quite know it, doesn't altogether accept himself in either aspect, a dying animal or eternal Brahman.

I thought of that sunset he watched up at Santa Barbara, in November 1944.[2] Little did he realize that almost everything good lay *ahead*, not behind him.

And now?

The Lawrence of Arabia Icelander on the beach asked me where Don was. He said, "I like to see you and your friend together."[3]

Two nights ago I had supper with Vidya before the Kali puja and we discussed *A Meeting by the River.* He has read the second draft and made notes throughout, with great care and a good deal of intelligence. Of course he is the best "technical adviser" one could wish for. Also, he had typed out an entry from his diary,

---

[1] Named after two Japanese astronomers who discovered it on September 18; it was nearest the sun and supposedly brightest on October 21.
[2] When his relationship with Vernon Old failed; see Nov. 16, 1944 in *D.1.*
[3] Isherwood never knew his name, but thought he looked Icelandic or like Peter O'Toole's T.E. Lawrence—slim, lanky, with pale blond hair, small clear blue eyes, a sharp jaw, and a strong chin. He, too, was sometimes with a friend, an older, heavier man with a moustache and goatee.

January 8, 1964, describing his period of taking sannyas. It ends, "Chris remained till the day of our glory, and rushed up to prostrate. Bless his heart."

There is really a deep affection between us, I believe.

November 13. I was just about to start writing this when something made me look into my little book of memorable dates, and there was [R.L.] Stevenson's birthday and Ben Masselink's. So I called and wished him a happy birthday. It's grey today and sad but rather snug. Don has gone out to draw the George Axelrods' moppet.[1]

On November 1, I started the first draft of my autobiographical book, which I plan to call *Hero-Father, Demon-Mother.* Am having a lot of difficulties with it already. Anyhow, work on it has been held up while I sorted out a lot of letters and manuscripts to go into a box which Peter Gowland made to fit exactly to the measurements of the rack in the safe-deposit vault of the bank. We took it in yesterday. So something will be saved if the house burns down.

On November 9, the Stravinskys, Bob Craft and Ed Allen came to supper. (This was the night of the great power failure and blackout over New England and New York. Gerald Heard says he is sure it was caused by UFO[s] (unidentified flying objects) and he points out how considerate they are: they chose the night of the full moon to do it, so people would be able to see their way around!)

Igor was very shaky, but as bright as usual. He told me he has been reading [Dostoevsky's] *Crime and Punishment* and that the language is marvellous, quite different from that of any other writer. He had also been rereading some of Tolstoy's later stories but didn't like them at all. He wanted to know how many words I write, on an average, per day and how many hours I work. I told him about one page and about three hours. This pleased him because it showed, he said, that I am very much concerned with technique. He believed that novelists didn't bother about technique a hundred years ago, though of course poets always did. He disliked the frame around Don's painting of Collin Wilcox (which now hangs over the fireplace) saying that it was only fit for a Renoir. (Don, so characteristically, took this as a rebuke to his inferior talent!) But Igor meant simply that it was fussy and overdecorated. We both

---

[1] Nina, Axelrod's daughter with his second wife, Joan Stanton. Axelrod (1922–2003) became famous for his plays *The Seven Year Itch* (1952) and *Will Success Spoil Rock Hunter?* (1955); his screen adaptations include *Breakfast at Tiffany's* (1961) and *The Manchurian Candidate* (1962).

refrained from telling him that it was Vera who gave it to Don originally, and we weren't sure if she remembered this! Bob said the painting reminded him of Schiele, approvingly.

He also remarked that Mondrian had abolished the picture frame. He has a superior, subtly embittered air nowadays. Ed Allen, who is now living in the house, has taken over his function as Igor's attendant. It was Ed who helped Igor down the steps from the carport. He also rose automatically when more drinks were needed and went into the kitchen to get them. Bob even remarked that they really must stop Ed from waiting on them so much. Bob hardly speaks to Igor directly.

I somewhat tactlessly asked Bob, "When are you publishing another Stravinsky book?" "I'm publishing another Robert Craft book," he answered. His piece about Aldous in this month's issue of *Encounter* has greatly offended Laura and Juliette Huxley.[1] I haven't seen it yet.

Ted is having another crazy fit. (So is Michael Leopold; he has gone to Camarillo for the second time.) Ted came into the house yesterday while we were out and left a would-be cute picture (framed) of a girl with pink hair playing a cello. He probably stole it from a drugstore. It was inscribed, "Merry Christmas in November with love from Ray DeLand." Ray DeLand is Ted's "mad" name. He left the picture on the mantelpiece, with a flower lying in front of it. This kind of behavior disturbs and infuriates Don, and Ted knows it only too well. His whole aggression is that Don is rejecting him.

This morning I talked on the phone to [an acquaintance], for the first time since his marriage. [...]. He is now teaching drama [...]. Living [nearby].

He began indirectly apologizing for his marriage at once—ah, how often I've heard that tone from backsliders into heterosexuality! "When you find your real goal in life, get on the right track, it's really refreshing, it's rather pleasant." And then they assure you that they're *happy*. (Happy, meaning safe, is the sure sign of one of these profoundly suspect relationships.) Oh yes, and then they tell you that the wife is a *wonderful cook*.

Yesterday I saw Gerald. An absurd feud has developed between Chris Wood and Michael. Michael insists that the little dog they gave to the Luces, and which the Luces almost instantly returned

---

[1] "With Aldous Huxley," Nov. 25, 1965, five diary entries, 1949–1963, presented Huxley as solitary and overworked after his first wife died and said that Huxley's second marriage separated Craft from him. Juliette Huxley was the wife of Aldous's brother Julian.

to them to look after while they were away, must never be left alone. So it was taken up to Chris Wood and his dog growled at it. So then Michael said they could never go there again, Chris Wood must come down and see *them*; without his dog, naturally. This Chris refused to do. So now Chris and Gerald can't meet. They exchange letters. It is understood that things will be all right again when Michael returns the dog to the Luces at the end of the year.

Gerald sanctions this bitchery by going along with it and not insisting on seeing Chris. Michael bitches me by always leaving the dog in Gerald's room when I come to see Gerald. (Don says that, in order to be a real bitch like Michael, you have to be unconscious of your bitchery or anyway refuse to think about it.) Gerald is perfectly well aware of how disturbing I find it, having the dog in the room.

Gerald suggested I ought to write about Peggy Kiskadden. I tried to explain to him that this would be almost impossible. Because Peggy is the kind of character you have to approach from what I regard as the wrong end; she is a noble self-sacrificing angel of mercy who turns out to be a devil-bitch. This is the sort of character which the debunker type of writer loves to describe; me it bores. Gerald said truly that you could make her interesting and touching by showing how she got to be that way. (How *did* she? I have no idea.) But even doing this wouldn't take the curse off her, from my point of view. She is still only the heroine of a bad Joan Crawford movie. You are supposed to feel sorry for her at the end of it—and you don't. There is only one thing to do about Peggy, forget that she exists.

November 14. It has rained all day, lightly. Good for the plants and bad for our spirits. Ted called this morning and said viciously, "Give my love to my dear brother." He told me he is attending a meeting of the Screen Actors' Guild tomorrow and plans to renew his membership. This made Don laugh when he heard it; it's so typical of Ted's crazy spells.

Today I tried hard to restart *Hero-Father*, but I can't. Something inhibits me. I think I must find another way of writing it. My instinct is against straight autobiography; suddenly it rather nauseates me. What I keep returning to is the idea of some kind of lecture form, but that seems so contrived. Perhaps I should attempt a sort of notebook. What I dislike is the prospect of writing everything out in sentences and paragraphs; it seems too literary, too *urbane*. Yes, perhaps something can be done with a notebook. Must consider this.

In any case, I must keep trying. It would be very bad to shelve this project now.

Last night, we had supper with Tom Dawson (who says he came to see me as a soldier during the war, while I was living up at Vedanta Place) and a friend of his named George Woodward. Dawson is the part owner of an apartment building on Olive Drive, called French Hill. He furnishes all the apartments himself, with French tapestries and chairs and chandeliers, the main effect is glass and satin. These places are absolutely terrifying. You can pay four hundred a month for a large one, and there isn't one spot in it where you could read a book or write a letter or have an intimate conversation. An ideal set for Sartre's *No Exit*. The people who mostly rent these apartments are men who have just been divorced; the wife has the kids and the house and the furniture. They have room service and they appeal (I guess) to the girls the men bring home with them at night. To think that you could live in one of them makes you sick with horror. They *sparkle* so!

I felt myself getting aggressive, so shut up. I had started to make ugly remarks about the dictatorship of the merchants.

November 18. Last night, Swami told me that he'd been astonished to find "some old *samskaras*"[1] which caused him to feel caste prejudice. An Indian had come to see him and he had realized, by various indications, that the man must be an untouchable. He had invited him to lunch, nevertheless, but had been "so relieved" when the invitation was refused!

Swami was so sweet, telling me this. And really, the prejudice seems grotesque from a western point of view. Because Swami hadn't (apparently) minded shaking hands with the man at all. It was just the idea of eating with him. "You see, Chris, they eat carrions." "But this man doesn't eat carrion, does he?" I asked. "Oh *no*—of course not!" Swami was shocked, "He is an educated man! He is sent by the Indian government!" But that didn't seem to be the point. The man would have been eating Vedanta Society food off Vedanta Society plates in the company of Vedantists—but, just the same. . . .

Jim Bridges reports another, much more drastic demonstration of the power of traditional folkways. He has been up in Utah on location for a Brando film he wrote.[2] The film company has been

---

[1] Subconscious imprints made by past actions or thoughts and forming the character.
[2] *The Appaloosa* (1966).

hiring a lot of Navajo Indian extras. The Indians elected two men to collect their pay for them and divide it among them. Then they found that the two men had been cheating them. So they cut one man's head off, and cut the other's tongue and legs off, and then they burned them in a trailer. According to Jim, nothing is being done about this by the authorities, "Because it happened on the reservation." Am not at all sure I believe this story. Jim just heard it.

Four days' rain in a row. Sunshine today, but it's raining in town and another storm is expected to break this evening.

Have given up wrestling with *Hero-Father* for the moment. Am now getting the notes lined up to prepare for starting the third draft of *A Meeting by the River*.

Don said this morning that maybe he should postpone his show. He feels he won't be able to sell any of the paintings he has done up to now.

Dr. Allen has been checking up on me. He says, "You'll live even longer than your mother." A marvellous opening for a film would be one of his entrances, wheeling the little cart on which he carries his drugs, instruments, medical books and other necessaries. You have to see it from the point of view of a new patient, who is already very nervous about himself. The cart appears in the doorway, stops. Then Dr. Allen's hand is seen, adjusting a clamp which holds some papers to its side. The hand withdraws. Then suddenly the cart is turned right around. A long pause. Again the hand appears and takes something from the cart. A long pause. Then slowly the cart begins to move. It has come three quarters into the room before Dr. Allen appears, pushing it. By this time the patient's nerves are shattered. (During the whole of the business there could be low cries of pain and cryptic medical asides between doctor and nurse, off-scene.)

November 30. On the 22nd, I started the third draft of *A Meeting by the River*. At present, most of the alterations I'm making are really just to set the stage, line up the reason for conflict and in general create more anticipation of trouble before the brothers actually meet. It is great fun doing this, almost pure play. But the fact remains, I still have to answer the basic question: in what way are the two permanently affected by their meeting?

Have just written to Wenzel Lüdecke,[1] who was here last week

---

[1] German film producer and screenwriter (1917–1989); in 1949, he founded Berliner Synchron, foremost dubbing studio in Germany.

and left with me a gruesome German screenplay called *General Frederik*, for my opinion. It's crude and corny beyond all description.

Ted is still semi-crazy. David Smith met him on the street the other day with a shopping cart from a market. In it, he had a framed drawing of Myrna Loy by Don. (Don had only lent it to him.) Ted offered to sell the drawing to David. Now Don has told his parents and they have persuaded Ted to give them the drawing, and also one of Dietrich.

December 1. Beautiful fall weather. Have spent most of the day working on *A Meeting by the River.* Now I begin more and more to realize what Edward meant; there must be a more direct struggle between the brothers—or rather, there must be more of a struggle inside Oliver, for I still don't agree that Patrick is such a consistent and conscious opponent. I have everything much better arranged up to Patrick's arrival at the monastery. After that, it gets confused and loses direction.

There must be definite *moves* (in the sense of chess moves) in this game. What moves do I have so far? Patrick's establishment of excellent and personal relations with the swamis; making Oliver jealous. Oliver's showing Patrick the swami's seat; this is really a move against himself. Patrick's involvement of Oliver with the journalist Rafferty; this is, as it were, an attempt to reduce Oliver's situation to a farce. Oliver's vision of the swami; this is a strong counter-move, putting Patrick in his place, though he doesn't know it. Patrick's confession to Oliver about Tom, and the ensuing "temptation scene" in which he leads around to suggesting that Oliver shall leave the monastery.

I'll only record one idea for the moment: what if Oliver's vision of the swami is delayed until right before sannyas? Must think this through carefully, pro and con.

December 10. Swami is still greatly distressed by Sarada's defection. She won't communicate, although she gets news of the center through her father. Before she left, she told the girls that she never had any spiritual emotion while doing the worship in the shrine. Swami recalls that he used to feel that the way she did the worship was "a little too sweet" (i.e. theatrical).

As for Vidya, he is still uncertain if he will go to Chicago, Paris or India. But, according to Swami, it seems doubtful if they want him either in Chicago or the French center, because he is "a troublemaker." His best bet would be to go to India and open a small center of his own, for visiting Vedantists from the West. Swami favors this because he thinks that Vidya can't stay long in any place unless he is running it. Meanwhile poor Vidya is terribly upset and at a loss what to do. He called me this morning, but of course I couldn't tell him all I know. I can only advise him to talk to Swami. Vidya obviously doesn't like or trust Swami and regards him as a devious character. But even I can't help feeling that the deviousness is all on Vidya's side. He really is a mystery of self-deception.

Things look very very bad on the international scene.[1] And I have no job. I suspect that both the Pakula movie and the Riverside professorship are falling through. Still, there is the novel to be finished and that will take a long time and be very enjoyable. Now Don is supposed to have his show early in January, less than a month from now. So he is panicky. But we have been happy together—at least I certainly have been—for a long time.

December 21. Two days ago I finished the first section of *A Meeting by the River.* It is six pages longer than the first section in the second draft, and I think really much improved because I'm setting the stage for the actual meeting much more elaborately.

On the 16th Maugham died, the same day as Denny Fouts, the day after Laughton. According to the *Los Angeles Times,* Willie said on his ninetieth birthday, "I have walked with death in hand, and death's own hand is warmer than my own. I don't wish to live any longer." Even allowing for misquotation, this seems curiously histrionic. What a masochist he was! But I wish I had kept in touch with him more. It would have been so easy, and it might have given a little pleasure. Today I absolutely must write to Alan Searle, I've kept putting it off.

It now seems that Vidya will definitely go to France and from there to India, for keeps, presumably. He calls me whenever he's in town, and I feel his affection—I must be almost the only person he's at all close to, now. I'm fond of him too, but there is always the embarrassment of knowing how he strikes other people and not being able to speak freely to him about this. The other day he told me that Swami had said, "Chris is the same inside as he is outside—I like that." Although this flatters and pleases me (I say flatters because alas it's a million miles from being true) I'm quite well aware that Swami said it—as he says so many things—in order

[1] Military activity in Vietnam was escalating amid fears of war with China; see Glossary under International Situation.

to deliver an indirect message to someone else; in this case, Vidya himself.

Last Wednesday in the reading there was a passage (page 305 of the Gospel) in which Ramakrishna says, "As a devotee cannot live without God, so also God cannot live without His devotee ... It is the Godhead that has become these two in order to enjoy Its own Bliss." This passage obviously moved Swami very much. In fact he seemed unable to explain fully to us *why* it meant so much to him. This must be something he has actually experienced for himself.

This is a period of maximum happiness with Don. Not that Don is happy—he is terribly worried about his show. Today we are to make a more or less final choice of the paintings. The other day he said, of one of his sitters, "He's so crushingly ordinary, and yet a kind of *person* with it, too, which is so irritating." How characteristic that is! He sees things with a subtlety which even I, who know him so well (and yet in some ways so little), can't always follow.

On the big lot at the end of Adelaide on Ocean Avenue, they have just started excavation for the dreaded apartment tower. It's said to be going to have thirty-two floors, which will make it one of the highest buildings in the city. I feel *threatened* by this and all other such demonstrations of the city's explosion—to an extent which is really neurotic. I find myself feeling exactly like M. as an old woman—her horror of "progress." I must examine this attitude of mine, I don't refer to it here as much as I should; if I bring it out into the open more, perhaps it won't torment me so. Undoubtedly there is a great deal of snobbery mixed up in it—a snobbery which goes back into my heredity. The snobbish horror of my upper-class family at the triumph of the caste of the merchants.

1966

January 1. The usual bright deadness of New Year's Day. Cold, sunny, clear, you can see right across the bay to the headland. Don in a flap about his show on the 7th, addressing hundreds of announcement cards. We seem to be inviting the whole earth. It's truly amazing, how many people we know, when you consider how few we see.

The Vietnam situation looks terrible, but then we don't know what is really happening or what anybody really intends.[1]

[1] During the Christmas bombing pause, Johnson launched a diplomatic initiative for peace talks, but it was viewed as an obligatory last effort before broadening the war.

I got this half-offer to work on a script called *The Defector* for a director named Raoul Lévy.[1] Clift is to play in it, and the suggestion was made through Jim Bridges. It will probably come to nothing because they don't want to pay. The fact remains I have no other job in prospect. Pakula is utterly vague, and no word comes from Riverside.

Have been working on the third draft of *A Meeting by the River* since November 22, but not nearly enough. By this time, on the usual basis of a page a day, I should have written forty-one pages; in fact I've done only thirty. Things are sorting themselves out, but much remains to be invented and I wouldn't be surprised if I have to do a fourth draft.

January 4. I have now finished the first two parts of *A Meeting* and I think they are greatly improved. The stage is set for a conflict.

But what exactly *is* this conflict? Now I come up against the problem and I feel I must examine it from the point of view of construction, even though Don rightly warns me against being too logical.

It's clear that the main action of the book is temptation—the temptation of any saint by any satan. It doesn't make any difference that Patrick's kind of temptation is really more to be described as teasing, and is only partly intentional—it is still a temptation, and in fact this kind of temptation is much more usual than the intentional kind, which seems nowadays a bit square and unmodern.

The key line is when Oliver says that he was inviting Patrick to come and judge the swami. He has to have Patrick's okay. He doesn't ever get it of course. What he does get is a spiritual intervention by the swami himself, proving to him that Patrick "belongs" whether he likes it or not, knows it or not. And this, in its turn, is sort of campily confirmed by Patrick's taking the dust of Oliver's feet.

But how am I going to show all this, and what is the big scene between the brothers to be about?

Suddenly all my foundations seem to be collapsing, except that I have plenty of good solid material with which to reconstruct them. But I have got to think hard before I write any more.

January 8. The opening of Don's show last night was really quite a big success. Nine pictures have been sold already, two drawings

---

[1] Producer of Bardot's first films and of *Moderato Cantabile* (1960) and *Marco the Magnificent* (1965).

and the rest paintings. Anne Baxter started the buying. She rushed across the room into Jo's arms screaming, with a kind of tearful triumph, "I've bought *two!*" Vidya was there, viewing the scene with the amused world-weariness of a swami about to depart forever into the depths of India; and Ronnie Knox seeming sad and lost in his new queer world, and Renate [Druks] who said of him that he is an ovary cutter; and Elsa Lanchester looking almost ladylike in a dark dress, gracious and bitchy-grand; and Jennifer Selznick in white, about to leave alone on a drive to Big Sur; and Jim Charlton drinking four times his share and telling me [about a woman who] is pregnant by Mark Cooper; and Dan[a] Woodbury quite drunk, saying it was a shame Rex didn't exhibit Don's nudes of him, and then taking a fancy to Jim and leaving with him; and Gerald Heard and Michael, bitchily arriving dead on time, and Chris Wood ditto; and old King Vidor being encouraged by his wife to paint again; and John Houseman, a little worried because he liked Don's work so much, almost more than he felt he should; and Cukor sly but friendly, planning a memorial supper for Maugham; and Antoinette and Jim Gill dramatically reconciled; and Jack Larson annoyed because his portrait wasn't among the exhibits though Jim's was; and Bill Inge terribly depressed about his life, sitting glum like a bankrupt on a couch, etc. etc. etc. etc.

On the 6th, Allen Ginsberg came to supper with us, bringing his friend Peter Orlovsky and a seventeen-year-old boy named Stephen Bornstein. We had prepared for them by inviting a "home team," Gavin and Brian Bedford. Everybody got high, and Ginsberg recorded our conversation and chanted Hindu chants, and Orlovsky took off his woollen cap and let his long greasy hair fall over his shoulders and kept asking me if I had ever raped anyone, and the boy Stephen unrolled a picture scroll he had made, under the influence of something or other, to illustrate the Bardo Thodol.[1] Brian, who is a very "good sport" on these occasions, joined enthusiastically in the chanting. Later, Ginsberg surprised me by showing great admiration for Hardy and introduced me to a poem I didn't know, "[A] Wasted Illness." While the chanting was going on, Don and Gavin watched in silence. Don sternly disapproving, Gavin demurely amused.

Don tells me he was possessed by paranoia and saw that Gavin is a witch, and when Ginsberg was out in the studio and remarked that Don's line was too "hard," Don says he really told him off.

---

[1] The Tibetan Book of the Dead, containing Buddhist instructions for the dying.

Also he actually shooed Orlovsky (who was leaving anyway) out of the house. But Don feels that his instincts in all this were correct, however hysterical, and I agree with him. Ginsberg came to the house prepared to show us up in one way or another (maybe he thinks he succeeded, I don't know). All three of them are to some extent demon guests, harpies who descend and wreck the homes of the fat bourgeoisie with self-righteous malice. They are quite tiresome, but also quite fun, and the evening was an occasion, though not necessarily one to be repeated.

At the beginning of this week they started excavating the site for the skyscraper at the corner of Adelaide and Ocean Avenue, and heavy trucks full of dirt have been roaring down our hill. When they are coming back up the hill empty they sometimes race each other abreast. It is hellishly noisy—how very noisy you only realize today, when there has been no work on the project. It is so tragic to look out over the Canyon, still so beautiful, and yet feel it is all ruined by the ever-increasing noise, jet planes, cars, motorcycles (the worst). I wonder how much longer we shall want to stay here and where we shall go if we leave. And again depression sets in, weariness at the prospect of this ever-crowding, ever-noisier future world.... Well, don't think about it, get on with your work. Be happy that you have what you have. And thank God for Don. He is so absolutely the companion I need. I even need his faults. And I am so proud of him.

Vidya said to me the other evening that the one thing he still feels is worthwhile in "worldly life" is to have someone you love who loves you.

I have decided to rough through parts 3, 4 and 5 of *A Meeting* before going on to the careful rewriting. I must first of all see how things fall into place and perhaps I shall discover some new problems in the process.

March 7. This long silence has been chiefly due to contentment, the contentment of being busy. I have just finished section 4 of *A Meeting by the River.* I don't know what I think of it, maybe it's all terribly stilted and contrived and literary, but at least it's as difficult as hell to do, and that's absorbing.

Don has been in New York since February 15 and at present he plans to return about the 21st of this month. I don't think he has accomplished much, though the engravings of his ballet drawings are ready at last or almost, but the trip was probably a very necessary holiday.

I'm now going to Riverside (UCR)[1] regularly on Tuesdays through Wednesdays, stopping the night at a motel, quite near the campus. It is fun and the students are both bright and beautiful. I only keep feeling a bit of a fake.

Now Alan Pakula wants me to do a picture about the history of art. When we talk about it, we both feel we *almost* know how it should be done. But we don't, not really. If only Don were here I know he'd be able to help me find an approach. The only other person who could do this might be Gerald, but Michael won't let me see him because he is still weak after his stroke (or whatever it was) and because I might give Gerald flu. No use raging against Michael and his bossiness—Gerald certainly encourages him in it.

When I do get *A Meeting* finished, I shall have to show it to Swami. I have no idea how he'll react and I don't want to think about this in advance. Some little bird had already told him that I was writing the novel, which I suppose is hardly surprising since I talk about it wherever I go.

March 19. The day before yesterday, they hung up a big street lamp from the telephone pole in front of this house. The lamp makes a great light, I woke last night and felt its oppression. I do hope I'm not going to let myself get obsessed by this, it is one of the ways in which I'm near to madness.

David Roth[2] has a friend who, when he was being examined by the psychologist at the draft board, was asked, "Could you kill a man?" and answered, "Yes, but it would take years."

Last night, I went up to Vedanta Place. Spent a little time in the temple, trying to get some sort of feeling of awareness, but all the lines seem to be disconnected. As I went into Swami's room I said to myself, I am going to visit a saint. I believe this was/is true, but I couldn't feel it. The only thing is to make japam and trust.

This week, on the night of [the] 15th, I went back as usual, to the motel in Riverside. I didn't feel particularly drunk. I woke up all in one piece. But then I absolutely couldn't find the galleys of Ned Rorem's *Paris Diary*. After searching the room, I went outside to look in my car and found to my astonishment that it was parked way over on the other side of the motel lot. Only later I found out (without ever being able to remember it) that I had been given another room, complained because there was a peculiar noise in the

---

[1] I.e., the University of California at Riverside.
[2] Then in his twenties; he often posed for Bachardy. Later he moved to San Francisco.

pipes, and moved to this room, leaving the proofs on a table! This sort of amnesia, when you aren't even drunk, alarms me rather. The least thing cuts me off from my memory, nowadays.

I wish I liked Rorem's diary better. I feel I should stand behind him, since he's sure to be attacked as a fag. And yet that's no good reason—

March 31. Beginning of my twelve-day Easter vacation. Fog. Now it looks as if Don may not come back for another two weeks, but I'm to talk to him tonight.

I have just read through the third draft of *A Meeting* as far as I've got, which is up to the beginning of the showdown scene between Patrick and Oliver. I must say, a lot of it seems boring, really chewy, and I feel the quality of the writing is poor. Granted that this letter form has to be written loosely, in a seemingly nonliterary manner, the way I have done it seems merely sloppy. However, the big thing is to push ahead to the end.

What is right and what is wrong about the showdown scene as I have it in the second draft?

I like the opening—Patrick saying to Oliver, I want to talk to you as I would to a priest, and then telling the story of his affair with Tom in deliberately gross sexual-fetishistic language, which disturbs Oliver.

But what then? I think Vidya was right when he said he disliked Patrick's asking, "What do you think I ought to do?" Patrick doesn't really want advice. His confession is actually an act of aggression. What he is saying, in effect, is *Here am I being utterly truthful about myself—dare you be equally truthful about yourself?* And when Oliver doesn't reply or react, Patrick prods him further, *I suppose you think I'm unfit to go on living with Penelope and our children.*

This disturbs Oliver even more, because of his slightly guilty conscience about Penelope. So Patrick continues to prod. He says how understanding Penelope is. This angers Oliver because he feels that Penelope is being insulted by being put on the same level as Tom. Perhaps Patrick now tries to get Oliver to say that Patrick ought to choose between Tom and Penny. Finally he maneuvers Oliver into saying that he should leave Penny, because that is a duty. The moment Oliver has said this, Patrick can insinuate that Oliver's motives are suspect, because he's in love with Penny himself.

That's part of the scene. But how does it join on to the other half which must now follow?

Don made the excellent suggestion that Patrick should say in effect, *If you really believe in God, you'll leave this place, because being a monk is just a way of hiding from your problems.* And he goes on to accuse Oliver of suffering from ambition. And then he comes right out and gets Oliver to ask him if he shouldn't leave the monastery at once, without taking sannyas.

I suppose the link is, in fact, that very line about Patrick having been frank and challenging Oliver to be equally frank.

Well, now I must try to get a rough draft.

Rereading Edward's letter about the second draft, I see one other thing which is very important. Oliver should turn on Patrick in this scene and ask him, either directly or indirectly, *Why do you want me to stop being a monk?* Perhaps I am wrong when I say, Oliver should actually *ask* Patrick this, but the question must somehow be raised. It must be shown, in other words, that Oliver's way of life makes Patrick feel insecure and challenged—just as Patrick's way of life makes Oliver feel challenged (as we have already been shown).

April 10. Easter Sunday, and according to my custom I managed to do quite a lot of work on the novel, in fact I have outlined the confrontation chapter very roughly to the end. Have reached page 100.

Brilliant sunny weather with strong wind. Don will return from New York at the end of the week, I hope. I miss him and need him badly. Clint [Kimbrough] and Gavin seem to be settling down rather successfully, with poor old Jim [Charlton] and Ronnie [Knox] out in the cold.

Anaïs Nin is really one of the best comic writers. This from her diary:

> June and I had lunch together in a softly lighted, mauve, diffused place which surrounded us with velvety closeness. We took off our hats. We drank champagne. We ate oysters. We talked in half tones, quarter tones, clear to us alone.

This afternoon, a newspaper man called to tell me that Evelyn Waugh is dead.

I have told Cukor to tell the people at USC that I won't come to the Maugham evening on Thursday because Mrs. Luce is going to be on the platform. I wasn't sure if I should refuse, it seemed childish, but then Don encouraged me when I told him about it over the phone. I know my motives are very mixed. I feel that Mrs

Luce—and Garson Kanin and Ruth Gordon for that matter[1]—will turn the meeting into a personal publicity stunt. But then aren't I resenting that partly because it won't be *my* stunt? Don's attitude is more realistic. He doesn't worry about my motives, he just doesn't want me getting myself into a senseless rage against this woman.

Virgil Thomson, who got into town today, says that Don is one of the most tactful people he has met in years. How strange this seems to me! I always think of him as wayward, impulsive, aggressive and passionate. In fact, I love him for those qualities. Jack Larson, Jim Bridges and Virgil had dinner with me at Chasen's, it was Jack's treat. Virgil, deaf and dogmatic but highly sympathetic with it, answered questions throughout the meal. When he said Camus was a phoney, I was so pleased that I shook hands with him. The Byron opera has been announced and will be performed at the Met in two years. Now Virgil merely has to compose it.

April 17. Yesterday I got a painful disappointment, Don isn't coming back yet—but the reason is a happy one, the portfolio of ballet drawings looks so good that Balanchine himself wants to be drawn and added to the collection. However—Missy[2] also wants three of the portraits done over again. So Don must stay at least ten more days, probably longer.

This morning I finished a draft of the big scene between Patrick and Oliver. I think it is fairly all right, but I shan't go at it again until after this visit to Riverside.

There is now really only one difficult bit to do, as far as I can see. That's Oliver's account of his vision of the swami. The two important points to be made there are: (1) that Oliver realizes that Patrick wanted to stop him becoming a swami because Patrick felt Oliver's monastic life was somehow a subversion of his own way of life. In other words, what Oliver confesses about himself earlier in the book is also true of Patrick—Patrick is just as insecure in his own way, or more so.

(2) that after the vision Oliver's problem about Patrick and what he himself is to do is automatically solved, because he suddenly no longer sees Patrick's world and his world as hostile opposites. It's

[1] Kanin (1912–1999), American actor, novelist, playwright, screenwriter, theater and film director, described his friendship with Maugham in *Remembering Mr. Maugham* (1966); Gordon (1896–1985), American actress, playwright, and screenwriter, was Kanin's wife and collaborator, for example, on *A Double Life* (1947), *Adam's Rib* (1949), *Pat and Mike* (1952), and *The Marrying Kind* (1952).
[2] I.e., Balanchine.

like the moment of *satori* in Zen, the *koan* becomes meaningless,[1] Oliver and Patrick are united within Swami's love.

Have been preparing a stew all afternoon, because Ronnie Knox is coming to supper. I really look forward to this, I am greatly at my ease with him and am really fond of him. But I do hope we don't get too drunk.

On Tuesday Gavin and Clint leave for Mexico, and Swami leaves for Chicago and New York. Swami is much disappointed in the two assistants, Sastrananda and Budhananda, because they have both requested permission to go back to India. They find the work here too hard. Also their manner towards the monks and nuns is superior. Swami puts it down to nationalism, a reaction to being snubbed for so long by the British.

April 23. A telegram from Don this morning to say he's returning tomorrow night. So now we enter a new phase. As usual I feel a lot of joy and some apprehension. The very thing that makes living with Don so unique is the constant uncertainty as to how he'll be from day to day—and the much greater uncertainty as to how he'll be after a longer separation than just sleep. And this makes me apprehensive, always. At the same time, I feel that this very apprehension is good—it is how life ought to be, one ought never to be certain, because certainty is maya.

Talking of maya, my spiritual life couldn't possibly be deader. I say my beads every morning in front of the *Life Magazine* photograph of Swami at the shrine. And now I wrap myself in the *chadar* he gave me. Does it help at all? Apparently not, but I must have faith in it like radiation treatment—because what's the alternative? For a longish time now I have been praying to Ramakrishna to be with me in the hour of death. Maybe he's waiting for that.

This weekend I haven't gotten around to working on the novel, because there's so much to be read—a great fat manuscript of part of a novel about China, by the divorced Chinese wife of a member of the Riverside faculty, a play by Jim Bridges called *The Papyrus Plays*, which I don't understand much of but must reread before I see him this evening, *three folders* of a book about Mexico by Tom Wright, which I haven't begun yet, and Gombrich's *Art and Illusion* which I have to return to the man who lent it to me at Riverside. (This last is for self-improvement, chiefly, because I

---

[1] *Satori* is the moment of illumination, like samadhi in Vedanta; *koans*, given by the Zen master, are sayings—typically fragmentary or puzzling—on which the disciple meditates in order to reach satori.

very much doubt if Alan Pakula's art film will ever materialize, at least with me doing it. *A Clockwork Orange* is off—Brian Hutton[1] begged to be excused.)

Well, no complaints, please. I should be thankful for a truly miraculous deliverance from a very serious accident. The day before yesterday, without the smallest warning, the clutch plate on the Volkswagen broke and the car stopped dead, just as I had backed up onto the road from this house. Imagine if that had happened the day before on the freeway! However, it did cost ninety-five dollars.

One of the Riverside students, Ken Day, invited me to have supper with him and his girlfriend before going to see him play Tom in *The Glass Menagerie*. The two of them live together in a charming little wooden house. Ken said, "I had to go and see her mother and tell her that we weren't going to get married, because that's not my nature." I think it was the girlfriend's nature, though. She said nothing, and meekly served a delicious dish of fried chicken.

Ken was excellent in the play, he is very Irish. He is also possibly the most talented writer on campus, though I can't be sure how much his screwball style is meant to mean and how much it is meant to shock the audience.

When I told Ronnie Knox about Ken's remark about his nature, Ronnie was delighted. I think this is because Ronnie himself has designs on a teenage girl he met on the beach. He asked her, "Would you be ready to live in sin?" She asked, "With a boy or a girl?" and he isn't sure if she meant this literally or is just dumb-innocent. The girl's mother has said she must see Ronnie before the girl may go out with him.

Ronnie is thirty-one. He says of himself. "I'm still a boy, really," but he also admits that there's less of the "tiger" in him than there used to be. I suppose most people would see him as a tragic figure, the ex-star, destined to end badly. But I'm not so sure. I hope a sort of fool's luck will pull him through.

Ronnie's stories of football are of course all about his exploits, but somehow he doesn't seem self-centered in a stultifying way. Even Renate says how wonderful he was when her son killed himself. He's certainly irresponsible and a bit mad, but at moments he seems surprisingly strong and gentle. When he speaks to people on the street, you see how utterly he charms them. As for his

---

[1] American actor and director (b. 1935), best known for directing *Where Eagles Dare* (1968).

writings, I'm not sure I showed him how to lay out this novel and the idea is really good and workable, and he has a lot of real humor—but he seems so undisciplined, his dialogue scenes are all over the place. A couple of times he has got me to read his stories aloud to him. When I did this he laughed wildly—either at his own jokes or my reading and funny accent.

May 31. Yesterday I finished the third draft of *A Meeting by the River.* It's more than twenty pages longer than the second draft and there's no question that it's much better constructed, as a piece of artifice I'm quite proud of it, but I still have an uneasy feeling about the writing, I fear it is flat and thin. And now I have got to show it to Swami, which makes me squirm inside. I hate the thought of him reading the parts about Tom—but *why* should I, actually? I'm not ashamed of them, I would never apologize for them artistically or morally, they are absolutely right for the book, I know. Furthermore, Swami has praised me for being myself, making no pretences about the way I live my life. Just the same, I squirm. Am going to take him the manuscript tomorrow.

Swami is very well, it seems, but he is very sad because of Sarada, who has now definitely told him that she doesn't want to return to the convent and that he can tell Belur Math to take her name off the roll of nuns. This he has done already.

Now I am confronted by this new job offer, Morris West's *The Shoes of the Fisherman.* At first glance the book seems corny beyond belief but I would like to do the job if I can possibly see how to, not so much for the money as for the sake of having something to occupy me—otherwise one feels so empty-handed after finishing a novel.

Riverside leaves very happy memories on the whole, and I plan to go back there next year. My favorite students—Ken Day with his prematurely wrinkled sweet Irish eyes, beautiful puzzled-looking Bob Edwards the blond rugby footballer who always seems to be in tears because his contact lenses bother him, carrot-haired Mickey Kraft the politician in his suede boots, Jeff Morehead the Salinger Kid with his spectacles and soft blurry face and gangling charm, Larry Johns the golfer with his tangled golden curls and creamy skin and long Jewish nose and sly smile. All of these have talent, to some degree. Bob's poetry is strangely aggressive, often directed against fat women in capri pants, Mickey will make a very intelligent political journalist, Jeff wrote a really moving and nostalgic little story about fishing, Larry writes strangely powerful nonsense verse. And there's a boy named Christopher MacDermott(?) who works in the steel mill at Fontana and who showed me one story which I thought really really good—but I hardly know him at all.

More about Riverside later, if I get around to it.

The amazing courage of Allan Carter (Michael Sean, Shawn?) lying there in Santa Monica Hospital after the car wreck. He still can't move one leg at all, but he jokes and chatters away. No hero could behave better.[1]

Ever since April 6 I have been going to Dr. Paul Macklin, a chiropractor in Santa Monica who was recommended by Jo Masselink and Bill Brown. I think he has done something for my chronically stiff neck, though nothing whatever for my thumbs. Am also on a nonfat diet for Dr. Allen, which is getting to be a bore. He found that my cholesterol count was up too high, though not much.

June 4. Yesterday, Swami rang me up to say that he'd finished reading my novel, and that, "As I finished reading the last scene there were two tears running down my cheeks." What an angel he is! He was obviously every bit as relieved as I was that he didn't have to say it would offend Belur Math. In fact, he went so far as [to] suggest that it ought to be sold at the Vedanta Center bookshop! I doubt if he really quite meant this, however. He did also say that there could be no question that the monastery in the novel is Belur Math, because there is no other monastery like that on that part of the Ganges.

So now I can relax and just let some time pass before rereading the manuscript for final changes. I can't help feeling that the slang used by the brothers could be improved. Also, Don is reading it right now and I await his verdict.

Dr. Macklin told me yesterday that my arthritis, such as it is, is incurable but can be prevented from becoming worse. He also told me arthritis is said to be often caused by aggression!

Have decided to give *The Shoes of the Fisherman* a try, if George Englund wants me.[2] I'll probably see him in a couple of days.

June 26. On the 23rd I mailed a typescript of *A Meeting by the River* to Vidya in Gretz. After reading it, Don said he thought Oliver is too sympathetic toward Tom in the big scene. Also he didn't like Oliver identifying with Patrick on the rock and Tom turning into Penny. I decided to leave the first passage as it is, for

[1] Sean was the passenger in a friend's small sports car. The back of his head was torn open, and he had to lie absolutely still so the wound would drain. The friend was not injured.

[2] American film and T.V. director (b. 1926); he produced the 1968 film.

the present; but I see that he is absolutely right about the second and I've already changed it. Also I slightly rewrote the passage in the last section, where Oliver says he feels that sannyas is an entering into freedom. (This was a remark made by Vidya.) I think I've now taken the sentimentality out of it.

Gavin has read the novel. He says it is "extraordinary" but I don't feel that he really likes it very much. Now Don Howard and Jack Larson are reading the two available copies. They were both to supper with us last night, along with Jim and Clint and Gavin and David Hockney, who has just returned here, to teach for a few weeks at UCLA.

Latest worry, a mysterious rupture of a blood vessel on the nose side of my left eye; it happened during the night, two nights ago. Don reminds me that the same thing happened about twelve years ago, only worse, and that it went away again soon.

This afternoon I visited Michael Sean at the hospital. He now has the neck brace off and he really does seem to be getting much better. The bang on the back of his head is a relatively undramatic pink healthy-looking scar. He has some small scars on his belly, from shooting one of the most dangerous beaches on Oahu and getting torn by the lava. In fact, what with his surfing and skiing and tobogganing, not to mention the crazy risks he took whizzing down our hill on his skateboard, that auto accident was long overdue! All the times Don and I have visited him, he has never once shown the least weakness, never stopped joking and laughing—although there were certainly times at the beginning when he was badly scared; the surgeons were frankly pessimistic. The only thing I dislike about him is his compulsive and almost sado-masochistic flattery. There are always other visitors present and he invariably tells them what great guys we are, what true friends, what geniuses, etc. etc. I can't shut him up by telling *him* (as I shall some day in private) how I have been moved by his courage—because, when I do say that, I shall mean it. The things he says he doesn't *really* mean. Oh yes, he means them up to a point but, because he says them in public, they are cheapened and devaluated, like advertisements.

A sign of the times: Lee Heflin saw a young man, bearded and with long hair, who goes around wearing a crown of ivy and roses. And there was that boy at Riverside who had a sweater inscribed, Jesus is Boss. He also wore a button: Draft beer not people.

On the 18th we had the customary Father's Day lunch at Vedanta Place and this time the boys, directed by Jimmy Barnett, put on a musical show consisting of Vedantically-revamped lyrics from *My Fair Lady*.[1] It was rather shocking, how slickly they went through their hoofer routines—linking arms, stepping forward to the mike with arms out-stretched, sidestepping, singing with legs planted firmly apart in the Al Jolson stance, etc. etc. Only the very stupid or pure in heart can listen to this kind of thing without squirming. I squirmed—all the more so because one of the songs was directed at me!

Some specimens:

Lectures and pujas, receptions for swamis,
Wednesday-night-living-room questions for Swami,
Saturday Ram Nam where everyone sings,
These are a few of our favorite things ...

Now the clock strikes, time for vespers!
There's no time to spare,
When you're in Vedanta your life is complete
With favorite things to share ...

I came to the hills and I found Vedanta
I know I am safe, never more to roam.
My life has been blessed with the sound of Brahman
Now at last I'm home ...

Aow, Mistafah Christafah Ishtafah,
Blimey, he's a blinking limey and a real fine bloke.
We call 'im Mistafah Christafah Ishtafah
Oh how I wish I were like 'im.

Now in his Harris tweed and denims
And his shock of bloomin' hair,
Every Wednesday night in livin' room
He's in his easy chair,
And when his voice rings out melodious,
What music fills the air!

Now he's a chappie mild and mannered
With a proper savoir faire
With his bushy-browed expressions
He will soon your heart ensnare.
And when he starts in tellin' stories, chum
He's quite beyond compare!
That's Christafah, our Christafah,
You'll never find another like him anywhere.

[1] And from *The Sound of Music*.

I squirmed, but I was touched of course and pleased. It *is* a family, even though I know so few of my relatives and can't honestly take much interest in them. Some woman came up to me afterwards and well-meaningly cooed, "Now you know how much we all love you!" Well, perhaps I am mildly liked—as an institution rather than a person—and that's quite sufficient.

I have always been and am now more than ever alien from the society as such. Only Swami's loyalty has forced them to accept me—for of course there must be all manner of lurid (and fairly accurate) rumors about my life. And then there are those dreadful novels of mine for the faithful to gag on. The fact that I've written a life of Ramakrishna and translated the Gita must only make my novels the less excusable in their eyes. It really is a very strange and comic situation—but I'm so accustomed to it that I seldom think about it.

Still this utter dryness when I sit down to meditate. I ought to mind about it, I know. But, if I minded, that in itself would mean that I wasn't dry.

Happiness with Don, by and large, since his return. And good health with even a slight loss in weight, down to just a speck under 150. Dr. Macklin really does seem to have made my neck much better, but my right thumb and left big toe are as bad as ever, most of the time.

Am still waiting to hear from George Englund about *The Shoes of the Fisherman*. Robin French thinks the job is definitely on. I don't really want to do it, but I do want to do something, and it's always nice to be earning.

July 5. Well, today I have just been told by Robin that the *Fisherman* job isn't on, because they have unwillingly got to let Morris West write a draft script, otherwise he won't let them renew the rights. Now that I can't have it, I'm disappointed, of course. Especially as I have nothing to do.

Never mind, cheer up, the gruesome Fourth of July holiday is over. As always I felt the terrible oppression of the crowding Folk, pressing in on our lives. Every year there are millions more of them.

Ben Masselink says that a whole gang of teenagers have occupied one of the apartments across the street from them. They play the radio at unearthly hours and shout at people on the street through a loudspeaker. "Will the boy with the blue surf board please come up?" etc.

Latest inscriptions in the tunnel to the beach and on the wall:

Overby can leap, also. Overby is alive. All is Overby.

This is a time of great happiness with Don. Yesterday evening, as we were having dinner at Ted's while the fireworks exploded outside—we found we simply couldn't be bothered to watch them—we began trying to think of marriages which we found "moving." After long efforts, we thought of four—the Masselinks, Michael Wilding and Margaret Leighton, the Stravinskys, Jimmy and Tania Stern. We couldn't think of any pairs of men!

I think Don Howard *really* likes *A Meeting by the River*. I was so much surprised to hear that he once seriously considered joining an Anglican monastery. Vidya likes it too and has now sent it on to Edward Upward. So far the necessary changes seem to be very few.

July 11. Have just heard from Edward. He likes the novel, or rather he likes the way it's constructed—that's what he stresses. Then he says that he's "uncertain" about Oliver's "motivation."

> Obviously the reader isn't meant to accept Patrick's view that Oliver is becoming a monk in order to escape the ambitiousness which is natural to him but which he knows their mother wouldn't approve of in him.... On the other hand the reader can't quite believe that Oliver becomes a monk solely because the social work he's been doing doesn't seem "real" enough to him. (It's to Oliver's advantage, of course, that the reader can't believe this.) ... Surely such a man becomes converted not only or even mainly because life doesn't come up to his expectations but because he comes to feel that the horrors of the world and the flesh are too great ever to be removed by any kind of social action, and that such action can only have meaning when it is performed sacramentally? I may be wrong in thinking this is how Oliver felt (particularly after being in the Congo) but if I'm not wrong, couldn't you add a sentence or two—possibly in Oliver's second letter to Patrick—which would indicate that Oliver wasn't insensitive to material horrors and that his becoming a monk was at least partly motivated by them?

I have copied this part of Edward's letter down because his writing is so tiny and untidy, and I want to be able to study it apart from the rest.

Edward goes on to tell me that Hector Wintle is dead. Now I'm sad that I didn't see more of him during my visits to England. He was, in later life, a mysteriously happy person—after being such a cheerfully gloomy young man. His happiness was something you felt immediately, and he never tried to describe it or explain it to me. His obituary notice never mentioned his novels—so I'm glad that I did at least speak of him as a writer in *Exhumations*.[1] Apparently he had always said he thought a coronary was the best thing to die of, and he died of one, at home in his garden.

July 18. This morning Don was putting something into one of the envelopes in the carton of photographs when an album fell on the floor, open. It was open at the two pictures of Hector Wintle.

Am depressed today. Partly (mostly) because I have what appears to be a cyst on the inside of my lip which Dr. Allen obviously doesn't like the looks of. He has told me to call him if it hasn't gone away by Friday, and then I fear he'll want to cut it out.

But also I'm depressed because we got terribly drunk for no reason the night before last, and I have been drunk far too often lately. It's such a boring vice and I have really no excuse for indulging in it, because it doesn't do anything for me except bring on black depression.

It now looks like I have a T.V. job, my first, a Christmas Spectacular about how "Silent Night" was composed, in 1818. The good thing about this, aside from the money, is that I think I could work quite harmoniously with Danny Mann.

On the 16th I had quite a long talk alone with Swami, after attending a lunch for Swami Sambuddhananda, from India. I told Swami that Vidya wanted my novel dedicated to him as Vidyatmananda rather than as John Yale. Swami evidently didn't like this, but said I should do it if Vidya wants it. It seems that Vandanananda brought back a bad report from Gretz; Vidya is said to be throwing his weight around already and making enemies.

Swami said that Maharaj had told him, morality is unimportant if you have devotion to God—"but of course we can't preach that," Swami added. I said that that was all very well, but I personally could feel no devotion at all. Swami said, "Anyone who says he has devotion or thinks he has devotion, doesn't have it. . . . People come to me every week and talk about their devotion to God, and I don't believe them." Also, he told me that, when he was fourteen, one of the swamis said to him, "Do you know what destructive means?" Swami said yes. "And do you know what constructive means?" Swami said yes. "Then be constructive, be constructive, be constructive." Swami went on to say that we

---

[1] In the preface to the final section "Stories"; see Glossary under Wintle.

must look on people's good qualities, not their faults, and I felt, as so often before, that he was saying this to me personally, because he observed my aggressions.

When I left Swami, I was full of good intentions, but already I'm back in the usual bad mental state. Never mind. The point is, Swami loves me—I don't care why and I can't possibly ever get to know why—but I ought to be able to feel that I'm under the protection of his love. Isn't that more than enough?

July 26. One month till my birthday. Why can't I try to spend it more fruitfully? I have wasted so much time—very nearly two months—since finishing the last draft of *A Meeting by the River.*

Above all else, I should dwell constantly on the thought of God. After the cancer scare last week— it proved to be non-malignant, merely a cyst—I had such a reaction of simple thankfulness and humble good humor. And then this horrible scene on the 21st. A shameful relapse to the mood of the bad old days of 1963. But never mind all that. The point is, I must keep praying to Ramakrishna to be able to love him. If only I could do that, then nothing else would matter, and in fact I should have a completely secure refuge from any outside disturbances—and, in addition, I should be of far more use to Don.

Lee Heflin reports that the latest word the kids use is "freaky." "Let's go into the woods and freak out" (take acid). About the squares who disapprove they say, "Listen to the freaks calling the freaks freaks."

July 28. Despite what I wrote two days ago I have frittered away this morning. What I need, when I'm not working, is a program. Even if the program included *deliberate* idling, that would be far less depressing than involuntary idling.

The evening before last, Betty Harford, Jim and Antoinette Gill, Jack Larson, Jim Bridges, Don and I had a picnic on the beach. It wasn't a success. The sun was setting on the hilltops, instead of out at sea, where we could have sat watching it—and that also spoiled the effect of the many surfboarders, which would otherwise have been magically beautiful. And then the beach itself was so dirty. And the food Betty and Antoinette had brought was wrong— nearly all cold, including raw tuna and beefsteak tartare. Betty's son Chris (who'd been surfing since six this morning!) came up with two wet water-shrunk friends and wouldn't touch it. And then Jack would talk only of the deaths of Monty Clift and Frank O'Hara, making a tremendous figure out of Frank as one of the half dozen who are running New York culture and whose loss will be a deathblow to the city. Jack was very manic and had been telephoning all over the place—to Salka [Viertel] in Switzerland, for example. He is determined to get Joe LeSueur to come out here, lest he should commit suicide as the result of brooding on Frank's death by himself.[1]

No word about the T.V. job.

The only achievement for me has been at the gym. The Air Force book rates the standard for forty-five to forty-nine years as twenty-three reps for exercise 2. I do thirty. Thirty-three reps for exercise 3. I do fifty-five. Twenty reps for exercise 4. I do twenty—though only just. Yesterday I weighed between 147 and 148. I still keep up the low-fat diet.

August 19. A week from today is my birthday, when I'll start a new volume, I think. But I am really very bored by diary keeping. I don't seem to be getting anything out of it at present. The last valuable entries were the ones about India.

Well, who knows, perhaps my entries about this trip to Austria will be worth something? It does seem now as if we really are going there quite soon. Danny Mann likes the outline I have done, and if ABC passes it then we are planning to go to Oberndorf before the teleplay is written.[2]

Am reading Hesse's *Steppenwolf* with great enthusiasm, after a sticky beginning. Then I'll try his [*Das*] *Glasperlenspiel*, though I doubt if I shall like it. Some other books I plan to read or re-read are: Cocteau's *Journals* and *Opium*, the rest of Dante's *Divine Comedy* (I never quite finished the *Inferno*!), the book of extracts from Ruskin edited by Rosenberg,[3] Gide's *Et Nunc Manet* and *So Be It*,[4] Moore's *Memoirs of My Dead Life*, Yeats's *Autobiography*, Anaïs Nin's *Diary*, Denton Welch's *A Last Sheaf* and *A Voice Through a Cloud*, Kerouac's *The Dharma Bums*, Samuel Butler's

---

[1] American poet and art critic Frank O'Hara (1926–1966) died in a sand buggy accident on Fire Island; Joe LeSueur (192[4]–2001), aspiring playwright, screenwriter, and critic, originally from California, had been his companion since 1955.

[2] "Silent Night" was first performed in the church of St. Nicholas in Oberndorf, Austria, near Salzburg; the melody was composed by the church organist, Franz Gruber, for words by the priest, Joseph Mohr.

[3] *The Genius of John Ruskin: Selections from His Writings* (1963), edited by John D. Rosenberg.

[4] *Madeleine, or Et Nunc Manet in Te* (And Now Dwells in Thee) was the title of Justin O'Brien's 1952 English translation. *So Be It, or The Chips Are Down*, was O'Brien's 1959 translation of *Ainsi soit-il, ou les jeux sont faits* (1952).

Notebooks, some at least of the Nietzsche *Portable*,[1] of Byron's and D.H. Lawrence's letters.

They have just reissued Glenway Wescott's *The Pilgrim Hawk*. I am quite horrified to see from the jacket that I praised it, saying that it was "truly a work of art" etc. Rereading it, it seems so stiff and mannered and empty. The first sentence starts to rustle already, like a lady novelist's (Elizabeth Rydal's?!) brocaded gown: "The Cullens were Irish; but it was in France that I met them and was able to form an impression of their love and their trouble." Again: "In the twenties it was not unusual to meet foreigners in some country as foreign to them as to you, your peregrination just crossing theirs; and you did your best to know them in an afternoon or so; and perhaps you called that little lightning knowledge, friendship." It is the rustle of a writer who's determined to write a truly elegant, sensitive novel. Those elegant pauses, while he visibly, in view of the audience, searches for the exactly right, perfect nuance! This is hairsplitting pretending to be truthfulness.

August 21. The day before yesterday, a girl named Jean Person came to see me. She had called first and asked if she could—so there was no reason for her to expect I'd be angry. However, when I opened the door, she jumped back a yard, covered her face with her hands, and exclaimed, "Oh, I'm so frightened!" Don says, rightly, that I should have told her to leave, then and there. I didn't, of course. She sat around, with bulging reddish-blue eyes which looked as if she'd been crying. She was fetchingly dressed, or meant to be, in a very tight little skirt, and wanted to know what she should do about her life.

Yesterday evening we went to Lee Heflin's and saw the film he has made, partly about his friend Duane [Hansen]'s body, partly about this other friend wandering along Hollywood Boulevard in a plastic bag. There were shots of really startling beauty, but the overall effect was too jerky and made me nervous.

So often I wake in the mornings with a sense of uneasiness. I feel that the situation is altogether out of my control. My life is out of my control. In fact, I get the feeling that I am my life. If I could sincerely say, "We are in God's hand, brother, not in theirs," that would be fine. But I feel, if anything, much more in "their" hand.

My character (that quaint old word) is simply awful. I am full of resentments, pushed this way and that by all manner of compulsions; and I am dull witted and unfeeling. Never mind all of that.

[1] *The Portable Nietzsche* (1954), selected and translated by Walter Kaufmann.

But I do wish I could just occasionally feel able to say, Chris is a mess but THOU ART.... I'm not writing any of this in a mood of self-dissatisfaction. I just want to try to describe what I am now. I mean, I want to register the resolve to start trying to describe it. This will be very difficult and at best something which can only be done piecemeal. But I really ought to start. (Some aspects I caught in *A Single Man*, but there it's all too simplified, because George is not me.)

August 22. I forgot to mention that we had supper with the Stravinskys on the 19th. Igor seemed pretty well. Poor Vera had hurt her leg. She was driving alone, and not attending, and she ran smack into a stationary car! You feel she is getting old and a bit vague. Bob called from Santa Fe, where he had been conducting *Wozzeck*.[1] When he is away one misses his rather brutal bossiness—which produces champagne and other goodies for the guests.

Igor told me that he has just finished a fifteen-minute piece which is some kind of a requiem and commissioned to be performed at Princeton.[2] He said that the music was so *dense* (I think that was the word he used) that it was equal to a whole symphony by Haydn.

Saw Ronnie Knox last night and we talked about his play. The other day he had another of his rows with his stepfather and his allowance was cut off. (It has been reinstated again since then.) Because of this he was depressed and complained to his French girlfriend, who replied, "Then why don't you kill yourself?" So the girl is in disfavor and Ronnie has started seeing Renate again. He says, "Whenever I get involved with a woman she starts trying to compete with me."

On the beach we saw Michael Sean, who comes out weekends from the clinic at Downey[3] and stays with the Gowlands. [...] He makes dates with all sorts of people and they have to take him around. Last Sunday they were humiliated because Michael made them arrive hours late at somebody's house for supper and the hosts were so furious they told Michael and the Gowlands to leave again. But Michael insisted on staying, and the hosts fed them contemptuously, "as if they were servants." Jo Masselink is strongly

[1] Alban Berg's 1922 opera.
[2] *Requiem Canticles* for alto and bass soloists, chorus and orchestra; completed August 13 and performed at McCarter Theater in Princeton October 8. It was also performed at his own funeral.
[3] Where he was continuing to recover from the car accident.

prejudiced against Michael. She also says quite bitter things about Ricky Grigg; that his behavior to Sandy is impossible, [...] etc.

[...] On the beach, [Michael] certainly seemed unable to stop talking, and it was all about his crazy fellow-patients at the clinic, including one who is masturbating himself to death, although he has to pull a catheter out of his penis in order to do it. [...] While we were talking to him, Peter Gowland dropped a hint that we should help with Michael's transportation to and from Downey. This we're certainly not about to do!

September 1. Don gave me three marvellous presents for my birthday, a Girard-Perregaux wristwatch, a pair of very powerful Japanese binoculars and an Uher tape recorder, which I have only just discovered how to work. The birthday was peaceful and happy, with a visit to Swami in the morning and supper with the Masselinks. But somehow it wasn't the psychically right moment to begin a new volume of this diary, so I'm waiting.

Swami is now in hospital, with visiting forbidden, and the doctor admits that he had a slight heart attack but doesn't seem alarmed. As always it is very hard to judge how sick Swami is, because he relaxes so completely toward his illness. For the same reason it is very hard to know how he feels about death. He seems nervous about his health, and never attempts to hide this by putting on a stoical act; and yet one suspects that he's quite relaxed about it, underneath. If he does worry it's because of his concern for all of us, and the Vedanta Society. Those idiots at Belur Math have chosen this time to suddenly start rocking the boat—demanding that the girls shall be separated from the boys. As though heterosexuality weren't actually the least of Swami's worries! Meanwhile Swami blames Vidya for mischief-making while in India, but at the same time (how characteristically!) repents of the severe letter he wrote Vidya and plans to write him another one since Vidya hasn't answered.

And then Pavitrananda—having first, of course, been told about this by Swami—said he felt definitely that my novel shouldn't be dedicated to Vidya as Swami Vidyatmananda. So I wrote this to Vidya, who replied that he didn't want me to dedicate it to John Yale, since John Yale was no longer alive—so the dedication is going begging! I am tempted to dedicate it to Gerald Heard, but that would raise the problem of hurting Michael Barrie's feelings. The way out would be to dedicate it to Gerald *and* Chris Wood, thereby excluding Michael on the grounds of an older relationship. But the whole business seems tricky and tiresome. Maybe better not dedicate it at all and have a sort of invisible dedication to Vidya! But Don is against this, he thinks it is encouraging Vidya in his sulks and sorriness for himself.

Discussion with Don about what happens to the mind during delirium or under the influence of drugs. Perhaps it can be described in the language of Patanjali's Yoga Sutras—*manas*, the recording faculty, continues to receive impressions through the sense organs, but *buddhi* and *ahamkar* are out of action,[1] so the impressions cannot be classified and the ego sense is not operating to decide what belongs to "I" and what doesn't. The result is a terrifying total onslaught of impressions which cannot be attended to individually because they can't be graded and arranged. So they make the mind feel that it is losing its own identity and being swept away in a flood—which is the terror of madness.

Tomorrow Danny Mann and I are supposed to confer with people from the ABC network and then decide what to do next about "Silent Night." I have a feeling that Danny, for some personal reasons, doesn't want to go to Austria yet. I do want to go, as soon as possible, because I have no other plans. More and more I feel that this visit to Europe will turn me on, somehow or other, and show me the way toward another book—not to mention *Hero-Father, Demon-Mother* for which I require all sorts of documentation from the papers at Wyberslegh.

September 2. Every morning (almost) I wrap myself in the chadar which Swami gave me and sit down on the couch in my workroom to make japam in front of the photograph of Swami doing the worship at the Hollywood shrine. I get almost nothing consciously from this, my mind takes off almost at once, skimming over all the current resentments, anxieties, wants and distractions. And yet I would feel a lack of something if I didn't do it. Am trying to get myself into the mood by first reading from Vidya's anthology of Vivekananda's writings.

Am rereading John Horne Burns's *The Gallery*, because I just found a hardcover copy at Needham's. (Getting a new or different copy of a book often stimulates me to reread it, because it becomes a slightly different book which you can open for the first time, as it were.) The parts I have read so far seem excellent, better than I'd remembered, and I can see now why Hemingway admired it

---

[1] *Buddhi* is the discriminating, classifying intelligence; *ahamkar* or *ahamkara*, the sense of self or individuality, lays claim to impressions as personal knowledge.

so much. What strikes me is the recurring theme of the unreality of the war, because I so constantly feel the same thing about my life nowadays. I wonder if one doesn't always feel this way when one is very much involved in the present day-to-day existence? It is so hard to feel the weight and depth of experience except in retrospect—and I seem to think so seldom about the past. The difference between me and the characters in *The Gallery* is that they were all presumably dissatisfied or dully miserable or acutely wretched and afraid—whereas I am living a life of contentment, by and large, and even very considerable happiness. How often I say to myself—especially with regard to Don—that this experience would be vividly and even poignantly beautiful if I could stand back and look at it!

Don said of his work, the day before yesterday, "My drawings are studies made under stress." The operative word is *stress*!

September 11. Now, suddenly, Austria is very much *on*. Danny Mann is prepared to leave about ten days from now, and we are to start getting tickets, passports, travel money immediately. My reaction to this is a bad back, for of course I now am unwilling to go, I cling to the pleasures of home—all the more strongly because I haven't been away since the beginning of 1965. So my back hurts and I am using Jo's hot pad on it.

The wonderful summer, the best in years, is holding on into a warm slightly foggy fall. I do so hate to leave it.

September 17. Well, now it's settled: Danny Mann and I are to leave next Wednesday, the 21st, by plane for New York, then on by another plane, Lufthansa[1], to Munich, then by car to Salzburg and Oberndorf. I realize already that Danny is a pusher and a penny-pincher (although the pennies are really not his but ABC's, he is no doubt anxious to prove to them that he can be an economical producer). So I shall have to show him, in the friendliest way, that I refuse to be either rushed or made uncomfortable.

The television, the other night, was giving forth some commercial which urged you to "accept your financial maturity" and buy the product in question.

Don and I were sitting at breakfast out on the deck—as always nowadays, because of this glorious and prolonged summer weather. We watched through the Japanese field glasses the elder blond brother in the garden of one of the houses below as he did

[1] I.e. Lufthansa as it is known now.

some carpenter work. He made a great play of doing this work in a grownup dead-serious way, frowning, taking measurements, regarding the wood with intense concentration before sawing it up. Meanwhile the younger blond brother hovered in the background, longing to be included. But the elder brother severely excluded him. Don exclaimed, "How badly people behave to each other!"

Ronnie Knox and I were on the beach. Ronnie took a piece of broken seashell which he had picked up (he pretended to me at first that he had had it with him for years) and put it into his mouth as a kind of horror tooth, the sort a werewolf would have. I said, "It ought really to have blood on it"—so Ronnie promptly scratched himself quite deeply in the thigh with the shell and got the blood. That's an example of his sort of craziness, tiny and minor but so characteristic.

A visit to Gerald Heard—the first Michael has permitted since their return from Hawaii. Gerald doesn't look nearly so fragile now, and he talks freely and brilliantly, without any impediment that I could detect. He does, however, hold himself crookedly when he stands up. Chris Wood hasn't been to see him at all, yet, and Gerald was making fun of Chris for this, saying how difficult he was, because he wouldn't come *exactly* when Michael told him he must.

When Michael had gone out shopping, Gerald told how he had written to one of the editors of *Life Magazine*, whom he knew, because of the caption *Life* printed underneath a photograph of a village in Vietnam being bombed. Gerald protested against the tone of the caption, which, he said, was characteristic of the increasing inhumanity of people in this country nowadays; they simply do not care about human suffering. So far he has had no answer from the editor. He says he hasn't told Michael about the letter to *Life* because Michael is very sensitive about anything to do with Mr. and Mrs. Luce, and would be hurt if he knew. (Don raised the question, how did Gerald get the letter off without Michael's knowing about it. The answer must be that there is an underground line to the outside world through Jack Jones!)

Gerald then talked about dolphins—how, it seems, they really do not want to communicate with us, although they have a language; all they want from us is love. The right whale has an even bigger brain than the dolphin, but if we don't do something quick, it will be wiped out by the whalers.

Gerald also spoke, with considerable self-satisfaction, about his behavior during his stroke. He says he wasn't in the very least alarmed when he lost the power to move his arm—and that Dr. Cohen had said later that he was "edified" by Gerald's attitude. Gerald also says that Cohen had been shaken (I think that was the word he used) by Aldous's attitude to death—i.e. his refusal to admit to himself that he was dying. The moral of this was that Cohen's faith in the value of spiritual disciplines had been shaken by Aldous's behavior but restored by Gerald's! Vanity on the very brink of the tomb.

I also saw Swami yesterday, at the convalescent hospital on San Vicente they have just moved him to. How utterly without pretension *his* behavior is! He seems much calmer now, though he still gets excited about Vidya, who is maintaining what seems a very bitchy silence. Apparently there is even a suspicion that Vidya told the Belur Math people that they should close the center altogether after Swami's death—but this is really more than I can believe. Swami is also sad about Sarada. "As long as she thought I was going to die," he says, "she was eager to come and see me. Now she doesn't care to."

However, he is seeing Maharaj in his dreams nearly every morning, early. Once they were sharing a bed. He also had an anxiety dream in which he was looking for Maharaj at Myavati[1] and couldn't find him, and meanwhile Krishna ran over someone in the car *and* cut off his leg![2]

---

[1] Himalayan location of the Advaita Ashrama of the Ramakrishna Order; usually spelled Mayavati.
[2] End of typescript; Isherwood labelled the next one, "Sep 21 1966–Jan 30 1970."

September 21. As Don was driving me to the airport, he advised me to dedicate *A Meeting by the River* to Gerald Heard only—not to Gerald and Chris Wood. If I just dedicated it to Gerald, he said, Michael and Jack Jones wouldn't have any excuse to feel offended. He's going to ask Gerald about this.

Danny Mann had come with Gigi (her surname is Michel, and I now discover to my surprise that she's thirty-one and has never yet been married, so I was probably all wrong about her being a gold digger.[1] Don and I parted discreetly at the car door. As for Gigi, I politely kissed her goodbye on the cheek. Danny took out ten dollars' worth of life insurance (which pays off three hundred thousand, I think he said). Danny spread his between his children and Gigi, I guess. So I took out the same amount in favor of Don—just to show Danny that we animals are every bit as valuable as humans.

In the first class (American Airlines) the television is a tactfully small box between each pair of seats. Danny didn't want to watch it either. So he read [Edmund Wilson's] *Memoirs of Hecate County* and I read Whitman's *Specimen Days.*

When we got near New York there was quite a long holdup, about two hours, circling around with occasional bumps and waiting to be allowed to land through rain and low fog and lots of air traffic. I was calm, thanks to Librium and drinks. Danny was pretty worried. "I'm an orthodox coward," he said several times—maybe this was a Jewish joke (almost all jokes made by Jews are Jewish, by queers queer). Anyhow I like all his behavior so far and really begin to think we shall get along.

The Kennedy Airport, when we did finally land, was in that state of yelling confusion and impotent haste which seems to be what I always encounter when I arrive there. A tall handsome German boy told us we'd missed the plane to Munich but could get on one to Frankfurt. Danny was much annoyed. I merely rejoiced that we didn't have to drive into New York—the only way one could have left the airport within an hour would have been in an ambulance. On the plane we were separated. I sat next to a manufacturer of rubber goods named Hans Christian Pauck, and practised my German on him until we were both thoroughly drunk. "You have not made any mistake for half an hour," he finally told me.

[1] She was an aspiring actress and played bit parts in several films, including *My Fair Lady.*

He gave me a jar of *gänseleber pastete*.[1] I slept snugly, wearing dear little blue slippers which the Luft Hansa gives its passengers.

September 22. Danny decided he couldn't face any more flying, so we came on to Munich by train. It didn't leave until the early afternoon however, so we had to wander around Frankfurt, which seemed stodgy and dull. On the train, I slept. Danny reported to me that, while I was asleep, some students in the carriage said to each other that only Americans would sleep on such [a] journey and never even open a book. Danny claims that he rebuked them, saying "We've been up two nights," and that this silenced the students. I say "claims" because Danny really speaks very little German, though he does make himself understood to an amazing degree in Yiddish.

We failed to get rooms with baths in Munich, because the Oktoberfest is on—the great beer-drinking festival. It makes the streets very loud at night. We are opposite the railway station, which is full of questioning eyes. But no opportunity or indeed inclination to explore. I woke suddenly in the night with a hunger that was like the sexual drive of a rapist. Quicker than it takes to tell, I'd plunged my hand in my bag, got out the jar of *gänseleber pastete* and eaten all, all. Was this the first symptom of loneliness?

September 23. We went out to the Bavaria Studios. We had no appointment, no letters of introduction, no particular proofs of identity, no guide except an ambitious young man who runs the Her[t]z Agency here and who had just rented us a car. And yet, within half an hour, we met the second in command and were shown round the studio, and within an hour we were having lunch with the head of the studio, Dr. Rosa(?). Imagine that happening in Hollywood!

Went to visit Herbert List after supper. Why was it so sad? Because he's now dry and lonely and looks like an old lady and is collecting drawings by minor Italian masters? Probably he thought of me as a sad little figure too.

October 8. What with work on the teleplay and idleness and the impossibility of doing *anything* private while one is with *any* director, I have left out the whole of our Salzburg visit and now I'm about to leave for London—this afternoon—and goodness knows how I shall ever be able to write anything at all while I'm there.

[1] Goose liver paté.

I'm to stay with Neil Hartley and Bob Regester. And Danny will be itching and chafing to get the teleplay revised and retyped because he is mad to get back to his Gigi. Well, anyhow—

Have been here just about two weeks, and this much I can say, Salzburg is one of my most favorite towns, along with San Francisco and London, I guess. Venice is too spooky and miragelike to come on the list and I don't know Rio enough. But Salzburg is so snug in the best medieval way, ratlike tunnels and alleys and cellars, and the setting is so beautiful, and the two ridges of mountain block off the old part so that all the new shit has to be built on the other side and can't be seen. The Austrians seem to be slightly sinister, with their unpleasant dialect and Disneylike unconvincing charm. The country is Disneylike too, so lawned and smooth and carpeted with green, and the improbable landscaped crags—but of course it is beautiful, beautiful like no other place, except maybe Japan, in its toylike way. After spending all these years in that junkyard, America, the sheer tidiness of Austria seems uncanny—everything, billboards, tin cans, neon signs, seems to have been swept under the rug.

All this refers to tourist Austria only. The extraordinary thing is, how tourist Austria stops near Salzburg and you get into this other land. At Salzburg, or near it, this other Austria begins. For example, Oberndorf, or Eugendorf which we are actually going to make the film Oberndorf, belongs to the atmosphere of East Prussia almost. There are little hills, but what you sense is the dreariness of the appalling Prussian plains, the squalor of the sad winter-stricken villages—a tall onion-dome church, rat-faced peasants, telephone wires, cowshit.

The weather has been beyond belief heavenly. So warm you could sunbathe, and the leaves all red and gold.

There is much to write about Danny Mann, and I will write it by degrees, I hope. I can say this at once, we have gotten along surprisingly well, considering what sort of people we individually are—neither very easy.

October 14. Here I am in London. Arrived on the 8th and this is my first entry. I doubt if it's going to be possible to write much while I'm here. Anyhow, I'll begin by copying out various notes I made while in Salzburg.

Walking with Danny in the town. His vice: buying sweaters. He simply cannot resist. Also, he is passionately fond of cheese. The way the foodshop windows here are displayed makes everything look rare and delicious. He bought me a silk scarf, saying,

"Perhaps this will remind you of old Danny." He let me pick it out and I got a very nice one, but I have to wear it in the tennis-anyone style, which is not mine. Danny says of himself that he was a Don Juan at one period of his life. If this is true (and why not) it makes you reflect on the nature of sexual success: his face is blasted with acne like a Hiroshima. All that is left of the acne now is a lumpy, unpleasantly shiny red surface.

Danny is sentimental about his own sentimentality—it seems wonderfully touching to him that he keeps photographs of his mother and father with him wherever he goes. He says that his marriage was a failure all along, and now he has so much alimony to pay that he must make $50,000 a year before he gets a cent. Gigi is the only love of his life, he says. (To Danny's love I oppose *my* love—which is like a merciless religion—it offers no salvation to anyone else and yet damns anyone who doesn't accept it.)

He suffers from vertigo. When we went up in the bus to see Hitler's house above Berchtesgaden, he couldn't look out through the window. This I found sympathetic, because I was hardly bothered by it at all.

Hitler's house is such a grim paranoid little dump. And most of the time it must have been shut in by wet clouds. The long torture-chamber passage through the mountain to the elevator, which is lined with brass, like a receptacle for boiling victims in.

The day Gottfried Reinhardt and I lunched at Mitteregg(?).[1] The two of us, such utterly urban creatures, sitting out of doors under the trees at a plain wooden kitchen table with a check cloth, stuffing ourselves with cheese and white wine as we looked out over a fairytale valley and talked international politics.

I love Gottfried as much as ever. He is really quite grotesquely ugly, with his great warts and spreading jowls; yet as soon as he begins talking you see only the beautiful intelligence and fun and sadness in his face. He dislikes everything about the Germans— they are utterly subservient, he says, and without any shame. They would support anybody who paid them. Their literature nowadays is worthless because it is written with complete cynicism. "I may be an idealist," Gottfried said, "but I like books to have a point of view." (And looking at him I saw that he really is an idealist.) He told me about an Australian manual for infantrymen which he was given while he was making training films for the army during the war. It told you that, if your best friend was badly wounded, you were to leave him lying there—and first you were to take away his

---

[1] Several Austrian towns have this name, including one near Salzburg.

water, because it would be more useful to you than to him!

We talked and ate and laughed until we were both drunk. I was so drunk that I returned to the hotel and slept all the rest of that day, woke up at one in the morning, went back to sleep again and slept till breakfast!

Salzburg notes: The surprisingly loud bangs of chestnuts dropping from the huge dark trees. There is nothing shorter than Austrian shorts. Every so often, on the right person, they are wildly erotic—but on most people so indecent that you quickly look away. The castle of Hohensalzburg need only be seen from below; inside, it is disappointing. But one must absolutely climb up to the Café Winkler; the view from it is the best in the city because it isn't spoiled, like every other view, by the café itself.

Salzburg is dobbintown. The mad marble dobbins on the Residenzplatz, vomiting water or squirting it through their noses, while the dear little live beige dobbins wait to pull tourist carriages through the city. The campy vain painted dobbins on the walls of the *Pferdeschwemme*.[1] And Pegasus being so silly, showing off his wings, obviously drunk. There is nothing on earth sillier than a silly horse when it is drunk.

The night before I left, I went into a cellar bar almost next to our hotel, where the *wirtin*[2] sang opera airs. A very drunk young man from Innsbruck talked his ugly unintelligible dialect, aggressively, to show me that it was no good my knowing High German. I felt strongly that I wasn't wanted and, being drunk myself, said sarcastically to a girl, "Bin nur ein Amerikaner." "*Nur!*"[3] she cried, with an indignant laugh, furious with me.

The morning I left I went up to the *festung*[4] and walked all along the cliffs to the Café Winkler. A beautiful walk on a beautiful morning. But I am incapable of taking walks. All physical acts become compulsive with me, as soon as they are consciously prolonged; and the walk is only around my head. (Oddly enough, I feared that a man who was apparently following me was a homicidal maniac. Even if he had been, it wouldn't have mattered; there were too many people about. What was odd was that I very seldom have fears of this particular kind.)

Gerhard Huber, my handsome young secretary, came to the hotel to say goodbye, with his fiancée, Irmi. They had nearly split up during my stay in Salzburg because Gerhard refuses to be bossed

[1] Horse pond, to wash and water the royal horses once stabled nearby.
[2] Hostess.
[3] "I'm only an American." "*Only!*"
[4] Fortress.

by her mother. Gerhard used to explain to me about the student corporation (the Catholic one) to which he belonged, and all its complex relationships. You have "fathers," "sons," "brothers," etc. Some student nicknames: Tristan, Orpheus, Taurus, Virus, Yogi, Rumba, Fouché, Dampf, Orplid.

October 24. Up at Disley, with Richard. No time yet to write about London, or Disley either. Am reading M.'s diaries, my father's war letters, etc. Will copy now some passages in M.'s diary relating to his death—because I don't want to take the book away with me.

1915. May 12. A telegram redirected on from Ventnor of last night "Lt.-Col. F.E.B. Isherwood reported wounded 9th of May, nature and degree not known." And this is Wednesday and nothing from him or any hospital. Wired to War Office but no news forthcoming.

May 27. Jack Isherwood came. His idea of comfort seems to be to catalogue and label the different degrees of dreadfulness of things that might have happened, then argue from the different hearsay reports that these things *haven't happened*, and therefore why be anxious? Just as if the *mere fact of not knowing* where or how he is, is quite enough in itself for anxiety, without speculating about anything else that can have happened, but I don't think the Isherwoods have much real feeling.

June 6. Aggravating note from Eric ... saying what a fearful and glorious fight they must have had on the 9th for so few to be left to account for the remainder.

June 8. A letter from Mr. Isherwood[1] in which Frank's being a prisoner (which alas we do not even *know*) he seems to think is quite a blessing in disguise. It all makes me feel *so* lonely.

June 10. A long letter with my fortune told by cards from Miss Cooke of Cappagh. My fortune all that could be wished with the return of my fair soldier.

June 21. To King George's Hospital just over Waterloo Bridge where Lady Wyman is working. Her son has been a prisoner in Westphalia since Mons, he was reported killed and she and her husband went to hospitals all over England to get news of his end. Three different soldiers and one a man of his own company told them in detail how he died.... Six weeks

[1] I.e., Frank Isherwood's father, John Bradshaw Isherwood.

later he wrote to them from Germany and said he had never even been wounded. Could not help feeling cheered. It seemed another ray of hope....

June 24. In the evening came a terrible letter from Arlington St., the British Red Cross Order of St. John. "We much regret to say that according to the Geneva list of June 12 received here on the 23rd it is intimated that a disc was found on a dead soldier close to Frezenberg[1] early in May with the following inscription: Isherwood, F.E.B. Y & L Regiment C. of E. We greatly fear this disc may have belonged to Col. Isherwood. I am faithfully, Louis Mallet." And so passes hope and life.

June 25. Wrote to Marple to Jack Isherwood as he and Esther [Isherwood], or at any rate the latter, continue to think it strange Henry [Isherwood] can see anything to be so worried about, they may still see hope.... I don't know how the rest of the day passed except that time seemed to have stood still and there appeared a vista of endless hopeless days of loneliness ahead.... Everything reminds me of him, the places we used to go together, our outings and the way he had of making everything nice and all his thousand kind and thoughtful ways.

July 11. Nine weeks today.... Now and then I have such hope and peace.... I think it *must* be well ... and if he has gone beyond, think he must have found the things he loved, and peace, and it may be, the beyond is near if one *could only know and see.*

July 16. I had a letter from Major Bayley yesterday ... it said it made them all the more determined to wage this war to the end—bitter to the Germans as we mean it to be. It is due to the memory of your husband and others like him that we out here should do all we can to avenge them. Rest assured we all shall do our best. Which makes one feel the lives that have been given have not been given in vain, but serve to stimulate by their example ... and that does so gladden one's heart.

July 27. A woman from Barker the mourning warehouse arrived between 10 and 11 laden with boxes of hats, bonnets, blouses and costumes and coats—all the latter designed for the very ample matron, and I was quite swamped. However we chose designs to have dresses made. M[ama] and I each a dull black silk and she a serge too. Hats, etc. Took till lunch.

---

[1] Belgium, northeast of Ypres.

July 29. Letter from Col. Clemson *quite sure it is all a mistake* about Capt. Purden having identified Frank, as he had promised all he heard to let him know at once.

July 30. Was fitted at William Barkers for my black silk coat and skirt ... feeling all the time so hopeful ... and yet there are alas no grounds for it.

September 8. In the paper today under *previously officially reported missing now unofficially reported killed* is Frank's name. A year and a day after he first went out. I think I miss him more every day and life seems harder and darker.

September 11. Heard from Graham he had replied for me to the Keeper of the Privy Purse: Please convey to Their Majesties with humble duty our grateful appreciation of their gracious sympathy.... I should never have dreamt of putting "humble duty!"[1]

September 27. In the evening I received from the War Office the disc bearing my husband's name—very polished up and torn from its string. It didn't seem real, somehow.

September 29. Met Mr. Robertson at the entrance to Somerset House at 11:45 and he took me into various bewildering departments in connection with presenting the will and valuing the property, which but for him would have been very tedious and distressing ... as it was it was after 1, and I had to return at 2:15 to take my oath and be asked more questions—felt so proud to say there were *no* debts ... few can say that.

October 5. Mr. Robertson from the Probate.... They cavil at the dates being given as the 8th or 9th and have inserted instead "between the 8th and 11th" though what difference it can make I can't imagine.

October 6. Heard from Mrs. Cobbold that Major Robertson was killed last week, this completes the casualties to those travelling on the 7th of September in the same carriage from Cambridge as Frank. Nothing has happened to any of those in the same carriage as Col. Cobbold....

---

[1] Sir Graham Greene, Permanent Secretary of the Admiralty through most of the war, a favorite cousin of Kathleen's on her mother's side; his nephew, the writer Graham Greene, was named after him. Their Majesties were then King George V and Queen Mary.

Have also read some of the letters which my father wrote to M. before they were married, in 1902. Here is one extract (November 12, 1902):

> I am reading a book on Esoteric Buddhism which is very interesting but rather difficult to understand. It doesn't seem to be incompatible with Christianity exactly either.... It appears that although we shall not really spend Eternity together we shall each think the other is with us! Do you think that satisfactory? Goodbye, my dear. I do hope we are going to have a *very* long happy time together, even though we don't get Eternity.[1]

The adjectives my father applies to M. are most often "kind" and "sweet." He says of himself that he is "early English," which seems to mean masculine, a bit crude and overdirect—as opposed to what he calls "party men"—smooth talkers, clever and oversubtle. Poor old Uncle Harry is put in this category. But that was how my father saw himself, not necessarily how he was.

October 29. I got back to London on the 26th. I'm to fly to California on November 2.

Richard was planning to drive down here with Dan and Mrs. Bradley (who wanted to see the motor show) and me. At the last moment he called it off, so I came on the train. Dan thought that perhaps Richard wanted me to himself for our parting. He gets jealous if he thinks other people are cutting in. But it may also have been that Richard simply wanted to slip away to Wyberslegh and spend the day quietly drinking. What is impressive about Dan is that he discusses Richard with real affection and unbitchy humor. I liked both of them greatly. Mrs. Dan is big and jolly and her cure for everything is to cook lots and lots of food. They are sort of ideal north-country working-class people—probably you don't find them like that any more; they are both in their fifties.

Richard's way of drinking is really most strange. He doesn't gulp or guzzle, he keeps quietly and prudently pouring small doses of beer into a glass; these he sips at. He started at breakfast some days and kept at it all day. He never seemed to get drunk, though he drooped toward evening and stopped speaking. If he throws up, Dan says, he will go right on drinking afterwards. Dan says that he usually only drinks at weekends; but while I was there

---

[1] Isherwood used all of these passages, with minor changes, in *Kathleen and Frank*, where they can be found by date.

he drank nearly every day. This surprised me, because I wasn't aware of any tension between us. Indeed, we have never gotten along so well. I spent a lot of time on the material I want for *Hero-Father, Demon-Mother*, and Richard was unfailingly bright, clear and helpful; he recalled names and dates and other facts without the smallest hesitation. He seems to have the past at his fingertips, and maybe this is partly because he can scarcely be said to live in the present at all. He does nothing in particular, except go over to Wyberslegh in the car, putter around and leave again. He says that walking uphill makes him breathless, so he doesn't go for walks any more. He eats almost nothing. He has a terrible cough and refuses to have his chest x-rayed. He no longer smokes, however. And he is much tidier. No more dribblings of food. Most of the time he wears a nice clean suit and a tidy tie—these are due to Mr. and Mrs. Dan, it's true. His only sloppiness is often to leave his fly unbuttoned. He doesn't strike one as in the least crazy or in any sense an invalid.

Several times, he told me how happy he was that I had come, and I think he meant this. Nevertheless, I discovered that he had really only expected me to come for the weekend—the time I had allowed for my visit was perhaps too much, at least from the Bradleys' point of view, since they had to clear out of their bed-room for me.

As for Disley, it seemed dripping with nostalgia, as usual. Those sad sodden hills under the low tragic sky. Lyme Cage[1] standing up black against the light, that enigmatic little structure, like some archetypal symbol one sees in a dream. The little pub called The Ploughboy, which I used in *A Single Man*. The telephone box on the corner with the ban-the-bomb sign daubed on it. The people in their raincoats, very sharp and distinct in the humid light. Actually, however, it hardly rained at all while I was there.

When I went down to Cambridge on October 16 to see Forster, I was overpowered by a sense of death. What seemed so terri-ble was that the buildings of King's hadn't changed at all. They were like merciless instruments which had whittled away at us human beings and worn us out. We had rubbed off our youth on them. Morgan was so gentle and faint and humble; thanking me for coming to see him until I wanted to weep. And look-ing down from his window through the yellow afternoon murk, I saw an undergraduate throwing up a bowler hat and another

---

[1] A sixteenth-century tower built as a hunting lodge in the 1,400-acre medi-eval deer park around Lyme Hall.

undergraduate trying to hit it with cracks of a long bullwhip—and it seemed that they, too, had only an instant in which to be young. And then we went into hall, passing down the lines of all those young men standing impatiently until we old fuddlers should be seated—and of course I was identifying myself with them, forgetting my own wrinkles, as I walked in the freak-show procession of dons, led by the poor arthritic provost who is bent down like a dromedary and had to have a younger American don support him so his chin shouldn't sink and rub against the floor. And no sooner had we sat down than a man opposite, whom I'd written off as an utter dotard, reminded me that we had both tried for an entrance scholarship to Charterhouse, the same year!

I see death too at Wyberslegh, which is slowly dying of neglect, and will almost certainly die, because repairing it would cost a fortune, and Richard doesn't even really want to repair it. He has become apathetic about the whole thing. He says he won't live in Wyberslegh again unless he can get someone to live there with him, which is surely next to impossible. The house is ghastly with damp and mould and black deathly dirt.

But on my last day at Disley, October 25, I insisted on taking a taxi and driving over to Marple Hall. You can hardly find either of the entrances to the park. What used to be called The Private Drive is now a road called Marina Drive, with a girls' and boys' school on it, big airy pleasing buildings, the Marple Hall Grammar School. All the times I had come here before, the place had seemed horrible, with the old disfigured house falling into ruin. But now the whole feeling was different. You could barely trace the foundations of the house, they were thickly grassed over. Only the stone over one of the doorways to the terrace lies there in the grass, engraved with the date 1658. And the two great beech trees are still standing—the one near the front gate and the one at the end of the terrace, in which I shot the wood pigeon with my air rifle (my most painfully remembered youthful "war crime"). It was a beautiful morning, and in the classrooms they were sitting at lessons, and over on what used to be the Barn Meadow a football game was going on. A new life had taken over. The Hall and its curse were forgotten, or remembered only as something romantic and mildly benevolent. And I felt as the narrator feels at the end of *Wuthering Heights*, when he sees that the graves are becoming overgrown by the vegetation of the moor and thinks that you could not imagine unquiet slumbers for the sleepers in that quiet earth.

November 17. (Back here on Adelaide Drive since November 2 and thank goodness no longer forced to use the Olivetti—though indeed it is a faithful little thing, infuriating and stubborn but strong as a donkey. I used to get so mad at it that I nearly tossed it out the window a couple of times; and I'd bang it so hard that it's a wonder I didn't wreck the keyboard.)

First, while I remember, a note on what's written above about my visit to the site of Marple Hall. I omitted to record what now seems almost the most significant detail—that my taxi driver was in a hurry and, in his ungracious north-country way, told me that he could only allow me fifteen minutes to walk around and recapture the past. Imagine Proust being goosed and speeded up like that!

I don't know if I shall ever have the memory or the energy to get my impressions of London written down, now. Will just relax and try to write every day for a week or so, and maybe something will come through. Up to now, I've been kept busy, or at any rate jittery, by having to rewrite the teleplay of "Silent Night." Now that's done and we await ABC's decision to go ahead or not to.

Projects ahead: Lamont Johnson asked me if I would be interested in adapting Shaw's *The Adventures of the Black Girl in Her Search for God,* to be produced at the theater next to the Music Center, downtown.[1] I've read this through once and can't make much sense out of it, but shall try it again. Then there's Tony Page's project of doing a condensed version of the two Lulu plays by Wedekind, for the English theater. Then there is the adaptation of Turgenev's *The Torrents of Spring* for the BBC, to be played by James Fox. Then there's the possibility that they may want me to work on *The Shoes of the Fisherman,* after all. They will if MGM is prepared to go on with the project, after all these false starts.

Then of course there is *Hero-Father, Demon-Mother.* Truman, whom we've seen twice in the past four days, doesn't like the title. Neither does Don. Truman thinks it too complicated and tricky.

Truman is curiously delightful to be with. We both felt this. Cecil Beaton had told me in London that success had spoiled him, but it wasn't apparent. He seemed the same as ever, except that this is one of his fat periods; he is quite gross, though in a compact powerful way, not at all flabby. He is giving a huge ball, "Black and White," shortly after Thanksgiving. I know Don would like to go to this. I absolutely loathe the thought of going to New York at all, but who knows what may not happen at the last moment! There would be the incentive of going to see *Cabaret,* which opens

[1] The Mark Taper Forum.

officially on the 20th—it's already on Broadway and in previews.

While we were driving to have supper with Truman Capote at the Bel Air Hotel we passed the Bracketts' house and wondered what the poor old things are up to, nowadays. Quite spontaneously a phrase came into my mouth, to describe their probable life: "The shuffle of slippers and cards."

November 23. Despite my resolutions, I've failed to make daily entries here and I begin to feel, more and more, that my memories of England are slipping away from me. Can't be helped.

I was just about to record, when I was interrupted while writing last, that Gerald had another stroke a short while ago, just before my return here. He lost the power of speech. Now he seems to be getting better. I talked to Jack Jones, who is my connection, this morning, and he told me that Gerald wrote several pages yesterday, in answer to some spiritual query from a woman he knows, but that he doesn't do much talking, on doctor's orders, although he is now able to.

Through Jack, I sent a message to Gerald asking if he would accept the dedication of *A Meeting by the River*. This seems to have pleased him and he did accept. The day before yesterday I finished correcting the American proofs and sent them back to Simon and Schuster. I still don't know what I think of the book. It seems thin and unconvincing at the beginning; later I think the fun of the psychological interplay starts to be felt. I am pleased with the final moves leading to the climax. But it's clever rather than emotionally powerful (whatever *that* means!)[.]

Talking about interplay, Don and I are engaged in one of our strange psychological wrestling matches. He wants to go to New York, attend Truman's ball and see *Cabaret*. I don't. That's to say, I don't want to go through all the trouble and discomfort involved. Also, I'm convinced, or I tell myself that I'm convinced, that Don would have a much better time there alone. Don, on his side, probably agrees with me and yet he feels that going to Truman's ball will be embarrassing unless I'm with him, so he wants me to come. Indeed this is very often his attitude, I think: he needs me, he would like my companionship, and yet he doesn't want me underfoot, because I can be a damned nuisance, with my slowness, my hatred of cold weather, my lack of enthusiasm for parties, etc. My tempo exasperates him, his exasperates me. We are always getting on each other's nerves and we love each other very much and so we wrestle emotionally. If I don't come, I'll have "deserted" him. If he puts off going because I won't go, I'll have deprived him of his fun. Similarly, if I stay behind I shall feel guilty, if I go I shall feel resentful because he dragged me over there against my will. Therefore we put off and put off the decision—it may well not be made until Saturday morning, when we have reservations on a midday flight. What I want, and am wrestling for, is an unconditional emotional "pardon," which will make me feel that I can stay here with a clear conscience. And what does Don want? If I knew that, there wouldn't be any problems.

Jerry Lawrence thinks *Cabaret* is a hit, but this seems to be largely hearsay. He hasn't got copies of the notices yet. *The New York Times* was very favorable, except for Jill Haworth, who was described as a hole in the production.[1] *Variety* and the *Los Angeles Times* unenthusiastic.

Gide's *So Be It* is one of the best things I have ever read about old age. I read it since I got home. Also Dougal Haston's book about climbing the Eiger,[2] which gave me vertigo even while lying down on the couch in the workroom—I felt I was on a ledge on the North Face. (I got the same sensation from reading Whymper's account of climbing the Matterhorn,[3] but this was much stronger.)

Now I'm wondering if I can use Gide's method of writing *So Be It* to do *Hero-Father*. I mean, I wonder if I can keep a sort of diary of my attempt to write the book—a diary which might turn out to be the book itself. After all, my theme is my archetypes, so it's to be presumed that these archetypes are present at all times within my consciousness—so their presence could be related to the happenings of today. I don't know exactly what I mean by this, but surely, if I watch out carefully, happenings will occur which will relate in one way or another to the archetypes? Well, for instance, the Eiger book raises the question, why does Dougal Haston appeal to me as a hero figure?

Perhaps I should loose-leaf all my material, without trying at first to connect it in any way. Just write it on separate pieces of paper and put them in a file. When I think about this problem I keep returning to the idea that the book ought to be written disconnectedly, or seemingly so. My first beginning of a draft of it seems too slick, largely because the narration is too orderly. I stopped because I was boring myself.

---

[1] *Cabaret*—book by Joe Masteroff, music by John Kander, lyrics by Fred Ebb—opened November 20 at the Broadhurst Theater; British actress Jill Haworth (b. 1945) starred as Sally Bowles.
[2] *Eiger Direct* (1966) by Haskell and Peter Gillman.
[3] *The Ascent of the Matterhorn* (1880) by Edward Whymper.

November 30. In bed this morning I had a very important thought about the book. It's this: this book is not about my father and my mother, it's about me. I mean, it is like an archaeological excavation. I dig into myself and I find my father and my mother in me. I find all the figures of the past *inside* me, not outside. I understand the importance of this approach when I think what it is that bores me about the conventional approach to autobiography. It's that the author says (in effect), "Before I tell you about me I shall tell you about them," and so the parents are forever separated from the child who is doing the telling—when, actually, the opposite is true, the parents are now only alive within the child and as a part of him. If I don't forget that I shall be able to write this book in a way which has some chance of pleasing me.

Don has been in New York since the 26th. He went to the ball, which he found tacky but very enjoyable, and he saw *Cabaret*, which he found merely awful. It is quite a hit, although the notices haven't all been good. It sounds Jewish beyond all belief and I now have scarcely any desire to see it.

Thick fog tonight. Am lazy, not wanting to work at anything. Not wanting to read, even—and certainly not to read the novels in manuscript I've been given; they bore me in advance. My belly is swollen up with gas and pressing hard and painfully against its ceiling, Paul Wonner and Bill Brown are coming to take me out to dinner and the movies, but I scarcely want that, even. Tonight I am very very old. I keep trying to fart, to relieve the gas pain. I wish, I wish ... what do I wish? That the man would come and fix the water heater.

December 17. Yesterday and the day before I worked on some polishings of the "Silent Night" teleplay. It was really an utter bore and yet I felt grateful for the compulsion to do *something*, anything. Now I'm at a loose end again. I ought to welcome this and relax, or else get on with reading the two manuscripts I am supposed to read.

Don got back from New York on the 13th and this much at least is clear, he had a much better time than he would have if I'd been along. So that's good. But his art and life problems remain. What *I* can see is that of course it would be a disaster if there weren't any such problems for him. Theoretically, I can see that about myself too, but only theoretically. Theoretically I'm sure Don sees that about *himself*.

Cabaret really does seem to be a hit, so I hope for steady drippings of money through the spring of 1967 at least.

December 18. Don said to me last night, at a dull party of dreary people, "Let's go—the night is younger than they are."

A proverb I thought of at breakfast: Miscarriages are made in heaven.

At the showing of student films,[1] the day before yesterday, was one of those men who are chronically and (as they say) unintentionally rude. His opening to me was, "When did you write The Novel?" He meant of course *Goodbye to Berlin*, which is "The Novel" simply because a second-rate play and a fifth-rate musical have been based on it. I was irritated but told him it was published in 1939. "That's a long long time," but said archly, implying that I'd better get busy and write something else. At this, even more irritated, I told him that I had published a book this year, last year, and the year before that, and that I am publishing another book next year. "You're a *machine!*" he exclaimed, mockingly. And, as I was leaving, he shouted after me, "Keep it up!" (I'm well aware how childish it is of me to let this sort of thing annoy me, even mildly.)

Swami was describing to me how he spends his days, now that he is on a convalescent routine. Instead of saying, "That's when I meditate," he said, "That's when I do my thinking." Apparently he called it thinking instead of meditating because he does it lying down. Such thinking one should be able to do!

Yesterday Don painted a really beautiful portrait of Collin Wilcox; one of his best. I keep hoping that he will go into another satisfactory work period. He denied that he really liked this picture, but no doubt that was partly out of superstition; he doesn't want to let the demons of the air hear him.

Jack [Larson] and Jim [Bridges] came to supper last Thursday. As usual Jim was silent and sleepy and seemingly bored in Jack's presence. This used to irritate me, but I have seen a good deal of Jim lately, while Jack and Don were both in New York, and now I think I understand his relation to Jack much better.

Jack, says Jim, is terribly worried about old age. He doesn't use a lot of cosmetics but he spends a great deal of time staring at his face in the mirror. Joan Houseman upset him by remarking (the bitch!) that it was good that he finally looked his age, because now one would take what he said seriously and not treat him like a child.

[...]

While Jack was away, Jim told me that he felt wonderfully free,

[1] At UCLA.

because when Jack is there he feels like a guest, he never feels that the house belongs to him. They are very formal about keeping separate accounts and each paying his share. After a quarrel it has always been Jim who has had to make the first advance and say he's sorry—but the other day, to Jim's amazement and delight, Jack said *he* was sorry.

Jack's too strict with Jim about clothes and disapproves if he dresses eccentrically. "I'm wild," said Jim, "sometimes I'd like to dress in velvet and put postage stamps all over my face."

Jim is terribly scared and worried by Jack's mental attacks. [...] But during these manic attacks, Jack also gets brilliant ideas—for example, it was during an attack that he saw how to rewrite the second part of the Byron opera.

Jim told me that at present he is "off" Gavin, Nellie Carroll and Clyde Ventura. He feels that Nellie is making use of him, particularly with regard to Miguel.[1] The other day, Nellie told Jim that Miguel had spent forty dollars on taxis. When Jim was astounded, Nellie said that Miguel's car had broken down and that of course he couldn't be expected to take a bus when he went to see his wife! Miguel has been building a workroom for Jim out in the garden. Jim had unwisely agreed to pay him by the day, and Nellie cut down Miguel's working hours by insisting that she and Miguel should make love in the mornings before he started out. As for Gavin, Jim feels that Clint [Kimbrough] has changed him—he is now so negative about everything. (Jim's specific charge was that Gavin had hated Beckett's *Happy Days*—which Nina Foch and Ted Marcuse performed at Pasadena the other day, very well—so violently that he refused even to discuss it.) Jim thinks that Gavin is terrifically impressed by show biz personalities, such as Natalie Wood. (This I don't believe.) Jim says that Gavin is such a careless writer, he never polishes anything or takes any real trouble. (This too I doubt.) As for Clyde, Jim complains that he's such a liar. I'm sure he is, but he's also truly charming, and if you ask him to the house he really sings (and dances) for his supper.

December 19. Another beautiful though slightly chilly day. Yesterday we went on the beach and something happened to my hip, I must have pinched a nerve or something. It is very painful if I make a wrong move. Christmas looms. Joe LeSueur is arriving tomorrow, Tony Richardson on Thursday—to spend a divorcee Christmas with Vanessa. Sometime in the next few days, Clint is

[1] Not his real name.

to have his screen test for *In Cold Blood*—astoundingly enough, he has made his way right up to the finals—and it is largely my doing! I dread his gratitude, if he wins—or rather, the gratitude which he will feel he must feel. How utterly strange it is, the way things of this sort come to pass! I met Richard Brooks all those years ago[1]—and what are the fruits of our meeting? That I took his house as a setting for the opening scene of *The World in the Evening*, and that I was able to call him and personally recommend Clint for the part in *Cold Blood*.

Some clinical notes: Y. and Z. have a lot of their sex in three-somes. Y. likes to jack off, but only with other people present. If Y. and Z. have had sex with someone together, then it's okay for either one of them to have that person separately later. But if either one goes with an outsider, the other is jealous.... Y. and Z. don't sleep together, and they never touch each other except when having sex....One time, when Z. had been away, Y. welcomed him home by fixing a very strong martini, taking a mouthful of it and then squirting it into Z.'s mouth. After they had done this a few times, Y. got Z. into the bath-tub and peed on him.[2]

The other day, Dr. Allen's wife committed suicide—this I heard from Gavin, who had happened to call his office. So I wrote Allen a note. Now we get a Christmas card, with a printed greeting from both of them. Inside this card there is another printed card, as follows: "Official reports of December 7, 1966 stated that my wife Jeri Selma Allen had died. However, Christmas Season and what it represents are strong reminders that the report is limited in what it is able to express. I am sure you will—in the true Spirit of Christmas—have no difficulty in understanding that this message and card which we picked out together some time ago comes to you now from both of us as always."

Danny Mann called me last night. It seems that "Silent Night" is really all set to be filmed—the only remaining decision is to be about the budget. Anyhow, my work on it is complete. And Danny is off to Austria either this week or right after Christmas. Now I have told Lamont Johnson that I want to work on Shaw's *Black Girl in Search of God*. And I am planning if possible to go to England next year and work on *The Torrents of Spring* and the adaptation of the Wedekind plays. This means that I won't be

[1] Around 1948, at MGM, as Isherwood tells in *Lost Years*. Brooks (1912–1992) was a novelist, screenwriter, director, and producer; his other films include *The Blackboard Jungle* (1955), *Cat on a Hot Tin Roof* (1958), *Elmer Gantry* (1960), *Sweet Bird of Youth* (1964), and *Lord Jim* (1965).
[2] Here Isherwood added by hand the names of Y. and Z., close friends.

going to Riverside to teach. That's partly laziness—it would be an awful sweat to give formal classes.

As for the *Hero-Father* book, I still don't know. I'm still searching for the tone of voice to tell it in. I seem to have lost my confidence in autobiography for autobiography's sake.

A mad thought—after seeing Antonioni's *The Blow-Up*[1]—that maybe I could write my first version of *Down There on a Visit* (the Mexican one) as a film script. But of course it would still have to mean something.

This morning I got a letter from Emilie Jacobson at Curtis Brown saying that *Vogue* wanted me to write a "profile interview" with Vanessa Redgrave (to go with some photographs they are publishing) "about how it feels to be in her shoes, her reaction to the limelight, the feelings of this young actress at peak fame about the theater in general, etc."

Here's my answer. Don advised me, quite rightly, not to send it, and I shan't. It is absurdly aggressive and, as Don points out, calculated to put Emilie Jacobson off even trying to find me any more job offers. I'm recording it here because it gives an all-too-true picture of the cantankerous side of me. Indeed, it is very unfair and bitchy. And yet, at the same time, it does say something I feel strongly.

".... Just for your information, not *Vogue*'s, I will explain why I'm refusing. To me, as a friend of Vanessa's, it would seem positively insulting to ask her how she feels about being a star, etc. You see, Vanessa is a professional, and the child of a professional family, and she has been an admired actress for years already. The only sense in which she is now a star is that she happens to have been hired by a big studio to engage in a somewhat dubious bit of pop art![2] If some unknown person has a lucky break, one can ask her or him how it feels—but not a professional whose importance has been belatedly recognized by a business concern, ages after everyone who cared for the theater was already aware of it. In Vanessa's case, this success is not a matter of luck, and it is insulting to suggest that it is. Of course, one doesn't expect *Vogue* to see that."

December 20. Yesterday was real winter solstice weather; thick white sea-fog, so that we could barely find our way down the Marina peninsula, to have supper with Michael [Leopold] and

---

[1] *Blow-up* (1966).
[2] I.e., *Blow-up.*

Henry [Guerriero]. The shack town was lost in fog, which made it snug to arrive and find lights and books and mobiles and drinks and a fire amidst the chilly whiteness. We talked about Costa Rica, which Michael had recommended to Gavin (without ever having been there) as the Switzerland of Central America, no Indian problem, political stability, the largest colony of North American residents south of the Mexican border, and a marvellous climate. Anyhow, Gavin is going there after Christmas, to work with the producer of the musical he is writing about Napoleon and Josephine.

With Michael and Henry there is much conversational ground which is usually avoided. One doesn't ask, "How crazy has Mikey been lately?" "How are you off for money?" "Are either of you in trouble with the police?" "How much longer do you think you can go on like this—and what do you plan to do when you can't?" Merely appearing gives both parties some reassurance, of course—this is the chief value of a visit.

I'm afraid Iris may be very seriously ill. I got a letter from her a bit less than a month ago, dated November 9, with a postscript dated November 24, in which she says:

> ... since writing it appears that I must go back to the stocks in Harley Street—I hope no slicing this time—so I shall be at Ivan's 2 Tregunter Road SW 10. Concerning death, life seems to become increasingly precious, which is unfair of it. At twenty death looks like a far, rather decorative portal, now a mean door in the hospital corridors, shut with a neat, inaudible click. Whereas one was at first so uneasy in it, with time one grows accustomed to life, like your description of driving on the freeway.★ "Now more than ever seems it rich to die." The nightingale has changed tune, and the old 'uns with ear trumpets, limping and groping, quaver "Darkling I listen"[1] with the hope that, after all, cockcrow will take over and breakfast coffee will bubble. I have just eaten an excellent fresh brown egg. The only thing is love, fresh brown love, that I miss, I never get used to the pangs of indifference. But you have a long way to go before you suffer such deprivations....

(★The reference is to *A Single Man* which Iris had just read and which had moved her to write the first part of the letter. After praising it, she said that the portrait of me in the book lacked the

---

[1] Both quotes are from Keats's "Ode to a Nightingale."

warmth which she felt towards me, as she looked back: "The feeling that we were together in one boat, becalmed, on flood, down waterfall....Your friendship was to me like a star sapphire shining above the lesser lights in California. I wear it still on my finger as I write with praise, with love....')

Some weeks ago, Geoffrey and Collin Horne decided, in their horrendous way, that it would be nice if each of the children sacrificed the belonging he valued most as a Christmas present to a member of a very poor family, preferably Watts-Negro. The problem was to find the right family. So we consulted Dorothy Miller, and she said she knew one. So the Hornes sent their presents over, and then paid the family a visit. To their dismay, Geoffrey and Collin found that this family owned a Cadillac (an old one, to be sure) and also a color T.V. set—which the Horne children had wanted but had been told they couldn't afford! However, the visit itself was quite a success.

December 21. Books I'm reading or dipping into just now: *Two Views*, Uwe Johnson (a good situation, but so stodgily written I hardly know if I can get through it), *The Last 100 Days*, John Toland (the sort of journalism that fascinates me. I feel such strong identification with so many of the characters, particularly on the German side—quite upset, for example, when Wenck is injured in the auto accident and prevented from leading the counterattack on the Russians![1] Because you know the Germans will lose, and indeed want them to lose, you get a kind of objectivity toward the whole struggle and are sad whenever the efforts of any heroic figure are frustrated.) *Elizabeth the Great*, Elizabeth Jenkins. (Although Jenkins writes well and Toland sloppily, I feel much the same about both books, strong identification. Fascinated by Elizabeth's hesitations, particularly the story of the signing of Mary Stuart's death warrant.)

Clint just called, in great excitement after taking his test for *In Cold Blood*. He thinks it went well. Once again, these hysterical warnings against telling anyone. Clint swears that somebody in the studio casting office told him Brooks actually signed someone else for the part, and then this someone talked to the press about it, and Brooks refused to have him and bought out his contract!

[1] As the Russians advanced on Berlin in early 1945, Walther Wenck (1900–1982), the youngest German general, initiated the counterattack, February 15–16, then was ordered to Berlin to update Hitler. Returning to the fighting at dawn, he fell asleep and had a spectacular crash. The attack was called off.

December 22. Am bothered by a pain through the right side of the groin, quite sharp; at times I can feel it in the hip and the back and down my right thigh. It started last Sunday, while I was lying on the beach with Don.

This morning, George Knox, the present head of the English department, called from U.C. Riverside. He wanted to know if I am coming to them for the spring quarter. I told him no. This decision is partly because of my vague intention of going to England next year, but also because of laziness and the probability of having enough money anyhow. It somehow depressed me, after I'd definitely made it, because it seems bad to start the post-solstice period with a negative decision and no particular plans. The *Black Girl* project is really postponed for a long while, because they have to get permission first from the Shaw estate. So I'm thrown back on *Hero-Father.*

A letter from Anne Geller. She has the thickest kind of Jewish skin. After having had the nerve to tell me I should give her this money from the *Cabaret* royalties (or rather, that the Frenches should) in memory of Jim, she now sulks because it isn't enough. "And this brings me to your enclosed check. Chris, I don't know what to say because my first thought was that it is a pretty bleak one. However, if you agree that it is all that your agents should rightfully release, then we'll let it go at that. And I do thank you dear, for thinking of me." Don says I should never have sent her any money in the first place.

Last night we had supper with Swami. The latest domestic crisis is that [one of the boys] has had a nervous breakdown. He came to Swami and said, "Swami, I know I have an ego and you want to destroy it, but please don't poison me." Swami told him he was crazy, but he began staying away from meals for fear of being poisoned. However, he doesn't seem violent. He has given up his gun, which he used to keep in his car, and Swami has it locked in the closet in his room until it can be handed over to [the boy]'s father, who is coming to take him away.

The fuss is still going on with Belur Math about the status of the nuns here. Are they part of the Ramakrishna Order, or are they part of the Sarada Convent in India? The girls mind this very much and no doubt it has political implications; spiritually speaking, it seems irrelevant. In one way, I must say it seems to me that Vidya's mischief making (if he really did make mischief, which I still slightly doubt) seems to have been heaven inspired; because it has indirectly caused the setting-up of this monastery in Hollywood, which may well become the safe retreat and stronghold of the boys against the bossiness of the girls. At least, in times to come, they'll have a household of their own which the girls can't interfere with.

After seeing Swami, we went on [to] Gavin's, where Jack and Jim were having supper. Clint, rather sententiously drunk, was holding forth about acting. When he was doing the test he had to read a speech from *Night Must Fall*[1] in which the word "murder" occurs. So Brooks had addressed him, and the crew, on the significance of the word, saying that murder is something extraordinary, something that no one present had ever done. He told Clint not to pause on the word, but to give it significance, etc. etc. Gavin was a bit scornful but not wanting to show this too much, lest Clint should be hurt or put off or shaken in his morale. Gavin's happiness about all this is truly touching. Jim seemed to be bitching Jack, meanwhile. We watched Truman Capote's T.V. show, "A Christmas Memory," which was quite astonishingly weak.

This morning it was chilly but fine, so we had breakfast out on the deck and opened letters and Christmas cards. A letter from Bob Regester to us both, written when drunk, and full of love. Don said, "Bob's love—where does it liveth?" and we both laughed wildly. I didn't really understand what he meant by saying "liveth" and I doubt if he did, either; but it touched off one of those subliminal flashes of understanding which are possible between intimates.

Have just finished *Two Views*. I don't think it works. And yet the idea is very good. The boy is attracted to the girl simply because he can't get at her, she is on the other side of the Berlin Wall. So he makes arrangements for her escape, and she does escape. But when she finally meets him again it means nothing to her. Either the boy and the girl should have been made far more interesting as people—so that you would give a damn about their relationship, which you don't—or else this should have been a short story. There is an evident lack of communication here, between Johnson as author and me as reader. I feel his intelligence and I am sure I am missing something of his intention. I would like to talk to him frankly about it, but of course we never could, even if we met.

December 24 [Saturday]. This pain in my hip and groin has got steadily worse. Lyle Fox massaged it, and seemed to locate it in my hip, towards the back, but it aches right through and makes me

---

[1] Emlyn Williams's 1936 play about a violent rural killer; it was twice adapted as a film.

feel sick. I have taken a codeine pill for it, but it's hardly any less. I can't see Dr. Allen about it until Tuesday, I guess, and meanwhile we have quite a social program to fulfil, including seeing Tony and Vanessa this evening and again at lunch tomorrow, going to Jennifer's with Joe LeSueur, then Swami's birthday on Monday and a "star" party for Joe LeSueur on Tuesday evening which we are giving with Gavin at our house because Gavin is leaving on Wednesday for Haiti.

Am depressed, because of the pain and because I got drunk two nights in a row, to ease it. On Thursday night I was very drunk. Don heard me say in my sleep, "I'm so happy....I like being happy." This somehow strikes me as a rather sick masochistic thing to say. It smells of self-pity.

December 28. The Christmas Eve gathering, which was to have been at that curious rather down-at-heel "English pub" The Mucky Duck, was transferred to Chez Jay's because it was too late at the Mucky Duck for them to serve fish and chips. So eleven of us ate and drank a dinner which must have cost Tony and Vanessa close on a hundred dollars—probably more, except that they saved on Rachel Harrison's meal because she wouldn't eat anything, only drink brandy and sing Welsh songs and quarrel semiseriously with Tony, to such an extent that I felt sure she was about to throw her drink at him across the table, and so kept guarding one of Don's precious mod ties which I happened to be wearing. Meanwhile, David Hemmings, looking treacherous smiling Sir Mo[r]dred to the life, in his wisps of beard, kept whispering with Guinevere-Vanessa or Lancelot-Franco Nero, as the case might be.[1] As for Rachel, when not quarrelling with Tony, she informed me that we were both geniuses, that she was a barren woman, and that she was Welsh; then turned to Franco and told him he was beautiful; then expressed her love-hate for Rex [Harrison], who smiled smugly. Don says that Rachel is his clown and that he encourages her to behave in this way.

The Christmas Day lunch was at Gladys Cooper's house, which Vanessa has rented. There really is something very sweet about her. She had taken the trouble to compose individual verses for each one of about a dozen grown-up guests—there were also swarms of children.

[1] Hemmings, the British actor (1941–2003), was appearing in Joshua Logan's film musical *Camelot* (1967) with Redgrave and Franco Nero, the Italian actor (b. 1941) who became her lover. Hemmings had just appeared in *Blow-up* with Redgrave.

[...]

The children fired rifles and guns and rushed about. After lunch, David Hemmings and Franco Nero engaged in a gun battle, trying who could draw fastest. This gradually became comically serious, though not malicious. Franco has played in Italian "Westerns" so he probably felt a professional obligation to win; and David is a compulsive competitor.

Later we took Joe LeSueur to see Elsa next door and then on to a party at Jennifer's. We are entertaining Joe because he entertained Don in New York. He is a faded prettyboy with a very thin veneer of talent. His manners are bad, he shakes hands with the left hand in his pocket, and is sulky in conversation. Jennifer had spent an alarming amount of money on unwantable presents. We got a huge candle glass on a fake marble stand and a glass bell for cheese on a fake antique pewter dish dated 1789.

On the 26th, I went up to Vedanta Place for Swami's birthday. This was the first day that Swami didn't eat in his room since his heart attack. All the monks and nuns were present, about thirty-five of them. Later I went into his room with the boys, and he talked—repeating what he'd said the other day, that one of the greatest signs of spiritual advance is when you think you aren't advancing. At the end of the talk he seemed suddenly very tired. (At three that afternoon there was a really violent earthquake jolt in the bay area; Don says it shook our house. But in Hollywood I didn't even feel it, although I was sitting on the floor.)

Yesterday evening, we had Joe LeSueur, Maggie and Michael Wilding, Gavin, and Tony Richardson in to eat some stew I'd cooked. Rex and Rachel Harrison had also been invited, but with characteristic theatrical rudeness they didn't show until later, and then of course Rachel didn't want to leave. Anyhow, today we had lunch with Tony and he told us how Gavin had driven him home and gotten drunk at his place and bored him terribly and sat up for hours—although he was leaving for Haiti in the morning. Tony said that he felt Gavin was at the end of his tether, that he had no more creative energy left and was in despair. Tony said he'd thought Gavin's last novel[1] was awful. He said he felt that Gavin was so aggressive and competitive ... Don is inclined to think that all this was an account of Tony's own attitudes and feelings, transferred on to Gavin. Tony certainly seems very unhappy on this visit, but that's probably because he is staying at Vanessa's and can't get any night fun. Yesterday he made some very strange

[1] *Norman's Letter* (1966).

statements at dinner—for example, that he worships Katharine Hepburn and Brigitte Bardot.

December 29. I have bought a humid-heat electric pad, so I'll try that on my hip for a few days before I decide to see Dr. Allen. I'm unwilling to see him, of course, because of the embarrassment of having to say something about his wife's death. If I knew him better or cared for him less, the embarrassment wouldn't be nearly as great.

Gavin has presumably left for Haiti, with or without Clint. No news of *In Cold Blood*, and now I have a growing fear that Clint won't get the part.

It is always very hard to write about such things in this diary, but I'll just record that relations with Don couldn't possibly be better as of this moment. Yesterday was a day of joy.

I have now completed nearly two months (since my return from Europe) of almost unbroken laziness. The result is that I have at least twenty letters to write, two manuscripts to read, plus the introduction which I must write for Swami's lectures.[1] And as for all the books I want to read—!

1967

January 1. Just after seeing the New Year in at Nellie's, I was hurrying to get into the Volkswagen with Don when I tripped over a gap in the sidewalk and hurt my toe. I don't think it's broken, but it's badly bruised. Furthermore, I'm starting a cold (unless the Contac stops it) and I still have the pains in my hip and groin. So the New Year seems to be starting badly. Never mind, says my superstitiousness, this will correct any possible hubris arising from the happiness of my life with Don just now, and the financial success of *Cabaret*.

What I must do, right away, is get to work. I did some token work this afternoon on the introduction to Swami's lectures. As for *Hero-Father* I still don't see how to write it but I must keep making motions of the will towards a solution. Once again I must observe the parallels between the life of the spirit and the life of art. In both cases one is saying, "Reveal yourself to me."

I was getting maudlin, because slightly drunk, the other day and talking about my death and its consequences for Don. He said, "Knowing you couldn't ever be a tragedy."

[1] *Religion in Practice* (1968).

David Sachs and Charles Aufderheide came to supper. David proposed a game to Don—it was actually a psychological test: David pretended to fall, to see if Don would move instinctively to catch him. Don didn't. I got the impression that Don didn't move because *he* knew instinctively that David was faking.

This caused Charles to confess how ashamed he was about a fight he had witnessed, some while ago. One man had hit another over the head with his crutch, and Charles said he was ashamed because he could so easily have run in and grabbed the crutch and prevented it.

Parties last night at the Rex Harrisons' and Jennifer's. Mike Nichols told me how much he admires *A Single Man*, Tony Richardson told me that I am one of his few real friends and that he loves me, Rex agreed with me that the new razor bands make it impossible for you to cut yourself. Drank enough to get depressed on, though not drunk.

January 2. Woke in the middle of last night, apparently because I wanted a glass of pineapple juice and some vitamin C tablets for my cold; actually because the Muse had a very important communication to make to me about *Hero-Father*.

The book will not be written in the first person, although it will of course be written from my point of view. The chief characters will be called Frank, Kathleen, Christopher, etc., and the words *father, mother, son* will never be used. Each of the chief characters will be observed as George is observed in *A Single Man*, by a disembodied observer, and subjective statements about any given character will be made by another of the characters, speaking in the third person. ("I used to hate her," Christopher says.) The whole book will be written in the present tense.

I have always longed to write my own version of Virginia Woolf's *The Waves*; this book will be it, I think. I see it as a collection of scenes, jumping back and forward in time. Woolf's "lifeday" progression will not be used. The end of the book is probably the visit I made to the site of Marple Hall, last year.

What is important to me in this plan is the idea of not insisting on the relationships. This sounds naive, since it will be made perfectly clear that Christopher and Richard are in fact the children of Frank and Kathleen; nevertheless I believe that this method will involve them all much more closely and organically with each other, as part of a process.

Well, enough said. Now I'll take a shot at it. As I decided back in November, I shall loose-leaf the material and not bother too much at present about the order in which the various episodes are presented.

Reading Garson Kanin's book about Maugham. He is rather a sneaky little worm, but is interesting, though depressing. Willie appears as an old bell (no, not belle!) which clangs so predictably when it is rung. Kanin keeps darting out like a naughty little boy and ringing it, and then running off again. Kanin isn't a nonvenomous worm, either. He has fangs, though he tries hard to avoid showing them.

January 12 [Thursday]. Just as I feared, Clint didn't get the part of Dick Hickock in *In Cold Blood*. It was given to a younger actor named Scott Wilson, who is said to resemble Hickock closely.

Two days ago, we had supper with Swami at Vedanta Place. He seemed much better. He told us that he has been going to the shrine in the mornings, lately, which is something he hadn't done in a long while. "I was there one hour," he told us with satisfaction—and both Don and I noticed the utter unself-consciousness with which he talks about himself. Before Don arrived, Swami also told me that, while he was ill in hospital, he had almost continuous awareness of the presence of Ramakrishna, Holy Mother, Maharaj and Swamiji. What he was saying was that they would always appear if his condition approached death, and that therefore he could never be fearful or even depressed, when he was sick.

Great happiness with Don—though he's mad at me right now for getting drunk last night. There's one scene which I can't describe here, I don't want to; we began to talk to each other in a way that was almost suprapersonal. If I write about it, I shall spoil something. I'll only record that I said something to him about our quarrels and how unnecessary they seemed, and Don answered, "*We have tiresome servants*"! It was a bit uncanny. We seemed to be talking from outside of ourselves.

Don has a name for David Roth: The Mouse of Rothschild.

Lyle Fox has got married, to a girl named Rez (I can't remember what that's an abbreviation of, except that it's a very unusual, biblical sounding name). They were married at a church in Westwood, last Saturday the 7th, and this wasn't a sudden decision, obviously. Yet, that very morning, while I was up at the gym, I talked to Lyle about Eileen and he said that it was sad and that he had really intended to marry her and that now he didn't suppose he'd ever marry anyone! That kind of secretiveness is the mark of very very stupid people, I think; it's so utterly pointless.

However, all credit to Lyle for curing, at least temporarily, the pain in my hip. He did it by giving me two exercises, which I only did once. The pain disappeared about half an hour later.

One of our gym regulars, John Hoyt,[1] has just been arrested in San Diego on the charge of having had sex with three boys from the Palisades, aged eleven and twelve. Lyle was quite decent about this. Poor man, he had been playing the local civic leader in a big way, which will mean of course that his fall will be all the heavier; probably he will have to leave town.

I must continue to record a disgusting inexplicable failure to do any work of any kind. I did make a start on *Hero-Father*, but stopped after one day's work.

January 21. Still stalled on *Hero-Father*. Something is wrong. But I'll only find out what it is by keeping on trying. Meanwhile I have corrected the British proofs of *A Meeting by the River*. Am stuck also on the introduction to Swami's lectures—not that that's so surprising. It will be almost impossible to write anything interesting about them.

On the 18th we went to a party up at Jinny Pfeiffer's, for Timothy Leary. He really is a fake. The smile on his face was so slimy that you could hardly bear to look at him. Leary is going round with a show he calls "A Psychedelic Religious Celebration"—sometimes it's about Jesus, sometimes about Buddha. Alan Watts, who was also at the party, was teasing him about this. Don overheard them.

Watts: "I don't like seeing on the marquee, Timothy Leary in the Incarnation of Jesus Christ—it sounds like Father Divine."[2]

Leary (with his coy smile): "But I *am* the incarnation of Jesus Christ."

Watts: "Oh yes, I know you are.... But there's a Zen story about the disciple who asked, 'Are not the lines of the hills like the body of the Buddha?' and his master answered, 'Yes, but it's a pity to say so.'"

(When we were talking about this at breakfast next morning, I said, "Maybe Leary really is an incarnation of Jesus—maybe it's a trick to test us," and Don retorted, "Just the sort of trick his Father *would* play!")

Then, the next evening, we went to Leary's show at the Santa

---

[1] American character actor (1905–1991), W.W.II veteran and former teacher; he appeared in *Julius Caesar* (1953), *The Girl in the Red Velvet Swing* (1955), *Spartacus* (1960), and others.
[2] Black Baptist preacher (*circa* 1880–1965), founder of a non-sectarian Christian movement, Peace Mission, which aimed to establish an interracial paradise on earth and attracted large sums of money. He claimed he was God.

Monica Civic Auditorium. This one was about the Illumination of the Buddha. To quote the ad: "Re-enactment of the world's great religious myths using psychedelic methods: sensory meditation, symbol overload, media-mix, molecular and cellular phrasing, pantomime, dance, sound-light and lecture-sermon-gospel." I wish I could describe it adequately, but I can't. There were projections on a screen by several magic lanterns, and there was singing and dancing by a group which calls itself The Grateful Dead; and Leary, all in white and barefoot, addressed us through a mike. A lot of it was asslicking the younger generation, telling them how great they were, and how free. Leary sneered at the oldlings and somehow tried to pass himself off as an honorary young man. He appealed to all the young to "drop out, turn on, tune in"—which means, as near as you could tell, drop all obligations imposed on you by your elders, take pot, acid or whatnot and thus tune in to the meaning of life. What was so false and pernicious in Leary's appeal was its complete irresponsibility. He wasn't really offering any reliable spiritual help to the young, only inciting them to vaguely rebellious action—and inciting them without really involving himself with them. But, of course, to many of the kids in the hall—and it was packed—the fact that Leary has a prison sentence hanging over him would be a quite sufficient guarantee of his sincerity.[1]

In the midst of this, a probably mad elderly woman threw several eggs at Leary and missed. A photographer (one of many who were there) marched her out. Whereupon she called the police and said he had assaulted her. So he came back down to where we were sitting—we had been very near to the woman—and appealed to Jack Larson and Don to testify that he hadn't assaulted her; which they agreed to do. (In fairness to the police, I must record that I'd expected the auditorium to be surrounded by them, but it wasn't. I saw only one officer, as we were leaving.)

January 22. We are in the midst of a rainstorm. It keeps beating in in waves, and the ocean is churned up brown near the shore.

This morning I asked Don to read the two bits of *Hero-Father*, the six pages I wrote in November 1965, and the two and a half pages of the new draft. I don't know why it is that I so often delay so long before asking Don's opinion; once again he was so

[1] Leary (1920–1996), former Harvard psychology professor, was sentenced to thirty years when he took responsibility for his daughter possessing marijuana in 1965. He appealed, on grounds of self-incrimination, and won in 1969, but meanwhile was arrested again for possession in 1968.

helpful and illuminating. Briefly, after our talk, I have come to the following tentative decisions.

The book should be written in a simple narrative style, in the first person; this third-person-present-tense style is too arty for my purpose. The book should ramble, or seem to ramble, switching from point to point in time as required. The book should be preoccupied with the concept of autobiography as myth, following Jung's remarks at the beginning of his autobiography, and there should be a lot of examples given of how myth is created out of the materials of experience. Therefore there will be quotations from my books, showing how different aspects of the myth were developed.

Obviously it will be very tricky, this "artless" arrangement of the reminiscent ramblings. For example, how soon do I reveal that my father was killed in the First World War? Probably such basic bits of information should be supplied right away, and then elaborated on and analyzed later.

One thing which Don found exciting was the idea that I really didn't know my father at all, and that the myth about him was created for my own private reasons—i.e. that I needed an anti-heroic hero to oppose to the official hero figure erected by the patriots of the period, who were my deadly enemies. Therefore it would be most interesting to show how certain aspects of my father had to be suppressed, because they were disconcertingly square; e.g. his references in his letters to "real men" etc.

Now I must make a really determined start; there is nothing to stop me from writing at least a rough draft of the whole book (it won't be a long one, I already know) before I go to England again and am able to do some more research among my mother's diaries and papers. However, before I start writing, I must plan a bit; and I fear I must reread at least parts of *The Memorial* and *Lions and Shadows*. (How deadly boring that prospect is!)

February 13. Will just start up this record again before going to the airport to meet Don, who has been staying the weekend in Santa Fe with Anthony Russo.[1]

Am still poking along with *Hero-Father* and haven't even finished the introduction to Swami's lectures.

Swami seems much better, but he has a good deal to worry him. Now, just as the new swami, Asaktananda, is about to arrive (the 17th) Swami has discovered that Vandanananda has been

---

[1] American hairdresser; longtime companion of John Goodwin.

applying to Belur Math for permission to leave here and return to India. He never told Swami anything about this. Swami says it must be because he wants to be put in charge of a monastery or center all of his own. It seems that Vandanananda had made some remarks about his dissatisfaction to one or other of the devotees. Gossip, gossip.... But what is really stunning in this situation is Vandanananda's not having told Swami when he must have known perfectly well that Swami would find out, in no seconds flat!

If Vandanananda leaves, that will mean, of course, another swami being sent out from India. (Swami says he has one picked, already); and then Swami would be left alone here with two newcomers to train. "Well, Swami," I said, "it just means you'll have to live another ten years at least!" Swami grunted humorously and protested, "Not *ten*!" But I got a feeling that he did indeed somehow accept the duty of not dying so soon.

(Which reminds me that Ronnie Knox rang me up in the middle of last night and told *me* not to die, because he needed my fatherly influence—though he didn't put it quite like that. He was very very drunk.)

A marvellous exit line: Dorothy, just as she was getting out of the car at the bus stop, two days ago: "Mis-tuh Isherwood, I hear they're going to teach Black Magic in the colleges.... but we'll talk about that next time—"

March 4. Three days ago, I finished my introduction to Swami's lectures—which means that it took me two whole months to produce eleven pages! Sheer tamas.

This afternoon, after another lapse, I have managed to get a page of *Hero-Father* sketched out. Now it's essential that I pile up pages as quickly as possible, so I reach a point of no retreat.

March 6. Resolved, to get on with the *Hero-Father* book quite recklessly, and meanwhile prepare a list of questions, which I'll keep firing at Richard—there's so much that I can't remember.

Last weekend, February 24–26, Wystan came to stay with us. He is on a reading tour, but he didn't have any reading dates in Los Angeles, so he was with us for the two nights and the whole of the 25th. He said he was tired and didn't want to see anyone, except the Stravinskys and Gerald Heard. The Stravinskys had only just returned from one of their concert trips, so Vera was exhausted and didn't even suggest our coming by for a drink—which surprised me, just a little. As for Gerald, we couldn't even reach Jack Jones, let alone Michael Barrie; anyhow I'm sure Wystan wouldn't have been let in the house, although Jack later told me that Gerald is getting better.

Wystan was wearing a sweater with the word GIMLI on it; Gimli is a dwarf who appears in the Tolkien trilogy, apparently. Both Don and I get the impression that Wystan is a little bit off Tolkien, though he didn't exactly admit this. He only remarked that the book on Tolkien which he has been writing has been held up, or maybe abandoned, because Tolkien didn't like having the sources of some of his material revealed. "That's because he's not a professional writer," Wystan said, "no professional would mind in the least."[1]

He told us, which I never knew before or had forgotten, that Virginia Woolf and Vita Sackville-West very nearly ran away together, and that there was a big scene with the husbands, Leonard [Woolf] and Harold [Nicolson], before they were persuaded not to.

He said that Benjamin Britten was the only friend he had ever lost. After we had had a lot of drinks, he said that Chester and Don and I were his only real friends—despite the fact that he knows so many people. He shed tears as he said this, which was curious rather than touching, because he recovered himself so quickly and because he apologized to me next morning, as I was driving him to the airport, for making a scene. Surely, I couldn't help thinking, a friend is hardly a real friend if you can't show emotion in his presence? Still, I believe in his emotion and in his loneliness, and I am indeed very very fond of him, and I think he is not only deeply affectionate but truly lovable and a great man. The power of his mind is absolutely amazing. Only I do wish he wouldn't say homosexuality is sinful; it is so terribly silly and unworthy of him to repeat this drivel, which was anyhow dreamed up by those beastly Old Testament Jews because they were so mad about breeding, so they could outnumber all the other tribes in the neighborhood and ultimately conquer them. Wystan was furious—quite rightly so—when "The Platonic Blow" was published in some magazine without asking his permission even.[2] And he made a very good

---

[1] Auden was writing about *The Lord of the Rings* (1954–1955) for an Erdman series, "Contemporary Writers in Christian Perspective." He destroyed the 1966 pamphlet, possibly already completed, because Tolkien angrily objected to it after Auden criticized the appearance of Tolkien's house.

[2] The pornographic poem, beginning "It was a Spring day, a day for a lay…," appeared in Ed Sanders's arts magazine *Fuck You*, March 1965; it was untitled but headed "A gobble poem snatched from the notebooks of W.H. Auden."

point when he said that the vice of modern journalism is that it tries to reveal everything, refusing to recognize any distinction between public and private—what I write for the public and what I write for my friends.

We talked about Stephen. I said that Stephen is envious of Wystan and he agreed. He thinks Stephen's real talent is as a prose writer, not a poet.

This visit of Wystan's was really a success; we both enjoyed it. Don drew him—that amazing checkered face; Don's best drawing of him so far, I think; and Wystan said he wanted to buy one of Don's nude drawings. I pointed out my favorite one to him, after he had almost decided on another one, of Mike Van Horn. (My favorite is of Larry Nichols.[1]) Am rather sorry I did this, as we shall be parting with it. For some reason, it reminds me of Puck.

Wystan seems to be drinking heavily on this tour (as well he may) but not drastically. We sent him off with a bottle of gin and a bottle of dry vermouth. He has to have his martinis.

Things-which-might-have-been-expressed-differently department: On February 16 we went with the Masselinks to have supper with Anne Baxter. She served a very rich oxtail dish and next morning both Don and I felt terribly sick. So later I talked to Ben on the phone, asking if they had felt sick too, and they hadn't. Ben remarked that this was surprising in Jo's case, because if anyone got sick she was almost always the one who did, and he added, "You can try things out on her—like they give poisoned food to the dog."

Last night, the sun set right on the end of the headland. Sad as always. The end of winter. No more ocean sunsets.

March 26. Easter Day, grey and sad. Don, feeling tense and wanting some action, has gone off to watch the hip "Love-in" at Elysian Park. I've been working. And I'm so happy to record that I have at last managed to get *Hero* going; fifty-four pages as of now. But I still have to find out what the book's about, or rather, whether all the material I have belongs to one book or two.

Saw Gerald Heard finally, on the 22nd. He speaks with great difficulty, and so indistinctly that you sometimes can't understand him. So you have to talk to him as much as possible. But he seems in very good spirits. I must say, I do see now why Michael doesn't want people around. I think he is wrong, but not all that wrong.

---

[1] A Los Angeles acquaintance, about Bachardy's age, who sat for him several times, once with his identical twin brother.

Gerald was obviously very pleased to see me.

Dodie Smith has read *A Meeting*. I feel she doesn't really like it, and Don agrees with me. She praises it of course, but says like Wystan that both characters are unpleasant—she puts it with characteristic tact, "One can't fully like them!"

There really isn't much else I want to say. I'm in a sad, slogging-along mood, but that's not necessarily bad. I feel old. I don't want to get sick. I keep praying and praying to Ramakrishna to let me feel his presence, so I can be strengthened by it to face whatever must come. Don is restless and unhappy, and we still haven't heard from the Redfern Gallery about a possible show. I do hope that will go through; sometimes I am so worried about him, and yet I know of course that one is never really helped through circumstances, Don has just got to pull himself together somehow and stop moping and act. There is nothing I can do to help him, except pray for him.

April 7. Have kept right on with the rough draft of *Hero-Father*, now there are eighty pages. But I realize more and more clearly that I am going to rely chiefly on my mother's diaries, which means that I'll have to read through and somehow make copious extracts from fourteen or fifteen volumes at least! If only Richard would let me bring them back here with me, but I doubt if he will, and the alternative of having them xeroxed may not be practical, the books are so small.

Our plans still uncertain. No word from the Redfern about a show for Don. I still don't know if I shall do the Shaw adaptation (*Black Girl*) for the people downtown, or this film, or what. The BBC, thank God, have cancelled *The Torrents of Spring*, which I didn't at all want to do anyway—or rather, they've given it to another writer to adapt. Willie Fox must have agreed to this.

As usual, what I really want is to stay here. But I know very well the book will make it absolutely necessary for me to go to England before too long. I should actually revisit Strensall, Frimley and Limerick,[1] before writing the final draft.

Don had a dream, two or three nights ago, that Swami was trying to help him; they were meditating together. Very auspicious!

From a bookseller's catalog: "CAPOTE. Autograph postcard from Brooklyn to an intimate friend in San Francisco, approx. 36 words, signed 'Truman.' The postcard is nicked around the edges

---

[1] Towns in Yorkshire, Surrey, and Ireland, where Isherwood lived in childhood while following his father's regiment before W.W.I.

due to the usual 'tender' handling the post office gives the mail these days. Also 3 or 4 small tack holes of unattributable origin. Autographic material of Mr. Capote is rare. $15.00."

May 2. Now it has finally happened; we have some plans, or at least I do. I'm to leave for England on Saturday next, the 6th. Objectives: to see that the Mercury Gallery does something definite about giving Don a show. (This is really Rex Evans's business, I just exert mild pressure from behind.) To give interviews to the press and television about *A Meeting by the River*, which will be published at the end of this month; this is Alan White's idea, rather to my surprise, he never suggested any such thing before. To see Richard and somehow persuade him to let me take away M.'s diaries and any letters I may need, so that I can study them at leisure back here at home. To talk to Anthony Page about the adaptation of Wedekind's Lulu. To apply pressure on the Shaw estate, if that is possible, to let us have the dramatic rights to *Black Girl*. To talk to Hugh French about the possibilities of a very vague job, to do with Anastasia.[1]

I'm to stay with Bob Regester for at least a week or ten days, which will be fun, at least, it will be a lot more fun than staying with anyone else who's available.

Meanwhile, the news is definite that Gertrude Macy is going to get her forty percent, which will mean no more earnings from *Cabaret* for about five months. However, Robin French says that he believes Macy won't be able to claim any percentage on the money from a probable film sale.

I have written 124 pages to date on the manuscript of *Hero-Father*, but a lot of that is just copying extracts from my father's letters. What stops me now is that I see these letters will have to be interleaved with extracts from my mother's diaries—or from her letters to Frank, if any of these have been preserved. More and more I see that this book will probably be a much straighter kind of narrative, a study of their marriage, primarily, without time jumps and other such tricks.

Sign of the times: extracted from a brochure of a new magazine, to be called *Avant Garde*. "*Not for Everyone.* Do you like Tom Wolfe? Edwardian haircuts? Simon and Garfunkel? *MacBird!*? Art Nouveau? Cassius Clay? Pot? Antique Clothes? Michael Caine?

---

[1] I.e., youngest daughter (1901–1918) of Tsar Nicholas II of Russia, rumored to have escaped execution with her family at Ekaterinburg. Ingrid Bergman had starred in the 1956 film, *Anastasia*.

Lord of the Rings? The Pill? Did you like Andy Warhol's films before they became fashionable? We have reason to believe that you can answer yes to most of the above. If so, you are a member of that small but influential group of tastemakers who set the cultural trends of the nation. And you are precisely the kind of person for whom a great new magazine called *Avant Garde* is edited. *Avant Garde* is a lavish, mirthful, daring new magazine of inordinate savoir faire and candor.... etc."

On May 6 I flew to London. Stayed with Bob Regester. On May 18 I moved to Chester Square (because Bob had to leave his place) to stay with Marguerite [Lamkin]. From May 19 to 21 I was at Coventry, staying with the Buckinghams. Forster and Joe Ackerley were both there. I returned to London until May 25, when I went up to Disley to stay with Richard. Came back to London on May 31. Joe Ackerley died, June 4; this was the day I went down to see the Beesleys in Essex. I returned to California on June 9.

As I didn't have my Olivetti with me there's no record of any of this trip. I'll write down a few memories as they occur to me, maybe.

June 15. I can at least say about the London trip that my missions were all more or less accomplished. I talked to the Mercury Gallery and at least produced a definite deadlock; they demand to see some actual drawings and paintings, as opposed to photographs, before they will give Don a show—and Don says to hell with them. I gave interviews in connection with *A Meeting by the River.* I saw Richard and, without any persuading, got all the letters and diaries I wanted to bring back here. I have taken a great liking to Anthony Page and let him know that I will definitely work on the Lulu adaptation. I have pushed the Shaw estate into a deadlock position—they demand script approval, we won't work in speculation. I have examined the material connected with the Anastasia project and decided I won't get involved in it.

Now here I am, faced with a long long grind of work; to read all the diaries and letters and see if they'll compose into a book. Richard was very helpful. His memory is excellent and he read through the stuff I've already written and corrected a lot of the facts and answered all my questions. So now I should go ahead.

Don has done some very exciting paintings while I've been away. He has his show with Rex Evans in November to work for.

Gifts I brought back from London: a flowered shirt and tie for Don, plus three other ties. A somewhat psychedelic shopping bag for the Masselinks, plus a rubber grapefruit which squeaks, plus a button inscribed "Keep the Pope off the moon."

Happiest day in England: June 3, when [Marguerite] took me to Cambridge and we had a picnic in a punt and got drunk. The Backs[1] were full of punts. Undergraduates jumped into the water or pretended to fall in, with all their clothes on. The trees were enchantingly green. It was almost like being young, but better; no anxiety. We went into Corpus Christi and saw the portrait of Christopher Marlowe.[2]

The weekend with Forster and Ackerley was happy too, but I had to work hard to keep up the *stimmung*.[3] It was strange and nurserylike, sharing a room with Morgan. He enjoyed it like a child, our talking in the morning from bed to bed. And he slept so peacefully—no grunts or groans or snores.

At Coventry I also met Harry Heckford who is writing this book about my work.[4] He is a pale tall gloomy young man who at one time wanted to be a Catholic monk. His interest in my writing must come from a suppressed area of his psyche. I invited him to meet Morgan and Joe. When he arrived he was very glum and all Joe's efforts to cheer him failed, even though he drank a bit. Then May Buckingham invited him to stay to supper and he did, but remained gloomy, earning bad marks from Morgan who likes bright young faces around him and why not.

June 16. This morning I heard from Heckford, enclosing an obituary of Joe by Maurice Ashley, rather chilly and omitting any reference to Joe's two marvellous later books, so that you got the impression that he had given up writing years and years ago.

Heckford wants to know if Edward Upward "was possibly the beginning of the Twin or Elder-Brother thing, if I may put it in that way?" And he continues, "He seems always to have gone just that much further than you; he believed in Mortmere to a greater degree than you, and he joined the Communist party whereas you kept your head. Yet he hasn't been half as creative as you. He was the out and out revolutionary and you were convinced by him as far as I can see, of the need for complaint and change. But why should you feel compelled to acknowledge his lead? It was as if he made you feel guilty about having a comfortable life."

---

[1] I.e., the back sides of the seven Cambridge colleges along the River Cam, with rear lawns running down to the water's edge and up the opposite banks.
[2] Isherwood's college, also attended by Marlowe.
[3] Mood.
[4] Henry Heckford's book was never published.

I tried to answer this in a letter to Heckford this morning, pointing out that what I really felt guilty about in relation to Edward was the charge of frivolity, unseriousness and the inclination to show off. *Now* I *know* that I am serious, even though my manner is unserious and even though I do admittedly show off. But I feel I somehow wandered away from answering the question.

It was actually from Edward that I heard about Joe's death. He came up from the Isle of Wight to see me in London on June 6; we had lunch with Olive Mangeot and then went for a walk. Edward talked about his new novel, the second volume of the trilogy, which he has almost finished—at least to the point where he will show it to me. It was only at the end of our long walk, just as we were approaching Cresswell Place again, to have tea with Olive, that Edward remarked that he had read of Joe's death that morning in the paper. I hadn't seen the notice and since Edward had never known Joe he hadn't thought it worth mentioning earlier.

Yesterday I started work on the first of M.'s diaries, 1883. It is going to be a long haul but fascinating, especially if I do a good deal each day so as to keep a sense of the continuity. I am typing out anything which seems even remotely of interest. This book will end up longer than a Dickens novel.

One of the things which strikes me about the first diary is that it isn't nearly as "period" as one would have expected. There are very few period expressions and few details which would seem dated, even twenty or thirty years later.

I ought at least to start messing a little with the Wedekind material; roughing out a translation just for my own use. Even if the job comes to nothing, it wouldn't be too much effort to do that, and it's the only way of finding out how I really feel about the plays. I have to discover a tone of voice.

Thick glum fog all today, though we had been promised afternoon sunshine.

June 20. On the 17th I went to Vedanta Place for the Father's Day lunch. Usually it's unpleasantly hot, but this year it was so cold that I was afraid Swami would get a chill, sitting outdoors. But he had put on thick long underwear and, he told me, taken a split of champagne in the bathroom before emerging to face the cameras of the devotees. He is looking wonderfully beautiful and calm and silver and happy nowadays, and his health seems quite good although he has to have a prostate operation in August. In the middle of lunch he turned to me and said quietly but with great emphasis, "Give Don my love," and he took my arm and pressed it, as though he were actually shooting it full of love for me to pass on to Don as a transfusion, later.

That evening we had Jennifer Selznick and Ivan and Kate Moffat to supper. Ivan gave a very funny imitation of Paddy Lee-Fermor[1] teasing Richard Burton, who was making a pompous speech. Ivan also recited a sort of prose poem about his return to America as a G.I. after the war and his feeling of joy in the innocent and un-damaged life of rural America and his feeling that it was just about to burst forth into a postwar flowering. This prose poem was false and embarrassing because it was so obviously a rehearsed effect. The evening seemed strained anyhow; Jennifer was restless and somehow excluded, and the conversation wasn't general although there were only the five of us. However, when I talked to Jennifer on the phone later she said they had all agreed on the way home that they had had a marvellous evening because Don and I were so full of "abstract love" towards them—whatever that may mean. We are in fact very fond of Ivan and Jennifer and prepared to be very fond of Kate when we know her better, but there's nothing abstract about that.

John Rechy came to supper on the 18th, and was very friendly and full of quite entertaining talk about himself—the way he smiles at himself in the glass after working out until his muscles are all pumped up; and the sleeping pills he has to take and the ear plugs he has to wear because of his neurotic sensitivity to barking dogs, he strains his ears to hear them, and finds out the phone numbers of their owners and calls them, etc., etc. He puts on a great show, in his tremendous and rather sympathetic eagerness to please. I do hope his new novel isn't as awful as he makes it sound.

Last night we had supper with Chris Wood, who seems to be seriously considering returning to England. He says however that he would never leave here until after Gerald's death (Gerald has had another stroke and couldn't be visited at all, last week). Also, he doesn't want to take Paul Sorel to England and he doesn't want Beau [his dachshund] to go through their quarantine. We talked a lot about Joe Ackerley, of course. It seems almost certain that Joe didn't suffer at all. He was certainly very lively at Coventry, drinking a lot but not excessively. He had complained to his sister Nancy of pains in the chest; but when she found him it was in the normal attitude of sleep, he didn't seem to have tried to get out of bed.

A letter from Richard; he says that he doesn't think the cute

---

[1] Patrick Leigh Fermor (b. 1915), British travel writer.

Danish son-in-law will leave the Bradleys' house for some time, so they are all very crowded. He was taking them over to Wyberslegh "where there is more room to move about"—but I'm quite unable to picture what they would actually do there. Richard tells amazingly little in his letters. For instance, I suppose he has gone back to drinking. All the time I was with him in Disley he didn't drink a drop, and neither did I; but not one word was said about this.

Speaking of Disley reminds me of another happy day, or rather morning. It was on May 29—there was a break in the rainy weather, just for a few hours, and I walked up to the Bowstones,[1] probably only a four or five mile walk but a long one for me after all these years in nonpedestrian Los Angeles. I was exhilarated by my own energy. "Such bloom rose around and so many birds' cries, / That I sang for delight as I followed the way."[2] But was I really attending to the scenery? No. I was mostly thinking about Don—wishing, in a way, that he was with me, but even more that he could see me striding along and laugh at old Dobbin feeling the Spring in his bones. I laughed to myself, picturing him laughing at me.

June 22. The prospect of these letters and diaries is quite discouraging, though I enjoy actually reading them. Today it was far more of an effort to get Granny Emmy's letters into (more or less) chronological order than to read through Kathleen's diary for 1891, as far as September 2, when her ex-beau, Anthony Thornhill, got married and Kathleen kept the leaf of the quotation calendar for that day: "Who seeks, and will not take when once 'tis offered, / Shall never find it more." No doubt this seemed extra tragically ironical to Kathleen because it's from *Antony and Cleopatra* (II.vii)!

As for the texture of Kathleen's life as a girl of 22-23, it seems to be made of dances, country house visits, gallery going, lecture going, church going, cricket watching, novel reading, letter writing, shopping—all somewhat compulsive. Most compulsive of all is a trip abroad, with catalogs of "sights." But then, who can write of such trips but D.H. Lawrence? Nevertheless, all this while, Kathleen was getting on with her drawing and sketching and watercolors.

Poor Bill van Petten is in St. Joseph's hospital again with his detached retina. He seems wonderfully cheerful, says that he sees visions with his bandaged eyes, a vast ebony dome scarred by fires,

---

[1] Two waist-high columnar stone markers on the Gritstone Trail, probably the shafts of a pair of Anglo-Saxon crosses destroyed during the Reformation.
[2] From "The Child's Grave" by Edmund Blunden.

perhaps sacrificial torches. But now he can't go on the trip he had been so greatly looking forward to, with Artie and another teenager, to Outer Mongolia.

Last night I read from the Gospel at Vedanta Place, for the last time till fall. A very big audience and the feeling of the end of the season and the hammy performance of a classic, like an old singer singing *Carmen*; everybody knows it by heart and enjoys it simply as an occasion.

Richard Charlton, Anita Loos's friend, took us to lunch yesterday and offered Don a show at the Phoenix Gallery and me two movie jobs, on a story about Suzanne Valadon[1] (for Bardot!) and an adaptation of C.S. Lewis's *The Screwtape Letters*. Does Charlton really mean business? We shall see. Am about to reread the Lewis book, which intrigues me, rather.

The checks from *Cabaret* are now stopped until Getrude Macy has been paid off. Robin French says this will only take three and a half months—provided the show keeps doing huge business, as at present.

A card from Bob Regester this morning. He is now in the South of France, apparently with Neil. He has already been in Turkey, which he says was "interesting." I suppose the shooting on *Light Brigade* must be over by this time.[2]

Being with Bob this time in London wasn't such a success. No doubt because Neil wasn't there, he was inclined to be bossy, and he showed a kind of possessiveness by harping on the theme that I am a hopeless alcoholic. If I did drink rather a lot it was because I was so depressed on first arrival—and of course the drinking led to deeper depression. I made a bad beginning by drinking heavily on the plane as well as taking Librium and Dramamine; in fact, I was so dazed on arrival that I walked right past the customs inspection, unchallenged! Then I drank all that day, but nevertheless woke up at 4:30 a.m. So I got a bus to Covent Garden and roamed around and ate three breakfasts, one of raw bacon and tea at the Garden, one of sausages at the West Kensington Air Terminal, and one of fish cakes at South Kensington Underground Station—all these from nostalgia rather than greed. The result was that I was so utterly exhausted that I slept right through a visit from Jeanne Moreau, whom I *would* like to have met, the next evening. She

---

[1] French painter (1865–1938), model and friend of Renoir, Degas, Toulouse-Lautrec and others. Salka Viertel had asked Isherwood to work on a Valadon film with her in 1958; see *D.1*.
[2] *The Charge of the Light Brigade* was filmed in Turkey, where Neil Hartley was assisting Richardson.

and Bob put on the record player in the living room and danced for hours without waking me.

At a dinner party at Patrick Woodcock's on the following evening, much fun was made of Stephen and his indignant statement about the CIA and *Encounter*; the general feeling is that he knew all the time.[1] Rosamond Lehmann said that Stephen "does everything double," he knows and he doesn't know, simultaneously. She also said that John Lehmann's autobiography is no good, because it isn't frank. Lionel Trilling said that he doesn't like Forster's work as much as he used to.[2] Rosamond, looking like a delicious suet dumpling covered with powder, said that Forster is not a great writer because he is prim and schoolmasterish and doesn't know about Love. She is such a cow. I like Mrs. Trilling better.[3] She had been to one of Timothy Leary's sessions and had reacted much the same way as we had. Patrick Woodcock (just off to watch them making *The Charge of the Light Brigade*) looked at my tongue and said there were no signs of cancer. Nevertheless, it bothered me all through my visit to England, and still does. Was rather surprised to find that Bob does not like Patrick, says he is a terrible bitch and gossip, never to be trusted.

July 8. We just heard that Vivien Leigh is dead. Don is very sad, and I am sad too. She was curiously lovable—I mean, as a mere acquaintance. I suppose it was because she seemed so vulnerable.

On one of the walls of the Canyon channel is written: "Good morning Sandy. Have a nice day today and every day." This part of the wall is right opposite the back of an apartment house. What makes the inscription charming [is] that it must surely have been written one night to surprise Sandy when he or she looked out of the window next morning.

When I saw Jo yesterday she said that she doesn't much care now if she does leave Ben, everything has been ruined between them. I wish I had written an account of the scenes on Wednesday and Thursday of last week, right after they happened; now I'm

---

[1] For years, leftist friends warned Spender that the CIA secretly funded *Encounter*. He resigned as co-editor May 7, 1967, telling *The New York Times* he had been unable to confirm the allegations until a month before. Patrick Woodcock's dinner party was May 8. See Glossary.

[2] Trilling (1905–1975), prolific literary critic and professor of literature at Columbia, wrote *E.M. Forster* (1943), which Isherwood read in November 1959. See *D.1*.

[3] Diana Trilling (1905–1996), also a literary critic. She reviewed *Prater Violet* with enthusiasm, as Isherwood tells in *Lost Years*.

disinclined to go into great detail. But it is a marathon sulk on Jo's part and a marathon drink-in on Ben's. Don says that of course if Jo had any sense she would allow Ben to spend two nights a week away from home, visiting Dee [Hawes] or any other girl he wanted. This would be the way to torpedo the Ben-Dee affair. But, as we both know, poor old Jo hasn't any sense. She is playing outraged American womanhood, at sixty-five (or more). She actually said to me, "What do guys think they are, that they can treat girls like that!" What is so dreadfully sad is that Jo is really being confronted with her own old age and oncoming death; Ben's behavior has merely caused this to happen and it's not what she is ultimately upset about. Naturally, Don identifies more with Ben and I with Jo. And yet Ben's act as the youthful prisoner of marriage in love with a girl of his own age is pretty grotesque too, in its own way; after all, he is well into the change of life. He is grotesquely self-conscious and literary about the affair, even when drunk (as he certainly was when he talked to us on Thursday). He has chosen, after months of reflection, apparently, two adjectives: *famous*, to describe Dee's personality, and *precise* to describe her physical movements in bed and elsewhere. (She is a dancer.) Jo was astonishingly indiscreet about her relationship with Ben—in a way I cannot imagine being to her—saying that they hadn't had any sex in years and that anyhow he was "a lousy lay."

Tomorrow they are coming to supper together. Will they talk about it? And if they don't talk about it, what *shall* we talk about?

July 11. We did talk about it, but only in twos, Jo and Don in the kitchen while he was cooking, Ben and I out on the deck while we were barbecuing swordfish steaks. As soon as Ben began to talk, it was obvious he was very drunk; I hadn't noticed it before. He burst out into a tirade against all the women, Dee and Jo included, who were nagging at him for drinking; "Hell, I've been drinking for twenty years, they never even knew I was drinking until I told them so myself!" He told me that there had been a terrible scene the night before, because Jo had found something he had written about her a year ago, discussing the difference in their ages. Ben's constant complaint was that non-writers don't understand "us writers"; "God damn it, when we write, that's how we keep ourselves from having to go to a psychiatrist and pay twenty dollars an hour."

Then, toward the end of supper, while I was talking to Jo, she suddenly began to tremble and then stiffen, and then, sobbing, she told me in a low voice, "Something's happening to me, I can't hear what you're saying." Ben jumped up and stood over her, with guilty concern. Then they went out on to the deck together. Jo came back very brave, smiling. It was nothing, she said. But soon they left. Next day, Jo called and said she must have been a little drunk, she thought; and she added that she really ought not to drink at all, "in the state I'm in now."

It is so horrible, being with them. Jo's jealousy and general despair is like a terminal disease. I guess it really is terminal, for she can't snap out of it. And [she has a friend], apparently, [who] is in pretty much the same state about [her husband]; the only difference is, [the friend's husband] has been having these affairs since they married.

I keep thinking, what a lesson to me. And I needn't preen myself on having cured myself of the disease. The Henry Kraft days aren't so far behind.

Total happiness in the Casa right now. Three nights ago, we had supper alone together, out on the deck, for the first time this year. (Maybe that's partly why I have a snivelly cold.) A very thin new moon showed itself over the hill for about fifteen minutes, just as we sat down to eat. "That's a good omen," Don said.

Last night we had supper with Jinny Pfeiffer and Laura Huxley. Laura had a whole lot of criticisms of her book[1] from the publisher, so I found myself promising to read it through again and consider each of them carefully. Jinny is concerned about Juan; he is bitter because he feels the anti-Mexican prejudice in his school. Jinny thinks he feels the lack of a father. Juan was even impressed by the political success of Julian Nava, just because he's a Mexican and has been elected to the L.A. Board of Education.[2] She says Juan murmured to himself, "He's my dad." So now she wants to get Nava to come to the house. This sentimentality makes Don indignant. "What the hell does it matter if someone has a father or not?"

We got home to find a red toy fire chief's hat in the mailbox. Round it was a paper ribbon inscribed: "Far bright. Star light. Bright land. Over hand. (When the moonlight falls upon the water). Ian Whitcomb's birthday tranquil, from Nancy." And another: "Happy birthday Ian Whitcomb. Happy birthday. You are a picnic. Happy birthday from Nancy Jane." This is "flower children" talk. On July 6, the day Ian Whitcomb was coming to

---

[1] *This Timeless Moment.*
[2] Nava, born in Los Angeles in 1927 to Mexican immigrant parents, became a history professor and was later ambassador to Mexico.

us to have supper and see [the movie] *The Kid from Spain* at Royce Hall, this girl Nancy called. (Ian *must* have told her that he was coming to us, but he denied this.) So I told her he was coming at seven, and she suddenly appeared, from the desert, with two or three other "flower children" including a middle-aged man. They had a skillet and they had come to cook hominy grits specially for Ian—because I suppose they thought that he would never have eaten them, being British. Ian was much embarrassed and told them to come and see his show and cook the grits on the stage, so it would be a happening.

July 19. Alarmist reports are issuing from the radio—10,000 or maybe 100,000 hippies are expected to descend upon California in the near future; and this, say certain doctors, may start a series of epidemics, because the hippies have syphilis, gonorrhea and hepatitis!

Have just written a letter to Glenn Porter (Chandala, as he now calls himself) saying that I won't endorse the Vietnam Summer antiwar project.[1] I rather hate to do this, but I have such a deep-down feeling that the whole Vietnam antiwar movement is something I must keep away from. Why do I feel this? On the rational level, I'm opposed to the kind of "just you hit me!" attitude which is designed to put the police more and more hopelessly in the wrong. (The latest idea is to use only mothers and children in these demonstrations.) I suppose it may have been better justified in India, where the population was subject to a foreign power, the British, and just dumbly obstructed that power as a last resort. But in this country the obstructors aren't the helpless victims of imperialism. If the police push them around they are damn well going to sue. God knows, I hate the police as much as anyone could; but I also know in my calmer moments that this is no way to deal with them, this will only alienate them more completely and drive them to identify themselves with the forces they represent—the money interests which actually want the war to go on. Whereas the police, and other such agents of authority, should ideally be seduced from their allegiance to the war-making forces; that's obvious.

Oh, I know very well the arguments on both sides of this. You can say, a peace demonstration of this kind is always used by those who merely wish to shake the foundations of the administration

[1] A nationwide movement started by Martin Luther King and Benjamin Spock in April 1967.

in any way possible—that is to say, by the extreme Right and the extreme Left, in all its aspects. And to this you can answer, what do I care *who* uses this demonstration? Its aims are aims I believe in, and if I suddenly find myself with allies I don't trust that doesn't invalidate the aims.

But, below the rational level, I am aware of another feeling; as a pacifist I must deny the rightness of every war, even the most apparently righteous ones. This war is too obviously unrighteous— indeed it is even politically deplorable, a mistake to be corrected as soon as possible, even from the point of view of the administration. Therefore objection to this war is primarily a political objection ... Is that mere dainty mindedness or indeed utter double-talk? I really don't know. I only feel ... And I will add this, I believe Aldous would have agreed with me. And Gerald Heard.

(Gerald is still alive but very weak. Chris Wood sees him, and Peggy Kis[k]adden. Michael never suggests my coming but maybe I ought to try to see Gerald again anyway; it is so hard to know if he really wants to be visited, any more.)

Jo is now concentrating on Ben's alcoholism. Dee has gone behind Jo's back and told a member of Alcoholics Anonymous to call Ben on the phone. Jo is naturally furious. [...]

We saw Cocteau's *Beauty and the Beast* and *Orpheus* last night again. No film creates a greater air of magic than *Beauty*; particularly the dining room where human hands hold the candles and the beautiful dark faces of the live statues on the fireplace breathe thick smoke and follow Beauty with their eyes. And then there is that unforgettable last long shot in *Orpheus*, with Death and her assistant Heurtebise being led away among the ruins by the two motorcyclists, under arrest.

July 31. Herb Compton, the young librarian from Sydney, Australia, who came to see me on the 28th, told me how happy I seemed. I said, I have every reason to be happy, and that is true. This time with Don, these last few weeks, have been among the happiest I have ever spent with him. Enough of that, I'm superstitious....

These are the projects which I might conceivably be involved in, sometime in the nearer or farther future: a film about Rimbaud and Verlaine for a producer named Kenneth Geist, an adaptation of *A Meeting by the River* which Jim Bridges thinks he wants to make and wants me to help him with, a film version of Tennessee's "One Arm" with Jim Bridges directing, *The Adventures of the Black Girl* (which now seems a little more likely to be okayed by the Shaw estate) and the expanded musical version of *Dogskin* made by Ray Henderson and to be directed by Burgess Meredith—oh, and I forgot the Lulu adaptation which I have now promised to do with Anthony Page, in London next year.

Dogskin is a bore, because Meredith and Ray want me to take the responsibility for it, and I don't want to because after all it is really Wystan's creation and I am very dubious about the lyrics which Ray has written and fairly dubious about the music. (Meredith seems to like it, however.) I have just written to Wystan trying to get him to at least listen to a tape or record of the music, so I don't have to be the one to say yes or no.

I do like Meredith. His story of how [Tallulah] Bankhead made a date with him at her apartment and received him naked. So he dutifully went ahead and "pumped her to the best of my ability" until she hastily warned him, "Don't come inside me, darling, I'm engaged to Jock Whitney."[1] He also says he told Michael Arlen, "There's nothing wrong with your stories that penicillin couldn't cure."[2]

Last night Don got a traffic ticket, the first in a long while. He was terribly upset, simply because it brought him into contact with a cop and he feels that cops are evil. At the same time he remembers that Vivekananda said that there is no such thing as evil. What can one do in order to understand this truth and live by it? Most of us, when bitten by a poisonous snake or spider, would be inclined not to feel hatred, saying that it was the creature's nature. Should one try to feel this about cops? But then one is denying their status as human beings.

August 2. Today at one we were supposed to visit Gerald, but this morning Michael called and said he was too weak for us to come. He told me that Gerald had recently remarked that, as you get to the end of your life, you realize how very seldom you have been kind to people. I'm not sure I would say that about myself; I think I have done quite a lot of "favors." With me, the trouble has been that I have usually done the favors without benevolence, because I felt myself pressured into doing them, emotionally blackmailed.

We both begin to wonder if we shall ever see Gerald again. Jack

---

[1] John Hay Whitney (1904–1982)—sportsman, financier, film producer, philanthropist, publisher, one of the richest men in America, from an old New England family—married twice but had only romantic liaisons with movie stars.
[2] Arlen (1895–1956), British-educated, Bulgarian-Armenian novelist, playwright, and screenwriter of cosmopolitan, debauched bohemia, wrote the 1924 bestseller *The Green Hat*; Bankhead starred in the stage adaptation.

Jones has become much more outspokenly critical of Michael and now gives us the impression that Michael has become really quite pathological about visitors, except for the ones he can't prevent from coming. Jack says that Gerald told him that he would like to see Don and me but that he didn't feel strong enough to go against Michael's wishes.

Don said at breakfast this morning that he is so happy with me and with our life together now. I feel the same way, but it is so important to remember that what is alive and flexible is also subject to change—change is a sign of emotional health. Therefore all statements and facts of this kind are merely to be recorded as one records the weather. Which doesn't make it any the less marvellous when the weather is fine!

A talk about the respective lives of the businessman and the artist, arising out of a slick but well-made film called *Hotel* we both saw last night. Is the artist really freer than the businessman? Is he really more creative? I said yes of course, and gave the illustration of two men being sent to work for the day; one returns with a painting and the other with a hundred dollar bill. But Don, who really thinks much more clearheadedly than I do, took this illustration to pieces and made me doubtful.

The night before last, we had supper with the Masselinks and [a friend], who has left [her husband] at least temporarily, moved out of the house and started looking for a job. So they all talked frankly and I must say Jo revealed truly shocking depths of self-righteousness and self-pity. Ben seems cowed but aggressive, and I think he's still seeing Dee. [The friend] has a lot more style. Of course, she has been living with the knowledge of [her husband's] sex excursions for many years. [...]

I have now been working on Kathleen's diaries and Frank's and Emily's letters, making excerpts from them and copying, for almost seven weeks and I've only worked through eight diaries, with twenty-three more to go. Not to mention all the mass of letters, which I have hardly begun to cope with yet because they belong to the later years of the period. I still have really no idea if this work will produce a book which can be published; maybe the material, when I have all of it, won't seem to be of sufficient general interest. I don't care. I'm really quite enjoying this work.

The point is made that the Detroit riots were not just Negroes "out to get whitey"; whites as well as Negroes were sniping at the police, the soldiers and the firemen. As J.D.R. Bruckner says in the *Los Angeles Times* Sunday opinion section: "It was full scale urban warfare conducted by the alienated and dispossessed against the society in which they live. And that society's government re-sponded to this threat with unbelievable violence.... A myth being propagated now about this war is that it was a kind of madness.... but it was not madness from the viewpoint of the dispossessed; to them it was a great happiness. Huge crowds of people gathered to watch and cheer the fires during the first two days ... and when a fireman tried to put out the blaze they threw rocks at him because he was interfering.... At least the snipers were selective.... The guardsmen were entirely indiscriminate; they shot up everything in sight."[1]

"As of mid-July, fifty-two percent of the American people said they disapproved of the way President Johnson is handling the situation in Vietnam. Only a third expressed approval." Gallup Poll.

August 4. John Rechy and Gavin to supper last night. It wasn't a satisfactory evening. We couldn't talk to John about his court case,[2] because Gavin wasn't to know about it, and we couldn't talk to Gavin about Clint's moving out, lest Gavin shouldn't want to do so in front of John. Gavin simply told us that Clint is look-ing for another place to live, and that was that. And, to make matters worse, my steaks on the barbecue turned out tough and Don's vegetables were undercooked and his dessert was frozen! If Don and I weren't in such a harmonious phase the evening would certainly have ended in gloom and recriminations; as it was, we laughed about it and went to bed and slept eight hours.

John talked at great length about his relations with his mother. He is apparently very frank [...].

[...] Jo says that she feels she and Ben are going to break up; he is still seeing Dee and Dee has "changed" him—for example, he now drives like a demon and curses other drivers!

The night before last, I had supper with Jim Charlton. His fate seems really to have taken a turn for the better. He's off to Honolulu to settle down in partnership with this friend he likes who's an architect, and there seems no reason why he shouldn't make lots of money and lead a happy life surfing and cruising and lapsing gradually into tropical inertia and middle age.

Reading M.'s diary gives a very strange sense of the inner nature

---

[1] A police raid on a black after-hours club, July 23, triggered five days of violence and looting; 2,700 businesses were sacked, 1,300 buildings burned, 5,000 people left homeless; and there were 41 dead, 347 wounded, and 3,000 arrested.

[2] For a sex offense; see Glossary.

of a life—just because it is written from day to day. Or perhaps it would be truer to say that it gives you a sense of the utter mystery of a life when it is viewed from very close to the surface. The mystery is in the lack of meaning. Everything Kathleen does seems compulsive. Does she really give a damn about all those art galleries and churches she visits? Does she give a damn about horse racing? And how about the dancing and the flirtations? She obviously finds some meaning in nursing Emily, partly because she loves her and perhaps even more because this represents self-control, self-sacrifice etc., which is her compulsive religion. But how much of the Christian ethic did she really believe in? And how squeamish she is about sex! Everything has to be *just so* in the relations between the sexes, otherwise Kathleen starts to get outraged or shocked. And yet at the same time she was an extremely perceptive person in many ways. Sometimes in later life she used to talk as if she viewed the whole of human activity as a sort of masquerade. Then, dismissing it all, she protested with genuine indignation that it would be "unfair" if there wasn't an afterlife in which she could be with Frank! At such moments she seemed to regard God as a hotel manager.

Talking of God, my morning japam gets more and more and more empty—or rather, fuller and fuller and fuller of myself and my daily preoccupations. My only grace is that I remain quite clearly aware of this emptiness. I still believe that it is only a phase; and therefore I wait apprehensively for the shock of fear or misfortune or loss which will bring me to my senses. It'll have to be a terribly big one!

August 8. Now at last the weather is quite beautiful, though windy. We lay on the beach on the 5th and saw Ted in the distance. He was lying reading (in itself unusual) and his legs were displayed in a way which would be perfectly normal for many people but seemed quite uncharacteristic of him—they were stuck up in the air, one crossed exhibitionistically over the other. I commented, "He'd never lie like that unless he was really absorbed in his book," and Don agreed, "Either he's absorbed or he's gone mad." The strangeness of this statement coupled with the fact that, in Ted's case, it was literally true or at least could be, made me laugh madly. But Ted didn't see us.

Yesterday morning we went to visit Gerald. He is very weak now and speaks very low, though his articulation seems better. You had the feeling that he was "high" for one reason or another. He seemed hugely amused by us, we were maya, absurd and trivial and yet absorbingly interesting to him, "and it doesn't alter one's affection," he added, "indeed it increases it." (I'm paraphrasing this because you couldn't properly understand everything he said.) He kept laughing, wildly but with a weirdly genuine amusement; it made me think of Ramakrishna. I have a feeling we shan't see him much more. He is very thin and tiny, but seems relaxed and perfectly happy. And yet—what an awful *toil* dying must be! Michael is a wizened little old man.

As I'd expected, Gerald was very eager to hear about Joe [Ackerley]'s death and he launched at once into some sort of scientific explanation of Joe's cheerfulness that last weekend, coming after his long years of depression. (Actually I don't think Joe *was* really cheerful, only extra stimulated by our company; but there was no point in saying that to Gerald.)

Charlie Locke showed up out of the past from New Jersey and made the afternoon very long, with the assistance of his wife and daughter. Not that I don't like them, but the meeting was symbolic, merely.

Yesterday evening we had a really very pleasant dinner with Horst Buchholz and his wife. Mrs. Buchholz (Myriam) seems a rather usual refined Frog but Horst has a proletarian jolliness, he is still a little Berlin hustler at heart which makes him quite charming. He showed us a couple of card tricks, showing us how to do them. One—"Posko Piati kennt alles"[1]—was his special secret and he said he would never have told it if he hadn't felt a "special affection" for us; Lee Thompson, the ex-alcoholic movie director[2] was there with his wife, and Thompson had been best man at the Buchholz wedding. Myriam encouraged her skinny Frenchified children to get Don to draw animals for them and to insist that he sign them! Just in case!

Ken McDonnell, Larry Paxton's half brother, whom I met just that once at Larry's funeral in 1963, writes again this morning out of the blue:

Dear Chris, I hope you do not think it too presumptuous of me to write you this letter after so long a time. The beauty and compassion of our one brief meeting has often filled my thoughts with fond memories of you.... It's funny how time

[1] "Posko Piati knows everything"; the sequence of ten letters—each appears twice—is a system for laying out twenty cards. The viewer chooses one, the cards are collected and laid out again, and the sharp can identify the card.
[2] John Lee Thompson (1914–2002), British actor, playwright, screenwriter; he directed *The Guns of Navarone* (1961) and two *Planet of the Apes* sequels.

would have me write this letter to you, but I suppose I could never forget the strength you gave me on that morning of sorrow. If, Chris, I may ever do anything at all for you, whatever it may be, please feel free to ask. With fond regards....

This makes me want to go to San Francisco. Maybe I will next month.

A brief quarrel with Don last night because he accused me, as so often, of "buttering him up," a phrase which never fails to annoy me. But we were soon friends again.

August 13. A beautiful day but we haven't been down to the beach, because it's Sunday and so crowded. As I bang away, copying out extracts from Kathleen's diary and Frank's letters, I realize more and more clearly what a gigantic job this is going to be. But it is quite fascinating, provided one doesn't press it—especially now that Frank and Kathleen are lovers, more or less; Kathleen is still so cagey one can't be sure of what she feels.

Yesterday we took Andee Cohen out to supper; she has just arrived home from Europe. She seemed foolish and pathetic and was maybe a bit high, at least at first. She has probably lost Willie,[1] [...] Andee was full of apprehension of world-doom and told us that one of her astrologers is so scared that he has "gone into hiding."

Andee also told us how her friends in Europe regard America as being already in a state of civil war. California is certainly no kindergarten, but both Don and I were surprised by something he saw the day before yesterday—he went to draw Mrs. Reagan (for this *Harper's Bazaar* assignment) and found their home absolutely unguarded;[2] this was the house just off Sunset, near Pacific Palisades. From the outside, Reagan himself could be seen lying by the pool. "I could have shot him," Don said, "without even ringing the doorbell." He found the family atmosphere completely false and dead; the Reagans and their children seemed to be giving a performance of "American family life." Then some visitors came and Reagan took them into the den. Don heard him tell a story using the word nigger.

August 25. From a brochure put out by the Egg and I Gallery: "The

[1] James Fox.
[2] Ronald Reagan was eight months into his first term as governor of California; Bachardy had been commissioned to do portraits of twelve Los Angeles celebrities.

formal arguments of this young Californian are direct, disarmingly simple. The clay forms of Carol Fumai are vaguely totemic and ambiguous, the surfaces warm and penetrating."

From an article in the *Los Angeles Times* of August 13 by Ian Nairn, a British architectural critic, called "The Sordid American Landscape":

> This is an environment of total confusion and mediocrity. And after a few miles the driver no longer cares whether he is in Maine or Texas. Yet stop the car and walk a few yards into the trees and you are in primeval America.... Yet this land is mostly inaccessible—and hence pointless. Urban wilderness and rural wilderness meet head-on without benefit to either; the environment has no structure at all.... The first thing that ought to be said about this disintegrated landscape is that it is not due to the pressure of population. In Western European terms the United States has no suburban problem at all; compared to the density of England and Wales—which still has some of the world's greatest rural landscape—the whole population of the United States could comfortably be fitted into Texas.

Today we went on the beach and in the water; the weather is glorious. I plod on with Kathleen's diaries and Frank's letters, which are very good indeed from South Africa; but don't get much other work done. Don is to leave on Monday for a few weeks in New York.

Gerald still very weak but no worse, it seems. Igor has been in hospital with a bleeding ulcer. Both Jo and Ben Masselink are going to psychiatrists. I keep on going to the gym but am still too heavy—153 lbs.—and my stomach bulges.

A correspondent writes this morning to ask what the pylorus is. He found it mentioned in *A Single Man* and insinuates that he doesn't believe there is really any such word.

August 29. Don left last night for New York, dashing for the plane laden with a portfolio of drawings, a covered hanger full of jackets and pants, and a couple of camera bags containing whatever else was needed. There was no time to park the car, much less to eat anything. I didn't even leave the driver's seat to say goodbye, because I was parked in a three-minute zone. This morning he called to say that he has got Maurice Grosser's apartment again, which is a huge advantage. Before he left, he had the usual acute doubts about the necessity for the trip. But this is a transition period for him and the only way to get through it is to keep knocking on doors, trying paths to see where they lead, etc. etc.

It is very hot. In town they have had a smog alert. Here i[t] is bearable and indeed wonderfully beautiful. I am typing this sitting out on the deck, a quarter past four and the sun is very hot on my back. At about six I shall go down to the beach and have a sunset swim. This house is so delightful now that all the trees and shrubs have grown tall and spread themselves. It is like a house in a wood. And the view from the windows of the ocean and the hills has not changed since the old days. I only wish I wasn't so obsessed by a sense of change—I suppose it is characteristic of growing old. But it is true that the noise has increased greatly. Daily sonic booms, and heavy backfiring trucks thundering down our hill from half past seven o'clock on.

We spent my sixty-third birthday going to a play (*Pantagleize*[1]) and two movies (*The Family Way* and *Hell's Angels on Wheels*). We also went to a very indifferent Mexican restaurant near Watts, recommended by Billy Al Bengston. But it was a happy day for me. And being sixty-three is certainly no burden as of this minute. My life with Don seems, as of this minute and indeed of the past couple of months in general, to be in a marvellous phase of love, intimacy, mutual trust, tenderness, affection, fun, everything. We have plenty of money and more to come, presumably, very soon from *Cabaret*, which has nearly paid off Gertrude Macy, according to Robin French. My health is good—am only concerned about the thickness of my waist; I seem to eat very little but it won't shrink any more. And I am very lucky to have work to occupy me for many many months ahead. What is bad, as of now, is my apparent spiritual condition—I say apparent because who knows what his spiritual condition really is. But I do "keep the line open" and try, throughout the day, to make acts of recollection. I am of course terribly uneasy about my "worldly" happiness; fearing to lose it and yet knowing that of course it will be necessary to lose it before I can find *ananda*. (Having said this, I suddenly ask myself if I'm not suffering from the delusions of puritanism; "sacred and profane love" and all that very unjazzy jazz. How *can* love be profane if it really is love? In my own case, hasn't my relation with Don now become my true means of enlightenment?)

Talk with Gavin on phone. He is off to Ingrid Bergman's birthday party. Shall he give her the new life of Sarah Bernhardt or a bottle of champagne? Does she read? Does it matter? She certainly

[1] A comedy by Belgian playwright Michel de Ghelderode (1898–1962).

drinks. But is that an argument? We decided on the Bernhardt.

Clint has found an apartment. Jo is going up to see her family in Oregon, without Ben. Ben suddenly has a lot of T.V. jobs.

August 30. The heat this morning was almost overpowering. (A breeze is getting up now, thank goodness.) After going out to get my cap put back on by Dr. Kurtzman I felt as lazy as a hog and lay drowsing on the couch in my workroom or in the sun on the deck.

Last night I had a dream in which Don reproved me for being so fat—and no wonder, after the delicious Mexican dinner I ate with Jim Bridges[,] plus four margaritas! Jim talked about our project, the *Meeting by the River* play, and again I had the impression that he really wants to do it.

Before going out with Jim I had the impulse to walk down to the beach shortly before sunset and swim. It's only at the height of the summer that this is really beautiful, the water creamy and bluish and the naked bodies shining-golden in the setting light. With Don it would have been a perfect moment, and even alone I was feeling nearly ecstatic, when who should appear but poor shriveled tearful old Jo. So we two old creatures greeted each other and hugged, eyed with bored faint distaste by the teenagers, and then Jo sat down and poured out her tales of Ben's cruelty; the horrible things he keeps saying to her all the time now. So the golden moment was lost. I do feel genuinely sorry for her, though, and I suppose I shall have to speak to Ben while she's away in Oregon.

Poor Gavin has had an abscess in his upper jaw and must have at least one tooth pulled tomorrow. But Ronnie Knox is staying with him—hiding out, in fact, because he owes five hundred dollars and some debt-collecting company is after him. The girl who called me the other evening, posing as a girlfriend of Ronnie's, was actually a spy of the company. How low can you sink? Gavin had been in agony all through the Bergman birthday party but had nevertheless enjoyed himself and had, as he said, "fallen in love with Cary Grant."

August 31. Shortly after ten in the morning. Am sitting in the gorgeous tiger-striped robe Don gave me for my birthday with a T-shirt under it to stop me from sweating coffee into its sleeves. (He also gave me a British queer novel called *The Ring*, which is no good I'm afraid—the sweetness of the gift was the trouble he must have taken to get it for me; I can't imagine how he did this unless he coaxed it out of a New York client of his who had a copy.) I am out on the deck. It isn't too hot yet and it may not be so hot today because there are a lot of high clouds. There are bees in the grape ivy and too many helicopters in the sky; they make more noise than the jets, with their motorbike clatter.

I wish I could describe how I feel. I'm trying to do so by setting the scene. But whenever I try to watch myself like this, the myself slips away from under and eludes me. (One of those slim graceful blond teenagers has just passed by along the road below. What is *he* thinking? Or rather, what is he feeling? It's the feeling I'm trying to get at.) In general, my feeling is daze, shot through with minor anxieties. For instance, when I hear voices below, especially young voices, I look down because I'm afraid that someone is about to come up the hill by our stairs. Why do I object to this so strongly? All very well to talk about trespass and not wanting to have people wandering around and maybe poking into Don's studio—that's not it. My deep objection is due to a fear of encroachment by every-one and everything which represents the external world; strangers, helicopters, high-rise buildings, the telephone company and its eyesore pole outside the window. I have such a sense of increasing pressure, the expanding population pushing us into the sea. Then why stay here? Not all places are like southern California; indeed, if you object to population-pressure you could hardly choose a worse one. I don't know why we should stay, except that we both love this house. We'll probably remain in it until circumstances pitch us out. (On T.V. last night there was a documentary about earthquakes, in which it is stated that southern California must expect a major quake any time!)

All right, enough for now. Must get on with Kathleen's diaries and Frank's letters. I made Jerry Lawrence laugh last night when I said, "I'm afraid this is going to be just another *War and Peace*." We swam on his beach which is full of rocks, both fixed and loose. One of the loose ones cut my foot twice, coming and going.

Jerry, as usual, had his exhibits: a young beachcomber-actor and a Hawaiian waiter at La Mer where we ate, who is a sword and fire dancer. What is Jerry's life all about? How good does he think his plays are? Does he see himself as a prophet of civil liberties? Is he a millionaire? Is he lonely? Does he have a religion? Does he really enjoy all this sex? Does he value objects? Are his friends just for display to other friends? Does he resent not being listed as a celebrated person in the *Information Please Almanac*? I find I can't answer any of these questions, and it is my fault that I can't. What am I doing, seeing someone even as often as I see Jerry and not finding out more about him? My lack of curiosity, however you look at it, is bad.

It's starting to get very warm and heavy. I'll move inside.

September 2 [Saturday]. It's two o'clock already. The whole morning has passed talking to Richard Pietrowicz, who came by to give me the revised manuscript of his novel *Patria*. He says he has improved the writing but I doubt this; I doubt if he knows how. And yet the story is really moving and the setting is impressively lifelike and the characters are quite adequately described; it would make a good movie, but a very expensive one. Richard is married now; a situation which one feels he has well in hand. He doesn't strike you as the sort of boy to let himself be pushed around by his wife. He is Polish by origin, which perhaps explains why he likes to drink and can drink a great deal and doesn't have hangovers or feel guilty about it. (Of course, this is what *he* tells me.) He said he would be drunk this evening, to celebrate having finished the novel. "I never drink when I'm feeling depressed." He is getting tired of his job with social security and is considering moving to Alaska. He can't get promoted here because the social security people are making it a policy to favor Negroes until they have readjusted the racial balance! He looks young for his age, twenty-seven, with a trim figure, quite goodlooking, very clean and soft-spoken, in neat dark pants and a very white T-shirt. I talked a lot, as I am supposed to, jumping from subject to subject; but it was always fundamentally about myself, and at the end of it all I felt rather bored with myself and my act. Rather but not very, and not really apologetic—because, after all, if you want to catch my act *and* drive clear across town to do it, then at least it's up to me not to disappoint you.

I met another ex-student yesterday evening on the beach when I went down for a late swim; the evening was cloudy but warm. I think his name is Ken Gross. Anyhow, he was at Los Angeles State College and he more or less appears in *A Single Man* as "Wally Bryant." He recognized me and came over to talk, in a matching beach outfit, blue shirt and trunks, grinning and sparkling with discreet indiscretion. He told me that he had a lot of stories about me which he wouldn't dare to tell me unless he'd had three or four martinis. This may have been a hint, and I did consider inviting him to have dinner with me—but no. Instead I went to the gym, then to eat at Johnson's Barbecue, then to see *You Only Live Twice* which consists almost entirely of explosions of various sizes, plus gunfire, karate and the pressing of buttons which open trapdoors to throw you into some compromising situation. One charmingly old-world touch, a villain who sits stroking a white cat. But even the cat got killed in the final explosion, I fear.

It now looks like I am definitely going to San Francisco next Friday, the 8th, to stay with Ben Underhill and see Ken McDonnell.

Swami, whom I talked to on the phone, says that Amiya was only here a few days. She returned to England because Swami's doctor told her her liver was in a very bad state, and since she had been able to bring very little money with her owing to currency regulations it seemed better for her to be treated at home. However, when she got back to London, her own doctor told her there was nothing the matter with her. But he said this without making an X-ray examination, and now Swami has told her to insist on having one. Swami is still waiting to hear the results of this. He says Amiya was drunk all the time, and had already started to make trouble, at Vedanta Place—something to do with Sarada, which I didn't follow. I am to see him on Wednesday.

Mike Steen has been in New York and talked to Tennessee about the "One Arm" project and Tennessee seems favorable. He and I talked on the phone. But I still have my grave doubts. Tennessee says (according to Mike) that he is delighted I want to work on the picture and that I'm the only one he would trust not to falsify it because I'm the only one who has never lied about homosexuality in my writing or my life, etc. But he sounds as if he could easily be talked out of the whole idea, or made suspicious of me; and I suspect that *he* may be afraid (grotesque as this sounds) of being compromised. Anyhow, I have said definitely that I'll do it, *if* it is clearly written into the contract that Jim Bridges is to be the director. Tennessee says (again according to Mike) that he would like to see the screenplay after I've done a rough draft and maybe make some suggestions. I know perfectly well that this means he will then get a half-credit—but why not?

The evening of the day before yesterday, the Masselinks, Peter, Alice, Ann and Marylee Gowland, along with Ann's daughter Tracy, and a girl friend and Bill Reid[1] and his [...] son all got together for a picnic (as they called it) in the park at Inspiration Point (as the city still dares to call it, after failing to stop those swine from putting up The Penthouse apartments to block off the hills, the ocean, the inspiration and the whole point of the point).

---

[1] The carpenter and contractor Isherwood and Bachardy hired to improve their first house, 434 Sycamore Road; son of silent film star Wallace Reid (1891–1923); mentioned in *D.1*.

Most of us wanted to go down on the beach but Peter Gowland insisted on staying there and eating at one of the long wooden tables, surrounded by dozens of elderly people at the other tables, many of them German or Jewish or both, who were sitting down to quite elaborate continental meals heated up on the barbecues. "I like to be among people," Peter said. Marylee said, "It's like communism," which I thought was a quite brilliant description of the atmosphere; it was that of a Workers' Park of Rest and Culture somewhere in East Europe. *Our* atmosphere, which we brought with us, was actually far worse—Jo about to leave home alone for the first time, Ben sulky-guilty-drunk [...], Marylee also sulking (she has moved out of the house), Alice sweetly sad. I do like her though. I don't like Bill Reid, never have. And he added to my depression by telling us that the projected skyscraper at the end of Adelaide isn't actually scrapped; all the metal has been bought and processed and the plans are complete—all some investor has to do is buy them and go ahead.

Books I'm reading: *The Dharma Bums, A Mummer's Wife,*[1] Sartre's *The Words, Turbott Wolfe,*[2] *The North Country* by Graham Turner, *The Ring* by Richard Chopping and Max Beerbohm's *Seven Men.*

September 7. Have just talked to Jo on the phone. Ben did leave home while she was away—she was the one who insisted on it—and has got a place of his own in Venice and is seeing Dee all the time. Jo has told very few people about it yet. I told her that she must; it will help if people know. But I quite understand how she hates to admit that Dee has taken Ben away from her—for that's how she privately sees it; the alternative would be to admit to herself that their dear little cozy play-relationship, with Jo so bossy and helpless, was a mess which was steadily getting messier and messier.

Swami goes into the hospital on Monday for his operation. I saw him last night. He is very cheerful and sweet and so marvellously sane. It seems odd and empty in the dining room without the boys; they are eating in the monastery now.

Dorothy, who came yesterday, praised a paper towel roll which has an orange pattern of leaves and curlicues printed on it. (As a matter of fact, I'd bought this roll by mistake instead of the plain white one we usually get, seen the curlicues with dismay and hoped to use the roll up before Don returns.) Dorothy said, "It's good you

[1] By George Moore.
[2] By William Plomer.

got a little bit of art in the house"; then, perhaps sensing that she hadn't been entirely tactful, she added, "other than the pictures." Now that Dorothy's nephew [...] has definitely split up with his wife, the wife has gone back east with the children. Dorothy went down with them to the depot. Just as they were about to get on the train, a man appeared in the crowd who looked rather like [her nephew]. His wife, who is an affected bitch, pretended to faint. Dorothy hit her as hard as she could, and she stopped pretending. The redcap was shocked and asked Dorothy why she had done it. Dorothy said, "If she pulls any more stunts like that I'll kill her."

Don Coombs, who has just got a job as a librarian at UCLA, had a friend who was dying of cancer. A woman faith healer claimed that she could cure him but only if he were her husband. So the friend married her and died anyhow, four months later.

I'm to leave for San Francisco tomorrow afternoon.

September 16. The visit to San Francisco was very enjoyable, largely because it was a holiday from every sort of a situation; I felt no emotional tension whatever between Ben Underhill and me, or between Ken McDonnell and me either, for that matter. We just had a good time—Sausalito, the Frank Lloyd Wright Marin County Civic Center (which I instantly named The Palaz of Hoon, after Wallace Stevens's poem[1]), the delicious margaritas at Señor Pico's and Trader Vic's, the Avalon Ballroom with its light effects like an old silent movie flickering, the topless boy dancers at 524 Union St., Sam's [Anchor Café] at Tiburon where I had a late lunch with Ben and Kin Hoitsma and the Muir Woods [redwood forest] which I visited with Ken.

Ken is powerfully built, plump, spotty, very intelligent and sweet. Despite the way he came on in his letters he is very reserved under his love-the-world manner. He kept repeating that this weekend was one of the most memorable of his whole life, because he had not only remet me but also a girl he'd taken acid with a year or two ago and "come closer to" than anyone he'd ever known, even Larry [Paxton]. He indicated that both these remeetings had been a huge success because they had both proved to him that he hadn't been wrong or deceiving himself about the girl or me the first time. But saying all this somehow didn't say very much.

Dear Ben's chief fault—and he is such an admirable person in so many ways—is that he cannot resist making social combinations. The first evening he stuck me with a dreary yakking drunk,

---

[1] "Tea at the Palaz of Hoon."

a real estate agent […]. Then, next morning, he had me come with him to the airport, to witness a strange kind of funeral scene; the boyfriend of a very rich man being extradited from the United States as an undesirable alien. Before boarding the plane (to England) he was taken into a room by a plain clothes detective and fingerprinted, so that his departure could be officially established. The boy had protruding teeth and a nose which was only half the normal length, but he was powerfully attractive. I didn't feel that he was going to have much trouble putting down new roots in England. (I did my bit to help by giving him Bob Regester's address.) The rich man was very drunk and the last moment of parting was too painful to watch—the two made an involuntary movement to kiss each other and then stopped short, exactly as if they had bumped their noses against glass, no doubt because they thought the detective might be somewhere around. But I couldn't help thinking why didn't they do it anyhow, just to show him and all his foul tribe; and why, for that matter, didn't the rich man drop everything and fly with the boy to England?

After these experiences I was very firm with Ben, telling him that I wanted to spend the time alone with him, or with him and Kin Hoitsma. We found Kin in a typically fixed-up apartment, decorated with sections of the American flag and large metal instruments and implements which looked as if they belonged in a torture chamber; actually they were harmless enough in their proper places—a rake, a scythe, a big hook, etc. With Kin I had a sense of distance and defensiveness; but this must have been due to Cecil [Beaton]. I tried, probably not successfully, to rebuild relations between them a little; each thinks the other has been cold, unreasonable, lacking in understanding.

Yesterday evening I chaperoned Jo to a dinner party at Bill [Brown] and Paul [Wonner]'s—Jim Gill and Antoinette were there too. It was a flop. Antoinette has been playing in T.V. all week and she was exhausted and silly. Paul, who had just been told by me about Jo's tragedy, was embarrassed and silent. Bill, who of course also knew, awkwardly tried to make things go. And, as always happens, there were these grotesque unintentional acts of tactlessness—Antoinette saying "everybody seems to be breaking up" and Bill loudly praising a group called The Doors and insisting we hear one of their numbers called "The End"! Jo got very weepy again as I was driving her home.

September 25. Jack Larson is back from New York and Don may be coming back Wednesday or Friday; I'll know when I talk to him tomorrow.

At long last I have worked through 1900. I shouldn't complain of the length, though (thirty-five pages compared with twenty, for 1899!) because there are these valuable drafts of Kathleen's letters in reply to Frank's, which reveal so much more than her cagey diary entries.

Now it really looks as if *Black Girl* may start late this week or next. Lamont Johnson wants us to begin.

An interesting evening with Byron Trott on the 23rd. His talk about the San Francisco scene. The Beats are an older generation. The Hippies are older than the Flower Children(?). The Diggers[1] look after the Flower Children, who can't look after themselves. David Roth made a social error when he came down here and met Byron and his UCLA friends, because David arrived dressed as a Hippie, long-haired, bearded, dirty. Byron and his friends were neat. I can see that Byron now rather looks down on David. Byron is very ambitious. He wants me to help him get a job in films. He is much impressed by McLuhan[2] and the fascination of computer programming. His most admirable quality is an apparent utter lack of self-pity, although he seems as badly crippled as ever. (When I asked him about his health on the phone he said he felt great. At the same time he quite objectively admits that he loses jobs because no one wants to hire a boy on crutches.) His body, aside from being bent double, is slender and well made. His face is becoming rather beautiful, but with a hint of ruthlessness.[3]

A capsule description of our epoch, which came to me as I lay in bed this morning: McNamara[4] or marijuana, Maharishi or McLuhan, Black Power or Flower Power, Reagan Now or Pay Later.

Jo and I went to hear the Maharishi Mahesh Yogi on the 21st. Jo had heard him a few days earlier and had been terrifically impressed—partly of course because this was practically her first

---

[1] A Haight-Ashbury movement providing free food and shelter. See Glossary.
[2] I.e., Marshall McLuhan (1911–1980), Canadian-born professor of English literature and communication, author of *Understanding Media: The Extensions of Man* (1964) and other bestsellers. San Francisco hosted the first McLuhan Festival in 1965.
[3] Trott (b. *circa* 1944) had rheumatoid arthritis and ankylosis spondylolisthesis, suffered chronic degenerative joint disease, and eventually required repeated hip replacements and other surgery. He later settled in San Antonio, Texas, where he founded the first Gay and Lesbian Community Center.
[4] I.e., Robert McNamara (1916–2009), secretary of defense under Kennedy and Johnson, 1961–1968, during the Vietnam build-up.

contact with ideas of this kind, and now she needs them. I was quite favorably impressed—I mean, I don't think he's an out-and-out fake. But his talk was most confusing. What he calls the Self can't be the Atman, for you are supposed to realize it right away, as a first step, almost. And yet if it isn't the Atman what is it? The suspect things are, his claim that "nowadays in the jet age" we can accomplish in five years what it took generations of holy men whole lifetimes to accomplish; his plugs for his schools of meditation, where all the teachers have been trained by himself—how is this possible? But he is fat and cheerful with a very attractive laugh or hiccup. He twiddles a flower, which must surely be artificial it seems so tough, all the time he is talking and he sits backed by a whole bank of flowers. Also he has an amusing mannerism of being seemingly unable to pronounce the word "relative" which he uses very frequently, with "absolute"; they come out as "relatew" and "absolew" approximately.

Swami's back home. I talked to him yesterday. I also talked to Gerald, though I couldn't understand his voice on the phone, it was so weak. This was a bit of bitchery on the part of Jack Jones; he let Gerald speak into the phone while Michael was out of the room. As before I got a great impression of Gerald's cheerfulness. When I told him how sick poor Igor is, he did manage to whisper fairly clearly, "Dying by inches," with some of his old relish. Gerald wanted to have Jack and me have supper together at his expense. I sidestepped this by having breakfast at Jack's house this morning; a very gracious meal.

For the first time in ages a very slightly better "sit" than usual this morning. Why? Largely because I did what I have been told one should do when alone, chanted the mantram aloud. Also a simple but helpful thought came to me, and I prayed for those I love *and for those I hate*. That's to say, instead of trying to think yourself into some sort of tepid benevolence, step outside yourself for an instant and pray, with no strings attached so to speak, for those whom Christopher hates. This made me feel good and, more important, truthful. I *can* pray for Peggy [Kiskadden] sincerely, but I can't sincerely say I love her. If I tried to, the prayer would automatically grow a tail on it—and, Oh Lord, please make her less of a bitch and turn her into the sort of person Christopher could approve of. Or—help me to love Peggy by making her more lovable.

October 17. Despite all my resolves I've neglected to write anything for nearly a month. So—

Am now into March 1901, and have sent off the first 107 pages of the carbon copy to Richard. The period just before Frank sailed for South Africa for the second time took a lot of time to cover because they corresponded such a lot, nearly every day.

Have also finished a rough draft or rather cut of Shaw's *Black Girl* in the form of a play. A girl named Kathy Delancey is taking it down from the tape I made.

Also Ray Henderson is working on the musical of *Dogskin* and we have sessions together. I am beginning to feel that maybe this will really turn into something, if Burgess Meredith keeps his enthusiasm and Elsa doesn't bitch it all up.

Jim Bridges on the other hand doesn't seem to be getting on with *A Meeting by the River*. If he really doesn't want to do it after all, I think I'll have a try at it myself, in consultation with Don. Because I do feel that something could be made of it.

It is very hot and dry, after a week of miserable cold fog.

Dorothy, who came yesterday to clean house, talked about [...] her nephew, and his wife [...]; they have split up. "My father he used to say, you can live without anything you wasn't born with, and he was *not* born with [his wife]." "[My nephew] says to me, if I was in hell and she was in heaven and could get me out, I'd rather stay in hell—so that means he *does not* care for her."

October 19. A surprisingly interesting meeting with Larry Holt yesterday. I only agreed to have tea with him because he had landed me this one thousand dollar lecture date at Long Beach State College, but now I realize that he is someone I would like to get to know better. I was simply being put off by externals, his bossiness and his beaky Jewish face, hollow-eyed air of tragedy and strutting pouter-pigeon walk. As a matter of fact he isn't really bossy or tiresomely tragic or even oppressively Jewish. He talked fascinatingly about Maupiti in the Leeward Islands northwest of Tahiti. The love relationship between men is officially recognized. It is called *tane-tane*(?).[1] You go to the *karuna*(?)[2] who is a psychic or seer and you drink from the same gourd and he tells you the word for love which must never be spoken by anybody else. Larry did this with the son of the chief, when they were both in their late twenties. The karuna had told this young man that someone like Larry would come to the island and enter into this relationship with him, but that he would leave again. Larry left after being on

---

[1] Literally male-male, evidently a pidgin-English formation from the Tahitian word *tane* for male.

[2] Possibly formed from the Polynesian *karu*, referring to part of the eye, with *na*, of or by.

the island for ten months, because the life there seemed "too easy" and "unreal," and because he was engaged to a girl in Austria. (Later she was shot by the Gestapo.) The son of the chief had been educated at the University of Hawaii before he met Larry. During World War II he joined the French navy and was drowned in a submarine.

Larry told me how, in 1948, he had been with Swami and I had called on the phone and Swami had turned to Larry with his eyes full of tears and said, "Chris's friend is dead." When Larry told me this I was completely at a loss; I just could not think of any friend who had died around that time. But when I mentioned this to Don he said at once "Denny,"[1] and of course that's who it was. It's really rather shocking that I shouldn't have remembered—but I mustn't think of this, because such thoughts are merely vanity. *I* am not my tired old clogged-up memory. I must not be vainly ashamed of its lapses.

Yesterday Jo went to see Swami and he gave her some instructions about meditation. She is still terribly weepy. One sees the appalling power of self-pity, the disgusting obsessive shamelessness of it. Jo hopes subconsciously that we shall report on her condition to Ben and tell him, "You *can't* go on doing this to her." Which is actually why I have such a terrific disinclination to call Ben. One time I did and he started going on about his guilt. "You have to live with the guilt." At first I thought he was saying "the Guild" and supposed he was mad at the screenwriters' guild! (God, what an object lesson poor old Jo is! How she keeps reminding me of my great self-pitying period in 1939-1940, when Vernon [Old] said to me, "I can't be sorry for you because you're so sorry for yourself.")

October 21. I called Ben Masselink yesterday—this was for the third time. He didn't answer, thank goodness. But I suppose I shall have to go on trying.

Last week I suddenly had the impulse to begin planned meditation again—that's to say, meditation according to the instructions Swami gave me all those years ago. I don't know how long it is since I dropped it, but ages certainly. Now I find it incredibly difficult; there is a terrific resistance. Making japam seems delightful by comparison; such a relief.

Blue movies last night at Jerry Lawrence's. Don, Jack [Larson], Jim [Bridges], Gavin and I were all rather disgusted by them,

[1] Fouts; see Glossary.

and I think a lot of the others were too. Sometimes they seemed positively absurd. You thought, why in the world would anyone want to do *that*? Improperly color-photographed, the cock can look really revolting, so raw-red, even a bit reptilian. But of course the truth is that these movies were inept beyond belief. The actors hardly responded to each other at all, and there was no buildup to the beginning of the act, and never any indication of an orgasm. The masturbation movie was the only one which hinted that the act might be even the least little bit pleasant; but it seemed such endless toil.

The evening before last was much more agreeable. I went to see Byron Trott and his friends at the house they have just moved into on Culver Boulevard. They all seem to have so much fun, with their psychedelic pictures which change colors under the black-light lamps, and their costumes, and their motorcycles. A photo of Bob Dylan with a real hypodermic needle stuck in its arm. The color-changing sign which says LOVE. The Buddha in which, after they had bought it, they found a coin slot. Bill Loskota, the medical student, and Bill Kincheloe, the art student,[1] have ridden down Sunset in costumes as Batman and Robin. Byron goes out in the jacket of a suit of evening tails.

This afternoon, Jack and Jim have just called in great excitement because they have bought a house; 449 Skyway, which was designed by Frank Lloyd Wright. I am almost sure this is the house which Jim Charlton took me to see in 1948 when I first met him. In those days the roof kept leaking, despite everything the owner did to stop it—and Wright is supposed to have answered his complaints by saying "Why don't you just relax and enjoy the house?" The place is said to be in bad repair, and Jack and Jim have got it for only $60,000 with all the furniture, including the furniture which Wright had made for it. (I seem to remember that this furniture was top-heavy and unpractical, and that the owner had had to fix extra supports on to the chairs to prevent them from tipping over. I didn't tell Jack and Jim this.)

October 30. This morning Jo says she is going to Hawaii at the end of the week, "to look around." She will see if she can find work for herself there. Alice Gowland is going with her for a short while, just to keep her company. Meanwhile Ben has written Jo a letter saying how beautiful their life together used to be. At the

---

[1] Undergraduates at UCLA; Loskota was later an anesthesiologist at USC School of Medicine, where he joined the faculty.

same time he assures me (over the phone) that he couldn't possibly go back to live with her, is in fact going to try living with Dee. His letter was apparently just written to relieve his guilt. But at the same time he is obviously enjoying his guilt quite a bit.

The day before yesterday I spent the whole day in bed, apparently as the result of eating some bad fish at Jack Allen's restaurant. Being sick has become quite a strange experience for me in these last years and I find I don't enjoy it any more. I am too old for it. Don was adorably sweet and kind and seemed to get real pleasure out of waiting on me, but I felt uneasy. This is no longer my scene. The days of Sudhira are over definitely. How one changes! Perhaps it's just because sickness could now so easily be something terminal, and I don't feel ready for that. I feel, in a way, less ready than ever before, despite all my prayers for help at the hour of death. A terminal sickness means the breaking of your will and the thawing of your heart, and my will has never been tighter nor my heart harder. It is only in my relationship to Don that I can still feel vulnerable, anxious, pliable. Well, that's a mercy at least. All the more so because, most of the time, he is so sweet and loving. His sweetness only makes me love him all the more—another mercy, certainly, that I'm not in the least afraid of being loved.

On the other hand, my meditation is dry as the desert. There is a kind of paralysis inside me. I just cannot keep my mind on formal meditation—the omnipresence of Brahman, goodwill toward all men etc. etc.—for even a couple of seconds at a time. Only making japam gives me a sort of cowlike contented ease. Provided I don't keep at it too long.

One is one's reverie, at any given time in one's life. Mine seems to get more and more possessive; money, comfort, reputation, possessions—and anxiety when these seem threatened, anxiety and rage. It is true that I am pretty contented with what I've got, but that's no great trick. I am awfully lucky, and correspondingly anxious because I know my luck can't hold.

A big brush fire, somewhere out beyond Malibu, is blowing smoke over the sea.

October 31 [Tuesday]. When it got dark you could see the flames clearly. They looked so near that I called Paul [Wonner] and Bill [Brown] to know if they were all right.[1] Today the fire is said to be almost out but the air is full of smoke-haze.

We had supper with Gavin. He praised Rechy's *Numbers* very

---

[1] They had bought a house in the hills above Malibu, on Rambla Orienta.

highly, but he hadn't yet read as far as the part about himself. We didn't mention this, but are both wondering how he'll take it.

On Sunday, after being sick the day before, I happened upon this, from Shaw's preface to *The Doctor's Dilemma*: "Use your health, even to the point of wearing it out. That is what it is for. Spend all you have before you die; and do not outlive yourself."

A man named Ronald Platt called this morning from New York, he wants to be the producer of "One Arm" but doesn't want Jim Bridges as director. He had broken his leg out fox hunting because he or his horse was kicked by the horses of Mrs. William Randolph Hearst Jr. He was indignant because she hadn't written him to apologize. I told him that was because she was afraid he'd sue her if she admitted it was her fault. I also told him I won't work on the film without Jim. In the morning mail came a letter from Mike Steen, evidently afraid I may make a deal with Platt before he gets back to New York.

Meanwhile Jim seems to be getting on with *A Meeting*, and Ray is nearing the end of composing for *Dogskin*, and nothing has been heard from the network about "A Christmas Carol."[1] In London tomorrow there will be a meeting of the representatives of the Shaw estate and we hope they will soon come to some decision about *Black Girl*.

I forgot to say that James Fox (in town for the weekend) came to supper on Sunday night with Andee [Cohen]. They are said by the press to be about to marry but they didn't tell us so. James was going back next day to play Gordon Craig opposite Vanessa Redgrave in a film about Isadora Duncan. He looks older, somewhat lined and ravaged, but still romantic. He was wearing an extremely mod shirt with ruffles from Mr. Fish (I think that's the name of the shop in London).

November 2. I had tea with Larry Holt again yesterday. He was concerned because Swami, when he appeared on Les Crane's show[2] on October 24, had publicly admitted that he has had the lower form of samadhi. True, this admission can have meant practically nothing to the vast majority of the viewers. Most of the few people who do vaguely know what samadhi is toss the word around as though the experience were quite usual. But Larry was worried because, as he rightly said, Swami used to be cagey

[1] Hunt Stromberg Jr. (1923–1986) was producing a T.V. adaptation of Charles Dickens's novel. Isherwood worked on it until Stromberg found he could not assemble the cast he wanted.
[2] A late-night T.V. talk show on ABC.

about mentioning spiritual experiences. He cited the example of Ramakrishna, who only revealed himself fully at the end of his life. Is Swami planning to leave the body, Larry wanted to know.

So I asked Swami this right out, when I went up there for supper later. And of course Swami simply laughed and then apologetically explained that he had felt obliged to speak of his experience in order to make a distinction between that and the experiences of a taker of lysergic acid or hemp. "But I didn't describe the experience," he added. People who keep imputing deep spiritual intentions to Swami understand him very little, I feel. As Don says, "He works on automatic, most of the time." Of course one may believe—in fact I think one has got to believe—that his actions are often inspired. But I don't think he is aware of this—I mean, I don't think he *chooses* to be aware.

Last night was Kali puja. It started at 10 p.m. We didn't stay, even for the beginning of it; we went to a movie instead. I felt guilty about this, but not very. Much later, when we were home again and I was making japam, I felt, "How blessed I am, to be praying," and I gave thanks, just simply because I was doing that, rather than anything else. This was a moment of insight for me; as a rule I take the act of prayer for granted and complain because I'm not able to pray "properly." But one should rejoice to be praying at all. Just as someone who has been deprived of food or sex can rejoice that he is eating or fucking—never mind the quality of the food or the sex partner. It is only by Grace that one is even able to perform the act.

Swami Turiyananda,[1] on pleasure: "You want always sunshine and a good time; but remember, all sunshine makes a desert."

November 5. Don and I have now stopped drinking for more than a month. We stopped immediately after celebrating Vera's name day on October 1, though we made an exception and drank a little on the 21st, when Bob Craft celebrated his birthday. We also drank one glass of wine apiece last night, because Tony Richardson was here.

Tony left this morning for Hawaii, on his way to Japan. He is travelling with a young engineer named Jeremy Fry[2] who is

---

[1] A direct disciple (1863–1922) of Ramakrishna; he visited America with Vivekananda in 1899 and founded Shanti Ashrama in the San Antonio Valley near San Francisco before returning to India in 1902.
[2] English industrialist and philanthropist (1924–2005), from the Quaker chocolate-making family; he invented an oil pipeline valve and amassed a fortune which he used for other projects. He was disqualified as best man

attractive and seems nice. Tony looked tired but was in very good spirits and we felt that he is really pleased with the film of *The Charge of the Light Brigade*, although he wouldn't quite admit it. His preoccupation seemed to be with the fun he was hoping to have in Waikiki and Tokyo. Nothing was said about the Verlaine-Rimbaud project (which Ken Geist apparently put forward to Woodfall a few days ago, in London) and nothing about the quarrel which James Fox told us Tony has had with John Osborne and nothing about the play Osborne is supposed to have written with Tony as the heavy.[1] The only person Tony did attack was Bob Regester; he obviously resents the fact that Neil Hartley is still living with him—all the more so because Neil is in Tony's good books. He is said to have done marvellous work during the shooting of *Charge*.

Igor has been in hospital for the past few days. They have discovered that his supposed gout is really the effect of a thrombosis. His condition is serious.

Jo left for Hawaii yesterday, as planned, with Alice Gowland.

On the 3rd I spent the day at Long Beach State College and gave three talks, or rather, question-answering sessions. I came away feeling more than usually a fake. This kind of encounter with the students seems so futile. And they obviously thought so too, although many of them quite enjoyed it, I think. One big red-headed boy, a bit like Kin Hoitsma, actually asked me how it felt to be in my position, if I believed I was really answering their questions, if I was interested in the questions at all, and so forth.

The wife of one of the professors told me that I looked exactly like her father, who was of German descent and who had died at the age of sixty-five in an automobile accident. He was a libra, however, not a virgo.

They told me that many of the students are older people and married; they commute from their homes. Long Beach State is noted for drug-taking; lysergic acid and pot. They have little interest in baseball or football, surf a lot, attend fewer dances than formerly. Chief interest is in politics, conservatives and liberals pretty evenly balanced. Plenty of protest against the draft.

As far as I was concerned, most of the questions asked me were about drugs, Hindu philosophy and the psychology (not the technique) of writing.

I came home depressed—by the ugliness of the place, the

at the wedding of Princess Margaret to Antony Armstrong-Jones over a homosexual offense. He married, had four children, and divorced in 1967.

[1] *The Hotel in Amsterdam*, in which the producer "K.L." is based on Richardson. See Glossary under Osborne for their quarrel.

multiplication of people and cars, the seeming hopelessness of their lives. But Don said I was really depressed because my audiences weren't larger! Anyhow, I haven't been feeling at all depressed in general, lately. And, much as I hate to admit it, laying off drink makes you far more energetic. People keep saying how well I look—by which they mean that my face is no longer puffy. As for Don, he is almost ready to decide to give up drinking altogether.

November 11. The following things have happened; I'll just mention them first because I'm not sure how much I shall have time to write this evening—I want to watch *Night People*[1] on television in fifteen minutes:

Ted has flipped again and is in prison. Jennifer made a suicide attempt, the night before last. Don has had a serious row with John Rechy because of the portraits of us in *Numbers.*

Ted started to go crazy on the 8th, right after he and Ted Cordes had moved into a new apartment together. The excitement seems to have done it. Then a cop stopped him and asked for his driver's license and he didn't have it and tried to walk away. (Not in order to escape, Ted says, but to go and see his mother—a short walk because this happened on Harold Way.) So then the cop came after him, and Ted said, "Don't you touch me," and threatened him with a ballpoint pen, and the cop did touch him, threw him down on the ground. So now he's in jail and no one will bail him out. Don is very anxious that he shall stay there until he is sane again, because if he's bailed out he'll most probably get into even worse trouble. Today there seems some possibility that his psychiatrist can at least get him moved into a mental home—but will he stay there?

Jennifer is in the Mt. Sinai Hospital and said to be quite out of danger and very much ashamed of herself. She was found in the water below Point Dume, after telephoning her doctor to tell him that she had taken a lot of sleeping pills and was going to throw herself off the cliff. We sent her a note, "Next time you go swimming *please* give us a call first." Hope she takes this in the spirit in which it was meant.

As for John Rechy, Don has told him that he wants John to take Don's drawing of John off the jacket of the next edition, if there is one, because the presence of the drawing and of Don's name on the book make it just that much more obvious that "Tony Lewis"

---

[1] Nunnally Johnson's 1954 East-West Berlin thriller starring Gregory Peck.

is supposed to be Don and "Sebastian Michaels" me.

(I can't concentrate sufficiently to go on writing this during the commercials, so will continue later, or tomorrow.)

November 12. Don as he got out of bed this morning, "The Old Cat—unclaimed baggage."

Bill Inge came in to see me this morning, to bring me a play and a story he has written. He seems depressed and a bit paranoid. He harps on betrayals of friendship, by Chancellor Murphy,[1] for example, and George Cukor. Talks of going to live in Australia— as most of us do nowadays from time to time.

A short while ago, Majl Ewing died. Yesterday the head of the English Department at UCLA, Bradford Booth, called to offer me the job of taking over Ewing's class on modern English literature. This may have been a routine matter of finding a replacement. Or it may have been a deliberate gesture—for Ewing, according to Evelyn Hooker, never forgave me for my remarks about him at the time of Dylan Thomas's visit in April 1950 and always blocked any invitation to me to lecture by the English department.[2]

I don't know if I have mentioned Marshall Bean, a man who has frequently written to me from 12 Emerson Avenue, Saco, Maine. He described himself as a schoolmaster who was dying of cancer. He said he loved my books and wanted to hear from me. He always wanted letters in handwriting not typescript, although I'd explained to him that my arthritis makes prolonged handwriting painful, and he always sent stamped and addressed envelopes. Don was the first to become suspicious. Now I hear that some letters which sound like the ones I wrote Bean are offered for sale by a dealer named Paul Richards, in Brookline, Massachusetts. Maybe this is because Bean has really died. Or maybe this is a racket for getting autographs and the cancer is only pretence. If so, as I remarked to the man who told me (his name is Bill Amboden(?)[3] and I met him in Needham's Bookshop) Bean must be dead to

---

[1] Franklin Murphy, an undergraduate friend of Inge's at the University of Kansas in Lawrence, became chancellor there and encouraged Inge to return as a writer in residence, but the idea made Inge anxious.

[2] Ewing (1903–1967), Professor of English at UCLA from 1930 and head of the department 1948–1955, married a wealthy southern Californian, Carmelite Rosencrantz, whose money allowed him to endow a lecture series in his name, collect books, found the Friends of the UCLA Library, and entertain the influential. Isherwood describes the Dylan Thomas episode in *D.1*, in the entry for Dec. 8, 1953.

[3] Amboen.

all superstition. The letters are being offered for around forty-five dollars each.[1]

(What follows, about my trip up to San Francisco and Santa Cruz, is based on notes I made from day to day in a pocket book.)

November 24 [Friday]. Am airborne, having taken off at 3:10 p.m. for San Francisco in a PSA[2] plane. The squeezed shore area of houses seems insignificant as soon as you turn in over the barren soot-brown mountains. To the east the valley beyond is pale blue, streaked with white skeins of cloud and absolutely featureless like sea or sky. All along the horizon the pinkish white snow of the Sierras. Every color in this view is pale and dirty and seemingly smog stained.

Don drove me to the airport. Roger Lind[3] will probably be coming to stay with him over the weekend. It isn't that I object to this, really, but I can't help wishing it were someone young and charming; he seems so cloddish.

Don loathes his classes in dynamic reading. Yet he admits that he can already read twice as fast and still understand as much or more.[4] He also admits that this technique may, in some utterly mysterious way, set him free to paint as he would like to—*if* he can bear to go on with it. This class and all it stands for—its sanction of normality, efficiency and the effort to make oneself able to play one's part in modern life—deeply offends and brutally challenges all that Don stands for, as an out-of-step individualist. When Don got back from the class last Tuesday (November 21) he became almost hysterical about all this and I suggested, only half kiddingly, that he had better drop the class before it made him go mad. Don said, "I may die, I may even attain enlightenment, but I promise you, I'll *never* go mad!"

He said this partly because Ted *is* mad again. He got into an argument with a cop, threatened him with a ballpoint pen, was

---

[1] Bean received autographed letters, poems, books, and photographs from James Baldwin, Samuel Beckett, Malcolm Cowley, Mary Hemingway, Randall Jarrell, and Stephen Spender, which can now be found in the collections of libraries and booksellers. He was alive to correspond with Cowley as late as 1971.

[2] Pacific Southwest Airways.

[3] Not his real name; New York painter, about ten years older than Bachardy, with whom Bachardy had a brief sexual relationship.

[4] Evelyn Wood Reading Dynamics, in Westwood. Bachardy attended about six sessions before realizing that he read only for pleasure and preferred to read slowly.

sent to jail, then moved to the prison hospital because of his crazy behavior, then bailed out by his parents—although Don begged them to let Ted stay in this time, in the hope it might teach him a lesson. Don is convinced that Ted's attacks are brought on voluntarily and that the only hope of stopping them is to make Ted feel that they aren't worthwhile. Yesterday Don and Ted were both at Thanksgiving dinner with their parents. Ted behaved so self-indulgently, obviously loving his own illness for the power it gives him over his father (who is tormented with guilt about Ted's condition and blames himself) that Don bawled Ted out on the way home in the car. He says that Ted was genuinely amazed. Don ended up with, "I don't think I like you very much," and Ted said, "I don't like you either," and he made Don stop the car and he got out.

We visited Gerald two days ago—Michael gave us such a bad report of his condition that I didn't want to postpone it until after my return from up north. He lies down all the time now—he can't walk—and his voice is weaker than ever, I could hardly understand anything he said, though Michael still seemed able to. But he is radiantly cheerful and one feels that he is absolutely aware of his predicament and yet, with one foot in the other world, still capable of feeling strong human affection, toward Don and me for instance.

November 25. I'm sitting on the couch in Ben Underhill's spare room writing this while we wait for Angus Wilson and Tony Garrett to come and have supper with us. It has been a beautiful warm day though hazy. We spent a lot of the morning in two queer bookshops, the Adonis and Rolland's, where we bought *Sex Life of a Cop*; *Go Down, Aaron*;[1] *Teleny*;[2] *Like Father, Like Son*; *A Fool's Advice*;[3] *The Beefcake Boys*. Buying such books is a sort of political gesture which is infinitely more satisfactory than actually reading them. It was also chiefly on principle, so to speak, that we watched two blue movies, one of which was called *Jack the Rimmer* and was literally blue because the color exposure was wrong. The other, a Japanese one, was better because the Japanese do these things with a ritual of seriousness and thoroughness; the very gradual penetration of the asshole by the penis was satisfactory because it was made visually important, it had, as they say, human dignity.

[1] By Chris Davidson, 1967.
[2] *Teleny or The Reverse of the Medal: A Physiological Romance of Today* (1893), sometimes attributed to Oscar Wilde.
[3] By Carl Corley, 1967.

Last night we roamed around Chinatown and ate at Sam Wo's after drinking two margaritas at The Empress of China, my first in all this long while. Ben showed me the building where he works on the Poverty Program.[1] He asked me if I thought he should get another job before the program is forced to fire all its employees, as may well happen if these conservative pressures continue. I said I didn't see any use in worrying about such possibilities when the entire social structure is so shaky. Who knows if a Black Power revolution may not fragment the United States into two or more separate countries? Ben told me how the Chinese are afraid of the Negroes and how they are tending more and more to vote conservative.

It seems awfully quiet here, in contrast to Santa Monica Canyon. In the night you hear only an occasional ship's siren. Ben amused me by complaining of the noise made by a bus going up Lombard, one or two of them an hour!

He is very sociologically minded and likes to go to seminars, rather than movies. He has just been to one by Buckminster Fuller,[2] who says that smog is caused by factories and *not* cars, as we are usually told.

Last night he dreamed that I had split into two people. One of them wanted to talk about literature and he and Benny went out to Marin County; the other was very cheerful and wanted to talk about boys. He was over in Berkeley, so Ben went across to him on the ferry (which doesn't exist any more). Ben couldn't make up his mind which Chris he wanted to be with. He says that this was an entirely pleasant dream, and that he had never had one like it before!

I woke this morning thinking that the real point of a householder's life is not simply that he is not a monk but that he loves a human being rather than God. So he must learn to love God through that human being. Very obvious and very important to remember.

November 26. Angus looks just the same, red-faced with wavy white hair[.] Tony hasn't changed either. Angus talked almost entirely about himself and "my new novel" (*No Laughing Matter*). As we were walking in the street and had momentarily paired off,

---

[1] Created under the 1964 Equal Opportunities Act. Chinatown was one of San Francisco's five Anti-Poverty Target Areas in President Johnson's War on Poverty.

[2] American architect and engineer (1895–1983); he invented the geodesic dome.

Tony told me almost guiltily how much he liked *A Single Man*. His praise of it was so strong that it seemed defiant and I felt that he must regard it as a sort of disloyalty to Angus. Angus didn't say one word to me about any of my books, so I didn't mention his new one which I anyhow find impenetrable. We ate at a restaurant called Gordon's which is queer but very respectable, except that the waiter addressed us as "child."

Angus was very good company, rattling away with vast good humor, and it's impossible not to like him. At Ben's apartment, before we left for supper, there was a rather dull man from Texas but Angus made the very most of him and rubbed him up the right way by asking about Galveston. The Texan disparaged it, saying it was run-down, but was pleased when Angus declared that he loved run-down seaside resorts! He also said he prefers plains to mountains. He was really funny describing the stuffiness of a graduate seminar he had recently had, at which they had discussed a novel by George Eliot (*The Mill on the Floss*, I think). Angus had remarked that George Eliot had been disappointed because her public hadn't liked the chief woman character as much as the man—at which a prissy student had said, "Mr. Wilson, we've only got three hours to devote to this book; why do we waste our time on what the author thought?" Another Wilson *mot* I remember: "You can always tell a homosexual from behind; he looks as if he's waiting for someone."

November 27. Am writing this on a bench in the sun in Washington Square, feeling very content to be in the warmth but like an old old person. I identify with that old Chinese lady for instance who has just shuffled past. Am waiting until it's time to go and have lunch with Thom Gunn. Ben will be away at work until late this evening.

Yesterday was perfect weather and Ben and I went flying. He drove me out to Buchanan Field which is an airport beyond the hills east of Oakland. We rented a four-seater Cessna with a re-assuringly middle-aged instructor named Dale. Ben sat beside him and had a first lesson which included taking off on his own: all you have to do is let the plane accelerate to sixty miles an hour along the runway and then pull back on the controls. We flew over a church which Ben said his father would be attending right at that moment, then out over the northern arm of the bay toward Mount Tamalpais, then down low over San Quentin because Ben works there, then south over the Golden Gate Bridge and the bare-ass beach and Golden Gate Park and right over the city and so back to the field. Forty-five minutes cost us twenty-five dollars. Again I was struck by the dull lifeless pallor of this landscape, and its terrible untidiness. Ben was distressed because there is so much man-made fill encroaching on the areas of water.

November 28. Am in a very noisy Italian café on Grant, drinking real English breakfast tea, until it is time for Ken McDonnell to come and pick me up and drive me down to Santa Cruz. Everybody is talking Italian and there is a scoreboard with the results of football games in Italy and the record player has started the champagne song out of *Traviata*. I have quite a bit of a hangover because of wine at lunch yesterday with Thom Gunn and scotch with Stanley Miron and drinks with Fred Kuh[1] at supper.

Thom has now become very thick with John Zeigel, who has been staying with him and just left. Thom looks and seems wonderfully healthy and cheerful. He has a black beard, and is still lean and vigorous. He has given up teaching so he can write all the time. He seems to have decided quite definitely in favor of living in America, after a final trial stay in England. I forget what we talked about, but I left him feeling really invigorated. He is one of the strongest people I've met for a long time.

Stanley Miron wasn't strong, but he didn't seem weak, either, or even depressed, as he told me about his marriage. He feels it was probably a ghastly mistake. Now he has a child coming. His wife is possessive, and they have almost no tastes in common. He has never told her about his past sex life. He doesn't know what to do. The only solution seems to be to keep her down in the country (which she loves) and spend a lot of time in the city, with his doctoring as an excuse. Despite all of this, Stanley still looks young and handsome and basically cheerful. I think he has a very hard heart.

The evening with Fred Kuh and a younger friend of his wasn't disagreeable but I only arranged it because I couldn't find anyone else to eat with. So I put the talk on automatic and got drunk. I also paid for the dinner which was unnecessary. Fred is rolling in success. I like him. He looks like a pig, an Edward VII pig.

November 29. Like an idiot I left my two suits behind in Ben Underhill's closet. No sooner had I pulled the downstairs door locked shut than I remembered this. Ken McDonnell clambered

---

[1] From Chicago, owner of The Old Spaghetti Factory, a restaurant in North Beach.

athletically around the back of the house, trying to find a way in, but he couldn't.

We drove down to Santa Cruz a roundabout way, through the beautiful woods, stopping for lunch at the Brookdale Inn—which, I now realize, is the same place I've been told of and have vaguely tried to find every time I've driven up this coast in the past twenty-eight years.

Ken is a sweet companion. He loves these mountains because he used to come to them on vacations in his childhood, with Larry. Since I saw him last he has fallen in love with a girl he met through the girl he had taken acid with. He even thinks they may get married soon. He is very insistent that he is wild and she is wild and that their life together will continue to be wild, after marriage. In other words, he isn't going to be trapped and pressured into taking a steady job and living like everybody else.... I kept my mouth shut.

Ken hates the Diggers. He says they're self-righteous.

We got to Santa Cruz early and walked along the front by the amusement park and I told him about the Marple ghost. Then we went up to the U.C. campus,[1] and reported to Miss Bartholomew at Crown College. When I introduced Ken to Dr. Thimann the provost,[2] I watched Thimann very carefully to see if he was jumping to any conclusions about us. I'm sure he didn't—and maybe the reason was because dear Ken looked so homely! He wanted to hear me talk, so I lent him one of my ties and he had supper with us first.

A student (psychology, Jewish) from Crown showed me to my guest room which is in Cowell College. It seemed very nice and clean and there was a lovely view stretching right away out to the ocean. The student hastened to remark that he didn't like the dormitories in Cowell and that the ones in Crown were far superior. To prove this, he took us on a tour of them and we were proudly shown a room in which one of the occupants had hoisted his bed right up to the ceiling, and another in which both beds were hung with thick curtains like four-posters, and another which had a draped cloth canopy like a tent.

After supper I talked a bit and then read a passage from *A Single Man*, about George arriving at college in the morning. The students seemed very enthusiastic, probably because they are nearly all undergraduates. Later I sat and talked to as many of them as

[1] I.e., University of California at Santa Cruz.
[2] Kenneth Thimann (1904–1997), English botanist.

wanted to listen. Lots of questions about Vedanta, particularly from a youngish man in a woollen cap, wearing a Zen beard. Later he told me that his name is Stanley Trout and that he isn't a student but lives nearby in a shack in the forest. He has written a manuscript which I am to read. It's my life, he said. He also said that he felt it was fated that we should meet here.

This morning, my room is still nice and clean but the view of the ocean is hidden by blank wet cloud and the rain is pouring down. I have soon got to slosh through the mud to Crown College where I have been given an office in which to receive students all morning and a classroom in which to hold a seminar all afternoon. The woods are full of bulldozers working on the foundations of future buildings. There is first to be college number four (at present called College Four) and then, as the years pass, a succession of colleges up to the number of about twenty![1]

One of the disadvantages of my room is that it is surrounded by rooms in which there are pianos. Life here is unusually permissive. The library is open all night, and apparently the pianos are too. Anyhow, there was piano (and drum) playing last night until at least eleven-thirty, and it began again this morning at six!

I have just had my breakfast; you have to eat it before 8:15.

Extract from *Sex Life of a Cop* by Oscar Peck:

"'Honey, take time to shower with me,' she begged, panting. He glanced down at the nude goddess lying there on the bed. Her suggestion sounded refreshing. 'Sounds great, baby. You lead the way,' he drawled."

December 1. We've just taken off (8:35 a.m.) from San Jose airport, homeward bound. This morning is brilliant, after two days of rain.

My chief concern was not to let my only pair of pants get wet and baggy-kneed and to keep the water out of my only pair of shoes. (Why on earth I brought a raincoat but no rubbers I will never know!) Mud everywhere, and seesaw planks laid across puddles. The builders go ahead as if they were ducks, although the excavations were all flooding and the newly made paths were streaming with water. There are said to be more than a million redwoods on this campus—it used to be a ranch—and of course it is very right and proper of the college authorities not to want to cut them down; but the alternative, it seems will be to build a lot

---

[1] The first college at UCSC, Cowell, was founded in 1965, followed by Adlai Stevenson in 1966, Crown in 1967, and Merril (College Four) in 1968. There are now ten colleges.

of the colleges down in hollows with trees all round and they'll be
as dark as tombs.

They have an institution which they call College Night (I suspect
it's a British idea, for two of the three provosts are Englishmen); one
night a week the entire student body of one of the colleges has a
sit-down dinner—the other nights it's self-service—and afterwards
they are entertained by some speaker or other type of clown. The
first night, College Night was at Crown, the night before last was
at Stevenson, last night was at Cowell. At Stevenson I was bad,
because dim with exhaustion from lack of sleep, so I consciously
forced it by acting merry and bright. (I must never again tell those
corny old show-biz stories.) Last night went much better, because
I read again, and because the Cowell dining room is better to
speak in than the Stevenson.

The piano playing the night before last was truly incredible. Just
slow chords struck at long irregular intervals but quite continuously
from 2:30 to 3:30! I'm sure whoever was doing it was high on pot
or something. Last night I made up my mind to sleep no matter
what, so I exhausted myself by watching the dress rehearsal of an
unthinkably ancient farce (by Sardou!) called *Let's Get a Divorce*,[1]
and then reading. And then to add to this there was a knock on my
door, well after midnight, and two girls wanted to know how to
write a play they'd had assigned to them; so I put on my bathrobe
and told them, for an hour. So then I did indeed sleep, until the
student who was to drive me to the airport banged on my door,
and I dove into my clothes and packed in the car, and here I am.

Further notes: Dr. Willson,[2] the provost of Stevenson, told me
that there is already an amazing amount of individuality in the three
colleges, so that they feel their experiment is a success. Stevenson
is the "activist" campus.

Saw two students who seemed definitely turned on. A girl who
joined a group of professors at lunch and didn't seem to know
where she was. And a boy who spent a lot of time with me yesterday, taking me to see the library etc.; in the evening, during the
rehearsal, he came up to me with a wildly scrawled bit of paper
saying how he had wanted to screw some girl and how she hadn't
wanted to because she said she only did that when she knew someone well and there was something more to it than just screwing. I

[1] *Divorçons* (1881) by Victorien Sardou and Emile de Najac; there was a new
English version by Angela and Robert Goldsby in 1967.
[2] F.M. Glenn Willson, professor of politics, and social sciences; he had degrees from the University of Manchester and from Oxford and had taught
in Rhodesia.

felt sure he was showing me this as a sort of test; but I didn't know how to react and he was disappointed and went away again.

Two boys who wanted to talk about Vedanta came to see me with the air of conspirators, waiting until the others had left the office. One of them, named Mark Bristow, had a lopsided but rather beautiful face. Oh the incoherence of their questions, which nevertheless puts me on the defensive; I feel apologetically that I ought to be able to understand them, if I were the real intuitive type!

The whole thing was a rat race, it always is; nevertheless, though depleted, I don't feel ashamed. I think I did impart a little something to a few of them, and anyhow I didn't fake it. I was, as far as it's ever possible to be, myself. Last night, Page Smith,[1] the provost of Cowell, held forth about the great revolution caused by the two world wars and the consequent vast gap between the generations nowadays. No doubt this is absolutely true, for squares. But I see more clearly now than ever before that our little gang in England in the thirties really dug all that stuff, although we used a different vocabulary. I've come away from this visit feeling pretty damn modern and not in the very least intimidated by the young. As far as I am concerned, the difference between the age groups is physical, far more than it is ideological. They want to stay up all night and make a noise. I don't. But this is merely due to a failure of energy and appetite on my part. (To take just one example—when Edward and I sat up drinking and writing down our sensations as we did so—making noise like "into the forests of the ether"—weren't we behaving psychedelically?)

December 15. It is very cold. Snow is even said to be possible later today. But inside, all is snug. Am so happy with Don.

Have been working on "A Christmas Carol" since I got back. Now the draft of a treatment is finished and submitted. So let's hope we get a director and go ahead.

It seems that the Shaw estate have given permission for our script of *Black Girl* to be performed; but Lamont Johnson won't be back until after Christmas. So I can get on with my book.

December 22. A resolution at the winter solstice: to keep this record more regularly and to be chatty.

---

[1] Charles Page Smith (1917–1995), American historian, prize-winning biographer and columnist, author of *A People's History of the United States* in eight volumes, previously at UCLA.

Hunt Stromberg Jr. seriously believes he can induce Queen Elizabeth to appear as a sponsor on our television movie of "A Christmas Carol"! This he plans to arrange through Douglas Fairbanks Jr.![1]

Jim Charlton, home for Christmas from Hawaii, is now delighted with his new life and says he has lots of work and is making money. He also says that Jo can have a far better business there, if she wants to, than she ever had in California.

A revival of Dodie's *Dear Octopus* is a big hit in London.

Olive Mangeot writes that she had a heart attack, but is all right now.

The Shaw estate has apparently okayed *Black Girl*.

Jack Larson and Jim Bridges have moved into their Frank Lloyd Wright house on Skyewiay (sic) and it is a very sacred shrine of art—at least to Jack. Jim is prepared to work in the cellar, at least for now. Later there may be a crisis about opening up another window, and thus vandalizing the shrine. (Which reminds me that the newspapers say the Christmas tableaux on Ocean Avenue have also been vandalized. Drove by them this afternoon but couldn't see anything wrong.)

On the 17th I at last finished working through the year 1901 in M.'s diaries and Frank's letters. If all the years take me this long, I won't get the book itself even started for about three and a half years!

December 25. All is peace on earth in this household, and the sun has just set after a beautiful mild gold-hazy winter beach day. Don has done two very striking paintings from photographs of women. Now he is with his parents. (He was obliged to take his father a terribly sickly Christmas card, the only one he could find, about how his dad had always been the best of dads, right from the beginning!) We're to meet up at a party given by Charles Aufderheide. Then see *The Roman Spring of Mrs. Stone.* Happiness!

Yesterday I began the teleplay of "Christmas Carol."

David Sachs, whom we saw yesterday evening, says that when Paul Goodman's son was killed the son's girlfriend put a cucumber in his coffin because it reminded her of his erection.

---

[1] American actor (1909–2000), intimate with the royal family since the 1930s, decorated by them for special service in W.W.II, knighted in 1949.

1968

January 1. Working hard on "A Christmas Carol." Hunt Stromberg is beginning to say he wants a draft of the teleplay this month. Rex Harrison is supposed to be interested. Hunt talks about bringing me to England to be on hand throughout the shooting of the film. This will fit in all right, unless the Mark Taper Forum people want to do *Black Girl* this spring, which I doubt.

Anyhow it's very good to have so much work. As for Don, he's equally hard at work painting.

No New Year's resolutions, except to make the bed every morning and try to avoid mooning about.

We spent midnight last night with Jack and Jim in their beautiful home. Looking into the living room from the deck, it was exactly like a stage set—in the sort of play in which the lights go up slowly to reveal an empty stage. I said this, and Don commented with his typical ambiguity, "It'd be the sort of play which ends exactly as it begins."

The first things I read this year, some of Ezra Pound's translations from Cavalcanti. How marvellous they are! The lines I woke up saying to myself, this New Year's morning:

> Language has not the power to speak what love indites:
> The Soul lies buried in the ink that writes.[1]

Don is listening a good deal now to the tapes I made for him of various poems. He likes Swinburne, Poe, Herbert. I asked him which New Year's Eve seemed most memorable to him. He said, "1961; I was sulking."

January 19. Swami has just telephoned this afternoon to say that Dick Thom is dead, apparently of a heart attack. Swami sounded very sad, and I suppose it was partly because he thinks of Richard as being a monk gone to waste—like me. I haven't seen Richard in such ages that I can't feel a great shock, and yet he's very vivid as he was up at the monastery in the old days.

I heard from Chris Wood last time I saw him that Peggy Kiskadden has had one of her breasts removed because of cancer. That did jolt me because it seems such a shockingly apt stroke of karma; I can never forget how Peggy gloated over Bill's brutal

[1] "Fragment" by John Clare (1793–1864), first published 1950 in *Selected Poems*, edited and introduced by Geoffrey Grigson.

frankness to her mother, when *her* breast had to be removed.

Have been working intensively on "Christmas Carol." I finished the teleplay on the 15th and Hunt Stromberg loves it, but Rex Harrison hasn't read it yet. Now I'm redoing some parts of *Black Girl*, trying to redistribute the material contained in Shaw's epilogue over the rest of the play.

The last three days have been beautiful, though the nights are cold. The day before yesterday I parked my car up on Ocean Avenue and went for a walk—I always find it hard to do this; I have to make myself. And yet there is so much to be seen when you're on foot; it's like a psychic world which exists, just over the threshold, invisible to the daily world of our auto-borne lives.

Up in the park, for example, there was a young man sitting under a palm; apparently he was a truck driver who'd got out of his truck, as one might get out to take a leak, in order to play a guitar! I went down the steps from the park and across the footbridge over the coast highway; it is fenced in now (I don't think it used to be) perhaps to stop people throwing things, or themselves, onto the road below. Some graffiti were: Pam please be my wife, Surfers rule, Veil, Silence is complicity, End war, Kill all pags (this must have originally been fags, and probably altered by a fag).

The old solarium is a parking lot now, but the bungalow court next door still exists, where one occasionally retired to repeat and prolong some hasty outdoor act. This section of the boardwalk is very nostalgic, it is the surviving remnant of a quiet little beach town. Sandy Bay House, where Jim Charlton used to live and where I so often spent the night and ran out early next morning to dip in the waves, far off across the wide stretch of sand. The Chalet, truly old world, with beach furniture on its tiny front lawn and a beach umbrella and a sleepy cat. Another building which looks quite empty, except for several bicycles and two mobiles made of sun spectacles. And the Synanon building, for sale now. And the miniature court of shacks covered with tar paper printed to resemble brick, where the old man used to take out his false teeth to give blowjobs. There is still some sort of clairvoyant doing business on the pier, in or near the place where I went with Vernon and Chaplin and the Huxleys to have our handwriting read.[1] I stood a long while watching a young long-haired surfer who seemed purely a native islander, quite unconnected with any of us ashore. And I remembered how I watched other boys on other boards here in this bay, twenty-nine years ago!

[1] Probably in early 1941; see *D.1*, pp. 144–145.

February 18. Back to this diary at last, now that I have more or less finished with *Black Girl* for the time being and have nothing more to do on "Christmas Carol" until something happens about casting, getting a director and deciding when to go to England.

All this time I have managed to creep ahead with the copying of diaries and letters for my book. Today I reached Kathleen's wedding day, March 12, 1903.

Will write a few things about the past month, but not today. Joe LeSueur is coming to supper, along with Jack Larson, Jim Bridges, Gavin and Basil Wright, whom we finally got around to inviting.[1] My relations with Jim are steadily becoming more and more strained, although he may not know this yet. I am getting irritated by his failure to restart work on the play of *Meeting by the River*, everything else seems to come first. I know very well from my own experience that, however busy you are, you can always sneak in some little extra effort on the side. Now I have talked to Robin French about it again, and he must surely have said something to Jim; but Jim never mentions it to me. Don says that this is because he is such a coward. Don also thinks that Jim is utterly under the thumb of Jack and that Jack maybe doesn't want him to do this job. We shall soon find out, I suppose.

Horst Buchholz wants me to work on a musical version of *Felix Krull*. He came to see me about this the day before yesterday. He thinks he can still play Felix, natch—on the stage, that is.[2]

February 21. Talked to Jo this morning. Her prospects for working as a designer in Honolulu are now the best; she has practically closed a deal with one of the biggest firms. But she is still very doleful. She hates to leave her apartment here ("it's my only security") and yet it makes her miserable most of the time by reminding her of the days with Ben. Ben finally told his father about their split-up, and about Dee. The father wrote back to Jo, "There used to be just the three of us, and now there's four." This sent Jo right up the wall, and no wonder. She has agonies of jealousy whenever she thinks of Dee being taken by Ben to any of their former friends, and received as Ben's wife. In our case she knows this won't

[1] Cambridge-educated British documentary director and producer (1907–1987); he began teaching at the University of California in 1960. Isherwood had met him during the 1930s, around the time when Wright co-directed *Night Mail* (1936) for the GPO Film Unit.

[2] The 1957 German-made film of Thomas Mann's last novel, *The Confessions of Felix Krull, Confidence Man* (1954), was among his greatest roles—more than a decade before.

happen, because we dislike Dee. But if Ben's father receives her as his "daughter" that will be worse than anything. This line about "now there's four" certainly sounds like an absolutely stunning bit of tactlessness. But I suppose the father, in the selfishness of his senility, merely thinks of Dee as one more person to look after him and be counted as "family." Except for the fact that he isn't in the least bit selfish *or* senile, according to Jo, and has never made any emotional demands on either of them.

The television cable trucks are making a maddening noise this morning. Everybody in the Canyon is getting the cable put in, as otherwise our reception is so poor. I have been trying to record more poetry for Don, and the mike keeps picking up their broadcast. Doing Pope, Cowper and Clare.

Richard Thom's funeral was on January 23, at Santa Barbara. I went up there for it; so did Webster Milam—and this seemed to please Mr. and Mrs. Thom very much. They were quite wonderful about the whole thing, appearing at the funeral in colors, no mourning. But Swami says that Mr. Thom was still reproaching himself because—when Richard didn't show up at his job and they went down to his apartment to find out the reason—Mr. Thom had said to Mrs. Thom, on the way there: "I hope he hasn't started drinking again—I'd rather see him dead." And then they found his dead body.

Swami says he is convinced that Richard died in spiritual ecstasy. Since he gave up drinking he had seemed very much changed, and full of love. And death came so easily, as he lay in bed; his hair was still tidily brushed and parted.

Talking of love—such a long period, lately, of happiness with Don. I always hate superstitiously to write about this, so I won't say more. But our visit to Truman Capote in Palm Springs (February 8–10) really made the best kind of fifteenth anniversary celebration. It was so pleasant being bossed auntishly by Truman, told when to come and see him, when to go off by ourselves, where to eat and how we were to take whirlpool mineral baths, steam and massage at the Spa Hotel.

February 24. A cliff slide this morning held up traffic in the Canyon. One car was buried but the driver not seriously hurt. A beautiful day. Have been working on Kathleen's diary, as usual. I am getting pretty bored with these uncommunicative entries and with Frank's letters about servants; but there is nothing to do but plod on. One unimaginable day, quite quite suddenly, it will all be finished—and then the real problem arises, what the hell does it all add up to, what does it *mean*? I simply have to have faith that the whole job is worthwhile, and I do, because even if a book doesn't hatch out of it I shall still have learnt such a lot about Kathleen and Frank.

We saw Emlyn and Molly Williams the night before last, at Marti Stevens's house. Judith Anderson was there too, and Emlyn teased her quite fiendishly. He looked the picture of fiendish sparkling good health, with his ruddy complexion and silk-white hair, and he was still bursting with energy although he had just given one of his Dylan Thomas performances at Royce Hall. For some rather mysterious reason, both he and Molly made a great fuss over us. Don thinks they have guilty consciences.

We left early but Gavin who was there too told me about the talk during supper. Molly got rather drunk and declared that she'd had "thirty-three years of perfect marriage," Emlyn pronounced that John Gielgud had spoilt his career by being too queer. He had liked *The Valley of the Dolls*[1] for its frankness(!) He was surprised and hurt because, at the home of some close friends of Frank Sinatra, his joke about Sinatra had been very coldly received. This was the joke: Emlyn was making fun of Gielgud's famous brick-droppings. He said, "If John were introducing the Sinatras, he'd say 'This is Frank Sinatra the crooner, and this is his daughter, Mafia Farrow.'"[2]

Judith looks marvellously young for seventy, but she sadly complains that no one wants her, she's too old. We suspect her of still harboring her corny old ideas about how women should be treated, etc.; she is as bad as Jo, or worse.

February 29 [Thursday]. A nice expression I just learnt: as queer as a treeful of ducks. (Good titles for a trilogy: *A Treeful of Ducks, Dick's Hatband, The Three-Dollar Bill.*)

Monday morning was foggy. I woke and immediately knew that, for some extraordinary reason, I wanted to reread *The Prisoner of Zenda*. So I did, right through, and most of *Rupert of Hentzau* too.[3] They are a very good demonstration of the supreme importance of narrative viewpoint. *Zenda* is far superior to *Hentzau*

---

[1] Jacqueline Susann's novel was published in 1966; the film was released in 1967.
[2] Sinatra (1915–1998) was thirty years older than actress Mia Farrow (b. 1945), his third wife, 1966–1968. In 1963, he lost his Las Vegas casino license for entertaining Sam Giancana in a casino in Lake Tahoe; his friendships with other Mafia bosses were monitored by the FBI.
[3] Anthony Hope's 1894 novel and its sequel, subtitled *From the Memoirs of Fritz von Tarlenheim* (1898), possibly sources for the film plot in *Prater Violet.*

chiefly because Rassendyll is so much more fun to be with than von Tarlenheim.

As regards plans, fog still predominates. I don't even know for absolute certain if it is Screen Gems[1] who owns "A Christmas Carol" now, or Aubrey-Stromberg. If it is Screen Gems, and I think it is, then I may not be sent to England after all. But Don will go, definitely; and I most probably shall, sooner or later.

Cabaret opened in London yesterday and Robin French has heard from Hugh that the notices are very good. *The Daily Mail* tracked down Jean [Ross] as the original of Sally Bowles and she has been interviewed and is being brought to the theater to meet Judi Dench.[2]

The day before yesterday, Jim Bridges spoke of *Meeting by the River* for the first time in weeks. He assured me that he will go to work on it again as soon as this screenplay he is doing now is finished.

Last night we watched Kubrick's *Paths of Glory* again, on T.V. It is a truly great picture. Aside from all the virtues of its direction, writing and acting, two things particularly struck me. The stylized sounds of firing, made (I suppose) by drums and strangely melodious, and thereby all the more sinister. And that marvellous last scene. For the first time in the whole picture we are shown a representative of "The Enemy," the pretty frightened German girl who sings "Es war einmal ..."[3] If the picture had ended just before this it would still have had a terrific impact but it would have been less than it now is; it would have been just another bitter brilliant exposure of The Way Wars are Run. But, after all this bitterness, Kubrick shows us a final scene which is like a shrug of his shoulders over the whole ghastly mess. No use staying furious. This is what human beings are. And, in the end, you have to forget your fury and weep or laugh (either is equally appropriate) and give them your reluctant blessing.

March 16. It looks as if we really are going to buy two duplex buildings from Paul Millard, as a tax write-off. I had violent fears of this, because I imagined all the fuss and expenses they might cause, but Don wants to do it, and I am feeling a lot less opposed

---

[1] Then the T.V. Division of Columbia Pictures.
[2] Starring as Bowles in the London premiere.
[3] "Once there was ..." from "The Faithful Hussar," sung by Kubrick's third wife Christiane Harlan (under the stage name Susanne Christian); the 1957 film, from Humphrey Cobb's best-selling 1935 novel, is about W.W.I.

after talking to Arnold Maltin[1] this morning.

Swami has been quite seriously sick, but is being moved out of the intensive care ward at Mt. Sinai Hospital either today or tomorrow.

Still no definite word about "A Christmas Carol." Don is anyhow leaving for England via New York around March 28. I suppose I shall stay on here and keep grinding away at Kathleen's diaries. Reached 1905 today.

Don remarked yesterday that the people he finds it fun to be with are Truman, Gore (who's coming here soon), Tony Richardson, Hunt Stromberg and his friend Dick Shasta[2] (with whom we had dinner the other evening and whom rather to our surprise we got along with very well). Also, says Don, it used to be fun with Doris Dowling—sad that this has come to an end because of Leonard's Jewish behavior. Of course, fun-to-be-with people are almost by definition not those one knows well. I said Igor and Vera are almost the only people I feel really snobbish about knowing; they are my royalty. I doubt if I could quite feel this about any writer, past or present.

March 22. A recent synchronicity: On the evening of the 16th, I went over to 147 to listen to a new song for *Dogskin* which Ray Henderson had composed. Elsa was there too. For some reason we got talking about quotations and how one sometimes can't hunt down their origins for years. Elsa gave as an example, "things that go bump in the night," which is a phrase from a Scottish prayer. That same evening, Don and I went to have supper with Jack [Larson] and Jim [Bridges], and one of the other guests was Terence McNally, who wrote a (very unsuccessful) play called [*And T*]*hings that Go Bump in the Night!*[3]

It now seems that the deal with Paul Millard will go through almost at once. Still no word from Screen Gems. Robin French thinks they will make "Christmas Carol" in England anyway, so I may well go there even if Stromberg and Aubrey are definitely out of the production.

Yesterday we went to see the Bracketts. Muff was just the same as ever. Charlie is now an invalid. His mouth is open most of the time, in a down-curving shark-grimace. However he seems

---

[1] Not his real name, an accountant.
[2] Not his real name.
[3] McNally, American playwright (b. 1939), later wrote *Love! Valor! Compassion!* (1994), *Master Class* (1995), *Corpus Christi* (1998), and the books for the musicals *Kiss of the Spider Woman* (1992) and *Ragtime* (1997).

to understand everything you tell him. They seemed delighted to see us, and the pathos of this was the only embarrassing feature of an otherwise really quite pleasant visit. Having to make an effort to entertain is actually much more amusing than not having to make any effort. Besides, Muff knew exactly what they wanted to hear; it was chiefly about Marguerite, her house and her life in London. They seemed much less interested in the doings of Dodie and Alec. Perhaps because Marguerite represents vitality, and that's what invalids need. They fed us caviar and iced white wine.

The day before yesterday we had supper with Jennifer, just the three of us. Jennifer has been doing a "marathon" (either at Big Sur, or one of their other centers[1]) during which between fifteen and twenty people are shut up together for twenty-four hours. Jennifer just loved it. Obviously she had had a truly significant experience. But neither Don nor I could quite envy her or want to do likewise. We asked her, of course, if many of the people had been aware that she was Jennifer Jones. She answered that only a few were and implied that this was unimportant. But surely it isn't? Surely being Jennifer Jones is a very important part of being Jennifer? It's all very well to talk about getting down to basic humanity, but, if you really can do such a thing, it must mean that the layer of being Jennifer Jones has to be broken through. Otherwise Jennifer remains a masked woman. She may be accepted by other members of a group as a woman, a mother, a middle-aged beauty, a hysteric, a would-be suicide, an elegant lady with money in the bank, etc. etc., but the mask is still on. And, as far as we could guess, this was what in fact happened. As for me, I don't believe I could feel I was "relating" to such a group until I'd read them at least half a dozen extracts from my books!

Jennifer admitted to us, with evident satisfaction, that her suicide attempt has made this doctor-lover[2] of hers much more in love with her. But he still won't leave his wife and family for her.

I forgot to mention a beautifully tactful act of Don's as we were leaving the Bracketts'. He kissed Charlie. So of course I did too. And Charlie was delighted. He said, "That's the nicest thing that's happened to me in weeks."

April 1 [Monday]. Three-thirty in the afternoon and raining hard in heavy gusty showers. The sea grey, the trees drinking, the slides sliding. Just the right weather for the situation in this house, which

[1] Evidently at the Esalen Institute.
[2] A fashionable psychiatrist then treating her.

is that Don took off at noon for London. We neither of us quite knew why he was doing this. Chiefly because David Hockney has lent us his apartment and since I still have no reason to go there it seemed as if Don had better use it.

After I'd seen him off I noticed that time seemed to slow down, to an uncanny extent. It apparently took me only twenty-five minutes to get home from the airport although I stopped at the bank on the way!

Ronnie Knox, seeming goofier than usual, came by to pick up Don's car, which Don has lent him. All he talks of are his quarrels with the manager of the hotel where he works.

The radio full of yesterday's big news: Johnson's resignation (which is called either nobility or tactics) and the bombing pause (which isn't expected to do much for peace).[1] Gore had supper with us last night. He is gleefully looking forward to the rat race of the convention and election, both of which he is to cover on television, with his enemy [William F.] Buckley.

While I was writing the above, Stephen called, en route to Santa Barbara. I shall see him on Friday-Saturday.

Rain, rain, rain coming down harder and harder. Jack and Jim have got Leslie Caron coming to stay with them, and now their roof will be leaking again, and they'll have to drive out to the airport in all this shit, poor bastards.

Am sort of lonely already and yet I don't really want to see anyone. I would just as soon watch telly, eat from the icebox (there's still some of the moussaka Don made) watch more telly, sleep.

April 4. So now they've shot Martin Luther King in Memphis and he may not live. Just heard this.

It's unfair, but I can't help remembering Jack Larson's patriotic carryings-on the other night (because of Johnson's speech and the bombing halt). He kept repeating excitedly, "This country *works!*"—meaning that the marvellousness of our constitution and the ultimate wisdom of our masses make everything work out all right in the end. Caron listened to this with French polite cynicism, as she sat eating a snack supper and taking sleeping pills in a mini skirt with her hair down her back, looking still so young. I do like her, but she is quite sour and no wonder. She found much

[1] On March 31, President Johnson announced he would not run again because he could not spare the time when the country was so divided and Americans were fighting overseas. He halted bombing in North Vietnam and renewed calls for peace talks.

fault with Vanessa Redgrave for her political attitudes. Since she had just arrived from London she was exhausted ...

Just before six, I heard that King has died. Oh fuck them all. How blood-horny this'll make the killers on both sides. The one good thing about this evening is that I'm going to spend it with Gavin, so shan't have to listen to a lot of phrases and duty attitudes.

April 7 [Sunday]. Such a beautiful day, though windy and a bit misty. Out on the bay, the flash of the water-skiers is like an appeal to get with it, to participate, to live in the moment, to make the scene, to be where the action is. Here I am, sitting up on the balcony with the typewriter plugged in, determined to witness, to record, rather than to run down to the beach and wander around looking for—what? Not sex. Just for the sense of being *there*. (But there, I remind myself, can just as well be here.) This isn't senile silliness even. I was just the same when I was young. The Beach and the Balcony—that's the story of my life.

Don just called from London, terribly disappointed and inclined to come right back home. He has left Marguerite's, where he quite liked it and where [she was] sweet to him, and moved into David Hockney's flat, which fills him with horror, it is so dirty and messy and cold and far away. [Marguerite] has suggested he shall come back to [her], but he doesn't like to accept. And nobody else has shown much enthusiasm on seeing him.... Well, I said of course come home any time, just play it by ear—but I hope very much he won't come home now because he will feel guilty for having wasted good money on this fruitless trip. God, how symbolic life is! Because why in hell shouldn't a trip be fruitless? And what *is* fruitless, anyhow? It's all equally a part of The Dance.

Perhaps, though, the very act of making this call to me has already relieved some of the tension. Perhaps Don will have a reaction and decide to wait a while and see what happens, secure in the knowledge that he *can* come home at any instant.... Even as I write this, I have a feeling that maybe I ought not to write it, because as soon as I have written it he will know. When two people are as psychologically interlocked as we are, telepathy becomes an absurdly inadequate word to describe what is probably going on between us.

Well, anyhow....

I enjoyed Stephen's visit. The morning of the day before yesterday, which was gloriously beautiful, I drove up to Santa Barbara to fetch him from the Center for the Study of Democratic Institutions, where he has been attending a conference (on communication—or

am I making that up, because it's such a typical nowadays theme?).
The center is a lordly building on a hilltop within a large estate
with noble trees and a vast view of the coast. The obvious sneer
would be to say that nothing could possibly look less democratic.
But, after all, this is a laboratory of sorts, and therefore demanding
some seclusion from the outside world. And I can well believe
that these conferences, futile as they may sound, have powerful,
long-term effects, after the participants have returned to their
respective campuses or other institutions. Stephen (white-haired,
paunchy, jowled, but looking essentially the same as when I first
met him) took me in to hear the end of the last conference. The
tone of it was an agreeable surprise to me; beneath the necessary
politeness there was real passion and the disagreement seemed
of intense importance. One delegate was attacking the views of
another and their confrontation had something at once theoretical and deadly about it, like the confrontation of prosecutor and
accused at a state trial for treason. Until Hutchins[1] stopped them
and made his smoothing farewell speech. (Stephen says Hutchins
is getting terribly bored by the center and its doings.) Afterwards
we had a nice lunch with wine, out on the terrace overlooking the
ocean. Elizabeth Mann Borgese[2] and Howard Warshaw ate with
us. Howard seemed a little out of his element and I didn't feel that
he and Stephen really hit it off anyway. Elizabeth was sympathetic
in a cranky European way; she told us she had a dog which typed
poems.

Then I drove Stephen back to Santa Monica. The weather was
still beautiful, the shore looked quite magical in spots between
billboards, and the drive, with not too much traffic, was a perfect
situation for talk. So what did we talk about? It is maddening
that I don't remember more—because, after all, here we were,
veteran intimates who understand each other's language and
tricks, and who had been separated a long time and had much
to tell each other, and much to communicate subliminally. Let's
see.... Stephen was worried of course about the riots and burn-
ings in Washington[3] because he had to go back there next day,
and because the house where he was staying was in the danger

---

[1] I.e., Robert Hutchins, who founded the center in 1959; see Glossary.
[2] German political scientist (1918–2002), fifth child of Thomas Mann; executive secretary to Hutchins's Board of Editors at the *Encyclopaedia Britannica*,
subsequently a Senior Fellow at the center and an expert in peaceful management of the world ocean. Later nominated for a Nobel Peace Prize.
[3] In reaction to King's assassination.

area, and because he had Lizzie[1] with him—though luckily she was in New York this weekend. Stephen seemed attached chiefly to Lizzie. About Natasha he said that she is the bravest person he has ever known; she was so brave about her cancer, which now seems definitely cured. (But in Stephen's anecdotes Natasha usually figures as the person who mustn't be allowed to suspect about his latest affair.) As for Matthew, I feel they have rather lost contact. He has married this boyhood girlfriend of his. [...] Stephen added that he hoped Lizzie would [...] marry someone rich. (According to him, she is having huge social success in Washington and has several admirers with money.) One thing I did get strongly from Stephen is his weariness of poverty and the need for doing wage-earning jobs. There is this Jewish combination of the desire for children—he said, "A marriage without children isn't a marriage at all"—and the underlying resentment against them as dependents.

He wasn't quite as bitchy as usual about Wystan. However he did say that Wystan is obsessed by his fame, and will go anywhere to make a public appearance or receive an honorary degree.

He said that being in Germany today is like being with a cured alcoholic; the bad part of them, which was also the source of their energy and individuality, has been removed and now they are rather empty and blank.

He told how Mary McCarthy[2] had said of one of his lectures that it was hopelessly above the heads of ninety-five percent of the audience and hopelessly below the heads of the other five. (Telling stories against himself has always been one of the modes of Stephen's bitchery—the most attractive and the most deadly.)

Must stop here and go over to see Ray Henderson and Elsa. Ray is probably in "a state" and well he may be. Burgess Meredith delayed so long in getting in touch with Wystan that they failed to make a date with him to listen to the *Dogskin* tape before he left for Europe. Burgess blamed this on the play he is producing, but actually he found time to go over to England in the middle of it, to watch the Grand National. Elsa exclaimed in exasperation, "Oh, these horselovers!"

April 8. To continue ... Stephen and I had supper that night with Gavin, David Hockney and Peter Schlesinger. I really do like David a great deal, he is so good-natured. He even seems quite

---

[1] Spender, his daughter.
[2] Leftist intellectual, novelist, theater critic, political journalist (1912–1989), author of *The Group* (1963).

affectionate, in a shy way—which I had never thought before. Stephen loves both David and Peter, but he and Gavin didn't get along. They just didn't connect. (Gavin admitted he had felt this, when I talked to him next day.)

After much watching of T.V. scenes of riots, Stephen finally decided to take off for Washington as planned. He had meanwhile called Lizzie in New York and told her to stay there for a day or two. As I drove back from the airport I put on the lights of the car, having heard on the radio that this was the way the Negroes were showing respect for King. But I saw only very few other white drivers doing this and several drivers warned me, thinking I had them on by mistake!

I have spent several hours of today on the beach. I decided to walk to the pier, and so I met Eddie Albert and his pretty son Edward who comes to our gym. And Edward told me all about the film *2001*, making me long to see it more than ever.

No work for days, now. I *must* get back to it.

April 9. I forgot to say that, at David's the night before last, I met a sculptor named Walter de Maria who wants to find a place in the desert where he can erect his artwork, two walls that run parallel for a mile.[1] He seemed more than somewhat square and somehow not of the party. However I think he may have made it with Andee [Cohen]—who told me that she is still madly in love with James Fox, although she sleeps around a lot. Today I got a postcard from James from Bolivia; he was going on to the Amazon.

Beautiful warm weather. Today I sat out writing letters on the deck. Yesterday afternoon, around six-thirty, we had an earthquake. I was in the car on my way back from the gym, and didn't feel it. But Larry Holt says it was so strong that he began to say his mantram. When I phoned Don in London last night and told him about the earthquake he was so disappointed that he hadn't been there. He still says that he's coming back soon but I sensed a slight brightening of mood, although he hasn't left David's flat.

King's funeral today. More cars with their lights on, but only a few. Mr. Garcia is disgusted with "The Black Man" because of the rioting. He feels that The Black Man missed his big chance to prove that he isn't a savage. Now Mr. Garcia doesn't want to live next to him.

[1] De Maria (b. 1935), educated at Berkeley, once a drummer for Lou Reed and John Cale, made *Mile Long Drawing* that year, two parallel chalk lines twelve feet apart running for two miles through the Mojave Desert. He also installed his first Earth Room, a room filled with dirt, in a Munich gallery.

Last night I had supper at Gavin's. Leslie Caron was there with Jack and Jim, and Camilla Clay and Linda Crawford. She is rather a magic person, so gay, almost affectionate, but with a welcome dash of lemon in it. Jack leaves for New York today. Jim told me that he has made up his mind to finish the play of *Meeting by the River* this month—or at least the first draft.

April 11. Out of respect for Dr. King, the casinos at Las Vegas (or at least many of them) were closed for two hours at the time of his funeral!

Saw Swami yesterday evening. He said that he had expected to die, the first day of his illness. I asked him if he had been worried about leaving the Vedanta Center. He said no. He said he had seen Maharaj twice during his illness, but he was unusually uncommunicative about this. He just said, "I saw him coming towards me." Then he said that he had decided, while he was in the hospital, that if he got better he would spend much more time in meditation; that was what he wanted to do now. He said that he had lost all desires. I thought he meant desires for the success of the work of the Ramakrishna Order, but he said no, he meant personal desires. This surprised me, because I never think of him as having any. I should have questioned him further, but I didn't. He looked absolutely marvellous—a little thinner in the face but not at all sickly; his face seemed to shine with love and lack of anxiety. I thought to myself, I am in the presence of a saint; and I asked Ramakrishna to help me "through the power of my guru." (I don't exactly know what I mean by this, though the phrase has come to me many times, but it's something like asking to have money paid to you through a particular bank, because you have absolute confidence in that bank.)

Today is almost chilly, after a very hot day yesterday. Just my luck. Yesterday I spent in Hollywood, and had to listen to poor old Larry Holt going on about Tommy [Thom]. How he bullies that boy and blackmails him emotionally is not to be believed. Oh, how sad.

Renewed interest in filming *A Single Man*, by a producer named Bruce Stark.[1] At first I rejected this idea, but it occurs to me that it just might be a good opportunity for setting up a collaboration with Don. I shall phone Don and ask him about this when I call him on Saturday evening to wish him a happy Easter. "The Christmas Carol" is said by Robin French to be about to roll again. And Ray

---

[1] Broadway stage manager and producer, in partnership briefly with Edgar Bronfman Jr., whose father was one of his backers.

Henderson and I are to have a talk with Burgess Meredith about the finale of *Dogskin*.

Today I got to the end of 1905 in Kathleen's diary.

A letter from Stephen. He met Senator McCarthy on the plane and they talked and Stephen was driven home in one of his cars, and has contributed one hundred dollars to the McCarthy fighting fund.[1] Stephen says, of his visit, "I don't think we've had a time which was so like old times that I can remember. Somehow Peter and David contributed a lot to the atmosphere. They are both quite new and at the same time something which we have known all those years reborn and able to carry on a conversation like equals, I mean in age." This is a perfect specimen of Stephen's literary incoherence, but I sort of know what he means and it makes me happy because I felt the same.

Yesterday I saw Gerald. He was pretty much the same; his voice may be a bit more distinct. The curious thing was, it seemed they'd been expecting me *last* week. But, when I didn't show up, Michael never attempted to get in touch with me. I had failed them, and that was that. This is the perfection of silent bitchery; it needs nothing added to its silence. I can appreciate it because I sometimes practise it myself.

April 15. Easter Monday. I didn't do much in the way of a resurrection yesterday as I was busy reading a sex paperback called *Caves of Iron*.[2] Also Don Coombs came to breakfast and then visited Elsa with me; she wanted him to help her find magazines from his department at the library with articles about Charles. And then in the afternoon Tito [Renaldo] came by and brought a friend named Nelson Barclay(?)[3] who wanted me to appear on a T.V. show—called, I'm afraid, "Boutique"! Said I would if I could do it with Laura Huxley (he suggested her). Tito again seemed terribly shaken up; his jaw worked nervously and it was an effort for him to get his words out. Before [Barclift] arrived, he spoke very movingly about his mental breakdown, and his loss and recovery of faith. He told me that he once burned his rosary and his copy

[1] Eugene McCarthy (b. 1916), Democratic senator from Minnesota since 1959, ran for president in 1968 calling for a negotiated peace in Vietnam.

[2] By Chris Davidson (1967).

[3] In a note, Isherwood corrected this to "Barclift." Edgar Nelson Barclift (*circa* 1917–1993), dancer and choreographer, was lead dancer in Irving Berlin's *This Is the Army* (1942) and afterwards worked in T.V. and films. He was a companion of Cole Porter, like Renaldo, and reportedly the subject of "Night and Day."

of the Gita! You do feel that religion matters to him passionately, to the point of agony. And that is a great grace, of course. A grace almost entirely denied to gross unspiritual fleshly old Dubbin.

The last three days, I've met with Jim Bridges and talked about his play version of *Meeting by the River*. We have really made quite a bit of progress, I think, and the dialogue seems to work, it is exciting. I was quite wrong to suspect Jim of dragging his heels on this project. Actually he has done an amazing lot of work, typed out practically one and a half drafts, despite all the movie writing he's had to do.

As always when we're alone together, he's quite a different person. He admits that Jack silences him. He adores Jack but feels he must get away from him. He says he is terrified of me, but I don't think he is, really. He says he knows he has no culture. Maybe not, but he is very quick and perceptive, and he knows this too. He isn't really a bit humble. In fact, he says he sees himself as a great producer, doing several jobs at once, writing, directing, making all sorts of decisions about music, sets, etc.

At the gym on Saturday, Lyle told me that he is probably giving up the business in June. For the last year they have been losing money steadily. I haven't the heart to tell him that this is because he has been so damned lazy. Anyhow it will be a terrible loss. I only hope someone else takes it on. Poor old Lyle, what will happen to him? All he seems to care about is making model racing cars.

Don called on Saturday and said he will be back on Thursday next, the 18th, unless I get a last-moment telegram. Well, I am longing to see him of course. If only he doesn't return disappointed and depressed!

April 22. This morning I got a call. Don postponed for the second time; he was to have come today and now it won't be till next weekend. I feel very sad. But it's good, actually, because he's done this job for the Royal Court and now the whole trip is justified. So I must stop moping and not be so lazy. The last four days I have had a bad stomach, but it's better now, apparently because of some pills Dr. Allen gave me. He has married again, a German girl. He said to me, so ingenuously, "As you get older you begin to find there isn't as much choice as there used to be." I don't [think] he can have meant this quite as crudely as it sounded. He added, "We're very happy."

Jim and I have finished a very rough draft of the first act of *A Meeting by the River*. Jim doesn't seem very inventive, but perhaps

I make him shy. Anyhow, once again, I find myself doing most of the work. He sits and types, agreeing with nearly everything I suggest. I took him up to Vedanta Place last Wednesday for the reading, and he went to the temple for the Sunday lecture, quite of his own accord. He has also been reading *An Approach to Vedanta*. I tell him we'll make a Hindu of him, but he says he is still fixated on his childhood religion, southern nonconformist, with lots of hymns. He has started to meditate, however!

April 24. The weather is fine but with a nasty cold searching breeze which chills my back and makes my fingers quite numb even in the sunshine. Am depressed and sick again, with gas and stomach pains and the shits, but I don't want to admit this to Dr. Allen for fear he starts a lot of tests. I feel I might get better if the wind dropped and it got warmer.

It is clear that Jim doesn't want to write this play, only direct it. He even said the other night that he wanted to give up writing altogether. He irritates me quite a bit, and yet he is truly good-natured and quite sensitive in many ways. He said, "I love Jack for his faults, the faults are what I love."

Am rereading Michael Campbell's *Lord Dismiss Us*. Am not sure how I feel about it yet. It is very well written and well constructed, but sometimes there's a nasty taste.

I long for Don to come back—so I can love and think and feel and be a human being again. At present I'm just a dull old dying creature. I really do not find my own company amusing, and yet I often prefer to be alone. I wish God were nearer, and yet, in a dull way, I know he's there.

April 26. Stomach still nervous but very little gas and none of that stuffed up feeling.

Having finished *Lord Dismiss Us*, I am impressed all over again; though the breakup of the love affair seems arbitrary. What the book chiefly leaves me with is a sense of the pricelessness of love. Anyone who shares it with anyone is to be envied beyond billionaires. And that brings me again to Don (whom I haven't heard from yet as to arrival). He wrote such a wonderful letter yesterday, and I realize more than ever that this is IT. Not just an individual, or just a relationship, but THE WAY. The way through to everything else. This seemed to be obliquely confirmed by Swami this morning. He is after me to find a title for a volume of Vivekananda's letters in English, and he called me because he thought he had found a clue to it in a quotation from one of the letters. Vivekananda writes, "Religion is the practice of one-ness with the infinite, the principle that dwells in the hearts of all beings, through the feeling of love."[1] If you are tuned in on personal love, then you are on the same wavelength as infinite love. There may be terrific interruptions from the static of egotism and possessiveness, but at least you *are* on the right wavelength and that's a tremendous achievement in itself.

Yesterday night, Gavin and I went to the opening of Marlene Dietrich's show here. It was really a wonderful experience. Not because of her singing, that was uneven and often spoiled by her mannerisms and too often repeated tricks; but because of the relationship which developed between her and us. The Ahmanson Theater (which is actually too big for an intimate show of this kind) was packed, the applause was like a thunderstorm which was always in the air throughout the performance and kept exploding and even interrupting the beginnings and ends of songs; and then there were several occasions when people rose and applauded standing. All this wasn't any the less moving because it was probably led by a claque. Because what one felt was that this was essentially a rite. A rite of enthusiasm. It was the enthusiasm itself, and our capacity for experiencing it, that we were celebrating; and she was the priestess. We were carried out of ourselves because we were moved, and moved because we were carried out of ourselves. We applauded our own applause. And Dietrich seemed to understand this so beautifully. There was so much humility in her reception of our emotion. She was so moved by it herself that she seemed quite beyond mere vanity.

Dorothy came to clean house this morning. She told me she had lived in a very bad neighborhood (in New York?) among West Indians and that she had learned to defend herself. The best weapon is a bottle. "A man will go up on a knife but he will *not* go up on a bottle." A bottle half full of hot water is good, because it breaks easily over your opponent's head. Or else a coke bottle with the bottom smashed off, because of its weight.

Am now reading through Frederick [Machell Smith]'s letters to Emily for my book. Otherwise the year 1906 is finished. (These will be included there because Emily read them through shortly after she moved into the Buckingham Street flat, and mentions this to Kathleen in a letter.) What a huge task this is! Yet, as a matter of fact, I have now worked through twenty of the diaries and have only(!) twelve more ahead of me.

---

[1] See *The Complete Works of Vivekananda*, vol. 7, pp. 498–499.

May 3 [Friday]. Something I forgot to mention about Dietrich. When we went round afterwards to see her—which was merely a formality, for she seemed hardly to know who we were—there was a man there from *Time.* He wanted to get some sort of statement from her. She refused. He giggled in that sick teasing bitchy way such creatures do, and asked, "Don't you want to be in *People?*" Dietrich cried out, so everyone around could hear her, "I am *not* People!"

I have been very sorry for myself and sad, these last few days. So terribly longing for Don, and also with an upset stomach, vagus nerve or whatnot. Dr. Allen thought perhaps it was the gallbladder, so yesterday I went through all the dreary jazz of X rays with gulps of barium. But now he says my gallbladder and stomach are functioning normally—as well they may, considering all the muscle massage I give them. So I will just have to ignore the whole thing. Probably it will get better as soon as Don gets home. He said, in a letter yesterday, that he'll most likely return on Monday.

Meanwhile, according to Robin French, George Schaefer[1] has definitely been hired to direct "Christmas Carol." He's supposed to call me this weekend to start our work together.

May 6. Don called last night, postponing his return. I talked to him again this morning, because I had made him feel guilty by telling him about my stomach pains and I wanted to reassure him. He *had* been worrying, and I'm glad I called again; we both felt very loving after our talk. It is so silly of me to mind his absence so much. I don't during the reasonable daylight hours, and yet I keep reverting to the rather horrible feeling that this is a kind of illness, his not being here, from which I may not recover. Suppose he died, over there. I wouldn't have any reason to go on living. That thought is simply terrifying.

My pains are much much better.

No call from Schaefer.

Shall I go to England if they want me to? Not if I can help it. Shall I fly to New York for this medal presentation ceremony at the Institute?[2] I may wriggle out of that, too.

Some personal ads from the *Los Angeles Press* (May 3):

---

[1] American director and producer (1920–1997), on Broadway and in movies, but mostly for T.V., notably "Hallmark Hall of Fame."
[2] Auden was to be awarded the Gold Medal for Poetry by the National Institute of Arts and Letters on May 28, 1968.

Virgin guy (23) good looks but shy lonely sensitive & sincere seeks beautifully minded and empathetic girl to learn ecstasy of love's ultimate experience.

Warm sincere gay seeks him who too is sick of casual affairs and meaningless moments.

AC/DC girl wanted by extremely sensitive couple for quiet (?) threesomes.

2 well built young guys want 3rd well built stud for fun and games.

Gay sincere goodlooking 27 with insatiable desire seeks same for fun.

ENNUI?? Discreet M.D. will be personal photographer to uninhibited couple.

Nice looking bi guy 21 like to hear from groovy people M & F no color hang-ups. Please pix if poss.

Goodlooking yng prof man 28 is just a country boy at heart. City living is tolerable and a practical necessity but I don't feel completely in touch with myself unless I'm grooving with the natural beauty and excitement of the great outdoors. I wonder if there isn't another physically rugged but sensitive and intelligent yng man who might want to join me some weekend and who wouldn't be afraid of a possible close friendship that sometimes develops in such an environment or of something even more beautiful if the magic is there.

Eccentric scientist 28, would like to meet a girl who can understand him.

Jim pointed out that at least eight of the ads in this issue are from the same address, 406 South 2nd Street, Alhambra!

May 13. Stomach pains have switched to back pains. I have been very miserable and sorry for myself, and the ache of not having Don has been acute. And yet I know I mustn't get him back by telling him too much about my ailments. How I hate being sick and lonely and old. No, I don't mind being old, only the other two. I am miserable for Don. And there is absolutely no substitute. Am reading the transcript of some of the talks Elsa and Ned Hoopes have been recording for their book on Charles,[1] and it's all so grim, the cancer and the keen-eyed wife, eternally on the lookout for Charles's loveboys.... At the same time, relations between Elsa and me couldn't be better—the night before last I

---

[1] Never published; see Glossary under Hoopes.

actually had supper with her and Ray and she fixed some delicious steaks!

May 15. The weather is warmer and there is less wind, but my back is still bad, although I pretended to Dr. Allen that it is better because I am getting so sick of it. My belly and groin are shot with nerve pains.

Nothing yet from Don and now I feel certain that he isn't coming home for his birthday and in a way I am glad because I don't want him to find me like this.

Am reading the Lanchester-Hoopes transcripts. Parts of them are truly horrible. I even find Elsa's frankness revolting; she has no business to be telling this to a gushy sob-sister like Hoopes, who is also a stranger. (Not that I'd be surprised if he has been making love to her.) Here are two extracts:

> They went out in cars ... As they went, they waved to me, Hulter[1] and Charles and Terry. Although I had no inkling of cancer, it didn't cross my mind, I knew that Charles was dying. You know, one can have flashes that one is not going to see a person again. He had a heart condition, had a gallbladder operation by then, and I knew that I was a "free woman." That's a terrible thing to say in a way. I haven't said it before, but that's the way I felt.

> It was really like a sort of Ariel, being freed from Prospero. I knew that I was free of something. I didn't know what I was in for, because missing a giant later is a terrible gap ... Whenever a phone rings you wonder, "Is it for Charles or me?" but he's not there anyway. And it's still there, you know. Not unpleasant. Sort of a compliment that he does remain.

May 18. Don's birthday and he isn't here. Talked to him after midnight last night—early this morning his time—and wished him a happy birthday. He said he had no date to go out to a birthday lunch or supper. His voice was so beautiful; it seemed full of tears and yet perfectly happy. (I don't know really what I mean by this, but it was the impression I got.) He says he will stay on for the opening of the John Osborne play in London[2] and then leave and be here this next weekend.

[1] Bob Hulter, Laughton's road man; he managed travel, hotel, and other arrangements on tour.
[2] *Time Present*, opening at the Royal Court, May 23.

I wish I understood life better—could see more clearly, I mean, what is happening to me. It is all a sort of vague dream, governed by compulsion. I go out to supper, like last night, and immediately know that I would rather be by myself at home. I want to be alone and just think about Don, brood on him, rather. And why do I like doing this? Because brooding on Don is love, and as long as I am tuned in on love I am happy, or at peace, which is infinitely better and the same thing. I wish I could brood on Ramakrishna in the same way—then I'd really have it made. But Don is love too, and perhaps I must somehow understand that this is enough, and indeed the same thing. But meanwhile I am grim and compulsionistic. I drive myself to perform chores, be it writing this diary, or working on Kathleen's diary, or copying bits out of Vivekananda's letters, or reading the Laughton transcripts, or whatnot. I hate writing this diary but I make myself do it because I know from experience that I will gradually relax while doing it and let something inside me speak. At least, I occasionally will. I ought to write every day, just making myself.

I still feel lousy and of course the thought arises that maybe I am perhaps seriously ill, that this indigestion and the pains around the groin etc. are beginnings of cancer. But I must keep reminding myself (instead of saying, Oh nonsense) that cancer is just another word. So I do have cancer? Lots of people have had it before, including poor wretched vulnerable terrified Charles, who had to face it with no one to help(?!) him but Elsa and Terry.

(Part of my compulsionism is the way I keep carefully erasing and making corrections while writing this. As if an inspector were about to arrive and look it over.)

May 19. The shits last night and now everything is out of whack again—my guts, groin, back, in addition to pains in the left knee and a sore throat on the left side. Had nothing for breakfast but tea and toasted honey-muffin without butter or jam. Am taking Festalan[1] and Maalox. Have given up alcohol, the past three days. This morning the pain in the pyloric region was so acute that it woke me. My head feels dull. When I got out of bed, my stomach was enormous with gas ... This isn't fussing, you understand, just a medical report. My next appointment with Dr. Allen isn't until the 23rd. Let's see if this yields to treatment.

Yesterday I did at least get through some work: a first draft of my speech to the Academy-Institute, presenting Wystan with his

---

[1] Atropine, to reduce secretions and relieve spasm in the digestive tract.

medal (I have now decided to call them tomorrow and say I can't come[1]), a first draft of the liner to be printed on the jacket of the Gita record for Caedmon,[2] plus my daily stint of Kathleen's diary, plus some copying into my quotation book[3] from Vivekananda's letters.

I forgot to record that, when I last saw Swami (no, it was the last time but one, May 1) I asked him if he was meditating a lot, as he had told me he would. He said yes. I asked him if he sat up in bed when he did this in the middle of the night. He said, "Sitting up or lying down, it makes no difference now." I spoke to him about my dullness and inability to feel God's presence. He assured me that this was all right, it didn't matter, I must just keep on trying. And then, referring to the presence of God, he looked at me and said with great emphasis and the obvious wish to reassure me: "It is a *fact*."

Have just been over to see Ray Henderson at 147 about *Dogskin*. The rough draft is nearly finished now. He told me a story about a Western singer and guitar player named Rusty(?) Draper.[4] One day, Draper was singing at some club and a girl came up to him and whispered something in his ear, one sentence. And immediately Draper cancelled the rest of his tour and they went straight to Las Vegas and got married. Neither one of them will tell what it was that she said. An American fairy story.

May 26. It was a painful disappointment yesterday when Don said he was staying on to do drawings for the second Osborne play,[5] because this collection of his drawings in the program was such a success. I don't think I showed how much I minded. I hope I didn't. I am really glad that he is getting a little recognition for a change, even on this small scale, and it is good for him to be able to enjoy it on his own. No ... it's only that I sometimes get the dreads. A dreadful mad silly voice says suppose he never comes back ... Actually he keeps assuring me—and I do believe him—

[1] Auden did not attend either. Isherwood's citation was read out by Glenway Wescott.
[2] *Selections from the Bhagavad-Gita (The Song of God), A New Translation in Prose and Poetry by Swami Prabhavananda and Christopher Isherwood.* Caedmon TC 1249 (1968). Isherwood read from the Gita translation made with Prabhavananda in 1944.
[3] Probably his commonplace book.
[4] Draper (1923–2003) recorded gold-selling singles throughout the 1950s, had stage roles in *Oklahoma* and *Annie Get Your Gun*, and appeared on T.V. in "Rawhide" and "Laramie."
[5] *The Hotel in Amsterdam* opening July 3 at the Royal Court.

that he's longing to return, loves me, longs to be with me, etc.

Well, anyway, shit. I'm not going to be like poor old Jo (who has got herself a kitten and now complains because it mews too much and bites her face). I am much better, temporarily at least, and yesterday, which was beautiful and warm, I went on the beach and jogged, even ran fast in tiny spurts, and then went in the water, the first time in ages. I am resolving to make a big stab at Kathleen's diaries. Have reached 1908.

Last night I went out with Paul Wonner. He meditates a lot—says Bill [Brown] does too, which rather surprised me—and says that he often gets a feeling of "euphoria." He also says that he has always known that there is a God, has known it from childhood. Listening to him, I thought how really naturally unspiritual I am—as Don has sometimes told me.

Paul asked me what my dearest wishes would be—this was because we'd seen a film called *Bedazzled* (quite brilliant in parts) about the devil granting wishes in exchange for a human soul. I said first that Don should have some big success, second that I should die a violent painless sudden death without anticipatory fear, outdoors. I later realized that both these wishes are wrong. Don must have a success all on his own, not through magic intervention. And I must die fully aware that I'm dying, so I can concentrate on Ramakrishna.... The worst thing was that, in my preoccupation with myself, I utterly forgot to ask Paul what *his* wishes were. He must have wanted to tell me, otherwise he wouldn't have asked that question.

May 31. The day before yesterday, Dorothy came to clean house and told me that she had dreamed the night before that Jo had gone mad and had been put away in an asylum and that I had asked Dorothy's help in getting Jo's things together and packing them! Dorothy took this dream seriously and was concerned about it, so I phoned Jo—without telling her anything, of course. Jo told me that she had been in a better frame of mind for some time, but that, that last night, she had woken up sobbing.

My symptoms are better, but I still have a slightly inflamed throat on the left side, plus early morning stomach spasms. Am now more inclined to believe that the whole thing was an infection, because one of the boys who comes to the reading at Vedanta Place said he had had the same ailment.

Loneliness for Don. I miss him nearly all the time; I mean, consciously. Fancy being able to "miss" God as often as that! Yesterday Jim [Bridges] told me that he and Jack [Larson] were going down to San Diego for the day, to see the production of *Hamlet* which Ellis Rabb has directed. So I went along and had an enjoyable outing. Ellis has produced some striking theatrical effects, but at the same time has neglected to make his Hamlet and his Horatio speak properly; their gabbling was incoherent, almost ducklike. However we saw the zoo first, and later there was a pleasant party at the house of Craig Noel who runs the theater[1]—pleasant because of the warmth and sweetness of the young actors, and the one young actress: Amy Levitt, who plays Ophelia and seems the only likely-to-succeed member of the cast.[2] After this we drove home, very very fast. Jim sat in the back with their stinky dog—I tried not to show my dislike of its presence—and Jack kept nearly falling asleep at the wheel; whenever he did this he unconsciously accelerated to almost a hundred. I had to talk, loudly and aggressively, to rouse him.

June 10. Don is to return today. I can hardly believe it. My throat is bad, and I am full of anxiety. A kind of dread—but of what? Of new developments? Silly old superstitious horse. Even the smell of happiness makes him tremble and twitch his nostrils.

I haven't been so depressed in years as I was this last week. And of course it was quite largely about Kennedy.[3] How nauseating these weepy liberals are, who cry that our society is sick with hate just because amidst millions one little foreign killer can be found![4] No—nobody is any sicker than usual (which *is* sick, all right); it's just that the pot is being stirred. What comes to the surface has always been there.

Julie Harris, whom I saw last night at Lamont Johnson's, seems sadder than ever—forlorn, guilty, talking about her "infidelities"—almost a fit sister for poor old Jo. But both of them are tough inside.

June 19. Don got home more than a week ago, on the 10th. I have never known him, or any other human being, manage to get through with so much baggage and not pay excess. At the Los

---

[1] The Old Globe.

[2] She played the role on Broadway in 1969, had a few small movie parts in the 1970s, and appeared in ABC T.V.'s "One Life to Live."

[3] Robert Kennedy was assassinated June 5 at the Ambassador Hotel in Los Angeles after winning the California Democratic presidential primary.

[4] Sirhan Sirhan, a Palestinian, was arrested and convicted after confessing to the murder.

Angeles Airport Customs Office a woman found the photos of his male nude drawings. She handed them to her superior, who said, "They're trash but they'll pass." Don asked, "Is it usual to get art criticism along with a customs inspection?"

Dorothy wanted to pray for Jo, or have her prayed for, I'm not quite clear which. In order that this might be done, I had to give Dorothy Jo's maiden name. How surprised Jo would be if she knew this!

Kubrick's *2001* is one of the very greatest artworks of our time. We saw it last night. The overwhelming feeling you get from it is fundamentally religious—a religious awe.

A strike has stopped performances of *Cabaret* everywhere except in England.[1] Soon there will be no more money coming in.

Papers and confusions and time wastings on account of the houses we have bought. I take out my irritation about this on Arnold Maltin. Don says he will deal with Arnold in future, so I won't lose my cool.

Don and I want to work together, but on what? *The Praying Mantises* for Anthony Page is no damn good and neither is *The Barford Cat Affair* for Hunt Stromberg.[2]

Have decided to begin my family book with a letter to Kathleen. Got this idea telepathically yesterday evening from Don while we were having dinner at Musso Frank's. Now that he's here I am functioning again. Back and stomach okay but throat still has something wrong with it.

The next hurdle to jump, a talk on Saturday for the ACLU[3] at Long Beach on "The Right to Dissent." My difficulty is, I don't give a damn whether or not I have the right to dissent, I just dissent or I don't. The only question which arises in any given instance is, is it prudent, is it strategic? Or shall I dissent and keep my mouth shut?

Have reached the beginning of 1909 in Kathleen's diary.

July 5. There was no problem with the ACLU talk because they sent me a paper telling me what to say. The day before yesterday I took part in a T.V. round table discussion (no table and a semi-circle) at Cal. State[4] with Ray Bradbury, Leon Surmelian, Bob Lee

---

[1] Actors' Equity struck June 17–21, mostly over pay, foreign actors, and chorus sizes. Nineteen Broadway shows and nine touring shows closed. *Cabaret* was then in Los Angeles.
[2] Both were novels to be adapted as films.
[3] American Civil Liberties Union; see Glossary.
[4] California State University, Los Angeles, where Isherwood taught when it was still Los Angeles State College.

and a man whose name I forget who produces Ray's plays. Ray talked nearly all the time but I rather warmed to him, he's so silly. At the beginning we each of us had to introduce ourselves, and I said that I was chiefly famous for the adaptations other people had made from my work, *I Am a Camera* and *Cavalcade*.[1] This *must* have been subconscious bitchery, but I wasn't even aware that I'd said *Cavalcade* until Bob Lee pointed it out!

Have been working full steam on Kathleen's diaries, determined to get the whole job finished if possible by the end of this month or soon after. Am in[,] now[,] February 1912.

Don is rereading *A Meeting by the River* and we shall work on this together if it seems at all possible, without saying anything to Jim Bridges who, luckily, is now busy with a musical of his own.

"Christmas Carol" is more or less finished until we shoot it. Lamont is still fussing with the script of *Black Girl*. He tried to get me to sneak *Saint Joan*[2] into it!

August 9. On August 5, I finished copying and cutting Kathleen's diaries up to the end of the year 1915. It was such a tiresome job that it was almost worth it, merely as discipline. I mean, the nervous effort of it was tiresome. Reading the diaries was often absorbingly interesting, especially toward the end. And I feel that I know Frank now for the first time.

I started this work on August 19, 1967, which means that I did 404 single-space pages (approximately) in 417 days, which doesn't sound like much, but then I had to read right through each of the thirty-two diary volumes and at least seventeen packets of letters, some of these in fairly difficult handwriting.

In a few days I shall begin reading through the typescript, making notes and hoping for a flash of inspiration as to how to present this material in a book.

On July 12, Don and I started working together on an attempted dramatization of *A Meeting by the River*. In a couple more days we should have finished the first act—assuming, as we do at present, that the play is to be in two acts and that the first act will contain, more or less, the material in the first four chapters of the novel.

Don is very much on the alert to prevent me from attributing ideas to him which I have actually had myself—this is part

---

[1] Noël Coward's 1931 play (filmed in 1933) chronicling the first decades of the twentieth century through the eyes of a well-to-do English family and their servants; widely regarded as a sell-out to middle-class values.
[2] The 1923 play also by Shaw.

of what he calls buttering him up. Nevertheless I do genuinely find it hard most of the time to remember which idea belongs to whom. I *think* it was Don who suggested putting Patrick and Oliver into actual structures, *cubes* we call them, which represent their subjective mental worlds. And we certainly agreed fifty-fifty that Mother and Penny must be live actresses and not photographs projected on the back wall, as in the version which Jim Bridges and I began to write. (Jim and Jack are now both in New York doing Jack's play which used to be called *The Queer Valentine*—I have a block against remembering its new title; and we hope to be able to present Jim with a rough draft of *A Meeting* by the time he returns.)

In any case, I am thoroughly enjoying this project. I love taking things apart and putting them together differently; and Don is always wonderful to work with. He will become better and better I know, as he gradually gets to know the material through and through by manipulating it.

August 13. On August 10 we finished the first act of *A Meeting*, and yesterday I started reading through the whole typescript of my cut version of Kathleen's diaries. I have also, incidentally, written a tiny foreword or whatnot for the paperback edition of *Journey to a War* which Faber is about to publish.[1] I only did this because Wystan said he wouldn't let his "Second Thoughts" be printed unless I wrote one too. Actually I don't feel this urge to revise things.

Peter Schlesinger's horoscope for the day, yesterday, told him to seek the advice of an older person, so he called me to know if he should take two trunks with him to England! Told him of course not; he was going to pack art books, records and all sorts of unnecessary ballast.

When Mark Bristow came to see me with his girlfriend Elise a couple of weeks ago, I was running around fixing them drinks, answering the phone, etc.; and then Mark said, with sweet graciousness, like a host, "Sit down, if you'd like." It was so funny and cute.

Names for books, gleaned from chance remarks, advertisements, etc.: *Lucky Blond, Pretty Boy with Big Feet, Food Smells from a Canyon*.

A synchronicity: yesterday Sharon Tate, Roman Polanski's wife, came to see Don about having her portrait drawn by him. And

---

[1] It appeared only in 1973 from Faber, and from Octagon books in New York in 1972.

yesterday morning I got from Ronnie Knox the current issue of a girlie magazine called *Knight* (which otherwise I never see, even on newsstands) with a story by him in it (called *The Hunk*)—and an interview with Sharon Tate!

Yesterday, being Monday, we saw Gerald and Michael; we have settled down into visiting them every Monday now. The last few times he has scarcely been able to speak at all. But he still seems perfectly aware of everything said in his presence. I notice that one can always get a reaction of some kind from him if one mentions Chris Wood. If you make fun of Chris a little, he laughs a noiseless laugh. It is, so to speak, Love's last faint signal.

Swami seems much recovered and he shows it by his renewed concern about the affairs of the Vedanta Society. He has lately reproved Vandanananda for seeing so much of certain women devotees. He hopes Vandanananda will decide to go back to India. Asaktananda has already said that, if Swami dies and Vandanananda becomes head of the center, he won't stay on! Swami is now considering getting a junior swami sent here as an assistant, so that Asaktananda would be the next in line to succeed him.

August 31. News of the day: On August 28 Don and I finished the first draft of our play of *A Meeting by the River*. Don is now going to type it and probably take a copy with him to New York when he goes there in about ten days, to show to Jim Bridges who is there directing Jack's play. (Jim says he's very optimistic at this point; they have a good cast.) Me, I'm nearly certain I shan't go, much as I hate being separated from Don.

Now I'm gradually working through a set of notes on the typescript of Kathleen's diaries. I reckon there will be about seventy-five pages of these. I don't exactly know why I am making them, except that it seems to be the only evident next step. I just hope that making them will somehow suggest what the book itself is to be about.

On the 29th, Dorothy Miller was bullied by Black Muslims while she was sitting waiting for a bus. They brought round pamphlets, and when the women who were waiting didn't respond enthusiastically and ask them questions they told them that Black Power was going to burn America down and that by 1975 it would be in control. Then they overturned the bench on which Dorothy and the others were sitting. A bus came by but didn't stop. Dorothy was later told by a cop that they have orders not to, if the driver sees there's a riot in progress at the bus stop. Dorothy was so upset she didn't come to work for us that day and didn't sleep at all that night.

David Burns[1] showed up late last night, after we'd gone to bed, and again today with a little Japanese girl he'd just met [...] Dave is going to be as tiresome as ever. Last night he slept in a shack on the building excavation at the end of the street. But I can't help feeling a kind of warmth toward him. Despite his crazy visions and pills he seems somehow sane and good. He doesn't appear to mind being arrested one little bit.

Tennessee, crazy in such a different way, is also here. And Gore has just arrived. Gore says that William F. Buckley called him a queer after losing his temper on their last T.V. appearance at the Democratic Convention on the 28th.[2] Gore says that newspapers from all over the country have been calling him to ask for his reactions to this public insult, and that he has replied, "I don't know what I did to deserve it. I always treated Mr. Buckley like the great lady he is."

At a party at Jennifer's on the 24th, Rex Reed[3] was rebuked by Richard Harris,[4] seconded by Rita Hayworth and Mia Farrow, for being a little bitch and uninvited to boot—he had been brought by Denise Minnelli.[5] We didn't witness this but were told about it by Gavin. And sure enough, one of the T.V. film columnists, a couple of days later, announced that Harris had been drunk and Rita ditto, and that Harris had insulted several major stars. No doubt this was Rex's revenge. The drunk part of it was certainly true. Rita, very sweet as usual, kept breaking in on Harris's conversation, probably because she had hot pants for him. He turned on her and said good-humoredly, "Shut up, you fat mischievous old thing!"

Ben Masselink married Dee a few days ago. Jo has been away, and we don't know if she knows yet.

On the eve of my birthday, at my request, Don arranged a dinner party at Chasen's for the Stravinskys, Bob and ourselves. Igor seemed much better. I was worried by Vera's exhausted appearance,

---

[1] Not his real name.

[2] See Glossary under Buckley.

[3] American film critic and arts columnist (b. 1938), he wrote for *The New York Post*, *The New York Daily News*, *The New York Observer*, *Vogue*, and *GQ* and published a number of books.

[4] Irish-born actor, singer, film star (1930–2002) made famous in *This Sporting Life* (1963) and as King Arthur in the screen musical *Camelot* (1967). Later, he played Dumbledore in two Harry Potter films.

[5] Third wife, 1960–1971, of film director Vincente Minnelli (1903–1986).

but she tried heroically to be gay and we had a beautiful evening. Perhaps one of the very last. The Stravinskys seem determined to leave Los Angeles for ever and settle in Europe. Dinner cost at least $175, maybe more—because, although it was our treat, Vera wouldn't tell us how much they tip the head waiter! My birthday itself we spent quietly. I went in swimming before breakfast, chiefly because I woke up with a hangover, having laid off all drink since July 7, which was the last time we had had supper with the Stravinskys! Since then I have been drinking a little, but it really does seem to depress me. Felt so sad this morning when I woke.

Part of a letter from Anne Geller, July 23: [She says she has been unhappy because she didn't say what she meant to about the check I sent her. She meant to comment not on the amount of the check but on Jim's respect for my abilities and his affection for me.]

September 2. Felt very low and blue yesterday, partly because Don told me the previous evening that he may go on to England to see [a friend] after his visit to New York. But this is something I have got to come to terms with, and I will. The other part of my blueness was due to David Burns who is really bugging us. He keeps showing up, always with "gifts," a flower, an empty cardboard cup with a bit of crumpled paper inside it, or whatnot. He starts talking astrology and you can't get a word in edgewise. He is always lurking around. He sits on the steps and writes poems which he keeps in a large paper bag. This afternoon Don almost trod on him before seeing him and let out a yell of shock. Dave left immediately but Don followed him in the car and gave him back his bag and told him never to come here again without phoning first. God knows if this will take. Anyhow Don is deadly determined to discourage him.

Michael called in the middle of the morning and told us that Gerald was after all not up to seeing us today. I wonder if this is the end at last.

Jo called. She knows about Ben's wedding. She seems quite calm about it.

Tennessee came to dinner with us last night. Also Bill Glavin[1] and Ronnie Knox, who is driving the Stravinskys' car to New York for them, starting today. It was a nice evening and Tennessee didn't seem unduly drunk or otherwise fuddled. The only awkward thing was, he brought a copy of his screenplay of "One Arm" for

---

[1] William M. Glavin, paid companion to Williams from 1965 to 1970.

me to read, more or less saying that I could work with him on it. This is of course out of the question because of the mix-up with Gavin, Jim Bridges and Ronnie Platt. Don called Tennessee a "situation queen."

September 5. Don and I had supper at the Vedanta monastery last night, Swami Swahananda was there on a visit. He's the one who has just joined the center in San Francisco. Swami now says he's glad that Swahananda isn't going to work here; he thinks only of social work, and only in terms of success, how many members does an organization have, how much money, etc. etc. Swami went on to say that the Belur Math itself is becoming just as bad. Gambhirananda is only waiting, Swami says, for the senior swamis who believe in meditation and the spiritual life to die off; then he'll start changing the Math's direction. Meanwhile, despite his indignation and pessimism, Swami looked wonderfully cheerful and well.

Don and I went on to see *2001* for the second time. It seemed even more wonderful than before, and there was so much I had missed the first time. As before, it dominated the audience completely. When it was over, about three-quarters of them remained seated throughout the credits and we felt they were simply too moved to want to leave the theater immediately.

On the night of the 2nd we had supper with Gore Vidal at the Bel Air. (Incidentally, his rooms there were quite like the eighteenth-century French suite in which the *2001* astronaut finds himself at the end of the film.) Gore looked slimmer than usual and very handsome. He says a fourth party will be formed, and when the election goes to the House of Representatives—as they believe it will—this party will force Humphey to change his policy on Vietnam, as the price of its support.[1]

Haven't heard any more from David Burns, knock on wood! Don asked me if my feeling of horror at having him around wasn't a little bit like what I felt about Guttchen. Yes, perhaps. But Guttchen made me feel guiltier and I disliked him. I don't dislike Burns, I'm very much touched by him, though deeply rattled and exasperated. He rattles me by making me feel that *I'm* the crazy one. His wandering life of meetings with strangers, his acceptance of whatever comes to him, including spells in jail or the funny farm, his ability to sit down anywhere and write poetry, his

[1] The House of Representatives decides a tie in the presidential race; see Glossary under Presidential Election 1968.

trustfulness and the little gifts he brings you—doesn't all this add up to something marvellous and even supersane? And doesn't my frantic demand to be left alone, my snarling at all invaders of my precious privacy, amount to a dreary sort of submadness?

Today, Don finished typing the first of the two acts of our play. It is obvious already that the speeches at the end of the act are far too long and that the whole thing will have to be rearranged. But it *is* an achievement and I'm already very pleased with parts of it. There are several distinct improvements on the novel, I believe. Don isn't so pleased with it, however. He still doubts if the material is suitable for the stage, at all.

September 11. Two days ago, Don left for New York, at 11:[1]5 p.m. That morning he finished typing a fair copy of our *Meeting by the River* play. (Whether it is to be ours or just mine isn't decided yet; he still has mysterious scruples about this and is going to consult Jack Larson—exactly why, I still can't figure out.)

As we were driving to the airport he said, "I think the Animals have been managing their life together much better lately," and he said that he had enjoyed this summer very much. I asked him if he hadn't been unhappy about his failure to work, or rather, to produce pictures he liked. He said yes, but he added that he keeps getting ideas all the time, "and that must mean that I want to be an artist, mustn't I?" He is planning to go to England, I think, though he won't say so in so many words; everything, as usual, has to be left open till the very last moment.

That morning we went to see Gerald, who amazingly could even talk a little in a relatively clear voice. But you couldn't follow him properly, he kept rambling off.

Poor old Larry Holt is going to have an operation under his tongue. I'm afraid it may be a malignancy. He called me and held a tedious conversation about the vulgarity of background music in films. I feel all these topics which he raises are somehow symbolic.

Ray Henderson has a tape of the whole rough version of the *Dogskin* musical and really it does seem to have something. Listened to it yesterday with Burgess Meredith, who seemed quite enthusiastic; but Elsa says he isn't to be trusted. He is a great raiser of false hopes.

Don called from New York last night. He had seen Jack's play and had liked it quite a lot. It is called *Cherry, Larry, Sandy, Doris, Jean, Paul*, which is one of the most off-putting names I have ever heard.

September 14. Yesterday I went down on the beach—I am hoping to go on the beach and in the water or to the gym, or both, every day that I am alone here; it's a valuable discipline and part of the survival technique which I am interested in following, after my disgraceful misuse of Don's last absence. I'm not indulging in self-pity or sentiment. This is strictly an experiment. Elderly people are apt to get left alone, anyhow, and they should know what to do about it. I am only trying to learn, rather late, what many of my friends seem to be experts at, already. I am bad at it, because I've been lucky and pampered. Well, anyhow—after that digression—I went down to the beach and the first thing I saw was a playing card, so I picked it up and it was the king of hearts. And then I saw an empty book of matches, with a drawing on the back of it of a sexy stud youth, advertising a bar on Melrose called The Tradesman; "a friendly bar" it said. Two good omens?

Also yesterday morning, a mild synchronicity. In Brother Lawrence I read, "So little time remains to us to live; you are near sixty-four...."[1] And, in *The Education of Henry Adams*, "To one who, at past sixty years old, is still passionately seeking education...."[2]

Talking of omens, last night when I came back into the house after supper, I found a cat. It must have got in by way of the balcony. This has never happened before. A messenger from Don? I keep hoping he'll call.

This afternoon, I've started roughing out the "letter" to Kathleen which is supposed to be the introduction to my book. I find it curiously exciting, writing "to" Kathleen in this way. It opens something up. It's a bit like meditation was, when I first tried it in 1939.

The thought came to me today that perhaps writing this letter to Kathleen will come easier to me *because* I am alone while writing it. Somehow being alone strengthens my sense of the confrontation between the two of us. We are both in states of waiting, as it were, and therefore have something in common. I'm not trying to think myself into the idea that Kathleen is alive somewhere or that she can hear me. Whether she is alive or not is neither here nor there. It's just how I feel.

---

[1] *The Practice of the Presence of God: The Best Rule of a Holy Life, Being Conversations and Letters of Brother Lawrence* (English translation, 1855); Frère Laurent (1611–1691) was a lay brother of the barefoot Carmelites in Paris.
[2] In "Twilight (1901)," chptr. XXVI.

September 16. The letter won't do. I find I have plenty to say in it, but it isn't the right form; it seems so artificial. Now I don't know. Should I bash on through it, hoping the annoyed unconscious will exclaim, as so often before, "No, *no*, idiot—here, let me—"? Or should I start again on the original draft of the book and try to write it better? Or should I aim at an annotated text; basically just Kathleen's selected diary extracts and Frank's letters, with my comments? I don't know. Maybe Don will be able to help me, when he gets back. Last night he called and said he would probably be returning in a day or two. He wasn't very enthusiastic about this, naturally; it doesn't seem to have been much of a visit for him and he hasn't accomplished anything beyond doing a couple of drawings. I kind of wish he had been able to go on to England and see [his friend]. But only kind of. I want him back but I want him back happy. If *only only only* he could somehow begin to paint!

However, I think Jim Bridges does truly like our play, and Jack has told Don he should have his name on it, so that's to the good. Jim wants to do it here, early next year, and then in England.

An American love story: Bob took his car into a filling station to get a tire fixed. A boy named Jack was on duty, a medical student studying dentistry at UCLA. They fell for each other instantly. Jack was so rattled that he couldn't do anything right. "I felt sorry for him," Bob says. It was nearly time for the station to close, but Jack worked on, doing various things inefficiently to the car, for two and a half hours! Yesterday evening Jim, who had just arrived back from New York, brought them both to have dinner with us at La Mer. It was Bob's twenty-first birthday (Jack is twenty-four). We drank Bob's health. I said, "I hope your life will be as happy as mine"—in other words, I was a little carried away. In the next booth was Henry Willson, the agent,[1] with some new tall beautiful client. As he left, he told the waiter that Bob was just the boy he was looking for, and he left his card. Much discussion, should Bob call him? Jim said to him, you're not an actor; then to Jack, but you are. And it turned out that Jack had played in summer stock. On the way home, as we rounded the traffic islands on Ocean Avenue, I murmured "the islets of Langerhans"[2] and Jack knew what they were—which was more than I did, exactly.

[1] He discovered Tab Hunter and Rock Hudson, who was strategically married to Willson's secretary, Phyllis Gates, from 1955 to 1958.
[2] Clusters of five cell types in the pancreas; each type secretes a different hormone to break down food; they are named after the German scientist, Paul Langerhans, who discovered them in 1869 and described the clusters as islets.

It seems Gavin didn't like Jack's play.

I have resolved to do more acts of recollection during the day, it's the only thing I *can* do; I seem incapable of meditation. So I'll try making japam at noon and again at sunset, with my beads whenever possible.

October 6. Today is one of the saddest greyest days we have had in a long while. Don did get back on September 17, but the day after tomorrow he's off again to England. More about that some other time.

Since he has been back from New York we have done quite a lot of rewriting on our play, building up the parts of Mother and Penelope and improving the end of act 1. We finished this on October 1. Jim is enthusiastic but obviously hasn't reread it properly; he is in the midst of rehearsals of this play *Niagara Falls* which he's directing. Gavin read and liked our play he said, though perhaps not madly. He calls it an anti-play. Actually I don't agree with this view at all. Just because the characters aren't always addressing each other directly, that doesn't mean that they are speaking unrelated soliloquies. They are addressing each other *indirectly*, which is something quite different; and I believe this will be apparent as soon as the play is actually performed—always provided that it is performed properly! Now the question arises, what's the next step? Who shall be shown it? Gavin suggests the APA.[1] Certainly we don't want to have anything to do with Gordon Davidson if we can help it, though the Mark Taper does seem the obvious first choice.[2] Don *may* show it to Tony Richardson.

(Speaking of the Mark Taper, it now seems that *Black Girl* will be done as the second production next season, that is, in April; because Gordon is determined to direct a play of his own choosing first. But he still hasn't signed a contract with Lamont Johnson. Lamont is trying to pressure him into doing this or at least into signing the two leads, which would amount to the same thing. Lamont wants Cicely Tyson and Stacey Keach.[3]

Chris Wood told me a strange thing. Last time he saw Gerald, about five days ago, he started talking about Joe Ackerley and his newly published last book, *My Father and Myself*. Gerald got intensely interested and suddenly began, according to Chris, speaking quite coherently and altogether behaving not only normally but

---

[1] Association of Producing Artists; see Glossary under Ellis Rabb.
[2] Where Davidson was Artistic Director; see Glossary.
[3] Tyson (b. 1933, in Harlem) and Keach (b. 1941, in Savannah, Georgia) had appeared together in *The Heart Is a Lonely Hunter* (1968), his first film.

"like a young man." Chris says Michael said he hadn't seen him like that in many many months. Chris didn't perhaps say everything he thought about this phenomenon. He seemed to me to be implying either that Gerald still playacts a bit for visitors or that Michael is such a powerful depressant that Gerald usually appears to be far more decrepit than he really is!

On October 4, Don got a letter from the Marquis Company, asking for his biographical data because they want to put him into *Who's Who in the West*!

Seth Finkelstein has just given me back my Berlin clock,[1] repaired and cleaned! He is such a sweet boy and I am racking my brains to think of something I can do for *him*. Talking about a date he had with his girlfriend that evening, Seth said, "First we'll have dinner and then come back here and enjoy the fruits of love."

Four nights ago Don and I were having supper at the Fuji Gardens, which we rather love, when I suddenly got and started to describe an idea for another novel. It's very vague, of course. Two old men—one of them more or less Swami and the other a writer who, a long time ago, seriously considered becoming a monk but then returned to the life of a householder. They have always kept in touch, however. They are neither of them at all solitary. Both have households, religious in the one case, secular in the other. In both households there are people who disapprove of or sneer at their friendship. The religious think the writer is a dirty old man who writes dirty old books. The secular think the Swami is an old imposter. I don't know any more yet. Except that the Swami gets very ill, nearly dies and then recovers. And that the writer then does actually die. All this wouldn't be quite as autobiographical as it sounds. I see the writer more as Henry Miller, though probably queer. It has some roots in Hesse's *Narziss und Goldmund,* some in [Mann's] *Death in Venice.* But very different, really. Essentially all about love—the only friction being between the supporting characters. As I was telling it, I felt that spooky thing, the sense of a vast terrain of almost virgin subject matter, waiting to be explored; and the gasp with which one recognizes and says to oneself, but *that*—that would *have* to be a masterpiece or nothing.... Don felt what I was feeling. His eyes filled with tears of excitement.

October 16. Don did leave for England on October 8 and he's there

[1] A present from Fräulein Thurau and described in *Goodbye to Berlin*: "a brass dolphin holding on the end of its tail a small broken clock" (p. 15). Finkelstein was a neighbor in the house directly below Isherwood.

now. One of the reasons why I haven't been writing in this book is that I've been writing letters to him, which has exhausted my urge to record. Also I've been so busy.

All I want to put down now is one thing, because it is worrying me. Yesterday morning, shortly before getting up, I had a very vivid dream. The dream was that Don was with me; I felt his presence intensely, and I saw him, I think he was lying beside me on the bed, or rather, sitting beside me, anyhow he was quite close. The terribly painful and disturbing thing about the dream was that we were actually in the act of parting, I felt he was slipping away from me; indeed that he had only partially come back to visit me and that, although he wanted to, he couldn't stay. He said something about the nerve ends not being connected—I think that was the expression he used. Why I am so worried about this is that it had all the atmosphere of an appearance before or at the moment of or just after death. It had that kind of frustration and poignancy. Immediately after it, I woke, and the time was exactly seven o'clock.

The last three days I have had a letter from him every day. But the latest one is dated last Saturday, the 12th. Until I get one from him written after the moment of the dream (which would be October 15, 3:00 p.m. London time) I shan't feel easy, and I shan't feel *quite* easy until he's home again. Of course I could call him on the phone but I don't want to because we agreed we wouldn't phone, and anyhow surely [his friend] would have let me know if anything bad had happened. Also, I was idiot enough to tell him about the dream in one of my letters, though of course I didn't say what I feared it meant. But he's so quick to sense things like this and it might upset him too.

November 5. Don got back last night safe, though looking tired. No more about this at the moment. I rejoice.

Today we voted and I feel sick, thinking of Nixon's most probable victory. I actually couldn't manage the voting contraption and had to be shown how, like several other gaga senior citizens. Which reminds me that Don brought back with him some capsules called K.H.3 because Ivan Moffat is taking them and recommending them strongly. This stuff—procaine hydrochloride—is supposed to improve mental alertness in the elderly and reduce irritability and depression, it also gives you more energy and makes your skin less yellow. Well, I'll try it—chiefly because Don took the trouble to buy it for me, it was a loving thought.

Dorothy Miller, last time she was here, said that Jackie "took the rag off the bush," referring to her marriage.[1]

Have run on the beach quite a lot lately and been in the ocean: did both today. I'm really in a very good state of health I guess, though fat bellied. And I have an almost constant pressure headache.

Don talks of our going to Australia while Tony Richardson is making his film about Ned Kelly.

The last time I saw Swami (October 23) he was in a wonderful exalted mood. He said, speaking of Maharaj, "Chris—to think of it! I *saw* him!" I said, "And we see you, Swami." He said, "You see the dust of his feet." His joy, and the *now*ness of it, was so beautiful.

November 22. We finally got Jim to read our play through again and make definite criticisms, so now we are rewriting it. We think it will be enormously improved. The idea is to make Oliver's moods more dramatic.

Today has been thick fog and so my car won't start and I have had to miss going to the gym. I try to exercise every day. What I prefer is trotting on the beach. It is empty now and often so beautiful, when the tide is far out and you can splash through the shallows. Have also started taking Alertonic,[2] which was recommended to me by the Masins. I wrote them to ask what they thought of K.H.3 and they said it is absolutely useless! Don and I discussed the advisability of telling Ivan Moffat this and decided not to. The Masins didn't say it was harmful, and if he believes in it, it probably will help him.

Have got into the second chapter of the Kathleen-Frank book. Very slow going but I'm quite pleased with it. Don loved the first chapter.

Here is a fan communication. No name or address on the envelope. Postmarked New York:

On Spanking a Friend

I did not know that I was strong or that
You would challenge me.
I was standing bored, unaware that I
Could so pleasingly
Defeat your shaft of impertinence and

---

[1] Jaqueline Kennedy married Greek shipping tycoon Artistotle Onassis on October 20, 1968.
[2] Vitamin B complex with trace minerals and a mild stimulant.

Turn you over to reward you.
But you were pleased and we did it again
And now I see that
Behind my sharp hand there exists an
Aspiration of hope
So strong that I pass it to you and wish
That from my strict message
You may receive my lesson of endearment.

SEIVAD DRAH[C]IR with admiration to a fine writer.

I feel pretty sure that the signature is looking-glass writing for RICHARD DAVIES.

Dave Burns is presumably now in jail. Someone called me a couple of days ago, saying that he himself had been in jail on a driving charge and Dave had told him to call me and get me to bail Dave out, or he would be removed from Santa Monica to the jail downtown before the friend who would pay the bail could be reached. I said no. Don says I was right, and I know I was from one point of view; but I still feel uncomfortable about it.

I lost the turquoise and silver Indian ring Don gave me and hunted everywhere for it without result. Then Jack Grinnich(?) who just happened to be down at Elsa's house next door,[1] miraculously noticed it lying beside our trash cans!

Have just had a long telephone talk with Vera Stravinsky. After leaving Los Angeles "for ever" they are quite glad to be home. Only Paris got good marks for beauty and food, but it was so terribly expensive.

No more news about *Black Girl*. It still hasn't been announced.

December 20. So windy and cold. I feel shrunken with the solstice. A withered little old thing. However I think Alertonic has helped, some. I do have quite a bit of energy and my joggings on the beach have become much longer.

Am only in the third chapter of *Kathleen-Frank*. It is terrible toil, especially copying out extracts from the diaries and letters, but I do see quite clearly how to write it. The main difficulty is to avoid cuteness.

Our revised version of *A Meeting by the River* is typed and Don has also xeroxed two copies, so now it must start out into the world.

---

[1] Grinnich was an actor who occasionally spent weekends with Lanchester.

Black Girl is definitely scheduled and the opening date is March 20 or thereabouts.

Dave Burns is still in jail or at least not yet heard from. Someone called about picking up his mail but never did so, and our box is getting slowly jammed with it.

The night before last, we saw Vanessa in *Isadora*. She was really wonderful in some of the dancing scenes: the good-natured silly girl, so full of shit and arty phrases, being *possessed* before your eyes by the god of the dance. It's uncanny, almost terrifying.

Violent last-minute attempts to get my name off "The Legend of Silent Night," which we finally saw on the 17th. It is unspeakable. Danny Mann is trying to get his name off too.

December 25. Wet Christmas. 10:00 p.m. Jim Charlton just called, back for the holidays from Honolulu. Dave Burns rang our doorbell yesterday morning and went off with his mail, except that he left the copies of *Life* behind; they are already billing him for them. He has again "escaped" from Camarillo. And Don came back from Christmas dinner with his parents with the news that Ted has definitely flipped again. I met Henry Kraft coming out of the movie I was about to go into this afternoon; *The Seagull*. He didn't exactly look older, only less interesting. He wasn't in the least pleased to see me.

Today I finished chapter 3 of *Kathleen and Frank*.

Yesterday I trotted on the beach. Very cold. An old seagull dying on the sand with wings flopped open and legs crumpled under it. It kept opening its beak in terrible fierce gasps. I've been thinking a lot about death lately.

Yesterday evening we and Gavin were invited to supper with Jack and Jim and watched the Apollo 8 rocket getting out of the lunar orbit and starting back to earth. Jack didn't like Gavin's attitude to it; not respectful enough. And he didn't like my saying that the crew's quoting Genesis was in poor taste—first because it was addressing the population of the earth, which doesn't entirely consist of Judeo-Christians, second because equating the exploration of the minor little moon by little us with the prime creation is surely a bit presumptuous.[1] But Jack has now definitely sponsored the moon, just as he sponsors the young American off-off-Broadway dramatists.

---

[1] The first manned lunar mission was launched December 21 and returned December 27; the crew read Genesis 1.1–12 to conclude their live T.V. broadcast from lunar orbit on Christmas Eve.

1969

January 7. The prospect of this year, the very look of its four digits, scares me. Nothing to be done about that.

Lots of Hong Kong flu around. Swami has just had it and I feel very slightly guilty because, the day I went to see him last, which was the 2nd, *I* felt as if I might be coming down with it. My throat was sore and I had chills but I took Coricidin and nothing happened. Swami showed me a letter he'd just got from Ritajananda about Vidya, saying how bossy he is and how he makes enemies. Ritajananda obviously didn't want to make a big fuss but Swami is sending a copy of the letter on to Belur Math and it will probably result in Vidya being recalled to India, or else being put into a position in which he is forced to resign from the order. I am glad I haven't written to Vidya, as I have had the impulse to do several times this winter, because then I should have to conceal from him what I know and I should feel part of the conspiracy against him. I suppose Swami is right about him, but that is Swami's affair.

Ted is having another breakdown. He came by yesterday with a bottle of liquor, gift wrapped. I guess he knows he is being "naughty." Don refused to let him stay even a moment, ordered him out. It's so awful for Don whenever this happens. He still minds, more even than he realizes.

A mysterious swelling on the middle joint of the small finger of my left hand. A tumor, seemingly, quite hard and without feeling.

Have got as far as Frank's proposal in the book, about halfway through chapter 4.

Gore Vidal is in town, his father is dying of cancer in a hospital. His greeting to me on the phone: "Mole? ... Toad."[1]

We have paid off the last of the mortgage, so now own this house.

A copy of our *Meeting* play has gone off to John Houseman. Jim Bridges is now to start passing it around.

January 8. We had supper with Vera Stravinsky and Bob yesterday; it was Vera's birthday according to the modern calendar reckoning. Don got her a white bag at Saks. She was pleased we remembered. They both looked worn out. Igor didn't come down; evidently he didn't feel like seeing us. He is terribly depressed and has

[1] Referring to the characters in Kenneth Grahame's *The Wind in the Willows* (1908).

all manner of things wrong with him, but Bob says his musical appreciation is still amazingly sharp; he really thinks of nothing else now. I felt Vera doesn't expect him to recover; they are all holed up together for the duration, including sweet little Hideiki (or how ever you spell it) who cooked the dinner last night and bowed when complimented, closing his eyes.[1] Bob had given him champagne a short while ago and he had been so excited he hadn't slept at all, he said. We drank champagne, lots of it; Dom Perignon and something else, very cold and calming and pure, like the waters of some magic spring. It is like going to see members of a royal family, either just before a revolution or in exile; a bit of both, because you feel their expectation of a debacle but at the same time the humorous quietness of resignation. Vera is, all in all, about the best woman I know and one of the best I have ever met in my life; truly kind and generous and strong. She complains but without the smallest self-pity and never feels herself a victim; what she does for Igor is done absolutely as a matter of course, out of love, not duty.

An unappetizing fat-faced ass named Sander Vanocur[2] interviewed Rita Hayworth on television last night. The whole interview was designed to show that poor Rita was all alone, unsuccessful, unhappy; and then he ended up by saying that the younger generation should be inspired to think that she could remain so marvellous and sexy and vital right up into her fifties! It's this not knowing what you're saying, and not giving a damn anyway, which is so hideously characteristic of nowadays.

A news item I forgot to record: "The Legend of Silent Night" was duly telescreened[3] on Christmas Day with our names on it, despite all our efforts—and *The Hollywood Reporter* and *Variety* both gave it very good notices!

The English have at last graciously condescended to give Forster the O.M.[4]—for being ninety.

January 9. Dr. Allen saw the swelling on my little finger, says it's a cyst. If it gets harder and larger he'll remove it but he doesn't

---

[1] Hideki Takami, houseboy and cook for the Stravinskys from 1967 until Stravinsky's death in 1971; later he opened a Japanese restaurant in Manhattan.
[2] American journalist (b. 1928) for *The New York Times*, *Washington Post*, and others; on T.V., he was Diplomatic Correspondent for ABC and a longtime White House Correspondent for NBC.
[3] I.e., televised.
[4] Order of Merit, for exceptional service, especially to the arts, learning, literature, and science. There are only twenty-four members at one time.

want to because it's near the nerve. As a sort of consolation prize he gave me a Hong Kong flu shot. While I was waiting to see him I tried to read Konrad Lorenz *On Aggression*. It is only just barely possible for me to make myself attend to this sort of writing, greatly as I am interested in the message of it. I think I got through three pages in nearly an hour!

Talk about [Don's friend] last night; Don had phoned him in London and he may just possibly be coming out here. The really important thing is that we do talk about him and similar problems. It's when we don't talk that the tension builds up.

Saw Swami yesterday evening, in bed but very cheerful. Swami Asaktananda presided at the reading after supper. He is quite a little schoolmaster; he points with his finger and gives curt orders. I'm not sure I should like to see him in command. Anyhow he has a lot to learn.

A nice letter from a woman named Rosemary Leonard with whom I've been corresponding for several years. She has just got a master's degree in philosophy and has been preparing a book on the enormous increase in the use of noun-noun sequences over the last two hundred years (Example: *school football field*, instead of *football field at school*). In "Mr. Lancaster," she informs me, I used 199 such sequences. In a passage of the same length, Meredith used only 121 in 1859, Johnson only 10 in 1759. Richardson and Fielding used 69 and 46 respectively. What does this prove? I must wait to read her book and find out.[1] But it *is* nice to have given someone so much occupation, and presumably pleasure.

January 10. Don has been suffering from stomach pains a lot lately. It frustrates him so, especially as he is working well. Rex Evans has again suggested he should have a show, this spring. Don says it's impossible; he can't get enough pictures finished.

Last night we had supper with Gore. He told us that he has had a researcher look into the early history of his family, chiefly to try to prove that he is racially Swiss, because he has been applying for a permit to reside in Switzerland and this would be helpful. Gore got his permit and has taken an apartment in Klosters. Meanwhile the researcher, going back to Austria in the thirteenth century, has discovered that the Vidals were then Sephardic Jews. The funny thing is, before Gore told us this, we had both independently

---

[1] Leonard was writing a Ph.D. thesis in linguistics at Lancaster University; she later published *The Interpretation of English Noun Sequences on the Computer* (1984).

thought he was looking curiously Jewish! He talked chiefly about success and money, his own and other people's. Don had the impression that he is deeply worried about his life.

January 11. This morning I got up at 5:30 and drove into Hollywood, to read the Katha Upanishad at Swamiji's breakfast puja. Don wanted to come but I dissuaded him, in view of the social ordeal of having Swami Ranganathananda there as a guest; that sort of thing is always easier to cope with alone.

After I'd read, I went in to see Swami, who is still officially sick but actually quite himself and looking absolutely radiant. The reading had turned me on more than usual; I think because of Krishna who somehow filled the shrine room with joy. Anyhow I wanted to talk about it to Swami but couldn't. I mean, I wanted to ask him to make me feel like this all the time instead of once a year. But, as so often, he seemed to know my thought and said, "Just think, Chris, Swamiji himself appeared in that shrine, and he liked your writing!" I suppose Swami was referring to his vision of Swamiji and Maharaj; but that hardly mattered to me. What mattered was that I knew Swami knew what I wanted.

As for Ranganathananda, he was hardly tiresome at all, beyond giving me a long mimeographed report on his lecture tour in the United States, twelve pages—but that I needn't read. He must be around sixty but he looks like an athletic coach of forty, big shoulders, erect walk, close-cropped hair. At breakfast he embarrassed us by taking only Horlick's Malted Milk. He says he never eats and drinks at the same time.

Got home and ran with Don on the beach. At least, we both ran, but separately because he runs so much farther and faster. The tide will be right for running now, for at least five days. So I'll try to run.

Jim Bridges has gone off to New York because he is so upset by the reception of his plays—although he knows that the only really bitchy notice, in *The Hollywood Reporter,* was written because of personal spite. Jack is urging him to write a *really important full-length modern* play. Don thinks this may mean that he'll lose all interest in *A Meeting by the River.*

January 15. We have just heard via Jack and Jim (who is still in New York) that John Houseman is very enthusiastic about our play and that we're to send him two more copies to show to people there. Jack seems terribly upset about Jim's general state of mind just now, but more of that later.

This afternoon, just as I was about to go down to the beach to run with Don, I somehow hurt a muscle in the calf of my left leg. It feels like a cramp. I can't rise on my toes without pain and this would make it awkward to drive my car, so someone is coming from the Vedanta Society to pick me up tonight.

This is a blissfully happy time with Don.

Very very slow progress on my book. And yet there really is no obstacle whatsoever. Just tamas.

Such a nice party on the 12th, lunch at the Vadims'. I do like both him and Jane, and Gore was there, about to retire next day to his slimming farm, and also beautiful broken-nosed Michael York and his wife Pat. We had them to supper last night.

Here is something I want to write about January 11 but didn't type out. I only scribbled it down in pencil, so now I'll copy it before continuing:

"Swami lying in bed, so snugly tucked in, and me prostrating before him. He looks out at me over the edge of the bed, as though he were in a boat and I was swimming alongside. Then he makes an effort and raises himself a little, in acknowledgement."

Gavin continues to go to seminars held by Krishnamurti at Mary Zimbalist's house up the coast. Krishnamurti says thought is bad, intelligence is good. Thought takes you round and round in circles, it doesn't solve anything. Intelligence is not personal or collective, it is something else.... Krishnamurti acts roles, in order to set problems for his students. "I'm an old man, I've lost my wife, I'm lonely, help me, feed me!" His tremendous energy. Gavin says he doesn't seem to love anyone or need anything. He doesn't read books—"that's an escape"—but he does watch T.V. He hates the word meditation.

January 16. Last night I saw Swami again. (I had to be driven in to Vedanta Place because I've somehow slightly strained a muscle in the calf; it's better this morning.) I asked Swami about his remark, on January 11, that Swamiji liked my writing. When I asked this, I noticed something I've noticed before—if you fail to ask Swami a question about one of his statements and you ask it some days later he always seems to have been expecting the question and he knows just what you're referring to, and he smiles, and seems pleased. He explained that when Ashokananda objected to a sentence in my introduction to *Vedanta for the Western World* (something about how Vivekananda could have become a great national leader if he had wanted to), he, Prabhavananda, had felt very unhappy; but then he had seen Vivekananda in the shrine and had felt reassured that Vivekananda approved of what I'd written and didn't find it offensive.

It is in cases of this kind that you really get a glimpse of the way Prabhavananda's mind works. The fuss with Ashokananda happened twenty years ago—and therefore, so did the vision. But the vision is at all times absolutely "now" in Prabhavananda's mind. Theoretically, of course, I can see that this should be so; but I still have to marvel that it *is* so!

Meanwhile, in the more usual "now," Vandanananda has been acting up again, taking advantage of Ranganathananda's presence to demonstrate Swami's tyranny and show that he, Vandanananda, can have no will of his own about anything. There was a ridiculous instance of this at supper, when Vandanananda absolutely refused to decide what time Ranganathananda should leave for the airport this morning; Swami had to be appealed to. And Ranganathananda was anyhow only flying to San Francisco, so he could have caught another plane if he missed one. The girls are very much in this act, which is disagreeable; sometimes they really seem to understand nothing but bitchery. Also, I seemed to detect signs of impudence toward Vandanananda on the part of Asaktananda, who now feels secure in Swami's favor. All this would be far worse than it actually is, if Swami seemed in the slightest degree gaga, but he sparkled with vitality. He still claims to be sick and exhausted but probably he is just instinctively conserving energy. Also, I'm sure, he didn't want to have to listen to Ranganathananda's nonstop table talk.

Meanwhile Don saw Jack Larson and agreed with me that he seems very worried about Jim. But there were no confidences.

January 24. Rainstorms on and off, more or less, since the 18th, and another five days of it are predicted! Am glad at least we didn't go down to stay with Truman Capote in the desert yet. Maybe by the 30th, when we're due to, it'll have let up. Because of the rain, my car has stalled twice. It had a new battery put in, but that has made no difference. Don's phone went out too; the rain leaked on the wire[,] which brought the ants who electrocuted themselves and short-circuited it. The young phone man told me he has been working sixteen hours a day. He said climbing poles in the rain wasn't as bad as going down manholes, because in the manholes you have waterfalls rushing all over you.

Igor is very ill. That is, his physical condition isn't worse than usual but he is in a state of terror, presumably about dying. Bob recommends getting a psychiatrist in. Vera feels that nothing will help. She says, "When I come in the room, his eyes are looking with such a fear. Never his eyes are quiet." What disturbs me is that Igor's religion, his cult of his icons, doesn't seem to strengthen him at all.

Don says he wants to go to England before we leave for Australia. I am sad about this, but it's the way things are, and who am I to complain, considering how things are for other people? I am the most pampered creature of my age I know of. It's just that I'm so happy with him.

The sprain in my calf seems to have cleared up. Maybe tomorrow I'll even run on the beach in the rain; unfortunately the tide isn't right. The cyst seems unchanged, neither harder nor softer, larger nor smaller.

Today I finished chapter 4 of *Kathleen and Frank*. Fifty-nine pages in all. It's a grim effort but I keep on.

Last night, Paul Wonner and Bill Brown took me to see *Fortune and Men's Eyes*. (Don had seen it in New York with Bob Christian in it and couldn't face it again.) It is terribly silly and woolly and well-intentioned. A cute blond boy of nineteen (Don Johnson) gets fucked by a prison bully (Sal Mineo) up against some bars, facing the audience. His jockey shorts are dragged down and the bully pretends to go up him from behind. He grimaces and yells.[1] I suppose this is a milestone on the road to freedom. But how wonderful when we have really explored fucks and can get on to the moments of postorgasm; *there* you have a whole almost unexplored dramatic territory!

Have seen Sybille Bedford about Aldous, for her biography. She is a hypochondriacal mess but intelligent, really perceptive. There's also a man who is writing about David Selznick, but Jennifer tells me neither she nor David's sons have seen him and they don't want that sort of book written; so I'll make that an excuse not to see him. His name is Bob Thomas.[2]

Ted still crazy, walked into the house unannounced yesterday. Don yelled at him and drove him out again. It is awful because it is partly pathetic, partly an attempt to be reconciled with Don, and mostly a quite conscious drive to trick Don into giving way, just for the sake of tricking him.

[1] The 1967 play was by Canadian playwright John Herbert. Mineo (1939–1976), a movie star since appearing in *Rebel Without a Cause* (1955), directed it, launching Johnson's (b. 1949) film and, later, T.V. career, which culminated in "Miami Vice" during the 1980s.

[2] Prolific Hollywood journalist and biographer (b. 1922), his *Selznick* appeared in 1970.

January 25. The rainstorms keep blowing in through the Canyon from the ocean. The waves are brown. A tremendous torrent is pouring out of the hills down the channels; on Rustic Road the water takes the curves like a racing car, so that the whole stream tilts sideways, spilling out over the road. An alarming lot of our hillside has gone down onto Ocean Avenue, including part of the wall and railing of our pie slice of property. It even looks as if the corner of the steps below Don's studio could be threatened if this keeps up. We walked down to the beach in a lull between storms and talked to Jo, alone in her gaudy little flat. She told us she can't ever see Ben's father again because he has had Ben and Dee to visit him.

Talk about death with Don after reading an ill-written but quite interesting piece by Gary Fisher, a weird boy we met at Laura Huxley's, who administers lysergic acid to some of the terminal cases at Mt. Sinai(?) Hospital to help them overcome their fear of death.[1] (We have even been wondering if he could help Igor.) Don says that his LSD experience was like being shut in a closet. When he came out of it, it was as if there were several hangers with suits of clothes revolving around him, and he chose the one which was Don Bachardy. "Since then," he says, "I sometimes feel as if, when I came out of the closet, I came out into a dream."

Gary Fisher says that we escape from identification with the self in the act of creativity, and that this helps us not to fear death, because fear of death is identification with the self. But, in that case, why is Igor so afraid? He, more than most people, has experienced the escape, surely, through his music?

February 2. On the 30th we drove down to Palm Springs to see Truman Capote, and got back yesterday. I spent yesterday afternoon and evening studying a book by Edgar Holt on the Boer War. I have read most of it now, looking up the places in the *Times Atlas*, and have really quite a good temporary grasp of what the war was all about. So I'll get on with chapter 5 before I forget. The history of the York and Lancaster Regiment still hasn't arrived. I sent chapters 1–4 off to Richard from Palm Springs.

Truman seemed as insecure as Gore and perhaps also sick or in fear of being, and quite obsessed by money. He says he has given up sex almost altogether. I got childishly angry the first day we

---

[1] Fisher, clinical psychologist and professor at the UCLA School of Public Health, treated schizophrenics, autistic children, and cancer patients; he published a number of articles on LSD and psilocybin therapy.

were there because we weren't alone with him at all. The Irving
Lazars and a director named Frank Perry and his wife Eleanor,
who writes scripts and is a bit of a fart.[1] But really it didn't matter
so much, except that I got extremely drunk (they all did a bit) on
margaritas and later fell down and mysteriously hurt my side. And
the evening before we left we *were* alone with Truman and he was
his fascinating self. Thank God, he at last is writing about the rich![2]

Still don't know if Don will leave for England or not; it depends
on whether Dicky Buckle can get him appointments to draw
the Harwoods (not sure if that is the right spelling of their royal
name).[3] And then of course [Don's friend] may have gone off to
Peru to see his sister. Anyhow, Don has been so angelically sweet
the last week or so that it is almost *literally* angelic; our relationship seems to exist in a better world than this one. When it's like
this—and, after all, it quite fairly often is—my only pain is a feeling
of poignancy; that it can't go on for ever, that I have to die. I think
of death very very often, much more often than of God, these
days. Which reminds me that *both* Don and I forgot to bring our
beads with us to the desert!

Last night, Don dreamt that he and I sat watching a show put
on by Elsa Lanchester, some kind of act in which she had a wolf.
And the wolf attacked her and she was terrified and ran away. Her
fear was horrible, Don says.

February 17. They say there will be more rain, but at present the
sun is shining. We have been soaked on and off all month. The
destruction of California by earthquake and flood is now predicted
by somebody for early in April.

Brian Bedford didn't like our play, thought the characters literary
and the whole thing more suited to radio. The offhand, superior
way he told us this offended both of us; we had to drag it out of
him, in Gavin's presence, at Matteo's. Now we both feel we don't
want him to direct *A Single Man*; he isn't the sort one wants to

---

[1] Perry (1930–1995) directed *David and Lisa* (1962), *The Swimmer* (1968),
Diary of a Mad Housewife (1970), *Play It as It Lays* (1972), *Mommie Dearest*
(1981), and worked on Broadway; he was to split in 1970 from his first wife,
Ellen Perry (1916–1981), who wrote his screenplays.
[2] In *Answered Prayers*.
[3] George Lascelles, 7th Earl of Harewood (b. 1923), a grandson of George
V and first cousin of Queen Elizabeth II, then director of the Royal Opera
House; his publications include Kobbé's *Complete Opera Book*. His second
wife was Patricia Tuckwell (b. 1926), an Australian violinist.

work with. (But Paul Bogarde,[1] the other contender, isn't either.)

With Lamont Johnson on the other hand, I get along fine. Rehearsals have started on *Black Girl*. The whole thing has practically turned into a spade musical with African singing and dancing, directed by native experts. Don is to have a drawing of me printed in the program, side by side with a drawing of Shaw (by [Feliks] Topolski); I can't help feeling that this is lese majesty on my part.

At the first rehearsal, which was on the 14th, it was amusing to see how the British and Canadian actors spoke out like trumpets while the black actors threw the lines away, in little squeaky method voices. But Monty will cure all that; he is admirably good-humored but firm.

Still no news from England. Don is unsettled, unhappy, aggressive, then sweet again. This is one of the cloudy periods. There is much we can't discuss. Earthquakes are in the air. As for me, I am horribly fat, gassy, worried about the cyst on my finger which sometimes seems to press on the nerve and pain me. And oh how bored I am with the Boer War part of my book! I am fat because of a psychopathic gluttony caused by my uneasiness. I sneak into the kitchen and stuff myself with dates.

February 21. Yesterday evening, [Don's friend] called from London. So now it looks like Don will go there quite soon. Everything is uncertain, otherwise. Shall we go to Australia? Will John Lehmann come here?

More rain is due. The house Dick Spencer built above the Pacific Coast Highway between Sunset and Topanga has slid down to the edge of the cliff and the highway is closed; so people have to drive downtown by way of Topanga or Malibu Canyon![2]

Am depressed and tamasik, but that's not important; just internal weather. As for *Kathleen and Frank*, there is nothing to do but get on with it.

Don is happy, now he smells England. I say I want nothing but his happiness. Well then, why aren't I rejoicing? Because I'm jealous. So what else is new?

We had Leslie Caron to supper last night. She is quite likable but not an utter darling, too cold-blooded. Her skinny cold ill-

---

[1] Paul Bogart (b. 1919), American director of T.V.—"The Defenders," "Get Smart," "All in the Family," "Golden Girls"—and movies—*Marlowe* (1969), *Cancel My Reservation* (1972), *Torch Song Trilogy* (1988).

[2] Spencer was an architect. He designed and built the house himself. Eventually it broke in half and had to be removed.

mannered husband[1] came in later. Why do we entertain people? We both hate it. And tomorrow, no less than five and maybe more people are coming in to watch Renate [Druks]'s films.

Robin French and his girlfriend got married on February 14; our sixteenth anniversary!

February 26. In six months I shall be sixty-five. I seem to understand less and less about life. I try to think about it, or indeed about anything, and it all blurs. Why aren't I wise, like it tells you you will be, toward the end? One is a dull-witted, gluttonous, timid, ill-natured—I was going to write animal, but let's leave the animals out of this; the thing I am isn't fit to touch their hooves or paws.... I do *not* write this in humility or even dismay, however. I know I have made a mess of my life but so do most of us. With the advantages I have had, the friends I have known and indeed all the happiness which still surrounds me—greater in some respects than ever before—I ought to have become a living wonder. And I, to put it very very mildly, haven't. So? Krishnamurti would tell me that self-improvement is a delusion and Vivekananda would tell me that duty is a snare. Perhaps I know this, deep down. Perhaps duty and self-improvement efforts are really just part of a game I play with myself—something I do to get "the click in my head."

I do have this angst, though; there's nearly always something to worry about, and when there isn't I just worry anyhow. Yesterday morning, a really big bit of the bank below our pie slice slipped down and entirely closed Ocean Avenue, including great chunks of the retaining wall. They had to break it up with drills before they could truck it away. Yesterday evening, wicked old Mrs. O'Hilderbrandt called me, trying to make my flesh creep with hints of how the house was doomed from the beginning to slide down the hill, how the first tenants had pulled out of it in terror after hearing the geologist's report—and then tempting me to buy the lot she owns across the street from us.[2] All very well to retort that the house has stood here since 1925; the fact remains that the pie slice embankment and the retaining wall did too, only because there had never been a steady downpour like this one we have just had—and with more of it predicted for the weekend.

Leslie Caron came by yesterday afternoon. Don drew her and she bought one of his new paintings.

Also Dennis Altman had supper with us; he's on his way back to

[1] Michael Laughlin.
[2] Where she lived in her own house; see Glossary.

Australia. Shall we go? Still don't know. Still don't know if Don will go to England; he is waiting for word from Dicky Buckle. John Lehmann is probably coming, around the 14th of March. The day before yesterday I finished chapter 5 of the book.

Dennis has been seeing a lot of Negroes while here. Two of his impressions; that those who talk loudest about Black Power are the ones who are most apt to have non-Negro lovers, they "talk black and fuck white"; that black homosexuals are queers first and blacks second. Dennis felt a terror of New York this time (he had been there before), said you didn't dare to look people in the face when you passed them on the street, for fear it would involve you in some sort of fight. He thinks Los Angeles is much more friendly and much less tense. He also remarked on the growing hostility between blacks and Jews.

There has been interracial hostility in the *Black Girl* company too, but Lamont thinks this was all to the good. Susan Batson and Douglas Campbell have had real open fights, and Susan has led the other black actors in complaining that this play isn't a proper protest play. But Lamont can handle them. The blacks respect him because he has done these two T.V. films, one in Watts and one in Stockton, about the ghetto life.[1] And I believe Susan may be really good in the part. She is a grotesque neckless little Hottentot of a thing but she is fascinating, sly, animal, childlike, campy, weird, tender, furious, wise; she can do all the moods and aspects which the different scenes demand. And the African drumming and singing are going to help a great deal, provided they don't become too much of a floor show on their own account. Lamont had me go down there last Sunday and radiate satisfaction and confidence; in fact, I put on a bit of an act for him, as I used to do for John [van Druten] during rehearsals of *I Am a Camera*.

Poor old Jo's Mexican money pig has been stolen from her apartment by two teenage youths (the barber saw them go up and come down); nothing else was taken. So now she has a new moan: "I feel so scared there, all alone—whenever I hear someone coming up the stairs, I feel scared." We have neglected her shamefully. I really would like to help her somehow, but when I'm with her this awful self-pity repels me every time.

February 28. More rain, and yesterday evening a mild earthquake (the radio says) which smashed some windows in the Los Angeles area—we didn't feel it.

---

[1] Probably "Losers Weepers" (1967) and "Deadlock" (1969); see Glossary.

Last night we went to see *Dr. Mabuse*[1] at UCLA, and found they'd moved it from Royce Hall to a much smaller theater and there was no room for most of the people who came. So we were fit to be tied. During supper Don broke the news to me that Jim Bridges had called to tell him our play has been rejected by the board at UCLA. A woman called Frances,[2] who has to report to the board on suggested productions, hadn't liked it ("They're only interested in their cocks") so naturally her account of it was most unattractive; no member of the board had bothered to actually read the play. So that's that. Maybe it is a blessing in disguise. But it has shaken Jim's morale, and Don's too, though not so severely.

After supper we went home and ran some of our old home movies, which were sweet and nostalgic but saddening. Don of course mourned over the disappearance of the enchanting kittenish bright-eyed boy. I marvelled to see how fat, grotesque, anxious-eyed, nervously grinning, stiff-jointed and joyless-laughing I already was—at that time when I still preened myself and imagined I was quite attractive! What phantoms such film-figures are; the bright flicker of activity and compulsive fun, gone in an instant and forever. And behind it, the mystery. What *is* Life really about?

This morning a cable from John Lehmann, who *is* coming for two or three days, and a cable from [Don's friend], who wants to know if Don is coming to England right away, as he'd like to go with him to Morocco. Much discussion forthcoming about this.

Jim Bridges tells me the latest expression current among the young: *heavy* meaning good, marvellous, inspiring—"It was a heavy experience." This expression was actually applied by Jim's friend Bob (see September 16, last year) to a performance by The Living Theatre of a show they call *Frankenstein*, at the Bovard Auditorium on the USC campus. Jack and Jim are both wild about this company and I must say I was thoroughly interested when Don and I went to see them the day before yesterday. I went with a good deal of preliminary hostility, because of Jack's uncritical ravings, because of the irritating claims implied by the title *Living* that all the company's competitors are dead, and because of the holy-superior note in their program: "These ensembles are participating groups of the RADICAL THEATRE REPERTORY, in the vanguard of a new phenomenon in theatrical and social history—the spontaneous generation of communal playing troupes, sharing

---

[1] *Dr. Mabuse der Spieler* (the Gambler), first of the two-part 1922 film.
[2] Probably Frances Inglis (b. 1910), a Stanford graduate, once concert manager and director of the Writers Guild of America West, who ran the Committee on Fine Art Productions and helped with the Professional Theater Group.

voluntary poverty, making experimental collective creations, and utilizing space, time, minds and bodies in manifold new ways that meet the demands of our explosive period." We're going again on Sunday, to see a different show, *Paradise Now*, so will write more about them after that.

March 6. Today we heard that the National Portrait Gallery has bought Don's drawing of Wystan. His first sale to a public gallery! And to add to the brightness of the day, Don's father was really pleasant to me (more so than ever before) when he came by with Glade to see if he could fix Don's car! *And* Nicholas Thompson (who is over here on a visit) seems to be really enthusiastic about our *Meeting by the River* play. He is coming to supper tonight. (Incidentally, he is going to settle here sooner or later and take over the literary department of Chartwell Artists and so presumably become our agent. Robin hasn't said a word about this to me.)

We never did see *Paradise Now* because the Living Theatre was closed down a couple of days previously. Some of the audience were encouraged to take off their clothes; one man allegedly took off his pants and shorts and then his tie, which he draped around his cock. The closing wasn't due to indecency but because the fire department said it couldn't be responsible for the safety of the audience!

There has been a flap over *Black Girl*. The black actors felt that the marriage of the Black Girl to the Irishman was a sellout. They denied this completely, after they realized how silly they sounded, but anyhow there were hot arguments. So Lamont devised a sort of happening in which the actors step out of character and discuss the play, and I typed it up as best I could and then they hated that, so the whole thing has been dropped (or so it seems as of this morning) and the play remains as is. I felt no resentment against the blacks or indeed against the whites, who got quite nasty, but I did feel that Lamont rather lost control—unless indeed his yielding can be regarded as a sort of judo. Douglas Campbell was really eloquent and made a lot of sense. Susan Batson was childish but dramatically impressive. Douglas told her she was wallowing in self-pity and refused to be intimidated by the ghetto-childhood references; he'd been raised in the slums of Glasgow, he said, and had to become a killer until he realized that violence is no good.

March 19. John Lehmann left yesterday morning. He had been staying in the house since the 14th and it seemed like two weeks. He is very big, he made the place seem tiny and there was nowhere for us to talk to each other where he couldn't hear, except the studio. We slept there, as we had given him the "basket."[1]

He didn't really want to see us. It was all symbolic. Now he can say we have entertained him and introduced our friends to him. But it is the symbolic aspect of anything which really impresses him—for instance, that Don has had a drawing bought by the National Portrait Gallery and that he is going to draw the Harewoods. And John himself thinks of his life in terms of his CBE[2] and meetings with the Queen Mum. I sound venomous but I am not; I ended up, as always, feeling simply sorry for him. He is quite stupid and thick-skinned and he expects to be waited on hand and foot, and he was scared lest he should somehow be maneuvered into having to buy us a meal. His talk[3]—actually a paper which he read aloud, badly—was so dull and dead that I was quite embarrassed, because I'd introduced him to the audience by saying that he was an absolutely unique authority on the thirties; the only person who had known its poets on three levels, as friend, fellow writer and editor-publisher, etc. etc.

I took him to see the Towers of Watts. They seemed more wonderful than ever—both as spires in the distance and as structures seen from below. They are an absolutely no-shit statement of individualism. You feel, everybody might do something like this in his backyard—and why the hell don't we all? But the purity of the whole thing consists in the fact that there wasn't anything else like it anywhere around; it has the purity of a monomaniac's hobby. (Actually, Simon Rodia *did* dream of being famous—"I had in mind to do something big, and I did"—but his way of going about it was so fantastic (although, actually, it succeeded) that it still seems like a hobby.)[4]

It was a beautiful afternoon and Watts itself looked anything but a sinister ghetto—so spacious and airy, with its little houses and wide roads; calm and rural, almost, after the teeming freeway.

Black Girl is as all right as it ever will be in this production. Susan is splendid; perhaps she will one day become a great actress. Douglas is so fat and old and ugly, but we should be lost without his voice. The black dancers are very exciting and one boy in particular, Fred Grey, is a brilliant mime and at the same time sexy

---

[1] Their own bed.
[2] Commander of the Most Excellent Order of the British Empire, for service to the nation.
[3] At UCLA on March 16.
[4] Sabato Rodia, a local workman, spent more than twenty-five years building the junk towers; see Glossary.

and sometimes beautiful in the grace of his movements. Gordon Davidson did everything possible to rock the boat; he nearly drove Lamont to resign. He is rude in a special Jewish-theatrical way; he doesn't know how to make suggestions or give advice without insulting people, and he keeps reminding them that he's the boss and that his word goes. He ought never never to be put in charge of a theater; he is a ruthless back-seat director.

The critics are coming to see the play tonight. The official opening is tomorrow.

March 22. The reviews (*Los Angeles Times, Herald-Examiner* and *Variety*) were all bad. The critic of *The Nation* says she is going to write a piece praising the play and attacking Los Angeles taste, but that won't appear until after we've closed. Monty talks of letting audiences of high-school and college kids in free, to fill the theater. It is tiresome, but I know the play isn't that bad—not nearly as bad as Shaw's own worst—and Susan is quite an experience. Probably we'll do wonderful business elsewhere later.

Don is now definitely booked on a plane to leave here for England on March 31. Everybody talks of the earthquake, which is predicted for either April 4, or 13, or 17, I think. Definitely April anyhow.

Am crawling steadily on with *Kathleen and Frank*.

An extract from a letter I got early this month, asking me to subscribe to an organization called Theater in the Street. How can such a truly noble cause be made to sound so funny? "Millions of dollars are being spent to find a solution to the strife in our cities. Yet it costs only three dollars to bring an evening of live entertainment to a child in a ghetto street. Instead of a night shattered by sounds of breaking glass, sirens and gunshots, this child could hear the plays of Molière and Chekhov."

March 29. On the morning of the 26th, when we got out of bed, and went out on the deck, there lay a beautiful four-masted schooner at anchor in the bay. Her flag (seen through the binoculars) looked foreign but you couldn't be sure, it was flapping about. Don said, "At last—our ship's come in!"

About an hour later the phone rang and it was Robin French, to say that a man named Sidney Beckerman[1] was about to buy the film rights of *Cabaret* for Tony Harvey to direct and that they

---

[1] Beckerman had a production company and worked with various studios; his best-known film was *Marathon Man* (1976) at Paramount.

wanted me to work on it and quite agreed to my having Don as a collaborator!

So now plans are altered or rather, dissolved. Don won't at any rate leave until Tony Harvey goes back to England, so we can talk about the film.

At present the whole thing seems too good to be true, because Tony is charming and our sort of person and he wants to get right away from the stage musical and shoot the picture on location in Germany. The only question is, will Beckerman back out at the last moment as Cinerama did? He must surely know about this copyright business,[1] so we hope he won't suddenly raise that as an excuse.

(Yesterday all day long and today until a couple of hours ago, there was a thick sea-fog; now the schooner has gone.)

Other good news: it looks as if our play will get a tryout at the Mark Taper, maybe in May. Ed Parone likes it very much. He wanted to direct it himself but Jim thinks he will give way about this. Also despite the bad notices, *Black Girl* is doing quite good business and the audiences are very enthusiastic.

Don is painting well and life with him is at its best. And Swami, when I saw him yesterday, was absolutely radiant. He filled me with joy.

April 5. Have just finished dusting my desk as a symbolic preparation for Easter tomorrow and because the desk was so dirty.

The *Cabaret* deal is still up in the air. Tony has gone off to San Francisco for the weekend and Mr. Beckerman's lawyers are still wrangling with Hal Prince over the number of songs which the musical will have to have in it. Also, we are not *quite* sure that Beckerman loves us or that Tony loves us either—enough, that is, to go to bat for us with Beckerman.

Forgot to mention that we went to see *The Lion in Winter* and really quite hated it. The picture stank literally; that is, the theater did; someone had thrown a stink bomb into it—a terrorist, it is believed, from the strikers against the *Herald-Examiner*; a protest because the film was advertised in the *Herald-Examiner*. Our clothes stank of it when we came out and indeed I had the same pants on when we went to see Tony Harvey next morning and they didn't lose the smell until I'd sat in the hot sun on his patio.

The earthquake scare is still very much on. Good Friday passed

---

[1] I.e., that Carter Lodge, as van Druten's heir, had rights in the material; see Glossary.

without incident—although there *was* quite a strong quake out at sea off the Mexican coast, which was felt on shore. The *Los Angeles Times* had a headline: *Mystics shaken up—not State*—the "mystics" being the people who had foretold the quake would be on the 4th. But actually the whole month will have to go by before the various dates set have all been proved false.

April 6. We heard this morning that Rex Evans died last Thursday, of a heart attack. One of the last things he did was to send a notice to the *Los Angeles Times* about Don's drawing of Wystan being bought by the National Portrait Gallery. It was in the Calendar, last week. Don didn't know this until he spoke to Jim Weatherford[1] this morning. I talked to Jim too. He says he's going to keep the gallery on, at least for the time being. I had the impression that it hasn't hit him yet. They were together twenty years.

On the 3rd we took Vera Stravinsky and Bob to see *Black Girl*. Igor of course couldn't come. And yet they are planning on taking him to New York soon! And Bob says he is composing all the time. I find it very disconcerting, being with him now, and I realize how much I used to depend on him in conversation. Now he is mostly silent, and stares at me with just a faint suggestion of suspicion. And I don't know what to say.

Bob and Vera told us how Mirandi Levy always used to call Igor "Pussycat" and how he'd always disliked it, and how this time, just as she was about to leave for Europe, she called him Pussycat and he said sharply, "Don't call me that!" and she was terribly shocked and asked, "What am I to call you?" And Igor said, "Either Maestro or Mr. Stravinsky." Vera had taken Igor to task for this later and he had been sorry, but it was awfully good for Mirandi, just the same.

Today has been cold and windy but beautiful, after the rain last night. To celebrate Easter, I have: tried to make a verse translation of an extraordinarily uninspiring poem written by Swamiji in Sanskrit (this for Swami, of course), worked on *Kathleen and Frank*, written a letter to a professor named Horst Jarka about an essay he did on "British Writers and the Austria of the Thirties" (this means Stephen, John Lehmann and me), cut my corns (and incidentally cut my leg with the razor blade) and sewn a button on the cuff of my green shirt. Had intended to restart the isometric exercises and write to tell Robin Maugham no, no, no, I will not and cannot see

---

[1] Evans's companion, also his silent partner in the gallery.

a film in his treatment of *The Second Window*.[1] But no time for this.

When I'm reading aloud in the temple I find myself (after I've finished reading and Swami is answering questions) looking from him to the shrine and back again to him and sort of creating a prayer triangle: calling upon them both to *make* me love God. It is very exciting and real, calling silently on Swami for his help *in the presence* of the shrine. Because his whole spiritual capital is invested right there, and Maharaj is there and so how can Swami refuse? I feel I am blackmailing him, in a good way. But why can't I ever tell him this?

April 7. This morning, in the middle of breakfast, there was a sudden nervous twinge in the palm of my hand and I touched the place with the fingers of my other hand, and there was a lump. It is down in the fleshy part of the palm of my left hand, right below the little finger and in line with the cyst on its middle joint. My first reaction was a sick panic, as though I'd been bitten by a snake. Mustn't this mean that the cyst is malignant and that the malignancy is spreading? What was shocking was its being so suddenly and instantaneously there. (Of course I may well not have noticed it before; it was the twinge which informed me.) Dr. Allen isn't in his office today; must wait till tomorrow. Now of course I begin to think more calmly and sensibly. If Dr. Allen wasn't disturbed by the first cyst, how can this one be something serious? But what I hate is the waiting, and the X rays, tests, etc. Don, as always on such occasions, is an angel of sweetness—which in a way makes the situation more painful, because I can see that he's a bit alarmed, too.

We have just been to see Gerald. Today he hardly spoke at all and regarded us with a stare, as though dazed. Michael rattled on about Atlantis, the Abominable Snowman; topics in books they are listening to.

April 10. Dr. Allen said he thought the lump in my palm is the beginning of a Dupuytren's Contracture; it isn't malignant and can be treated by X rays or surgery. He doesn't think it's connected with the cyst. I'm to go to a hand surgeon (or whatever you call them) and have it examined. Hands, says Allen, are "like watches," very complicated, and hand surgery is far more delicate than heart surgery.

---

[1] Maugham (1916–1981)—barrister, soldier, author of fiction, plays, and screenplays, and a nephew of Somerset Maugham—published the novel in 1968.

Yesterday evening, Swami asked me about a quotation from *All's Well That Ends Well* (Act I, Scene iii, 230–33):

> Thus, Indian-like,
> Religious in mine error, I adore
> The sun, that looks upon his worshipper,
> But knows of him no more.

Swami is convinced that this refers to the Gayatri mantra: "May we meditate on the effulgent Light of him who is worshipful ... etc."[1] Because the Sanskrit word used also means the sun. I find it very hard to believe that Shakespeare knew about this, but will look it up.

April 12. Poor Jo, she was all set to go to Rio yesterday—actually it was I who suggested it, and it *was* a good idea. Jo was thrilled and she arranged it all, including getting Alice Gowland to go with her. And then, on the evening of the 10th, Alice called the trip off because of some work she has to do!

Dorothy came to clean house yesterday and asked about Jo. Dorothy had had a very sinister dream about her: Dorothy was down in Coronado—we lived down there and she had been working for us and was just coming away from our house, and crossing by a bridge over a river which ran out to the ocean. The river was in flood and obviously very dangerous. Just as Dorothy reached the middle of the bridge, Jo came floating down the river clinging desperately to a large log. She was being washed out to sea. Dorothy yelled for us to come and help but she couldn't make us hear.

An expression of Dorothy's I never heard before: "I was up late watching television last night, oh I was real late, I sat there till they'd tied up the last dog."

Jim has gone to join Jack in New York, to be there when Jack's opera is tried out for the Met. So no news about a possible performance of our play at the Mark Taper. And no word from Tony Harvey about the *Cabaret* film. So Don has decided to leave on the 15th and just come back early if he has to. When he gets there, he can go and stay at the Harewoods' place in the country and draw them while he's in the house, Buckle says. [Don's friend] is back from Peru. As for Tony Richardson and Australia, we've heard nothing more.

---

[1] The most sacred mantra of the Hindu tradition.

Today I finished chapter 6 of *Kathleen and Frank*—more than thirty-five days to a chapter!

April 20. Don left for England at noon on the 17th. When we woke up that morning he said, "Pan Am will have the taking of me up!" (He loves Melville's "Billy in the Darbies.")[1] On the way to the airport, he said, "If anything happens, you'll know I have my beads with me all the time."

I had supper that night with Jim Bridges. He described the drastic scenes of emotion in New York between Virgil Thomson and Jack, after the opera had been successfully auditioned for the Met. Virgil told Jack, "You've given me a live baby." They both wept. Virgil also told Jack that he is a great poet. Jim says he's never jealous of Jack's success.

On the 18th, I had a visit from a wonderful pair, they seemed more like angels or Venusians than eighteen-year-old Californians: Peter Schneider and Jim Gates. Peter is little, curly-headed and probably Jewish. He's a magician, really performs at parties and on the stage; he has a card: "Esmereldo the Mediocre Magician." He said, "I'm making magic now." Jim is tall and skinny and blond, with a thin blond mustache; he studies Vedanta and has been up to see Swami. When I showed him the D.H. Lawrence candlestick[2] they bowed down before it, mockingly. They give imitations of me, Alan Watts and others in imaginary dialogues. One could hardly call them fans; their admiration seems boundless but the axe is hanging over your head every instant and they are watching you like doctors. Peter's father is a psychologist. Peter said, "Up to a few years ago he was my best friend, but I can't stand his arrogance any more," and Jim agreed, "Yes, he's the most arrogant man I ever met." Nevertheless, they are both staying with him, and I talked to him on the phone as he was going to drive them over to see me and wanted directions. He told me, "I'm a colleague of yours"; he teaches at Cal. State.

Impossible to decide if they are "lovers." The word seems impertinent, anyhow. Their relationship, like everything else about

---

[1] Isherwood also loved this ballad, from the end of *Billy Budd*, which contains the line, "Heaven knows who will have the running of me up!" "Up," meaning heaven, was a special habitat of Bachardy's Kitty persona.

[2] Made by Lawrence from a twisted branch fixed to a hand-cut, oval, wooden base possibly from the same tree; given to Isherwood by Frieda Lawrence, Dorothy Brett, or Mabel Dodge Luhan in Taos in 1950. During the same visit, Georgia O'Keeffe gave him a cedar root she found at the Lawrence ranch; he mentions the root in *D.1*.

them, seems extraterrestrial. I want to go and watch Peter perform, at some hippie theater in Venice, next Saturday. Then maybe I'll see his father.

Saw Swami yesterday. He told me how Krishna said that he gave liberation easily but was niggardly about giving devotion. This seemed like a telepathic answer to me, for I've felt more and more dull and dead when I try to meditate. I always ask for devotion and never feel a spark of it. *And yet*—it's so strange, but I am not in the least bothered by this; because then I think, I've got Swami, and how can anything really bad happen to me as long as I'm under his protection? I *know* it can't. Here, in this life, everything seems fogged in solid; but when I think of life after death I suddenly realize that I really do have faith, a great deal of it!

Supper with Chris Wood. He told me that Paul Sorel has suddenly become absorbed in teaching a mentally retarded child how to read. It makes him so happy, Chris says, that the boy's face lights up whenever he comes into the room.

May 1. Well, the *Cabaret* film is on. We stand to win at least ten thousand dollars, for a treatment; then, if that's accepted, ninety thousand for the screenplay; then, if the picture is made, a bonus of twenty-five thousand if we're the sole credited authors and of ten thousand if we share the credit!

Tony Harvey is going away on Tuesday, leaving us unbugged, thank God, to work on the treatment. Don will either return next Tuesday or Thursday.

Yesterday I saw the hand doctor, Dr. Ashworth; he says it is a contracture and that the "cyst" (which isn't one) is part of it. These conditions are found chiefly among northern races, British and Scandanavian, never in Asia. They are associated with epilepsy and diabetes in some cases but no link has been discovered. They are never malignant. If mine gets worse he'll operate and will be able to cure it; but often they arrest themselves for years. Many patients, he says, don't come in until their hands are half clenched! The clinic looked more like a palmist's, with these doctors sitting and gazing into the outspread hands of their patients.

Peter Schneider and Jim Gates came up to Vedanta Place last night and I drove them back. Peter was a bit suspicious and hostile; Jim was beaming with devotion for Swami. But they really both are sweet and so amusing. They now behave as if they had known me for years. They only became distant and embarrassed when we ran into Peter's father while getting something to eat afterwards. He and I talked father talk, but I kept catching Jim's eye. The boys told me that they feel a difference between themselves and other people of their age because they no longer feel "cool," that is, detached, objectively amused. At their house on the Sherman Canal there is a shed where they meditate; at least Jim does.

May 5. It's bitterly cold and Doug Walsh[1] has never shown up to fix the electric control on the heater, so am freezing. Dorothy is sick and a much younger friend of hers, Mrs. Marie Jackson has come over to clean the house; she seems very nice. Am rather worried about Dorothy; on the phone she sounded so depressed and even a bit scared, as if she didn't expect to get better. However she cheered up when we talked about Jo. I told her that Jo's Siamese has had six kittens and how Jo spends hours watching the cat feeding them. (Bill Brown told me how Jo had said to the cat in his presence: "You're the *only* thing I love in the whole world!" Which Bill had taken a bit personally, after all he and Paul have done to make her feel wanted!) So Dorothy said, "It sort of takes the place of that other"; I loved Ben being referred to as "that other." And then I told her [I think] that Jo is still hoping to get Ben back. Dorothy said that she didn't believe Ben would ever go back to Jo, because it wasn't really Dee who had taken Ben away from Jo, he must have wanted to leave her anyway. "He must have been thinking far back how he could dump her."

Have just seen Gerald. He had another stroke about two weeks ago and has now been fitted with a catheter. Michael says this is really a good thing because Gerald doesn't wake up (and thus doesn't wake Michael up) in the night to pee. Gerald was cheerful and quite articulate; we talked a lot about volcanoes.

(Which reminds me that there *was* an earthquake in California in April. It happened on April 29, during the afternoon, on the north shore of the Salton Sea. A lot of windows were broken in Indio. The shock was quite considerable (six on the Richter Scale, I think) and it was felt in Las Vegas; and the girls in the French office in Beverly Hills were quite scared, the building rocked so. The extraordinary thing is, I didn't feel it at all down here; I had just come in from jogging on San Vicente with Jim Bridges.)

Haven't seen Peter Schneider and Jim Gates since I took them to *Black Girl* on May 1. I suppose they won't call me, they are waiting for me to call them. And now I'm embarrassed to because I want to; I would love to see them. There is something strangely delightful in being with them. I find I haven't recorded our visit

---

[1] Not his real name.

to the so-called theater in Venice, The Walrus. It's really a dance hall, part of what must once have been a very big apartment in an old building in the hippie section of Washington Boulevard. All sorts of people come, all ages, black and white, and they dance and there is a light show. Everybody dances, it seems, in the way he or she feels; there is no question of "knowing how to" and in this sense it is like the kingdom of heaven, everyone is equal and happy. Quite young boys flutter around like moths, flickering their arms; and old women do a sort of hula; and young couples perform various rituals. Cop cars are cruising below and two narcotics squad plainclothesmen came upstairs and mingled with the dancers; they looked disarmingly benevolent. But Peter didn't perform after all. He felt the atmosphere wasn't right for his magic, the audience wasn't in the mood for it. And I think his instinct was correct. Anyhow we hung around there for about four hours, and Jim gave me a letter, saying, "We know so much about you, I think you should know something about me." From the letter I infer that Peter is definitely not queer and that Jim isn't in love with him. But the funny thing is, I haven't had one moment alone with Jim since then to ask him all the questions which arise from the letter. The evening was most enjoyable (though Peter was sulky and morose) as far as I was concerned—this was on April 26. One thing which Peter and Jim said to me that night I remember because it flattered me: "You're not like a writer." They meant that I seemed to them to have a life apart from writing. Their little house is in a part of Venice I've never visited before, with very steeply arched bridges over the canals, which are surprisingly clean and have ducks swimming on them. At the back of their lot, near the canal bank, is the shed where they meditate; it is much tidier and cleaner than their house, with mats spread on the floor and a little shrine with a picture of Ramakrishna. They have a girlfriend whom they jokingly decided to both of them marry, calling themselves "Schnates." Her name is Allyn [Nelson], she was there when I visited them. I thought she was pregnant but it turned out that she just had a huge stomach. The boys thought this wildly funny when I told them.

Don is supposed to come back tomorrow. I have a sort of sick-ish premonition that he won't. Not that it really matters, and if he has got some extra work that's all to the good; but now the *idea* of postponement in itself upsets me, because of the other times. It's a sort of death image.

Igor has had an operation in New York, the blood clot was removed from his leg. His condition is very serious but he is still alive, or was when the last telephone call came through; Igor's secretary keeps in touch with them and phones Bill Brown who phones me. He is so tough that you never know.

Swami called this morning to say that they all loved *Black Girl*. It's too bad that it has to come off, because there has been a real word of mouth success for it, these last few weeks. And they have been having discussions after the show. Susan Batson answers questions with a good deal of intelligence. Of course the blacks simply cannot get it through their heads that Shaw wrote this more than thirty-five years ago and that anyhow he was definitely not writing about the Negro Problem as such—much less the Negro Problem in America in 1969. There was almost a minor riot when a white man in the audience objected to the end of act one, the black dancers sneering at the audience as whites. The blacks told him he was full of shit, and a lot of whites took their side, and a lot more whites didn't. The final performance was last night.

I got a sudden fit of irresponsibility and spent most of yesterday out at shows. Saw Cocteau's *The Testament of Orpheus*, which I liked very much except for the long yakky part during the trial. And *Fortune and Men's Eyes* for the second time, because I was curious about the changes they've made. In the rape scene, Sal Mineo takes all of his clothes off before going into the shower room and shows his cock; and then in the shower-room he pulls Don Johnson's shorts off and you see his cock too. And at the end of the play, when Mona is being beat up by the guard offstage and you hear his cries, Don Johnson gets sadistically excited, puts his hand inside his pants and mimes jacking himself off, with gasps of lust. When he is supposed to have come, he slowly brings out his hand, evidently full of semen, and stares at it in horror like Lady Macbeth. Blackout! The audience was mixed and predominantly hetero. A few went out but many frankly loved it. A much younger boy played Queenie and a much prettier boy Mona; Don Johnson's body is really beautiful, so was the Queenie-player's, and even Sal Mineo still looks pretty good with his clothes off. This helped a lot. But oh dear most of the play is so hysterically silly.

May 6. Woke up with an absolute certainty that Don was going to postpone, but he hasn't, and it's twenty of twelve this morning and he's scheduled to arrive (early) at 4:55 p.m.

Forgot to mention that, when I was about to go into the theater to see *Fortune and Men's Eyes*, Don Johnson arrived in a sport car driven by a girl whom he proceeded to embrace and very publicly and repeatedly kiss, as if to warn us all not to get any wrong ideas about him.

Two earthquakes yesterday, very mild and I felt neither of them. But one had its epicenter in downtown Los Angeles!

Also forgot to mention Gerald's astonishing violence when I happened to refer to Napoleon. It seemed so odd that he would still feel such a very long-range resentment in his present condition.

Had supper with Bart Johnson last night. A boy was there who collects desert junk—for example, a can of Log Cabin syrup in the form of a log cabin, no longer marketed, he says. It was completely rusted over. He claims that you can actually sell this stuff. Tried to see it aesthetically but couldn't. The boy's name: John Ingraham.

On April 27 I had supper with Jennifer (another "forgot"); a very nice evening, sort of Persian. She wore a brocaded robe and we sat on a divan in beds of cushions, eating our supper out of doors by candlelight. While I was waiting for her I looked through her books and discovered that Jung's essay on synchronicity[1] which I've always wanted to read is in volume 8 of his collected works. Yesterday she sent me a copy and I was so delighted; how seldom one gets a really thoughtful gift.

May 31. This is a madly compulsive period. We finished the rough draft of the *Cabaret* film treatment on the 27th and are now well into the rewrite. I think it's really very well constructed and quite a bit different from novel or play—you can't be different from the musical because you can't be different from nothing. But the instant our half day of work is done on *Cabaret* there's Don's painting for him (it looks like he will have a show before long, after all) and for me *Kathleen and Frank.* (On May 19 I finished chapter 7; this one took thirty-seven days! Since then, with immense toil and the aid of three books, Churchill's,[2] Rayne Kruger's[3] and Kearsey's,[4] I have managed to write *just under two pages,* describing the battle of Spion Kop!!) And then, if any time remains over, there's letter writing to be done and this diary to be kept (which is why it hasn't been) and the guilty urge to work out, either at the gym

---

[1] "Synchronicity: An Acausal Connecting Principle."

[2] Winston Churchill's war correspondence—*London to Ladysmith via Pretoria* (1900) and *Ian Hamilton's March, Together with Extracts from the Diary of Lieut. H. Frankland, a Prisoner of War at Pretoria* (1900)—appeared in a one-volume edition in 1962.

[3] *Goodbye Dolly Grey: The Story of the Boer War* (1959).

[4] A.H.C. Kearsey, *War Record of the York and Lancaster Regiment 1900–1902* (1903).

or jogging or at least doing my minimum schedule of isometrics; haven't been too bad about this.

On the 28th, Tony Richardson called, wanting me to do a rewrite on the script he already has for *I, Claudius*. I said yes, *we* would; but am still not sure he really understands about my partnership with Don. He is leaving very shortly for Australia to make the Ned Kelly film, that is, if Mick Jagger can be sprung; he is to play Kelly and has just been arrested again on some narcotics charge.[1]

Saw Swami this afternoon. He is still in this marvellous state; not merely in excellent health and spirits but able to convey, as almost never before to the same degree, an absolute spiritual guarantee: *this thing is true*. He tells the same stories, that's beside the point; because, when he's like this, the story is spiritually fresh each time. Today, he told how Maharaj had commanded him three times: *love me*. And today this made my flesh creep and my eyes water. Swami wouldn't have needed to explain, even to an outsider, that it didn't mean "love me, Brahmananda"; his tone and manner made that obvious.

Yesterday afternoon I went down to Peter and Jim's shack on the canal. Jim is away up north in Washington with his sister. When I came in, Peter was playing his guitar and mouth-organ combination (it has become somehow protocol not to greet one another). He likes to get into a conversation and then discuss *why* we are talking about what we're talking about. He also wanted to know, did I think him attractive. (Leading up to making me answer this never spoken question took him about an hour.) Told him he was attractive and that Jim was beautiful. None of this was flirting, exactly. Flirting means trifling with somebody and we were neither of us doing that; we were sincerely curious and our motives weren't ulterior. I still feel that they are both of them altogether unusual creatures; so does Don. He has already drawn them.

When I said my motives weren't ulterior, I must confess that I am beginning to wonder if Jim and Peter can't somehow be brought into my projected novel about the two old men, the Swami and the Writer!

June 1. Another thing Swami told me yesterday was that he has

---

[1] On May 24, 1969 in London for possession of cannabis. The trial was delayed until September 29 so he could make *Ned Kelly*. He was previously arrested on May 10, 1967 at Redlands, the home of Rolling Stone Keith Richards, and convicted of possessing amphetamines.

come more and more to realize that Maharaj, Swamiji, Holy Mother and Ramakrishna are all the same. He says it was a long time before he could feel the presence of Ramakrishna but that now this comes to him very strongly.

Talked to Peter Schneider this morning. He is getting terrible stage fright about his engagement to do his magic act in Bel Air next weekend before an audience of six hundred people. He had a nightmare about it last night in which he told the audience he was going to vanish and then simply walked right out of the hall, and walked and walked until he got to the freeway—and it was on fire! He is also puzzled and worried because Jim has never written to him since he went up north, and Jim is such an ardent letter writer; wondered if he could be sick, dead, or what.

Last night Don and I had supper with Jo. The six kittens of her cat are now quite big and running about everywhere. Jo's latest self-pity ploy is to ask us about our movie collaboration and then murmur tearfully that it must be so wonderful to do things *together*. The curry she had cooked was too watery and not nearly hot enough, and this she excused by saying how she hates to cook *alone*. Poor dreadful old Jo! How she degrades herself by talking like this, and degrades us too by drying up our compassion (such as it is!) and making me write this bitchy note about her. The best thing about the evening was that we trotted all the way down to her house and I wasn't the least out of breath. That's what stopping smoking does for you! As Don said, I couldn't have done that when he first met me.

June 12. Just as an exhibit of my "public relations" self, here's a postcard I wrote to a college boy named Robert Cullinane who asked me "what or who it was that inspired you to become a writer and how you made headway in your calling":

My father told me stories, my grandmother talked about the theater, so it seemed the most natural thing to me, as a child, to act and write—it was play, as opposed to work. It has remained play, more or less, except for a few breadwinning chores. Certainly I have never thought of it as a profession; it is far too important to me. As for the acting, that turned into lecturing which is much more fun because you speak your own lines and hog the stage! As for getting ahead, that's a matter of luck and not stopping. Ambitious people are actually more likely to give up, they get so discouraged. If it's play, why stop?

The various shades of falseness in this are material for a whole autobiographical essay—maybe something I could work into *Kathleen and Frank*. And yet, overall, it's an approximately true statement.

We sent the treatment of *Cabaret* off to Tony Harvey in England on the 9th; haven't heard yet. Robin French says he likes it. Meanwhile we are playing around with ideas for *I, Claudius*. Shall we do both of them, or neither? Shall we go to Berlin, England, Australia? All is uncertain.

There's also a production of *A Meeting by the River* which Nicholas Thompson is "putting together" in London. We suspect that he and Robin are in favor of pushing Jim Bridges out of directing this, and he is making it easy for them by taking on all this other work; directing Jack's *Cherry, Larry, Sandy...* for the Edinburgh Festival, writing a film for Leslie Caron and Michael Laughlin and directing his own screenplay, *The Babymaker*. Nevertheless, Jim keeps saying how excited he is about our play, how he wants to go to India before he does it, etc. He simply has the rush bug, like all of us nowadays.

Jim Gates is coming back from Washington tomorrow, recovered from his appendicitis operation. Peter's magic-making on the 7th was a fiasco because they made him perform to an audience which was mostly drunk and didn't particularly want to hear him and couldn't anyway, because the mike didn't work and the tables at which they were sitting were back inside a sort of porch, looking through arches into the room where he was. Don and I went. It was horrifying to see all these Others, nine hundred of them, most fairly elderly and compulsively determined to enjoy themselves; when they danced it was macabre. The party was given by a rich man named Jack Ryan who has a huge house and big grounds on a hilltop in Bel Air.[1] He gives lots of parties. While they are going on he retires to a tree house and drinks with his friends. Peter behaved very well. He stolidly went through with his act and didn't complain. The lady who had invited him decided that he had better not appear again; he was originally supposed to give three performances that evening. (I met her two evenings later when I had to be on a round table discussion for the ACLU at the Stage Society Theater. I was with Susan Batson, who has just married Kaye Dunham from our *Black Girl* cast, and Susan was quite shocked because I acted so "coldly"; but it was good I did,

---

[1] Ryan (d. 1991), an electronics engineer, helped design Mattel's Barbie and Chattie Cathy talking dolls; he was later a husband of Zsa Zsa Gabor. The party was a UCLA alumni fundraiser.

because next morning, the lady phoned Peter and apologized for having involved him in this mess!)

June 24. Yesterday we got this note from Tony Harvey: "Thank you for the treatment received today. It is interesting but I am afraid it only confirms my original feelings about this subject working for an audience today. I realize you have devoted much time to it and I am grateful. I wish I could be more cheerful about the whole thing but I must be honest and tell you that I have grave doubts about repeating something which has been done many times." The whole tone of this silly cunt's note is infuriating of course, but what really matters is, will he try to blame us and our treatment for his decision not to go ahead with the picture? This will be seen within the next few days, as he's on his way to New York by boat and will meet Beckerman and Joe Wizan[1] there.

This next exhibit needs no comment. It is a *Life Magazine* ad published in *The New York Times* on June 10. Jack Larson organized a letter of protest against it signed by a lot of the younger dramatists and he himself wrote a poem abusing *Life* which does more honor to his heart than his Muse. The ad is full page and includes a photograph of Tennessee which is obviously chosen for its expression of cocky idiocy:

"Played out? Tennessee Williams has suffered an infantile regression from which there seems no exit. Almost free of incident or drama ... nothing about *In the Bar of a Tokyo Hotel* deserves its production." That's the kind of play it is and that's the kind of play it gets in this week's *Life.* From a theater review that predicts the demise of one of America's major playwrights to a newsbreaking story that unseats a Supreme Court judge, we call it a bad play when we see it. And that's the kind of strong stuff in *Life's* pages that gets us a major play from 36.5 million adults. Every week.

And now, just to take the taste away, here's a couple of lines from the play about the life of Buddha which the children performed at the Father's Day lunch at the Vedanta Center on the 21st. Siddhartha's favorite horse Kanthaka and his charioteer Channa are waiting to say goodbye to him.[2]

Channa: "I did not know a horse could cry."

---

[1] American talent agent, movie producer, and studio executive at CBS and Twentieth Century-Fox (b. 1935).
[2] When Gautama renounces the world to wander alone.

Kanthaka: "Do not be surprised when my heart breaks. It will make a very loud noise, because it is bigger than yours."

Dobbin wept a few tears at this.

July 4. We are just back from visiting Paul Wonner and Bill Brown at their new house in Montecito. God knows why they chose to move into it; it's dark and beset by neighbor noises, but it was nice seeing them, I miss them both around here and next time they'll probably move even farther off. Soon they are leaving on a trip to South America and will meet up with Jo in Rio.

Two cute boys from the Royal Ballet, friends of David Hockney, have come out with us to spend an afternoon on famed fag beach, their names are Wayne Sleep and Graham Powell. They tell us that Christopher Gable has left the ballet to become an actor but that he hasn't had much success so far—this is of interest to us because we liked Gable so much in the BBC television film about Delius and thought he might be good to play Oliver in our *Meeting* play.[1]

They've just got back from the beach, so no more now.

July 5. These two boys have beautiful manners; they thanked us so nicely and they have arranged for tickets for tomorrow night, so we can see *Romeo and Juliet*. A contrast to Peter and Jim, who think politeness is "insincere"! So it is, but that's not the point and something which you simply cannot explain to an American hippie.

On the 30th, Irving Blum, who has a gallery here, came to see Don's work. He doesn't want to show any of his painting, thinks it's primitive and that he hasn't found a style, but does want to show his drawings and seems really prepared to get Don a lot of publicity and commissions. So this was both a blow and an encouragement. Don on the whole seemed encouraged, or at least stimulated; he had been fearing this confrontation so much, and it was a relief to be frankly rejected. The worst of Don's attitude is that he really refuses to trust *anything* but rejection. Even something like getting a picture into the National Portrait Gallery is dismissed because it's only been chosen as a portrait, not as a drawing! I absolutely see Don's point of view and yet it sometimes makes me so furious I could slap his face; it's so *sulky*.

On July 2 I got a ticket on the Strip for going through a red

[1] Gable (1940–1998), a star at the Royal Ballet, played Eric Fenby, the composer's amanuensis, in Ken Russell's *Delius: Song of Summer* (1968); later he founded the Central School of Ballet and was artistic director of Northern Ballet Theatre.

light. This upset me ridiculously, but at least I *had* just completed four whole years without one (June 30).

Clement Scott Gilbert, the manager who's interested in *Meeting by the River*, is here now and coming to see us tomorrow. According to him, we may quite possibly open in August, which would mean rushing Australia or going there later. Tony Richardson expresses great interest in the Claudius project but unfortunately won't suggest paying our fares because we'd already told him we were wanting to come there anyhow! As for that unspeakable fart Tony Harvey, not one word from him all this while! Next week we shall have to light a fire under him and Beckerman somehow—at least get them to say we're off the job.

Last time I was up at Vedanta Place, I got Annamananda(?)[1] to give me the key to the shrine and I went inside and sat there a few minutes. It seemed extraordinarily important to do this, though I didn't "feel" anything much while I was in there.

Last night we saw *Citizen Kane* and *The Magnificent Ambersons*, which are being shown in an Orson Welles festival season. Seeing *Kane* makes me question my impatience with Don when he behaves what seems to me negatively about his painting. Isn't it perhaps because I'm being possessive, just as Kane was with his opera-singer wife? That I want Don to succeed at painting because I've already made it *my* success? Yes, it's possible, maybe probable, and that would account for my getting so angry with him when "my" success is thwarted. Just the same, I believe Don's painting *is* good *and* original. So fuck Blum.

July 8. Have just heard from Robin that Tony Harvey *is* going to direct *Cabaret* but that we are out; he wants to do it in some other way. This is the one possibility we had ruled out and of course it is irritating because it amounts to saying that we have failed, when the truth is that we did this treatment according to what Harvey himself had told me he wanted. Harvey now as good as says that he thinks we would be incapable of coming up with any other kind of story line and he doesn't even ask for our suggestions. Maybe there has been some mischief-making here. If there has we shall probably find out, before long. Meanwhile, Robin assures me that we'll be paid for our work.

No news about the play yet. So Australia is still up in the air. I haven't talked this over with Don; he's out.

The day before yesterday we went to the ballet but at the last

---

[1] *An*amananda, previously Arup Chaitanya and before that, Kenny.

moment they couldn't give us seats, so we stood in the wings, which was tiring but far more interesting. [Rudolf] Nureyev and [Margot] Fonteyn were so near sometimes that you could see every line on their faces, which was plenty. As Don said, there was a terrific pathos in Fonteyn's absolute determination to look nineteen—which she doubtless succeeded in doing, from the viewpoint of the audience. Nureyev seemed very cold to me, and his smile reminded me so strongly of Cyril Connolly, it is subtle and mocking and essentially hostile. Don found him so extremely feminine. Don was altogether convinced that this sort of richly costumed mime isn't ballet at all but acting;[1] he says it's ridiculous even to speak of it in the same breath as Balanchine. Anyhow, we both enjoyed ourselves greatly; and Wayne especially was so sweet and considerate, very carefully inserting us into the best vantage spots and steering us out of the way when props had to be moved on to the stage. We ate with him and Graham and had drinks with Freddy Ashton afterwards. And Freddy was charming. We didn't get home till nearly four which is late even for us.

On July 2, I went to Dr. Ashworth who said my contracture is no worse, so no operation is needed right now. There is some trace of its beginning in the right hand as well!

One of the best dramatic touches in the ballet was that, when Juliet recovers consciousness lying on top of the tomb, she thinks she is still in bed and tries to pull the bedclothes up over her. Fonteyn was also truly marvellous in her power of making that old tired face express the most brilliant, innocent joy.

July 12. Three days ago, I drove down to Trabuco and returned after supper. The San Diego freeway goes right through now, fusing with the Santa Ana, all you have to do is get off it at the road to El Toro. But when I did, I didn't even know where I was at or which way I should turn, all was tract houses, gas stations, wires, wires, wires, and smog. You couldn't even see the hills. However, within a mile or two, the old countryside reappeared and the view from the monastery itself has hardly changed yet; there's only the church and one house overlooking you from the ridge above.

The whole place looks beautiful, better cared for than I've ever seen it, no doubt because there are now so many monks to look after it. The new building is rather like a glorified motel, certainly, but it will look all right when it's overgrown with creepers.

[1] Choreographed by Kenneth MacMillan, 1965.

Swami, Vandanananda, Asaktananda were all three there. Vandanananda was very much muted, as always in Swami's presence. We kept exchanging glances but he didn't say anything at all confidential to me; how could he? He knows I am on Swami's side. Swami says Vandanananda is very sad to be going. Certainly he now seems much more sympathetic, because he is the naughty one. Asaktananda seems much less sympathetic for the same reason; he is the good little pig. I really am beginning to get bad vibrations from Asaktananda. Probably he doesn't like me. He seems haughty and even potentially spiteful. What is shocking is that nice big cute Bob[1] has left the monastery. According to Swami, he has been undermined by Vandanananda so that he has taken a dislike to Asaktananda. So he left in protest against Vandanananda's going back to India. At the same time, Swami says Bob was "in very bad company" before he came to the monastery, "That's why I took him in at once." "Bad" must mean queer. I suppose good old sex was just reasserting itself, disguised as righteous indignation—don't I know its tricks!

While I was at Trabuco I went into the shrine all by myself and sat in the dark, saying, "speak to me." It didn't. In fact I couldn't possibly feel drier than I do now. And yet, somehow, it's all right. I ought to be frantic, I know. But I am not. How can I be? I have known Swami. He sits there and shines. He is the beacon which shows the way out through the reef. I know I ought to steer for it but I don't. It's enough to know that there is a way. What I keep wondering is, when I get gravely ill or am in some great danger, shall I suddenly be able to steer towards the beacon?

We are now definitely planning to go to Australia in about ten days' time. As soon as the visas arrive we shall get reservations.

Une vie:[2] J.B. has been a lawyer all his life in a small town where everybody knows him. So he has never dared to have a friend to live with him. Now he's my age and he has met a doctor aged sixty-nine and they like each other a lot, and J.B. feels he *must* experience domestic love before he dies. *But* the doctor is living with an eighty-year-old friend who has a couple of million and if the doctor leaves him (as he wants to) he won't inherit the money. I said, why doesn't the doctor kill him? J.B. replied, perfectly seriously: "You're the second person who's said that. But, you know,

[1] Robert Hoffman (1921–1997), twice-married monk who also lived at the Hollywood Vedanta Society. He founded the Hoffman Institute to promote his 1967 insight that unconditional love is a universal birthright and published *Getting Divorced from Mother and Dad* (1976).
[2] See Aug. 12, 1961.

Chris, if he was the sort of person who was capable of doing a thing like that, then I wouldn't like him the way I do." So I had to hasten to explain that I was kidding! J.B. is a very good simple innocent person. His sexuality is that of a fourteen-year-old boy. This in itself makes him almost attractive, or at least nonrepulsive.

Incidentally, J.B. has had Dupuytren's Contractures in both hands! The doctor cured him with cortisone, not surgery.

Thundershowers yesterday, today beautiful weather. Jim Gates came and went on the beach with us; then Don drove him into town as he had to try to sell his violin to raise some money. Jim and Peter seem to be drifting apart, rather. Oddly enough, Peter has changed from an agnostic into an ardent Vedantist, but this makes him disinclined to talk about religion to Jim. Maybe he has this Jewish possessive thing; he has to disown something or own it personally. Furthermore, Peter is getting much involved with girls again and this brings up the sexual difference between him and Jim. They have had several different people staying with them in their tiny shack—they entertain as if they lived in a palace!—and in each case Jim has felt that Peter and the guest were getting together and that he was left out. (I must admit that this is more an impression I have received than what Jim has told me in so many words.) Because of all this, Jim feels strongly that he must get a job as soon as possible, so he can contribute to the household; he doesn't like the fact that Peter's mother is paying their rent (seventy a month). Putting on swimming trunks revealed that Jim has terribly thin legs, which is a real shame, when he is otherwise so beautiful. I said, "Maybe he'd better become a monk, after all." Don laughed and said, "Worldly old Drubbin!"

July 17. Now suddenly it's upon us, we're to fly down to Tahiti at midnight, next Sunday the 20th. Thus we skip out from under Nixon's Moon Day,[1] though that wasn't intended.

A terrible fuss is on with Beckerman, who wants to pay us less or nothing for our treatment of *Cabaret* because Tony Harvey has told him it isn't what he told us in advance he wanted. This is such a barefaced lie that it makes Tony seem psychopathic.[2]

Meanwhile we hear that Christopher Gable loves our play and longs to be in it, thinks it the best play he ever read and a perfect expression of how he himself feels about life! Anthony Page is

---

[1] Apollo 11 was launched July 16; Nixon declared July 20 a national holiday so that Americans could watch Neil Armstrong and Edwin "Buzz" Aldrin walk on the moon.

[2] Isherwood and Bachardy eventually received satisfactory payment.

getting together with Jim Bridges socially but at the same time (Don thinks) dimly hinting that he himself would like to direct the play.

As for Tony Richardson, down there, unimaginably far off in the Australian outback, we have absolutely no guarantee that he seriously wants us to work on the Claudius film. So our attitude must be that we are off to enjoy a holiday and that any resulting benefits are fringe benefits.

The worst of these projected travels is that they utterly alienate me from even the vaguest thoughts of God. Those travel jitters! They get stronger, not weaker, with experience. It was quite inspiring to talk to Jim Gates, who is humbly happy to have found a job as busboy at a bar-restaurant on Washington called The Black Whale. How truly sweet he is.

July 20. I'd better sign off now, although it's only three in the afternoon, because there may well be distractions later.

Anthony Page called this morning to tell Don that *Cherry, Larry* was a disaster at its London opening and that Jim and Jack are both crushed and that he, Anthony, doesn't feel Jim is much of a director and that the Royal Court wouldn't want to do our play with Jim directing it. We have decided not to get involved in any of this. Let them all fight it out.

The moon landing took place a couple of hours ago. Oh the horrible falseness of the commentator we heard, and the way the thing was built up into being somehow a rebuke to the indifference of the Young. The emotion they tried to whip up was so false, and this in spite of the moment being genuinely dramatic. The English, with their tight-lipped terseness, understand so much better how to create drama.

It still seems quite unthinkable that we shall either be in Tahiti or in the drink by tomorrow morning. Don says this is the first trip he has really looked forward to in years, because we're making it together. Yesterday we went and got Swami's blessing; he shone upon us.

No more, unless there are last-moment bulletins.

We left for Tahiti by UTA plane[1] shortly before midnight, July 20. Arrived 4:50 a.m., July 21. Stayed at Hotel Maeva. Drove around island.

July 22. Took a boat over to Moorea.

---

[1] I.e., Union de Transports Aériens, a French airline.

July 23. By plane to Bora Bora, 10–11 a.m. Stayed at Hotel Maitai.

July 24. Waited all day on airfield for plane back to Papeete. Left 7:20 p.m. Got Pan American plane out of Papeete at 9:45 p.m., arriving Pago Pago, Samoa, 11:55 p.m. Stayed Intercontinental Hotel.

July 25. Left by plane to Western Samoa, 4:45 p.m., arriving 5:30 p.m. Stayed at Aggie Grey Hotel, Apia.

July 26. Saw R.L. Stevenson's house and his grave.

July 27. At Apia.

July 29. Left Apia 9:30 a.m., arrived Pago Pago soon after 10 a.m. Left by Pan American plane 3:30 p.m. (Crossed International Date Line.)

July 29. Arrived Auckland 5:55 p.m. Stayed at Great Northern Hotel.

July 30. Drove around Auckland harbor.

July 31. In Auckland.

August 1. Left Auckland by Air New Zealand plane at 9:00 a.m. for Sydney. Arrived 10 a.m. Flew to Canberra at 11:50 a.m., arriving 12:45 p.m. We drove with Neil Hartley to *Ned Kelly* location, where we met Tony Richardson and Mick Jagger. We stayed with them at Palerang.

August 2, 3, 4, 5, 6. At Palerang. Worked on outline of *Claudius*. Visited location, etc. Marianne Faithfull and her mother arrived on 6th.

August 7. Left Canberra 3:25 p.m. by plane for Sydney, arrived 4:00 p.m. Stayed at the Chevron Hotel. Saw Dennis Altman and his friend Reinhardt Hassert.

August 8. We moved to the Florida Motor Inn on MacDonald Street. We had supper with James Fairfax, a friend of Robin Maugham. To Manly Beach.

August 9, 10. At Sydney. On 10th, Dennis Altman drove us out to Thirroul (where D.H. Lawrence lived). Don called [his friend] in London and arranged to go over there.

August 11. We left by QANTAS plane at 8:00 p.m., via Nandi, arriving Honolulu at 10:30 a.m. (It was still August 11 because we had recrossed International Date Line). Don went on by United Airlines plane to Los Angeles at 12:15 p.m. I saw Jim Charlton. Stayed at Park Shore Hotel.

August 12. Spent the day with Jim. Stayed the night at his apartment.

August 12. Left Honolulu by United Airlines plane at 8:30 a.m. Arrived Los Angeles 4:15 p.m. Don met me.

August 20. Above is the time scheme of our trip. I've written it down because I realize that I can't (don't want to) carry out my original intention, which was to fake an account of the trip from my notes and memory. That's such a bore and so untrue, because you can't put yourself back in thought and feeling, even a few weeks.

Don is now in London. He left two days ago. Haven't heard from him yet. He is staying with [his friend] and will probably go to Algeria with him next week, then return here to work on the screenplay of *Claudius*; we settled this with Tony Richardson while we were in Australia. But there's also the possibility that our *Meeting by the River* play may be done in London quite soon; then of course I should have to go over there.

Meanwhile, Ray Henderson and Burgess Meredith are rehearsing parts of the *Dogskin* musical at a theatrical workshop in Hollywood. I went for the first time yesterday. The lead is played by a black actor, Booker Bradshaw; he's clever and effective but a bit slick. A boy named Ricky Drivas[1] seems really good. And there is a sexy young man with a terrific voice named Peter Jason, who is very funny as the lover in Paradise Park. The feminine lead girl is good too: Julie Gregg. She has something touching about her, a bit like Alice Gowland. Altogether it's an astonishingly good group.

Last night I had supper with the Stravinskys. Igor is "better" but he looks terribly thin and his eyes are really tragic with anxiety. He hardly speaks. Vera has just discovered that they have relatively very little money. All their investments have been sold by the lawyer without asking their permission, to pay debts. Their house is devaluated because the hippies on the Strip have made people unwilling to live anywhere in that neighborhood. Their chief resource would be to sell Igor's manuscripts at auction. Otherwise they just have his royalties. And his nurses alone cost more than forty thousand dollars a year! Vera and Bob are determined to get out of Los Angeles and find some small place where they can live. Vera is resentful against Igor's children, who have behaved ungratefully, she feels.

After supper I wanted to get a copy of Genet's *Funeral Rites* which has lately been published in translation. So I stopped off at a bookshop on Ventura Boulevard which advertises itself as the only all-gay bookstore in town. It looked very snug, with light shining through its screened windows (so as not to give offense, I suppose; this was close on midnight) but when I got inside they didn't

---

[1] Richard *Dreyfuss*; see Glossary.

have the Genet and indeed had very few books—mostly beefcake photos, magazines and some huge pink plastic dildos. I would have examined the stock at some length but was embarrassed by the cool eyes of the goodlooking boy in charge and left feeling like a very dirty old nag.

I fear I notice that my contracture is increasing; I can no longer completely straighten the little finger of my left hand. It'll be too tiresome if I have to have this operation just now.

Heard this morning that Rudolf von Strachwitz, Barbara Greene's husband, is dead. One of Hitler's deadliest enemies; and he has been living and will be buried in Berchtesgaden![1]

Leslie Caron told me on the phone that the murder of Sharon Tate and the others in Benedict Canyon, followed by the two other murders at Silver Lake and Marina del Rey,[2] created a tremendous panic. She and Michael actually moved to the Beverly Hills Hotel for a few nights, feeling that a murder epidemic was about to break out.

Psychologically speaking, the flight to Tahiti was incomparably the most significant part of our journey. Ben Masselink said in an article that he had thought of Tahiti every day of his life. I probably had too. I had thought of it as utterly alone (disregarding the thousands of other islands) on a vast blue empty map; the most distant point. And the fact that nowadays you can take a plane to it several times a week from our huge roaring mobbed airport only made it more remote, more romantic. There was, so to speak, amidst the tiresome popular vulgarity of departing for London, Rome, New York or Tokyo, this one unobtrusive alternative. To make a genuine departure, to the one place on earth which wouldn't be exactly like all the others.

Then it seemed (though we hadn't planned that) the perfect night to depart—right after the moon rape. (Oh, how sad it was to look up at the poor violated thing and know that it was now littered with American junk and the footprints of the trespassers!) It was right to be going.

The flight took place entirely in darkness, which added to the effect of its being a teleportation rather than a gradual progress. All those thousands of miles entirely over water! I woke a few hours later, somewhere near the equator, and how awesome and all encompassing the star dome was, millions more of them out

---

[1] Rudolph Graf Strachwitz von Gross-Zauche und Camminetz helped plan the assassination attempt against Hitler in 1944; see Glossary.
[2] The murders in Benedict Canyon and Silver Lake were committed by the Manson gang; the third incident was unrelated. See Glossary.

there, great burning misty blobs of light right down to the ring of the ocean. And then I kind of thought or felt that now we had entered the other world, within which Tahiti really exists; we had almost already arrived.

The darkness eased us through the crowded Tahiti airport and the French and the customs and our arrival at the too-luxurious hotel (the Maeva). We got right into our room and it was still dark and we lay down on our beds and waited for the sun to rise and it did, and here we were in Tahiti, looking out at a grove of tall palms surrounding a parking lot. But that didn't spoil anything. We moved at once to a more expensive room on the beach side; and there was Moorea, towering up with its jagged pinnacles under clouds beyond the lagoon. It was absolutely there and we were beholding it, and for this fact alone the entire trip was justified in advance.

August 22. It's silly to say that Tahiti has been spoilt, just because there are hotels and tourists. The life in Papeete was probably corrupt long before Gauguin's arrival. Quinn's[1] kind of loudness isn't really that much more romantic than an old-fashioned rough trade bar in San Francisco; it's just that a lot of foreigners find it kicky to get drunk and pick up Polynesians in their native surroundings. The waterfront appears to be in the process of rebuilding and I personally wouldn't be bothered if it developed luxury shops like Palm Springs.

Because there is the rest of the island and the interior of it which is still virgin jungle and there are all the surrounding islands, some of them nearly a thousand miles away and still only to be reached by schooner after a week of sailing. And anyhow I felt that the charm of Tahiti isn't at all a sense of having entered another culture, an innocent Garden of Eden. To me the most romantic spot was the tomb of the last king, Pomare V, with a brandy bottle on top of it, a monument which has the gravity of a nineteenth-century European graveyard amidst the untidiness and irresponsibility of tropical vegetation. Another thing Ben Masselink said came back to me often: the sense of the ocean all around you, brimming full and seemingly ready to flood you at any moment. How beautiful it was to sit at the restaurant near the Gauguin museum and look out across the pale leaf-green water of the lagoon to the reef on the horizon, where you see the waves bursting huge into the air and the dark blue dangerous ocean beyond.

[1] A bar.

I'm glad we didn't fly over to Moorea but went by boat. The violently rough trip across the channel was the right preparation for the overwhelming experience of entering Cook's Bay; the waves outside the reef create a dreamlike sense of gliding calm within the lagoon. This *is* a Garden of Eden but a very sophisticated one. The fantastic garlanded pinnacles seem contrived and campy and you gasp as if the curtain had gone up on a supreme theatrical spectacle; this is literally one of the most *enchanting* places I have ever seen in my life. We swam about in the warm water, gazing at it all and trying not to mind the other tourists. We did however miss seeing the Bay of Opunohu (which some say is even more beautiful) because we should have had to go there with a party, packed into a minibus.

This morning I drove Jim Gates and Peter Schneider to Acres of Books at Long Beach. All these years I have been meaning to go there, so in a minor way it was an accomplishment like getting to Tahiti. It was terribly hot and the whole area reeked of oil. I found a novel of Masefield's which I didn't even know existed, *The Street of Today*; it seems to be a sequel to *Multitude and Solitude*. Peter bought books by Stephen Leacock, Benchley, Thurber and [Heard's] *Is God Evident?* Jim also bought a book by Heard, *The Five Ages of Man*, and Tagore's *Gitangali*.

August 23. Bora Bora stands up like a little monument in the middle of the sea. It is beautiful but it doesn't seem at all mysterious and for this reason one might well prefer it for a South Sea holiday of quietness to Tahiti or Moorea. We arrived in a heavy rainstorm after a rough flight in a small battered plane. Don lost his cool because of the delays and the crowding and the heat of the closed launch which took us from the airstrip across the lagoon and began one of his fuck fuck fuck outbursts. A woman behind him was sincerely surprised and whispered to her husband, "What's he so mad about?" However we found a dear little hotel of grass huts called the Maitai and were quite happy, the tropical rain fell warm and soothing on the dark snug huts. Next day, the launch took us over to the airstrip. Having been deposited there we were told the plane would be delayed owing to engine trouble. It was delayed for ten and a half hours, during which time they kept us out at the airstrip rather than bring us back to the island and have the expense of giving us lunch at the hotel. Everybody got angry in different ways and formed groups. We were angry too but we didn't want to belong to the groups. So we bathed in the lagoon and strolled about the beaches which have sand in very large yellow grains. We also ate coconuts, more as a protest than because we wanted them. We finally got back to Papeete just in time to board a big plane for American Samoa. It was full of the members of some huge tour, mostly Jewish. Pago Pago airport seemed awful and we resolved not to go to the Intercontinental Hotel where we had been booked in by our travel agent. So we found a driver and asked to be taken somewhere cheap and quiet. There was literally nowhere. He drove us to a house full of sleeping people, probably it was his home, and was told they couldn't take us in. So we went to the hotel after all and got in a fight with the management because the air-conditioning made such a noise and the breakfast was practically uncooked. We had left Bora Bora without ever having seen Maupiti, the island which was the scene of Larry Holt's long-ago love story. It lies over on the other side of Bora Bora and we had no opportunity to make the trip. But the fact that it was there, all the time, created a brooding tragic background which was just right as a contrast to Bora Bora's sweetness and shining calm.

Last night I talked about all this to Jo, who has just got back from Brazil. Paul Wonner was sick in Chile and Buenos Aires and so he and Bill decided they couldn't wait for her and went straight back home. So Jo decided to go alone, and she met some people who entertained her and all was well—rather grimly and heroically so; she has proved to herself—and to Ben too, of course—that she can make out on her own. Not that she *enjoyed* the trip but it *taught* her a lot. Poor old Jo, now her kitties are sick with a fever and have had to go to the vet.

August 24. I forgot to mention how, when we left Acres of Books and were on our way home, we noticed one of the two very impressive bridges—much more impressive, *I* think, than Sydney Bridge, and outrageous looking too, like roller coasters—between Long Beach, Terminal Island and San Pedro. I had never seen them before. Neither had the boys. So of course we had to drive over them. All around was this almost comically ghastly junk-landscape, reeking of oil. Jim delivered himself of a pensée which was so sententious and yet perfectly natural in the mouth of someone very young that I made him write it down on the notepad in the glove compartment of the Volkswagen, to his great embarrassment: "Man has finally created a situation in which it's necessary for him to bypass his own colossal blunders!"

Talked to Don in London this morning. He now feels that we should sign a contract with Jim Bridges without delay, binding him to direct our play right after he has finished this film or else step down. Don seems to be disgusted with the London agents and backers. I don't quite understand the inwardness of this but maybe I'll find out more from Jim who is returning from England by plane this evening.

Pago Pago harbor is impressive but oppressive, with its very high walls of greenery all around. You feel you recognize the tacky-tropical wooden architecture of "Rain" and indeed the house where Maugham set the story is still standing, though transformed into a market.

We managed to get seats on the plane for Western Samoa that afternoon. It's less than an hour's flight. After Pago Pago we wanted to like it but the experience was much more than liking; joy and delight. The long road to Apia from the grass-covered airstrip runs along the coast through a succession of villages, and it was like entering a better land and Beatrice saying, "Don't you know that here man is happy?"[1] The cooking ovens were smoking and the open, pillared fales[2] were full of people, and all the boys and youths, so handsome and gay with their beautiful golden bodies, were laughing and shouting and running about. It was a perfect evening, though, as always in the Pacific, there were great clouds piled on the horizon. I asked our driver if it was going to rain and he answered, with what seemed a good-humored irony, "Your flag is on the moon—how can it rain?"

August 25. Jim Bridges arrived and we are to talk to Robin French together this afternoon. There's something about Jim that irritates me, he is so weak and so childishly eager to be a big shot; but let's hope directing this film will stiffen him up a bit. Meanwhile, they are running into difficulties with *Dogskin*; the black actor, Booker Bradshaw, has had to go off to do a part in a film so they're rehearsing a replacement. I keep thinking about *Claudius*, rather negatively. I do hope I'm not going to start hating it. Much depends on Don's attitude. He has just written to say he thinks Graves is odious and his book a bore. But of course this can easily turn into an inspiration to do better.

Apia isn't particularly picturesque but it's open and cheerful and not dominated by a luxury hotel like Pago Pago. Aggie Grey's hotel is comfortable and untidily built, a whole lot of buildings have been gradually added to it. We had a room without air conditioning; a fan was quite sufficient. And, at least at this time of year, this

[1] Cf. Dante, *Purgatorio*, Canto XXX, l. 75.
[2] Samoan houses.

island (and all the other places we visited) appeared to be nearly bugless. We had no insect bites at all. Aggie Grey is a character and as such to be avoided; she belongs to the category of the gracious imperious lady-madam. Her place was a whorehouse during the war, it's said.[1] One evening we had a fiafia[2] at the hotel—mostly very amateur—and she danced, with the airs of a very great retired actress condescending. Don't know why I'm being so nasty about her. Surely I didn't want her to make a fuss over me?

Stepping outside the hotel after dark we got involved with a bunch of girls—they were supposed to show us where the cable station was, as we wanted to let Tony Richardson know we'd be late arriving in Australia. Two of the girls turned out to be boys in drag. It seems that some boys "decide" they don't want to live as boys in the Samoan culture, so they help with the housework and wear wom[e]n's clothes. This doesn't stop them from marrying later. Several girls and some little boys accosted us. It was all lively, half-mocking. Sex seems very cheerful, uncomplicated and ambivalent here—sort of Andy Warhol primitive.

Of course I had to see the R.L. Stevenson house and tomb—particularly the tomb; it was sort of a funeral rite on Frank's behalf, he would have so loved to go there and in a sense he sent me; the very last of his South African letters I copied out before leaving is about the Stevensons and how he wishes he and Kathleen could have visited them.

August 26. So far, I've spent most of my birthday recording poems on tape for Don; Fulke Greville, Vaughan, Dryden, Pope, Cowper, Coleridge. Also I went in the ocean for the first time in months—except for our bathes during our trip.

Don called this morning, says he'll come back early next week he thinks. Last night, we ran through the *Dogskin* selections at Theater West in front of an audience. Everybody was quite good—except for Burgess, who made a mess of the introduction. Julie Gregg was really touching and beautiful and sweet as Elaine; she has a kind of untiresome sadness which is just right. Peter Jason did the Dog with terrific sexiness. Ricky [Dreyfuss] was extremely good in the last speech of the chorus; he got a big laugh with "I am the nicest person in this room" and he said "act" in exactly the

---

[1] Agnes Genevieve Grey (1897–1988), of British-Samoan descent, ran a bar, once called the Cosmopolitan Club, which flourished when the U.S. military arrived in Samoa in 1942; she is rumored to have been a model for "Bloody Mary" in James Michener's *Tales of the South Pacific* (1948).
[2] Traditional Samoan feast with music and dancing.

right tone. Elsa was quietly turning up her nose at the whole affair. She bitches Burgess because she says he will cast any cute young girl in anything, regardless of her talent.

August 28. Am feeling more than somewhat harassed; so much to be done either at once or in the near future. I had a charming evening on my birthday with Igor and Vera and Bob and Ed Allen—Igor very distant and deaf and nervous but at least he knew me and was friendly. However I got a bit drunk, not really but more than I like to be, and that depressed me all yesterday and stopped me from working. The only good thing was supper with Jeanne Moreau. She has style; in fact, she is very like herself on the screen. Probably she's fat, but that was all covered by a flowing Hawaiian-type garment, and what you saw was quite beautiful in its own way—the tired restless face and the graceful arms and hands. An agent named Mike Medavoy[1] was there too, a young not bad-looking Jewish wheeler-dealer, he seemed to me, rude and insecure and really quite anxious to please. I wondered if maybe he hadn't been to bed with her; she might well have found him sexy. During dinner, cooked by her French maid, we heard noises in the trees and bushes beyond the patio—this was quite a lonely house, away up in the heights of Bel Air. All three of us were extremely conscious of the noises—indeed, Moreau asked me a couple of times, "Do you hear them too?" as though she thought they might be an hallucination brought on by murder fears. We kept repeating that they were obviously deer; and I'm sure they were. But it brought home to me how the ghost of Sharon Tate haunts people nowadays, especially in this part of town. (Incidentally, Moreau told me that she can't smoke pot, it makes her paranoid.)

To get back to Robert Louis Stevenson. It was a very hot morning when we went to see his house—they only showed us the outside—and the policeman on duty advised us not to go up Mount Vaea. When our driver saw that we meant to, he produced a guide, a young girl. She was soon joined by at least nine others, an oldish woman, youths, male and female children. The beginning of the path was easy, then quite abruptly they told us, "This way," pointing up a slope so steep that one could hardly have climbed it if there hadn't been lots of bushes and vine roots to hang on to; the hill is entirely covered with trees and undergrowth. I struggled up

---

[1]Medavoy (b. 1941) became a powerful studio executive at United Artists, was a founder of Orion Pictures in 1978, and later chairman of Tri-Star Pictures. Isherwood typed Midavoy.

somehow, panting like an animal and streaming sweat; even Don was soon exhausted and we kept stopping. Our "guides" pushed us from behind and dragged us from in front, laughing and encouraging us; when we rested they examined my face like doctors who are considering whether or not the patient will be able to stand the operation. So of course I reacted by showing as much energy as I possibly could. Don filmed my struggles with our movie camera. Actually, the absurd effort of this trip didn't surprise me unduly because I remembered reading that, when they took Stevenson's coffin up to the summit, one fat mourner collapsed and died soon afterwards! The campy fun of the climb was increased by our feeling of instant intimacy with the Samoans; they seemed to know everything about us at a glance and probably they did. "Is he your son?" "No." "Oh, I am sorry!" "No need to be sorry." "You are half boy, laugh like woman." They had tremendous jokes about us. When we got to the top, there it was. It doesn't look like much. But the view is tremendous—Vailima[1] right below, and the mountains rising behind and, on the other side, the corrugated iron roofs of Apia, rusty red or painted blue, and the ocean. In a show-off mood because pleased with myself for not having collapsed, I recited Stevenson's "Requiem" aloud; but this was also a ritual act offered to Frank[2]—not so much to Kathleen, who would have hated it all, I'm sure. On the way down we discovered what we might have guessed at once—this brutal climb is a trick to squeeze money from tourists; actually you can get to the top by a quite easily graded path winding slowly through the woods. The "guides" wanted a dollar each. We ended by giving them about six dollars eighty, which was absurd, but after all the climb was psychologically worth it. After this, we drove on around the island and picnicked on a beautiful beach where we could swim in the lagoon. (The only serious hazard on such beaches are the coconuts, which fall quite frequently and could lay you out or even kill you.) A beautiful Samoan youth appeared while we ate and rode into the water on a horse, which seemed symbolically appropriate.

Our first day on Tahiti, I said to Don that it had been one of the happiest days of my life. This was another. You get such a feeling of joy, in these islands anyhow. Don was feeling it too and our being here together made it perfect. One of those rare and perfect passages in the life symphony when one's interior monologue, one's

---

[1] Stevenson's house.
[2] The last two lines of "Requiem" appear on the tomb; Isherwood first read Stevenson in childhood with Frank.

psychological duet with the companion and one's subjective sense-poem about the surroundings all relate and make harmony, and yet continue individually. One of them doesn't drown out the other two, as usually happens. That's what makes the passages so rare.

September 1. Was disappointed at first and rather mad at Don for deciding to stay in England another week. (The real reason why he wants to do this I still don't know, but no doubt he'll reveal it after he gets back.) However, I now feel that maybe it's a good opportunity to get things squared away before his return. It is certainly a punishment for my idleness so far. What have I accomplished? I did get quite a lot of Swami's manuscripts on Narada's Bhakti Sutras[1] edited (but he promptly gave me a lot more!). And I have done quite a bit of poetry recording for Don on the tape recorder in his studio. (But that's actually play rather than work.) I have *not* done one page of chapter 9 of *Kathleen and Frank.* I have *not* started the article on David Hockney to be published in the book of his reproductions.[2] I still have thirteen letters to answer. I still haven't finished my notes on our trip in this diary. But what I'm really worried about is *Claudius.* I should never have told Don it didn't matter his going to England, in the first place, because we really do need all the time we can get and he can't possibly realize this since he has no experience of this kind of work. I'm now seriously doubtful if we can finish the rough draft in time—except in the sense of just fudging it. Well, maybe we can get Woodfall to let us have another couple of weeks; only I hate that too, I who have always surprised them by getting the work done quicker than they expected. The whole problem is, how to tell the story—how to get into it at all. That's what I don't see at present.

On the 30th we had the official farewell lunch for Vandanananda and it was a real success. A very large portion of the congregation showed up, the weather was perfect, not too hot, and the food was managed excellently. Swami rose to the occasion and made a nice little speech and put a garland around Vandanananda's neck and Vandanananda made a genuinely touching, humble and dignified speech in which, without giving the show away, he in fact said he was sorry for all his indiscretions and involvements with these various women. I do see now that it would have been practically impossible for him to stay on here—because, however careful he became, these bitches would never have left him alone; they are

---

[1] Published as *Narada's Way of Divine Love* (1971), translated and with a commentary by Prabhavananda, introduced by Isherwood.
[2] *72 Drawings: Chosen by the Artist* (1971); Isherwood's introduction was not used.

worse than any vice squad operatives when it comes to entrapment and enticement. Swami, when he and I were alone together, giggled and said, "It's a terrible profession being a swami, even I, in my old age, a woman wrote me the other day and said, you are the star in my blue sky, imagine!" Jimmy [Barnett] and his fellow songsters sank even lower than usual to the occasion and produced some adapted farewell songs, based on "Auld Lang Syne," "Dixie," etc., which must have made all hell blush. (I talked to Jimmy later and he told me that Bob [Hoffman] didn't leave the monastery so much for sexual reasons as because he couldn't adjust to group living.) As for me, I made Vandanananda a very public pranam—the first time I've ever done this to him—and we embraced. I knew he would like it that I did this in front of everybody and it wasn't that I didn't "mean" it—I sincerely respect his attitude nowadays and was furthermore much softened up by his flattering remarks about me in his speech—but of course there was an element of playacting in it. I wondered if he'd read and remembered the end of *A Meeting by the River*!

About Auckland: you might walk quite a long way down one of the main streets before you realized that you weren't somewhere in England—or Scotland, for the population seems overwhelmingly Scots; the girls fresh faced and homely, the boys often strikingly beautiful, hero types with narrowed eyes who seem to be facing antarctic blizzards. The bars are almost exclusively for men and one gets the impression that only lesbians and whores come out at night. The bar beneath our ponderously respectable hotel, the Great Northern, had a few queers also heroic, of the screaming pioneer fag variety. We saw quite a few hippies on the streets.

When we arrived, about 7 p.m., we were told that we must go instantly to the dining room, if we wanted any dinner; it didn't serve after 7:30. In the mornings, they try to make you sit at an already partly occupied table although there are plenty of empty ones, and when you refuse they raise their eyes to heaven and sigh in true British grumble-style. About seven, the maid arrives with the tea. We had locked our door, so she battered on it as though this were a police raid until it was opened, and then entered saying indignantly, "You locked your door!" One evening I went to a cafeteria and asked for a meat sandwich. The waitress told me severely, "We *never* serve meat *at night!*" At six on Friday evening the shops snap shut like mousetraps until Monday morning. In an Auckland newspaper there was a letter from a New Zealand wife who had been in California, studying at Stanford. She said, quite innocently, that the two things she would miss, back here at home, were that you could shop at night and on Sundays and that in California husbands do everything with their wives, "They even help them decorate the Christmas trees!"

There is a very cozy homely charm behind all this, however. In a movie theater I sat down on a piece of chewing gum. Without complaining, I pointed it out to the manager, so he would stop others sitting on it. "The swines!" he exclaimed. "That's why we won't sell chewing gum in this cinema. We'll be able to get it out again but the stuff we use burns like buggery, wouldn't want to risk it on you, not with your trousers on." The girl at the hotel produced some cleaning fluid which, much to our surprise, got the gum off my pants. When we told her this, she said, "Then nothing's wounded but your pride."

The harbor is naturally beautiful and enormous, a whole water-world of bays and beaches and islets, around which you could sail for days. The city is mostly suburbs of dull little wooden houses; and all around it the bush. If there is no building it comes right up to the road, great tropical tree ferns growing amidst this northern respectability and the vegetation so dense that you could get lost within fifty yards. After seeing this, it no longer seemed strange that the government puts out a booklet on "How to Survive in the Bush, in the Mountains and on the Beach."

We saw a film called *The Taking Mood*, made strictly by New Zealanders for New Zealanders, which was quite revealing. It's about a race from the North to the South Island and back between two fishermen; they have to catch fish in various specified locations. One is an old-fashioned expert, a Scot, in old clothes, driving an old beat-up car. He is the idol of the South Island. The other is an amateur, a young Auckland lawyer, English rather than Scots, handsome, slick, elegant, a girl chaser, a sort of younger James Bond. Both of them cheat and they end by dividing the prize. You got such an overwhelmingly provincial atmosphere from it; the lawyer, despite his imitative role-playing, was just as provincial as the old fisherman. One sees them down there, much influenced by the outside world and yet pretty satisfied with themselves and really, if the northern hemisphere is wiped out with bombs, quite capable of keeping things going, with their fishing and yachting and rugger. The idea of actually settling there is full of horror yet fascinating; it's all so marvellously self-contained and self-sufficient. A surprising fact: The southern end of the South Island is only about two hundred miles nearer to the South Pole than to the equator. A conclusion: Being in bookshops in Auckland (and in

Sydney too) made me realize what a tremendous and admirable cultural role is played by Penguin Books. They make Culture seem chic, modern and relevant in a way that no U.S. paperbacks seem able to. You feel suddenly—oh well, all right, I can face this place if I must, as long as there's Penguins.

September 5. Swami told me the day before yesterday that, despite his touching speech, Vandanananda has quietly made arrangements to meet some of the people who have involved him in this scandal after he has arrived in India; they are going out to join him there! It seems as if he is really very childish in some ways; which makes him endearing, of course, although a security risk. And Swami seems just a little bit amused by it all.

I lost my turquoise ring again, this time in the restaurant, The Black Whale, where Jim Gates works as a busboy. But the waitress found it. I went there to see Jim (who happened to be away) but also because I had to take Lee Prosser somewhere and try to raise his morale: poor little thing, he is being drafted, and this after having split up with Mary,[1] who says they have outgrown each other and she doesn't want to live with him any more. To my astonishment, I succeeded quite a bit—although it may well have been that Lee was chiefly depressed because he hadn't had a proper meal in so long plus a chance to talk to *somebody*. Anyhow, he talked himself out of his angst and wrote me a sweet little note to thank me for it: "In some inexplicable way, you gave me the strength I needed to go on. I'm not afraid now, and thank you. I can't explain."

Gavin told me that not long ago, after walking on the beach at night, he went into the men's room over the life guard station, say around ten o'clock, and found two men screwing on the floor, a third going down on a fourth and a fifth watching and jerking off. They didn't seem bothered when he walked in. He said "hi" to reassure them, peed and went out again. As he left, a young man and a girl approached. The girl wanted to use the public phone and the young man went into the john. He came out again with a face of horror and hurried the girl away. The question is, are these people insane? They might almost as well go and do it right in the police station, you'd think. I suggested that maybe it's like a professional robbery; all based on exact timing between the janitor's rounds. But I doubt it.

Yesterday the man who is producing a film Gavin is to write,

---

[1] His first wife.

The Woman Who Rode Away, called me and told me that he and his associates are quite probably buying *Cabaret,* so his wife, Anne Haywood(?) can play in it. He wanted to see our treatment and guaranteed that anyhow he would see that we get paid for it. He says Manny Woolf, the head of Allied Artists, hasn't even been shown a copy! Oh yes, this man's name is Raymond Stross.[1] Gavin has been told that he is a "monster." He seemed a little too friendly, wanted me to call him Raymond right away.

When we got to Sydney, a man from Ajax Films,[2] John Daniel, met us and put us on the plane for Canberra, so we never left the airport. Neil Hartley was waiting for us there. The airline had sent our bags on into Canberra while we were talking, so Neil got furious and said something about fucking incompetence. The airline clerk said severely that they didn't have to listen to obscene language, and when Neil said he'd fucking well say what he fucking well pleased, he said, "We have laws to deal with that sort of thing"—all of which wasn't a very reassuring introduction; besides which, Neil kept going on about how he loathed the country and everybody in it. But when we got outside Canberra, which one does very easily, it was all so strange and empty, the hills and the grey woods of gum trees, whole groves of them deliberately killed by ringing the trunks, and the vast stretches of pasture land and the brilliant pure clear light. White and black cockatoos and a kind of magpie which makes liquid noises like a myna bird. And the silence—"the noiseless antipodes" as Lawrence says[3]—and the feeling of emptiness. It can't be described in words, only remembered as a feeling. Waking in the night—utter silence, nothing to make a sound. And then the sun rising, very brilliant, though there's a mist steaming along the ground. The first morning, Don and I got up and ran down the long dirt road from the house which stretched away into the distance. Don shouted, "We're in Australia!" Some dogs joined us, and we went crashing uphill into the woods, splintering the dead fallen branches, making a terrific noise in the silence.

But first, that first afternoon, we went out to where they were filming, a stony riverbed. Ned Kelly and the rest of the gang

---

[1] Anne Heywood, Miss Britain 1949 (b. Violet Pretty, 1932), appeared in the film of another D.H. Lawrence story, *The Fox* (1968), and was being promoted by her producer husband, Stross (1916–1988). Their other films include *A Terrible Beauty* (1960), *Midas Run* (1969), and *Good Luck Miss Wyckoff* (1979), but they did not make the two Isherwood mentions.
[2] An Australian studio.
[3] In *The Boy in the Bush* (1924), chptr. 4, "Wandoo."

are trading for horses in the rain; a hose is spraying them. Tony Richardson looking like the Duke of Wellington, in a kind of Inverness mackintosh cape; we embraced in front of the whole crew and the actors, including Mick Jagger. It was such an improbable encounter, after these thousands of miles, like Stanley and Livingstone, rather. Mick Jagger, very pale, quiet, good-tempered, full of fun, ugly-beautiful, a bit like Beatrix Lehmann; he has the air of a castaway, someone saved from a wreck, but not in the least dismayed by it.

Whenever you meet Tony after a separation it always seems to be the same situation; he tells you with a gleeful conspiratorial air that something absolutely terrible has just happened. ("Just" is the operative word.) What had happened this time was that an army of students from Canberra were celebrating "Bush Week"—they go out into the countryside and get drunk in small town hotels and then often wreck them or set fire to them, either by accident or for fun. "Bush Week" was now coming to an end and three hundred students were said to be heading for Palerang, the house where Tony and Neil and Mick Jagger were staying; they had vowed to kidnap Mick, or, failing him, Tony, and hold him for a thousand dollars' ransom, to be given to charity.

So Tony and Neil had called in the police. Our first night, or was it maybe our second, at Palerang there were ten policemen sitting up in the kitchen all night, waiting for the students, who never showed. Incidentally, without knowing it, they were guarding a pot party which was going on in the living room!

Palerang was a ranch house standing in the midst of a big ranch, with other ranches (they call them "stations") adjoining it. It belonged to a family named Sykes. The young Mr. Sykes was a daredevil-type skier, rider and auto racer, not really much interested in farming, who had married a rich Jewess who believed herself to have taste as a decorator. His sister Annabelle cooked for us; she was a very nice girl. But the general atmosphere was that the Sykes[es] hadn't quite vacated the house; they had moved away while Tony was there but they were keeping an eye on it and there would probably be a big bill for breakages when he left. Mrs. Sykes, the Jewess, had overdecorated that place and made it townish, but it still had a certain magic, chiefly because it was so solitary. This being the Australian winter, the nights were cold, with frost on the ground, and we had big fires. Annabelle told us that the summers are very hot and that life is made unpleasant by the many varieties of poisonous snake, the tiger snake chiefly, which you find all around the garden and in outside privies.

This situation, a somewhat beleaguered household, surrounded by protectors, infiltrated by spies (Mrs. Sykes looked in every day to check up, and there were two villainous cleaning women) is probably a standard situation which Tony sets up for himself whenever he makes a picture. Of course his headquarters is bound to be a focus for gossip and scandal. And having Mick there—not to mention Marianne Faithfull, who emerged from hospital with her dreadful little Austrian baroness of a mother and descended on Palerang just before we left—made this without doubt the most fascinatingly wicked house in Australia. But Tony managed to create a very similar setup when he came to Los Angeles in 1960 with John Osborne and Mary Ure, and took that house which belonged to one of the Gabors and had the colored maid who used to report all our conversations to a gossip columnist.[1]

As guests, Don and I were in a very awkward position the moment Tony and Neil had left for work in the mornings. The cleaning women tolerated us because we made our own beds and tried to be helpful; they called us "chappies." But they were fiercely on the lookout for any attempt to load them with extra work. The more aggressive of the two had emigrated from England and she grumbled like a Londoner. When she thought the soup was unfit to eat she described it as "on the nose."

On the Sunday there was a beard-judging get-together at a pub in the nearby town of Braidwood. (All these places seem half deserted.) Tony, Jocelyn Herbert[2] and I were the judges and we gave prizes for the best, the most sinister, the most sexy beards. The contestants got parts in the film, if they wanted them. The beards seemed entirely natural here. Although this country seems relatively law abiding, far more so than the States, there is much more of a frontier feeling. These little empty towns with their wooden houses and ironwork balconies and corrugated iron roofs give you the impression of something missing—it's the body of a man lying shot dead in the middle of the street. Crowds of beer-drinking men. Men everywhere. The women much less in evidence.

I only got to talk to Mick alone on the last day. He made a great impression on me. On Don too, but Don didn't see as much of him. Mick seems almost entirely without vanity, for one thing. He hardly ever refers to his career or himself as a famous and successful

---

[1] Passages about the visit appear in *D.1*.
[2] British production director and costume designer (1917–2003) for *Ned Kelly*; production designer for the English Stage Company since painting scenery for its first season at the Royal Court. She lived with company co-founder George Devine and worked on other Richardson films.

person and you might be with him for hours and not know what it is he does. Also, he seems equally capable of group fun, clowning, entertaining, getting along with other people, and of entering into a serious one-to-one dialogue with anybody who wants to. He talked seriously but not at all pretentiously about Jung, and about India (he has a brother who has become a monk in the Himalayas[1]), and about religion in general. He also seems tolerant and not bitchy. He told me with amusement that the real reason why the Beatles left the Maharishi[2] was that he made a pass at one of them: "They're simple north-country lads; they're terribly uptight about all that." Am still not sure if I believe this story. And indeed I am still not sure what I think about Mick. I would have to see him again. It can be that I was carried away—I certainly *was* when I suggested he should come and stay with us when next he comes to Los Angeles; we'd have the press and the police and the public on our backs for six months afterwards!

We talked to Tony about *Claudius* and did an outline. Admittedly he was tired and full of *Ned Kelly* problems, but he seemed so languid and bored and yet dogmatic during these discussions that we were on the point of telling him to forget all about it. Then suddenly he said it was fine and we should go ahead. Now I kind of wish we had got ourselves out of it because I really am unsure if we can deliver anything. We shall see.

Other memories of our stay at Palerang: The nice suppers in the dining room with all of them, including usually three or four of the young actors and members of the crew, very noisy and sexy. Walks with Don in the woods and along the river shore; this is such a marvellous place to be together in, sometimes you feel you are the only two people in the world. The discomfort of the shared bathroom which could only be reached through the kitchen full of people. The thrill of watching the members of the Kelly Gang come thundering down the creek side on their horses, splash through the river, gallop uphill past the camera; the wild cocky little boy who was Mick's double, lounging in the saddle as if it were a galloping easy chair. The sheep grazing in the frosty fields in the early morning; the strange sparkling pallor. But I must repeat, this country can't be described. It is a feeling, it's your own predicament in being there. The sense of being so far, far. And of nature being alien. The harsh weird cries of the birds—they are

---

[1] Christopher Jagger, musician, songwriter, bandleader.
[2] Maharishi Mahesh Yogi (b. 1917), who devised and promoted Transcendental Meditation. He initiated the Beatles in 1967, and they went on a retreat at his Himalayan ashram; Jagger and Faithfull were also there.

so *foreign*. You feel they are actually using a foreign language. Yes, that's how I remember the country round Palerang: the endless empty silence, the pistol-sharp crack of grey dead gum-tree wood, the foreign squawking of birds.

September 7. The first night we spent in Sydney was in a tall glass tower called the Chevron Hotel. The windows came down to the floor and gave me vertigo, but it was a marvellous place for our first view of Sydney Harbor at sunset, the bridge and the scallop shells of the opera house and the dark cliffs of the harbor gateway. It seemed indeed at that moment worthy of being called one of the most beautiful harbors of the world. Later daylight impressions were much less thrilling; the surrounding heights seemed too regular, the atmosphere suburban, the scallop shells merely silly, the bridge just another bridge ("lumpish," Alan Moorehead[1] calls it). No one who has seen the Golden Gate and San Francisco Bay can think much of this. The most attractive places we visited were Manly, which is a real Victorian English beach town with sunshine and sharks added, and the zoo and Thirroul, down the coast south of Sydney, where D.H. Lawrence and Frieda stayed. (I have only just discovered that Wyewurk[2] still stands and is preserved as a Lawrence memorial; I so regret that we didn't visit it.) Altogether, you still get an impression of transplanted English—despite all the other European immigrants—with their dreadful parody cockney accent; English and northern and bred to a small island, and yet here they are, with all the wastes of wilderness behind them and the wastes of the ocean in front. There's also a kind of British rigor which is displayed by the wearing of shorts, almost exhibitionistic it seems, by street cleaners and garbage-truck drivers and many others; we noticed this also in Auckland.

Other memories of our stay in Sydney aren't particularly pleasant. A boring evening with a rich man named James Fairfax who was descended from the seventeenth-century general; we got into a royalist-republican tiff because I'd told him about Bradshaw and my approval of his political career![3] Dennis Altman who meant

---

[1] Australian journalist (1910–1983), celebrated for his coverage of W.W.II; his books include the prize-winning *Gallipoli* (1956).

[2] The bungalow where Lawrence and Frieda lived in 1922 when he was writing *Kangaroo*, in which he used it.

[3] Isherwood's ancestor Judge John Bradshaw (1602–1659) sentenced Charles I to death. Thomas Fairfax (1612–1671) led the Parliamentary forces to victory against the king but stayed away from the trial and tried to get the execution postponed.

well but is so Jewishly thick-skinned that he unintentionally insulted Don. And my resentment when Don decided to go off to
North Africa with [his friend], leaving me alone on my birthday.
I write this down because it is there in my mind; but it makes
no difference to the fact that this was one of the best trips of my
whole wandering life. Indeed, there was more to come, for I spent
two days in Honolulu with Jim Charlton which were beautiful
and happy and brought back so many memories of our past times
together.

Jim, as usual, has made himself a magic little home, full of psychedelic posters and black lights, in a tropical semi-slum above the
"bad" Hotel Street, looking out over the lights below and the dark
shoulders of forest above. This time, I saw that the monster hotels
of Waikiki aren't really all that important, because Jim took me up
into the rain forests below the Pali[1] and to a waterfall in the deep
woods, where he stripped off his clothes and swam—extraordinary,
how boyish his body still is, at, I guess, nearly fifty!

He seems lonely, but then he always did. He is building quite
a lot of houses. He has few people he can talk to, plenty he can
go to bed with. He is engaged in one of his satirical projects, it's
called Autopia, a scheme for turning the entire United States into
stripes of superhighway; he describes it with that deadpan matter-
of-factness which sometimes seems a little mad but is probably his
safety valve to relieve paranoid pressures.

We talked about things we were ashamed of. Jim was ashamed
of having failed to defy the cops that evening long ago when we
were taken to the Santa Monica police station because I'd asked
them why the café we were in was being raided.[2] I'm ashamed of
having consented to answer the loyalty questions which Marvin
Schenck(?) put to me at MGM, when I was about to work on
Diane in 1954. (This always rankles, but, on the other hand, it gave
me a chance to put in a good word for Salka and for Virginia, and
I said nothing that wasn't true. The reason I hate having done it is
that my prospective job was at stake, *maybe*, plus the possibility of
being put on an "uncooperative" list.[3] Well, anyhow—I'm sure I
have done worse things which I refuse to remember at all, much
less tell Jim.)

Jim talked about his first meeting with Mark Cooper—whom
he now chooses to regard as one of the chief heavies in his personal

---

[1] I.e., cliff; it is well over a thousand feet high.
[2] The raid on the Variety, December 1, 1949, mentioned in *D.1* in Isherwood's
"Outline" and in his entry for December 6, 1949.
[3] Schenck was a studio executive; see Glossary.

drama: "I was driving back home along the coast highway and there he stood, holding out his arms—he was like a man on a flat-top, signalling in a plane. And then, when I'd got him home, I said to him, 'I want to go all the way with you'—and I sure did, *Jesus!*"

Another thing I now feel about Honolulu—or rather, don't feel—is that it's "the gateway to the South Seas." All kinds of places used to be a gateway to them, and Tahiti in particular—even San Pedro![1] But now, because I've been down there, Honolulu is Honolulu simply. Which makes me like it better.

Yesterday I went over to Venice to collect my turquoise ring from Jim Gates, who had got it back from the waitress. He told me how he and Peter had gone into Westwood Village to see a movie (*If . . .*), and how it had seemed very exciting and disturbing, like going to the great city; their life in Venice is so pastoral! The other day, when I left, he hugged me and then felt embarrassed, because Peter was there. So yesterday, as we were alone, I hugged him. He was pleased and said, "You're out of sight!"

September 21. Don got home on the 8th. We started work on *Claudius* next day, so tomorrow we shall complete two weeks on the script. We have got as far as the marriage to Messalina, not bad, but still I am really worried because there are so many problems still to be solved and then the whole thing has to be written out in detail—and all this within another five-and-a-bit weeks! I know already that it can't be done properly, only fudged, and I loathe fudging from the bottom of my soul.

Jeanne Moreau told me that both she and her father have Dupuytren's Contracture. They will have to have the operation soon. I met her again, with Don. The evening wasn't quite a success, she seemed chilly.

A cute marine in Vietnam was being interviewed on T.V. the other evening. Asked what he would say if he heard they were to be returned home, he grinned and answered, "It'd be outstanding. We look on the States as a dream world."

Got a letter from *Avant Garde* (magazine) asking me (as one of "one hundred notable Americans") "who qualifies as the most hated man in America?" They explain: "Ever since the retirement of Lyndon B. Johnson, America has been left with a hate vacuum. That is, Johnson was almost universally despised, the object of all our scorn and frustration." The people they suggest are: Roy

[1] The San Pedro Channel, between the Los Angeles coastline and Santa Catalina Island.

Cohn,[1] Mayor Daley,[2] J. Edgar Hoover, General Hershey,[3] George Wallace,[4] Richard Nixon. But surely some newspaper editors are far more despicable than any of these?

October 21. We're still at it, and I doubt if the screenplay can be finished by the end of the month, though probably soon after. It has meant a most drastic reconstruction of the story and really I do think we have done a quick and quite good job; but ideally it should all be rewritten before Tony sees it.

For the past ten days or so, I've been trying a new method of meditation, which is to imagine myself sitting alone in front of the shrine at Vedanta Place. To do this properly, so that I really feel what I feel (sometimes) when I *am* sitting in front of it, is very difficult but wonderfully effective. As long as I can hold the feeling I have no trouble at all with distracting thoughts.

The shrine is a piece of furniture that is *alive*. I have to be absolutely alone with it to feel this; it's a sense of radiation. You don't do anything, don't even pray, just expose yourself to the radiation. The radiation is "safe" as long as you expose yourself to it without any conditions; it would be dangerous to ask it for anything. (Except, of course, to be able to experience the radiation.) It is a confrontation, with the implication that the shrine is really inside me—for otherwise I couldn't get this feeling at all, even occasionally. Now I am wondering, should I try to keep visiting the shrine at times when I can be alone in it? Will that help me to meditate on it?

A couple of weeks ago, when I came to see Swami, I took the dust of his feet. It was just an impulse. Swami said quickly, "What's the matter, Chris? You aren't ill?"—which made me laugh. Did he mean that he found my behavior artificial? Perhaps so.

He says that the Belur Math people won't allow Vidya to come

---

[1] Manhattan lawyer (1927–1986); he helped prosecute Alger Hiss and the Rosenbergs for the New York District Attorney and then became chief council to Joseph McCarthy.

[2] Richard J. Daley (1902–1976), mayor of Chicago from 1955 to his death and president of the corrupt Cook County Democratic Central Committee; he issued a shoot-to-kill order against rioters following Martin Luther King's assassination and appeared to encourage police violence against demonstrators outside the 1968 Democratic convention.

[3] Lewis Blaine Hershey (1893–1977), director of the Selective Service System (the U.S. draft) 1940-1970.

[4] Pro-segregation, four-time governor of Alabama (1919–1998); he ran for president four times.

to India because he made conditions—told them where he would and would not go, etc.

November 21. Well, we finished the *Claudius* screenplay and sent it off on November 3. Since then we have been waiting to hear something. At last, yesterday, we did hear—by way of Peter Schlesinger on a card and Bob Regester in New York—that Tony Richardson has only just returned to England. So Robin French sent a cable to Neil to ask him if he ever got the copy of the screenplay and today we hear that Neil cabled back that Tony is writing us about it and that he is sending Robin the money for it. This sounds frigid, but we mustn't jump to conclusions till we get Tony's letter. Anyhow, I still feel obstinately that we did a really good job, allowing for the untidiness of first-draft dialogue.

As for our play, we are still waiting to hear if this Clifford Williams, the suggested director, really and truly wants to do it, rather than a gruesome-sounding musical about Oscar Wilde in prison, which is another "property" Clement Scott Gilbert wants to produce.

Since finishing the screenplay I have spent *eleven days* laboriously writing and rewriting two and three-quarter pages as a foreword to David Hockney's book of drawings and etchings! Am only just now restarting chapter 9 of *Kathleen and Frank*.

November 22. This morning a note arrived from Tony Richardson: "I have read the script very quickly and liked a lot of it, especially the first part which I liked enormously.... as soon as I have had time to think about it clearly I shall write to you again. It is a super job and I am very grateful to you both." This may still be a brush-off, but at least it raises our morale greatly. Now we have to wait for a further word from Tony or some news from Gilbert before we can decide about going to England.

Meanwhile, Irving Blum continues to talk in the most definite way about Don's show at his gallery, early next year. Don of course is still skeptical and I am bound to admit that Don's kind of work does seem very far from that of Blum's other artists. Why does he want to show it? As a sort of offbeat joke, Don says. I have a theory that it is a mad attempt at an art putsch.[1] If it succeeds, the bottom will drop out of the market for a lot of his competitors' clients; if it fails, Don will be sacrificed as the figurehead of a sunk revolt and Blum will get an E for effort.

[1] Of figurative art over abstract.

We were discussing possible alternative titles for *Kathleen and Frank*. One is *For Life*.

December 10. Tony Richardson never wrote, so I called him in London, the night before last and he told me that he had decided our script wasn't what he wanted and that he had already made a script of his own, during a trip to the South of France. He was embarrassed and slightly apologetic and said that he wished we had been with him, implying that, in that case, we might have worked together. He also asked if we'd be in New York next week because he was probably coming over. I said that we might be and asked him to call us when he knew for certain.

As far as we can guess, this leaves the door open a little way. But maybe we won't want to open it further. It's so difficult to know what to do. I think we may go to New York, because Don has anyway some drawings he could do there for his show, and we should see Wystan and the Stravinskys and even, who knows, manage a reconciliation with Lincoln. As for going to London, it seems we have got to, if we are to see Clifford Williams and decide if we want him to direct *A Meeting by the River*. He is directing another play there now and couldn't come over and meet us in New York—I suggested this to Clement Scott Gilbert on the phone last night. But if I go to London that means visiting all around or offending everybody by dashing off again in a couple of days. Which seems senseless, considering that we will presumably have to be in London for several weeks in February and March for the rehearsals.

Meanwhile I plug along with *Kathleen and Frank* and should soon finish chapter 9. I now incline to the idea that the book will change its character at the end of chapter 10, when Kathleen and Frank are married. The rest of the material will be summarized as much as possible and there will be much more about my own feelings, fantasies, myths and memories with an increasing number of flash-forwards. I estimate that the book will consist of twenty-four chapters, so I'll soon have done three-eighths of it.

Two synchronicities:

I remarked to Jennifer Selznick last time we saw her that I think Jung is one of the greatest men of our age. I was actually referring to his autobiography, but Jennifer reacted to this by sending me a copy of his *Psychology and Alchemy*, saying she was sure I'd be interested in it. Shortly before it arrived, I got a book from a reader in England who had seen me there in an interview on T.V.; it is his autobiography, called *The Diary of a Mystic* (his name is Edward

Thornton). In it he writes of his great preoccupation with Jung's ideas and he particularly quotes from *Psychology and Alchemy*!

I was copying a passage from one of Frank's letters to Kathleen about the two of them leaning over Johns bridge together during a visit to Cambridge. I wondered, does one write Johns (as Frank did) or John's, so I looked up Cambridge in *The Everyman Encyclopaedia* and there was a photograph of John's[1] seen from the Backs—the only illustration to the Cambridge article!

December 23. Yesterday was the solstice (I checked on it this year by phoning the Griffith Park Observatory) and there's the usual feeling of starting the upward climb again. This is a period of unanswered questions which I always rather enjoy recording, it gives one a sense, at least, that *something* has *got* to happen. (So often it actually doesn't, however!)

Robin French sent us this morning a copy of a note he got from Neil Hartley about our *Claudius* screenplay: "This letter is to inform you that we have decided not to go forward with their draft and to thank you very much for your cooperation. Please give my great devotion to both Christopher Isherwood and Don Bachardy for the work they have done."

This sounds so weird that I almost begin to wonder if Tony isn't having some quite other problems, either financial or psychological. And Neil also says he has written to us. He hasn't.

We await news about Don's opening at Irving Blum's gallery. Don is lunching with him on Christmas Day to arrange this. Then we can settle about going to England. Don thinks we should both go and interview Clifford Williams together, otherwise we can't observe him properly. The bore of it is, we should presumably have to make two trips: See Williams, come back for Don's show, go back again to England for the rehearsal and opening of the play.

Working like a donkey I have revised all the muddled-up manuscript of Swami's translation of Narada's Bhakti Sutras. When I told him this in triumph on the phone just now, he asked sweetly, "And you have written the introduction?" which of course I haven't. Also he asked me to paraphrase a whole chapter of the Gita which he has quoted verbatim! To be ready for his birthday on the 26th!

Finished chapter 9 of *Kathleen and Frank* on the 17th.

Jim Gates has gone to Trabuco for the holiday week. This is the one truly satisfactory thing I've managed to arrange. Swami

---

[1] I.e., St. John's College.

really knows who he is now and likes him and Jim got good marks by waxing the library floor soon after arrival, after hitchhiking all the way down there! Jim's lump is not satisfactory, however. The doctor now says it should be removed immediately as it may be malignant. Jim doesn't seem much worried about this. Peter Schneider is coming to see me this afternoon—very shortly, in fact.

December 24. Peter did show up. His face is spotted with tiny scars because he dropped a container of sulphuric acid at the restaurant where he works—they use it for cleaning pots—and it splattered all over him and he got some up his nose, which still burns. I was worried about this and called him this morning but he won't go to the doctor, says it's getting better. He really is quite sweet, though moody, envious and a bit of a poseur. I asked him why had he come to see me and he said because he liked to hear me talk about God (which I hadn't been doing—I wish now that I had because that might have flushed out some ideas from the unconscious which I could use in writing this gruesome Narada introduction). This morning he asked me what he should read, and I read him the ending of *Ulysses* over the phone, old Dobbin show-off.

First thing this morning a telegram arrived. Hoped of course it would be from Tony Richardson but no, Tony Page, wanting the books and papers he gave me when I was to adapt Wedekind's Lulu plays; he has evidently got someone else to do it now. No apologies—he just asks for them as though I owed him something. Well, never mind. Forget it. This is, ha, Christmas.

Talking of Christmas, I have been playing Scrooge by proxy; demanding that one of our tenants at the Hilldale duplexes shall pay or get out. His name is Harold M. Evans and he is a Negro, which of course makes me feel guilty but only very slightly. Anyhow, he has paid, now. Arnold Maltin, who had to do the demanding, claims that he couldn't sleep on account of it. He is hysterical but maybe not without purpose; he is trying to prepare us for his fee as manager. There will be a fuss about this, I'm pretty sure. As for Mr. Evans, it seems likely that he was just probing to see if we'd get tough or not; he has a tailoring business of his own.

Have just looked through my papers and it seems that all I have of Lulu material is just one paperbound book; the cut version of the two plays in German, made by Kadidja Wedekind. So am feeling less put-upon and less indignant with Tony Page.

A beautiful pearly evening with flamingo sunset. Just back from the gym where my weight is exactly 155 lbs.; it has come down just a bit from an alltime high of about 157-8. Something drastic must be done. I blame it largely on eating dates. Nicked my hand on a barbell; it bled.

Last night we had supper with David Sachs. He bores me terribly but that's because he is still awed by me and makes professor talk, carefully weighed and measured generalizations about life, etc. He's turning into a little old Jewish professor of philosophy; last time I saw him, he was still a brash cute little Jewboy grotesquely masquerading as a professor.

December 25. We talked to Vera Stravinsky in New York this morning. She was very pleased to be called, I think. She said that they have got possession of the manuscripts of Igor's works which are now their only important assets, but that she's afraid the lawyers will charge a lot for having got them. Igor is wonderfully better. Vera is all right, only so tired—"morally tired" the doctor had told her. It was all touching and sad, because you got a feeling that, no matter how much money came out of the sale of the manuscripts, it would be frittered away on these astronomical hotel and medical bills which they keep running up. And Vera seems so defenseless now and an old lady.

Ben Underhill talked to me too, phoning from the airport on his way back to San Francisco. He is now in charge of a school for children up to fourteen in a little town called Paicines, south of Hollister, where they have vineyards. Very cheerful as usual and with his slightly amused, calm, self-sufficient air.

Went to see the new James Bond movie, *On Her Majesty's Secret Service*, with a much less good Bond[1] and little to recommend it except scenes of skiing and an avalanche in Switzerland. Going to a performance at half past twelve in the morning was chiefly significant as a symbolic act of schedule breaking. But I had no business to do it. Should have started chapter 10 of *Kathleen and Frank*. I did fudge my way to the end of a rough rough prefirst draft of a foreword to the Narada Bhakti Sutras, however.

Now I'm waiting for Don to call me and tell me what happened when he had lunch with Irving Blum today and discussed his show. He agreed to do this because we can't very well talk about it at the two parties we're to go to tonight—at Charles Aufderheide's and at Leslie Caron's.

Don just called, and Irving Blum seems really to mean business; the show is to be late February or early March. So that's a good solid Christmas present—the best we could have.

[1] I.e., George Lazenby instead of Sean Connery.

December 29. On Christmas Eve, they reran "The Legend of Silent Night," and it had my name on it, not "Magda Bergmann,"[1] despite all the fuss I made last year. However, it seems that my residual won't be shared with the man who wrote the additional material, so I'll get the full ten thousand, minus taxes etc.

We have sent a cable to Clement Scott Gilbert, trying to get a definite statement from him about the deadline for deciding on a director and the possibility of opening our play in March. Have heard nothing from Tony Richardson.

Jim Gates is just back from a week at Trabuco; a great success. His description of it consisted of *wow*-s. He liked Mark and Krishna the best of the monks.

Don is busy drawing people for his show.

Terrific winds, such as I've hardly ever experienced before in this town. Driving was really quite dangerous even in Beverly Hills.

Ray Henderson is getting married, but the girl has to get divorced first, so they've gone to Maryland to do it. I think Elsa is more upset about this than she'll admit.

After a dip, during which it seemed that this might be the end at last, Gerald is a tiny bit better. We saw him today.

December 31. Chilly but beautiful. Don is in Pasadena, drawing two people, both for Irving Blum's show. There has been a big earthslide on the Golden State Freeway; hope he wasn't held up by it.

I've been to the gym; was there yesterday too and the day before. Weight still just over 150. Did some more work on the introduction to Narada this morning and will now try to start chapter 10 of *Kathleen and Frank*.

Nothing from Clement Scott Gilbert, nothing from Tony Richardson and nothing from [Daniel] Selznick about our play; he asked us for a copy to show to Irene Selznick in New York.[2]

---

[1] Isherwood devised the pseudonym as a comment on the T.V. show. "Bergmann" is the character in *Prater Violet* whose wife tells him, "Go and write your poems. When I have cooked the dinner, I will invent this idiotic story for you. After all, prostitution is a woman's business." "Magda" is indeed a woman's name, borrowed from one of the Gabor sisters, at Bachardy's suggestion.

[2] Isherwood wrote David Selznick, but almost certainly meant Daniel (b. 1936), Harvard-educated writer and producer who was the younger son of David with his first wife Irene Mayer Selznick (1907–1990). She became a theatrical producer when they separated—of Williams's *A Streetcar Named Desire* (1947) and van Druten's *Bell, Book, and Candle* (1950) among others.

Don't know if we'll spend this evening with Jack and Jim or just watching T.V. in bed. Jim is full of rehearsals; his film starts almost at once.[1]

Ted may be starting another attack.

Swami is mildly sick, at Santa Barbara, but it's thought to be no more than the usual congregation fatigue.

[1] *The Babymaker.*

Glossary

Abedha. American disciple of Swami Prabhavananda, born Tony Eckstein. He spent many years at Trabuco and at the Hollywood Vedanta Society, but never took sannyas and eventually left to work for Parker Pens.

Acebo, Eddie (b. 1940). American Vedanta devotee, born in Los Angeles of Mexican parents. When he was just fifteen, he saw Gerald Heard moderate a television program called "Focus on Sanity"; later he read essays by Heard, Huxley, and Isherwood, which guided him to Vedanta, and he spent six years living as a monk at the Hollywood Vedanta Society and at Trabuco studying Indian philosophy. In 1968, he moved to Mexico and settled there for many years, but eventually he returned to Trabuco.

Ackerley, J. R. (Joe) (1896–1967). English author and editor. He wrote drama, poetry, fiction, and autobiography, and is well known for his intimate relationship with his Alsatian, described in *My Dog Tulip* (1956) and *We Think the World of You* (1960). Other books include *Hindoo Holiday: An Indian Journal* (1932) and *My Father and Myself* (1968). He was literary editor of *The Listener* from 1935 to 1959 and published work by some of the best and most important writers of his period; Isherwood contributed numerous reviews during the 1930s. Their friendship was sustained in later years partly by their shared intimacy with E.M. Forster. Ackerley was also close to his sister, Nancy West, who was a great beauty and a drunk, as Isherwood records. Ackerley appears in *D.1* and is mentioned in *Lost Years*.

ACLU. American Civil Liberties Union, nonpartisan, nonprofit organization founded in 1920 to protect and preserve individual liberties guaranteed by the U.S. Constitution and its amendments.

Alan. See Campbell, Alan.

Albert, Eddie (1906–2005). American actor, on Broadway from the mid-1930s. his films include *Carrie* (1952), *Roman Holiday* (1953), *Oklahoma* (1955), and *The Heartbreak Kid* (1972), and he played Oliver in the T.V. series "Green Acres," which first aired in 1965. His wife of forty years, the actress and dancer Margo (1917–1985) was born in Mexico City as Maria Marguerita Guadalupe Teresa Estel Bolado Castilla y O'Donell. She had a brief first marriage. Her films include *Winterset* (1936), *Lost Horizon* (1937), and *Viva Zapata!* (1952). She sat for Bachardy in 1973.

Albert, Edward (1951–2006). American actor; son of Eddie and Margo Albert. He appeared in his first film when he was fourteen. Later he went to UCLA and Oxford. He starred in *Butterflies Are Free* (1972) and eventually became a photographer. He sat for Bachardy in 1973.

Aldous. See Huxley, Aldous.

Alec. See Beesley, Alec and Dodie Smith Beesley.

Allen, Alan Warren. Isherwood's general practitioner from April 1961; he was then about forty, tall, handsome, soft-spoken, easygoing, and married. As Isherwood tells in this diary, his first wife committed suicide and Allen later remarried.

Allen, Edwin (Ed). American librarian, at Wesleyan College; once a salesman for Oxford University Press. He met the Stravinskys in 1962, when he was about thirty, and cataloged Igor Stravinsky's library. He also helped with errands and domestic tasks in California and New York, becoming a weekend fixture in the Stravinskys' Fifth Avenue apartment when they moved permanently to New York.

Altman, Dennis (b. 1943). Australian academic, author, gay activist. He did graduate work at Cornell in the mid-1960s and published one of the first accounts of the gay liberation movement in the U.S., *Homosexual: Oppression and Liberation* (1971), followed by many books on sexuality and political culture. Later, he became a politics professor at La Trobe University in Melbourne and President of the AIDS Society of Asia and the Pacific.

Amiya (1902–1986). English Vedanta devotee, born Ella Sully, one of ten daughters of a handsome Somerset farm laborer whose upper-class wife chose scandal and poverty in order to marry him. Amiya travelled to California in the early 1930s with an older sister, Joy, who married an American artist named Palmerton. She was hired by Swami Prabhavananda and Sister Lalita as housekeeper at Ivar Avenue. By the time Isherwood met her at the end of the decade, she had received her Sanskrit name from Swami and become a nun. She became a particular friend of Isherwood's when he lived at the Vedanta Society during the 1940s. She had married in the late 1920s, becoming Ella Corbin, but the marriage failed; in 1952 she met George Montagu, 9th Earl of Sandwich (1874–1962), when he visited the Vedanta Society. A few weeks later Swami gave them permission to marry, and Amiya returned to England, divorced Corbin, and became Countess of Sandwich. She grew close to Isherwood's mother and brother. She was also close to her own younger sister, Sally Hardie (1906–1990), over whom she tried to hold sway with her social position, with lavish gifts, and by financing her company, Sphinx Films, which, in the late 1950s and early 1960s, made several travel films about Italy. Bachardy drew Amiya twice. She appears in *D.1*.

Amohananda. American monk of the Ramakrishna Order. Until he took sannyas in 1971, he was called Paul Hamilton.

Anamananda. See Arup Chaitanya.

ananda. Sanskrit for bliss or joy; an aspect of Brahman. It is used as the last part of a monk's sannyas name in the Ramakrishna Order, for example, Vivek*ananda*, "whose bliss is in discrimination."

Anandaprana or **Ananda.** See Usha.

Anderson, Judith (1898–1992). Australian-born actress; she made her first appearance on the New York stage in 1918 and played major roles throughout the 1930s and 1940s, including the lead in *Mourning Becomes Electra* (1932), Gertrude to Gielgud's Hamlet in 1936, Lady Macbeth twice, and Medea twice. She also had many movie roles, including Mrs. Danvers in Hitchcock's *Rebecca* (1940) and other often chilling parts. She took the lead in the brief Broadway run of Speed Lamkin's play *Comes a Day* at the end of the 1950s. In 1961, Isherwood records that he saw the "snippets" from her most famous shows, which she performed with Bill Roerick on a U.S. tour that included Los Angeles. She appears in *D.1*.

Anderson, Phil. A big, dark, good-looking American whom Isherwood and Bachardy sometimes ran into on the beach in Santa Monica in the early 1960s. Isherwood often told Bachardy he found Anderson attractive. He had particularly nice legs.

Andrews, Oliver and Betty Harford. California sculptor, on the art faculty at UCLA; American actress, his wife until the 1970s. They are mentioned as a couple in *D.1* and in *Lost Years*. Oliver knew Alan Watts well and travelled with him to Japan. Betty acted for John Houseman in numerous stage productions and appeared in a few movies, including *Inside Daisy Clover* (1965). Also, she was a close friend of Iris Tree and acted at Tree's High Valley Theater in the Upper Ojai Valley. They had a son, Christopher, born in the 1950s and named after Isherwood. After they separated, she lived with Hungarian actor Alex de Naszody until he died in the early 1980s. Oliver died suddenly of a heart attack in 1978, while still in his forties.

Angus. See Wilson, Angus.

Animals, The. Isherwood and Bachardy. Also, homosexuals in general, as against human beings or heterosexuals. Isherwood and Bachardy called their Adelaide Drive house La Casa de los Animales. See also Dobbin for Isherwood in his identity as a horse and Kitty for Bachardy.

Arizu, Betty. The daughter Jo Masselink had with Ferdinand Hinchberger. She married Fran Arizu, a Mexican, with whom she had two children.

Arup Chaitanya. American disciple of Swami Prabhavananda, born Kenneth (Kenny) Critchfield. He arrived at the Vedanta Society towards the end of the 1940s and lived there and at Trabuco. He took his brahmacharya vows in 1954, becoming Arup Chaitanya; then in 1963, on taking sannyas, he became Swami Anamananda. He worked for many years in the Vedanta Society Hollywood bookstore, and he died at Trabuco in the early 1990s. He appears in *D.1*.

Asaktananda, Swami (1931–2009). Indian monk of the Ramakrishna Order. He was groomed by Swami Prabhavananda to take over the Hollywood Vedanta Society, until Prabhavananda unexpectedly decided that Asaktananda had the wrong personality for the role and sent him back to the Belur Math in India against his wishes and amid much controversy. Asaktananda later headed the Narendrapur Center, an enormous educational establishment of the Ramakrishna Order outside Calcutta.

asanas. Yoga postures, or the mat on which they are performed.

Ashokananda, Swami. Indian monk of the Ramakrishna Order. Head of the Vedanta Center in San Francisco, where Isherwood first met him in 1943. He appears in *D.1*.

Ashton, Frederick (Freddy) (1904–1988). British choreographer and dancer; born in Ecuador, raised in Peru, and educated in England from 1919. He studied with Léonide Massine and Marie Rambert; Rambert was the first to commission a ballet from him, in 1926. In the late 1920s, he worked briefly as a choreographer in Paris; then, in 1926, he joined the Vic-Wells (later Sadler's Wells) Ballet, where he spent the rest of his career. The company gradually evolved into the Royal Ballet, and, in 1963, Ashton succeeded Ninette de Valois as director. He appears in *Lost Years*.

atman. The divine nature within man; Brahman within the human being; the self or soul; the deepest core of man's identity.

Aubrey, James (1918–1994). Film and T.V. executive; born in Illinois, educated at Princeton. He was a fantastically successful president of CBS television, dominating ratings and doubling profits, until he was fired in 1965. In 1969, he took over MGM Studios, but meanwhile, he briefly headed Aubrey Productions, joined by Hunt Stromberg, who had worked closely with him at CBS. Isherwood refers to their producing partnership as Aubrey-Stromberg. Aubrey was said to be the model for the ruthless Robert Stone in Jacqueline Susann's 1969 novel *The Love Machine*.

Auden, W.H. (Wystan) (1907–1973). English poet, playwright, librettist; perhaps the greatest English poet of his century and one of the most influential. He and Isherwood met as schoolboys towards the end of Isherwood's time at St. Edmund's School, Hindhead, Surrey, where Auden, two and a half years younger, arrived in the autumn of 1915. They wrote three plays together—*The Dog Beneath the Skin* (1935), *The Ascent of F6* (1936), *On the Frontier* (1938)—and a travel book about their trip to China during the Sino-Japanese war—*Journey to a War* (1939). A fourth play—*The Enemies of a Bishop* (1929)—was published posthumously. They also wrote a film scenario "The Life of an American," probably in 1939. As well as several stints of schoolmastering, Auden worked for John Grierson's Film Unit, funded by the General Post Office, for about six months in 1935, mostly writing poetry to be used for sound tracks. He and Isherwood went abroad separately and together during the 1930s, famously to Berlin (Auden arrived first, in 1928), and finally emigrated together to the U.S. in 1939. After only a few months, their lives diverged, but they remained close friends; Auden settled in New York with his companion and, later, collaborator, Chester Kallman. Auden's librettos include *Paul Bunyan* (1941), for Benjamin Britten, *The Rake's Progress* (1948) with Kallman for Stravinsky, and *Elegy for Young Lovers* with Kallman for Hans Werner Henze. As this diary records, Isherwood and Bachardy attended the premiere of *Elegy for Young Lovers* at Glyndeborne in July 1961. Auden is caricatured as "Hugh Weston" in *Lions and Shadows* and figures centrally in *Christopher and His Kind*. There are many passages about him in *D.1* and in *Lost Years*.

Aufderheide, Charles. American technician, from the Midwest. He moved to

Los Angeles with Ruby Bell and the From twins in the 1940s and worked on cameras at the Technicolor laboratories for about thirty years. He was an amateur poet, read widely, liked to entertain, and was a crucial unifying personality in the Benton Way group. In the early 1970s, he moved to San Francisco. He appears in *D.1* and *Lost Years*.

Austen, Howard (Tinker) (1928–2003). Companion to Gore Vidal from 1950. He worked in advertising in New York and studied singing, then devoted most of his time to Vidal, managing his business and social life. He appears in *D.1*.

Ayer, A.J. (Alfred, Freddie) (1910–1989). British philosopher, educated at Eton and Oxford. He married four times and had many affairs. His second (and fourth) wife was American journalist Dee Wells, née Chapman (b. 1925), author of the best-selling novel *Jane* (1973). They married in 1960 and again in 1989. Ayer's third wife was Vanessa Lawson, formerly wife of Nigel Lawson; she and Ayer married in 1982 but were involved with each other from 1968; she died of cancer in 1985. Another long affair, in the early 1950s, was with Jocelyn Rickards, who remained a friend. Ayer was also a close friend of Tony Bower, whom he met in New York during World War II, and with whose half-sister, Jean Gordon-Duff, he was briefly involved around the same time.

Bachardy, Don (b. 1934). American painter; Isherwood's companion from 1953. Bachardy accompanied his elder brother Ted to the beach in Santa Monica from the late 1940s, and Isherwood occasionally saw him there. Ted first introduced them in November 1952. They met again in early February 1953 and, on February 14, began an affair which quickly became serious. Don was then an eighteen-year-old college student living at home with his brother and his mother. He had studied languages for one semester at UCLA, then transferred at the start of 1953 to Los Angeles City College in Hollywood, near his mother's apartment. He studied French and Spanish but dropped French for German as a result of Isherwood's influence. He had worked as a grocery boy at a local market, and, like Isherwood in youth, spent most of his free time at the movies. In February 1955, Bachardy went back to UCLA to begin his junior year and almost immediately changed his major to theater arts. In July 1956, he enrolled at the Chouinard Art School, supplementing his instruction by taking classes with Vernon Old, and within a few years got work as a professional artist, drawing fashion illustrations for a local department store and then for newspapers and magazines. During this period he also began to do portraits of Isherwood, close friends, and favorite film stars, and to sell his work. He drew a set of Hollywood personalities to accompany an article in the *Paris Review* in 1960, but his first major portrait commission, from Tony Richardson, was to draw the cast of the 1960 stage production of *A Taste of Honey*. During 1959 and 1960, Bachardy worked a few days a week in a West Hollywood studio loaned to him by Paul Millard. In 1961 he attended the Slade School of Fine Art in London, supported partly by his patron Russell McKinnon and partly by *Women's Wear Daily* which, since he had no work permit, paid him generously in cash to be their London fashion illustrator. His work at the Slade led to his first solo shows, in London in 1961 and in New York in 1962. Since then, he has done countless portraits, both of the famous and the little known, and exhibited in many cities. His work is held in numerous public and private collections, including the Smithsonian Museum of American Art in Washington, D.C. and the National Portrait Gallery in London, and he has published his drawings in several books, including *October* (1981) with Isherwood, *Last Drawings of Christopher Isherwood* (1990), and *Stars in My Eyes* (2000). Together, Isherwood and Bachardy wrote several stage and film scripts, including their award-winning screenplay for the T.V. film "Frankenstein: The True Story" (1973). He figures centrally in *D.1*.

Bachardy, Glade De Land (1906–198[8]). Don Bachardy's mother, from Ohio. Childhood polio left her with a limp, resulting in extreme shyness. Her father was the captain of a cargo boat on the Great Lakes, and she met her husband, Jess Bachardy, on board during a summer cruise with her sister in the 1920s. They married in 1928 in Cleveland, Ohio, and travelled to Los Angeles on their honeymoon, settling there permanently. The Bachardys divorced in 1952, but later reconciled; once Don and his brother Ted Bachardy had moved out of their mother's apartment, their father moved back, in the late 1950s. An ardent movie-goer, Glade took Don and Ted to the movies from their early childhood because she could not afford babysitters, thus nurturing an obsession which developed differently in each of them. According to Don, Glade did not know what homosexuality was until her elder son Ted had his first breakdown in 1945. She appears in *D.1*.

Bachardy, Jess (1905–1977). Don Bachardy's father, born in New Jersey, the youngest of several brothers and sisters in an immigrant German-Hungarian family. Jess's mother, who never learned to speak English, was pregnant with him when she arrived in the U.S.; his father drowned accidentally shortly before. Jess was an automobile enthusiast and a natural mechanic and took several jobs as a uniformed chauffeur when he was young. Afterwards, he worked on board a cargo boat on the Great Lakes, where he met his future wife. They moved to California, and he turned his mechanical skills to the aviation industry, working mostly with Lockheed Aircraft for the next thirty years. His progress was limited by the fact that he never finished high school, but he advanced to the position of tool planner before he retired in the 1960s. He never allowed his sons to learn Hungarian, and they barely knew their Bachardy grandmother or any of her family. For fifteen years, he refused to meet Isherwood, but he finally relented and came to like him. He was a lifelong smoker and died of lung cancer in 1977. He appears in *D.1*.

Bachardy, Ted (1930–2007). Don Bachardy's older brother. Isherwood spotted him on the beach in Santa Monica, probably in the autumn of 1948 or spring of 1949, and invited him to a party in November 1949 (Ted's name first appears in Isherwood's diary that month). Isherwood was attracted to Ted, but did not pursue him seriously because Ted was involved with someone else, Ed Cornell. Around the same time, Ted experienced a mental breakdown—about the third or fourth he had suffered since 1945, when he was fifteen. Eventually he was diagnosed as a manic-depressive schizophrenic. He was subject to recurring periods of manic, self-destructive behavior followed by nervous breakdowns and long stays in mental hospitals. Isherwood continued to see Ted intermittently during the early weeks of his affair with Don, but a turning point came in February

1953 with Ted's fourth or fifth breakdown. Isherwood sympathized with Don and intervened to try to prevent Ted from becoming violent and having to be hospitalized; nevertheless, Ted was committed on February 26. He had another breakdown, in March 1955, and was again committed to the Camarillo State Mental Hospital for a number of weeks, until April 7. When well, Ted took odd jobs: as a tour guide and in the mail room at Warner Brothers, as a sales clerk in a department store, and as an office worker in insurance companies and advertising agencies. Isherwood writes about him in *D.1* and *Lost Years*.

Bacon, Francis (1909–1992). Irish-born English painter. He worked as an interior decorator in London during the late 1920s and lived in Berlin in 1930, around the time that he taught himself to paint. He showed some of his work in London during the 1930s, but came to prominence only after the war, when his *Three Studies for Figures at the Base of a Crucifixion* made him suddenly famous in 1945. His paintings present anguished, distressed figures in vague, nightmare spaces, often with deliberately smudged paintwork and blurred outlines; he urged that art should expose emotions, rather than simply represent, and expressed his intention to leave evidence of his human presence and experience on his work. Isherwood records some of his remarks on art in *D.1*. He also appears in *Lost Years*.

Balanchine, George (1904–1983). Russian-born choreographer, son of a composer. He studied ballet at the Maryinsky and piano at the St. Petersburg music conservatory. In 1924, he emigrated via Berlin and spent a decade working in Europe, mostly for Diaghilev and the Ballet Russe. In 1933, Lincoln Kirstein persuaded him to emigrate again, to New York, and together they founded the American School of Ballet, struggling off and on for another decade to finance and house the company that would eventually become the New York City Ballet. Balanchine made over four hundred ballets and is known for his modernist approach—abstract, technically demanding, and based on a committed understanding of music. He was to twentieth-century ballet what Picasso was to painting and Stravinsky to music, and he collaborated with Stravinsky a number of times. He married five times.

Barbette (1899–1972). American tightrope walker, born Van der Clyde Broodway, in Texas. As a young member of Ringling Brothers' Circus, he filled in for a woman tightrope walker who fell ill, and he afterwards began to perform as a woman, though he concluded his act by removing his wig to reveal his gender. He became well-known in Paris, where he was photographed by Man Ray in 1926 and appeared in Cocteau's first film, *Le Sang d'un poète* (1930). At the start of World War II, he returned to the U.S. but a fall from the high wire in 1942 ended his performing career. He continued as a circus producer and choreographer; in Hollywood, he choreographed *The Big Circus* (1959). In old age, he was twisted and painfully stiff as a result of his injuries. He sat for Bachardy twice.

Barnett, Jimmy. American monk of the Ramakrishna Order, also known as Sat and as Swami Buddhananda. He lived at Trabuco during the 1960s and later at the Hollywood Vedanta Society. Eventually, he left the order and settled in Sedona,

Arizona, where he became a Native American chieftan and worked as an artist, counsellor, and medicine man. Isherwood mentions him in *Lost Years*.

Barrie, Michael. A one-time singer with financial and administrative talents; friend and secretary to Gerald Heard from the late 1940s onward. He met Heard through Swami Prabhavananda and lived at Trabuco as a monk until about 1955. He was friendly with Isherwood and Bachardy throughout the 1950s, and they rented Barrie's house, at 322 East Rustic Road, for roughly two months in 1956. Barrie nursed Heard through his five-year-long final illness until Heard's death in 1971. He appears in *D.1*.

Batson, Susan (b. 1944). American actress, teacher, director, producer. She was in the original off-Broadway cast of *Hair* (1967), appeared in T.V. serials, and was later acting coach to Tom Cruise, Spike Lee, Jennifer Lopez, Nicole Kidman, and Sean "P. Diddy" Combs. She won the 1969 Los Angeles Drama Critics Circle Best Performance Award for her Black Girl in Isherwood's adaptation of George Bernard Shaw's *The Adventures of the Black Girl in Her Search for God*.

Baxter, Anne (1923–1985). American actress, a granddaughter of Frank Lloyd Wright; educated in New York private schools. She studied acting with Maria Ouspenskaya, debuted on Broadway at thirteen, and made her first movie by seventeen. Her films include *The Magnificent Ambersons* (1942), *The Razor's Edge* (1946) for which she received an Academy Award as best supporting actress, *Yellow Sky* (1949), *All about Eve* (1950) for which she received an Academy Award nomination, *The Outcasts of Poker Flat* (1952), *The Blue Gardenia* (1953), *The Ten Commandments* (1956), *Cimarron* (1960), and *Walk on the Wild Side* (1962). From 1971, as Isherwood records, she returned to Broadway, replacing Lauren Bacall in *Applause*. She also acted on T.V., including, from 1983 to 1985, "Hotel." Her first husband was the actor John Hodiak, with whom she had a daughter; the second, from 1960 to 1968, was Randolph Galt, an outdoorsman and adventurer with whom she had two daughters; the third was David Klee, an investment banker. With Galt, Baxter went to live in the Australian outback on a cattle station; after the marriage failed, she published a book about her experience there, *Intermission: A True Story* (1976). She was a client and friend of Jo Masselink, and she appears in *D.1*.

Beaton, Cecil (1904–1980). English photographer, theater designer, author, and dandy. He photographed the most celebrated and fashionable people of his era, beginning in the 1920s with the Sitwells and going on to the British royal family, actors, actresses, writers, and others. From 1939 to 1945 he worked successfully as a war photographer. Isherwood and Beaton were contemporaries at Cambridge but became friendly only in the late 1940s when Beaton visited Hollywood with a production of *Lady Windermere's Fan* and was helpful to Bill Caskey, then trying to establish himself as a photographer. Returning later to Hollywood, Beaton designed costumes and productions for *Gigi* (1958) and *My Fair Lady* (1964) and both times won the Academy Award for costumes. He collected many of his photographs into books and travel albums, often with commentary, and he published five volumes of diaries. He appears in *D.1* and *Lost Years*.

Beckman, Mathilde von Kaulbach (Quappi) (1904–1986). German Vedanta devotee at the Ramakrishna-Vedanta Center in New York on the Upper East Side. She trained as a violinist and studied voice and acting in Vienna. In 1925, she became the second wife of German painter Max Beckman (1884–1950) and she was a subject of some of his paintings. They fled to Amsterdam in 1937, and after the war they settled in St. Louis and later in New York.

Bedford, Brian (b. 1935). British stage actor and, later, director; an American citizen from 1959. He trained at the Royal Academy of Dramatic Art (RADA) and starred in the West End and on Broadway in Shakespeare and other classic dramas as well as new plays by Stoppard, Shaffer, and others. During 1969, he appeared in revivals of *The Cocktail Party* and *The Misanthrope* in Ellis Rabb's APA-Phoenix Theater repertory program on Broadway. In 1971, he won a Tony Award for his role in *The School for Wives*. He appears regularly at the Stratford Festival in Ontario, Canada, and on T.V. and in films.

Beesley, Alec (1903–1987) and Dodie Smith Beesley (1896–1990). She was the English playwright, novelist and former actress, Dodie Smith. He managed her career. They spent a decade in Hollywood because he was a pacifist and a conscientious objector during World War II. She wrote scripts there for Paramount and her first novel, *I Capture the Castle* (1949). Isherwood met them in 1942 through Dodie's close friend John van Druten, and when Isherwood left the Vedanta Society in August 1945, his first home was the Beesleys' chauffeur's apartment. Dodie encouraged his writing, and he discussed *The World in the Evening* with her extensively. It was Dodie Beesley who challenged John van Druten to make a play from *Sally Bowles*, leading to *I Am a Camera*. In the summer of 1943, the Beesleys mated their Dalmatians, Folly and Buzzle, and Folly produced fifteen puppies—inspiring Dodie's most famous book, *The Hundred and One Dalmatians* (1956), later filmed by Walt Disney. Her plays include *Autumn Crocus* (1931) and *Dear Octopus* (1938). In California, the Beesleys lived on Tower Road in Beverly Hills from the autumn of 1943, then on the Pacific Coast Highway in Las Tunas from the spring of 1945; in November 1945, they moved further out on the old Malibu Road, beyond the Malibu Colony. They returned to England in the early 1950s and settled again in their cottage, The Barretts, at Finchingfield, Essex. They appear in *D.1* and *Lost Years*.

Behrman, S.N. (1893–1973). American playwright, producer, screenwriter, short story writer, journalist. His successes on Broadway include *The Second Man* (1927), *End of Summer* (1936), *No Time for Comedy* (1939), the book (with Joshua Logan) for *Fanny* (1954), and *Lord Pengo* (1962). He also adapted work by others, including *Serena Blandish* and Maugham's short story "Jane." He worked for the Hollywood studios off and on from 1930, specializing in dialogue, and was known for his contributions to Garbo's films *Queen Cristina, Conquest,* and *Two-Faced Woman*. He also wrote for *The New York Times* and *The New Yorker*. He is mentioned in *D.1*.

Ben. See Masselink, Ben.

Bengston, Billy Al (b. 1934). American artist, born in Kansas, educated at the California College of Arts and Crafts in Oakland, at Los Angeles City College, and at the Los Angeles County Art Institute (now Otis Art Institute). He had his first one-man show at the Ferus Gallery in Los Angeles in 1958, followed, from the 1960s onward, by shows and public and private commissions throughout the United States, Canada, Germany and Japan. His work includes painting, sculpture, textiles, lithography, and architectural design. He has been a guest artist and a professor at the Chouinard Art Institute, UCLA, and elsewhere, and has held numerous fellowships and grants, including a Guggenheim. Based for years in Venice, California, he moved in 2004 to Victoria, British Columbia, with Wendy, his Japanese-American wife of many years, but they returned in 2007. Isherwood met him through Bachardy who was commissioned to do Bengston's portrait, along with other prominent Los Angelinos, for *Harper's Bazaar* in 1967.

Bennett, Alan (b. 1934). English actor and playwright, born in Yorkshire, educated at Oxford, where for a time he pursued a graduate degree in medieval history. He has written for stage, film, T.V., radio, and print, revolutionizing the possibilities of comic satire and winning many awards. His works for one medium have frequently been presented later in at least one other; they include *Beyond the Fringe* (1960, with Peter Cook, Jonathan Miller, and Dudley Moore), *Forty Years On* (1968), *An Englishman Abroad* (1982), *Talking Heads* (1987), *A Question of Attribution* (1988), *The Madness of George III* (1991), *Writing Home* (1994), *The History Boys* (2004), and *The Uncommon Reader* (2007).

Berlin crisis. At the Vienna talks in early June 1961, Khrushchev advised Kennedy that he would soon transfer Soviet authority in Berlin to East Germany, thereby ending agreements made among the four victors at the end of World War II which guaranteed Britain, France, and the U.S. access to Berlin across the East German territory surrounding it. The Western Allies would be forced to renegotiate with the new communist East German state, a state they did not formally recognize, in order to get food, supplies, and military personnel and equipment into Berlin, 110 miles from the western border. Kennedy replied to Khrushchev that the Allies would not give up the right of access won in the war and that the West had a moral duty to the 2,000,000 people in West Berlin.

On June 8, the USSR protested the meeting of the Upper House of the Bonn Parliament planned for June 16 in West Berlin. The Parliament had been meeting in Berlin for years, and the USSR had been protesting since 1959; nevertheless, the renewed protest was seen as the first move in Khrushchev's attempt to force the Western Allies out of West Berlin. Isherwood first mentions the crisis two days later, on June 10, and he mentions it again on June 16, the day after Khrushchev recapped the Vienna talks on Soviet T.V. and set the end of 1961 as a deadline for a German peace agreement. On T.V., Khrushchev reiterated his plans for a peace conference, warning that if the West did not attend, other countries would sign a treaty with East Germany without them. He also warned that any fighting over access to West Berlin could bring nuclear holocaust.

By June 26, when Isherwood mentions the crisis a third time, some British papers were suggesting that the West should recognize communist East Germany in exchange for a guarantee of self-determination for West Berlin, but in England the crisis was heavily shadowed by the disaster of appeasement in September 1938. By mid-July, the West was preparing military resources, including building up troops. Macmillan, de Gaulle, Adenauer, and former U.S. President Eisenhower, as well as President Kennedy, were publicly advocating a tough position against Russian plans.

Meanwhile, more and more East Germans were fleeing to West Berlin, and they were increasingly students and young professionals needed in the work force. Whereas at the end of the 1950s, over 100,000 refugees a year had been crossing to West Berlin, during the crisis, the numbers doubled and, in bursts, tripled. The East Germans closed the border between East and West Berlin on August 13 and began to build the Berlin Wall. The British and the Americans responded by sending in more troops, and Vice President Johnson visited, promising West Berlin would not be forgotten, but these moves were widely seen as symbolic. By September 23–24, when Isherwood records signs that the West would sell West Germany down the river, the U.S. was advising West Germany to accept the reality of two German states, even while proclaiming U.S. policy was unchanged. Kennedy reassured Khrushchev that the U.S. would not pursue reunification of Germany, and on October 17, Khrushchev rescinded his December 31 deadline for a peace settlement.

But tension rose all over again only a few days later, when East German police at Checkpoint Charlie stopped the American Chief of Mission in West Berlin and asked to see his passport as he was travelling to the theater in East Berlin. His car bore an occupation forces license plate, entitling him to travel throughout the city without being stopped. The Americans sent another diplomat in a similar car to test the police procedure at the border; he, too, was asked for his passport. The Americans sent him again, backed by tanks and infantry. The Russians answered by sending tanks of their own, evidently because they thought the Americans might attempt to break through the new wall. On October 27 and 28, U.S. and Soviet tanks faced each other about 50–100 yards each from the wall, with live ammunition and orders to fire if fired upon and with tactical nuclear weapons in the vicinity, until Kennedy and Khrushchev finally agreed how to back down from the face-off.

Blanch, Lesley (1904–2007). English journalist and author. She studied painting at the Slade, designed book jackets, and from 1937 to 1944 was an editor at *Vogue*. Her books include *The Wilder Shores of Love* (1954), *The Sabres of Paradise* (1960), *The Nine-Tiger Man* (1965), *Pavilions of the Heart* (1974), biography, travel essays, cook books, and an autobiography titled *Journey Into the Mind's Eye* (1968). She married twice, the second time to the Russian-born French novelist, diplomat, and film director, Romain Gary (1914–1980), whom she met in England during World War II. She was posted with him to Bulgaria and Switzerland, and she travelled widely elsewhere before they divorced in 1962 in Los Angeles, where he was the French consul. Blanch was a close friend of Gavin Lambert who introduced her and Romain Gary to Isherwood and Bachardy during the 1950s; Isherwood also tells about their friendship in *D.1*. Later she settled in France.

Blum, Irving (b. 1930). American art dealer. As a salesman for Hans Knoll, purveyor of modernist furniture, he helped Knoll's Cranbrook-trained wife, Florence, carry out corporate decorating assignments, which often included paintings, and he frequented the Manhattan art scene before joining the Ferus Gallery on La Cienega Boulevard in 1957. His efforts to create a clientele included organizing classes with co-owner Walter Hopps to educate West Coast collectors. In 1967, he opened his own gallery, where he continued to show contemporary Californian artists including Ed Moses, Billy Al Bengston, and Don Bachardy, and more widely known talents like Diebenkorn, Stella, Lichtenstein, Warhol and Johns. Later he opened a New York gallery with Mark Helman.

Bob. See Craft, Robert.

Bopp, Bill (b. 1932). A friend of Bachardy; he worked in administration for Burroughs Corporation, the data processing company, and lived in an apartment in Hollywood.

Bower, Tony (1911–1972). American editor and art dealer; educated in England at Marlborough College and University College, Oxford. His real name was Albert Kilmer Bower. His mother became Lady Gordon-Duff through a second marriage, and his accent and manners gave the impression he was English. He is said to have made a living by playing bridge for money and was a spectacular gossip. He worked briefly for *Horizon*, was drafted into the U.S. Army twice during World War II, and trained on Long Island and later in San Diego. After the war he worked at New Directions, and in 1948 he became an editor at the New York magazine *Art in America*. Eventually, he became an art dealer. Isherwood met Bower in Paris in 1937 through Jean and Cyril Connolly. He appears in *D.1* and *Lost Years*, and, as Isherwood tells, was the model for "Ronny" in *Down There on a Visit*.

Bowles, Paul (b. 1910). American composer and writer, best known for his novel *The Sheltering Sky* (1949), filmed by Bertolucci. In addition to fiction, he wrote poetry and travel books and made translations. Isherwood first met Bowles fleetingly in Berlin in 1931 and used his name for the character Sally Bowles without realizing that he would later meet Bowles again and that Bowles would become famous in his own right. Bowles and his wife, the writer Jane Bowles (1917–1973), lived in George Davis's house in Brooklyn with Auden and others during the 1940s. They later moved to Tangier, where they lived separately from one another but remained close friends. As Isherwood tells in *D.1*, he and Bachardy visited them there in 1955.

Brackett, Charles (1892–1969) and Muff. American screenwriter and producer and his wife. He was from a wealthy East Coast family, began as a novelist, then became a screenwriter and, later, a producer. He often worked with the Austro-Hungarian writer-director Billy Wilder. He was one of five writers who worked on the script for Garbo's *Ninotchka* (1939); he won an Academy Award as writer-producer of *The Long Weekend* (1945); and he produced *The King and I* (1956), as well as working on numerous other films. When Isherwood knew him best during the 1950s, Brackett worked for Darryl Zanuck at Twentieth Century-Fox, where he remained for about a decade. His second wife, Lillian, was called Muff; she had been the spinster sister of Brackett's first wife, who died, and Muff was already in her sixties when Brackett married her. Brackett also had two grown daughters, and one, Alexandra (Xan), was married to James Larmore, Brackett's assistant. The Bracketts appear in *D.1* and *Lost Years*.

Bradbury, Ray (b. 1920). American novelist, poet, playwright, and screenwriter;

he finished high school in Los Angeles and never went to college. He is best known for his science-fiction classics *The Martian Chronicles* (1950) and *Fahrenheit 451* (1953). Other works include *Something Wicked This Way Comes* (1962) and, among his collections of stories, *I Sing the Body Electric* (1969). From the mid-1980s, he adapted his short stories for his T.V. series, "The Ray Bradbury Theater." He married and had four daughters. He appears in *D.1* and *Lost Years*.

Bradley, Alan. English farm laborer. He worked at Wyberslegh Farm in Cheshire and befriended Richard Isherwood there after World War II. He and his wife, Edna, looked after Richard when Kathleen Isherwood died. Richard called the Alan Bradleys "the Alans."

Bradley, Dan. Younger brother of Alan Bradley; with his wife, Evelyn, he took over from Alan the care of Richard Isherwood, partly because Dan Bradley was not fully employable after an accident at work. Richard referred to these Bradleys as "the Dans." After Marple Hall was torn down, the Dans lived next door to Richard in one of the new houses built on the estate. Richard left much of his property and money to the two Bradley families in his will.

Bradshaw, Booker (1940–2003). American actor. He appeared in "Star Trek" during 1967 and in several films, including *The Strawberry Statement* (1970) and *Skulduggery* (1970).

brahmachari or **brahmacharini.** In Vedanta, a spiritual aspirant who has taken the first monastic vows. In the Ramakrishna Order, the brahmacharya vows may be taken only after five or more years as a probationer monk or nun.

Brahman. The transcendental reality of Vedanta; the impersonal absolute existence; infinite consciousness, infinite being, infinite bliss.

Brahmananda, Swami (1863–1922). Rakhal Chandra Ghosh, the son of a wealthy landowner, was a boyhood friend of Vivekananda with whom, ultimately, he was to lead the Ramakrishna Order. Later he was also called Maharaj. Married off by his father at sixteen, he became a disciple of Ramakrishna soon afterwards. Like Vivekananda, Brahmananda was an *Ishvarakoti*, an eternally free and perfect soul born into the world for mankind's benefit and possessing some characteristics of the avatar. He was an eternal companion of Sri Krishna, and his companionship took the intimate form of a parent/son relationship (thus reenacting a previously existing and eternal relationship between their two souls). After the death of Ramakrishna, Brahmananda ran the Baranagore monastery (two miles north of Calcutta), made pilgrimages to northern India, and in 1897 became president of the Belur Math and, in 1900, of the Ramakrishna Math and Mission, founding and visiting Vedanta centers in and near India.

Brando, Marlon (1924–2004). American actor; raised in Nebraska. He was kicked out of a military school in Minnesota, studied acting for a year in New York, debuted on Broadway in *I Remember Mama* in 1944, and became a star as Stanley Kowalski in Tennessee Williams's *A Streetcar Named Desire* in 1947. He continued to explore Method Acting as a professional and joined the Actors Studio in the late 1940s. Once he arrived in Hollywood, his stardom became phenomenal; his blend of defiance and charisma gave him the stature of Garbo and few others, and the violent eccentricities of his private life (including three failed marriages and murderous and suicidal offspring) did not diminish his fame. His films include: *The Men* (1950, Academy Award nomination), *Viva Zapata!* (1952), *Julius Caesar* (1953, Academy Award nomination), *The Wild Ones* (1953), *On the Waterfront* (1954, Academy Award), *Guys and Dolls* (1955), *The Young Lions* (1958), *Mutiny on the Bounty* (1952), *Reflections in a Golden Eye* (1967), *The Godfather* (1972), *Last Tango in Paris* (1972), *Apocalypse Now* (1979). Isherwood first met Brando when Tennessee Williams came to Hollywood to polish the film script for *A Streetcar Named Desire* (1951), for which Brando received an Academy Award nomination; he tells about this in *Lost Years*, and he also mentions Brando in *D.1*.

Breese, Eleanor and Vance. She was a novelist and secretary; he was a pilot. They were divorced but remained close and made an attempt to renew their marriage in 1956. She worked for Isherwood at Twentieth Century-Fox starting in September 1956. Her novel *The Valley of Power* appeared in 1945 under her pen name Eleanor Buckles, but a second novel, about her marriage to Vance, was evidently never published. Later she co-wrote a memoir for Wynne O'Mara, *Gangway for the Lady Surgeon: An Account of W. O'Mara's Experiences as a Ship's Surgeon* (1958). Vance Breese also became a friend of Isherwood and Bachardy. They appear in *D.1*.

Bridges, James (Jimmy, Jim) (1936–1993). American actor, screenwriter and director; raised in Arkansas and educated at Arkansas Teachers College and USC. He was frequently on T.V. in the 1950s and appeared in a number of movies, including *Johnny Trouble* (1957), *Joy Ride* (1958), and *Faces* (1968). He lived with the actor Jack Larson from the mid-1950s onward, and through Larson became close friends with Isherwood and Bachardy. In the early 1960s, he was stage manager for the UCLA Professional Theater Group when John Houseman recommended him as a writer for a Hitchcock suspense series on T.V. He turned out plays constantly, some of which were shown only to Larson, and many of which were never staged, among them *The Papyrus Plays* mentioned by Isherwood. Bridges came to prominence in the 1970s when he directed and co-wrote screenplays for *The Babymaker* (1970), *The Paper Chase* (1973), *The China Syndrome* (1979), *Urban Cowboy* (1980), and, later, *Mike's Murder* (1984), *Perfect* (1985), and *Bright Lights, Big City* (1988). He directed the first production of Isherwood and Bachardy's play *A Meeting by the River* for New Theater for Now at the Mark Taper Forum in 1972, and he directed the twenty-fifth anniversary production of *A Streetcar Named Desire* at the Ahmanson in 1973. He appears in *D.1*.

Brown, Bill (b. 1919). American painter, educated at Yale and Berkeley. He has used the professional names W.T. Brown, W. Theo. Brown, W. Theophilus Brown, and Theophilus Brown. He was born in Illinois and made his career on the West Coast with his longterm partner, Paul Wonner, also a painter. Brown and Wonner were companions from the late 1950s until the mid-1990s, sharing apartments and houses in Santa Monica, Malibu, New Hampshire, Santa Barbara, and finally San Francisco, where, after twenty years, they settled into separate apartments in the same building. Along with Wonner, Richard Diebenkorn, Wayne Thiebaud, David Park, Nathan Oliveira and others first perceived as a group in the early 1950s, Brown has been characterized as an American or a Californian Realist, a Bay Area Figurative Artist, a Figurative Abstractionist. Isherwood met Brown and Wonner in August 1962, when they attended Don Bachardy's first Los Angeles show at the Rex Evans Gallery with Jo and Ben Masselink.

Brown, Harry (1917–1986). American poet, playwright, novelist, screenwriter. He was educated at Harvard and worked for *Time Magazine* and *The New Yorker*. His first novel, *A Walk in the Sun* (1944), was filmed in 1945, and afterwards he worked on numerous Hollywood scripts, especially war movies. He won an Academy Award for co-writing *A Place in the Sun* (1951), and he also wrote *Ocean's Eleven* (1960) among others. In the early 1950s, he worked at Twentieth Century-Fox and MGM, and he was married for a few years to Marguerite Lamkin. Later he married June de Baum. His other novels are *The Stars in Their Courses* (1968), *A Quiet Place to Work* (1968), and *The Wild Hunt* (1973); he published five volumes of poetry. He appears in *D.1*.

Buchholz, Horst (1933–2003). German stage and screen actor, son of a shoemaker. He starred in European films in the 1950s, then achieved Hollywood fame as a gunslinger in *The Magnificent Seven* (1960). His wife, Myriam Bru (b. 1932), was an actress, and, later, a talent agent in Paris.

Buckingham, Bob and May. British policeman and his wife, a nurse. E.M. Forster met and fell in love with Bob Buckingham in 1930. In 1932 they made a radio broadcast together, for a BBC series "Conversations in the Train," overseen by J.R. Ackerley who had introduced them. When Buckingham then met and married May Hockey, it threw his relationship with Forster into turmoil, but the three eventually established a lifelong intimacy. Forster even gave the Buckinghams an allowance as they grew older. In 1951, Buckingham retired from the police force, joined the probation service, and settled with May in a new post in Coventry in 1953. They had one son, Robin, who married and had children of his own, before dying in the early 1960s of Hodgkins Disease.

Buckle, Christopher Richard Sanford (Dicky) (1916–2001). British ballet critic and exhibition designer; educated at Marlborough and, for one year, Oxford. He was ballet critic for *The Observer* from 1948 to 1955 and for *The Sunday Times* from 1959 to 1975. He designed an influential exhibition about Diaghilev and the Ballets Russes in 1954, a less successful one about Shakespeare in 1963, and a Cecil Beaton exhibition at the National Portrait Gallery in 1968. He also created the 1976 exhibition "Young British Writers of the Thirties" at which Isherwood spoke at the Portrait Gallery. He published biographies of Nijinsky (1971), Diaghilev (1979), and Balanchine (1988), as well as three volumes of autobiography.

Buckley, William F. (1925–2008). Right-wing, Roman Catholic columnist and founder of *The National Review*. Gore Vidal and Buckley were hired by ABC News to cover the 1968 Democratic and Republican conventions on T.V., and they argued bitterly on the air about the violence between police and anti-war demonstrators outside the Democratic convention in Chicago. Their host, Howard K. Smith, asked Vidal whether the attempt of some protestors to raise a Vietcong flag was inflammatory, like raising a Nazi flag during World War II.

Vidal asserted that some people in the U.S. believed the Vietcong had the right to organize their country in their own way; Buckley asserted that raising the Vietcong flag encouraged the killing of American soldiers and marines, and called Vidal pro-Nazi. Vidal responded, "The only pro- or crypto-Nazi here is yourself," whereupon Buckley said, "Now listen, you queer, you stop calling me a crypto-Nazi or I'll sock you in the jaw and you'll stay plastered."

Buddha Chaitanya (Buddha). An American disciple of Swami Prabhavananda; born Philip Griggs. He lived as a monk both at the Hollywood Vedanta Society and at Trabuco during the 1950s and took brahmacharya vows with John Yale in August 1955, becoming Buddha Chaitanya. In 1959 he left Vedanta for a time, but eventually took sannyas and became Swami Yogeshananda. Later he led a Vedanta group in Georgia. He appears in *D.1*.

Bürgi, Maria. Swiss academic and Vedanta devotee. She taught about Hinduism at the University of Lausanne and as Maria Bürgi-Kyriazi published several books, including one about Ramana Maharshi.

Burton, Richard (1925–1984). British actor, born Richard Jenkins in a Welsh coal-mining village; he took the surname of his English master and guardian, Philip Burton. He made his professional stage debut in the early 1940s, then served in the air force and briefly studied English at Oxford, where he acted in Shakespeare. He starred in *Hamlet* in London and New York in 1953 and 1954, followed by other Shakespearian roles and, later, the musical *Camelot* (1960) on Broadway and *Equus* (1976). His films include *My Cousin Rachel* (1952), *The Robe* (1953), *Alexander the Great* (1956), *Look Back in Anger* (1959), *Becket* (1964), *The Night of the Iguana* (1964), *The Spy Who Came in from the Cold* (1965), *Who's Afraid of Virginia Woolf* (1966), *The Taming of the Shrew* (1967), *Where Eagles Dare* (1969), *Equus* (1977), and *California Suite* (1978). He was nominated seven times for an Academy Award. His first wife was a Welsh actress, Sybil Williams, with whom Isherwood met him in the late 1950s, and with whom he appears in *D.1*. He began a tempestuous public romance with Elizabeth Taylor when he played opposite her in *Cleopatra* (1963); they married twice, in 1963 and 1975, and divorced both times. His third wife, whom he married in 1976, was an English model, Susan Hunt, and he met his fourth wife, Sally Hay, when she worked as an administrator on his television film "Wagner"; they married in 1983. His legendary drinking hastened his death. In September 1960, Isherwood began to work for Burton on a screenplay of "The Beach of Falesá" by Robert Louis Stevenson; Burton had bought the rights to the story and wanted to produce it himself. Nothing came of the screenplay drafts by various authors although the project limped on for a few years. Around the same time, the Burtons loaned their Hampstead house to Don Bachardy when he studied at the Slade, and Isherwood lived there with Bachardy during 1961. Richard Burton's brother, Ivor Jenkins, lived adjacent with his wife, Gwen, and looked after the property; the Jenkinses also appear in *D.1*.

Calley, John (b. 1930). American film producer and studio executive, born in New Jersey. He began his career in T.V. in the 1950s, worked in advertising, and then became an executive at Filmways, Inc. From 1968 to 1981, he was a senior executive at Warner Brothers and president for a time. During the 1980s, he was an independent producer, working closely with Mike Nichols, and then he ran United Artists and Sony Pictures. Isherwood worked for him when Calley was co-producer with Haskell Wexler of *The Loved One* (Neil Hartley was associate producer). Calley's other films include *Ice Station Zebra* (1968), *Catch-22* (1970), *Postcards from the Edge* (1990), *The Remains of the Day* (1993), *Goldeneye* (1995), *Closer* (2004), and *The Da Vinci Code* (2006).

Campbell, Alan (1904–1963). Actor and screenwriter, second husband of Dorothy Parker. They first married in 1933, divorced after the war, and later remarried, eventually settling on Norma Place, West Hollywood, the heart of Boys Town, as it was known among the gay community (Campbell was rumored to be homosexual). They worked on more than a dozen screenplays together, including, with Robert Carson, *A Star Is Born* (1937) and Lillian Hellman's film adaptation of her play *Little Foxes* (1941). Campbell was involved with the Hollywood Anti-Nazi League and other leftist causes; he was blacklisted by the House Un-American Activities Committee during the 1950s.

Campbell, Dean. American musical comedy actor, singer, dancer, singing coach; he appears in *D.1.*

Campbell, Douglas (b. 1922). Scottish actor and director, mostly of classic stage plays, including Shakespeare; later, he appeared in small T.V. roles and made a few movies. He spent much of his career in Canada, where he performed at the Stratford Festival from 1955 onward and founded The Canadian Players, of which he was artistic director. He played Bernard Shaw in Isherwood's version of *The Adventures of the Black Girl in Her Search for God.*

Capote, Truman (1924–1984). American novelist, born in New Orleans; his real name was Truman Persons. In *Lost Years*, Isherwood describes meeting Capote in the Random House offices in May 1947 shortly before the publication of Capote's first novel, *Other Voices, Other Rooms*. They quickly became friends, and Capote also appears in *D.1.* Capote wrote for *The New Yorker*, where he worked in the early 1940s, and for other magazines. His books include *The Grass Harp* (1951), *Breakfast at Tiffany's* (1958), and the non-fiction novel *In Cold Blood* (1966). He never finished his last novel, *Answered Prayers*, though a chapter, "La Côte Basque, 1965," was published in *Esquire* magazine in 1975, forever alienating rich and powerful friends who were portrayed in it. The rest of what he had written of the novel was published posthumously. Capote's companion for many years was Newton Arvin, a college professor; afterwards, he lived and travelled with Jack Dunphy, and then later picked up new boyfriends with increasing frequency, including John O'Shea. Drink and drugs hastened his death.

Caron, Leslie (b. 1931). French dancer and actress. Her father was a chemist, her American-born mother a dancer. Caron studied ballet from childhood, performed in Paris as a teenager, and was discovered by Gene Kelly, who made her a star in *An American in Paris* (1951). Isherwood first met her during the 1950s, when she was appearing in Hollywood musicals such as *The Glass Slipper* (1955) and *Gigi* (1958). She received British Film Academy Awards and was nominated for Academy Awards for *Lili* (1953) and *The L-Shaped Room* (1962). Later, she appeared in *Damage* (1992), *Jean Renoir* (1993), *The Reef* (1997), *Chocolat* (2000), and *Le Divorce* (2003). She also worked on the stage in New York, London, and Paris. She was married briefly to George Hormel in the early 1950s, then for ten years to British director Peter Hall, with whom she had two children. Her third husband, from 1969 to 1980, was American producer Michael Laughlin. She is mentioned in *D.1*.

Carroll, Nellie (d. 2005). American artist, born Jean Dobrin; she designed and drew greeting cards. She was a close friend of Jim Bridges and Jack Larson. Bachardy drew and painted her many times after they met in 1963. She married once and had a daughter, Amy, who died of cancer in the early 1990s. For the last forty or so years of her life, she lived with a Mexican man about fifteen years her junior, who also had a wife and son.

Carter. See Lodge, Carter.

Caskey, William (Bill) (1921–1981). American photographer, born and raised in Kentucky; a lapsed Catholic of Irish background, part Cherokee Indian. Isherwood met him in 1945 when Caskey arrived in Santa Monica Canyon with a friend, Hayden Lewis, and joined the circle surrounding Denny Fouts and Jay de Laval. They became lovers in June that year and by August had begun a serious affair. Caskey was briefly in the navy during World War II and was discharged neither honorably nor dishonorably (a "blue discharge") following a homosexual scandal in which Hayden Lewis was also implicated. Caskey's father bred horses, and Caskey had ridden since childhood; he had worked in photo-finish at a Kentucky racecourse, and in about 1945 he took up photography seriously. He took portraits of his and Isherwood's friends, and he took the photographs for *The Condor and the Cows*, which Isherwood dedicated to Caskey's mother, Catherine. Caskey's parents were divorced, and he was on poor terms with his father and two sisters. He and Isherwood split in 1951 after intermittent separations and domestic troubles. Later, he lived in Athens and travelled frequently to Egypt. As well as taking photographs, he made art objects out of junk, and for a time had a business beading sweaters. There are many passages about him in *D.1*, and he is a central figure in *Lost Years*.

chaddar. A length of cloth worn on the upper body, often draped on the shoulders as a shawl, by monks and nuns of the Ramakrishna Order and by many other Hindus. Some Western Vedantists meditate in it, to keep warm, and to conceal their rosary.

Chamberlain, Richard (b. 1935). American actor and singer; born in Beverly Hills and educated at Pomona College before serving in Korea for a year and a half. He became famous in the series "Dr. Kildare," in which he starred from 1961 to 1966, and he never shook off the role, despite ambitious appearances on the London stage (for instance as Hamlet) and in a number of films, including *The Madwoman of Chaillot* (1969), *Julius Caesar* (1970), *The Three Musketeers* (1974), and *The Four Musketeers* (1975). He returned to T.V. successfully in the mini-series "Shogun" (1980) and "The Thorn Birds" (1985).

Chapman, Hester (1895–1976). Novelist and biographer of royal figures such as Anne Boleyn, Caroline Matilda of Denmark, and the Duke of Buckingham. She was a cousin of Dadie Rylands, a habitué of Bloomsbury, and a longtime friend of Rosamond Lehmann. With her first husband, she ran a boys' prep school in Devon. Her second husband, Ronnie Griffin, a banker, died in 1955.

Chapman, Kent (b. 1935). A student acquaintance of Isherwood, first mentioned in *D.1* in July 1957. He was an aspiring writer, followed artistic developments among the California poets and painters in West Venice where he lived during the late 1950s, and told Isherwood about the scene there, including his own first attempt to smoke pot. Isherwood also met a girlfriend, Nancy Dvorak. In 1958, Chapman was drafted into the army and served briefly and unhappily in Korea. The night before he received his induction letter, he ran across Los Angeles to Venice from a friend's house in Hollywood and was stopped by a policeman in Beverly Hills. He no longer recalls whether the policeman drove him the rest of the way home, but Isherwood perhaps drew on the episode for *A Single Man*. After Korea, Chapman threw away a novella he had completed and which Isherwood admired, about Vivekananda. In 1963, he moved to San Francisco where he developed a serious drug problem. Five years later, he gave up drugs, got married, and became a Roman Catholic. Eventually, he divorced, moved to France in 1979, and, in 1982, entered a Benedictine monastery, the Abbey of En Calcat, in southern France where, as Frère Laurent, he continued to write and where he also took up painting. He was also a close friend, in France, of Swami Vidyatmananda.

Charles. See Laughton, Charles.

Charlton, Jim (1919–1998). American architect, from Reading, Pennsylvania. He studied at Frank Lloyd Wright's Taliesin West in Arizona and also at Wright's first center, Taliesin, in Wisconsin. He joined the air force during the war and flew twenty-six missions over Germany, including a July 1943 daylight raid. Isherwood was introduced to him by Ben and Jo Masselink in August 1948 (Ben Masselink had also studied at Taliesin West), and they established a friendly–romantic attachment that lasted many years. Towards the end of the 1950s, Charlton married a wealthy Swiss woman called Hilde, a mother of three; he had a son with her in September 1958. The marriage ended in divorce. Afterwards he lived briefly in Japan and then, until the late 1980s, in Hawaii, where he wrote an autobiographical novel, *St. Mick*. Charlton was a model for Bob Wood in *The World in the Evening*. He appears in *D.1* and *Lost Years*.

Cherry, Budd. Creative assistant on *The Loved One* for Tony Richardson. He sat for Bachardy around this time and signed his portrait using a double "d" for his first name. He had an apartment in New York, on East 68th Street, which he once loaned to Bachardy. He was also a dialogue director on *Faces* (1968).

Chester. See Kallman, Chester.

Christian. See Neddermeyer, Christian.

Claxton, Bill and Peggy Moffitt. He was a photographer known for his work with musicians and actors. His wife, Peggy Moffitt, was a model and actress. She was muse to fashion designer Rudi Gernreich, modelling his topless bathing suit in the mid-1960s, and she had a small role in Antonioni's *Blow-up* (1965). In 1991, they published *The Rudi Gernreich Book*, full of Claxton's photographs of

Gernreich's designs worn mostly by Moffitt in her signature white pancake make-up with heavily blacked eyes. One shot shows Bachardy drawing her portrait while Gernreich looks on. Isherwood met Bill Claxton through Jim Charlton, and he appears in *D.1*.

Clay, Camilla (d. 2000). American stage director. She assisted Ellis Rabb at the APA Repertory Company in 1966 and occasionally later. In 1967, she assisted José Quintero when he directed O'Neill's *More Stately Mansions* with Ingrid Bergman and Colleen Dewhurst at the Ahmanson before bringing it to New York. From 1967 to 1972 she rented a house in Malibu with writer Linda Crawford and before that, briefly, they lived at the Chateau Marmont. In 1972, the pair moved back east where Clay directed *Cabaret* at a community theater on the North Fork of Long Island in 1974 and *Stuck* by Sandra Scoppettone in 1976. She lived in Los Angeles again for a few years from 1979 onward before finally settling in New York, where she died of cancer. Isherwood met her through Gavin Lambert.

Clift, Montgomery (1920–1966). American actor, born in Nebraska. He began his career on Broadway at fourteen and appeared in Moss Hart and Cole Porter's musical comedy *Jubilee* (1935), Robert Sherwood's Pulitzer Prize-winning *There Shall Be No Night* (1940), Thornton Wilder's Pulitzer Prize-winning *The Skin of Our Teeth* (1942), Lillian Hellman's *The Searching Wind* (1944), and Tennessee Williams and Donald Windham's romantic comedy *You Touched Me* (1945). He was an early member of Actors Studio in the late 1940s before becoming a Hollywood star in Fred Zinnemann's *The Search* (1948). His other films included *A Place in the Sun* (1951), *From Here to Eternity* (1953), *Raintree County* (1957), *The Young Lions* (1958), *Suddenly Last Summer* (1959), *The Misfits* (1961), *Judgment at Nuremberg* (1961), and *Freud* (1962). Isherwood met him at the end of the 1940s through Fred Zinnemann. As Isherwood records in *D.1* and *Lost Years*, Clift had a drinking problem and was insecure about his looks. He had a car crash in 1957 which badly affected his face, and while he was making *Freud*, in which he appeared as the young doctor, he had cataracts removed from both eyes. He died of a heart attack when he was only forty-five.

Clint. See Kimbrough, Clint.

C.O. Conscientious objector.

Cockburn, Jean. See Ross, Jean.

Cohen, Andee (b. *circa* 1946). American photographer. She began taking pictures of her friends—actors, artists, and rock-and-roll musicians in London and Los Angeles—when her boyfriend, James Fox, gave her a camera in 1966. Her work appeared on album covers for Frank Zappa, Joe Cocker, Tom Petty, and others. Later she married Rick Nathanson, a film producer.

Coldstream, William (Bill) (1908–1987). English painter and teacher; educated at the Slade School of Fine Art. He exhibited with the New English Art Club and the London Group in the late 1920s and became a member of the London Group in 1934. During the mid-1930s, he worked for John Grierson's General Post Office film unit with Auden, already an acquaintance, and Benjamin Britten; Coldstream and his first wife, the painter Nancy Sharp, took in Auden as a lodger. (Nancy later married Stephen Spender's older brother Michael Spender.)

Coldstream painted mostly portraits, including of Isherwood, Auden, Auden's mother, and some landcapes. In 1937, he founded the Euston Road School of Drawing and Painting with Victor Pasmore. During the war, he joined the Royal Artillery and in 1943 was made an Official War Artist. Afterwards he taught at Camberwell School of Art and became a professor at the Slade, where he remained until 1975. He chaired the National Advisory Council on Art Education from 1959, producing the Coldstream Report in 1960, which called for art students to study art history as a requirement and led eventually to degree status being awarded to recognized art courses.

Colloredo-Mansfeld, Countess Mabel (Nishta) (1912–1965). American secretary of the Ramakrishna-Vedanta Center in New York, on the Upper East Side, from 1956. She first visited the center in 1951. She was born Mabel Bayard Bradley in Boston and educated at Foxcroft School, Virginia. In 1933 she married Franz Colloredo-Mansfeld, an Austrian count raised partly in New York and educated at Harvard. They had two sons and a daughter before he was killed flying for the Royal Air Force during World War II.

Connolly, Cyril (1903–1974). British journalist and critic; educated at Eton and Oxford. He was a regular and prolific contributor to English newspapers and magazines, including *The New Statesman*, *The Observer* (where he was literary editor in the early 1940s), and *The Sunday Times*. He wrote one novel, *The Rock Pool* (1936), followed by collections of criticism, autobiography, aphorisms, and essays—*Enemies of Promise* (1938), *The Unquiet Grave* (1944), *The Condemned Playground* (1945), *Previous Convictions* (1963), and *The Evening Colonnade* (1973). In 1939, he founded *Horizon* with Stephen Spender and edited it throughout its publication until 1950. He was perhaps the nearest "friend" of Isherwood and Auden who publicly criticized their decision to remain in America during World War II. He blamed them for abandoning a literary-political movement that he was convinced they had begun and were responsible for. Connolly married three times: first to Jean Bakewell, who divorced him in 1945, then to Barbara Skelton from 1950 to 1956, and finally, in 1959, to Deirdre Craig with whom he had a son, Matthew, and a daughter, Cressida. From 1940 to 1950 he lived with Lys Lubbock, who worked with him at *Horizon*; they never married, but she changed her name to Connolly by deed poll. He appears in *D.1* and *Lost Years*.

Connolly, John. Secretary to Bill Inge. They met at a party given by Glenway Wescott in 1952; in 1957 Connolly left a job assisting Carson McCullers with a play script and began working for Inge—filing, typing, paying bills, answering letters, organizing travel and domestic arrangements. He understood and coped well with Inge's depressive mood swings. According to Connolly, they were never lovers; he maintained his own apartment and social life. He advised against Inge's move to California in 1964, remained behind in New York, and was replaced by Mark Minton. Connolly was friendly with George Platt Lynes and took a particular interest in Don Bachardy after meeting Isherwood and Bachardy at Lynes's New York apartment on New Year's Eve 1953.

Coombs, Don. Instructor of English at UCLA during the 1940s and, as Isherwood tells, a librarian there from 1967. He was a sex friend of Isherwood's beginning in 1949. He appears in *Lost Years*.

Cooper, Gladys (1888–1971). British stage and film star; she was a teenage chorus girl, World War I pin-up, and silent film actress before establishing her reputation on the London stage. As Isherwood tells in *D.1*, he first met her in Los Angeles in 1940 when she was past fifty and had made few films. She had a supporting role in *Rebecca* that year and afterwards appeared in *The Song of Bernadette* (1943), *Green Dolphin Street* (1947), *The Secret Garden* (1949), *Madame Bovary* (1949), *The Man Who Loved Redheads* (1955), *Separate Tables* (1958), and *My Fair Lady* (1964), among many others.

Cooper, Wyatt (1927–1978). Actor, screenwriter, editor, from Mississippi; educated at Berkeley and UCLA. He appeared on stage and T.V., had a small role in *Sanctuary* (1961), and wrote the screenplay for *The Chapman Report* (1962). In *D.1*, Isherwood describes meeting Cooper when he was involved with Tony Richardson. He became the fourth husband of Gloria Vanderbilt (b. 1924), the only granddaughter of Cornelius Vanderbilt and in girlhood the subject of a headline-making custody battle between her widowed, reportedly lesbian mother and her forceful, richer aunt, Gertrude Vanderbilt Whitney, who raised her in lonely splendor on Long Island. At seventeen, Gloria married Pasquale "Pat" DiCicco, a Hollywood agent; at twenty-one she inherited four million dollars. Her two other husbands were conductor Leopold Stokowski and film director Sidney Lumet. She again made her name a household word with her designer jeans in the 1980s. She had two sons with Cooper, the younger one, Carter, committed suicide in 1988; the older one is CNN anchor Anderson Cooper. Cooper also wrote *Families: A Memoir and a Celebration* (1978).

Cordes, Ted. A companion of Ted Bachardy, and the last steady partner he had. He was a few years younger than Bachardy and worked in advertising or publicity. They lived together for several years in the late 1960s, and Cordes weathered at least one of Bachardy's breakdowns. Eventually the relationship collapsed over Bachardy's mental health, and Cordes asked Bachardy to move out of the apartment which they shared. Don Bachardy did several portraits of Cordes.

Coricidin. A brand-name cold remedy. Some versions contain a cough suppressant (dextromethorphan) attractive to recreational drug users and dangerous in combination with the antihistamine ingredient (chlorphenamine maleate). When Isherwood used it, Coricidin contained a decongestant (pseudoephedrine) and not an antihistamine. Some versions also contain an analgesic (acetaminophen), for fever and pain.

Cotten, Joseph (1905–1994). American actor. He worked on Broadway from the early 1930s and played the lead opposite Katharine Hepburn in *The Philadelphia Story* in 1939 and 1940. He was also a member of Orson Welles's Mercury Theater from 1937 to 1939, and Welles brought him to Hollywood to appear in *Citizen Kane* (1941). He went on to star in *The Magnificent Ambersons* (1942) and *Journey Into Fear* (1943) for Welles and then, for Hitchcock, in *Shadow of a Doubt* (1943). His many other films include *Portrait of Jennie* (1948), *The Third Man* (1949), and *Hush ... Hush, Sweet Charlotte* (1964). He appears in *D.1* with his first wife, Lenore Kipp; she was wealthy in her own right and a friend to Isherwood and Bachardy until her death from leukemia in 1960. The year Lenore died, Cotten married British actress Patricia Medina (b. 1920), who was in *The Three Musketeers* (1948) and Welles's *Mr. Arkadin* (1955) and starred with Cotten on Broadway.

Coulette, Henri (1927–1988). American poet; born in Los Angeles, educated there and at the University of Iowa where he taught for several years in the Writers' Workshop. He contributed to *The New Yorker*, *The Paris Review*, *The Hudson Review*, various anthologies, and published two volumes of verse, including *The War of the Secret Agents* (1965), which won several prizes. He was a professor of English at L.A. State from 1959 until his death, and, briefly, Isherwood's colleague.

Craft, Robert (Bob) (b. 1923). American musician, conductor, critic, and author; colleague and adopted son to Stravinsky during the last twenty-three years of Stravinsky's life. Isherwood first met Craft with the Stravinskys in August 1949 when Craft was about twenty-five years old and had been associated with the Stravinskys for about eighteen months. Craft was part of the Stravinsky household, and travelled everywhere with them, except when his own professional commitments prevented him. Increasingly he conducted for Stravinsky in rehearsals and supervised recording sessions, substituting entirely for the elder man as Stravinsky's health declined. In 1972, a year after Stravinsky's death, Craft married Stravinsky's Danish nurse, Alva, who had remained with Stravinsky until the end, and they had a son. Craft published excerpts from his diaries as *Stravinsky: Chronicle of a Friendship 1948–1971* (1972; expanded and republished 1994), edited three volumes of *Selected Correspondence* by Stravinsky, which appeared in 1981, 1984 and 1985, and produced other books arising from his relationship with the Stravinskys as well as articles, essays, and reviews on musical, literary, and artistic subjects. He appears in *D.1* and in *Lost Years*.

Crawford, Linda (b. 1938). New York writer. She shared a house in Malibu with Camilla Clay from 1967 to 1972, then returned to New York and began to publish novels in the mid-1970s; they include *In a Class by Herself* (1976), *Something to Make Us Happy* (1978), *Vanishing Acts* (1983), and *Ghost of a Chance* (1985).

Cuban Missile Crisis. Isherwood first mentions this on September 12, 1962, a week after President Kennedy revealed on September 4 that the Soviets had shipped surface-to-air missiles and large numbers of military personnel to Cuba; the Russians insisted they were supplying only defensive weapons. Isherwood mentions it again on October 23, the day after Kennedy announced on October 22 at 7 p.m. that there were Soviet nuclear missile launch sites and nuclear-capable bombers in Cuba. Kennedy explained that he was establishing a blockade 500 miles off Cuba's coast to prevent the arrival of more weapons, and he warned that nuclear weapons launched against any country in the Western Hemisphere would be regarded as an attack on the U.S. and would bring full retaliation.

Cukor, George (1899–1983). American film director. Cukor began his career on Broadway in the 1920s and came to Hollywood as a dialogue director on *All Quiet on the Western Front* (1930). In the thirties he directed at Paramount, RKO, and then MGM, moving from studio to studio with his friend and producer

David Selznick. He directed Garbo in *Camille* (1936) and Hepburn in her debut *A Bill of Divorcement* (1932) as well as in *Philadelphia Story* (1940); other well-known work includes *Dinner at Eight* (1933), *David Copperfield* (1934), *A Star is Born* (1954), and *My Fair Lady* (1964). Isherwood met Cukor at a party at the Huxleys' in December 1939; later they became friends and worked together. Cukor appears in *D.1* and *Lost Years*.

Curtis Brown. Isherwood's first literary agency in London and in New York, from the mid-1930s. In September 1935, Curtis Brown's London office oversaw the contract committing Isherwood to deliver his next three full-length novels to Methuen; 1935 is also the year Isherwood first appears on the books of Curtis Brown in New York. (At the time, Isherwood was still being published by the Hogarth Press, but Methuen began publishing him with *Prater Violet* after the war.) Isherwood evidently formed a relationship with Curtis Brown's play department for *The Ascent of F6* in the mid-1930s, and Curtis Brown also represented Auden from about this time. In the New York office, Alan Collins was Isherwood's agent until 1959 when Perry Knowlton took over and continued as Isherwood's American agent until 1973. Cindy Degener, head of the dramatic department (film, television, stage), worked with Isherwood and Auden on plans for a musical based on the Berlin stories; at first, in 1959, Frank Taylor wanted to produce it, but he wasn't confident Auden had the popular touch, and so he tried to match Isherwood with a professional musical comedy writer; Isherwood writes about this in *D.1*. At the end of 1960, Oscar Lewenstein—who had worked with Charles Laughton and Tony Richardson and shared offices with Richardson in Curzon Street, London, around this time—became interested in the musical, and Isherwood, Auden, and Chester Kallman again considered the project in 1961. In the mid-1970s, Isherwood left Curtis Brown, New York, for Candida Donadio. But he stayed with Curtis Brown in London, where he was represented by John Barber, James McGibbon, Richard Simon, Peter Grose, and Anthea Morton-Saner.

Dan and Mrs. Dan. See Bradley, Dan.

Dangerfield, George (Geo) (1904–1986). British historian. He emigrated to New York City in 1930 and worked as a publisher's editor and then literary editor at *Vanity Fair* until 1935. By 1968, he was a lecturer in the history department at the University of California at Santa Barbara. His best known books are *The Strange Death of Liberal England* (1935) and *The Era of Good Feelings* (1952)—about the shift in America from Jeffersonian to Jacksonian democracy, 1812–1829—which won the Pulitzer and Bancroft Prizes in 1953. He co-authored, with David Gebhard and Larry Ayres, *Howard Warshaw: A Continuing Tradition* (1977), which was the catalog produced by the UCSB Art Museum for the Warshaw exhibition staged there in August and September 1977. Dangerfield's wife, from 1941, was Mary Lou Schott. Isherwood evidently borrowed his nickname for the main character in *A Single Man*.

Danquah, Paul (b. 1925). British actor and barrister. His father, a philosopher and also a barrister, was founder of the United Gold Coast Convention, an African nationalist party, and helped create the Republic of Ghana, but he fell out of favor with its first president, Kwame Nkrumah, and died in prison there in 1965. Paul Danquah's mother was white, and Danquah was raised in England. While he was still a law student, he played Jimmy in the film version of *A Taste of Honey* and had several other film and T.V. roles during the 1960s. His companion was Peter Pollock, an industrial heir who had once been Guy Burgess's lover. From 1955 until 1961, Danquah and Pollock shared their Battersea flat with Francis Bacon. Later, in the 1970s, they moved to Tangier where Pollock ran a beach bar, The Pergola. Danquah was a consultant to the World Bank in Washington until 1986.

David. See Hockney, David.

Davidson, Gordon (b. 1933). Theater director, raised in Brooklyn. He worked as stage manager and director at the American Shakespeare Festival, where he met John Houseman who invited him to UCLA to work with the Theater Group in 1964. In 1967, the company moved into the new Mark Taper Forum, and Davidson became artistic director of what is now called the Center Theatre Group. He continued in the job for thirty-eight years, opening the new Ahmanson Theater in 1989 and the Kirk Douglas Theater in 2004. By the time he retired, he had won eighteen Tony Awards, three Pulitzer Prizes, and sent thirty-five productions to Broadway. He also won a Margo Jones Award for his contribution to the development of American regional theater.

Davies, Marion (1897–1961). The Ziegfeld Follies chorus girl taken up by newspaper magnate William Randolph Hearst, who made her into a romantic star and financed her films. Her relationship with Hearst is sketched in Orson Welles's *Citizen Kane* (1941). She lived with Hearst at San Simeon and at houses in Beverly Hills and Santa Monica until he died in 1951. Ten weeks after his death, she married Captain Horace Brown, whom she first met during the war and who was previously a suitor of her sister Rose. Speed Lamkin introduced Isherwood to Davies in 1950, and she appears in *D.1* and *Lost Years*.

Davis, Vince. A boyfriend of Ted Bachardy with whom Ted lived for about four years at the end of the 1950s (after Ted lived with Bart Lord). The relationship broke down over Ted's mental health problems; later Davis became a born-again Christian and married. He appears in *D.1*.

Day-Lewis, Cecil (1904–1972). Irish-born poet, novelist, translator, editor; educated at Sherborne School and Oxford, where he became friends with Auden and, through him, met Isherwood. He was a schoolmaster during the 1930s, wrote for leftist publications and joined the Communist party in 1936. His poetry from the period reflects his political involvement, but he later returned to personal themes and abandoned his radical opinions during World War II. He wrote roughly twenty detective novels under a pseudonym, Nicholas Blake, as well as three autobiographical novels, and he published verse translations of Virgil and of Paul Valéry. He was Professor of Poetry in Oxford from 1951 to 1956 (just before Auden) and became Poet Laureate of Britain in 1968. He married twice, the second time to actress Jill Balcon (b. 1925), with whom he had three children (one is the actor Daniel Day-Lewis); he had two children with his first wife. He also had a long affair with Rosamond Lehmann during his first marriage. He appears in *D.1*.

"de Laval, Jay" (probably an assumed name). American chef; he adopted the role of the Baron de Laval. In the mid-1940s he opened a small French restaurant on the corner of Channel Road and Chautauqua in Santa Monica, Café Jay, frequented by movie stars seeking privacy. In 1949, he opened a second restaurant in the Virgin Islands, and in 1950 he was briefly in charge of the Mocambo in Los Angeles before opening a grand restaurant in Mexico City. There, he also planned interiors with Mexican designer Arturo Pani and created a menu for Mexico Air Lines and crockery for Air France. Isherwood met de Laval through Denny Fouts. He was a lover of Bill Caskey before Isherwood and a friend of Ben and Jo Masselink. He appears in *D.1* and *Lost Years*.

Dench, Judi (b. 1934). British actress, educated at The Mount School, York, and Central School of Speech and Drama. She made her stage debut as Ophelia in *Hamlet* for the Old Vic in Liverpool and London in 1957 and was Juliet in Franco Zeffirelli's 1960 production of *Romeo and Juliet*. She joined the Royal Shakespeare Company in 1961, played classic roles for them and for other companies, left for a time, then returned in 1976. In 1968, she was Sally Bowles in the London production of *Cabaret*, her first singing role. Her fame as a movie star grew later in her career; her films include *Four in the Morning* (1965), *A Room with a View* (1986), *A Handful of Dust* (1988), *Mrs. Brown* (1997), *Tomorrow Never Dies* (1997) and subsequent James Bond films, *Shakespeare in Love* (1998, Academy Award for Best Supporting Actress), *Tea With Mussolini* (2000), *Chocolat* (2000), *Ladies in Lavender* (2004), *Iris* (2001), and *Notes on a Scandal* (2006).

DePry, Bert and Bess. Vedanta devotees. He was a successful businessman and President of the Vedanta Society of Southern California throughout the 1960s and into the 1970s.

Depuytren's Contracture. A disease of the hand in which the connective tissue underneath the skin of the palm and fingers develops fibrous bumps or cords. The cords gradually shorten, contracting the fingers into a bent position so they cannot be straightened. It usually affects only the third and fourth fingers. Cortisone injections can alleviate the condition and wearing a splint at night can slow its progress, but eventually, the bumps and cords may have to be removed surgically, in particular to prevent the middle joint of the fingers from becoming fixed in a bent position and in severe cases, where nerve and blood supplies are cut off, to prevent the fingers from requiring amputation. The cords can grow back after surgery and are more difficult to remove the second time. The disease is more frequent in men than in women, and more common in middle age.

Dexamyl. Dextroamphetamine, an antidepressant or upper, combined with amobarbitol, a barbiturate to offset its effect. Isherwood was introduced to it by Bachardy, and they both used it when tackling big swathes of work; for many years they shared Bachardy's prescription since neither of them relied on it habitually. Bachardy was first given it at eighteen by a friend, Alex Quiroga. (Dexedrine, which Isherwood also mentions, is a brand name for a preparation of dextroamphetamine without the barbiturate.)

Diggers. A sixties movement which grew out of the radical underground theater and arts scene in San Francisco and spread to other cities. It took its name from the mid-seventeenth-century English Diggers who opposed private property and asserted the poor man's right to dig on and cultivate common land. The Diggers organized Free Fairs during 1966, bringing avant garde art, poetry, and rock-and-roll to the streets of San Francisco. Later they gave away food in the parks, ran a bakery and stores where the goods were free, and established a free medical clinic. Hippies and Flower Children slept in their shelters for twenty-five cents a night and dressed themselves at their clothing exchanges.

Dobbin. A pet name for Isherwood, known in his lifetime only to himself and Bachardy. Other names included Dubbin, Dub, Drubbin, and Drub, all associated with his private identity as a reliable, stubborn old workhorse.

Dodie. See Beesley, Alec and Dodie Smith Beesley.

Don. See Bachardy, Don.

Donna. See O'Neill, Donna.

Doone, Rupert (1903–1966). English dancer, choreographer and theatrical producer; founder of The Group Theatre, for which Isherwood and Auden wrote plays in the 1930s. His real name was Reginald Woodfield. The son of a factory worker, he ran away to London to become a dancer, and then went on to Paris where he was friendly with Cocteau, met Diaghilev, and turned down an opportunity to dance in the corps de ballet of the Ballets Russes. He was working in variety and revues in London during 1925 when he met Robert Medley, his longterm companion. He died of multiple sclerosis after years of increasing illness. He appears in *D.1* and *Lost Years*.

Dorothy. See Miller, Dorothy.

Douwe. See Stuurman, Douwe.

Dowling, Doris (1923–2004). American actress. She moved from stage to screen with a Paramount contract in the 1940s. Her handful of films include *The Lost Weekend* (1945), *The Blue Dahlia* (1946), and *Bitter Rice* (1948; made in Italy). Isherwood met her in the early 1950s through Shelley Winters and Ivan Moffat, and she appears in *D.1*. Her marriage to the musician and bandleader Artie Shaw (his seventh of eight) was then breaking up; in 1960, she married Leonard Kaufmann, her third husband. In the 1970s, she appeared on T.V., returned to Broadway in a 1973 revival of Clare Booth Luce's play *The Women*, and toured in *Follies* during 1974. Her son Johnno, from her marriage to Shaw, settled with his wife and children in South America. Bachardy often drew Doris Dowling and once drew her sister, Constance Dowling (1923–1969), also an actress.

Dreyfuss, Richard (b. 1947). American actor, writer, producer, and director. His films include *American Graffiti* (1973), *Jaws* (1975), *Close Encounters of the Third Kind* (1977), *The Goodbye Girl* (1977), *Whose Life Is It Anyway?* (1981), *Always* (1989), *Postcards from the Edge* (1990), *What About Bob?* (1991), and *Mr. Hollands's Opus* (1995), but he was still unknown when he appeared in the musical of *The Dog Beneath the Skin* in 1969.

Druks, Renate (1921–2007). Austrian-American painter, actress, film director, scenic designer; born in Vienna, where she studied at the Vienna Art Academy for Women. Later, she studied at the Art Students League in New York. She settled in Malibu in 1950. Her paintings were mostly allegorical portraits of women friends—Anaïs Nin, Joan Houseman, Doris Dowling—in naïve, magic-surrealist style. Druks had a role in Kenneth Anger's experimental film *Inauguration of the Pleasure Dome* (1954) and in other underground films. She often sat for Don Bachardy.

Dub, Dub-Dub, Dubbin. Isherwood; see under Dobbin.

Dundy, Elaine (1921–2008). American actress and writer; born in New York, educated at Mills College in California, at Sweet Briar, and at Jarvis Theater School. Her real name was Elaine Rita Brimberg. She acted intermittently, mostly for T.V., then published a best-selling novel *The Dud Avocado* (1958) and wrote a play, *My Place*, which was staged successfully in 1962. She was the first wife of theater critic Ken Tynan, whom she married in 1951 and with whom she had a daughter, Tracy. The marriage became unstable and increasingly belligerent, and they divorced in 1964. Afterwards she lived largely in New York. Her later books include two more novels, a biography of Peter Finch, a book about Elvis Presley and his mother, and her memoir *Life Itself!* (2001). She appears in *D.1*.

Dunphy, Jack (1914–1992). American dancer and novelist; born and raised in Philadelphia. He danced for George Balanchine and was a cowboy in the original production of *Oklahoma!* He was married to the Broadway musical-comedy star Joan McCracken, and from 1948 he became Truman Capote's companion, although in Capote's later years they were increasingly apart. He published *John Fury* (1946) and *Nightmovers* (1967). He appears in *Lost Years*.

Easton, Harvey and June. He ran probably the first gym in the Los Angeles area, on Beverly Boulevard in Hollywood. He aspired to be a lyricist and had some talent, but died in his early forties of cancer. June ran a dress shop. Diane Easton, their daughter, was at art school with Bachardy briefly during the 1950s, then acted for T.V. Harvey and June appear in *D.1*.

Eckstein, Tony. See Abedha.

Edward. See Upward, Edward.

Elsa. See Lanchester, Elsa.

Emily, or Emmy. See Smith, Emily Machell.

Evans, Rex (1903–1969). British music hall comedian; he moved to Broadway, then became a Hollywood character actor, playing small film roles from the late 1930s through the 1950s. He ran an art gallery on La Cienega Boulevard. He appears in *D.1*.

Fairfax, James (b. 1933). Australian art collector and philanthropist, educated at Oxford. He was the last family chairman, from 1957 to 1987, of the Fairfax newspaper empire.

Faithfull, Marianne (b. 1946). English singer, songwriter, actress. Her recording of "As Tears Go By," by Mick Jagger and Keith Richards, was a Top Ten U.K. hit in 1964, and she became Jagger's girlfriend in 1966. She was to play Ned Kelly's girlfriend in Tony Richardson's film, but as she was arriving in Sydney, she tried to kill herself by swallowing 150 Tuinals (a barbiturate), and went into a coma for six days. Isherwood mentions that she was accompanied on location in Australia by her mother, Eva, Baroness Apollonia and Erisso, a half-Jewish Viennese dancer, actress, and, after she emigrated to England, teacher, who was divorced from Major Glynn Faithfull, a British intelligence officer during World War II.

Falk, Eric (1905–1984). English barrister, raised in London. Falk, who was Jewish, was a school friend from Repton, where he was in the same house as Isherwood, The Hall, and in the History Sixth. He helped Isherwood edit *The Reptonian* during Isherwood's last term, and they saw one another during the school holidays and often went to films together. Falk introduced Isherwood to the Mangeots, whom he had met on holiday in Brittany. Later, he lived in The Temple, a group of mostly late seventeenth-century, college-like buildings in which barristers have offices and also keep residential apartments on the Thames Embankment at the western edge of the City of London. He appears in *Lions and Shadows*, *D.1*, and *Lost Years*.

Faye, Alice (1915–1998). American actress, singer, comedienne; born and raised in New York, where she went on the stage at fourteen. She starred in Hollywood musicals from 1934 to 1945—including *Every Night at Eight* (1935), *Poor Little Rich Girl* (1936), *Alexander's Ragtime Band* (1938), *The Gang's All Here* (1943)—but quit movies over conflicts with Darryl Zanuck at Twentieth Century-Fox and focused back on radio and stage. She made only a few further films. In 1972 and 1973, she revived the musical *Good News* on Broadway, then toured in it for a year, including to Los Angeles. She was a childhood favorite of Bachardy.

Finney, Albert (b. 1936). English actor, trained at RADA; son of a bookie. He acted in Shakespeare from the mid-1950s for the Birmingham Repertory Theatre and came to prominence on the London stage in *Billy Liar* (1960). Afterwards, he appeared in several John Osborne plays directed by Tony Richardson, receiving great praise for *Luther* in 1961, and taking the role to Broadway in 1963. In 1965, he joined the National Theatre Company and appeared in Peter Shaffer's *Black Comedy* (1965) and Peter Nichols's *A Day in the Death of Joe Egg* (1967); then, after a hiatus, he returned to the company to star in *Hamlet*, *Tamburlaine*, *Macbeth*, and others. His film career was launched with *Saturday Night and Sunday Morning* (1960), and he became an international star in Richardson's *Tom Jones* (1963). His other films include: *The Entertainer* (1960), *Night Must Fall* (1964), *Scrooge* (1970), *Murder on the Orient Express* (1974), *The Dresser* (1983), *Under the Volcano* (1984), *The Browning Version* (1994), *Erin Brokovich* (2000), and *Traffic* (2000).

Foch, Nina (1924–2008). American actress, born in Holland and raised in Manhattan. Her films include *The Return of the Vampire* (1944), *Johnny Allegro* (1949), *An American in Paris* (1951), *Scaramouche* (1952), *Executive Suite* (1954), *The Ten Commandments* (1956), *Spartacus* (1960), and *Mahogany* (1975). She had roles on Broadway, was a member of the American Shakespeare Festival, and appeared regularly on T.V. in John Houseman's "Playhouse 90," "The Outer Limits," and others. She also directed and, from the 1960s, taught acting at USC and at the American Film Institute. Her third husband, from 1967 to 1993, was stage producer Michael Dewell (b. 1931).

Fonda, Jane (b. 1937). American actress, born in New York, raised in Hollywood and Greenwich, Connecticut, educated at Vassar; daughter of actor

Henry Fonda and his socialite second wife, Frances Seymour Brokaw, who committed suicide in 1950. She worked as a model before studying with Lee Strasberg at the Actors Studio. There she met Andreas Voutsinas, a Greek would-be actor and director born and raised in Africa and educated in London; he directed her in a disastrous Broadway comedy and coached her in films at the start of the 1960s. Attracted to the vanguard of cultural trends, she opposed the Vietnam War during the 1960s, toured American G.I. camps with Donald Sutherland and other actors as the Anti-War Troop and, in 1972, travelled through North Vietnam followed by press and making radio broadcasts. In 1988, she apologized publicly for supporting the enemy and allowing herself to be photographed at the controls of a North Vietnamese anti-aircraft gun. Her films include *Walk on the Wild Side* (1962), *Period of Adjustment* (1962), *The Chapman Report* (1962), *Cat Ballou* (1965), *Barefoot in the Park* (1967), *They Shoot Horses, Don't They?* (1969), *Klute* (1971, Academy Award and New York Film Critics Award), *Julia* (1977), *Coming Home* (1978, Academy Award), *California Suite* (1978), *9 to 5* (1980), and *On Golden Pond* (1981). She married three times: in 1965 to Roger Vadim, who directed her in *La Ronde/Circle of Love* (1964) and *Barbarella* (1968), then from 1973 to 1990 to political activist Tom Hayden, and from 1991 to 2001 to CNN tycoon Ted Turner. She had one child with Vadim and another with Hayden.

Foote, Dick. American actor and singer. A longtime lover of Carter Lodge. Isherwood and Bill Caskey first met him in early 1949, and saw him regularly over the years with Lodge and van Druten, sometimes at the AJC Ranch. He appears in *D.1* and *Lost Years*.

Forbes, Bryan (b. 1926). English actor, director, producer, screenwriter, novelist; born in London and educated at RADA. He worked on the stage from seventeen, had film roles during the 1950s, and appeared in *The Guns of Navarone* (1961) and *A Shot in the Dark* (1964). From the 1960s, he turned mostly to directing—including *The L-Shaped Room* (1962), *King Rat* (1965), *The Madwoman of Chaillot* (1969), and *The Stepford Wives* (1975)—and contributed some of his own screenwriting and producing. His second wife is the English actress Nanette Newman (b. 1934), with whom he has two daughters, Emma, an actress, and Sarah, a fashion journalist.

Ford, Glenn (1916–2006). Canadian-born actor raised in Santa Monica. He was already making movies by 1939, served in the marines during World War II, and afterwards became a star opposite Rita Hayworth in *Gilda* and opposite Bette Davis in *A Stolen Life*, both in 1946. Among his many other films are *The Big Heat* (1953), *The Blackboard Jungle* (1955), *Cimarron* (1961), *The Courtship of Eddie's Father* (1963), and *Midway* (1976). He also acted in T.V. films and in the series "Cade's County" (1971) and "The Family Holvak" (1975). When Isherwood met him with Hope Lange in July 1960, he was working on *The Four Horsemen of the Apocalypse* (1962), Vincente Minnelli's remake of the 1921 silent film about a family fighting on opposite sides in World War I; Minnelli's film was set during World War II. From 1943 to 1959 Ford was married to the American tap dancer, Eleanor Powell (1910–1982); they had a son, Peter. Later, Ford was married to actress Kathryn Hays, from 1966 to 1968, and then to actress Cynthia Hayward from 1977 to 1984 and to Jeanne Baus from 1993 to 1994.

Forster, E.M. (Morgan) (1879–1970). English novelist, essayist and biographer; best known for *Howards End* (1910) and *A Passage to India* (1924). He was an undergraduate at King's College, Cambridge, and one of the Cambridge Apostles; afterwards he became associated with Bloomsbury and later returned to King's as a Fellow until the end of his life. He was a literary hero for Isherwood, Upward, and Auden from the 1920s onward, and Isherwood regarded Forster as his master. They were introduced by William Plomer in 1932. Forster was a supporter when Isherwood was publicly criticized for remaining in American during World War II. He appears in *D.1* and *Lost Years*. He left his papers and copyright to King's College with a life interest to his literary executor, the psychologist and translator of Freud, Professor W.J.H. "Sebastian" Sprott (1897–1971). Isherwood was also named in his will, as the heir to the American rights of Forster's unpublished homosexual novel *Maurice*, written 1913–1914 and heavily revised 1959–1960; it was published posthumously in 1971 under Isherwood's supervision. Forster and Isherwood shared an understanding that any proceeds would be used to help English friends in need of funds for U.S. travel; Isherwood assigned the proceeds to the National Institute of Arts and Letters where an E.M. Forster Award was created to support English writers on extended visits in the U.S.

Forthman, William H. (Will). American professor of Philosophy of Religion. Isherwood met him at the start of the 1940s, when Will and his brother Bob were teenage parishioners of Allan Hunter, the Congregational minister who participated in the La Verne Seminar in 1941 and involved himself with Gerald Heard's spiritual pursuits. Bob Forthman attended one of Heard's Trabuco seminars in 1942, and Will Forthman continued for many years to attend events sponsored by Heard or by the Vedanta Society. In the 1950s, Will lived on Spoleto Drive in the house of Margaret Gage, Heard's patroness. He became an instructor at California State University, North Ridge in 1958 while still working on his Ph.D.; later, he was a full professor and taught there for many years. He appears in *D.1*.

Fouts, Denham (Denny) (*circa* 1914–1948). Son of a Florida baker; he worked for his father as a teenager then left home to travel as companion to various wealthy people of both sexes. Among his conquests was Peter Watson, who financed *Horizon* magazine, and Fouts helped solicit some of the magazine's earliest pieces. During World War II, Watson sent Fouts to the U.S. with Jean Connolly, and she and Tony Bower introduced Fouts to Isherwood in mid-August 1940 in Hollywood. Fouts determined to begin a new life as a devotee of Swami Prabhavananda, but Swami would not accept him as a disciple, so, after a spell in the East, Fouts moved in with Isherwood in the early summer of 1941, and they led a spartan life of meditation and quiet domesticity. Isherwood describes this in *Down There on a Visit* where Fouts appears as "Paul," and there are many passages about Fouts in *D.1* and *Lost Years*. In August 1941, Fouts was drafted into Civilian Public Service camp as a conscientious objector; on his release in 1943, he lived with a friend from the camp while studying for his high-school diploma; afterwards he studied medicine at UCLA. In 1945 and 1946, Isherwood and Bill Caskey lived in Fouts's apartment at 147 Entrada Drive while Fouts was mostly away; eventually, when Fouts returned, Caskey quarrelled with him, ruining Isherwood's friendship. Soon afterwards, Fouts left Los Angeles for good.

He became an opium addict in Paris, and Isherwood saw him there for the last time in 1948 before Fouts died in Rome.

Fox, James (Willie) (b. 1939). British actor, from childhood; his real name is William; he is a younger brother of actor Edward Fox. Tony Richardson gave him a small role in *The Loneliness of the Long Distance Runner* (1962), and he later starred in *The Servant* (1963), *The Chase* (1966), *Isadora* (1968), *Performance* (1970), *A Passage to India* (1984), *Absolute Beginners* (1986), *The Remains of the Day* (1993), and other films. He left acting for Christian evangelism for a time during the 1970s.

Fox, Lyle. He ran the gym in Pacific Palisades attended by Isherwood, Bachardy, and the Masselinks from the start of the 1960s. He was blond and muscular. In 1967, he married an attractive younger woman called Rez. Eventually he became personal trainer and masseur for Gregory Peck and travelled with Peck on location all over the world.

Frandson, Phillip (Phil) (1925–1981). American educator, from the Midwest. He studied geography, geology, and economics in Paris and Mexico and earned a doctorate in adult education from UCLA. He worked for the Adult Education Association in Washington and Chicago, and from 1956 onwards, for UCLA Extension, where he became Associate Dean in 1970 and Dean in 1973 and developed the Extension into probably the largest continuing education program in the U.S. He was a consultant to the U.S. Office of Education and the National Endowment for the Humanities and travelled in the U.S. and abroad to advise other continuing education administrators. He also collected and lectured about American antiques and received an Emmy Award for producing and hosting a nationally televised T.V. series on American folk art. He also helped plan the Los Angeles Zoo and was a member of its Board of Trustees.

Frank. See Isherwood, Frank Bradshaw.

Franklin. See Knight, Franklin.

French, Hugh (1910–1976). Hollywood agent and former actor. He first approached Isherwood with project ideas in the late 1950s after opening his own agency, Chartwell Artists, with his son, Robin, in about 1956. The Frenches took over from Jim Geller as Isherwood's film agents in 1963. At the start of the 1970s, Hugh French left Chartwell Artists to produce films. He appears in *D.1.*

French, Robin (b. 1936). Hollywood agent and, later, producer and T.V. syndicator; educated at boarding school in England and briefly at college in California. He worked with his father, Hugh, and increasingly represented Isherwood in the film business, taking over entirely in about 1970. He presided over the "incredible rights mess" of the play *I Am a Camera* and the musical and film *Cabaret*, securing Isherwood a substantial income for many years. By 1974, he had left the agency business to become head of domestic production at Paramount Pictures. He later produced a few films, but worked primarily as a T.V. syndicator; eventually he operated and part-owned several T.V. stations before retiring in the late 1990s. French is mentioned in *D.1.*

From, Isador (Eddie, Isad). American film technician and, later, psychotherapist.

Isherwood first met him in 1944, though he became closer to Eddie's identical twin, Sam, who was among the first to answer one of Evelyn Hooker's questionnaires. (The Froms did not look alike because Sam had his nose bobbed.) Sam became wealthy as a businessman, but was a frequent drunk driver and died in a car crash in the mid-1950s. The Froms were at the center of The Benton Way Group which began when Ruby Bell, a librarian from the Midwest, inherited some money and encouraged a group of friends, mostly homosexuals and including the Froms and Charles Aufderheide, to move with her to Los Angeles where she bought a house for them downtown on Benton Way. Later, the group moved to a bigger house, above the Sunset Strip behind the Chateau Marmont; the new house looked like an Italian villa and became known as The Palazzo. It was the scene of many parties and also of serious discussions about homosexual love. Their third home was a large apartment above some shops in a two-story building on Melrose Place. According to Alvin Novak, Eddie was once picked up by the police for an offense relating to his homosexuality, and Isherwood made a great impression on him by coming to his aid. The Froms appear in *D.1* and *Lost Years*.

Frost, Ron (Ronny). American musician, writer, teacher, and registered nurse. He was accepted as a private piano student by Elizabeth O'Neil De Avirett, director of the Los Angeles Conservatory, when he was still a teenager, but he abandoned his studies and became a surgical technician in the army. Afterwards, he settled in Hollywood and worked at Mount Sinai Hospital, but had difficulties adjusting to civilian life. He heard Swami Prabhavananda lecture at the Hollywood temple in 1957, and the following year he became a monk. By the start of the 1960s, he felt able to devote himself to his music again, and he also studied for a master's degree in English. Eventually he returned to Texas, where he taught English Composition at the Community College in El Paso, gave private music lessons, and was the organist at Unity Church.

Gage, Margaret. A rich, elderly patroness of Gerald Heard; she loaned him her garden house on Spoleto Drive in Pacific Palisades, close to Santa Monica, from the late 1940s until the early 1960s. She also provided Will Forthman with a room in her house during the same period. She appears in *D.1.*

Gain, Richard (Dick). American ballet dancer. He danced in the original Broadway chorus of *Camelot* (1960) and in Martha Graham's company and later worked as a choreographer and teacher. He became friendly with Richard and Sybil Burton during *Camelot* and shared their Hampstead home with Isherwood and Bachardy in 1961, while touring with Jerome Robbins's "Ballets: USA." Gain's friend, Richard (Dick) Kuch, also danced in *Camelot*. In 1972, the pair moved to East Bend, North Carolina, to teach dance at the North Carolina School of the Arts where Kuch became an assistant dean. They resigned in 1995.

Gambhirananda, Swami (1899–1988). Indian monk of the Ramakrishna Order; philosopher, scholar, translator. A powerful General Secretary of the Ramakrishna Order for many years, in charge of its practical daily operation. Then, from 1985 to 1988, he was the eleventh president of the order, the spiritual leader, with no role in temporal matters.

Garrett, Anthony (Tony) (b. 1929). Companion to Angus Wilson from the late 1940s; son of a bank clerk; born and raised in London. He left school at sixteen and in 1945 began work as an assistant librarian at the British Museum, where he met Wilson who advised him what to read and took him abroad. From 1948 to 1950, Garrett served in the Army Intelligence Corps in occupied Austria. He returned briefly to the British Museum, then tried the Foreign Office, and, from 1952 to 1954, studied at the London School of Economics for a Social Science Certificate which enabled him to become a probation officer. In 1960, Garrett sacrificed his career when the probation service, evidently responding to gossip about his homosexual ménage with Wilson, requested that he move out of Wilson's cottage in Sussex to a separate residence at least forty miles distant. Thereafter, Garrett devoted himself to Wilson's career, serving as secretary, typist, research assistant, driver, and photographer. He also acted in amateur dramatic productions.

Gates, Jim (1950–*circa* 1990). American non-conformist, violinist, monk of the Ramakrishna Order; born in Washington state and raised in Claremont, California; his father taught Latin and English. He moved out of his parents' house before the end of his sophomore year in high school and after high school went with Peter Schneider to Los Angeles, where they shared various living arrangements in Venice, San Marino, and Hollywood, and where he eventually revealed to Schneider that he was gay. He was obsessed with Isherwood and hoped to run into him on the beach; Schneider looked up Isherwood's telephone number in the phone book and called him to explain this, and Isherwood invited them to Adelaide Drive. Gates attended Santa Monica College briefly and worked as a busboy, library clerk, and live-in assistant to the husband of Marlene Dietrich. He also joined the Hollywood Vedanta monastery for a time. He died of AIDS.

Gavin. See Lambert, Gavin.

Gaynor, Janet (1906–1984). American film star. She appeared in her first movie in the 1920s, and by 1934 was the biggest box office attraction in the U.S. Her films include *The Johnstown Flood* (1926), *Sunrise* (1927), *Seventh Heaven* (1927), *Street Angel* (1928), *State Fair* (1933), *A Star is Born* (1937), *The Young in Heart* (1938), and *Bernadine* (1957). For many years, she used the name of her second husband, fashion designer Gilbert Adrian (known as "Adrian," d. 1959), with whom she appears in *D.1*. In 1964, she married producer Paul Gregory. Gaynor was also an accomplished painter and showed her still lifes in New York in 1976.

Geist, Kenneth (Ken). American writer; educated at Haverford College, the London Academy of Music and Dramatic Arts, and Yale School of Drama. He worked as an actor, stage manager, and director in New York, and became assistant to the director of the Theater Group based at UCLA. During 1965 he worked at CBS T.V. in Los Angeles. Then he turned to writing, first as a film critic in New York, and later as a biographer of Joseph Mankiewicz.

Geller, Jim (d. 1963). Isherwood's Hollywood film agent. He was a story editor at Warner Brothers during the 1940s and expressed interest in Isherwood's work, especially the script written with Aldous Huxley, *Jacob's Hands*. Isherwood was employed with him briefly on *The Woman in White* in 1945. Later, Geller abandoned his studio career, and by the early 1950s he had become Isherwood's agent. Isherwood moved on to Hugh and Robin French when Geller died. His wife, later widow, was Anne Geller. Geller appears in *D.1* and *Lost Years*.

Gerald. See Heard, Henry Fitzgerald.

Gerda. See Neddermeyer, Heinz and Gerda.

gerua. Hindi for ocher, the color of the cloth worn by monks and nuns who have taken their sannyas vows and symbolizing their renunciation.

Gielgud, John (1904–2000). British actor and director; born and educated in London, trained briefly at RADA. He achieved fame in the 1920s acting Shakespeare, Wilde, and Chekhov, and as a director, he had his own London company from 1937. From the 1950s, he also worked with contemporary British playwrights, including Peter Shaffer, Alan Bennett and David Storey. He won three Tonys for his work on the New York stage. His movies, in which he often majestically played supporting and character roles, include: *Hamlet* (1939), *Julius Caesar* (1953), *Richard III* (1955), *Becket* (1964), *Hamlet* (1964), *The Loved One* (1965), *The Charge of the Light Brigade* (1968), *Oh! What a Lovely War* (1969), *Lost Horizon* (1973), *Murder on the Orient Express* (1974), *Joseph Andrews* (1977), *A Portrait of the Artist as a Young Man* (1979), *The Elephant Man* (1979), *Arthur* (1981, Academy Award), *Chariots of Fire* (1981), *Gandhi* (1982), *The Shooting Party* (1985), *Plenty* (1985), *Prospero's Books* (1991), *Hamlet* (1996), *The Portrait of a Lady* (1996), *Shine* (1996), *Elizabeth* (1998). During the 1970s and 1980s, he also worked in television, notably as Charles Ryder's father in the series "Brideshead Revisited." His companion in the 1950s was Paul Anstee, an interior decorator. In 1960, he met Martin Hensler at an exhibition at the Tate Gallery in London; Hensler, Hungarian by background, moved in with him about six years later, and they remained together for the last thirty years of Gielgud's life. Isherwood tells in *D.1* that he first met Gielgud in New York in 1947 and didn't like him; they met again in London in 1948 and became friends. Gielgud also appears in *Lost Years*.

Gill, Jim and Antoinette. American painter and his girlfriend, Canadian actress Antoinette Bower. They lived together for a time but never married. He exhibited his primitivist representational work at the Felix Landau Gallery in the 1960s. One day, he staged an exhibition of his entire store of work, sold it off cheaply, gave up being an artist, and went off with a new girlfriend. Antoinette Bower appeared on T.V., for instance in "Star Trek," and had a small role in the film *Mephisto Waltz* (1971). They both sat for Bachardy a number of times.

Gilliatt, Penelope (1932–1993). English critic, novelist, screenwriter; born Penelope Connor in London and briefly educated at Bennington College in Vermont. She was a staff writer for British *Vogue* and later for *Queen*, and by 1961 she was film critic for *The Observer*. Later she became widely known in America as film critic for *The New Yorker*. Her first husband, Roger Gilliatt, was a London neurologist and best man at the wedding of Princess Margaret to Antony Armstrong-Jones. In 1963, she married John Osborne and had a daughter with him; the marriage broke down by 1966, and Gilliatt settled in New York in 1967, where she had a relationship with the stage and film director Mike Nichols which lasted until 1969, followed by a brief affair with Edmund Wilson. She became an alcoholic, and her career at *The New Yorker* ended when she fabricated an interview with Graham Greene for the magazine. She wrote the prize-winning original screenplay for *Sunday, Bloody Sunday* (1971), more than ten volumes of fiction including short stories, and several volumes of film history and criticism.

Ginsberg, Allen (1926–1997). American poet, born in New Jersey; educated at Columbia University; a member of the Beat scene in New York, San Francisco, and Paris in the 1950s and early 1960s, and a central figure in 1960s counterculture. He was a Zen Buddhist and campaigned against the Vienam War and in favor of drugs, communism, and homosexuality. He invented the phrase "flower power" and claimed he was the first to chant Hare Krishna in North America, through his friendship with Swami Prabhupada who launched the movement in the West. He was a poetic disciple of William Blake and of Walt Whitman and is best known for his early works *Howl and Other Poems* (1956) and *Kaddish and Other Poems* (1961). He published numerous volumes of poetry and also lectures, letters, and journals telling about his friendships with William Burroughs, Jack Kerouac, Neal Cassady, and with his longterm lover, American poet Peter Orlovsky (b. 1933), whom he first met in San Francisco in 1954. Orlovsky dropped out of high school and served as a U.S. Army medic before becoming Ginsberg's secretary; he published several volumes of his own poetry. With Ginsberg and Orlovsky, Isherwood mentions Stephen Bornstein, another close friend of theirs.

Glade. See Bachardy, Glade.

Goetschius, George (1923–2006). American sociologist, educated at New York University and Columbia. In 1954, he settled in London where he met Tony Richardson and moved into Richardson's flat in Hammersmith. He worked for the London Council of Social Service and the Ford Foundation and later taught at the London School of Economics; he also contributed his progressive views to the founding of The English Stage Company at the Royal Court, urging Richardson and George Devine to put on John Osborne's *Look Back in Anger* which he was among the first to read. His relationship with Richardson ended around 1959; later his partner was the playwright Donald Howarth. In 1966, he published an article about the social importance of The English Stage Company, and he wrote two books, *Working with Unattached Youth: Problem, Approach, and Method* (1967) and *Working with Community Groups* (1969). His health was never stable and he suffered a mental and physical breakdown in the 1970s.

Gokulananda, Swami (d. 2007). Indian monk of the Ramakrishna Order, once Swami Prabhavananda's personal attendant in India. He spent some years in the hills of Cherrapunji, and then became head of the Vedanta Center in Delhi.

Goldwyn, Samuel (1882–1974). Polish-American film producer; his real name was Samuel Goldfish. He was partner in several early film companies before forming Goldwyn Pictures with the Selwyn brothers in 1916; from this partnership he took his new name. He was bought out of Goldwyn when it merged in 1924 with the Metro and Mayer production companies to form Metro-Goldwyn-Mayer, and by 1925 he had founded his own Samuel Goldwyn Studios. He remained a top Hollywood producer for thirty years, producing many celebrated and award-winning films—*Wuthering Heights* (1939), *The Best Years of Our Lives* (1946), *Guys and Dolls* (1955), *Porgy and Bess* (1959)—and Samuel Goldwyn Studios continued in business after he retired. In *D.1*, Isherwood describes his first Hollywood job as a writer at the Goldwyn Studios beginning November 1939 for a few weeks; he found Goldwyn difficult. Goldwyn's wife was called Frances.

Goodman, Paul (1911–1972). American novelist, poet and, critic. He wrote works of social commentary on subjects as various as city planning, psychology, political theory, juvenile delinquency, and education. Three early novels—*The Grand Piano* (1942), *The State of Nature* (1946), *The Dead Spring* (1950)—appeared together in 1959 as *The Empire City*, and he published *The Break-up of Our Camp: Stories 1932–1935* as well as many other short stories, but Isherwood was unimpressed by his work until the publication of Goodman's autobiographical novel *Making Do* (1963). Goodman also published literary, film, and T.V. criticism and an autobiography; he was a frequent contributor to *The New York Review of Books*. He is mentioned in *D.1* and appears in *Lost Years* as a member of the Benton Way Group.

Goodwin, John (1912–1994). American novelist. A wealthy friend of Denny Fouts; Isherwood met him probably during the first half of 1943 and often mentions him in *D.1* and *Lost Years*. He owned a ranch near Escondido, a house in New York, and later a modern house in Santa Fe. His intermittent companion was Anthony Russo, a hairdresser of Italian background, much younger than he. Isherwood and Bachardy were invited to spend Christmas 1964 with Goodwin and Russo in Santa Fe, but since Bachardy was then living mostly in New York, they cancelled the trip to be at home together in Santa Monica; this ended the friendship between Isherwood and Goodwin. Goodwin published *The Idols and the Prey* (1953) and *A View of Fuji* (1963).

Gore. See Vidal, Gore.

Gowland, Peter (b. 1916) and Alice. Photographer and camera maker and his wife, director of his photo shoots. He is known for his photographs of celebrities and his nudes, many of which have appeared as *Playboy* centerfolds. His Gowlandflex camera, designed in 1957, is still on the market and is widely used by professionals. The Gowlands were among the Masselinks' closest friends, and Isherwood met them through the Masselinks in the early 1950s. They have two daughters, Ann and Marylee Gowland. The Gowlands appear in *D.1*.

Granny Emmy. See Smith, Emily Machell.

Green, Henry. See Yorke, Henry.

Gregg, Julie (b. 1944). American actress; she was nominated for a Tony Award for her role in the Broadway musical *The Happy Time* (1968), had film parts in *The Godfather* (1972) and *Man of La Mancha* (1972), and many T.V. roles.

Gregory, Paul (b. 1920). American film, T.V., and theater producer; born and raised in Iowa, and, briefly, in London. He acted in two films, then turned to booking and management. In 1950, he persuaded Charles Laughton to be his client and arranged tours for him, then T.V. appearances and stage and film productions for which Laughton acted, directed, and sometimes wrote material. Gregory's Broadway shows include *John Brown's Body* (1953) and *The Caine Mutiny Court-Martial* (1954); his movies, *The Night of the Hunter* (1955) and *The Naked and* *the Dead* (1958). At the suggestion of his girlfriend, Janet Gaynor Adrian, he hired Bachardy to draw Charles Boyer for a poster to promote the Broadway opening of *Lord Pengo* (based on S.N. Behrman's *The Days of Duveen*) in November 1962, but Boyer was difficult and the drawing was never used. Gregory married Gaynor a few years later.

Grigg, Richard (Ricky). Research oceanographer. He grew up in Santa Monica, where he learned to surf in 1950. In 1953, he moved to Hawaii and pioneered big-wave surfing from 1958 onward, first in Waimea Bay and later on the north shore of Oahu. He also free-dived all over the Pacific and later windsurfed. He got a Ph.D. in marine biology at the University of Hawaii and was a professor there until 2004 when he retired.

Griggs, Phil. See Buddha Chaitanya.

Grosser, Maurice (1903–1986). American painter and writer; raised in Tennessee and educated at Harvard. He wrote about art for *The Nation*, and published a number of books including *The Painter's Eye* (1956), *Painting in Our Time* (1964), and *Painter's Progress* (1971). He was the longtime companion of Virgil Thomson; both were close friends of Paul and Jane Bowles, and Grosser lived partly in Tangier. He had a Manhattan apartment which he often loaned to Bachardy, on 14th Street between Seventh and Eighth Avenues.

Guerriero, Henry (b. 1929). American painter and sculptor; from Monroe, Louisiana, where he had known Marguerite and Speed Lamkin and Tom Wright, who were roughly his contemporaries in age. In Los Angeles, he moved in different circles with his companion Michael Leopold. Isherwood met him in the early 1950s, and he is mentioned in *D.1*. He changed his name to Roman A. Clef in 1978.

Guinness, Alec (1914–2000). English actor, born in London; his mother's name was de Cuffe; he never knew who his father was. He trained at the Fay Compton School of Dramatic Art and began his career on the London stage in the mid-1930s, appearing in Shakespeare, Shaw, and Chekhov at the Old Vic. During World War II, he served in the navy, and afterwards began making films, generally in comic or character roles. These include *Great Expectations* (1946), *Oliver Twist* (1948), *Kind Hearts and Coronets* (1949, in which he played eight parts), *The Lavender Hill Mob* (1951, Academy Award nomination), *The Bridge on the River Kwai* (1957, Academy Award), *The Horse's Mouth* (1958, for which he wrote the screenplay, nominated for an Academy Award), *Our Man in Havana* (1959), *Lawrence of Arabia* (1962), *Doctor Zhivago* (1965), *The Comedians* (1967), *Scrooge* (1970), *Brother Sun, Sister Moon* (1973), *Hitler, The Last Ten Days* (1973), *Murder by Death* (1976), *Star Wars* (1977, Academy Award nomination), *The Empire Strikes Back* (1980), *Return of the Jedi* (1983), *A Passage to India* (1984), *Little Dorrit* (1987, Academy Award nomination). He also played the lead in the television mini-series of John Le Carré's *Tinker, Tailor, Soldier, Spy* and *Smiley's People*. He was knighted in 1959, and he was awarded an honorary, career Academy Award in 1980. His wife—Lady Guinness, once he was knighted—was Merula Salaman, an actress and later a painter and children's author.

guna. Any of three qualities—sattva, rajas, tamas—which together constitute

Pakriti, or nature. When the gunas are perfectly balanced, there is no creation or manifestation; when they are disturbed, creation occurs. Sattva is the essence of form to be realized; tamas, the obstacle to its realization; rajas, the power by which the obstacle may be removed. In nature and in all created beings, sattva is purity, calm, wisdom; rajas is activity, restlessness, passion; and tamas is laziness, resistance, inertia, stupidity. Since the gunas exist in the material universe, the spiritual aspirant must transcend them all in order to realize oneness with Brahman.

Gunn, Thom (1929–2004). English poet. Thomson Gunn was educated at University College School, Bedales, and Cambridge. His father edited the London *Evening Standard*; his mother, also a journalist, committed suicide when he was fifteen. He contacted Isherwood in 1955 on his way from a creative writing fellowship at Stanford to a brief teaching stint in Texas; Isherwood invited him to lunch at MGM and they immediately became friends. Gunn later taught at Berkeley off and on from 1958 until 1999. His numerous collections of poetry include *Fighting Terms* (1954), *My Sad Captains* (1961), *Moly* (1971), *Jack Straw's Castle* (1976), *The Man with Night Sweats* (1992), and *Boss Cupid* (2000). He appears in *D.1*.

Gurian, Manning. Stage manager and, later, producer. In the late 1940s, he was the companion and professional partner of Margo Jones (1911–1955), founder of America's first nonprofit professional theater, the Dallas Civic Theater, which she envisioned as part of a nationwide network of community theaters. She was producer and director of many young American playwrights, notably Tennessee Williams. In 1951, Gurian was company manager for *I Am a Camera*, and a few years later became the second husband of Julie Harris, the play's star. He produced Joe Masteroff's *The Warm Peninsula*, in which Harris took a leading role, but it ran for only a few months, from October 1959 until January 1960. He and Harris had one son, Peter Gurian, and divorced in 1967.

Guttchen, Otto. German refugee. Isherwood met him in Hollywood during World War II and writes about him in *D.1*. Guttchen was tortured in a Nazi concentration camp, and his kidneys were badly damaged. He left his wife and child in Switzerland. He struggled to find employment in Hollywood, was often too poor to eat, and became suicidal late in 1939. Isherwood found it difficult to help him adequately and felt intensely guilty about it. In the mid-1950s, they met again and Guttchen appeared to have regained his hold on life.

Hackett, Albert (1900–1995) and Frances Goodrich (1891–1984). American stage and screen writers. He was the son of actress Florence Hackett and had been a child actor; she was a former actress, educated at Vassar, with two previous husbands. They were married for many years and collaborated on plays and numerous filmscripts, including *The Thin Man* (1934), *It's a Wonderful Life* (1946), *The Virginian* (1946), *Father of the Bride* (1950), and *Seven Brides for Seven Brothers* (1954). They won the Pulitzer Prize and a Tony Award for their 1955 Broadway play *The Diary of Anne Frank*, later adapted as a film. They are mentioned in *D.1*.

Hall, Michael. American actor and, later, antique dealer; he appeared in *The Best Years of Our Lives* (1946). In *Lost Years*, Isherwood describes how he met Hall at a party in the winter of 1945–1946 and began a friendship which lasted for twenty years and included occasional sex. Eventually, Hall left the West Coast and settled in New York.

Halsey, Edwin (Ed) (d. 1964). American professor of religion; educated at Dartmouth and Harvard, where he obtained a Ph.D. A naval officer during World War II, he then became a monk at Trabuco while Gerald Heard was running it. During the 1950s, he taught at the Claremont Colleges, where he met John Zeigel and, in 1956, began an affair with him while Zeigel was still an undergraduate. The relationship brought disapproval from the college and community, so Halsey resigned and travelled to the Caribbean and Mexico in search of somewhere he could settle with Zeigel. The pair spent two years in Ajijic, Mexico, Halsey writing a book. In 1962, Zeigel returned to California to take his Ph.D. qualifying exams and teach part-time. Halsey visited him, then drove back to Mexico in October to await Zeigel's homecoming at the end of the academic year. On the way to Mexico, he was killed in a crash in Yuma, Arizona. Isherwood, then at work on *A Single Man* and perhaps struck by parallels between his relationships with Heinz Neddermeyer and Don Bachardy and Halsey's relationship with Zeigel, used a fatal car crash to establish the situation in the novel, although the younger lover dies in his story rather than the older one.

Hamilton, Gerald (1890–1970). Isherwood's Berlin friend who was the original for Mr. Norris in *Mr. Norris Changes Trains*. His mother died soon after his birth in Shanghai, and he was raised by relatives in England and educated at Rugby (though he did not finish his schooling). His father sent him back to China to work in business, and while there Hamilton took to wearing Chinese dress and converted to Roman Catholicism, for which his father, an Irish Protestant, never forgave him. He was cut off with a small allowance and eventually, because of his unsettled life, with nothing at all. So began the persistent need for money that motivated his subsequent dubious behavior. Hamilton was obsessed to the point of high camp with his family's aristocratic connections and with social etiquette, and lovingly recorded in his memoirs all his meetings with royalty, as well as those with crooks and with theatrical and literary celebrities. He was imprisoned from 1915 to 1918 for sympathizing with Germany and associating with the enemy during World War I, and he was imprisoned in France and Italy for a jewelry swindle in the 1920s. Afterwards, he took a job selling the London *Times* in Germany and became interested there in penal reform. Throughout his life he travelled on diverse private and public errands in China, Russia, Europe, and North Africa. He returned to London during World War II, where he was again imprisoned, this time for attempting to promote peace on terms favorable to the enemy; he was released after six months. After the war he posed for the body of Churchill's Guildhall statue and later became a regular contributor to *The Spectator*. He appears in *Lost Years* and *D.1*, where Isherwood tells that at the start of the war, he sent Hamilton a letter which was quoted in William Hickey's gossip column in the *Daily Express*, November 27, 1939, without permission. In the letter, Isherwood mocked the behavior of German refugees in the U.S. His remarks, frivolously expressed for Hamilton's private amusement but fundamentally serious, seemed to Isherwood to have triggered the public criticism which continued into 1940 in the press and in Parliament, of both his own and Auden's absence from England.

Harford, Betty. See Andrews, Oliver and Betty Harford.

Harris, Bill (d. 1992). American artist, raised partly in the USSR and Austria. Harris painted in the 1940s and later made art-objects and retouched photographs. Isherwood met him through Denny Fouts in 1943, while still living as a celibate at the Hollywood Vedanta Society; early in 1944 they began an affair which helped weaken Isherwood's determination to become a monk. Harris was a beautiful blond with a magnificent physique, and Isherwood found him erotically irresistible; the relationship soon turned to friendship, and Harris later moved to New York. Isherwood refers to Harris as "X." in his 1939–1945 diaries (see *D.1*), and he calls him "Alfred" in *My Guru and His Disciples*. Harris also appears in *Lost Years*.

Harris, Julie (b. 1925). American stage and film actress, born in Grosse Pointe, Michigan; educated at finishing school, Yale Drama School, and the Actors Studio. She became a star in the stage adaptation of Carson McCullers's *The Member of the Wedding* (1950), a status she confirmed when she originated the role of Sally Bowles in *I Am a Camera* (1951). She received a Tony Award for *Forty Carats* (1969) and for *The Last of Mrs. Lincoln* (1972), and toured with a one-woman show on Emily Dickinson, *The Belle of Amherst* (1976). Altogether, she has won five Tony Awards, more than any other actor, and she has been nominated ten times. She moved to the screen with early stage roles, receiving an Academy Award nomination for her film debut in *The Member of the Wedding* (1952); later Hollywood movies include *East of Eden* (1955), *Requiem for a Heavyweight* (1962), *The Haunting* (1963), *Harper* (1966), *Reflections in a Golden Eye* (1967), *The Bell Jar* (1976), and *Gorillas in the Mist* (1988). She has also been nominated for nine Emmy Awards, and won twice. During the 1980s, she appeared in the television series "Knots Landing." Isherwood first met her in 1951 after she was cast as Sally Bowles, and their close friendship is recorded in *D.1*. She was married to Jay Julien, a theatrical producer, and then to Manning Gurian, a stage manager and producer, with whom she had a son, Peter Gurian. She divorced Gurian in 1967 during a long affair with actor James (Jim) Murdock. In 1977 she married the writer William Carroll.

Harrison, Rex (1908–1990). English stage and film star, educated at Liverpool College. He made his stage debut in Liverpool at sixteen and was successful in the West End, on Broadway, and in films by the mid-1930s, especially in black-tie comedies. He married six times: to Marjorie Colette Thomas (1934–1942), to actresses Lilli Palmer (1943–1957), Kay Kendall (1957–1959), and Rachel Roberts, whom Isherwood mentions both as Rachel Harrison and as Rachel Roberts (1962–1971), to Elizabeth Harris, ex-wife of actor Richard Harris (1971–1975), and to Mercia Tinker (1978 until his death). His affair with another actress, Carole Landis, was presumed to have contributed to her suicide. He won a Tony Award as Henry Higgins in *My Fair Lady* (1956) on Broadway, and the 1964 film brought him an Academy Award. His other films, many of which also reprised stage roles, include *Blithe Spirit* (1945), *The Rake's Progress* (1945), *Anna and the King of Siam*

(1946), *The Ghost and Mrs. Muir* (1947), *Cleopatra* (as Julius Caesar, 1963), *The Agony and the Ecstasy* (as Pope Julius, 1965), and *Doctor Doolittle* (1967).

Harrity, Rory. American actor. Second husband of Marguerite Lamkin, from 1959 until 1963. Harrity began on the stage and had a film role in *Where the Boys Are* (1960). He also had writing ambitions, but died young of alcoholism.

Hartley, Neil (1916–1994). American film producer, from North Carolina; Tony Richardson's collaborator in Woodfall Productions from 1965. He was production manager for Broadway impresario David Merrick, who imported several of Richardson's stage plays, and he met Richardson in 1958 at the Boston try-out for *The Entertainer*. The pair worked together for the first time on *Luther* when it opened in New York in 1963. *The Loved One* was the first film that Hartley produced for Richardson, and the partnership lasted until Richardson's penultimate film, *Hotel New Hampshire* (1984). Hartley also produced for T.V., including "The Corn Is Green" (1979) and several Agatha Christies. He was a semi-closeted homosexual and died of AIDS. His companion for a long time was Bob Regester.

Harvey, Anthony (Tony) (b. 1931). English actor turned film editor, then director. He directed his first film in 1967. When he was mooted to direct *Cabaret* in 1969, he had just had a popular success with *The Lion in Winter* (1968) and an Academy Award nomination.

Harvey, Laurence (Larry) (1928–1973). Lithuanian-born actor, educated in South Africa and briefly at RADA. He played the Christopher Isherwood character in the film version of *I Am a Camera* in 1955, and Isherwood first met him in London in 1956. He appears in *D.1*. He worked on stage and in films in England from the late 1940s and through the 1950s before going to Hollywood; other films include *The Good Die Young* (1954), *Storm Over the Nile* (1955), *Three Men in a Boat* (1956), *Room at the Top* (1958), *Expresso Bongo* (1959), *Butterfield 8* (1960), *Walk on the Wild Side* (1962), *The Manchurian Candidate* (1962), *Of Human Bondage* (1964), *Life at the Top* (1965), *Kampf um Rom* (1966), and *The Magic Christian* (1968). He lived with the actress Hermione Baddeley for a number of years and was married three times: to actress Margaret Leighton from 1957 to 1961, to American heiress Joan Cohn from 1968 to 1972, and to model Paulene Stone from 1972 until his early death from cancer. With Stone he had a daughter, Domino (1969–2005), subject of the eponymous 2005 film. He also reportedly had male lovers, including Jimmy Woolf, who boosted his career.

Hayworth, Rita (1918–1987). American movie star and World War II pinup. Her parents were dancing partners in the Ziegfeld Follies, and she was a professional dancer at twelve. Her career swelled and faded like her love life, but she was Hollywood's sex goddess during the 1940s. Her films included *Only Angels Have Wings* (1939), *Blood and Sand* (1941), *You Were Never Lovelier* (1942), *Cover Girl* (1943), *Gilda* (1946), *Miss Sadie Thompson* (1953), *Pal Joey* (1957), and *Separate Tables* (1958). She had five husbands—including Orson Welles and Aly Khan, the son of the Aga Khan—and many lovers. She appears in *D.1* with her last husband, producer James Hill.

Heard, Henry FitzGerald (Gerald) (1889–1971). Irish writer, broadcaster, philosopher, religious teacher. Auden took Isherwood to meet him in London in 1932 when Heard was already well known as a science commentator for the BBC and author of several books on the evolution of human consciousness and on religion. A charismatic talker, he associated with some of the most celebrated intellectuals of the time. One of his closest friends was Aldous Huxley, whom he met in 1929 and with whom he joined the Peace Pledge Union in 1935 and then emigrated to Los Angeles in 1937, accompanied by Heard's friend Chris Wood and Huxley's wife and son. Both Heard and Huxley became disciples of Swami Prabhavananda. Isherwood followed Heard to Los Angeles and through him met Prabhavananda. Then Heard became an ascetic, rejecting association with women and criticizing Swami's insufficient austerity; he broke with Swami early in 1941, straining his friendship with Isherwood, and set up his own monastic community, Trabuco College, the same year. By 1949 Trabuco had failed, and he gave it to the Vedanta Society of Southern California to use as a monastery. In the early 1950s, Heard's asceticism relaxed, and he warmed again to his friendship with Isherwood and, later, Don Bachardy. During this period, he shared Huxley's experiments with mescaline and LSD.

He contributed to *Vedanta for the Western World* (1945) edited by Isherwood, and throughout most of his life he turned out prolix and eccentric books at an impressive pace, including *The Ascent of Humanity* (1929), *The Social Substance of Religion* (1932), *The Third Morality* (1937), *Pain, Sex, and Time* (1939), *Man the Master* (1942), *A Taste for Honey* (1942; adapted as a play by John van Druten), *The Gospel According to Gamaliel* (1944), *Is God Evident?* (1948), and *Is Another World Watching?* (1950) published in England as *The Riddle of the Flying Saucers*. For a number of years, he obsessively documented sightings of flying saucers, which he believed were either ultra-fast, experimental aircraft kept secret by the U.S. government or, more exciting to him, visitors from Mars.

Heard is the original of "Augustus Parr" in *Down There on a Visit* and of "Propter" in Huxley's *After Many a Summer* (1939). His role in Isherwood's conversion to Vedanta is described in *My Guru and His Disciple*, and he appears throughout *D.1* and *Lost Years*.

Heinz. See Neddermeyer, Heinz.

Henderson, Ray. American musician, educated at USC. As a pianist, he accompanied Elsa Lanchester for many years, performing in her nightclub act at The Turnabout, a Los Angeles theater, and on tour, notably, in the autobiographical revue Laughton created for her, "Elsa Lanchester—Herself," for which he was billed as Musical Director, and on her T.V. show. He was also her friend and lover, although he was much younger than Lanchester. He composed the music for some operettas which she recorded privately, and he scored and wrote lyrics for a musical version of *The Dog Beneath the Skin*, which was never produced. He died young, of a heart attack.

Herbold, Mary. A member of Allan Hunter's Congregational church. Isherwood met her in the early 1940s. She was a typist and a notary public whose services Isherwood evidently used over the years. On Isherwood's recommendation she typed *Time Must Have a Stop* for Huxley in 1944. She appears in *D.1*.

Hockney, David (b. 1937). British artist, educated at Bradford Grammar School in West Yorkshire, Bradford School of Art, and the Royal College of Art. By 1961, he was identified with his friend R.B. Kitaj and others as leaders of a new movement in British art. Versatility and an appetite for new projects and techniques continually energized his career, in oils, acrylics, photography, photo-copying, drawing, printmaking, faxing, computer images, watercolor, stage and opera design, as well as commentaries about art and the historical development of artistic technique. Hockney's early success allowed him to travel to the USA, Europe, and Egypt; in 1964 he settled in Los Angeles, and met Isherwood soon afterwards. During the 1960s, he taught at the University of Iowa, the University of California at Berkeley, and the University of Colorado, as well as at UCLA where, in 1966, he met Peter Schlesinger, a student in his class who became an important subject. Hockney and Schlesinger rented an apartment on 3rd Street in Santa Monica, a few minutes' walk from Isherwood and Bachardy. Hockney returned to England towards the end of the 1960s and then worked in Paris for a time, near Gregory Evans who became his lover by 1974 and another important subject. Later, he moved around among studios in the Hollywood Hills, Malibu, London and eventually, Bridlington in Yorkshire.

Hoitsma, Kinmont (b. 1934). American college instructor. He fenced for Princeton and for the U.S. in the 1956 Olympics, studied art history as a post-graduate, and for thirty years taught humanities, philosophy, and religious studies in a community college in the San Francisco Bay area. He was photographed by Cecil Beaton in 1964 and 1965, and he appears in Beaton's diaries. He published a booklet, *The Real Mask—Albee's Tiny Alice* (1967).

Holt, Larry. Dr. Hillary Holt, a Hollywood devotee of German or Austrian background. He lived in Hollywood and was friendly with the American monk, Swami Anamananda, who assisted him with his work and errands. He died of cancer in the 1970s.

Holy Mother. See Sarada Devi.

homa fire. Prepared in an ancient Vedic ceremony according to scriptural in-structions, the fire is a visible manifestation of the deity worshipped. Offerings to the deity are placed in the consecrated fire. The homa ritual aims at inner puri-fication; at the end of the ritual, the devotee mentally offers his words, thoughts, actions, and their fruits to the deity.

Homolka, Oscar (1901–1978) and Florence Meyer (1911–1962). Viennese-born actor and his wife, a photographer. He moved from stage to screen in Germany at the end of the 1920s and went to Hollywood in the 1930s. Isherwood met him in 1941 during the filming of *Rage in Heaven*. Homolka was in countless other movies, including *The Seven Year Itch* (1955), *War and Peace* (1956, as General Kutuzov), *A Farewell to Arms* (1957), and *Funeral in Berlin* (1966). Isherwood knew Homolka's first wife, the actress Grete Mosheim, and he remained friendly with Homolka's second wife, Florence, after her marriage to Homolka ended. Florence was wealthy in her own right; her father, Eugene Meyer, was publisher of *The Washington Post*, later succeeded by her younger sister, Katharine Meyer Graham. The Homolkas appear in *D.1*.

Hooker, Evelyn Caldwell (1907–1996). American psychologist and psycho-therapist, trained at the University of Chicago and Johns Hopkins; professor of psychology at UCLA. She was among the first to view homosexuality as a normal psychological condition. She studied homosexuals in the Los Angeles area for many years, through questionnaires, interviews, and discussion in various social settings, accumulating many file drawers of notes which she referred to as "The Project." She first presented her research publicly at a 1956 conference in Chicago, demonstrating that as high a percentage of homosexuals were psycho-logically well-adjusted as heterosexuals. (Her paper was titled "The Adjustment of the Male Overt Homosexual." Isherwood mentions another 1961 paper, "The Homosexual Community," published in the *Proceedings of the XIV International Congress of Applied Psychology* in Copenhagen. There were many more.) Born Evelyn Gentry, she took the name Caldwell from a brief first marriage then changed to Hooker at the start of the 1950s when she married Edward Hooker, a Dryden scholar and professor of English at UCLA, who died of a heart at-tack in 1957. Isherwood met her in about 1949, possibly through the Benton Way Group. In 1952, Jim Charlton refurbished the Hookers' garden house on Saltair Avenue in Brentwood, and Isherwood stayed there until tension developed over the arrival of Don Bachardy in 1953. After an uneasy period, the friendship resumed, as Isherwood tells in *D.1* and *Lost Years*. In 1961, Isherwood refers to a young woman Hooker counselled who suffered complications after an abortion and had to be hospitalized. The young woman's father pressed charges against Hooker, two other psychiatrists who had each advised a therapeutic abortion, the obstetrician who performed the operation, and his daughter's boyfriend. The five were indicted by a grand jury. After six months, the judge ruled there was not enough evidence to indict Hooker, who, as a personal friend of the young man in question, had only referred the couple to one of the two psychiatrists. Both psychiatrists and the young man were eventually declared innocent, but Hooker came to believe that the police had pursued the charges against her because of her research on homosexuality, about which she was questioned at her university office. Anxiety led her to remove all personally identifying data from her notes and records, a task which took her and her secretary nearly a year.

Hoopes, Ned (1932–1984). Teacher, children's anthologist, and, during 1962–1963, host of "The Reading Room," a CBS T.V. series about children's books. He worked on a biography of Charles Laughton, at first with Elsa Lanchester's approval and later without. The book, *A Public Success–A Private Failure: The Unauthorized Biography of Charles Laughton*, was never published.

Hope. See Lange, Hope.

Horne, Geoffrey (b. 1933). American actor and acting coach, born in Argentina. He appeared in *The Bridge on the River Kwai* (1957) and worked in T.V. from the 1950s through the 1980s. He taught acting at the Lee Strasberg Theater and Film Institute and, later, at New York University. One of his numerous wives was Collin Wilcox, with whom he looked after several children evidently born to other parents.

Houseman, John (1902–1988) and Joan. Romanian-born writer, director, producer, and actor; his real name was Jacques Haussmann. His mother was British and he was educated in England, then travelled to Argentina and the U.S. as an agent for his father's grain business which collapsed during the Depression. He worked as a journalist and translated plays, then in 1934, directed Virgil Thomson's opera of Gertrude Stein's *Four Saints in Three Acts*, a Broadway hit. Afterwards, he collaborated with Orson Welles with whom he founded the Mercury Theater in 1937. He produced Welles's film *Citizen Kane* (1941), but after a disagreement over who developed the story, he went on to work for David Selznick in Hollywood and was responsible for a string of widely admired films. He returned often to direct on Broadway, taught at Vassar, was Artistic Director of the American Shakespeare Festival in the late 1950s, and later of the UCLA Professional Theater Group, and he ran the drama division at Juilliard from 1967. He took his first of many movie roles in *Seven Days in May* (1964), and he won an Academy Award for his supporting role in *The Paper Chase* (1973), which he reprised in the T.V. series. He divorced his first wife and, in 1950, married a beautiful and stylish French woman, Joan Courtney, with whom he had two sons. He appears in *D.1* and *Lost Years*.

House Un-American Activities Committee. A four-member subcommittee of the anti-communist HUAC held hearings in Los Angeles in April 1962, and the hearings were heavily picketed. On the second day, a handful of supporters of the HUAC marched against the picket. As Isherwood records in his diary on April 27, one of his students compared the scene to the stand-off at Little Rock Central High School in September 1957, when nine black students, testing the Supreme Court's 1954 ruling against segregation in schools, tried to attend the previously all-white school. They were turned back day after day in front of a threatening mob of about 1,000 white antidesegregationists.

Howard. See Austen, Howard.

Howard, Donald (Don) (1927–1987). American literary scholar and university professor; he was born in St. Louis, raised in Boston, and educated at Tufts, Rutgers, and the University of Florida, where he wrote his dissertation on fourteenth-century English literature. He published books on Langland, the Gawain poet, and most notably on Chaucer, and he also wrote essays about many other aspects of Christian Europe in the Middle Ages and Renaissance. He taught at Ohio State and Johns Hopkins before becoming an associate professor at the University of California at Riverside in 1963. Afterwards he taught at UCLA and, from 1974, at Stanford. He died of AIDS.

Hoyningen-Huene, George (1900–1968). Russian-born photographer, also known as George Huene; son of an American diplomat's daughter and a Baltic baron who had been chief equerry to Tsar Nicholas II. By the end of World War I, he was an exile in Paris, where he studied art and sold drawings to a fashion magazine. Eventually he became a regular photographer for *Vogue* and *Vanity Fair*, and, after 1936, for *Harper's Bazaar*. He published books containing his photographs of Greece, Egypt, North Africa, and Mexico. After the war, he settled in Hollywood where he taught photography and was color consultant on films for his longtime friend George Cukor. He also made several amateur documentaries. Isherwood met him in the late 1940s or early 1950s, through Gerald Heard and the Huxleys. He appears in *D.1*.

Huston, John (1906–1987). American film director, screenwriter, actor. Son of actor Walter Huston and father of actress Anjelica. As a young man, he was California lightweight boxing champion, served as an officer in the Mexican cavalry, worked briefly as a reporter in New York, and lived down and out in Paris and London. He wrote a number of successful scripts in the 1930s and 1940s before his directing debut with *The Maltese Falcon* (1941). During World War II, he filmed documentaries in battle conditions as a member of the army signal corps and was awarded the Legion of Merit for his bravery. Afterwards, he directed many further celebrated films, including *The Treasure of Sierra Madre* (1947, two Academy Awards: Best Director, Best Screenplay; his father won a third: Best Supporting Actor), *The Asphalt Jungle* (1950), *The Red Badge of Courage* (1951), *The African Queen* (1951), *Beat the Devil* (1954), *The Misfits* (1960), *The Man Who Would be King* (1975), and *Prizzi's Honor* (1985). In 1952, he moved with his fourth wife, Ricki Soma, and their family to Ireland. Isherwood was friendly with Huston by 1950, possibly through the Huxleys or through Gottfried Reinhardt who produced *The Red Badge of Courage*. Huston appears in *D.1* and *Lost Years*.

Hutchins, Robert (1899–1977). American educator, born in New York, educated at Oberlin and Yale. He was Dean of Yale Law School while still in his twenties, President of the University of Chicago at thirty, and later Chancellor there. He was also chairman of the board of editors of the *Encyclopaedia Britannica*, an associate director of the Ford Foundation, founder and president of the Center for the Study of Democratic Institutions in Santa Barbara, and author of books on education in modern democratic society. He was a long-time friend of Aldous Huxley.

Huxley, Aldous (1894–1963). English novelist and utopian; educated at Eton and Oxford; a grandson of Thomas Huxley and brother of Julian Huxley, both prominent scientists. In youth, he published poetry, short stories, and satirical novels such as *Crome Yellow* (1921) and *Antic Hay* (1923) about London's literary bohemia and Lady Ottoline Morrell's Garsington Manor, where he lived and worked during World War I and where he met his first wife, Maria. The Huxleys lived abroad in Italy and France during the 1920s and 1930s, partly with D.H. Lawrence—who appears in Huxley's *Point Counter Point* (1928)—and Lawrence's wife, Frieda. In 1932 Huxley published *Brave New World*, for which he is most famous.

An ardent pacifist, Huxley joined the Peace Pledge Union in 1935, and his *Ends and Means* (1937) was a basic book for pacifists. In April 1937, he sailed for America with his wife and son, accompanied by Gerald Heard and Heard's friend Chris Wood. Plans to return to Europe fell through when he failed to sell a film scenario in Hollywood, became ill there, and convalesced for nearly a year. He was denied U.S. citizenship on grounds of his extreme pacifism. California benefited his health and eyesight—he had been nearly blind since an adolescent illness. *After Many a Summer* (1939) is set in Los Angeles, and Huxley wrote many other books there, including *Grey Eminence* (1941), *Time Must Have a Stop* (1944), *The Devils of Loudun* (1952), *The Genius and the Goddess* (1956).

Not long after he arrived in Los Angeles, Isherwood was introduced to Huxley by Gerald Heard. Huxley and Isherwood collaborated on three film projects together during the 1940s: *Jacob's Hands*, about a healer, *Below the Equator* (later called *Below the Horizon*), and a film version of *The Miracle*, Max Reinhardt's 1920s stage production. Like Heard, Huxley was a disciple of Prabhavananda, but subsequently he became close to Krishnamurti, the one-time Messiah of the theosophical movement. Huxley's study of Vedanta was part of his larger interest in mysticism and parapsychology, and beginning in the early 1950s he experimented with mescaline, LSD, and psilocybin, experiences which he wrote about in *The Doors of Perception* (1954) and *Heaven and Hell* (1956).

In May 1961, the house he shared with his second wife, Laura, on Deronda Drive in the Hollywood Hills, was consumed in a brush fire. Huxley saved the novel he was writing, *Island* (1962), and three suits; otherwise all his books and papers were lost. Laura preserved only her Guarneri violin and a few clothes. On May 26, *Time Magazine* reported, "While firemen restrained the nearly blind British author from rushing into the blaze, Huxley wept like a child." Huxley's letter describing the scene and stating that there were neither tears nor any need for him to be restrained appeared in *Time* on June 16.

In 1960 he found a malignant tumor on the back of his tongue. He refused surgery in favor of radium needle treatment at the site of the tumor, a procedure recommended by his surgeon friend, Max Cutler. The tumor went and Huxley retained his power of speech. Eventually a new tumor appeared in his neck; Cutler removed it surgically, but then a third grew in the same location. Huxley died of cancer on the day John F. Kennedy was assassinated. He appears throughout *D.1* and *Lost Years*, and Isherwood helped the novelist Sybille Bedford with her *Aldous Huxley: A Biography, Volume 1 1894–1939* (1973) and *Aldous Huxley: A Biography, Volume 2 1939–1963* (1974).

Huxley, Laura Archera (1911–2007). Italian second wife of Aldous Huxley. Isherwood met her in the spring of 1956 at the Stravinskys' after she and Huxley married secretly in March. She was the daughter of a Turin stockbroker, had been a concert violinist from adolescence, and worked briefly in film. She became a psychotherapist, sometimes using LSD therapy on her patients, and she published two popular books on her psychotherapeutic techniques. Her 1963 bestseller, *You Are Not the Target*, was an early self-help book. She also published a memoir about Huxley, *This Timeless Moment* (1968), and a children's book. She first befriended Aldous and Maria Huxley in 1948 and used her special method of therapy on Huxley to help him recapture lost parts of his childhood. He incorporated some of her psychotherapy results into his utopian novel, *Island*. Before marrying Huxley, Laura lived for many years with Virginia Pfeiffer; after the marriage, she and Huxley settled in a house adjacent to Virginia's. After Huxley's death, she eventually became a children's rights campaigner. She appears in *D.1*.

Huxley, Maria Nys (1898–1955). First wife of Aldous Huxley; eldest daughter of a prosperous Belgian textile merchant ruined in World War I. Her mother's family included artists and intellectuals, and her childhood was pampered, multilingual, and devoutly Catholic. She met Huxley at Garsington Manor where she lived as a refugee during World War I; they married in Belgium in 1919 and their only child was born in 1920. Before her marriage, Maria showed promise as a dancer and trained briefly with Nijinsky, but her health was too frail for a professional career. She had little formal education and devoted herself to Huxley. Her premature death resulted from cancer. Isherwood met her in the summer of 1939 soon after he arrived in Los Angeles. She appears in *D.1* and *Lost Years*.

Huxley, Matthew (1920–2005). British-born only child of Aldous and Maria Huxley. He was brought to America in adolescence and Isherwood met him in Santa Monica in 1939. He attended the University of Colorado intending to become a doctor, served in the U.S. Army Medical Corps during World War II, and was invalided out of the army in 1943. He worked briefly as a reader at Warner Brothers, and, as a militant socialist, was involved in a strike there in 1945. During the same year he became a U.S. citizen. He took a degree from Berkeley in 1947 and later studied public health at Harvard. This became his career, and for many years he worked at the National Institute of Mental Health in Washington, D.C. He also published a book about Peru, *Farewell to Eden* (1965). He married three times, and had two children with his first wife. He appears in *D.1* and in *Lost Years*.

Igor. See Stravinsky, Igor.

Inge, William (Bill) (1913–1973). American playwright, born and educated in Kansas and at the University of Kansas; he earned a teaching degree in Tennessee and taught high-school and college English, then became the music and drama critic for the St. Louis *Star-Times*. In 1944, he interviewed Tennessee Williams, who befriended him and took him to Chicago to see *The Glass Menagerie*; afterwards, Inge accepted another university teaching job and wrote his first play, *Farther Off from Heaven*, which was produced in 1947 by Margo Jones at the Dallas Civic Theater. His next play, *Come Back, Little Sheba*, opened on Broadway in 1950 to great praise, and he won a Pulitzer Prize and two Drama Critics Awards for *Picnic* (1953). *Bus Stop* (1955) and *The Dark at the Top of the Stairs* (1957) were equally acclaimed, but in the 1960s his stage work failed repeatedly. His plays were adapted for film mostly by others, but he received an Academy Award for *Splendor in the Grass* (1961), which he co-produced; his other screenplays are *All Fall Down* (1962) and *Bus Riley's Back in Town* (1965, under the pseudonym Walter Gage). In 1963, he moved from New York to Los Angeles, and in the late 1960s, he briefly returned to teaching at the University of California at Irvine. He wrote two novels, *Good Luck Miss Wyckoff* (1970) and *My Son Is a Splendid Driver* (1971). He was depressive and had problems with alcohol. Isherwood and Bachardy first met Inge in New York in 1953 during the original run of *Picnic*. He is mentioned in *D.1*.

International situtation, December 1965. December 9 and 10 saw the most intensive bombing so far of North Vietnam and further marine landings in the Quanting foothills, a Vietcong stronghold. There were already about 180,000 U.S. troops in Vietnam. U.S. forces were also bombing Laos to disrupt supply routes. The Soviets criticized the U.S. aggression and asserted that North Vietnam should begin peace talks only on its own terms. Also on December 9, Secretary of State Dean Rusk told the press that China might decide to engage the U.S. in a wider war, and President Johnson asked for civilian spending to be kept down in 1966 budget proposals to allow for increasing military needs in the new year.

Iris. See Tree, Iris.

Isherwood, Esther (1878–1944). Youngest sister of Isherwood's father, fifth child of John and Elizabeth Bradshaw Isherwood. She married a clergyman, Joseph Toogood, against the wishes of her family, and had a long and happy marriage which produced a son and a daughter.

Isherwood, Frank Bradshaw (1869–1915). Isherwood's father; second son of John Bradshaw Isherwood, squire of Marple Hall, Cheshire. He was educated at Sandhurst and commissioned in his father's old regiment, the York and Lancasters, in 1892 at the age of twenty-three. He left for the Boer War in December 1899, caught typhoid, recovered, and served a second tour. In 1902 he left his regiment and became adjutant to the Fourth Volunteer Battalion of the Cheshire Regiment, based locally, in order to be able to offer his wife a home despite his meager income. He married Kathleen Machell Smith in 1903, and they settled for a time in a fifteenth-century manor house, Wyberslegh Hall, on the Bradshaw Isherwood family estate. In 1908, Frank rejoined his regiment and the family, now including Christopher, followed the regiment to Strensall, Aldershot, and Frimley; in 1911 a second son, Richard, was born and the family moved again to Limerick, Ireland, early the following year. Frank was sent from Limerick via England to the Front Line almost as soon as war was declared in the summer of 1914, and he was killed probably the night of May 8, 1915 in the second battle of Ypres in Flanders, although the exact circumstances of his death are unknown. Isherwood felt that Frank was temperamentally unsuited to the life of a professional soldier, though he was dutiful and efficient. He was a gifted watercolorist, an excellent pianist, and he liked to sing and take part in amateur theatricals. He was also a reader and a story-teller. He was shy and sensitive, but mildly good-looking, and a keen and agile sportsman. He was conservative in taste, in values, and in politics, but, unlike Kathleen, he was agnostic in religion and was attracted to theosophy and Buddhism. Isherwood wrote about his father in *Kathleen and Frank*.

Isherwood, Henry Bradshaw (1868–1940). Isherwood's uncle and his father's elder brother. In 1924, Uncle Henry inherited Marple Hall and the family estates on the death of Isherwood's grandfather, John Bradshaw Isherwood. Though Uncle Henry married late in life (changing his name to Bradshaw-Isherwood-Bagshawe in honor of his wife, Muriel Bagshawe), he had no children; Isherwood was his heir and, for a time after his twenty-first birthday, received a quarterly allowance from his uncle. The two had an honest if self-interested friendship, occasionally dining together and sharing intimate details of their personal lives. When Uncle Henry died in 1940, Isherwood at once passed on the entire inheritance to his own younger brother, Richard Isherwood. Uncle Henry is mentioned in *D.1*.

Isherwood, John (Jack) Bradshaw (1872–1962). Isherwood's Uncle Jack; he was the youngest son of Isherwood's grandfather, also named John Bradshaw Isherwood, and the younger brother of Frank Isherwood. He trained as a lawyer and joined the civil service, dealing with death duties and property deeds at Somerset House in London.

Isherwood, Kathleen Bradshaw (1868–1960). Isherwood's mother, often referred to as "M." in the diaries. Only child of Frederick Machell Smith, a wine merchant, and Emily Greene. She was born and lived until sixteen in Bury St. Edmunds, then moved with her parents to London. She travelled abroad and helped her mother to write a guidebook for walkers, *Our Rambles in Old London* (1895). In 1903, aged thirty-five, she married Frank Isherwood, a British army officer. They had two sons, Isherwood, and his much younger brother, Richard. When Frank Isherwood was lost in World War I, it was many months before his death was officially confirmed. Isherwood's portrait of his mother in *Kathleen and Frank* is partly based on her own letters and diaries. She was also the original for the fictional character Lily in *The Memorial*. Like many mothers of her class and era, Kathleen consigned her sons to the care of a nanny from infancy and later sent Isherwood to boarding school. Her husband's death affected her profoundly, which Isherwood sensed and resented. Their relationship was intensely fraught yet formal, intimate by emotional intuition rather than by shared confidence. Like her husband, Kathleen was a talented amateur painter. She was intelligent, forceful, handsome, dignified, and capable of great charm. Isherwood felt she was obsessed by class distinctions and propriety. As the surviving figure of authority in his family, she epitomized everything against which, in youth, he wished to rebel. He deemed her intellectual aspirations narrow and traditional, despite her intelligence, and she seemed to him increasingly backward looking. Nonetheless, she was utterly loyal to both of her notably unconventional sons and, as Isherwood himself recognized, she shared many qualities with him. There are many passages about her in *D.1* and *Lost Years*.

Isherwood, Richard Graham Bradshaw (1911–1979). Christopher Isherwood's brother and only sibling, younger by seven years. He was reluctant to be educated and never held a job in adulthood, although he did National Service during World War II as a farmworker at Wyberslegh and at another farm nearby, Dan Bank. In childhood, he saw little of his elder brother, who was sent to boarding school by the time Richard was three. Both boys spent more time with their nanny, Annie Avis, than with their mother. Richard later felt that Nanny had preferred Christopher; she made Richard nervous and perhaps was cruel to him. When Richard started school as a day boy at Berkhamsted in 1919, he lodged in the town with Nanny, and his mother visited at weekends. Isherwood by then was at Repton. The two brothers became closer during Richard's adolescence, when Isherwood was sometimes at home in London and took his brother's side against their mother's efforts to advance Richard's education and settle him in a career. During this period Richard met Isherwood's friends and helped Isherwood with his work by taking dictation. Richard was homosexual, but he seems to have had little opportunity to develop any long-term relationships, hampered as he was by his mother's scrutiny and his own shyness.

In 1941, he returned permanently with his mother and Nanny to Wyberslegh—signed over to him by Isherwood with the Marple Estate—where he eventually lived as a semi-recluse. Nanny died in 1948, and after Kathleen Isherwood's death twelve years later, Richard was looked after first by a married couple, the Vinces, and then by a local family, the Bradleys. He became a heavy drinker, Marple Hall fell into ruin and became dangerous, and he was forced to hand it over to the local council which demolished it in 1959, building several houses and a school on the grounds. Eventually, Richard moved out of Wyberslegh into a new house on the Marple Estate; the Dan Bradleys lived in a similar new house next door. When he died, he left most of the contents of his house to the Dan Bradleys and the house itself to their daughter and son-in-law. Richard's will also gave money bequests to the Dan Bradleys, Alan Bradley, and other local friends. Family property and other money were left to Isherwood and to a cousin, Thomas Isherwood, but Isherwood himself refused the property and passed some of his share of money to the Dan Bradleys. Richard appears in *D.1* and *Lost Years*.

Ivan. See Moffat, Ivan.

japa or **japam.** A method for achieving spiritual focus in Vedanta by repeating one of the names for God, usually the name that is one's own mantra; sometimes the repetitions are counted on a rosary. The rosary of the Ramakrishna Order has 108 beads plus an extra bead, representing the guru, which hangs down with a tassel on it; at the tassel bead, the devotee reverses the rosary and begins counting again. For each rosary, the devotee counts one hundred repetitions towards his own spiritual progress and eight for mankind. Isherwood always used a rosary when making japa. Japam is a Tamil form which came into use among Bengali swamis of the Ramakrishna Order—including Prabhavananda, Ashokananda, Akhilananda—because they spent varying periods of time in the Madras Math.

Jason, Peter (b. 1944). American actor. He played small roles in T.V., films, and commercials.

Jenkins, Ivor (1906–1972). An older brother of Richard Burton. He helped run the family home in Wales after Burton's mother died in her early forties, and he worked in a local coal mine. When Burton became successful in acting, Ivor and his wife, Gwen, often travelled with him, and Burton bought them a house adjacent to his own in Squire's Mount, Hampstead. In the late 1960s, Ivor broke his neck in a fall; he spent the last four years of his life in a wheelchair.

Jenkins, Terry. British model and aspiring actor. Isherwood met him when Charles Laughton brought Jenkins to Hollywood in 1960. Jenkins was then in his twenties. Laughton was in love with him, coached him and got him a screen test, but Jenkins had no real talent for acting. He was heterosexual, but admired Laughton and entered into a sexual relationship with him in an untroubled manner. When Laughton was dying, Jenkins looked after him with great care and sensitivity. Later Jenkins married a nurse of Laughton's. He is mentioned in *D.1*.

Jennifer. Also Jennifer Selznick and, later, Jennifer Simon; see Jones, Jennifer.

Jim. See Charlton, Jim.

Jo. See Masselink, Jo.

Joe. See Ackerley, J.R.

The John Birch Society. Founded in 1958 by millionaire candy-manufacturer Robert Welch to promote personal freedom, limited government, and traditional Christian values. It was a bastion of anti-communism in the 1960s.

Johnson, Lamont (Monty) (b. 1922). American actor and, especially, director; born in Stockton, California, educated at UCLA and the Neighborhood Playhouse. He worked in radio as a teenager to pay his way through college, then moved to New York where he continued in radio soaps and directed an off-Broadway production of Gertrude Stein's *Yes Is for a Very Young Man*. In 1959, he was a founder of the UCLA Theater Group, and during the 1960s and 1970s he won numerous Emmys and Screen Directors Guild Awards for his T.V. miniseries and made-for-T.V. movies. He also directed episodes of popular shows like "Have Gun Will Travel," "The Rifleman," and "The Twilight Zone," and a few feature films, including *The Last American Hero* (1973) and *Lipstick* (1976). He acted on T.V. and in a number of movies. "Losers Weepers," broadcast February 19, 1967 as the premiere to NBC's series "Experiments in Television," was about the desperation of a poor black family in Watts; "Deadlock," a made-for-T.V. movie broadcast February 22, 1969 as a pilot for NBC's "The Protectors" (1969–1970), explored the tensions between a white police officer and a black district attorney during ghetto violence triggered by the police killing of a black youth. The first was a play written in the Watts Writers Workshop; it was filmed in Watts and framed by clips from the workshop and comments by the cast and by Budd Schulberg, who founded the workshop, and his brother Stuart Schulberg, who produced the segment.

Jonathan. See Preston, Jonathan.

Jones, Jack. American painter; a disciple and, later, close friend of Gerald Heard. Some of his best work was of Margaret Gage's garden on Spoleto Drive, where Heard lived until 1962, and he shared Heard's interest in clothing and costume. Jones was about the same age as Don Bachardy and lived nearby in Santa Monica Canyon, so they sometimes sat for each other.

Jones, Jennifer (Phylis Isley) (1919–2009). American actress, born in Tulsa, Oklahoma; in childhood, she travelled with her parents' stock stage company and spent hours in her father's movie theaters. After a brief stint at Northwestern University, she attended the Academy of Dramatic Arts in New York, left for a radio job in Tulsa, and began her Hollywood career in B-movies in 1939. She was discovered in 1941 by David Selznick, who changed her name and took control of her career with spectacular results. She won an Academy Award for *The Song of Bernadette* in 1943, followed by *Since You Went Away* (1944, Academy Award nomination), *Love Letters* (1945, Academy Award nomination), *Duel in the Sun* (1946, Academy Award nomination), *Portrait of Jennie* (1948), *Madame Bovary* (1949), *Carrie* (1951), *Love Is a Many-Splendored Thing* (1955, Academy Award nomination), *A Farewell to Arms* (1957), *Tender is the Night* (1962), *The Towering Inferno* (1974), and others. Her 1939 marriage to the actor Robert Walker, with whom she had two sons, ended in divorce in 1945, and Selznick left his wife, Irene Mayer, for Jones; they married in 1949. His obsession with Jones combined with her own emotional instability (including suicide attempts) made a melodrama of their careers and their private lives. In 1965, Selznick died, leaving huge debts. In 1971, Jones married a third time, to Hunt Foods billionaire and art collector Norton Simon. Later that decade, her only child with Selznick, Mary Jennifer, committed suicide; partly as a result, Jones created the Jennifer Jones

Foundation for Mental Health and Education and trained as a lay therapist and volunteer counsellor. Isherwood first met her when he worked with Selznick on *Mary Magdalene* in 1959, and he took her to meet Swami Prabhavananda in June that year. He writes about the friendship in *D.1*.

Julie. See Harris, Julie.

Kali. Hindu goddess; the Divine Mother and the Destroyer, usually depicted dancing or standing on the breast of a prostrate Shiva, her spouse, and wearing a girdle of severed arms and a necklace of skulls. Kali has four arms: the bleeding head of a demon is in her lower left hand, the upper left holds a sword; the upper right hand gestures "be without fear," the lower right confers blessings and boons on her devotees. Kali symbolizes the dynamic aspect of the godhead, the power of Brahman: she creates and destroys, gives life and death, well-being and adversity. She has other names: Shakti, Parvati, Durga. Kali was Ramakrishna's Chosen Ideal, and for a number of years, he devoted himself to worshipping her image in her temple at Dakshineswar. Kali puja is usually celebrated in November.

Kallman, Chester (1921–1975). American poet and librettist; Auden's companion and collaborator. They met in New York in May 1939 and lived together intermittently in New York, Ischia, and Kirchstetten for the rest of Auden's life, though Kallman spent time with other friends, often in Athens as he grew older. He published three volumes of poetry and with Auden wrote and translated opera libretti, notably *The Rake's Progress* (for Stravinsky), *Elegy for Young Lovers*, and *The Bassarids* (both for Hans Werner Henze). He appears in *D.1* and *Lost Years*.

Kaper, Bronislau (1902–1983). Polish-born composer, trained at Warsaw Conservatory; he wrote music for German films in Berlin, then, fleeing Hitler, emigrated to Paris and Hollywood where he continued his career at MGM. His film scores include the Marx Brothers' *A Night at the Opera* (1935); *Green Dolphin Street* (1947), from which the theme became a jazz favorite; *The Great Sinner* (1949), for which Isherwood wrote the script; *Invitation* (1952) directed by Gottfried Reinhardt; and *Lili* (1953), for which Kaper won an Academy Award.

Kaplan, Abbot (1912–1980). University administrator and professor, born in New York, educated at Columbia and at the Jewish Theological Seminary. He worked as a high school principal, served in the navy during World War II, and in 1946 began lecturing in labor economics at UCLA's Institute of Industrial Relations. He became a professor of adult education and director of the continuing education program, UCLA Extension, for which he hired Isherwood to lecture in 1961. He was also a founder, with William Melnitz, of UCLA's Professional Theater Group, which became the resident company at the Mark Taper Forum. Later, he was a dean and a professor in the College of Fine Arts. He left UCLA in 1967 to be founding president of the State University of New York at Purchase, but after retirement returned to the Graduate School of Management where he taught Arts Management in the 1970s.

Kaplan, Abraham (1918–1993). American philosopher, born in Odessa, son of a rabbi. He was raised in Minnesota and educated there and at the University of Chicago and UCLA, where he later taught. He asked Isherwood to be a public signatory, with Aldous Huxley and others, to a telegram in the *Los Angeles*

Times, December 6, 1961, protesting Soviet treatment of Jews. Three leaders of the Leningrad Jewish community had been given prison sentences ranging from four to twelve years for passing information about the USSR to a foreign embassy, possibly the Israeli Embassy. Their trial, from October 9 to 13, was made public only in mid-November. In addition, three leaders of the main synagogue in Moscow were said to have been tried and convicted in early October, and, reportedly, synagogues and schools were being closed down. Tass denied anti-Semitic persecution and said Western criticism insulted the hundreds of thousands of Jews working in all parts of Soviet life.

Kaplan, Al. American doctor. He was wealthy, interested in theater, and backed at least one West End show, *Keep Your Hair On*, by John Cranko. He lived in West London and died fairly young.

Karapiet, Rashid. Anglo-Indian actor; he played small roles on the British stage and in T.V. and films, including *Bombay Dreams*, "Miss Marple," and *A Passage to India*. Isherwood met him through his lover, John Lehmann's friend Jeremy Kingston, and as Isherwood mentions, Karapiet typed some of *Down There on a Visit*.

Kathleen. See Isherwood, Kathleen Bradshaw.

Kaufmann, Leonard (Len). Beverly Hills agent and publicist, third husband of actress Doris Dowling.

Kennedy, Paul. A young man with whom Isherwood had an occasional sexual relationship towards the end of the 1950s. As Isherwood tells in this volume, Kennedy developed cancer suddenly and died in August 1962. Isherwood drew on his visits to Kennedy in the hospital for the episode of *A Single Man* in which George visits Doris in similar circumstances. Kennedy is mentioned in *D.1*.

K.H.3. An anti-aging preparation developed by a Romanian doctor, Ana Aslan, and promoted in the 1960s and 1970s; the active ingredient is procaine, the local anaesthetic widely used by dentists and typically known by its brand name, Novocaine. K.H.3 possibly has a mild antidepressant effect, but it never persuaded the medical establishment and is not approved by the Food and Drug Administration in the U.S.

Kidd, Kap. American actor. Isherwood met him towards the end of the 1950s through Bill Jones, a writer with whom Kidd shared an apartment for many years. He appears in *D.1*.

Kimbrough, Clinton (1933–1996). American actor and director, sometimes credited as Kimbro. He appeared at the American National Theater and Academy at Washington Square in New York during the first half of the 1960s; he also had small roles in a few films and narrowly missed being cast as one of the killers in the film of Capote's *In Cold Blood*. He had affairs with both sexes. Bachardy did many drawings and paintings of him and of his wife, Frances Doel. She was a script girl on low-budget films, then began to work on her own screenplays and original stories, including *Big Bad Mama* (1974) and *Crazy Mama* (1975), in which her husband had a small part. Later she became a producer and a script development consultant.

Kingston, Jeremy (b. 1931). British writer. He was trying to establish himself as a playwright when he met John Lehmann in 1956 and began an affair with him that cooled after some months into a lifelong friendship. Lehmann hired him as an editorial assistant at *The London Magazine*, and Kingston also typed some of Lehmann's books. In the late 1950s, Kingston had some success with a radio drama and a play, *No Concern of Mine*, staged in 1958 and with another play, *Signs of the Times*. But his novel, *The Prisoner I Keep*, mentioned by Isherwood, was never published. He was theater critic for *Punch* for eleven years. Later he became a drama critic for *The Times*.

Kirstein, Lincoln (1907–1996). American dance impresario, author, editor, and philanthropist, raised in Boston, son of a wealthy self-made businessman. He was educated at Berkshire, Exeter, and Harvard, where he was founding editor of *Hound and Horn*, the quarterly magazine on dance, art, and literature. He also painted, and he helped found the Harvard Society for Contemporary Art. In 1933, Kirstein persuaded the Russian choreographer George Balanchine to come to New York, and together they founded the School of American Ballet and the New York City Ballet. Kirstein was also involved in founding the Museum of Modern Art. His taste and critical judgement combined with his entrée into wealthy society enabled him to recognize and promote some of the greatest artistic talent of the twentieth century. In 1941, he married Fidelma (Fido) Cadmus, sister of the painter Paul Cadmus. He served in the army from 1943 to 1946. Isherwood's first meeting with Kirstein in New York in 1939 was suggested by Stephen Spender, who had met Kirstein in London, and Kirstein appears in *D.1* and *Lost Years*. He is a model for Charles in *The World in the Evening*. His poetry was admired by Auden, and Isherwood mentions reviewing his *Rhymes of a PFC* (1964) when Kirstein reissued it as *Rhymes and More Rhymes of a PFC* (1966). But their friendship ended that year, when Kirstein commissioned Bachardy to do portraits of the New York City Ballet stars without consulting Balanchine. Kirstein, who suffered from depression and paranoia, could not bring himself to ask Balanchine's permission to sell the drawings as a ballet souvenir, even though he had already had them printed, because he became irrationally afraid that Balanchine and others associated with the company would assume Bachardy was a boyfriend he was trying to promote. Balanchine's red pencil markings on a copy of the portfolio returned to Bachardy by Kirstein reflect Balanchine's personal attitudes to some of the dancers—he wrote, for instance, "I hate her," on one portrait, and "He can't dance," on another—but there are no criticisms of Bachardy's work. Bachardy was disappointed when the project was scrapped, and he later blamed himself for the fact that Kirstein thereafter refused to see Isherwood even though Auden tried to reconcile them.

kirtan. Hindu devotional singing or chanting.

Kiskadden, Peggy. Thrice-married American socialite from Ardmore, Pennsylvania; born Margaret Adams Plummer, she was exceptionally pretty and had an attractive singing voice. From 1924 until 1933, she was married to a lawyer and (later) judge, Curtis Bok, the eldest son of one of Philadelphia's most prominent families. In the early 1930s, she accompanied Bok, a Quaker, to Dartington, England, where she met Gerald Heard and Aldous and Maria Huxley. Her second marriage, to Henwar Rodakiewicz, a documentary filmmaker, ended in 1942, and she married Bill Kiskadden in July 1943. She had four children, Margaret Bok (called Tis), Benjamin Bok, Derek Bok (later president of Harvard University), and William Kiskadden, Jr. (nicknamed "Bull"). Isherwood was introduced to her by Gerald Heard soon after arriving in Los Angeles; they became intimate friends but drew apart at the end of the 1940s and finally split irrevocably in the 1950s over Isherwood's relationship with Don Bachardy. By the mid-1960s, he had forgotten how to spell her name, sometimes typing Kiscadden. She died in the 1990s. There are numerous passages about her and her family in *D.1* and *Lost Years*.

Kiskadden, William Sherrill (Bill) (1894–1969). American plastic surgeon; third husband of Peggy Kiskadden. He was born in Denver, Colorado, the son of a businessman, studied medicine at the University of California and in London and Vienna in the late 1920s, and eventually established his practice in Los Angeles. He was the first clinical professor of plastic surgery at UCLA and founded the plastic surgical service at UCLA County Medical Center in the early 1930s as well as holding distinguished positions at hospitals in Los Angeles—teaching, administering, practicing—and writing articles on particular procedures and problems. He became interested in the population problem and with Julian and Aldous Huxley and others founded Population Limited in the early 1950s. He served in both world wars, the second time in the Army Medical Corps. In February 1958, he nearly died after a cardiac operation. He appears in *D.1* and *Lost Years*.

Kitty. Bachardy's pet name, once known only to himself and Isherwood, and denoting his identity as an exotic, temperamental feline creature in the private myth world they shared.

Knight, Franklin (Frank) (*circa* 1924–2005). American monk of the Ramakrishna Order; he was given the name Asima Chaitanya when he took his brahmacharya vows, probably in 1965 or 1966, and came to be known as Asim. He first joined the Trabuco monastery in about 1955 and settled there permanently, but Swami Prabhavananda never allowed him to take sannyas because of an episode—referred to by Isherwood in his diary entry for December 16, 1963—in which he behaved inappropriately towards a woman. Knight was a cousin of Webster Milam, and is mentioned in *D.1*.

Knox, Ronnie (1935–1992). American football player and aspiring writer, born in Illinois. His real name was Raoul Landry. He was a football All American for Santa Monica High School, a star freshman quarterback for the University of California, and widely considered to be the most talented college football player in the country when in 1953 he suddenly transferred to UCLA, sacrificing a year of eligibility and generating a scandal. He was also written up as a glamor boy because of his physical beauty. As a professional, he joined the Canadian Football League and played for Montreal. Later he wrote fiction, and published at least one of his stories. He lived with Renate Druks from 1960 to 1964 and afterwards had a French girlfriend called Véronique. All three sat for Bachardy several times. Knox died virtually homeless in San Francisco.

Kolisch, Dr. Joseph. Viennese physician practicing in Hollywood; a follower of

Prabhavananda. Aldous and Maria Huxley, Gerald Heard, several monks and nuns at the Vedanta society and even Greta Garbo followed his advice and were on his vegetarian diets during the 1940s. On Heard's recommendation, Isherwood first saw Kolisch in January 1940 for what appeared to be a recurrence of gonorrhea, but as Isherwood tells in *D.1*, Kolisch attributed Isherwood's symptoms to his psychological makeup. Kolisch also appears in *Lost Years*.

Krishna. One of the most widely worshipped Hindu gods, a hero of the Mahabharata and the Bhagavatam. Krishna was also the Sanskrit name given to George Fitts, an American monk of the Ramakrishna Order, from New England. He joined the Vedanta Society in Hollywood in 1940 and was living there as a probationer monk in 1943 when Isherwood moved in. He was then about forty years old, had some private wealth, and spent his time obsessively tape recording and transcribing Swami Prabhavananda's lectures and classes. He took his brahmacharya vows in 1947, and early in 1958, he took sannyas and became Swami Krishnananda. He lived in Hollywood, but usually accompanied Swami on trips to Santa Barbara, Trabuco, and elsewhere. He appears in *D.1*.

Krishnamurti (1895–1986). Hindu spiritual teacher. As an impoverished boy in India, he was taken up by the leaders of the Theosophical movement as the "vehicle" in which their Master Maitreya would reincarnate himself. He was adopted and educated in England, then in 1919 sent to an orange ranch in Ojai, California for his health. In 1929, he renounced his messianic role and rejected the guru–disciple relationship along with the devotional and ritual aspects of Hinduism. Although he broke with the Theosophists, he went on speaking to devotees, sometimes in huge numbers, for the rest of his life all around the world. He was extremely handsome and charismatic and had many secret sexual liaisons which introduced tension among his followers and led to a series of lawsuits with a colleague and rival, Desikacharya Rajagopalacharya, whom he cuckolded. Mary Zimbalist, widow of Hollywood producer Sam Zimbalist (1904–1958), eventually became one of his closest companions, hosting him at her house in Malibu, travelling with him, and helping him with his lectures and books. She was a founder, trustee, and vice-chairman of the Krishnamurti Foundation of America, set up in 1969 to wrest control of assets from Rajagopal. Isherwood first met Krishnamurti in 1939 through Aldous and Maria Huxley and later went to hear him speak in Ojai. He appears in *D.1*.

Lambert, Gavin (1924–2005). British novelist, biographer and screenwriter; educated at Cheltenham College and for one year at Magdalen College, Oxford. He edited the British film magazine *Sight and Sound* before going to Hollywood in 1956. He was working for Jerry Wald at Twentieth Century-Fox on *Sons and Lovers* (1960) when Ivan Moffat introduced him to Isherwood; he appears often in *D.1*. His novel *The Slide Area: Scenes of Hollywood Life* (1959), which Isherwood read in manuscript in 1957, was influenced by Isherwood's Berlin stories. He and Isherwood worked on a television comedy project "Emily Ermingarde" for Hermione Gingold and later for Elsa Lanchester, but the series was never produced. Lambert also helped Isherwood revise the film script of *The Vacant Room*. During the 1950s and early 1960s, he planned a musical version, never produced, of Thackeray's novel *Vanity Fair*. He wrote and directed an independent film,

Another Sky (1956), wrote the screenplay for his own 1963 novel *Inside Daisy Clover* (1965), and scripted *Bitter Victory* (1957), *The Roman Spring of Mrs. Stone* (1961), *I Never Promised You a Rose Garden* (1977), and others. His books include *On Cukor* (1972); *The Dangerous Edge* (1975), a study of nine thrillers; *The Goodbye People* (1977); *Running Time* (1983); *Norma Shearer: A Life* (1990); *Nazimova: A Biography* (1997); *Mainly About Lindsay Anderson* (2000); and *Natalie Wood: A Life* (2004). During the 1970s, he settled in Tangier for a time, returning to Los Angeles in the early 1980s.

Lamkin, Hillyer Speed (b. 1928). American novelist; born and raised in Monroe, Lousiana. Isherwood met him in April 1950 when Speed was twenty-two and about to publish his first novel, *Tiger in the Garden*. He had studied at Harvard and lived in London and New York before going to Los Angeles to research his second novel, *The Easter Egg Hunt* (1954), about Hollywood, in particular Marion Davies and William Randolph Hearst; he dedicated the novel to Isherwood who appears in it as "Sebastian Saunders." Lamkin was on the board, with Isherwood, at the Huntington Hartford Foundation. With a screenwriter, Gus Field, he tried to adapt *Sally Bowles* for the stage in 1950–1951, but Dodie Beesley criticized the project and encouraged John van Druten to try instead. In the mid-1950s, Lamkin wrote a play *Out by the Country Club* which was never produced, and in 1956, he scripted a T.V. film about Perle Mesta, the political hostess who was Truman's ambassador to Luxembourg. During 1957, he wrote another play, *Comes a Day*, which had a short run on Broadway, starring Judith Anderson and introducing George C. Scott. Eventually, when the second play failed, Lamkin returned home to Louisiana. He appears in *D.1* and *Lost Years*.

Lamkin, Marguerite (b. *circa* 1934). A southern beauty, born and raised in Monroe, Louisiana, and briefly educated at a Manhattan finishing school. She followed her brother Speed to Hollywood, and married the screenwriter Harry Brown in 1952, but the marriage broke up melodramatically in 1955 as Isherwood records in *D.1*, where Marguerite is frequently mentioned. Bachardy had a room in the Browns' apartment during the early months of his involvement with Isherwood, and Marguerite was an especially close friend to him. In later years she also became close to Isherwood. She assisted Tennessee Williams as a dialogue coach during the original production of *Cat on a Hot Tin Roof*, and afterwards she worked on other films and theatrical productions on the East and West coasts and in England when southern accents were required. She was married to Rory Harrity from 1959 to 1963, and later settled in London, where she had a successful third marriage, became a society hostess, and raised large sums of money for AIDS and HIV research and care.

Lanchester, Elsa (1902–1986). British actress; she danced with Isadora Duncan's troop as a child, then began acting in a children's theater in London at sixteen. In 1929, she married Charles Laughton and went with him to Hollywood in 1934, settling there for good in 1940 and becoming an American citizen. Lanchester began making films before Laughton and they acted in several together—for instance *The Private Life of Henry VIII* (1933) and *Witness for the Prosecution* (1957) for which she received an Academy Award nomination. Her most famous film was *The Bride of Frankenstein* (1935), but she was in many more, including *The*

Constant Nymph (1928), *David Copperfield* (1935), *Lassie Come Home* (1943), *The Razor's Edge* (1946), *The Secret Garden* (1949), *Come to the Stable* (1949, Academy Award nomination), *Les Misérables* (1952), *Bell, Book and Candle* (1958), *Mary Poppins* (1964), and *Murder by Death* (1976). She also worked in television and for many years she sang at a Los Angeles theater, The Turnabout, on La Cienega Boulevard. She toured with her own stage show, *Elsa Lanchester—Herself*, during 1960 and opened at the 41st Street Theater in New York on February 4, 1961 for seventy-five performances. She met Isherwood socially in the late 1950s, was greatly attracted to him and introduced him to Laughton, afterwards vying with Laughton and Bachardy for Isherwood's attention. She appears in *D.1*.

Lane, Homer (1875–1925). American psychologist, healer, and juvenile reformer. Lane established a rural community in England called The Little Commonwealth where he nurtured young delinquents with love, farm work, and the responsibility of self-government. For Lane, the fundamental instinct of mankind "is the titanic craving for spiritual perfection," and he conceived of individual growth as a process of spiritual evolution in which the full satisfaction of the instinctive desires of one stage bring an end to that stage and lay the ground for the next, higher stage; he believed that instinctive desires must be satisfied rather than repressed if the individual is to achieve psychological health and fulfillment. In practice, Lane identified himself with the patient's neurosis in order to allow it to emerge from the unconscious; personally loving the sinner and the sin, he freed the patient from his sense of guilt. Auden discovered the teachings of Homer Lane through his Berlin friend, John Layard, a former patient and disciple of Lane's, and in late 1928 and early 1929, became obsessed with Lane, preaching his theories to his friends and in his poems.

Lange, David (d. 2006). American film producer; brother of Hope Lange. He was educated at Principia and Harvard, acted a little, and produced several films, including, with Alan Pakula, *Klute* (1971). He later taught screenwriting. He appears in *D.1*.

Lange, Hope (1931–2003). American actress, born and raised in Connecticut; she was twelve years old when she debuted on Broadway in *The Patriots* (1943). As a teenager, she waitressed in her mother's Greenwich Village restaurant, modelled, and continued as a stage actress and in live T.V. drama until she was brought to Hollywood with her first husband Don Murray to appear in *Bus Stop* (1956). Afterwards she played in numerous other films including *Peyton Place* (1957), for which she was nominated for an Academy Award, *The Young Lions* (1958), *The Best of Everything* (1959), *Deathwish* (1974), *Blue Velvet* (1986), and *Clear and Present Danger* (1994). During her love affair with Glenn Ford, she co-starred with him in *Pocketful of Miracles* (1961) and *Love Is a Ball* (1963). She made a number of T.V. films and won two Emmy Awards for her role in the television comedy series "The Ghost and Mrs. Muir" (1968–1970); she also appeared on "The New Dick Van Dyke Show" (1971–1974). In 1977, she returned to Broadway in *Same Time Next Year* opposite Don Murray. Lange had two children with Murray, but the marriage ended in 1960. In 1963, she married the director and producer Alan Pakula; they divorced in 1969. In 1986, she married Charles Hollerith, a theatrical producer. Isherwood first met Lange with Murray in the late 1950s; she appears in *D.1*.

Lansbury, Angela (b. 1925). British star of stage and film; granddaughter of pacifist labor politician George Lansbury and daughter of actress Moyna Macgill, who brought her with her twin brothers Edgar and Bruce to Hollywood to escape the Blitz. Lansbury was making feature films before the end of the war and went on to appear in *National Velvet* (1944), *The Picture of Dorian Gray* (1945), *The Three Musketeers* (1948), *The Dark at the Top of the Stairs* (1960), *The Manchurian Candidate* (1962), *Something for Everyone* (1970), *Bedknobs and Broomsticks* (1971), *Death on the Nile* (1978), and *Nanny McPhee* (2006) among others. Isherwood first mentions her at the time she made a hit in Tony Richardson's *A Taste of Honey* on Broadway in 1960; other Broadway successes include *Mame* (1966), *Sweeney Todd* (1979), and *Deuce* (2007). She has also had a T.V. career, especially in "Murder, She Wrote" (1984–1996). She has won eleven Tony Awards, six Golden Globes, and been nominated repeatedly for Academy Awards. She married twice, the second time, in 1949, to Peter Shaw with whom she had two children.

Lansbury, Bruce (b. 1930). British-born T.V. writer and producer; brother of Angela Lansbury and twin brother of stage and film producer Edgar Lansbury. Bruce Lansbury produced "The Wild Wild West," "Mission Impossible," and many other shows in a long career at the Hollywood studios and for the major networks there and in New York. Bachardy drew Bruce and Edgar Lansbury as well as Angela and also Bruce Lansbury's two daughters.

Laos crisis. Power struggles among communist-dominated Pathet Lao, neutralists, and right-wing militarists brought down government after government in Laos following independence from France in 1953. Isherwood first mentions a crisis on December 31, 1960, when a newly installed pro-Western, right-wing government came under threat internally from communist-led rebels. On December 30, this new government asked for U.N. help to counter a reported invasion across its northern border by 2,000 North Vietnamese troops; they hinted the Chinese were also involved. Hanoi denied any such invasion and also denied supporting the rebels, as did the Soviet Union. But over the next few weeks the North Vietnamese, the Soviets and the U.S. all supplied arms, men, or transport to the localized conflict, and Laos—poor, undeveloped, but strategically positioned as a corridor to the rest of Southeast Asia—became the focus of possible generalized war between the West and the Communist bloc.

Larmore, James and Alexandra (Xan). He was assistant to the film producer Charlie Brackett, her father. They were married during World War II, when he was a soldier. Previously he had been a chorus boy, and, according to rumor, Brackett's lover. Xan became an alcoholic, and their relationship was turbulent. They appear in *D.1*.

Larson, Jack (b. *circa* 1933). American actor, playwright and librettist; born in Los Angeles and raised in Pasadena. His father drove a milk truck, and his mother was a clerk for Western Union; they divorced. At fourteen, Larson was California bowling champion for his age group. He attended Pasadena Junior College, where he was discovered acting in a college play and offered his first film role by Warner Brothers in *Fighter Squadron* (1948) (also Rock Hudson's first film.) He is best known for playing Jimmy Olsen in "The Adventures of Superman," the original T.V. series aired during the 1950s. He lived for over thirty-five years with the director James Bridges and co-produced some of Bridges's most successful films, *The Paper Chase* (1973), *Urban Cowboy* (1980), and *Bright Lights, Big City* (1988). He wrote the libretto for Virgil Thomson's opera *Lord Byron*. Larson and Bridges were close friends of Isherwood and Bachardy from the 1950s onward and appear in *D.1*.

Laughlin, Leslie. See Caron, Leslie.

Laughlin, Michael. American film producer, and later, director and screenwriter; educated at Principia College in Illinois and at UCLA, where he studied law. He produced *Two Lane Blacktop* (1971) among others. He was the third husband of French actress Leslie Caron from 1969 to 1980.

Laughton, Charles (1899–1962). British actor. He played many roles on the London stage from the 1920s onward and made his first film, *Piccadilly*, in 1929; other films include *The Private Life of Henry VIII* (1934, Academy Award), *Les Misérables* (1935), *Mutiny on the Bounty* (1935), and *The Hunchback of Notre Dame* (1939). His last film was *Advise and Consent* (1962), which Isherwood mentions. Laughton also acted in New York and Paris and gave dramatic readings throughout the U.S. from Shakespeare, the Bible, and other classic literature. He became an American citizen in 1950. Isherwood met him in Hollywood in 1959 through Laughton's wife, Elsa Lanchester, and Laughton proposed various projects; in particular, he asked Isherwood to help him write a play about Socrates. Isherwood first mentions the project in March 1960, and in *D.1* he describes their sessions reading through Plato together and devising the script. He also tells how, that summer, Laughton bought 147 Adelaide Drive, next door to Isherwood, so that he could spend time with male friends away from his wife in their house on Curson Avenue.

Laura. See Huxley, Laura Archera.

Laurents, Arthur (b. 1918). American playwright, director, screenwriter; educated at Cornell. He is probably best known as the author of the musicals *West Side Story* (1957) and *Gypsy* (1960); his other plays include *Home of the Brave* (1945), *The Time of the Cuckoo* (1952), *A Clearing in the Woods* (1957), and *Invitation to a March* (1960). He directed the last of these and two later hits, *I Can Get It for You Wholesale* (1962) and *La Cage aux Folles* (1983). He rewrote several of his musicals and plays for the movies, and he wrote the screenplays for, among others, *Rope* (1948), *Anastasia* (1956), *Bonjour Tristesse* (1958), and *The Turning Point* (1977), which was nominated for an Academy Award. He also turned his novel, *The Way We Were* (1972), into a screenplay for Sidney Pollack, who directed Barbara Streisand and Robert Redford in it. Isherwood and Bachardy first became friends with Laurents and his longterm companion Tom Hatcher in the mid-1950s. He appears in *D.1*.

Lawrence, Jerome (Jerry) (1915–2004). American playwright; born in Ohio, educated at Ohio State University and UCLA. He was a reporter and editor for small daily newspapers in Ohio, then a continuity editor for a Beverly Hills Radio station. By the time he joined the U.S. Army during World War II, he was a senior staff writer for CBS radio. In the army, he worked as a consultant to the secretary of war, then as a correspondent from North Africa and Italy, and he co-founded Armed Forces Radio with Robert Lee. They continued their partnership as playwrights after the war. Among their best-known plays are *Look, Ma, I'm Dancin'!* (1948), the prize-winning *Inherit the Wind* (1955) about the Scopes monkey trial, the stage adaptation of *Auntie Mame* (1956), and *The Night Thoreau Spent in Jail* (1971). Lawrence and Lee also wrote the book and the lyrics for the musical *Mame* (1966), adapted James Hilton's novel *Lost Horizon* as the book and lyrics for *Shangri-La* (1956), and were involved in adapting much of their work for film. Lawrence taught play writing at several universities and was an adjunct professor at USC. Isherwood often went to parties at his house, especially to meet good-looking young men, mostly actors, whom Lawrence knew through his theater connections. Lawrence often claimed that he had introduced Isherwood and Bachardy to each other because Bachardy and his brother Ted attended a party at Lawrence's house on February 14, the date Isherwood and Bachardy marked as the start of their romance, but, in fact, Isherwood and Bachardy met earlier. Lawrence appears in *D.1*.

Layard, John (1891–1974). English anthropologist and Jungian psychoanalyst. He read Medieval and Modern Languages at Cambridge and did field work in the New Hebrides with the anthropologist and psychologist W.H.R. Rivers. In the early 1920s, he had a nervous breakdown and was partially cured by the American psychologist Homer Lane. Lane died during the treatment, leaving Layard depressed and seeking further treatment, first unsuccessfully with Wilhelm Stekel and eventually more productively with Jung. Auden met Layard in Berlin late in 1928 and introduced him to Isherwood the following spring; for a time all three were obsessed with Lane's theories recounted by Layard. During this period, Layard had a brief and tortured triangular affair with Auden and a German sailor, Gerhart Meyer, whereupon he tried to kill himself. Isherwood used the suicide attempt in *The Memorial*, and Layard appears as "Barnard" in *Lions and Shadows*. Layard eventually recovered his psychological health so that he was able to work and write again, and he married and had a son. Like Auden, he also returned to the Anglican faith of his childhood.

Lazar, Irving (Swifty) (1907–1993). Agent and deal-maker for movie stars and authors such as Lauren Bacall, Humphrey Bogart, Truman Capote, Noël Coward, Ernest Hemingway, Vladimir Nabokov, Cole Porter, Diana Ross, Irwin Shaw, and Tennessee Williams. He practiced bankruptcy law in New York during the Depression, then relocated to Hollywood in 1936. His wife was called Mary.

lectures 1960. Isherwood gave eight lectures at UCSB that year. The first two were "Influences" and "Why Write at All." "What Is the Nerve of the Interest of the Novel," which he mentions in his entry for September 17, became the third and fourth lectures. The next four were "A Writer and the Theater," "A Writer and the Films," "A Writer and Religion," "A Last Lecture." The series was titled "A Writer and His World," and it was broadcast the following year by KPFK. Much later, the lectures were published in *Isherwood on Writing* (2007), edited by James J. Berg.

Ledebur, Christian (Boon) and Henrietta. Iris Tree's second son (by Austrian actor Friedrich Ledebur) and his wife. They lived in Santa Monica intermittently, in a corner apartment in the merry-go-round building on Santa Monica Pier where, until 1954, Iris also lived. Boon was a psychologist. The Ledeburs had a son, Marius, before divorcing, and Boon later remarried and had another family in Switzerland.

Lederer, Charles (1910–1976). American screenwriter and director, born in New York and educated at Berkeley. His mother was a sister of Marion Davies, the actress and second wife of William Randolph Hearst; his father was a theatrical producer. He briefly worked as a journalist but was writing films by the time he was twenty-one, often collaborating with Ben Hecht. His screenplays include *His Girl Friday* (1940), *I Love You Again* (1940), *Kiss of Death* (1947), *I Was a Male War Bride* (1949), *Monkey Business* (1952), *Gentlemen Prefer Blondes* (1953), *Can-Can* (1960), *Ocean's 11* (1960), and *Mutiny on the Bounty* (1962). He also co-wrote and produced *Kismet* on Broadway (1953–1954), later adapting it for film. During the 1940s he was married to Virginia Nicholson (once married to Orson Welles); in 1949 he married the actress Anne Shirley. Isherwood knew Lederer through studio writing jobs, possibly they met at MGM; he mentions Lederer in *D.1.*

Lee, Robert (Bob) (d. 1994). American writer. During World War II, he co-founded Armed Forces Radio with Jerry Lawrence, and together they produced the official military radio programs for D-Day, V-E Day, and V-J Day; afterwards they became longtime collaborators on radio, stage, T.V. and film projects. In addition to their plays and films, they wrote numerous radio plays for CBS and the series "Columbia Workshop." They also co-founded the American Playwrights Theater, The Jerome Lawrence and Robert E. Lee Theater Research Institute at Ohio State, and the Margo Jones Award, in honor of the producer-director who produced their prize-winning play, *Inherit the Wind*, and who died young.

Lehmann, Beatrix (Peggy) (1903–1979). English actress; youngest of John Lehmann's three elder sisters. She met Isherwood in Berlin in 1932, and they remained close friends. She had a London triumph in O'Neill's *Mourning Becomes Electra* in 1938 when Isherwood was in China, and he returned in time to see her in the Group Theatre's performance of Cocteau's *La Voix Humaine* in July. During 1938 she had an affair with Berthold Viertel.

Lehmann, Helen (1899–1985). Eldest of the talented, beautiful siblings who made their mark on English literary life from the 1930s through the 1970s; educated at Cambridge. She had no public reputation, although she was the model for "Kate" in Rosamond's novel *Invitation to the Waltz*. Soon after leaving Cambridge, Helen married a soldier, Montague Bradish-Ellames, and devoted herself to family life. During World War II, she was a driver for the U.S. Army, and when her husband divorced her for another woman, she moved to London and worked for the Society of Authors until 1960.

Lehmann, John (1907–1988). English author, publisher, editor, autobiographer; educated at Cambridge. Youngest child and only son of a close family; his mother was an American from New England; his father trained as a barrister and wrote for *Punch*. Isherwood met him in 1932 at the Hogarth Press where Lehmann was assistant (later partner) to Leonard and Virginia Woolf. Lehmann persuaded the

Woolfs to publish *The Memorial* after it had been rejected by Jonathan Cape, publisher of Isherwood's first novel *All the Conspirators*. Isherwood helped Lehmann with his plans to found the magazine *New Writing* and obtained early contributions from friends like Auden. He tells about this in *Christopher and His Kind* and also writes about Lehmann in *D.1* and *Lost Years*. When Lehmann left the Hogarth Press, he founded his own publishing firm and later edited *The London Magazine*. He wrote three volumes of autobiography, *The Whispering Gallery* (1955), *I Am My Brother* (1960), and *The Ample Proposition* (1966). For many years he shared his house with the dancer Alexis Rassine.

Lehmann, Rosamond (1901–1990). English novelist, educated at Cambridge, second-eldest sister of John Lehmann. She made a reputation with the sexual and emotional frankness of her first novel *Dusty Answer* (1927), and her later works—including *Invitation to the Waltz* (1932), *The Weather in the Streets* (1936), *The Echoing Grove* (1953)—also shocked. Her first marriage, in 1923, was to Leslie Runciman, son of a Liberal Member of Parliament, and from 1928 to 1944, she was the first wife of the painter Wogan Philipps, with whom she had a son and a daughter. Afterwards, she had a nine-year affair with Cecil Day-Lewis. Her daughter with Philipps, Sally, died suddenly of polio in 1958 when she was twenty-four; Rosamond described her continuing spiritual relationship with Sally in *The Swan in the Evening: Fragments of an Inner Life* (1967). She appears in *D.1* and *Lost Years*.

Leigh, Vivien (1913–1967). English stage and film star; born Vivian Hartley in India, and educated in convents and finishing schools in England and Europe. She trained at RADA and in 1932 married a barrister, Leigh Holman, with whom she had a daughter. She made her first film, *Things Are Looking Up*, in 1934 and became a theatrical sensation the following year in *The Mask of Virtue*. When she starred opposite Laurence Olivier in *Fire Over England* (1937), they fell in love, divorced their respective spouses and, in 1940 married each other. Her stage roles, frequently opposite Olivier, included many from Shakespeare, as well as Shaw, Thornton Wilder, Tennessee Williams, Rattigan, Coward, Dumas (*fils*), Chekhov, and a musical comedy adaptation of Sherwood's *Tovarich* in 1963. She won an Academy Award as Scarlett O'Hara in *Gone with the Wind* (1939) and another as Blanche Du Bois in *A Streetcar Named Desire* (1951), already played to acclaim on the stage. Other films include *Waterloo Bridge* (1940), *That Hamilton Woman* (1941), *Caesar and Cleopatra* (1945), *Anna Karenina* (1948), *The Roman Spring of Mrs. Stone* (1961), and *Ship of Fools* (1965). She suffered from tuberculosis and exhaustion, had two miscarriages during her marriage to Olivier, and became incurably manic-depressive. Olivier divorced her in 1961. Isherwood met her in Los Angeles in the summer of 1960, when she was touring with Mary Ure in Jean Giraudoux's *Duel of Angels*. She appears in *D.1*.

Leighton, Margaret (Maggie) (1922–1976). English actress. She made her London debut as a teenager and established her reputation in the Old Vic Company in the late 1940s. From the mid-1950s until the late 1960s, she also played on Broadway, where she won Tony Awards for *Separate Tables* (1956) and *The Night of the Iguana* (1962). Her films include *The Winslow Boy* (1948), *The Sound and the Fury* (1959), *The Loved One* (1965), and *The Go-Between* (1971). Her second marriage was to Laurence Harvey, from 1957 to 1961, and her third, in 1964, to actor Michael Wilding. She had a small role in Bachardy and Isherwood's "Frankenstein: The True Story."

Len. See Worton, Len.

Lenya, Lotte (1900–1981). Austrian actress and singer. She became famous in pre-war Berlin for her roles in *The Threepenny Opera* and other musicals created by Bertolt Brecht and the composer, Kurt Weill, whom she married. With Weill, she fled the Nazis, settled in the U.S., and for a time gave up her career. After Weill died in 1950, she starred in a long-running off-Broadway revival of *The Threepenny Opera*—which Isherwood and Bachardy saw in 1960 when it toured to Los Angeles—in *Brecht on Brecht*, and, later, in *Cabaret*. She also made a few movies, including *Die Dreigroschenoper* (The Beggar's Opera, 1931), *The Roman Spring of Mrs. Stone* (1961), *From Russia with Love* (1963), and *Semi-Tough* (1977). Her second marriage was to the writer and editor George Davis. Bachardy drew her portrait in 1961 and 1962.

Leopold, Michael. Aspiring writer, from Texas; he was about eighteen when Isherwood met him at the apartment of a friend, Doug Ebersole, in December 1949. They began a minor affair soon afterwards. Leopold was interested in literature, admired Isherwood's work, and later wrote some stories of his own. During the 1960s, he lived with Henry Guerriero in Venice, California. He appears in *D.1* and *Lost Years*.

Levant, Oscar (1906–1972). American composer, pianist, and actor. He was a close friend of George Gershwin and became famous as an interpreter of his music. His film appearances include: *Kiss the Boys Goodbye* (1941), *Rhapsody in Blue* (1945), *Humoresque* (1946), *You Were Meant for Me* (1948), and *An American in Paris* (1951). Levant wrote the music for several popular musicals and had a live talk show in Hollywood, "The Oscar Levant Show," broadcast out of a shed on a minor network. His show was shut down by the sponsors in the early 1960s despite its popularity, because he insulted their products for laughs and encouraged his guests to do the same. Isherwood appeared on the show in the mid-1950s, sometimes reading poetry; this led to his occasionally being recognized in the street. In 1958, he argued with Levant about Churchill and then refused to return to the show for a time because Levant attacked him for remaining in Hollywood during the war. He appears in *D.1*.

Levy, Miranda Speranza (Mirandi). Sister-in-law of the Italian-American jewelry designer, Frank Patania (d. 1964) whose Native American-influenced work in silver, turquoise, and coral was bought by Mabel Dodge Luhan and Georgia O'Keeffe and can be seen in museums. She ran Patania's Thunderbird shop in Santa Fe and married Ralph Levy, a Hollywood film director. She appears in *D.1* under her maiden mame, Mirandi Masocco.

Lewis, Jack. Los Angeles doctor. A colleague of the endocrinologist Jessie Marmorston; he began to treat Isherwood in about 1957 and became his main doctor for some years.

Lincoln. See Kirstein, Lincoln.

List, Herbert. German photographer. Probably introduced to Isherwood by Stephen Spender who, in 1929, became friends with List in Hamburg, where List was working as a coffee merchant in his family's firm. List appears as "Joachim" in Spender's *World Within World* and as "Joachim Lenz" in Spender's *The Temple*. Isherwood writes about him in *D.1*.

Locke, Charles O. (1896–1977). American journalist, novelist, screenwriter. He wrote for his family's newspaper in Toledo, Ohio, while still in college, then had a career in New York working for several major papers and in advertising and publicity. He wrote poetry and song lyrics as well as five novels. In 1957, he wrote the screenplay from his most famous novel, *The Hell-Bent Kid*, in the office next door to Isherwood's at Twentieth Century-Fox. Isherwood describes their friendship in *D.1*, where Locke's wife and his daughter, Mary Schmidt, are also mentioned.

Lodge, Carter (d. 1995). American business manager of English playwright and novelist John van Druten. Lodge was van Druten's lover in the late 1930s and early 1940s; afterwards he began a long-term relationship with another man, Dick Foote, but remained close to van Druten. Isherwood first met Lodge in November 1939. He lived mostly in the Coachella Valley at the AJC Ranch, which he and van Druten purchased in the early 1940s with Auriol Lee, a British actress. Lodge managed the ranch, where they grew corn and tomatoes, and handled his own and van Druten's financial affairs very successfully. When van Druten died in 1957, Lodge inherited his property and rights in his work, including *I Am a Camera*, which later entitled him to a percentage in *Cabaret*. He appears in *D.1* and *Lost Years*.

Lord, Bart. Amateur actor, avid movie and show-business fan. He was a boy-friend of Ted Bachardy, with whom Ted lived for a few years during the 1950s. They eventually fell out of touch because of Ted's mental breakdowns. He appears in *D.1*.

Loy, Myrna (1905–1993). Hollywood star; born in Montana, raised there and in Los Angeles, where she became a chorus girl at eighteen. She was an exotic screen vamp in over sixty films until the mid-1930s, when she began to appear as Nora Charles in movies based on Dashiell Hammett's *The Thin Man*, and became Hollywood's number one woman star. Her later films include *The Best Years of Our Lives* (1946), *The Red Pony* (1949), and *Midnight Lace* (1960). During World War II, she worked for the Red Cross and, later, for UNESCO. She was also active in the Democratic party.

Luce, Henry R. (1898–1967) and Clare Booth (1903–1987). He was born in China and educated at Hotchkiss, Yale, and Oxford. He worked as a journalist before he co-founded *Time Magazine* in 1923. In the 1930s, he launched *Fortune* and *Life*, and then *House and Home* and *Sports Illustrated* in the 1950s. She was the illegitimate, peripatetically educated daughter of a dancer mother and a violin-ist father who deserted them. She worked as an actress briefly, was editor of *Vogue* and *Vanity Fair*, a successful Broadway playwright—*Abide with Me* (1935), *The Women* (1936), *Kiss the Boys Goodbye* (1938)—a Republican congresswoman and U.S. Ambassador to Italy (1953–1957). Each was married once before. She converted to Roman Catholicism in 1946 after her daughter, an only child by her first husband, was killed in a car accident. The Luces were both vehement anticommunists.

Luckenbill, Dan (b. 1945). American writer and librarian. From 1970, he worked at the UCLA library in the manuscripts division. Later he curated exhibitions and wrote catalogs on gay and lesbian studies. He published short stories from the 1970s onwards.

Ludington, Wright (1900–1992). Art collector and philanthropist; raised in Pennsylvania, educated at the Thacher School in Ojai, at Yale, at the Pennsylvania Academy of Fine Arts, and at the Art Students League in New York. In 1927, he inherited a fortune from his father, a lawyer and investment banker who worked with the Curtis Publishing Company. He also inherited an estate in Montecito—Val Verde—which he spent decades improving with the help of a school friend, the landscape architect Lockwood de Forest. Val Verde included an art gallery for Ludington's collection of modern paintings and outdoor settings for his ancient sculpture. In 1955, he sold Val Verde and built a new house off Bella Vista Drive—Hesperides—which was designed by Lutah Maria Riggs especially to display his art collection. De Forest's widow landscaped Hesperides. Ludington was a founder and board member of the Santa Barbara Museum of Art and gave the museum many pieces from his collection.

Lynes, George Platt (1907–1955). American photographer; educated at the Berkshire School, where he met Lincoln Kirstein, and, briefly, at Yale. Lynes photographed Auden and Isherwood during their brief visit to New York in 1938. In the spring of 1946, he photographed Isherwood again and encouraged Bill Caskey in his efforts to become a professional photographer. Later, in 1953, Lynes befriended and photographed Don Bachardy. He made his living from advertising and fashion photography for magazines like *Town and Country*, *Harper's Bazaar*, and *Vogue*, but he is also known for his photographs of the ballet, male nudes, and surrealistic still lifes. He did many portraits of film stars and writers. He appears in *D.1* and *Lost Years*.

M. Isherwood's mother. He called her "Mummy" and began letters to her with "My Darling Mummy," and later, "Dearest Mummy," but he invariably wrote "M." in his diaries. See Isherwood, Kathleen.

Isherwood sometimes uses "M." for Mahendranath Gupta, the schoolmaster who became Ramakrishna's disciple and recorded Ramakrishna's conversations and sayings in his diaries, later compiling them in *Sri Ramakrishna Kathamrita or The Gospel of Ramakrishna*.

MacDonald, Madge. Nurse. She worked at UCLA hospital and lived in Isherwood's neighborhood. She appears in *D.1*.

Macklem, Francesca (Jill). Secretary to Fred Shroyer, who was responsible for Isherwood's appointment to teach in the English Department at Los Angeles State College. Isherwood became friendly with her in 1959 when he began the teaching job, and she appears in *D.1*. She had a life-threatening heart condition. She was married three times, the third time to her first husband, Les Macklem, again; they had two children.

Macy, Gertrude. New York stage manager and producer; secretary and biographer of actress Katherine Cornell. She co-produced the 1951 stage version of *I Am a Camera* with Walter Starcke and thereby had a substantial financial stake in *Cabaret*, which reduced Isherwood's earnings and which he several times refers to in his diaries. She is mentioned in *D.1*.

Madhavananda, Swami (1888–1965). Hindu monk of the Ramakrishna Order and scholar; a disciple of Sarada Devi. He translated a number of Vedanta texts, and his versions are still widely used, notably Shankara's commentary on the Brihadaranyaka Upanishad. He was General Secretary and President of the Ramakrishna Math for many years and an influential supporter of the women's Math, Sarada Math, founded in 1954.

Maharaj. See Brahmananda, Swami.

mahasamadhi. Great samadhi, usually referring to the moment of death, when the illuminated soul leaves the body and is absorbed into the divine. See samadhi.

Mailer, Norman (1923–2007). American writer, raised in Brooklyn and educated at Harvard. He fought in the Pacific during World War II and became famous with the publication of his first novel, *The Naked and the Dead* (1948), about an American infantry platoon invading a Japanese-held island. He twice won the Pulitzer Prize: for *The Armies of the Night* (1968) and *The Executioner's Song* (1979). Other novels and works blending fiction with non-fiction and personal commentary include *The Deer Park* (1955), *An American Dream* (1965), *Why Are We in Vietnam?* (1967), *Of a Fire on the Moon* (1970), *The Prisoner of Sex* (1971), *Ancient Evenings* (1983), *Tough Guys Don't Dance* (1984), *Harlot's Ghost* (1991), *Oswald's Tale: An American Mystery* (1995), *The Gospel According to the Son* (1997), and *The Castle in the Forest* (2007). He also wrote screenplays and directed films. He co-founded *The Village Voice* in 1955, and in 1969 he ran for mayor of New York. He married six times. In 1960, he stabbed his second wife, Adele Morales, with a penknife, leaving her in critical condition. They had been married for six years and had two daughters together; she did not press charges. In *Lost Years*, Isherwood tells of his first meeting with Mailer in 1950.

Mallory, Margaret. Art collector and philanthropist. She lived with Alice Story, known as Ala, and they travelled and collected together. Mallory was a benefactor of UCSB, endowing fellowships in music and art history and donating parts of her collection.

Mann, Daniel (Danny) (1912–1991). American film and television director, born and educated in Brooklyn. He trained as a musician and acted in the theater from childhood, then made his name directing movies, including *Come Back Little Sheba* (1952, for which Shirley Booth won an Academy Award), *The Rose Tattoo* (1955, for which Anna Magnani won an Academy Award), *Butterfield Eight* (1960, for which Elizabeth Taylor won an Academy Award), *Our Man Flint* (1966), *For Love of Ivy* (1968), *Willard* (1971), and *How the West Was Won* (1977). Isherwood first met him in 1954 during the filming of *The Rose Tattoo*, when he took Bachardy to visit Tennessee Williams and Frank Merlo in Key West.

Manning. See Gurian, Manning.

Mangeot, Olive. English wife of Belgian violinist André Mangeot; mother of Sylvain and Fowke Mangeot. Isherwood met the Mangeots in 1925 and worked for a year as part-time secretary to André Mangeot's string quartet which was organized from the family home in Chelsea. He brought friends to meet Olive when he was in London. She is the original of "Madame Cheuret" in *Lions and Shadows*, and Isherwood drew on aspects of her personality for "Margaret Lanwin" and "Mary Scriven" in *The Memorial*. Olive had an affair with Edward Upward and through his influence became a communist. Later she separated from her husband and for a time shared a house with Jean Ross and Jean's daughter in Cheltenham. Hilda Hauser, the Mangeots' housekeeper and cook, also moved with Olive to Cheltenham, where, together, they raised Hilda's granddaughter, Amber. As Isherwood tells in *D.1* and in *Lost Years*—in which the Mangeots also appear—Hilda's daughter, Phyllis, was raped by a black G.I. during World War II and Amber resulted.

Mangeot, Sylvain (1913–1978). Younger son of Olive and André Mangeot. Isherwood's friend, Eric Falk, initially introduced Isherwood to the Mangeot family because Sylvain, at age eleven, had a bicycle accident which confined him to a wheelchair for a time, and Isherwood had a car in which he could take Sylvain for outings. They grew to know each other well during the time that Isherwood worked for Sylvain's father, and together they made a little book, *People One Ought to Know*, for which Isherwood wrote nonsense verses to accompany Sylvain's animal paintings (it was eventually published in 1982, but one pair of verses appeared earlier as "The Common Cormorant" in Auden's 1938 anthology *The Poet's Tongue*). Sylvain is portrayed as "Edouard" in *Lions and Shadows*. Later he joined the Foreign Office and then became a journalist, working as a diplomatic correspondent, an editor, and an overseas radio commentator for the BBC. He appears in *D.1*.

mantram or mantra. A Sanskrit word or words which the guru tells his disciple when initiating him into the spiritual life and which is the essence of the guru's teaching for this particular disciple. The mantram is a name for God and includes the word *Om*; the disciple must keep the mantram secret and meditate for the rest of his life on the aspect of God which it represents. Repeating the mantram (making japam) purifies the mind and leads to the realization of God. With the mantram, the guru often gives a rosary—as Swami Prabhavananda gave Isherwood—on which the disciple may count the number of times he repeats his mantram.

Marguerite, previously Marguerite Brown, also Marguerite Harrity. See Lamkin, Marguerite.

Markovich, John (Mark) (193[2]–2008). American painter and monk of the Ramakrishna Order, born in Detroit. He was known as Mark, and later as Brahmachari Nirmal and then Swami Tadatmananda. He became the official abbot of Trabuco.

Marple Hall. The Bradshaw Isherwood family seat; see entries for Frank Bradshaw Isherwood and Richard Bradshaw Isherwood.

Masocco, Mirandi. See Levy, Miranda.

Mason, James (1909–1984). British actor, educated at Marlborough and Cambridge. He was a conscientious objector during World War II. He joined the Old Vic Theatre Company in the 1930s, soon began making films, and became a star in *The Seventh Veil* (1945), before moving on to Hollywood where he often played villains. Later films include *Julius Caesar* (1953), *The Desert Rats* (1953), *20,000 Leagues Under the Sea* (1954), *North by Northwest* (1959), *Lolita* (1962), *Lord Jim* (1965), *Georgy Girl* (1966), and *The Verdict* (1982). On T.V. he played Franz Gruber in Isherwood and Danny Mann's "The Legend of Silent Night" and Polidori in "Frankenstein: The True Story." He was married to actress Pamela Kellino from 1941 to 1965 and, from 1971, to Clarissa Kaye.

Masselink, Ben (1919–2000). American writer, born in Michigan and educated at DePauw University. Probably Isherwood and Bill Caskey met Ben Masselink with his longtime companion Jo Lathwood in the Friendship Bar in Santa Monica around 1949; they appear often in *D.1* and *Lost Years*. During the war, Masselink was in the marines; one night on leave, he got drunk in the Friendship and Jo Lathwood took him home and looked after him. When the war was over he returned and stayed for twenty years. Isherwood alludes to this meeting in his description of The Starboard Side in *A Single Man*. Masselink had studied architecture, and Isherwood helped him with his writing career during the 1950s. His first book of stories, *Partly Submerged*, was published in 1957. He then published several novels: two about his war experience—*The Crackerjack Marines* (1959) and *The Deadliest Weapon* (1965), the second of which Isherwood greatly admired— and *The Danger Islands* (1964), for teenage boys. He also wrote for television throughout the 1950s and in 1960 worked at Warner Brothers on the script for a film of *The Crackerjack Marines*. As Isherwood tells, Masselink left Jo in 1967 for a younger woman, Dionyse (Dee) Humphrey, the wife of their friend, Bill Hawes. Dee had an adopted daughter, Heather, from her first marriage.

Masselink, Jo (*circa* 1900–1988). Women's sportswear and bathing suit designer, from Northville, South Dakota; among her clientele were movie stars such as Janet Gaynor and Anne Baxter. In youth, she worked as a dancer and was briefly married to a man called Jack Lathwood whose name she kept professionally. Also, she had a daughter, Betty (see Arizu), and a son with a North Dakotan, Ferdinand Hinchberger. From 1938, Jo lived in an apartment on West Channel Road, a few doors from the Friendship, and by the late 1940s she knew many of Isherwood's friends who frequented the bar—including Bill Caskey, Jay de Laval, and Jim Charlton. She never married Ben Masselink, though she used his surname while they lived together. She and Masselink figure through much of *D.1* and *Lost Years* as Isherwood's closest heterosexual friends.

Maugham, William Somerset (Willie) (1874–1965). British playwright and novelist. In 1917, he married Syrie Wellcome, but he never lived with her. His companion was Gerald Haxton, eighteen years younger, whom he met in 1914 when they both worked in an ambulance unit in Flanders. They travelled and entertained on Cap Ferrat at the Villa Mauresque, which Maugham bought in 1926. Haxton died in 1944, and Maugham's subsequent companion and chosen heir was Alan Searle. Isherwood met Maugham in London in the late 1930s and saw him whenever Maugham visited Hollywood, where many of Maugham's works were filmed; with Bachardy, Isherwood later made several visits to the Villa Mauresque. In 1945, Isherwood worked for Wolfgang Reinhardt on a screenplay for Maugham's 1941 novel *Up at the Villa* (never made), and he enlisted Swami Prabhavananda to advise Maugham on the screenplay for *The Razor's Edge* (1944). Although Maugham did not follow their advice, Isherwood and Swami again helped him in 1956 with an essay "The Saint," about Ramana Maharshi (1879–1950), the Indian holy man Maugham had met in 1936 and on whom he had modelled Shri Ganesha, the fictional holy man in *The Razor's Edge*. Maugham appears in *D.1* and *Lost Years*.

maya. In Vedanta, maya is the cosmic illusion, the manifold universe which the individual perceives instead of perceiving the one reality of Brahman; in this sense, maya veils Brahman. But maya is inseparable from Brahman and can also be understood as the manifestation of Brahman's power, god with attributes. Maya has a double aspect encompassing opposite tendencies, toward ignorance (avidya) and toward knowledge (vidya). Avidya-maya involves the individual in worldly passion; vidya-maya leads to spiritual illumination.

McKinnon, Russell. Californian patron of Don Bachardy. He commissioned a sanguine (red pencil) drawing of himself in about 1960, when he was in his early forties and married to a wealthy woman called Edna, ten or fifteen years older. Possibly he also had money of his own. Bachardy felt McKinnon was attracted to him, but McKinnon was shy and expressed his interest mostly in a brotherly way. He urged Bachardy to gain exposure to European art and culture and offered to pay for art study in Europe. Bachardy admired Walter Sickert, so he suggested the Slade where Sickert had studied. McKinnon visited London after Bachardy's show at the Redfern Gallery and took him out to dinner. They met again in Los Angeles, then eventually fell out of touch.

Mead, Mr. Isherwood and Bachardy's mechanic.

Medley, Robert (1905–1994). English painter. He attended Gresham's School, Holt, with Auden, and they remained close friends after Medley left for art school at the Slade. In London, he became the longtime companion of the dancer Rupert Doone and was involved with him in 1932 in founding The Group Theatre, which produced *The Dog Beneath the Skin*, *The Ascent of F6*, and *On the Frontier*. He also worked as a theater designer and teacher and founded the Theatre Design section at the Slade in the 1950s before becoming Head of Painting and Sculpture at the Camberwell School of Arts and Crafts in 1958. He appears in *D.1* and *Lost Years*.

Meredith, Burgess (1907–1997). American actor and director. He distinguished himself in the theater in the early 1930s, moved to film in 1936 with his stage role in *Winterset*, and appeared in many subsequent films, including *Of Mice and Men* (1939), *The Story of G.I. Joe* (1945), *Advise and Consent* (1961), and *Rocky* (1976). He also played the Penguin in the Batman T.V. series. His third of four wives was actress Paulette Goddard, from 1944 to 1949. Meredith was blacklisted in 1949, and disappeared from movies for nearly a decade. In *D.1*, Isherwood tells how his plan to play Ransom in *The Ascent of F6* was interrupted by the war; in this volume, Isherwood records Meredith's interest in directing *The Dog Beneath the Skin*.

Merlo, Frank (1921–1963). Italian-American companion of Tennessee Williams; raised in New Jersey by his Sicilian immigrant parents. He was about twenty-five years old when he met Williams in 1947. He had served in the navy and for a time continued to work as a truck driver. He was handsome and capable, kept house, cooked, and made travel, social and business arrangements for Williams. Isherwood first met him in Los Angeles when Merlo accompanied Williams on a visit there in 1949, and Merlo appears in *D.1* and *Lost Years*. The relationship grew less stable during the late 1950s, and the pair were often apart during Merlo's fatal illness with lung cancer; but it was the most lasting romance of Williams's life, and Williams was at his bedside when Merlo died.

Methuen. Isherwood's English publisher from the mid-1940s. His cousin, Graham Greene, recommended *Mr. Norris Changes Trains* to E.V. Rieu, a managing director at the firm, but Isherwood's first book published by Methuen was *Prater Violet* in the spring of 1946 (well after the U.S. publication because of the war). In September 1935, Isherwood had signed a contract for his "Next three full-length available novels" and accepted half of a £300 advance on the first (styled in the contract *Prata Violet*); he had already promised his next novel to his current publisher, The Hogarth Press, and at Leonard Woolf's insistence, *Sally Bowles*, *Goodbye to Berlin*, and *Lions and Shadows* were published by Hogarth. After the war, the contract with Methuen was honored, the second novel being *The World in the Evening* and the third, *Down There on a Visit*, which he delivered to Alan White in 1961. White joined Methuen in 1924, became a director in 1933, and retired as Chairman in 1966. When White retired, John Cullen became Isherwood's editor, and after Cullen, Geoffrey Strachan. Methuen remained Isherwood's U.K. publisher for the rest of his life and posthumously until 1998, when Random House attempted to take over the imprint which by then belonged to a larger group, Reed Books. Methuen achieved independence through a management buy-out, but agreed in the negotiations to let Isherwood go to Chatto & Windus at Random House.

Michael. See Barrie, Michael.

Milam, Webster. Vedanta devotee from Arizona and, as a seventeen-year-old high school student, an aspiring monk. He was among the handful of men who lived with Isherwood in Brahmananda Cottage at the Vedanta Society in 1943. By 1949, he had left the society, and he soon married. He appears in *D.1*.

Millard, Paul. American actor. He lived with Speed Lamkin in West Hollywood for a few years during the 1950s. He briefly called himself Paul Marlin, then later changed to Millard; his real name was Fink. He was good-looking and relatively successful on the New York stage and on T.V., but eventually joined his mother's real estate business and invested in property. During 1959 and 1960, he loaned Bachardy a little house in West Hollywood to use as a studio, and around this time, the two had an affair which Isherwood apparently did not know about. Millard appears in *D.1*; he died in the 1970s.

Miller, Dorothy (d. 1974). Cook and cleaner to Isherwood and Bachardy from 1958 until the early 1970s. On their recommendation she later kept house for the

Laughtons as well, both in Hollywood and in Charles Laughton's house next door to Isherwood and Bachardy in Adelaide Drive.

Miller, Harry Tatlock (1913–1989). Australian-born writer, critic, and curator. He met his longtime companion, Loudon Sainthill, in the mid-1930s, travelled to London with him in 1939 and served with him in the Australian Army Medical Corps during World War II; after the war, they shared a room at Merioola, a Sydney boarding house which gave its name, the Merioola Group, to the artists living there. Miller curated two shows at the David Jones Gallery in Sydney and wrote about the tours of European dance companies and theater companies to Australia and New Zealand during the 1930s and 1940s, especially Ballet Rambert. In 1939, when Sainthill's paintings of the Ballet Russes de Monte Carlo were shown at the Redfern Gallery in London, Miller began an association with the gallery and in 1949 settled in London permanently and was made a director. His books with Sainthill include *Royal Album* (1951), *Undoubted Queen* (1958), and *Churchill* (1959); he also edited *Loudon Sainthill, With an Appreciation by Bryan Robertson* (1973).

Miller, Henry (1891–1980). American writer, born in New York, son of a Brooklyn tailor. He dropped out of college, drove a cab, worked at odd jobs and for Western Union, then lived in Paris from 1930. His first novel, *Tropic of Cancer* (1934), about an American artist abroad, was published there but banned in English-speaking countries for its sexual explicitness. In the U.S. the ban was lifted in 1961 and Grove Press brought out an edition, but the book was labelled obscene and booksellers were prosecuted for selling it. Grove Press and the ACLU fought more than sixty cases, eventually reaching the U.S. Supreme Court, which ruled in December 1963 that the book was not obscene. A volume of Miller's stories, *Black Spring* (1936), and his second novel, *Tropic of Capricorn* (1939), were also banned until the early 1960s. Other works include *The Colossus of Maroussi* (1941) about Greece, *The Plight of the Creative Artist in the United States* (1944), *The Air-Conditioned Nightmare* (1945), *The Rosy Crucifixion*—a trilogy comprised of *Sexus* (1949), *Plexus* (1953), and *Nexus* (1960)—*The Time of the Assassins* (1956) about Rimbaud, *The Intimate Henry Miller* (1959), and *Just Wild about Harry* (1963), a play. He also painted, contributed to *Why Abstract?* (1945) with Hilaire Hiler and William Saroyan, and assembled *To Paint Is to Love Again* (1960) with reproductions of his work. During World War II, he returned to America and settled in Big Sur in 1944; in 1962, he moved to Pacific Palisades. Partly because of his long battle with censorship, he became a guru for the Beat generation. He was married five times. His triangular love affair with his second wife, June Mansfield, and Anaïs Nin, who encouraged and assisted him financially during his years in Paris, was recorded by Nin in her diaries and became the subject of the movie *Henry and June* (1990). With his third wife, Janina Martha Lepska, to whom he was married from 1944 to 1952, he had Valentine Miller (b. 1945) and Henry Tony Miller (b. 1948); they lived with their mother in Los Angeles. His fourth wife was Eve McClure, whom he divorced in 1962.

Miltown. A tranquilizer widely used in the 1950s and 1960s, generically called meprobamate. In the mid-1960s, it was discovered to be a habit-forming sedative and became a controlled substance available only by prescription.

Mishima, Yukio (1925–1970). Japanese author, of novels, short stories, poems, plays, and essays; born Kimitake Hiraoka; educated at Tokyo University. He was already famous in Japan when Alfred Knopf published *The Sound of Waves* in English translation in 1956 and invited Mishima to the U.S. the following year. Isherwood first met him during that visit, which he tells about in *D.1*, and they met again in November 1960 when Mishima returned to the U.S. for the staging in New York of three of his Noh plays. By then Mishima had married Yoko Sugiyama, a student of English literature and daughter of the painter Yagushi Sugiyama, and the couple had had the first of their two children. More of Mishima's work had been translated into English: *Five Modern Noh Plays* appeared in English in 1957, then *Twilight Sunflower* (also a volume of plays) and *Confessions of a Mask* (1949) in 1958, followed by *The Temple of the Golden Pavilion* (1959) and many others. *Confessions of a Mask* addressed Mishima's discovery of his homosexuality. His masterwork, *Sea of Fertility*, conceived in 1962 and completed in 1970, is a tetralogy about Japan in the twentieth century. Mishima was obsessed by the warrior traditions of Imperial Japan and was expert in martial arts. In 1968, he founded a military group, the Shield Society, to revive the Samurai code of honor. Disillusioned when the young did not answer his call for a return to nationalist ideals, he committed Seppuku; in his diary entry for November 25, 1970, Isherwood transcribed an account of the ritual suicide from the *Los Angeles Times*. Mishima was twice nominated for the Nobel Prize.

Moffat, Ivan (1918–2002). British-American screenwriter, educated at Dartmouth; son of Iris Tree and her American husband Curtis Moffat. He worked on government-sponsored documentaries for Strand Films in London, served in the U.S. Army Signal Corps Special Coverage Unit under the American film director George Stevens during World War II, and returned to Los Angeles in 1946 as Stevens's assistant. He worked with Stevens on *A Place in the Sun* (1951), was Stevens's associate producer for *Shane* (1953), and co-wrote *Giant* (1956), before going on to work for David Selznick on *Tender Is the Night* (1962). Other screenplays include *Bhowani Junction* (1956), *Boy on a Dolphin* (1957), and *Justine* (1969). Moffat's first wife was Natasha Sorokin, a Russian, once part of a ménage à trois with Simone de Beauvoir and Jean-Paul Sartre. Their marriage broke up at the start of the 1950s, leaving a daughter, Lorna. Moffat then had a number of beautiful and talented girlfriends, including the writer Caroline Blackwood with whom he fathered a daughter, Ivana, born in 1966 during Caroline's second marriage to the composer and music critic Israel Citkowitz; Ivana's paternity was kept secret until Caroline Blackwood's death in 1996. In 1961, Moffat married Katherine Smith, known as Kate, a well-connected English heiress. The marriage ended in 1972. Moffat appears often in *D.1* and *Lost Years*. Although Moffat was always heterosexual, Isherwood identified with him, as an expatriate and as a romantic adventurer, and based both the main character in the first draft of *Down There on a Visit* and "Patrick" in *A Meeting by the River* partly on him.

Moody, Robert L. (1910–1973). British psychotherapist; raised in Surbiton and educated at Bromsgrove School. He began the Bachelor of Medicine course alongside Isherwood at King's College Medical School, London, in October 1928 but left in 1930 without taking any exams. Later, he became one of the first directors of the Jungian organization, the Society of Analytical Psychology, when it was formally registered under that name in 1945. He was also an editor of the *Journal of Analytical Psychology* and wrote various articles. He married three times, last to Louise Diamond. He appears as "Platt" in *Lions and Shadows*.

Moreau, Jeanne (b. 1928). French stage and screen star and singer; daughter of an English chorus girl. She was educated at the Paris Conservatory of Dramatic Art and became a leading actress for the Comédie Française and the Théâtre National Populaire before coming to international prominence in Louis Malle's *Frantic* (*Ascenseur pour l'Echafaud*, 1957, in France; released as *Lift to the Scaffold* in the U.K. and later retitled *Elevator to the Gallows* in the U.S.) and *The Lovers* (1958). Her many film roles after that, for a range of celebrated directors, include *Les Liaisons Dangereuses* (1959), *Moderato Cantabile* (1960), *La Notte* (1961), *Jules and Jim* (1961), *Eva* (1962), *The Trial* (1963), *Diary of a Chambermaid* (1964), *Chimes at Midnight* (1966), *The Bride Wore Black* (1968), and *Going Places* (1974). She was married twice, briefly both times, and had many love affairs, including a complicated one with Tony Richardson while she was working with him on *Mademoiselle* (1966) and *The Sailor from Gibraltar* (1967).

Morgan. See Forster, E.M.

Morris, Phyllis (d. 1982). English character actress and author. She was a student with Dodie Smith (later Beesley) at the Academy of Dramatic Art (precursor to RADA) in London during World War I, and they sometimes shared lodgings. She came from a wealthy family and had a lifelong income; in youth, she was briefly married to a doctor thirty years her senior. She published a volume of poems, *Dandelion Clocks* (1917), two children's books, *Peter's Pencil* (1920) and *The Adventures of Willy and Nilly* (1921), and wrote plays, including *Made in Heaven* and *The Rescue Party* staged in London in the 1920s. She appeared in a number of Dodie Smith's plays and had minor roles in a few Hollywood films, for instance, *That Forsyte Woman* (1949). Isherwood met her in Hollywood after the war, and he mentions her in *Lost Years*. Eventually, she settled in a retirement apartment at Gosfield Hall, a restored Tudor manor house near the Beesleys' Essex cottage. She wrote for T.V., took up painting, and continued her stage career until near the end of her life.

Morrow, Vic (1932–1982) and Barbara. American actor and his wife, an actress and T.V. writer. He was born and raised in the Bronx, joined the navy, then attended college in Florida on the G.I. Bill. Afterwards, he studied acting in Mexico City and at the Actors Workshop in New York where he appeared off Broadway and was discovered by MGM and cast in *The Blackboard Jungle* (1955). His other films include *God's Little Acre* (1958), *Cimarron* (1960), and *Dirty Mary Crazy Larry* (1974), and he directed *Deathwatch* (1976), which he adapted with his wife from Jean Genet's play. Morrow starred in the T.V. series "Combat!" from 1962 to 1967 and wrote and directed some episodes. He married Barbara Turner in 1958, and they had two daughters, one the actress Jennifer Jason Leigh. Barbara Morrow worked on several projects with her husband, but the marriage ended in 1965. She later married T.V. director Reza Badiyi. In later years, Vic Morrow acted in made-for-T.V. movies and mini-series, including "Roots" (1977). He died in a helicopter accident on the set of *Twilight Zone—The Movie* (1983).

Mortimer, Raymond (1895–1980). English literary and art critic; he was writer and editor for numerous magazines and newspapers and wrote books on painting and the decorative arts as well as a novel. He was at Balliol College, Oxford, with Aldous Huxley and later became a close friend of Gerald Heard, introducing Heard to Huxley in 1929; he was also intimate with various Bloomsbury figures and an advocate of their work. From 1948 onward, he worked for *The Sunday Times*, spending the last nearly thirty years of his life as their chief reviewer. He appears in *D.1*.

Mortmere. An imaginary English village invented by Isherwood and Edward Upward when they were at Cambridge together in the 1920s; the inhabitants were satires of generic English social types and were all slightly mad. As part of their rebellion against public school and university, Upward and Isherwood shared an elaborate fantasy life which was described by Isherwood in *Lions and Shadows*. The fragmentary stories the two wrote for each other about Mortmere were eventually published as a collection in 1994; Upward's *The Railway Accident* appeared on its own in 1949.

Murdock, James (1931–1981). American actor; born David Baker. He worked mostly in T.V. Westerns in the late 1950s and 1960s: "Gunsmoke," "Rawhide" (as Harkness "Mushy" Mushgrove, the cook's assistant), "Have Gun, Will Travel," and "Cheyenne." Isherwood tended to misspell his name as Murdoc*h*.

namaskar. Salutation; see pranam.

Natasha. See Spender, Natasha Litvin.

Narendra, also Naren. See Vivekananda, Swami.

National Institute of Arts and Letters. The National Institute of Arts and Letters had 250 members chosen in recognition of their individual achievements in art, literature, and music. In 1976 it amalgamated with the American Academy of Arts and Letters and was called the American Academy and Institute of Arts and Letters, then later simplified its name to the American Academy of Arts and Letters. Members are chosen for life, and they confer several awards of their own, including the E.M. Forster Award. The organization maintains a library and museum in Manhattan. Isherwood was elected in 1949.

National Portrait Gallery, London. After acquiring Bachardy's 1967 portrait of Auden in 1969, Roy Strong left the NPG to become Director of the Victoria and Albert Museum in 1974 and went on to a career as a writer and broadcaster. He was succeeded at the NPG by the art historian John Hayes, who was director from 1974 to 1993. Hayes was an expert in the paintings of Thomas Gainsborough. It was not until 1996, when Charles Saumarez Smith was director, that the NPG acquired Bachardy's portraits of Ackerley (1961), Forster (1961), John Osborne (1968), and Thom Gunn (1996) and commissioned portraits of James Ivory, Ismail Merchant, and Ruth Prawer Jhabvala. In 1998, the NPG purchased a ninth portrait, of Dodie Smith (1961).

Neddermeyer, Heinz, Gerda and Christian. Isherwood's German boyfriend and his wife and son. Heinz was about seventeen when he met Isherwood in

Berlin, March 13, 1932. Their love affair, the most serious of Isherwood's life until then, lasted about five years. Hitler's rise forced them to leave Berlin in May 1933, and they lived and travelled in Europe and North Africa. In a traumatic confrontation with immigration officials at Harwich, Heinz was refused entry on his second visit to England in January 1934, so Isherwood went abroad more and more to be with him. In 1936, Heinz was summoned for conscription in Germany, and Isherwood scrambled to obtain or extend permits for Heinz to remain in the ever-diminishing number of European countries that would receive him. An expensive but shady lawyer called Salinger failed to obtain a new nationality for him. Heinz was expelled from Luxembourg on May 12, 1937 and returned to Germany, where he was arrested by the Gestapo and sentenced—for "reciprocal onanism" and draft evasion—to a three-and-a-half-year term combining imprisonment, forced labor, and military service. He survived and married Gerda in 1938, and Christian was born in 1940. Isherwood did not see Heinz again until 1952 in Berlin, though he corresponded with him both before and after this visit. It was Heinz's conscription that first turned Isherwood towards pacifism. Their shared wanderings are described in *Christopher and His Kind*, and their friendship also serves as one basis for the "Waldemar" section of *Down There on a Visit*. Heinz is also mentioned in *D.1* and *Lost Years*.

Neil. See Hartley, Neil.

Nelson, Allyn L. A girlfriend of Jim Gates and Peter Schneider. She lived in Oxnard, California, and attended Claremont College, where she met Gates and Schneider. Schneider later recalled she may have been studying nursing.

Nichols, Mike (b. 1931). American actor, director, producer; born Michael Igor Peschkowsky in Berlin. He emigrated to New York at seven and was educated at the University of Chicago. He studied acting with Lee Strasberg and became famous with Elaine May in a comedy duo which they took to Broadway as *An Evening with Mike Nichols and Elaine May* (1960). He directed many Broadway hits, including *Barefoot in the Park* (1964), *The Odd Couple* (1965), *The Little Foxes* (1967), *Plaza Suite* (1968), *The Prisoner of Second Avenue* (1971), *The Real Thing* (1984), *Death and the Maiden* (1992), and *Spamalot* (2005). His Hollywood successes were just as numerous, among them: *Who's Afraid of Virginia Woolf?* (1966), *The Graduate* (1967), *Carnal Knowledge* (1971), *The Day of the Dolphin* (1973), *Silkwood* (1983), *Working Girl* (1988), *Postcards from the Edge* (1990), *Primary Colors* (1998), *Closer* (2004), and *Charlie Wilson's War* (2007). When Isherwood knew him, his wife was Pat Scot. His fourth wife, since 1988, is newscaster Diane Sawyer.

Niem, Jan (d. 1973). Chauffeur to Tony Richardson, for twenty years. He was born in Poland and sent to a camp in Siberia by the Russians during World War II. After the war he came to the U.K. on a training scheme Churchill offered Stalin and was made a British citizen. He married an English woman with whom he ran a car service for the film industry. According to rumor, Tony Richardson "won" him in a poker game with Cubby Broccoli. He died on top of a prostitute while on location in the South of France.

Nikhilananda, Swami. Indian monk of the Ramakrishna Order; a longtime head of the Ramakrishna-Vedanta Center in New York, on the Upper East

Side; author of numerous books on Vedanta and translator of *The Gospel of Sri Ramakrishna* from Bengali into English with the help of Joseph Campbell, Margaret Woodrow Wilson, and John Moffitt who put Nikhilananda's translations of the songs into poetic form. He appears in *D.1*.

Nin, Anaïs (1903–1977) and Rupert Pole (1919–2006). American writer and her second husband. She was born in Paris, raised in New York after the outbreak of World War I, and spent the 1920s and 1930s back in Paris seeking out the company of writers, intellectuals, and bohemians. She became a psychoanalyst as well as writing novels, short stories and literary criticism. Her six-volume *Diary* began to appear in 1966, and tells, among other things, about her friendship with Henry Miller. Her other books include *Children of the Albatross* (1947)—which Isherwood read and admired before he met her—*The Four-Chambered Heart* (1950), and *A Spy in the House of Love* (1954). Some of her work, like Miller's, was published in Paris years before it appeared in the U.S. She had many love affairs and married twice. Pole, much younger than she, was a stepgrandson of Frank Lloyd Wright. He had a Harvard music degree and acted before studying forestry at UCLA and Berkeley. He worked as a forest ranger, and, as Isherwood tells in *Lost Years*, Nin lived with him at his forest station in the San Gabriel Mountains among all the other rangers in defiance of the rules. By the 1970s, the pair settled in the Silver Lake District of Los Angeles in a house designed by Wright's grandson, Eric Lloyd Wright.

Nirvanananda, Swami. Hindu monk; he was a direct disciple of Brahmananda (Maharaj), initiated as a brahmachari in 1914, and became Brahmananda's personal attendant. His original name was Surya, pronounced Surja in Bengal, and this was the basis for the affectionate name, Sujji Maharaj, by which he was mostly known. He was manager of Belur Math and became a vice-president of the Ramakrishna Order in the 1960s. He died in the early 1980s.

Oberon, Merle (1911–1979). British film star, raised in India and discovered in London by her first husband Alexander Korda who made her internationally famous during the 1930s. Her films include *The Private Life of Henry VIII* (1933), *The Scarlet Pimpernel* (1935), *Wuthering Heights* (1939), *The Lodger* (1944), and *A Song to Remember* (1945). She divorced Korda in 1945 and married cinematographer Lucien Ballard, whom she divorced in 1949. In 1957 she married again, to Bruno Pagliai, an Italian industrial tycoon with vast holdings in Latin America and especially Mexico, where they went to live until she divorced him in 1973. Her fourth marriage was to actor Robert Wolders, her co-star in her last film, *Interval* (1973), which she produced herself.

O'Hilderbrandt, Mrs. Isherwood's neighbor. As Mary Miles Minter, she was a teenage star of silent films. Her career ended in scandal before she was twenty when her director, and, by rumor, her lover, William Desmond Taylor, was murdered, possibly by her mother. The crime was never solved, and she once offered to tell Isherwood what had happened if he would write a book about it on her behalf.

"Old, Vernon" (not his real name). American painter. During Isherwood's first visit to New York in 1938, George Davis introduced him to Vernon Old at an establishment called Matty's Cell House. Blond, beautiful, and intelligent, Vernon matched the description Isherwood had given Davis of the American boy he'd like to meet, and Vernon featured in Isherwood's decision to return to New York in 1939. They lived together in New York and Los Angeles until February 17, 1941, when they split by mutual agreement. Vernon then lived unsteadily on his own, painting, drinking, and being sexually promiscuous, until a suicide attempt later that year. During World War II, he tried to become a monk, first in a Catholic monastery in the Hudson Valley and later at the Hollywood Vedanta Society and at Ananda Bhavan in Montecito. Eventually, he turned to heterosexuality, married "Patty O'Neill" (not her real name) in November 1948, and had a son before divorcing. His painting career was increasingly successful, and in the late 1950s he tutored Don Bachardy. He appears in *Christopher and His Kind* and in *My Guru and His Disciple* (as "Vernon," without a surname) and throughout *D.1* and *Lost Years*.

Olivier, Laurence (1907–1989). British actor, director, producer; celebrated as the greatest Shakespearian actor of his time. He became a Hollywood star by the start of World War II in *Wuthering Heights* (1939), *Rebecca* (1940), *Pride and Prejudice* (1940), and *That Hamilton Woman* (1941), and was appointed co-director of the Old Vic with Ralph Richardson near the end of the war. In 1963, he became director of the National Theatre in Britain. He directed and produced himself in a number of movies, beginning with *Henry V* (1944) and *Hamlet* (1948), which together won him several Academy Awards for acting and directing, and he appeared in more than fifty other films and over a hundred stage roles in London, New York, and elsewhere. He was married three times, to actress Jill Esmond from 1930 to 1940, to Vivien Leigh until 1960, and then to Joan Plowright until his death. Isherwood became friendly with him during 1959 when Olivier was in Los Angeles filming *Spartacus* (1960). He appears in *D.1*.

One, Incorporated. Homosexual advocacy and support group founded in 1953; publisher of *One* magazine. *One* was the subject of a legal struggle with the U.S. Post Office during the 1950s; in 1958 the Supreme Court ruled that gay publications were not *a priori* obscene and could be sent and sold by mail. The business manager at One, Inc., and its driving force for over a decade, was Bill Legg. He had several names: Dorr Legg, William Lambert, and Marvin Culter.

O'Neill, Donna (d. 200[1]). A companion of Ivan Moffat. She was beautiful, an accomplished horsewoman, and married to a wealthy man who objected to her involvement with Moffat. She remained with her husband, and she spent many years in analysis. She often sat for Don Bachardy in the late 1950s and early 1960s. She appears in *D.1*.

Osborne, John (1929–1994). English playwright, born in Fulham, West London. He worked as a journalist briefly and then acted in provincial repertory until his third play, *Look Back in Anger* (1956), established him as the center of a generation of working-class realist playwrights called "the angry young men." During the 1950s, his work was mostly produced by George Devine and Tony Richardson's English Stage Company at the Royal Court. Other plays include *The Entertainer* (1957) starring Laurence Olivier, *Luther* (1961), *Inadmissable Evidence* (1964), *A*

Patriot for Me (1965), *West of Suez* (1971), *A Sense of Detachment* (1972), *Watch It Come Down* (1976), and *Déjàvu* (1991), a later sequel to *Look Back in Anger*. Several of his plays were filmed. Osborne also wrote the screenplays for Richardson's *Tom Jones* and *The Charge of the Light Brigade*. Their collaborations ended with the latter because Osborne was sued for plagiarizing Cecil Woodham-Smith's *The Reason Why*. The rights to her historical account belonged to Laurence Harvey, who agreed to sell them and abandon the suit if he could be in the film. So Richardson gave Harvey a small role previously promised to Osborne. Osborne and Richardson quarreled and never worked together again.

In *D.1*, Isherwood records that he met Osborne in Hollywood in 1960, when Osborne came to join Mary Ure (his second wife) and Richardson, both working there. In September 1961, as Isherwood tells in this diary, he and Bachardy were guests at La Baumette, the house which Osborne rented from Lord Glenconner during August and September in Valbonne in the South of France. About three weeks before the visit, Osborne's "A Letter to My Fellow Countrymen," sometimes called his "Damn You, England" letter, appeared in the *Tribune* sparking enormous controversy in the British press. Isherwood copied into his diary on August 20 a few phrases from Osborne's letter quoted by J.W. Lambert in *The Sunday Times* that day (p. 3), and he paraphrased reactions the paper printed on the same page from the Angry Young like Shelagh Delaney (b. 1939), playwright of *A Taste of Honey*, John Braine (1922–1986), whose first novel was *Room at the Top* (1957), and Arnold Wesker, and from more established figures like Oxford Regius Professor of Modern History, Hugh Trevor-Roper (1914–2003), known for *The Last Days of Hitler* (1947), and critic, broadcaster, novelist, and playwright J.B. Priestley (1894–1984).

Osborne married five times; first to Pamela Lane, an actress, then to Ure, then to Penelope Gilliatt. His fourth wife, from 1969 to 1978, was Jill Bennett (1931–1990), the British actress, who starred in several of his plays and died a suicide. Finally, he married Helen Dawson (d. 2004), drama critic and arts editor at *The Observer* during the 1960s, and remained with her until his death. He wrote three volumes of autobiography, *A Better Class of Person* (1981), *Almost a Gentleman* (1991), and *Damn You, England* (1994).

Page, Anthony (Tony) (b. 1935). Oxford-educated British actor and director, born in India. He was Artistic Director at the Royal Court from 1964 to 1973 and directed five plays there by John Osborne, three during 1968, when Bachardy contributed drawings to the programs. In 1966, he asked Isherwood to adapt Wedekind's Lulu plays, *Erdgeist* (*Earthsprite*, 1895) and *Die Büchse der Pandora* (*Pandora's Box*, 1904), but the project wasn't completed. He also directed productions of Ibsen, Tennessee Williams, Albee and others in the West End and at the National Theatre and in New York. He made a few movies, including *I Never Promised You a Rose Garden* (1977) and later became known for his television documentaries, biographies, and mini-series.

Pagli. Vedanta nun; her real name was Amelia Monsour. Her nickname, Pagli, means madwoman; she had a reputation for well-meaning loopiness, but was not mad. Isherwood struggled with the spelling of her name, but usually got it right. After sannyas she became Amitaprana. She left the convent to care for her sick father and became proprietor of Travels with Amelia, a Hollywood travel agency, but remained a Vedanta devotee.

Pakula, Alan (1928–1998). New York-born producer and director; educated at Yale School of Drama. He began his Hollywood career in the cartoon department at Warner Brothers in 1949. His first big success as a producer was *To Kill a Mockingbird* (1962). He also produced and directed *Klute* (1971), produced *Inside Daisy Clover* (1965), directed *All the President's Men* (1976), and directed, co-produced and scripted *Sophie's Choice* (1982). He was Hope Lange's second husband, from 1963 to 1969, and later married writer Hannah Boorstin. Isherwood habitually misspelled his surname as Pacula. He appears in *D.1*.

Paley, William (Bill) (1901–1990). American media mogul; son of a Ukrainian cigar manufacturer; educated at the University of Chicago and the University of Pennsylvania. He bought CBS in 1928 when it was a small radio network and developed it into the radio and T.V. giant which he ran for over fifty years. During World War II, he was deputy chief of psychological warfare for the Allies. He was a figure in American cultural and intellectual life, devoting time to the Museum of Modern Art in New York, to hospitals, universities, and think tanks. His second wife, Barbara (Babe) Cushing Mortimer (1915–1978), a Boston-born society beauty, was one of Truman Capote's closest friends from 1955 until 1975. She epitomized the glamor of the rich women Capote called his Swans, but she ended their friendship when she discovered that Capote was using her private life as material for his fiction.

Parker, Dorothy (1893–1967). American poet, short story writer, journalist, and literary critic; born in New Jersey. Celebrated for her wit and associated with the Algonquin Hotel in New York where for years she lunched with writer friends. Her brief first marriage was to a New York stockbroker, Edwin Parker. She contributed to *The New Yorker* from its debut and to many other American magazines. Her 1929 short story "The Big Blonde" won the O. Henry Prize. She wrote plays—*Close Harmony* (1929) with Elmer Rice and *Ladies of the Corridor* (1953) with Arnaud d'Usseau—and screenplays—notably *A Star is Born* (1937) and Hitchcock's *Saboteur* (1942) with her second husband, Alan Campbell. She protested the execution of Sacco and Vanzetti in 1927, covered the Spanish Civil War for *The New Masses*, and was involved in the founding of the Hollywood Anti-Nazi League and other Hollywood committees opposing fascism; she also supported the Civil Rights movement and willed her estate to the National Association for the Advancement of Colored People. She was blacklisted at the end of the 1940s and later testified before the House Un-American Activities Committee where, in contrast to many of her colleagues, she cited the First Amendment (freedom of speech) instead of the Fifth (the right not to serve witness against oneself). She took over Isherwood's teaching position at L.A. State College when he left, during the 1960s. Bachardy drew her portrait a number of times during the same period.

Parone, Edward (Ed). American stage director. He assisted Gordon Davidson with the professional Theater Group at UCLA, where he directed *Oh! What a Lovely War*. In 1967, he moved with Davidson to the Mark Taper Forum and ran New Theater for Now to develop new plays, including *A Meeting by the River* in 1972, directed by Jim Bridges. He stayed at the Mark Taper Forum for about twelve years, and was a director in residence and eventually associate artistic director, turning his hand to producing, writing, and editing. He was also assistant to the producer on *The Misfits* and directed for T.V.

Pavitrananda, Swami. Indian monk of the Ramakrishna Order; head of the Vedanta Society in New York, on the Upper West Side, and a trustee of the Ramakrishna Math and Mission. He spent many years in the order's editorial center, Advaita Ashrama, at Mayavati in the Himalayas. He often paid a month-long visit to Swami Prabhavananda during the summers. Other than Prabhavananda, he was Isherwood's favorite swami.

Peggy. See Kiskadden, Peggy.

Pfeiffer, Virginia (1902–1973). A sister-in-law and friend of Ernest Hemingway; born in St. Louis to a wealthy pharmaceutical manufacturer and raised on her family's 60,000 acre farm in Piggott, Arkansas. During the 1920s, she travelled with her older sister Pauline Pfeiffer in Europe; they befriended Hemingway and his first wife Hadley in Paris, and Pauline married Hemingway in 1927. Virginia was often with the Hemingways in Paris, Key West, Cuba, and Bimini during the 1930s, and she renovated a barn at the family property in Arkansas for Hemingway to work in. When it burned down, she renovated it again. She sometimes took charge of their two sons and of Hemingway's son from his first marriage. Hemingway left Pauline for Martha Gellhorn in 1940, and Pauline died at Virginia's house in Hollywood in 1951. Laura Archera lived with Virginia for many years before marrying Aldous Huxley. Virginia adopted two children, Juan and Paula, whom Laura helped her raise. After they married, the Huxleys settled in a house in Deronda Drive in the Hollywood Hills a few hundred yards from Virginia's house, and Virginia and Laura continued to be frequent companions. In May, 1961, both houses burned down in a bush fire; Virginia moved, in the autumn, to a nearby house in Mulholland Drive and invited the Huxleys to join her there. Aldous Huxley died in that house two years later.

Phipps, William Edward (Bill) (b. 1922). American actor, born in Indiana. He made his first Hollywood film in 1947, appeared in a number of science fiction movies in the 1950s, and continued to act in films and T.V. until 2000. He was a friend of Charles Laughton, who introduced him to Isherwood and Bachardy. He is mentioned in *D.1*.

Plomer, William (1903–1973). British poet and novelist born and raised in South Africa. He met Isherwood in 1932 through Stephen Spender after Spender showed Isherwood Plomer's poems and stories about South Africa and Japan. Plomer was a friend of E.M. Forster and soon introduced Isherwood. In South Africa, Plomer and Roy Campbell had founded *Voorslag* (Whiplash), a literary magazine for which they wrote most of the satirical material (Laurens van der Post was also an editor). Plomer taught for two years in Japan, then, in 1929, settled in Bloomsbury where he was befriended by the Woolfs who had published his first novel, *Turbott Wolff* (1926). In 1937, he became principal reader for Jonathan Cape where, among other things, he brought out Ian Fleming's James Bond novels. During the war he worked in naval intelligence. He also wrote libretti for Benjamin Britten, notably *Gloriana* (1953). A 1943 arrest for soliciting a sailor in Paddington station was hushed up, but led Plomer to destroy correspondence with homosexual friends and to practice extreme circumspection in his private life. He lived with Charles Erdmann, who was born in London of a German father and Polish mother, raised in Germany from about age five, and returned in 1939 to England where he worked as a waiter and pastry-cook. He appears in *D.1*.

Plowright, Joan (b. 1929). British actress; trained at the Old Vic Theatre School. She first appeared on the London stage in 1954 and joined Tony Richardson and George Devine's London Stage Company in 1956. Isherwood met her when she was in the New York production of *A Taste of Honey* in 1960. During the same year, she starred with Laurence Olivier in *The Entertainer* at the Royal Court; she then appeared with Olivier in the film, and later in the stage and film versions of Chekhov's *Three Sisters*. She became Olivier's third wife in 1961, and they had two children. In the early 1960s, she and Olivier joined the National Theatre Company when it was founded at the Old Vic in London, and she played leads in Chekhov, Shaw, and Ibsen. Later, she played Shakespeare and had a long West End run in De Filippo's *Saturday, Sunday, Monday* (1973) followed eventually by his *Filumena*. She also appeared in the stage (1973) and film (1977) versions of *Equus*. Among her other movies are *Moby Dick* (1956), *Uncle Vanya* (1963), *101 Dalmatians* (1996), and *Tea with Mussolini* (1999). She appears in *D.1*.

Prabhaprana (Prabha) (d. 1998). Originally Phoebe Nixon, she was the daughter of Alice Nixon (Tarini) and after sannyas became Pravrajika Prabhaprana. The Nixons were wealthy southerners. Isherwood first met Prabha in the early 1940s at the Hollywood Vedanta Society, where she handled much of the administrative and secretarial work, and he grew to love her genuinely. By the mid-1950s, Prabha was manager of the Sarada Convent in Santa Barbara. She appears in *D.1*.

Prabhavananda, Swami (1893–1976). Hindu monk of the Ramakrishna Order, founder of the Vedanta Society of Southern California based in Hollywood. Gerald Heard introduced Isherwood to Swami Prabhavananda in July 1939. On their second meeting, August 5, Prabhavananda began to instruct Isherwood in meditation; on November 8, 1940 he initiated Isherwood, giving him a mantram and a rosary. From February 1943 until August 1945 Isherwood lived monastically at the Vedanta Society, but decided he could not become a monk as Swami wished. He continued to be closely involved with the Vedanta Society, travelled twice to its headquarters in India, and remained Prabhavananda's disciple and close friend for life. Their relationship is described in *My Guru and His Disciple*, and Prabhavananda appears in *D.1* and *Lost Years*, as well as providing inspiration for *A Meeting by the River*.

Prabhavananda was born Abanindra Nath Ghosh, in a Bengali village northwest of Calcutta. As a teenager he read about Ramakrishna and his disciples Vivekananda and Brahmananda and felt mysteriously attracted to their names. By chance he experienced an affecting meeting with Ramakrishna's widow, Sarada Devi. At eighteen he visited the Belur Math, the chief monastery of the Ramakrishna Order beside the Ganges outside Calcutta. There he had another important encounter, this time with Brahmananda, and abandoned his studies for a month to follow him. When he returned to Calcutta, he became involved in militant opposition to British rule, and joined a revolutionary organization for which he wrote and distributed propaganda. At one time, he took charge of some stolen weapons, and some of his friends who engaged in terrorist activities met with violent ends. Because he was studying philosophy, Abanindra attended Belur Math regularly for instruction in the teachings of Shankara, but he regarded the monastic life as escapist and put his political duties first, until he had another compelling experience with Brahmananda and suddenly decided to give up his political activities and become a monk. He took his final vows in 1921, when his name was changed to Prabhavananda.

In 1923 he was sent to the U.S. to assist the swami at the Vedanta Society in San Francisco; later he opened a new center in Portland, Oregon. He was joined there by Sister Lalita and later, in 1929, founded the Vedanta Society of Southern California in her house in Hollywood, 1946 Ivar Avenue. Several other women joined them. By the mid-1930s the society began to expand and money was donated for a temple which was built in the garden and dedicated in July 1938. Prabhavananda remained the head of the Hollywood society until he died; he frequently visited the Ramakrishna monastery in Trabuco and the Sarada Convent in Santa Barbara and also stayed in the home of a devotee in Laguna Beach.

Isherwood and Prabhavananda worked on a number of books together, notably translations of the Bhagavad Gita (1944) and of the yoga aphorisms of Patanjali (1953). Prabhavananda contributed to two collections on Vedanta edited by Isherwood, and Isherwood also worked on Prabhavananda's translation of Shankara's *Crest Jewel of Discrimination* (1947). Prabhavananda persuaded Isherwood to write a biography of Ramakrishna, *Ramakrishna and His Disciples* (1964); this became an official project of the Ramakrishna Order and was subject to chapter-by-chapter review by a high authority at the Belur Math.

pranam. Greeting of respect made by folding the palms, by taking the dust of the feet (i.e., touching the greeted one's foot and then touching one's own forehead), or by prostrating. Namaskar, from the same Sanskrit root "nam" for a salutation expressing love and respect, can also be made in a variety of ways depending upon local tradition and social situation: verbally, by nodding, by folding the palms, by bending at the knees to touch the ground with the forehead, or by lying flat on the ground.

prasad. Food or any gift consecrated by being offered to God or a saintly person in a Hindu ceremony of worship; the food is usually eaten as part of the meal following the ritual, or the gift given to the devotees.

Prema Chaitanya (Prema) (1913–2000). American monk of the Ramakrishna Order, originally named John Yale and later known as Swami Vidyatmananda; born in Lansing, Michigan, educated at Olivet College in Illinois, Michigan State College, and later at the University of Southern California where he obtained a doctorate in education. He taught high school before moving to Chicago in 1938 and working as a schoolbook editor. In 1941, he tried unsuccessfully to join the navy, and then in 1942 he joined Science Research Associates (SRA), a publishing house that specialized in teaching tools and psychological tests, and which was later bought by IBM. He ran SRA during the war and eventually became a director. While he was there, the entire male staff was interviewed by Alfred Kinsey for his work on *Sexual Behavior in the Human Male*, so Yale recorded his sexual history with one of Kinsey's assistants and was made ill by recounting it. He decided to give up sex, evidently because he was homosexual.

In 1948, after reading Isherwood and Prabhavananda's translation of the Bhagavad Gita and other works on Vedanta, he moved to Los Angeles, where he began instruction with Prabhavananda that November and moved into the Vedanta Society in Hollywood in April 1950. Isherwood met him at the Vedanta Society in the spring of 1949. In August 1955, Yale took his brahmacharya vows at Trabuco and was renamed Prema Chaitanya. He continued to live at the Hollywood society and briefly at Santa Barbara but never at Trabuco. He developed the Vedanta Society's bookshop, building a mail-order business. He also edited the Vedanta Society magazine, *Vedanta and the West*, collaborating with Isherwood on the magazine's chapter-by-chapter publication of Isherwood's biography of Ramakrishna. Prema's own 1961 book, *A Yankee and the Swamis*, describing his journey in 1952–1953 to the Ramakrishna monastery and various holy places in India, was also published serially in *Vedanta and the West* and caused a scandal with its remarks about the residents of the various places Prema visited; offending passages were deleted by Swami Prabhavananda from the final text. Prema also edited *What Religion Is: In the Words of Swami Vivekananda*, for which Isherwood wrote the introduction, and *What Vedanta Means to Me* (1960), to which Isherwood also contributed. He appears in *D.1* and *Lost Years*, and Isherwood drew on his sannyas experience in India in 1963–1964 for *A Meeting by the River*, originally intended to be dedicated to him.

In his unpublished memoir "The Making of a Devotee," Vidyatmananda tells that he became estranged from Prabhavananda while living at the Belur Math in India, where he discovered that the family model of religious life adopted by Swami Prabhavananda for the Hollywood Vedanta Society was not approved by the Ramakrishna Order, which advocated a strict monastic model with the sexes separated—officially only men could join the order. Swami had initiated many women and allowed them to live alongside the men because there were not enough devotees in southern California to populate two separate orders; Vidyatmananda began lobbying to have the nuns turned out of the Hollywood Vedanta Society. He also expressed surprise over his discovery in India that despite being a Westerner, he might still hope to rise through the hierarchy of the order, and even be permitted to transfer away from the Hollywood center. He remained in India for nearly a year, hoping to live there permanently, but he failed to find useful work, and returned to Hollywood after falling severely ill with paratyphoid. Swami Prabhavananda felt Prema's discussions with the elders of the Belur Math were disloyal, and when in 1966 Prema was invited to transfer to the Centre Védantique Ramakrichna, east of Paris in Gretz, France, Swami wrote saying he wanted nothing more to do with him. They met again, however, in 1973, and Prabhavananda gave Vidyatmananda his blessing.

Vidyatmananda remained in Gretz for the rest of his life. When he arrived the center was in decline; the first generation of devotees had died or left following the death of the founding Swami, Siddheswarananda, and only a few new devotees had appeared. Vidyatmananda, as manager, saw to reorganizing, rebuilding, and modernizing the property; he learned how to speak French and how to farm. The center was run as an ashram, and eventually thrived on the physical labor of young spiritual trainees who generally returned to secular life after a period of retreat there.

presidential election 1960. As Isherwood records in his diary entry for November 15, the result was not confirmed for some weeks. Kennedy won by just over 100,000 popular votes out of more than 68 million cast. His margin in California was about 40,000 until California's more than 150,000 absentee ballots were counted on November 17, and it turned out that Nixon, not Kennedy, had won California with its thirty-two electoral votes. Six of Alabama's eleven electors were "unpledged" and could vote as they liked when the Electoral College convened. Hawaii's three votes, which eventually went to Kennedy, were still under dispute in January, and the Republicans challenged the vote in Illinois (twenty-seven electoral votes), Texas (twenty-four electoral votes), and nine other states. But they did not press for a recount. When the Electoral College met in December, Kennedy won 303 of the 522 votes.

presidential election 1968. The Twelfth Amendment to the U.S. Constitution provides for equal votes in the Electoral College to be decided by the House of Representatives. The Democrats voted down the minority peace plank at their 1968 convention and nominated Johnson's vice-president, Hubert Humphrey, whose position on the war was much like Johnson's. This left antiwar voters with no candidate, so, as Gore Vidal evidently told Isherwood, McCarthy considered running as a fourth-party candidate with a view to extracting a peace promise from Humphrey in the event of a tie. The third party was George Wallace's American Independent Party.

Preston, Jonathan. A young Englishman introduced to Isherwood by Phil Burns when Preston came to live in Los Angeles towards the end of 1958 with a Canadian companion, John Durst. Isherwood found him attractive and they became friends. Later, Preston returned to England, where he became a publicist, and Isherwood occasionally met him there. He appears in *D.1.*

Prince, Hal (b. 1928). American theatrical producer and director; born in New York, educated at the University of Pennsylvania; winner of more than twenty Tony Awards for his many Broadway hits. He expressed an interest in 1959 in an Auden-Isherwood-Kallman musical based on *Goodbye to Berlin*, but the project didn't progress. In 1966, he produced and directed the spectacularly successful version which he commissioned from Joe Masteroff (book), Fred Ebb (lyrics), and John Kander (music).

Prosser, Lee (b. 1944). American author, painter, musician, Vedanta devotee, student of ancient religions, shamanism, witchcraft, and Wicca. He composed "The Ramakrishna Waltz" and wrote a memoir, *Isherwood, Bowles, Vedanta, Wicca, and Me* (2001). He married three times: first to Mary, from whom he was divorced in 1970, then to Grace, with whom he had two daughters, and, much later, to Debra.

puja. Hindu ceremony of worship, a watch or vigil; usually offerings—flowers, incense, food—are made to the object of devotion, and other ritual, symbolic acts are also carried out depending upon the occasion.

Rabb, Ellis (1930–1998). American actor, director, and producer, from Memphis, Tennessee. He appeared in and directed Shakespeare, Chekhov, Shaw, Pirandello and contemporary drama, including Tennessee Williams. In 1959, he founded a group called the Association of Producing Artists (APA) of which he was Artistic Director. The APA worked in affiliation with The Phoenix Repertory Company, and many of their productions had Broadway runs in the 1960s, usually as a group of plays in repertory, including *You Can't Take It With You*, *Right You Are If You Think You Are*, *The Wild Duck*, *War and Peace*, *The Show Off*, *The Cherry Orchard*, *Pantagleize*, *The Cocktail Party*, *The Misanthrope*, and *Private Lives*. Rabb starred in some of the productions. His wife from 1960 to 1967, actress Rosemary Harris, was also in the company. The APA-Phoenix lasted until 1969. In their penultimate Broadway run, Rabb himself played Hamlet (March 3, 1969–April 26, 1969), evidently replacing the actor Isherwood saw in the role in San Diego in May 1968. Afterwards, Rabb continued to act and direct, and he won a Tony Award for his 1975 production of *The Royal Family*.

rajas. See guna.

Ramakrishna (1836–1886). The Hindu holy man whose life and teachings were central to the modern renaissance of Vedanta. He was widely regarded as an incarnation of God. Ramakrishna, originally named Gadadhar Chattopadhyaya, was born in a Bengali village sixty miles from Calcutta. He was a devout Hindu from boyhood, practised spiritual disciplines such as meditation, and served as a priest. He was a mystic and teacher, and in 1861 he was declared an avatar: a divine incarnation sent to reestablish the truths of religion and to show by his example how to ascend towards Brahman. Ramakrishna was also initiated into Islam, and he had a vision of Christ. His behavior was sometimes highly unconventional, in keeping with his beliefs and with the extreme spiritual practises which he undertook. For instance, in youth, he put his tongue to the flesh of a rotting corpse as part of his Tantric discipline, and in order to emulate the gopis, he undertook *madhura bhava*, identifying himself as a female devotee of Krishna, assuming a feminine attitude, and actually dressing in women's clothes. He several times danced with drunkards because their reeling reminded him of his own when he was in religious ecstasy. His followers gathered around him at Dakshineswar and later at Cossipore. His closest disciples, trained by him, later formed the nucleus of the Ramakrishna Math and Mission, now the largest monastic order in India. Ramakrishna was worshipped as God in his lifetime; he was conscious of his mission, and he was able to transmit divine knowledge by a touch, look, or wish. Isherwood's biography, *Ramakrishna and His Disciples* (1964), was written with the help and encouragement of Swamis Prabhavananda and Madhavananda.

Ram Nam. A sung service of ancient Hindu prayers which invoke the divinities Rama, his wife Sita, and the leader of Rama's army, the monkey god, Hanuman. In Ramakrishna practise, Ram Nam is sung on Ekadashi, the eleventh day after the new or full moon, generally observed with worship, meditation and fasting.

Ranganathananda, Swami (1908–2005). Hindu monk, born in Kerala.

President of the Ramakrishna Order from 1998 until his death and Vice-President from 1989; before then he had prominent roles at other centers, including Secretary of the Delhi center, which he made a gathering place for Indian intellectuals, from 1949 to 1962; President of the Ramakrishna Mission Institute of Culture in Calcutta from 1962 to 1967; and President of the Hyderabad Center for a long period beginning in 1967. He was a self-taught Sanskrit scholar and an orator, and he lectured extensively in over fifty countries. Many of his lectures were recorded for sale and he also published a number of books. He suffered extreme privation in Burma during World War II; afterwards, in order to be able to travel, he kept a very strict personal diet, which Isherwood mentions, of bananas and milk made up with Horlick's, a powder of dried milk and malted cereals.

Rattigan, Terence (1911–1977). British playwright and screenwriter; educated at Oxford. He wrote mostly comedy at the start of his career, including *French Without Tears* (1936). After World War II, during which he served as an air-gunner, he turned to social and psychological drama, achieving repeated acclaim with *The Winslow Boy* (1946), *The Browning Version* (1948), *The Deep Blue Sea* (1952), *Separate Tables* (1954), *Man and Boy* (1963), *A Bequest to the Nation* (1970), *In Praise of Love* (1973), *Cause Célèbre* (1977), and others. He adapted a number of these plays for the screen, and he wrote numerous further films, among them *The Day Will Dawn* (1942), *Brighton Rock* (1947; based on Graham Greene's novel and called *Young Scarface* in the U.S.), and the musical remake of *Goodbye, Mr. Chips* (1969). Isherwood and Bachardy were introduced to Rattigan in London in 1956, and he appears in *D.1*.

Ray, Andrew (1939–2003). British child actor. His first movie was *The Mudlark* (1950), and he played Geoffrey in Tony Richardson's Broadway production of *A Taste of Honey*. He appeared in a made-for-T.V. "Great Expectations" (1975) and as George VI in the mini-series "Edward and Mrs. Simpson" (1978). His father, Ted Ray, was a violinist and a music hall and radio comedian who also appeared in a number of movies.

Rechy, John (b. 1934). American writer; born in El Paso, educated at the University of Texas and then at the New School for Social Research in New York. He served in the U.S. Army in Germany and for many years taught creative writing at the University of Southern California. His novels about the homosexual communities of New York and Los Angeles generally explore marginal themes of violence, drugs, and crime. *City of Night* (1963), which includes his story "The Fabulous Wedding of Miss Destiny," tells about his experiences as a hustler. Later novels are *Numbers* (1967), with thinly disguised portraits of Isherwood, Bachardy, and Gavin Lambert, *The Day's Death* (1970), *The Vampires* (1971), *The Fourth Angel* (1973), *Rushes* (1979), *Bodies and Souls* (1983), *Marilyn's Daughter* (1988), and *The Miraculous Day of Amalia Gómez* (1991). Rechy also wrote *The Sexual Outlaw* (1977), a documentary study of urban homosexual sexual practices. In his memoir, *About My Life and the Kept Woman* (2008), he tells about his trial, mentioned by Isherwood August 4, 1967. He was arrested in Griffith Park for oral copulation (a felony) with a partner who had a previous conviction for the same offense, but he was found guilty only of a misdemeanor and fined $1,000.

Redgrave, Vanessa (b. 1937). English star of stage and screen; she trained at London's Central School of Speech and Drama, made her stage debut in 1957, and established her reputation with the Royal Shakespeare Company in the early 1960s. Her films include *Morgan* (1966, Academy Award nomination), *Blow-Up* (1966), *A Man for All Seasons* (1966), *The Sailor from Gibraltar* (1967), *Camelot* (1967), *The Charge of the Light Brigade* (1968), *Isadora* (1968, Academy Award nomination), *Oh! What a Lovely War* (1969), *Mary, Queen of Scots* (1971, Academy Award nomination), *Murder on the Orient Express* (1974), *Julia* (1977, Academy Award), *The Bostonians* (1984, Academy Award nomination), *Prick Up Your Ears* (1987), *The Ballad of the Sad Café* (1991), *Howards End* (1992), *Mission Impossible* (1996), *Mrs. Dalloway* (1998), *Girl, Interrupted* (1999), *The Cradle Will Rock* (1999), *Running with Scissors* (2006), *Atonement* (2007). Her stage roles are too numerous to name, and she has often appeared on T.V. Much of her work during the 1960s was for Tony Richardson, whom she married in 1962 and with whom she had two daughters, actresses Natasha Richardson (1963–2009) and Joely Richardson (b. 1965), before divorcing in 1967. In 1969, she had a son with actor Franco Nero, whom she married in 2007. She is well-known for her leftist political activism and has unsuccessfully run for parliament as a member of the Workers' Revolutionary Party.

Regester, James Robert (Bob) (19[32]–1987). American theatrical producer and advertising executive, from Bloomington, Indiana. He met Tony Richardson in Los Angeles in the 1960s and worked for him in Europe as a member of the production team for *Mademoiselle* (1966) and *The Sailor from Gibraltar* (1967). He became a longtime companion of Neil Hartley, and they shared a house in Maida Avenue. With financial backing from a friend, Louis Miano, he co-produced *Design for Living* with Vanessa Redgrave, Jeremy Brett, and John Stride at the Phoenix Theatre in 1973; *The Seagull*, in 1985, starring Vanessa Redgrave and Natasha Richardson; Gerald Moon's *Corpse*, with Keith Baxter and Milo O'Shea in 1984; and *Legends*, starring Mary Martin and Carol Channing, which toured in the U.S. in the mid-1980s. He died of AIDS.

Reinhardt, Gottfried (1911–1994). Austrian-born producer. He emigrated to the U.S. with his father, Max Reinhardt, and became assistant to Walter Wanger. Afterwards he worked as a producer for MGM from 1940 to 1954 and later directed his own films in the United States and Europe; his name is attached to many well-known films, including Garbo's *Two-Faced Woman* which he produced in 1941 and *The Red Badge of Courage* which he produced in 1951. He was Salka Viertel's lover for nearly a decade before his marriage to his wife, Silvia, in 1944. Through Salka and Berthold Viertel, Reinhardt gave Isherwood his second Hollywood film job in 1940, and he remained Isherwood's favorite Hollywood boss. During the war, he enlisted and wrote scenarios for films on building latrines, preventing venereal disease, cleaning rifles, etc. Reinhardt and his wife eventually returned to Germany and settled near Salzburg. He appears in *D.1* and *Lost Years*.

Reinhardt, Wolfgang (1908–1979). Film producer and writer; son of Max Reinhardt, brother of Gottfried Reinhardt. He produced *My Love Come Back* (1940), *The Male Animal* (1942), *Three Strangers* (1946), *Caught* (1948), and

Freud (1962), for which he was nominated for an Academy Award as co-writer. Isherwood probably met Wolfgang Reinhardt through Gottfried soon after arriving in Hollywood, and as he records in *D.1* and in *Lost Years*, he and Wolfgang tried to work together several times during the 1940s when Wolfgang was a producer at Warner Brothers. With Aldous Huxley in 1944, they discussed making *The Miracle*, a film version of the play produced by Max Reinhardt in the 1920s, and in 1945 Wolfgang hired Isherwood to work on Maugham's 1941 novel *Up at the Villa*, but neither film was ever made. Much later, in June 1960, Wolfgang approached Isherwood to write a screenplay based on Felix Dahn's four-volume 1876 novel, *Ein Kampf um Rom* (*A Struggle for Rome*), about the decline and fall of the Ostrogoth empire in Italy in the sixth century, but Isherwood turned the project down. Wolfgang's wife was called Lally.

Rembrandt. The drawing Isherwood mentions in his entry for August 11, 1962, of the angel leaving Manoah is probably "Manoah's Offering" (*circa* 1639, Berlin, Kupferstichkabinett; cataloged as Number 180 and shown as Figure 210 in Otto Benesch, *The Drawings of Rembrandt* (1954), vol. 1, p. 49). It was exhibited in Berlin in 1930, and Don Bachardy possesses a small book of forty-eight Rembrandt drawings, *Rembrandt Handzeichnungen* (Leipzig), which includes it under the title "Der Angel verläßt Manoah und sein Weib" ("The Angel Departs from Manoah and His Wife"); the book is inscribed "Christopher from William / Christmas 1936" in, evidently, the hand of William Plomer.

Renaldo, Tito. Mexican actor; he played the first son in *Anna and the King of Siam* (1946). Isherwood met him in 1947, apparently through Bill Caskey, who first came across him when Renaldo was a companion of Cole Porter. Renaldo took up Vedanta as a disciple of Swami Prabhavananda and for a time lived at Trabuco as a monk; he left and returned to Vedanta many times. He was an exceptional cook and in the late 1950s and 1960s worked in Carlos McClendon's shop in West Hollywood. He suffered from severe asthma. In the 1970s, in frail health, he returned for good to his family in northern Mexico. He appears in *D.1* and *Lost Years*.

Renate. See Druks, Renate.

Reventlow, Lance (1936–1972). American Grand Prix driver, son of Barbara Hutton, the Woolworth heiress, and her second of six husbands, Count Curt Haugwitz-Reventlow, a Danish aristocrat. Reventlow was born in London, and his mother built Winfield House in Regent's Park, now the official residence of the U.S. Ambassador to Britain, to protect him from kidnapping threats when he was a baby. He was brought back to the U.S. during World War II when his parents divorced, though his mother continued to travel constantly. He began racing in California at nineteen and shared his hobby with the actor James Dean, whom he saw on the day of Dean's death. His first Grand Prix start was in the Belgian Grand Prix in 1960. After racing Maseratis and Coopers, he began building his own cars, and produced the Scarab sports car during the 1950s. His Formula 1 model was less successful, and after producing a third car during the 1960s he eventually lost interest in racing. He was briefly married to the actress Jill St. John. Reventlow died in a small-plane crash in Colorado. He appears in *D.1*.

Richard. See Isherwood, Richard Graham Bradshaw.

Richardson, Tony (1928–1991). British stage and film director; educated at Oxford where he was president of the Oxford University Dramatic Society. During the 1950s, he was a T.V. producer for the BBC, wrote about film for *Sight and Sound*, and was a founder of the Free Cinema movement, collaborating with Karel Reisz on a short, *Momma Don't Allow* (1955). He co-founded The English Stage Company with British actor and director George Devine (1910–1966) and under its auspices directed John Osborne's *Look Back in Anger* at the Royal Court in 1956. Then he and Osborne formed a film company, Woodfall, and Richardson went on to make movies, many adapted from his stage productions. In 1960, when Isherwood first mentions him in *D.1* (he also appears in *Lost Years*), Richardson was involved with Wyatt Cooper, a young actor, and he was directing for screen and stage virtually simultaneously. He was filming *Sanctuary* (1961)—amalgamated from Faulkner's *Sanctuary* (1931) and its sequel, *Requiem for a Nun* (1951), which he had already staged separately at the Royal Court in London in 1957—and he was also directing Shelagh Delaney's *A Taste of Honey* in New York with a mostly English cast brought over from London. As Isherwood tells in this diary, he worked for Richardson on film scripts of Evelyn Waugh's 1948 novel *The Loved One* (1965), Carson McCuller's *Reflections in a Golden Eye* (later directed by John Huston with a different script), *The Sailor from Gibraltar* (1967) based on Marguerite Duras' novel, and, with Don Bachardy, adaptations of Robert Graves's *I, Claudius* and *Claudius, the God*, though much of the work was never used. Richardson's other films include *The Entertainer* (1960), *A Taste of Honey* (1961), *The Loneliness of the Long Distance Runner* (1962), *Tom Jones* (1963, Academy Award), *The Charge of the Light Brigade* (1968), *Hamlet* (1969), *Ned Kelly* (1970), *Joseph Andrews* (1977), *The Hotel New Hampshire* (1984), and *Blue Sky* (released posthumously, 1994). He was married to Vanessa Redgrave from 1962 to 1967 and had two daughters with her, and he had a long affair with Grizelda Grimond, producing a third daughter in 1973.

Rickards, Jocelyn (1924–2005). Australian-born artist and costume designer, she attended art school in Sydney, lived at the artists' boarding house Merioola with Alec Murray, and became friends there with Loudon Sainthill and Harry Tatlock Miller. At the end of the 1940s, she moved to London, where she became a lover of the philosopher A.J. Ayer who introduced her to his London literary friends during the early 1950s. She painted decorative murals, designed theater costumes as Sainthill's assistant, then assisted Roger Furse on the film *The Prince and the Showgirl* (1957). She worked on several of Tony Richardson's films—*Look Back in Anger* (1959), *The Entertainer* (1960), *The Sailor from Gibraltar* (1967, Academy Award nomination)—and also on *From Russia With Love* (1963), *Blow-Up* (1966), *Ryan's Daughter* (1970), and *Sunday, Bloody Sunday* (1971). From early 1960 until the autumn of 1961, she lived with John Osborne who was then still married to Mary Ure and who, when Isherwood met them, was leaving Rickards for his second wife, Penelope Gilliatt. Graham Greene was another lover. In 1963, she married the painter Leonard Rosoman; later, she moved in with the film director Clive Donner and married him in 1971.

Ritajananda, Swami. Indian monk of the Ramakrishna Order; chief assistant to Swami Prabhavananda at the Hollywood Vedanta Society from 1958 to 1961. He then went to France to run the Vedanta Center at Gretz, near Paris, until his death in 1994. As his assistant at Gretz, he later took on Prema, by then called Swami Vidyatmananda.

Robbins, Jerome (Jerry) (1918–1998). American dancer, choreographer, theater and film director, born and raised in New York where he studied dance and appeared on Broadway before joining Ballet Theater in 1940. He collaborated with Leonard Bernstein on *Fancy Free* and *On the Town* in 1944 and went on to choreograph many Broadway hits, including *The King and I* (1951), *The Pajama Game* (1954), *Peter Pan* (1954), *West Side Story* (1957), *Gypsy* (1959), and *Fiddler on the Roof* (1964). In Hollywood, he worked on *The King and I* and *West Side Story*, for which he shared an Academy Award for Best Director in 1961. He worked simultaneously in classical ballet, sometimes with George Balanchine, and became a Ballet Master with The New York City Ballet in 1972. He ran a touring company, "Ballets: USA," from 1958 to 1962, which he brought to London in 1961 when Isherwood and Bachardy were there. The company performed many of his own works, including *Afternoon of a Faun* (1953, to Debussy's *Prélude à l'après-midi d'un faune)*, and *The Cage* (1951), both of which Isherwood mentions.

Roberts, Rachel (1927–1980). Welsh-born actress, educated at the University of Wales and RADA. She had many stage roles, beginning in 1951. Her films included *Saturday Night and Sunday Morning* (1960), *This Sporting Life* (1963, Academy Award nomination), *O Lucky Man!* (1973), *Murder on the Orient Express* (1974), and *When a Stranger Calls* (1979). She also appeared regularly on American T.V. in "The Tony Randall Show" from 1976 to 1978. Her second marriage, in 1962, was to Rex Harrison; they divorced in 1971.

Robinson, Bill. A young man with whom Isherwood became friendly during 1958. Robinson was then in analysis, and soon settled into a successful longterm relationship. He appears in *D.1*.

Rodia, Sabato (1879–1965). Italian-born construction worker, tiler, and telephone-line repairman. He worked from 1921 to 1948 in a lot beside his house in Watts, building seven towers of steel rods, reinforced cement, and wire mesh, which he decorated with tile fragments, broken dishes and bottles, sea shells, car parts, corncobs, fruit, and other objects, and titled "Nuestro Pueblo" (Our Town). His remark, which Isherwood quotes in his entry for March 19, 1969, about doing something big was recorded by William Hale, a student, who made a 1952 documentary about Rodia.

Roerick, Bill (1912–1995). American actor. Isherwood met him in 1943 when John van Druten brought Roerick to a lecture at the Vedanta Society. He was in England as a G.I. during World War II and became friends there with E.M. Forster, J.R. Ackerley, and others. His companion for many years was Tom Coley. In 1944, Roerick contributed a short piece to *Horizon* defending Isherwood's new way of life in America after Tony Bower had made fun of it in a previous number. He appears in *D.1* and *Lost Years*.

Ronnie. See Knox, Ronnie.

Rorem, Ned (b. 1923). American composer and writer, born in Indiana and raised in Chicago. His mother was a civil rights activist and his father a medical economist whose work formed the basis for Blue Cross; both were Quakers. Rorem attended the Music School of Northwestern University before winning a scholarship to the Curtis Institute in Philadelphia; he continued his studies at Juilliard. He was Virgil Thomson's copyist in the late 1940s and in 1949 moved to France where he entered Parisian aristocratic circles and pursued his own eccentric, dissipated adventures. He chronicled this life in *The Paris Diary* (1966) and published four further volumes of diaries as well as books of lectures and criticism. His musical compositions, which are widely performed and recorded, include three symphonies, four piano concertos, many other orchestral and chamber works, six operas, a variety of choral works, ballet and theater music, and the hundreds of songs and song cycles for which he is best known.

Rosen, Bob. Assistant director and production manager on various T.V. serials, and, later, a producer. He was a casual friend who lived near Isherwood in Santa Monica and was often on the beach.

Ross, Jean (1911–1973). The original of Isherwood's character Sally Bowles in *Goodbye to Berlin*. He met her in Berlin, possibly in October 1930, but certainly by the start of 1931. She was then occasionally singing in a night club, and they shared lodgings for a time in Fräulein Thurau's flat. Ross's father was a Scottish cotton merchant, and she had been raised in Egypt in lavish circumstances. After Berlin, she returned to England where she became friendly with Olive Mangeot, lodging in her house for a time. She joined the Communist party and had a daughter, Sarah (later a crime novelist under the name Sarah Caudwell), with the Communist journalist and author Claud Cockburn (1904–1981), though Ross and Cockburn never married. She appears in *D.1* and *Lost Years*.

Russell, Bertrand Arthur William, 3rd Earl Russell (1872–1970). English philosopher, mathematician, social critic, writer; educated at Trinity College, Cambridge, where he was a Cambridge Apostle; afterwards he worked as an academic. He published countless books and is one of the most widely read philosophers of the twentieth century. Chief among his awards and honors was the Nobel Prize for Literature in 1950. Throughout his life, Russell expressed his convictions in social and political activism. When he opposed British entry into World War I and joined the No-Conscription Fellowship, he lost his first job at Trinity, and he was fined and imprisoned more than once for his role in public demonstrations as a pacifist. Partly as a result, he became a visiting professor and lecturer in America and returned to Trinity as a Fellow only in 1944. Isherwood first met him through Aldous and Maria Huxley in late 1939 in Hollywood, as he tells in *D.1*. By December 1939, Russell had renounced pacifism because of the evils of fascism. In 1949, he began to champion nuclear disarmament. In 1958, he helped to found and was elected president of the Campaign for Nuclear Disarmament, but he resigned in 1960 to launch his more militant Committee of 100 for Civil Disobedience Against Nuclear Warfare, which sponsored several public protests in 1961. A week before the September 17 demonstration which Isherwood mentions, Russell and more than thirty other committee members (including three Royal Court playwrights, Arnold Wesker, Robert Bolt and Christopher Logue) were summoned to Bow Street magistrates' court and enjoined not to breach the peace by participating; they refused and were sentenced to a month in prison. Because he was eighty years old, Russell served only seven days, in Brixton.

Sachs, David (1921–1992). American philosopher and poet, born in Chicago, educated at UCLA and Princeton where he obtained his doctorate in 1953. He lectured widely and taught philosophy at a number of American and European universities, longest at Johns Hopkins. His essays on ethics, ancient philosophy, philosophy of the mind, literature, and psychoanalysis were published in many journals, as were his poems, and he edited *The Philosophical Review*. He appears in *Lost Years* as a participant in the Benton Way Group.

Sahl, Mort (b. 1927). Canadian-American comedian and political satirist, raised mostly in Los Angeles and educated at UCLA. As he tells in *D.1*, Isherwood first saw him perform in Los Angeles in July 1960; that year, Sahl was pictured on the cover of *Time Magazine* as the father of a new kind of comedy. He appeared in films and on T.V. and made many recordings.

Sainthill, Loudon (191[8]–1969). Australian-born theatrical designer. He painted Colonel De Basil's Ballet Russes de Monte Carlo on tour in Australia in the late 1930s and was invited to return with the company to London where he had a show at the Redfern Gallery in 1939. Afterwards, he designed sets and costumes for Helene Kirsova when she left the Monte Carlo Russian Ballet to set up her own company in Sydney, and he served in the Australian military during World War II. After the war, he lived with Harry Tatlock Miller at Merioola, and they produced books about Ballet Rambert and the Old Vic Theatre Company on tour. In 1949, they settled in London, where Sainthill designed sets and costumes for the Royal Shakespeare Company and others, including *The Tempest* (1951), Shaw's *The Apple Cart* (1953), *A Woman of No Importance* (1953), Rimsky-Korsakov's *Le Coq D'Or* at the Royal Opera House (1954), *Othello* (1955), *Tiger at the Gates* (1955), *Expresso Bongo* (1958), *Orpheus Descending* (1959), *Aladdin* (1959), *Belle or the Ballad of Dr. Crippen* (1961), *Canterbury Tales* (1967), and a handful of films. He collaborated on several books with Miller, and he taught stage design at the Central School of Arts and Crafts in London during the 1960s.

St. Just, Maria (1921–1994). Russian-born actress, Maria Britneva. She was educated in England and in 1956 married Peter Grenfell, 2nd Baron St. Just (1922–1984) the heir and putative son of English banker and politician Edward Grenfell. She was known for her obsessive friendship towards Tennessee Williams more than for her acting. Williams made her his literary executor, a role she played with ferocity until her death in the early 1990s. Isherwood met her through Williams, certainly by the mid-1950s, and she is mentioned in *D.1*.

samadhi. The state of superconsciousness, in which an individual can know the highest spiritual experience; absolute oneness with the ultimate reality; transcendental consciousness.

Sambuddhananda, Swami. Indian monk of the Ramakrishna Order. He was a disciple of Holy Mother and was inspired to become a monk by Swami Premananda, a disciple of Ramakrishna. Sambuddhananda was head of the Ramakrishna Math in Bombay from 1935 to 1965. He also served as secretary on several centenary committees. He died in the mid-1970s.

Sandwich, George. See Amiya.

sannyas. The second and final vows of renunciation taken in the Ramakrishna Order, at least four or five years after the brahmacharya vows. The sannyasin undergoes a spiritual rebirth and, as part of the preparation for this, renounces all caste distinctions. In *D.1*, Isherwood's entry for March 13, 1958 refers to the way in which Krishna (George Fitts) had first to join the Brahmin caste in order to have a caste to renounce; then Krishna had to imagine himself as dead, and to become a ghost in preparation for being reborn. At sannyas, the spiritual aspirant becomes a Swami and takes a new Sanskrit name, ending with "ananda," bliss. Thus, the new name implies "he who has the bliss of" whatever the first element in the name specifies, as in Vivekananda, "he who has the bliss of discrimination." A woman sannyasin becomes a pravrajika (woman ascetic), and her new name ends in "prana," meaning "whose life is in" whatever is designated by the first element of the name.

Sarada (d. 2009). American nun of the Ramakrishna Order; of Norwegian descent, originally called Ruth Folling. She studied music and dance and, while at the Vedanta Society, learned Sanskrit. Her father lived in New Mexico. Isherwood met her when he arrived in Hollywood in 1939; in the mid-1940s, she moved to the convent at Santa Barbara where he occasionally saw her. He tells about her in *D.1* and *Lost Years*. She was a favorite of Prabhavananda, but she suddenly left the Vedanta Society in October 1965 when she was forty-three years old. She married a few years later and became a painter. Prabhavananda was distressed by her abrupt departure and for a long time afterwards forbade her to be mentioned by her Sanskrit name, insisting she be called Ruth.

Sarada Convent, Montecito. In 1944, Spencer Kellogg gave his house at Montecito, near Santa Barbara, to the Vedanta Society of Southern California. The house was called "Ananda Bhavan," Sanskrit for Home of Peace. Kellogg, a devotee, died the same year, and the house became a Vedanta center and eventually a convent housing about a dozen nuns. During the early 1950s, a temple was built adjacent to the grounds.

Sarada Devi (1855–1920). Bengali wife of Ramakrishna; they married by arrangement when she was five years old. After the marriage, she returned to her family and he to his temple, and their relationship was always chaste although she later spent long periods of time living intimately with him. She became known as a saint in her own right and was worshipped as Holy Mother, the living embodiment of Mahamaya, of the Divine Mother, of the Goddess Sarasvati, and of Kali herself. Isherwood was initiated on Holy Mother's birthday, November 8, 1940.

Satprakashananda, Swami. Indian monk of the Ramakrishna Order. In *D.1*, Isherwood records meeting Swami Satprakashananda for the first time in 1957. He was head of the St. Louis Vedanta Center from 1938 until his death in the late 1970s.

Schary, Dore (1905–1980). American actor, writer, film producer, studio executive. He ousted Louis B. Mayer to run MGM in 1951, then formed his own production company. *Lonelyhearts* (1958) was his adaptation of Nathanael West's novella, *Miss Lonelyhearts*, and *Sunrise at Campobello* (1960) was based on his award-winning play about Roosevelt. He appears in *D.1*.

Schenck, Marvin (1897–1993). Studio executive, at MGM; a nephew of Hollywood bosses Joseph Schenck (Loews, United Artists, Twentieth Century-Fox) and Nicholas Schenck (Loews, MGM). He began work aged fifteen as an office boy for Marcus Loew Vaudeville Booking Agency in New York, joined the navy during W.W.I, then managed Loew's theaters in New York and New Jersey before joining MGM and, later, moving to Los Angeles. In 1954, when MGM hired Isherwood to write the script for *Diane*, Schenck formally questioned him about his loyalty to the U.S., and as he mentions in his diary entry for September 7, 1969, Isherwood always felt ashamed of answering. He had signed a petition requesting a review of the case against the "Hollywood Ten," who were jailed without appeal and blacklisted by the studios for refusing to answer questions before the House Un-American Activities Committee. The studios adopted the uniform position that they would not knowingly employ a Communist or a member of any group proposing to overthrow the U.S. government. Isherwood implied to Schenck that, upon reflection, he might not sign the petition again. He assured him that he had never been a member of the Communist party and affirmed his loyalty to the U.S. He also explained that his leftist sympathies in the 1930s, like those of so many of his contemporaries, had primarily reflected his strong opposition to fascism. He reiterated most of this in a subsequent letter written at the request and on behalf of Salka Viertel, but the letter denies any knowledge of the political activities or attitudes of Salka or her daughter-in-law, Virgina, who had divorced leftist writer Budd Schulberg to marry Peter Viertel.

Schenkel, John. A novice monk at the Hollywood Vedanta Society from about 1949; by 1952 he was living at Trabuco. He became Ananta Chaitanya in 1954 when he took his brahmacharya vows and then, around 1960, he fell in love and left Vedanta to get married. Eventually, he settled in India with his wife, an Indian woman he met in Mexico City. He appears in *D.1*.

Schlesinger, Peter (b. 1948). American painter, photographer, and, later, sculptor; born and raised in Los Angeles. In the summer of 1966, Schlesinger studied drawing with David Hockney at UCLA and became Hockney's lover and model. In 1968, they travelled together to England, where Schlesinger studied at the Slade. He remained in England for the better part of a decade, also travelling in Europe. Some of the photographs which he took during this period later appeared in his book, *A Checkered Past: A Visual Diary of the 60s and 70s* (2003), including many of Eric Boman, the photographer who became his lover and companion in 1971 and with whom he moved to New York in 1978.

Schneider, Peter (1950–2007). American writer, editor, and teacher, raised in Claremont, California, where he became friends with Jim Gates who first introduced him to Vedanta. He and Gates moved to Los Angeles together when they were about eighteen, at first living briefly with Schneider's father, Dr. Leonard Schneider, a Gestalt therapist and a psychology professor at L.A. State, and afterwards in a cabin on the canals in Venice. Around this time, Schneider attended Santa Monica College; he later got a degree in English at UCLA. He introduced himself and Gates to Isherwood over the telephone; Isherwood began driving them to Wednesday night meetings at the Vedanta Society and soon introduced them to Swami Prabhavananda, whereupon Schneider joined the society, toying with the idea of becoming a monk. Instead, he married Sumishta Brahm, a Vedanta devotee; the marriage was short, but he stayed away from the Vedanta Society for some years. Afterwards, he lived with Anya Cronin, also known as Anya Liffey, with whom he had two children in the 1980s. With Cronin, he wrote and produced musicals and other theater events for the Vedanta Society and elsewhere. He contributed to Vedanta publications and served as the society librarian in Hollywood. His Sanskrit name, given to him by Vivekaprana, was Hiranyagarbha. As P. Schneidre and later P. Shneidre, he published poems in *Paris Review*, *Rolling Stone*, *Antioch Review*, and others. He also founded and ran a literary press, Illuminati, publishing work by James Merrill, Charles Bukowski, Viggo Mortensen, and a book of Don Bachardy's drawings of artists, *70 x 1*.

Schorer, Mark (1908–1977). American literary critic, novelist, biographer, anthologist; educated at the University of Wisconsin. He taught at Dartmouth and Harvard before becoming Professor of English at Berkeley in 1945; he was department chair from 1960 until 1965. He lectured around the country on modern English and American literature, held distinguished research fellowships (including three Fulbrights, four Guggenheims, and a Bollingen), and served on the boards of various learned societies. His short stories appeared in *The New Yorker*, and he wrote three novels, *House Too Old* (1935), *The Hermit Place* (1941), and *The Wars of Love* (1954). He also published *William Blake: The Politics of Vision* (1946), *Sinclair Lewis: An American Life* (1961), and edited two widely used books for students of English, *The Harbrace College Reader* (1959, with Philip Durham, Everett L. Jones, Mark Johnston) and *Modern British Fiction: Essays in Criticism* (1961). He had a son and a daughter with his wife, Ruth.

Schreiber, Taft. American entertainment executive. He worked at the Hollywood talent agency MCA, and when Isherwood met him, he represented Charles Laughton, among others. MCA—Music Corporation of America—was founded in 1924 to book bands and actors but eventually expanded into all areas of entertainment. In the early 1960s, MCA bought Universal Pictures and its parent company Decca Records, started Universal Television, and dropped out of the talent business to avoid antitrust violations. Its subsidiaries eventually included publishers, record and video companies, cable T.V., real estate, and many other businesses. Schreiber and fellow MCA executives became extremely powerful (he was close to Ronald Reagan) and wealthy, and he became known, with his wife Rita, as an art collector and philanthropist. He is mentioned in *D.1*.

Schubach, Scott. A wealthy doctor who lived with Michael Hall for some years in West Hollywood.

Schwed, Peter (1911–2003). Isherwood's editor at Simon & Schuster; raised on Long Island and educated at Princeton. Eventually he became editorial chairman of the firm. He wrote several books himself, mostly about golf and tennis, and published a volume of his editorial correspondence with P.G. Wodehouse. Isherwood never genuinely felt that Schwed liked or understood his work although they worked together for about fifteen years. He appears in *D.1*.

Scott Gilbert, Clement. British would-be theatrical producer, he was wealthy and evidently became the owner of at least one theater. With Ernest Vadja, he created the characters for "Presenting Charles Boyer," an NBC radio show which ran a handful of times in 1950. He backed two 1961 productions staged in Croydon, Surrey: *Mother*, with David McCallum in the cast, and *Compulsion*. And he backed the proposed London production of *A Meeting by the River* in 1970.

Searle, Alan (1905–1985). Secretary and companion to Somerset Maugham from 1938; he was the son of a Bermondsey tailor and in youth had a cockney accent. Lytton Strachey was a former lover. When he first met Maugham in London in 1928, Searle was working with convicts—visiting them in prison and helping them to resettle in the community on release—but he told Maugham he wanted to travel. Maugham reportedly invited him on the spot to do so, but for a decade they met again only when Maugham was in London. Eventually, Searle devoted his life to Maugham and became his heir. He appears in *D.1*.

Sellers, Dr. Alvin. One of Isherwood's doctors. He had an office in Beverly Hills and was on the staff at Cedars of Lebanon Hospital. Isherwood first saw him in the mid-1950s, and he appears in *D.1*, where his name is spelled incorrectly as Sellars. (In 1961, Cedars of Lebanon merged with Mount Sinai Hospital and became Cedars-Mount Sinai Medical Center.)

Selznick, David O. (1902–1965). American movie producer, most famous for *Gone with the Wind* (1939). He also brought Alfred Hitchcock to Hollywood to direct *Rebecca* (1940). Among Selznick's many other movies are *King Kong* (1933), *David Copperfield* (1934), *Reckless* (1935), *Anna Karenina* (1935), *A Tale of Two Cities* (1935), *A Star Is Born* (1937), *The Prisoner of Zenda* (1937), *Intermezzo* (1939), *Spellbound* (1945), *Duel in the Sun* (1946), *Portrait of Jennie* (1948), and *The Third Man* (1949). He worked for his father's movie company until Lewis Selznick went bankrupt in 1923; in 1926, his father's former partner, Louis B. Mayer, hired him as an assistant story editor at MGM. Selznick soon moved to Paramount, then RKO, then back to MGM until 1935 when he formed Selznick International Pictures with John Hay Whitney. Selznick's aspirations were monumental, and he tried to control every detail of his pictures. Despite his box-office success, he went into debt, and by the end of the 1940s he had to close his companies. He married Louis B. Mayer's daughter Irene in 1931, and they had two sons, Jeffrey and Daniel, before separating in 1945. When their divorce was finalized in 1949, he married Jennifer Jones. During the 1950s, he took Jones to Europe to work, and her career absorbed him at the end of his life. He traded his rights in *A Star is Born* to get Jones the lead in *A Farewell to Arms* (1957); the film failed, and it proved to be his last. Isherwood worked for Selznick in 1958, developing a script for a proposed film, *Mary Magdalene*, and they became friends, as Isherwood records in *D.1*.

Selznick, Jennifer. See Jones, Jennifer.

Shawe-Taylor, Desmond (1907–1995). Dublin-born critic and author, educated at Oxford. He was assistant editor for the magazine of The Royal Geographical Society before starting to write book reviews for *The Times*, *New Statesman*, and *The Spectator*. During the war he served with the Royal Artillery and worked in intelligence. When he returned to his career, he concentrated on music, notably opera, and became the chief music critic for *The Sunday Times*, where he remained until 1983. He published a book about Covent Garden and, with Edward Sackville-West, edited *The Record Guide* (1951) and *The Record Year* (1953). He also broadcast for the BBC on music and records. He owned a house in Dorset with Raymond Mortimer and Edward Sackville-West.

Shroyer, Frederick B. (Fred) (191[7]–1983). Professor in the English department at Los Angeles State College, where he was responsible for Isherwood being hired to teach in 1959. He wrote novels—*Wall Against the Night* (1957), *Wayland 33* (1962), *There None Embrace* (1966)—and produced a number of college English books and anthologies—*College Treasury: Prose Fiction, Drama* (1956) edited with Paul Jorgensen, *Informal Essay* (1961), *Art of Prose* (1965) both with Paul Jorgensen, *Short Story: A Thematic Anthology* (1965) edited with Dorothy Parker, *Types of Drama* (1970) with Louis Gardemal, and *Muse of Fire: Approaches to Poetry* (1971) compiled with H. Edward Richardson. He appears in *D.1*.

Sino-Indian War. China invaded India across a disputed boundary, the McMahon Line, between Tibet and India, and also into Kashmir on October 20, 1962. As Isherwood mentions, Vengalil Krishnan Krishna Menon (1897–1974), a London-trained barrister and diplomat, Minister of Defense since 1957, lost his post, and after the war the Indian military was widely reformed. Nehru followed a policy of nonalignment throughout his time as Prime Minister of India, but he requested help from the West when this long-simmering border dispute erupted. A ceasefire was declared on November 20, when the Chinese evacuated back across the McMahon Line. But they retained territory in Kashmir.

Sleep, Wayne (b. 1948). British ballet dancer; educated at the Royal Ballet School; in 1966, he joined the Royal Ballet Company, where he became a Principal with numerous roles choreographed on him. He also appeared as a guest dancer with other ballet companies and starred in West End musicals, including *Cats* (1981). He is a choreographer and teacher and created his own review of dance, Dash, in which he toured world wide.

Smith, David. A young admirer of Isherwood's work; he occasionally paid court at the house in Adelaide Drive and once sat for Bachardy.

Smith, Dodie. See Beesley, Alec and Dodie Smith Beesley.

Smith, Emily Machell (Granny Emmy) (1840–1924). Isherwood's grandmother on his mother's side. Emily's husband, Isherwood's grandfather Frederick Machell Smith, was a wine merchant in Bury St. Edmunds; they married in 1864 and in 1885 moved to London with Kathleen, their only child, due to Emily's unpredictable health. She was beautiful, passionate about the theater, and liked to travel with Kathleen, who helped her to prepare a book of guided walks, *Our Rambles in Old London* (1895). Emily's maiden name was Greene; her brother Walter Greene was a prosperous brewer in Bury St. Edmunds, went into politics, and became a baronet. Walter Greene entertained lavishly at his country house, Nether Hall, and Kathleen enthusiastically attended house parties and dances there as a young woman. Through Emily's family, Isherwood was related to the novelist Graham Greene.

Smith, Katharine (Kate) (1933–2000). English second wife of Ivan Moffat, from 1961 until 1972. She was a daughter of the 3rd Viscount Hambleden whose family fortune derived from the book and stationery chain, W.H. Smith, and Lady Patricia Herbert, daughter of the 15th Earl of Pembroke, elder sister of Isherwood's Tangier friend, David Herbert, and lady-in-waiting to Queen Elizabeth the Queen Mother. Kate Smith was a bridesmaid to Princess Alexandra and a close friend of Princess Margaret. She had two sons with Moffat, Jonathan (b. 1964) and Patrick (b. 1968), a godson of Princess Margaret. In 1973, she married thriller-writer Peter Townend, author of *Out of Focus* (1971), *Zoom!* (1972), and *Fisheye* (1974).

Smith, Margot. An art student at Chouinard with Bachardy; she was about seventeen when they met in 1956, but she dressed in a sophisticated manner and occasionally modelled for a fashion illustration class. She later married a screenwriter and editor, Sam Thomas. Bachardy drew her many times over the years.

Sorel, Paul (b. 1918). American painter, of Midwestern background; born Karl Dibble. He was a close friend of Chris Wood and lived with him in Laguna in the early 1940s, but moved out in 1943 after disagreements over money, living intermittently in New York. Wood continued to support him, though they never lived together again. Sorel painted portraits of Isherwood and Bill Caskey in 1950; he appears in *D.1* and *Lost Years*.

Southern, Terry (1924–1995). American writer, born in Texas and educated at Southern Methodist University, Northwestern, and the Sorbonne. He was a satirist with a dark, hip, comic touch. He published journalism, short stories and novels—including the erotic parody *Candy* (1957) and *The Magic Christian* (1959)—and co-wrote the screenplay for *Dr. Strangelove* (1964) with Stanley Kubrick and Peter George. Later films included *Barbarella* (1968) and *Easy Rider* (1969).

Spender, Elizabeth (Lizzie) (b. 1950). British actress and writer; educated at North London Collegiate School; daughter of Stephen and Natasha Spender. She had small parts in Isherwood and Bachardy's "Frankenstein" and in Terry Gilliam's *Brazil* (1985), and she worked in publishing. In 1990, she married Australian actor and satirist Barry Humphries—best known for his caricature "Dame Edna."

Spender, Matthew (b. 1945). British sculptor and author; educated at Oxford and the Slade; son of Stephen and Natasha Spender. In 1968, he married the painter Maro Gorky, a daughter of the Turkish-Armenian painter Arshile Gorky, and settled with her in Italy. He exhibited from the late 1980s at the Berkeley Square Gallery in London, and his large stone, clay, and wood sculptures appear in Bernardo Bertolucci's 1995 film *Stealing Beauty*, made at his property in Tuscany. His books include *Within Tuscany: Reflections on a Time and Place* (1992) and *From a High Place: A Life of Arshile Gorky* (1999). He appears in *D.1*.

Spender, Natasha Litvin (b. 1919). Russian-born concert pianist; she married Stephen Spender in 1941 and had a son and a daughter with him. In his entry for April 7, 1968, Isherwood alludes to the fact that, in 1964, she underwent two operations for breast cancer followed by radiotherapy. She appears in *D.1* and *Lost Years*.

Spender, Stephen (1909–1995). English poet, critic, autobiographer, editor. Auden introduced him to Isherwood in 1928; Spender was then an undergraduate at University College, Oxford, and Isherwood became a mentor. Afterwards Spender lived in Hamburg and near Isherwood in Berlin, and the two briefly shared a house in Sintra, Portugal, with Heinz Neddermeyer and Tony Hyndman. Spender was the youngest of the writers who came to prominence with Auden and Isherwood in the 1930s; after Auden and Isherwood emigrated, he cultivated the public roles they abjured in England. He worked as a propagandist for the Republicans during the Spanish Civil War and was a member of the National Fire Service during the Blitz. He moved away from his early enthusiasm for communism but remained liberal in politics. His 1936 marriage to Inez Pearn was over by 1939; in 1941 he married Natasha Litvin, and they had Matthew and Lizzie. He appears as "Stephen Savage" in *Lions and Shadows* and is further described in *Christopher and His Kind*, *D.1*, and *Lost Years*. He published an autobiography, *World Within World*, in 1951, and his *Journals 1939–1983* appeared in 1985.

Spender was co-editor with Cyril Connolly of *Horizon* and later of *Encounter*, and in 1968, he helped to found *Index on Censorship* to report on the circumstances of persecuted writers and artists around the world. As Isherwood mentions in his diary entry for June 22, 1967, it was widely presumed that Spender knew for years that *Encounter*, of which he was co-editor from 1953 to 1967, was funded by the CIA through the Congress for Cultural Freedom. But Spender's biographer, John Sutherland, argues he did not. Spender made the first of several formal enquiries into the matter in 1963, because the press had grown interested, and he was evidently told reassuring lies by his backers. Nevertheless, in 1964, he found new funding for the magazine (from Cecil King at the *Daily Mirror*), and he distanced himself from it by spending more time teaching in the U.S. But even the new funding may have been corrupt. Frank Kermode, who resigned as editor with Spender, was also lied to, as he tells in his memoir *Not Entitled* (1995). And Kermode's reputation, like Spender's, was cleverly employed to defend the magazine. On May 20, 1967, *The Saturday Evening Post* published an article in which Thomas Braden, who was known to have headed the CIA's division of international organizations in the early 1950s, stated that one editor at *Encounter* was a CIA agent. Spender told *The New York Times*, "I can't imagine anyone believing I was a CIA agent." He also doubted it was his American co-editor, Irving Kristol, and he said it could not have been Kermode who had been at the magazine only two years. He did not comment on whether it may have been Melvin Lasky, the American co-editor who had succeeded Kristol and who had supposedly pointed the finger at Spender. As the truth emerged (it never fully did), Spender was threatened and attacked by both adversaries and "friends" who were evidently continuing to try to control him. When Melvin Lasky attacked him in *The Observer*, Spender warned, on June 27 in *The New York Times*, that he would sue *Encounter* for libel if the attacks did not stop.

Starcke, Walter (b. 1922). American actor and theatrical producer, and later, spiritual teacher and author. He altered the spelling of his last name to Starkey, but returned to Starcke, when he gave up acting. Isherwood first met him in January 1947 through John van Druten. Starcke starred in an unsuccessful play of van Druten's in the late 1940s, then became van Druten's producer and his longterm boyfriend. Van Druten's previous lover, Carter Lodge, and Lodge's new lover, Dick Foote, never liked Starcke, resulting in complicated rivalries among the four of them; van Druten and Starcke finally split up in 1957, the year van Druten died. Isherwood enjoyed Starcke's company and remained in touch with him; he appears in *D.1* and *Lost Years*. His books include *Joel Goldsmith and I: An Inside Story of a Relationship with a Modern Mystic*, *It's All God: The Flowers and the Fertilizer*, and *Homesick for Heaven: You Don't Have to Wait.*

Steen, Mike (1928–1983). American stuntman, actor, author, born Malcolm H. Steen. He was from Louisiana, where he was friendly with Speed Lamkin, Tom Wright, and Henry Guerriero. Lamkin introduced him to Isherwood in the early 1950s. Gavin Lambert became romantically involved with Steen during 1958, and Steen also had relationships, perhaps sexual, with Nicholas Ray, William Inge, and Tennessee Williams. He worked as a stuntman in Ray's *Party Girl* and did stunts or played bit parts in other movies in the late 1950s and 1960s, including a tiny part in the 1962 film of Williams's *Sweet Bird of Youth*. He published two books: *A Look at Tennessee Williams* (1969) and *Hollywood Speaks: An Oral History* (1974). He appears in *D.1.*

Stephen. See Spender, Stephen.

Stern, James (Jimmy) (1904–1993) and Tania Kurella (1906–1995). Irish writer and translator and his German wife. He was educated at Eton and, briefly, Sandhurst. In youth, he worked as a farmer in Southern Rhodesia and a banker in the family bank in England and Europe, then travelled until settling for a time in Paris in the 1930s, where he met Tania, the daughter of a psychiatrist. She was a physical therapist and exercise teacher, exponent of her own technique, the Kurella method. She fled Germany in 1933 to escape persecution for the left-wing political activities of her two brothers, already refugees. The Sterns married in 1935. Isherwood met Jimmy in Sintra, Portugal, in 1936 through William Robson-Scott and introduced him to Auden with whom the Sterns became intimate friends, later, in America. In 1956, they returned to England and settled, eventually, at Hatch Manor in Wiltshire. His books include *The Heartless Land* (1932) and *Something Wrong* (1938)—both story collections—and *The Hidden Damage* (1945), about his trip with Auden to survey bomb damage in postwar Germany for the U.S. Army. Tania collaborated on translations of Mann, Kafka, and Freud. The Sterns appear in *D.1* and *Lost Years*.

Stevens, Marti (b. 1931). American singer and actress; she appeared in a few films, including *All Night Long* (1961), several times on Broadway, and often on T.V.

Stewart, Donald Ogden (1890–1980) and Ella Winter. American actor, novelist, playwright, screenwriter and his wife, an author and translator. He was educated at Yale, served in the navy during World War I, and appeared on Broadway during the 1920s. His most famous screenplay was *The Philadelphia Story* (1940), for which he won an Academy Award. She wrote a bestselling book about the Soviet Union, *Red Virtue: Human Relationships in the New Russia* (1933), and translated from German *The Diary of Otto Braun With Selections from His Letters* *and Poems* and Wolfgang Koehler's *The Mentality of Apes*. Her first husband was Lincoln Steffens (1866–1936), the journalist and author, and she edited his letters. The Stewarts were neighbors of Salka Viertel in Santa Monica, and they remained close to her when Salka's career foundered during the McCarthy period. Stewart was blacklisted for his involvement with the Hollywood Anti-Nazi League, and the couple left Hollywood for London in 1951 and settled in a house in Hampstead near the one loaned to Don Bachardy by the Burtons in 1961. Ella is mentioned in *D.1*.

Strachwitz von Gross-Zauche und Camminetz, Graf Rudolf (1896–1969) and Barbara Greene (1907–1991). German diplomat and his wife, Isherwood's cousin. Each of them had close ties to the conspirators in the July 20, 1944 plot to assassinate Hitler. Strachwitz's name was removed from the conspirators' list for a proposed future German government because he was married to an English woman, officially an enemy. This saved his life and his wife's when the plot failed and the list was found. After the war, he served as a German ambassador abroad and also made his home in Berchtesgaden, Bavaria, where Hitler once had his summer retreat. Barbara Greene's books include *Land Benighted* (1938) about her trip through Liberia with another Isherwood cousin, Graham Greene. Isherwood's 1948 meeting with the Strachwitzes in Buenos Aires, where Strachwitz was then a university professor, is mentioned in his introduction to *The Condor and the Cows*.

Stravinsky, Igor (1882–1971). Russian-born composer; he went to Paris with Diaghilev's Ballets Russes in 1920 and brought about a rhythmic revolution in Western music with *The Rite of Spring* (1911–1913), the most sensational of his many works commissioned for the company. He composed in a wide range of musical forms and styles; many of his early works evoke Russian folk music, and he was influenced by jazz. Around 1923 he began a long neo-classical period responding to the compositions of his great European predecessors. During the 1950s, with the encouragement of Robert Craft, he took up the twelve-note serial methods invented by Schoenberg and extended by Webern—he was already past seventy. After the Russian revolution, Stravinsky remained in Europe, making his home first in Switzerland and then in Paris, and he turned to performing and conducting to support his family. In 1926, he rejoined the Russian Orthodox Church, and religious music became an increasing preoccupation during the later part of his career. At the outbreak of World War II, he emigrated to America, settled in Los Angeles, and eventually became a citizen in 1945. Although he was asked to, he never composed for films. His first and most important work for English words was his opera, *The Rake's Progress* (1951), for which Auden and Kallman wrote the libretto. Isherwood first met Stravinsky in August 1949 at lunch in the Farmer's Market in Hollywood with Aldous and Maria Huxley. He was soon invited to the Stravinskys' house for supper where he fell asleep listening to a Stravinsky recording; Stravinsky later told Robert Craft that this was the start of his great affection for Isherwood. He appears throughout *D.1* and *Lost Years*.

Stravinsky, Vera (1888–1982). Russian-born actress and painter, second wife of Igor Stravinsky; she was previously married three times, the third time to the painter and Ballets Russes stage designer Sergei Sudeikin. In 1917, she fled St. Petersburg and the bohemian artistic milieu in which she was both patroness and muse, travelling in the south of Russia with Sudeikin before going on to Paris where she met Stravinsky in the early 1920s; they fell in love but did not marry until 1940 after the death of Stravinsky's first wife. Isherwood met her with Stravinsky in August 1949. She painted in an abstract-primitive style influenced by Paul Klee, childlike and decorative. She appears throughout *D.1* and in *Lost Years*.

the Strip, also Sunset Strip. A once-glamorous section of Sunset Boulevard between Laurel Canyon Boulevard in the east and Doheny Drive in the west. Home to nightclubs like Ciro's, Trocadero, Mocambo, Crescendo and expensive restaurants including La Rue and The Players.

Stromberg, Hunt, Jr. (1923–1985). American T.V. executive; son of Hunt Stromberg (1894–1968), who was one of MGM's most profitable and powerful film producers from the mid-1920s until he retired in 1951. Stromberg Jr. began his career as a theater producer and moved to T.V. early in the 1950s. He worked with James Aubrey at CBS on the original idea for the series "Have Gun, Will Travel" (1956) and during the 1960s, he supervised "The Beverly Hillbillies," "Green Acres," and "Lost in Space." He was fired from CBS at the time of Aubrey's downfall in 1965 and for a time ran a production partnership with Aubrey until Aubrey took over MGM Studios in 1969. Later, Stromberg worked at Universal Studios. He produced "Frankenstein: The True Story" (1973), written by Isherwood and Bachardy, and "The Curse of King Tut's Tomb" (1980), adapted by Herb Meadow from Barry Wynne's book.

Stuurman, Douwe (1910–1991). American classicist and professor of literature, raised in Washington state, where his Dutch immigrant parents were farmers. Of seven children, he alone completed his formal education, at Calvin College in Michigan, at the University of Oregon, and at Balliol College, Oxford, where he was a Rhodes Scholar and studied philosophy, literature, and Greek. He taught at Reed College and Santa Barbara College before serving in U.S. military intelligence during World War II. At the end of the war, as part of a Library of Congress military team, he retrieved over 100,000 Nazi books, pamphlets, and files of unpublished chancellory correspondence, transporting them in army trucks and storing them in an army warehouse for later historical analysis. He taught in the English department at the University of California at Santa Barbara from 1949 to 1975, arranged for Huxley to lecture there in 1959, and was responsible for Isherwood's appointment in 1961. He married twice, first to a blind poet, later, briefly, to Phyllis Plous, director of the UCSB art museum. He appears in *D.1*.

Sudhira. A nurse of Irish descent, born Helen Kennedy; she was a probationer nun at the Hollywood Vedanta Society when Isherwood arrived to live there in 1943. In youth, she had been widowed on the third day of her marriage. Afterwards, she worked in hospitals and for Dr. Kolisch and first came to the Vedanta Society professionally to nurse a devotee. In *D.1*, Isherwood tells how he relished being ill under her care; she also appears in *Lost Years*. She enlisted in the navy in January 1945 and later married for a second time and returned to nursing.

Sujji Maharaj. See Nirvanananda, Swami.

Surmelian, Leon (1905–1995). Armenian-American writer, popular historian, teacher. His books include *I Ask You, Ladies and Gentlemen* (1945), a memoir about his orphaned childhood in Turkey during the Armenian genocide. He was a professor at Los Angeles State College, later called California State University. He appears in *D.1*.

Sutro, John (1903–1985). Oxford-educated British film producer and screenwriter. He was a friend of Evelyn Waugh and worked as a rubber merchant at the family firm in London before starting a production company, Ortus Films. He produced *49th Parallel* (1941), *The Way Ahead* (1944) directed by Carol Reed, and *Men of Two Worlds* (1946), among others; later, he translated scripts for Roman Polanksi's *Cul-de-Sac* (1966) and *The Fearless Vampire Killers* (1967). Another close friend was Graham Greene, and, in 1960, Isherwood mentions Sutro's proposed film of *England Made Me*, but it was produced only in 1973 by five others: C. Robert Allen, David Anderson, Jerome Z. Cline, Jack Levin, and Zika Vojcic. Sutro's wife was Gillian Hammond, a fashion journalist.

Swahananda, Swami. Monk of the Ramakrishna Order, from India; born near Habiganj, now in Bangladesh, and educated at Murari Chand College, Sylhet, and at the University of Calcutta where he got an M.A. in English. He was initiated in 1937 by Swami Vijnananda, a direct disciple of Ramakrishna, joined the order in 1947 and took his final vows in 1956. He was head of the Delhi Ramakrishna Mission, and then in 1968 was sent to the San Francisco center as assistant to Swami Ashokananda. Later, he became head of the Berkeley center, and in 1976 he was moved to Hollywood to replace the late Swami Prabhavananda. As a young monk, he edited *Vedanta Kesari*, a scholarly publication of the Ramakrishna Order, and he went on to publish numerous articles and books of his own.

Swami. Used as a title to mean "Lord" or "Master." A Hindu monk who has taken *sannyas*, the final vows of renunciation in the Hindu tradition. Isherwood used it in particular to refer to his guru, Swami Prabhavananda, and he pronounced it *Shwami*; see Prabhavananda.

Swamiji. An especially respectful form of "Swami," but also a particular name for Vivekananda towards the end of his life.

tamas. See guna.

Tate, Sharon (1943–1969). Texas-born model and starlet; she appeared in *The Fearless Vampire Killers* (1967) directed by her husband, Roman Polanski, and in *The Valley of the Dolls* (1967). On August 9, 1969, Tate, eight months pregnant, was murdered with four friends by Charles Manson and his followers in the Polanskis' rented house in Bel Air, Benedict Canyon. The next day, August 10, the Manson gang murdered Leno and Rosemary LaBianca in the Los Feliz/Silver Lake area. On August 12, Chet Young, a deranged fan of the singing Lennon Sisters, shot their father, William Lennon, in the parking lot of a golf course in Marina del Rey where Lennon was an instructor.

Ted. See Bachardy, Ted.

Teller, Edward (1908–2003). Hungarian-born nuclear physicist. He worked on the atom bomb at Los Alamos under the direction of Robert Oppenheimer (1904–1967), then left to start the Lawrence Livermore Laboratory at Berkeley, where, with Stanislav Ulam, he developed the hydrogen bomb, first tested in 1952. He had a reputation for disdaining the moral implications of his appetite for research, and colleagues found him difficult. In 1954, he failed to support Oppenheimer in public testimony after Oppenheimer's security clearance was revoked over youthful links to the Communist party. Oppenheimer had opposed fast-paced development of the H-bomb, and he worked for international control of nuclear arms. The rivalry between the two scientists may lie behind Isherwood's resentment dream about Teller, recorded in his diary April 19, 1962, at a time when Isherwood felt that he, like Oppenheimer, was being professionally persecuted, or at least misunderstood, by colleagues, partly for his liberal convictions. And Isherwood shared not only Oppenheimer's liberalism but also his religious inclinations. Oppenheimer studied Sanskrit and Hinduism and publicly quoted the Baghavad Gita to express his awe when Hiroshima and Nagasaki were destroyed by atom bombs. In his entry for December 28, 1963, Isherwood observes that in a photograph taken in India, he resembles Oppenheimer.

Tennant, Stephen (1906–1987). British poet, aesthete, eccentric; youngest son of Lord Glenconner, a Scottish peer, and Pamela Wyndham, a cousin of Oscar Wilde's paramour, Lord Alfred Douglas. He studied painting at the Slade and worked for decades on a novel which he never completed. He was known for his extravagantly camp dress and manners, his interior decorating, and spending the last seventeen years of his life mostly in bed. He was an intimate friend of Cecil Beaton and was photographed by him several times, and he had a long affair with Siegfried Sassoon. He is said to be a model for Sebastian Flyte in Waugh's *Brideshead Revisited* and for Cedric Hampton in Nancy Mitford's *Love in a Cold Climate*.

Tennessee. See Williams, Tennessee.

Terry. See Jenkins, Terry.

Thom, Richard (Dick) (d. 1968). American disciple of Swami Prabhavananda. His parents were among Swami's devotees in Portland, Oregon in the 1920s, and Thom began preparing to be a monk while still in high school. He was also Northwest weightlifting champion in adolescence. He lived at the Vedanta Society in Hollywood with Isherwood and the other probationer monks briefly in 1943, until he got into trouble and was expelled from school. He took various jobs, then joined the marines, later returning to the monastery intermittently until the mid-1960s. He had trouble with alcoholism, was a loner, and never settled into the monastic life, although he remained close to the nuns in Santa Barbara through his parents who were gatekeepers at the temple. He appears in *D.1*. His cousin, Tommy Thom, was also at the Hollywood center for a time.

Thompson, Nicholas. American theatrical agent, based in London. He was introduced to Isherwood and Bachardy by Robin French and became the London agent for *A Meeting by the River*. Isherwood records in March 1969 that Thompson planned to return to Los Angeles eventually and work with French at Chartwell Artists, but French does not recall offering him a job.

Thomson, Virgil (1896–1989). American composer, music critic, author; born in Kansas and educated at Harvard. During the 1920s, he lived off and on in Paris, where he studied with Nadia Boulanger and met the young French composers, including Poulenc and Auric who were known as "Les Six" and who were, like him, influenced by Satie. He became friendly with Gertrude Stein, who supplied libretti for his operas *Four Saints in Three Acts* (1934) and *The Mother of Us All* (1947). In 1940, he returned permanently to New York. He was the music critic for the *Herald Tribune* for a decade and a half, published eight books on music, including *American Music Since 1910* (1971), and composed prolifically. His third and last opera, *Lord Byron*, on a libretto by Jack Larson, was planned for the Metropolitan Opera in New York, but instead premiered at the Juilliard Opera Center in April 1972. His film scores include *The Plow That Broke the Plains* (1936) and *The River* (1938), both directed by Pare Lorentz, and *Louisiana Story* (1948), directed by Robert Flaherty. His companion for many years was Maurice Grosser.

Tony. See Richardson, Tony.

Trabuco. Monastic community sixty miles south of Los Angeles and about twenty miles inland; founded by Gerald Heard in 1942. Heard anonymously provided $100,000 for the project (his life savings), and he consulted at length with various friends and colleagues as well as with members of the Quaker Society of Friends about how to organize the community. In 1940, he planned only a small retreat called "Focus," then he renamed the community after buying the ranch at Trabuco. Isherwood's cousin on his mother's side, Felix Greene, oversaw the purchase of the property and the construction of the building which could house fifty. By 1949 Heard found leading and administering the group too much of a strain, and he persuaded the Trustees to give Trabuco to the Vedanta Society. It opened as a Vedanta monastery in September 1949.

Tree, Iris (1896–1968). English actress and writer; third daughter of actor Herbert Beerbohm Tree. She published three volumes of poetry (two before 1930, a third in 1966) and wrote poems and articles for magazines such as *Vogue* and *Harper's Bazaar*, as well as *Botteghe Oscura*, *Poetry Review*, and *The London Magazine*. In youth, she travelled with her father to Hollywood and New York and married an American, Curtis Moffat, with whom she had her first son, Ivan Moffat, born in Havana. Until 1926 she lived mostly in London and in Paris where she acted in Max Reinhardt's *The Miracle*; she toured with the play back to America where she met her second husband, an Austrian, Count Friedrich Ledebur, with whom she had another son, Christian Dion Ledebur (called Boon), in 1928. Iris Tree had known Aldous and Maria Huxley in London, and the Huxleys introduced her to Isherwood in California during the war. With Allan Harkness, she brought a troupe of actors to Ojai to start a repertory theater, The High Valley Theater, and she adapted, wrote and acted in plays for the group including her own *Cock-a-doodle-doo*. Many of the actors were pacifists like herself. She moved often—from house to house and country to country—and in July 1954 left California for good, settling in Rome where she worked on but never finished a novel about her youth. Her marriage to Ledebur ended in 1955. Isherwood modelled "Charlotte" in *A Single Man* partly on Iris Tree, and she appears in *D.1* and *Lost Years*.

Tutin, Dorothy (1931–2001). English actress; she played Sally Bowles in the original London stage production of *I Am a Camera* in 1954. As he records in *D.1*,

Isherwood became friendly with her in London and visited her several times on her houseboat. *The Hollow Crown*, with which she toured in 1963 in the U.S., was a "Royal Revue," devised by John Barton for the Royal Shakespeare Company, about the kings and queens of England from William the Conqueror to Victoria. Afterwards, she appeared in another RSC production, John Gay's 1728 play *The Beggar's Opera*. Her films include *The Beggar's Opera* (1953) opposite Olivier and *A Tale of Two Cities* (1958) opposite Dirk Bogarde.

Tynan, Elaine. See Dundy, Elaine.

Tynan, Kenneth (Ken) (1927–1980). English theater critic, educated at Oxford. During the 1950s and 1960s, he wrote regularly for the London *Evening Standard*, then for *The Observer*, *The New Yorker*, and other publications. He was literary adviser to the National Theater from its inception in 1963, but his anti-establishment views brought about his departure before the end of the decade. His support for realistic working-class drama—by new playwrights such as Osborne, Delaney and Wesker—as well as for the works of Brecht and Beckett, was widely influential. Many of his essays and reviews are collected as books. At the end of stage censorship in 1968, he devised and produced the sex revue *Oh! Calcutta!* which opened in New York in 1969 and in London in 1970. His first wife, from 1951, was the actress and writer Elaine Dundy, with whom he had a daughter, Tracy Tynan, later a costume designer for films. In 1963, he began an affair with the newly married Kathleen Halton Gates (1937–1995), a Canadian journalist and, later, novelist and screenwriter, raised in England; they married in 1967 and settled for a time in the U.S. with their children, Roxana and Matthew. Isherwood first met Ken and Elaine Tynan in London in 1956, and they are mentioned in *D.1*.

UCLA. University of California at Los Angeles.

UCSB. University of California at Santa Barbara, where Isherwood taught during the autumn of 1960.

Upward, Edward (1903–2009). English novelist and schoolmaster. Isherwood first met him in 1921 at their public school, Repton, and followed him (Upward was a year older) to Corpus Christi College, Cambridge. Their close friendship was inspired by their shared attitude of rebellion toward family and school authority and by their literary obsessions. In the 1920s they created the fantasy world Mortmere, about which they wrote surreal, macabre, and pornographic stories and poems for each other; their excited schoolboy humor is described in *Lions and Shadows* where Upward appears as "Allen Chalmers." Upward made his reputation in the 1930s with his short fiction, especially *Journey to the Border* (1938), the intense, almost mystical, and largely autobiographical account of a young upper-middle-class tutor's conversion to communism. Then he published nothing for a long time while he devoted himself to schoolmastering (he needed the money) and to Communist party work. From 1931 to 1961 he taught at Alleyn's School, Dulwich where he became head of English and a housemaster; he lived nearby with his wife, Hilda, and their two children, Kathy and Christopher. After World War II, Upward and his wife became disillusioned by the British Communist party and left it, but Upward never abandoned his Marxist-Leninist convictions. Towards the end of the 1950s, he overcame writer's block and a nervous breakdown to produce a massive autobiographical trilogy, *The Spiral Ascent* (1977)—comprised of *In the Thirties* (1962), *The Rotten Elements* (1969), and *No Home But the Struggle*. The last two volumes were written in Sandown, on the Isle of Wight, where Upward retired in 1962. They were followed by four collections of short stories. Upward remained a challenging and trusted critic of Isherwood's work throughout Isherwood's life, and a loyal friend. He appears in *D.1* and *Lost Years*.

Ure, Mary (1933–1975). British stage actress who appeared in a few Hollywood films, including *Sons and Lovers* (1960), for which she received an Academy Award nomination. She grew up in the suburbs of Glasgow, the daughter of a prosperous engineer. She played the female lead in John Osborne's *Look Back in Anger* at the Royal Court in London in 1956, then in New York, and afterwards in the film, and she became Osborne's second wife. In *D.1*, Isherwood tells of meeting her with Tony Richardson in 1960 when she was appearing (as Clare Bloom's replacement) in the touring stage production of Jean Giraudoux's *Duel of Angels* with Vivien Leigh. She divorced Osborne and, in 1963, married Robert Shaw (1927–1978), the English actor and writer, with whom she appeared in Tony Richardson's production of *The Changeling* (1961) and several other plays and films, and with whom she had four children. Their first child was born while Ure was still married to Osborne. Her other films include *Storm Over the Nile* (1955) and *Where Eagles Dare* (1969). She died from a combination of barbiturates and alcohol.

Usha. A nun at the Vedanta Society and later at the convent in Santa Barbara; originally called Ursula Bond and, after sannyas, Pravrajika Anandaprana. She was a German Jew, educated in England, and came to the U.S. as a young refugee during the war. Until the war ended, she worked for the U.S. government as a censor. She had been married before taking up Vedanta, and she had a daughter, Caroline Bond; she left three-year-old Caroline with the child's father, but Caroline later joined the Hollywood convent where she was known as Sumitra, took brahmacharya vows, and remained for about ten years before leaving to join an ashram, which emphasized Sanskrit and scriptural study. Later, Sumitra did graduate work in Sanskrit and became a freelance editor. Usha appears in *D.1*.

Vadim, Roger (1928–2000). French actor of stage and screen, journalist, screenwriter, director; he dropped his Ukrainian surname, Plemiannikov. Vadim helped launch the New Wave in movies when he directed his wife, Brigitte Bardot, in *Et Dieu créa la femme/And God Created Woman* (1956), the first of many erotically charged films, mostly in French. Later work included *Les Liaisons dangereuses* (1959), *Et mourir de plaisir/Blood and Roses* (1960), *La Ronde/Circle of Love* (1964), *Barbarella* (1968), *Don Juan 1973 ou si Don Juan était une femme/Ms. Don Juan* (1973), *La Jeune fille assassinée/Charlotte* (1974), *Une Femme fidèle* (1976), and three U.S. films, *Pretty Maids All in a Row* (1971), a remake of *And God Created Woman* (1988), and *The Mad Lover* (1991). He collaborated on many of his screenplays and continued to appear as an actor, for instance in *The Testament of Orpheus* (1959), *Rich and Famous* (1981), and *Into the Night* (1985). His second wife was Danish actress Annette Stroyberg, with whom he had his first child; he had another child with Catherine Deneuve before marrying Jane Fonda in 1965. He and Fonda also had a child, then divorced in 1973. He later married twice more, to Catherine Schneider, with whom he had a fourth child, then to Marie-Christine Barrault.

Vandanananda, Swami (1915–2007). Hindu monk of the Ramakrishna Order. He arrived at the Hollywood Vedanta Society from India in the summer of 1955 and eventually became chief assistant, replacing Swami Asheshananda. In 1969 he returned to India, was head of the Delhi Center, and later, at Belur Math, became Assistant Secretary of the order, and then General Secretary. He appears in *D.1*.

Vanderbilt, Gloria. See Cooper, Wyatt.

van Druten, John (1901–1957). English playwright and novelist. Isherwood met him in New York in 1939, and they became friends because they were both pacifists. Van Druten's family was Dutch, but he was born and educated in London and took a degree in law at the University of London. He achieved his first success as a playwright in New York during the 1920s, then emigrated in 1938 and became a U.S. citizen in 1944. His strength was light comedy; among his numerous plays and adaptations, many of which were later filmed, were *Voice of the Turtle* (1943), *I Remember Mama* (1944), *Bell, Book, and Candle* (1950), and *I Am a Camera* (1951) based on Isherwood's *Goodbye to Berlin*. Van Druten acquired a sixty percent share in material which he used in *I Am a Camera* from *Goodbye to Berlin*; Isherwood retained a forty percent share. Anyone acquiring rights after van Druten needed his agreement or, after his death, the agreement of his estate. In 1951, van Druten directed *The King and I* on Broadway. He also wrote a few novels and two volumes of autobiography, including *The Widening Circle* (1957). He usually spent half the year in New York and half near Los Angeles on the AJC Ranch, which he owned with Carter Lodge. He also owned a mountain cabin above Idyllwild which Isherwood sometimes used. A fall from a horse in Mexico in 1936 left him with a permanently crippled arm; partly as a result of this, he became attracted to Vedanta and other religions (he was a renegade Christian Scientist), and in his second autobiography he describes a minor mystical experience which he had in a drug store in Beverly Hills. He was a contributor to Isherwood's *Vedanta for the Western World*, and there are numerous accounts of him in *D.1* and *Lost Years*.

Vanessa. See Redgrave, Vanessa.

Van Horn, Mike. American artist and top fashion illustrator, especially for Lord & Taylor. A close friend of Bachardy. In the early 1970s, he left Los Angeles to continue his career in New York State where he now renovates country properties, paints, and sculpts.

Van Petten, Bill (d. mid-1980s). An Ivy Leaguer who divided his time between La Jolla and Santa Monica, where he lived at the bottom of the Canyon in an apartment once inhabited by Jim Charlton. He contributed to the *Los Angeles Times* and was an early friend of Tom Wolfe. He had a small independent income and often travelled abroad, especially to the Middle East. During the 1960s he had radiation treatment for cancer, and the treatment damaged his face and eyes.

Van Sant, Tom (b. 1931). Californian sculptor, painter, conceptual artist, and kite maker; he produced the first satellite map of Earth, the Geosphere Image, and founded the Geosphere Project to model issues of earth resource management.

Vaughan, Keith (1912–1977). English painter, illustrator, and diarist. He worked in advertising during the 1930s and was a conscientious objector in the war; later he taught at the Camberwell School of Art, the Central School of Arts and Crafts, and the Slade where he was Bachardy's tutor in 1961. He also taught briefly in America. Isherwood met him in 1947 at John Lehmann's and bought one of his pictures, "Two Bathers," a small oil painting, still in his collection. Vaughan's diaries, with his own illustrations, were published in 1966. He appears in *D.1* and *Lost Years*.

Vedanta. One of six orthodox systems of Hindu philosophy, Vedanta is based on the Upanishads, the later portion of the ancient Hindu scriptures known as the Vedas. More generally, Vedanta is the whole body of literature which explains and comments on these texts. Probably first formulated by the philosopher Badarayana (second or first century BC), Vedanta teaches that the object of existence is not release but realization—that we should learn to know ourselves for what we really are. This realization is not obtained through logic, but through the direct experience of the inspired sages recorded in the Upanishads. Vedanta is uncompromisingly non-dualistic. Everything exists through Brahman; Brahman is consciousness, joy, bliss absolute. He is the Ultimate Principle and the Final Reality and the Indivisible One. Brahman within the individual is the divine self or Atman; Atman is eternal, unchanging, and one with Brahman. The phenomenal world of nature and man has merely a temporary, changing existence; it is the result of maya, illusion. Ignorance of this leads to a belief that things exist apart from the absolute. Ignorance is responsible for samsara, the continuous cycle of death and rebirth and death which lasts as long as the individual remains in the toils of maya. The search for Reality is a mystical search, pursued by introspective means such as meditation and spiritual discipline. Vedanta honors all the great spiritual teachers and impersonal or personal aspects of Godhead worshipped by different religions, considering them as manifestions of one Reality. Vedanta had a broadreaching revival in the nineteenth century under the leadership of the Indian holy man Ramakrishna, and it spread to the West through his disciples, especially Vivekananda, and the disciples of his disciples. Isherwood began his study of Vedanta in July 1939 under the guidance of Swami Prabhavananda, a second-generation disciple.

Vedanta Place. The Vedanta Society of Southern California was able to adopt the address Vedanta Place when, in 1952, the Hollywood Freeway cut off the tail end of Ivar Avenue where the society was located, leaving it in a cul-de-sac. The property, previously number 1946 Ivar Avenue, had been the home of Carrie Mead Wyckoff, later known as Sister Lalita. As Isherwood tells in *D.1*, he lived at the Hollywood society as a novice monk during World War II.

Ventura, Clyde (1936–1990). American actor and stage director, born in New Orleans. He was in the 1963 Broadway cast of Tennessee Williams's *The Milk Train Doesn't Stop Here Anymore*, and he was Artistic Director of Theater West in Los Angeles, where he directed a number of Williams plays. He was also an acting coach at Actors Studio on the West Coast, and he had a few small movie roles.

Vera. See Stravinsky, Vera.

Vidal, Gore (b. 1925). American writer. He introduced himself to Isherwood in a café in Paris in early 1948, having previously sent him the manuscript of his third novel, *The City and the Pillar* (1948). Vidal's father taught at West Point, and Vidal was in the army as a young man. Afterwards, he wrote essays on politics and culture, short stories, and many novels, including *Williwaw* (1946), *Myra Breckinridge* (1968, dedicated to Isherwood), its sequel, *Myron* (1975), *Two Sisters: A Memoir in the Form of a Novel* (1970), *Kalki* (1978), *Duluth* (1983), *The Judgement of Paris* (1984), *Live from Golgotha* (1992), *The Smithsonian Institute* (1998), ancient and medieval historical fiction such as *A Search for the King* (1950), *Messiah* (1955), *Julian* (1964), *Creation* (1981), and the multi-volume American chronicle comprised of *Burr* (1974), *Lincoln* (1984), *1876* (1976), *Empire* (1987), *Hollywood* (1989), and *Washington, D.C.* (1967). He also published detective novels under a pseudonym, Edgar Box. During the 1950s, Vidal wrote a series of television plays for CBS, then screenplays at Twentieth Century-Fox and MGM (including part of *Ben Hur*), and two Broadway plays, *Visit to a Small Planet* (1957) and *The Best Man* (1960). His adaptation of Friedrich Duerrenmatt's *Romulus the Great*, about Romulus Augustulus, ran on Broadway from January to March 1962. In 1960 Vidal ran for Congress, and in 1982 for the Senate, both times unsuccessfully. He bought Edgewater, a Greek Revival mansion on the Hudson River north of New York, and lived there off and on with Howard Austen, from 1950 until he sold it in 1968; later he settled in Rapallo, Italy, and finally in Los Angeles. Over the years, Vidal campaigned to become a member of the National Institute of Arts and Letters; with pushing from Isherwood, he was finally elected in 1976 and turned it down. He describes his friendship with Isherwood in his memoir, *Palimpsest* (1995) and in *Point to Point Navigation* (2006). There are many passages about him in *D.1* and *Lost Years*.

Vidor, King (1894–1982). American film director, screenwriter, producer; born and raised in Texas. He began directing in the silent era with *The Turn in the Road* for Universal in 1919, then formed his own studio, and afterwards worked for MGM. His movies include *The Big Parade* (1925), *The Crowd* (1928), *Show People* (1928), *Billy the Kid* (1930), *The Champ* (1931), *Northwest Passage* (1940), *Duel in the Sun* (1947), and *The Fountainhead* (1949). In *D.1*, Isherwood describes meeting him in Italy in 1955 on the set of *War and Peace* (1956), which Vidor was directing with the assistance of Mario Soldati. Vidor retired when his next film, *Solomon and Sheba* (1959), failed. During the 1960s he taught at UCLA film school, and in 1979 he received a special Academy Award as a creator and innovator in film. His first wife was silent screen star Florence Vidor, who came to Hollywood with him in 1915 as a newlywed; they divorced in 1924. He was married to Eleanor Boardman, also a film star, for a few years in the mid-1920s. His third wife was a writer, Elizabeth Hill.

Vidya, also Vidyatmananda, Swami, previously John Yale. See Prema Chaitanya.

Viertel, Salka (1889–1978). Polish actress and screenplay writer; first wife of Berthold Viertel with whom she had three sons, Hans, Peter, and Thomas. Sara Salomé Steuermann Viertel had a successful stage career in Vienna (including acting for Max Reinhardt's Deutsches Theater) before moving to Hollywood where she became the friend and confidante of Greta Garbo; they appeared together in the German language version of *Anna Christie* and afterwards Salka collaborated on Garbo's screenplays for MGM in the 1930s and 1940s (*Queen Christina*, *Anna Karenina*, *Conquest*, and others). Isherwood met her soon after arriving in Los Angeles and was often at her house socially or to work with Berthold Viertel. In the 1930s and 1940s, the house was frequented by European refugees, and Salka was able to help many of them find work—some as domestic servants, others with the studios. Her guests included the most celebrated writers and movie stars of the time. In 1946, Isherwood moved into her garage apartment, at 165 Mabery Road, with Bill Caskey. By then Salka was living alone and had little money. Her husband had left; her lover Gottfried Reinhardt had married; Garbo's career was over; and later, in the 1950s, Salka was persecuted by the McCarthyites and blacklisted by MGM for her presumed communism. In January 1947, she moved into the garage apartment herself and let out her house; in the early 1950s, she sold the property and moved to an apartment off Wilshire Boulevard. Eventually, she returned modestly to writing for the movies, but finally moved back to Europe, although she had been a U.S. citizen since 1939. She published a memoir, *The Kindness of Strangers*, in 1969. She appears in *D.1* and *Lost Years*.

Vince. See Davis, Vince.

Vishwananda, Swami. Indian monk of the Ramakrishna Order. Isherwood met him in 1943 when Vishwananda visited the Hollywood Vedanta Society and other centers on the West Coast. Vishwananda was head of the Vedanta Center in Chicago. He appears in *D.1*.

Vivekananda, Swami (1863–1902). Narendranath Datta (known as Naren or Narendra and later as Swamiji) took the monastic name Vivekananda in 1893. He was Ramakrishna's chief disciple. He came from a wealthy and cultured background and was attending university in Calcutta when Ramakrishna recognized him as an incarnation of one of his "eternal companions," a free, perfect soul born into maya with the avatar and possessing some of the avatar's characteristics. Vivekananda was trained by Ramakrishna to carry his message, and he led the disciples after Ramakrishna's death, though he left them for long periods, first to wander through India as a monastic, practising spiritual disciplines, then to travel twice to America and Europe, where his lectures and classes spawned the first western Vedanta centers. In India he devoted much time to founding and administering the Ramakrishna Math and Mission. His followers hired a British stenographer, J.J. Goodwin, to transcribe his lectures; these transcriptions, which Goodwin probably edited into complete sentences and paragraphs, along with Vivekananda's letters to friends and to his own and Ramakrishna's disciples, were published as *The Complete Works of Vivekananda*. Isherwood wrote the introduction to one volume, *What Religion Is: In the Words of Swami Vivekananda* (1960), selected by Prema Chaitanya.

Wald, Jerry (1911–1962) and Connie. American screenwriter and producer, born in New York and educated at New York University, and his wife. He worked as a journalist, produced film shorts for Warner Brothers in the early 1930s, and then became a screenwriter before starting to produce his own features.

In 1950, he formed a production company with Norman Krasna and afterwards joined Columbia Pictures as a vice-president and executive producer. Then in 1956, he formed Jerry Wald Productions, releasing his films through Twentieth Century-Fox where he had a high reputation and a great deal of power. As a producer, his name is associated with a long list of films from the 1940s and 1950s, including *All Through the Night* (1942), *Mildred Pierce* (1945), *Key Largo* (1948), *Johnny Belinda* (1948), *Adventures of Don Juan* (1948), *The Glass Menagerie* (1950), all at Warner Brothers, and *Miss Sadie Thompson* (1953), *Peyton Place* (1957), *The Long Hot Summer* (1958), *Sons and Lovers* (1960), and *Let's Make Love* (1960). He also did screenwriting work at Warner Brothers on *In Caliente* (1935), *Brother Rat* (1938), *The Roaring Twenties* (1939), *Torrid Zone* (1940), *They Drive by Night* (1940), and others. He was rumored to be a model for Budd Schulberg's *What Makes Sammy Run?* Isherwood was hired by Wald in 1956 to work on *Jean-Christophe* (never made), and Gore Vidal and Gavin Lambert also worked for Wald. He appears in *D.1*. Only months after his death, on July 13, 1962, his wife, Connie, married a successful Beverly Hills doctor called Myron Prinzmetal. Prinzmetal was a drug addict, and the marriage did not last. She went back to calling herself Connie Wald and continued in her role of Hollywood hostess.

Warshaw, Howard (1920–1977). American artist, born in New York City, educated at Pratt Institute, the National Academy of Design Art School, and the Art Students League. He moved to California in 1942 and worked in the animation studios of Walt Disney and later Warner Brothers, then taught briefly at the Jepson Art Institute where he was influenced by a colleague, Rico Lebrun. In 1951, he began teaching at the University of California at Santa Barbara, where he remained for over twenty years. He painted murals in the Dining Commons at UCSB and at the Santa Barbara Public Library. At UCLA he painted the neurological mural, and he also did murals for U.C. San Diego and U.C. Riverside. He was blind in one eye. His wife Frances had money of her own. She had been married previously to Mel Ferrer with whom she had a daughter called Pepa and a son. Isherwood and Warshaw became friendly when Isherwood began teaching at UCSB, and Isherwood usually slept at their house when he stayed overnight in Santa Barbara. In 1954, Isherwood encouraged Bachardy to attend drawing classes with Warshaw in Los Angeles, but Bachardy found the class too theoretical and left. Years later, Warshaw sat for Bachardy.

Watts, Alan (1915–1973). English mystic, religious philosopher, teacher, and Dean of the American Academy of Asian Studies in San Francisco, where he settled. He became a Zen Buddhist while still a schoolboy at King's College, Canterbury, and was ordained as an Anglican priest in 1945. He was a friend of Aldous Huxley, experimented with LSD in the 1950s, and became a figure on the Beat scene. He opposed the Hindu emphasis on asceticism, asserting that sex improved spiritual presence, and he married three times. His many books and pamphlets include *An Outline of Zen Buddhism* (1932), *Behold the Spirit: A Study in the Necessity of Mystical Religion* (1947), *The Supreme Identity: An Essay on Oriental Metaphysic and the Christian Religion* (1950), *The Wisdom of Insecurity* (1951)—which, in *D.1*, Isherwood records had a profound effect on the young Bachardy—*Nature, Man and Woman: A New Approach to Sexual Experience* (1958), and *Psychotherapy East and West* (1961). Isherwood first met Watts in 1950, and he is mentioned in *D.1* and *Lost Years*.

Weidenfeld, George (b. 1919). Viennese-born British publisher; trained as a lawyer and diplomat. He emigrated in 1938 and worked for the BBC during World War II, mostly commentating on European Affairs; he also wrote a foreign affairs column for the *News Chronicle*. After the war, he began publishing a magazine, *Contact*, and then founded Weidenfeld & Nicolson with Nigel Nicolson. Their first books appeared in 1949, focusing initially on history and biography. In the 1950s, they turned to fiction as well; Nabokov's *Lolita* (1959) was their first best-seller. Weidenfeld & Nicolson also became known for its books by world political leaders and for diaries, letters, and memoirs of public figures. In 1949, Weidenfeld spent a year in Israel as a political adviser and Chief of Cabinet to President Chaim Weizmann, and afterwards he sustained close ties there. He wrote *The Goebbels Experiment* (1943) and an autobiography; for a time he had a column in *Die Welt*. He married four times and had one daughter with his first wife; his second wife, Barbara Skelton, had previously been married to Cyril Connolly.

Wescott, Glenway (1901–1987). American writer, born in Wisconsin. He attended the University of Chicago, lived in France in the 1920s, partly in Paris, and travelled in Europe and England. Afterwards he lived in New York. Early in his career he wrote poetry and reviews, later turning to fiction. His best-known works are *The Pilgrim Hawk* (1940) and *Apartment in Athens* (1945). Wescott was President of the American Academy of Arts and Letters from 1957 to 1961. His longterm companion was Monroe Wheeler, although each had other lovers. In 1949, Wescott went to Los Angeles expressly to read Isherwood's 1939–1945 diaries. While he was there, Isherwood introduced Wescott to Jim Charlton with whom Wescott had an affair. Wescott appears in *D.1* and *Lost Years*.

Wesker, Arnold (b. 1932). English playwright; his early plays were staged at the Royal Court, where *Roots* (1959) was a hit. In 1961, when Isherwood first met him, he was running Centre 42 at the Roundhouse in Chalk Farm, a project to make the arts accessible among the working classes. His other plays include *The Kitchen* (1957), *Chips with Everything* (1962), and *Shylock* (1976).

Wheeler, Monroe (1899–1988). American arts publisher, from the Midwest. He met the writer Glenway Wescott in Chicago in 1919, and they became lifelong companions. During the 1920s and early 1930s, they lived in Paris, where Wheeler designed and published deluxe-edition books, mostly with California millionare Barbara Harrison, who married Wescott's brother, Lloyd Wescott, in 1935. He returned to New York in the mid-1930s, joined the staff of the Museum of Modern Art, and in 1941 became director of exhibitions and publications. He oversaw 350 books during his career and became a trustee of the museum in 1967. He appears in *D.1*.

Whitcomb, Ian (b. 1941). British pianist, singer, ukulele player, record producer; educated at Bryanston and at Trinity College, Dublin. He moved to Los Angeles in 1965 when his song "You Turn Me On" reached number eight in the American Top Ten, and he performed there with The Beach Boys, The Rolling Stones, and The Kinks, among others. Later, he hosted a Los Angeles radio show and published books about the history of popular music, including *After the Ball: Pop Music from Rag to Rock* (1972), *Tin Pan Alley: A Pictorial History (1919–1939)*, and *Rock Odyssey: A Chronical of the Sixties* (1983).

White, Alan J. See Methuen.

Wilcox, Collin (1937–2009). American actress, also known as Collin Horne and sometimes credited as Wilcox-Horne or Wilcox-Paxton. She was educated at the University of Tennessee, the Goodman School of Drama in Chicago, and Actors Studio. She attracted praise on Broadway in *The Day the Money Stopped* (1958) and appeared in a few films, including *To Kill a Mockingbird* (1962), *Catch-22* (1970), and *Midnight in the Garden of Good and Evil* (1997). She also worked in T.V. She was married for a time to actor Geoffrey Horne. Later, she taught acting on the East Coast.

Wilding, Michael (1912–1979). British actor of stage and screen. His films include *In Which We Serve* (1942), *Dear Octopus* (1943), *English Without Tears* (1944), *An Ideal Husband* (1947), *Torch Song* (1953), and *Waterloo* (1970), and he played Sir Richard Fanshawe in the Isherwood-Bachardy "Frankenstein." He married four times. He was Elizabeth Taylor's second husband, from 1952 to 1957, and father of two of her children. His last marriage was to actress Margaret Leighton.

Williams, Clifford (1926–2005). British actor and director, educated at Highbury County Grammar School in London. He worked in mining, served in the army, and acted and directed in South Africa during the 1950s. In 1961, he became a staff producer at the Royal Shakespeare Company, and from 1963 until 1991 a prolific associate director. His many independent productions, some of which went on to successful Broadway runs, include an all-male *As You Like It*, Anthony Shaffer's *Sleuth*, both in 1967, and Kenneth Tynan's *Oh! Calcutta!* in 1970. He was married twice and had two daughters with his second wife.

Williams, Emlyn (1905–1987). Welsh playwright, screenwriter, and actor. He wrote psychological thrillers for the London stage, including *Night Must Fall* (1935), and is best known for *The Corn is Green* (1935), based on his own background in a coal-mining community in Wales and in which he played the lead; both of these plays were later filmed. His many other stage roles included Shakespeare and contemporary theater, and from the 1950s, he toured with one-man shows of Charles Dickens and Dylan Thomas. Among his films are *The Man Who Knew Too Much* (1934), *The Dictator* (1935), *Dead Men Tell No Tales* (1938), *Jamaica Inn* (1939), *The Stars Look Down* (1939), *Major Barbara* (1941), *Ivanhoe* (1952), *The Deep Blue Sea* (1955), *I Accuse!* (1958), *The L-Shaped Room* (1962), and *David Copperfield* (1970). His wife, Mary Carus-Wilson, known as Molly (d. 1970), was an actress under her maiden name, Molly O'Shann. Isherwood first met Williams and his wife in Hollywood in 1950; they appear in *D.1*.

Williams, Tennessee (1911–1983). American playwright; Thomas Lanier Williams was born in Mississippi and raised in St. Louis. His father was a travelling salesman, his mother felt herself to be a glamorous southern belle in reduced circumstances. *The Glass Menagerie* made him famous in 1945, followed by *A Streetcar Named Desire* (1947). Many of his subsequent plays are equally well known—*The Rose Tattoo* (1950), *Cat on a Hot Tin Roof* (1955), *Sweet Bird of Youth*

(1959), *The Night of the Iguana* (1961)—and were made into films. He also wrote a novella, *The Roman Spring of Mrs. Stone* (1950). When he first came to Hollywood in 1943 to work for MGM, he bore a letter of introduction to Isherwood from Lincoln Kirstein; this began a long and close friendship which Isherwood tells about in *D.1* and *Lost Years*. *Period of Adjustment*, for which, as Isherwood records, Bachardy designed the poster, opened in New York in November 1960, directed by George Roy Hill, with James Daley, Barbara Baxley and Robert Webber and was moderately successful, running to 132 performances. Later George Roy Hill directed Tony Franciosa and Jane Fonda in the film version.

Willie. See Maugham, William Somerset.

Wilson, Angus (1913–1991). British novelist and literary critic, educated at Westminster and Oxford. He worked as a decoder for the Foreign Office during World War II and otherwise made his career in the library of the British Museum before turning to writing full time in 1955. His novels include *Hemlock and After* (1952), *Anglo-Saxon Attitudes* (1956), *The Middle Age of Mrs. Eliot* (1958), *The Old Men at the Zoo* (1961), *No Laughing Matter* (1967), and *Setting the World on Fire* (1980). He also published short stories and literary-critical books about Zola, Dickens, and Kipling. From 1966, he was a professor of English literature at the University of East Anglia, where, with Malcolm Bradbury, he oversaw the first respected university course in creative writing in Britain. In *D.1* Isherwood tells that he met Wilson at a party in London in 1956. Wilson's companion, from the late 1940s onward, was Anthony Garrett.

Wilson, Colin (b. 1931). English novelist and critic. Author of psychological thrillers, studies of literature and philosophy, criminology, the imagination, the occult, and the supernatural. He became well known with his 1956 book, *The Outsider*—which Isherwood mentions twice in *D.1*—about the figure of the alienated solitary in modern literature. Isherwood met him in London in 1959.

Winslow, Al (d. 200[6]). American doctor and monk of the Ramakrishna Order, roughly five or six years younger than Isherwood. He was later known as Swami Sarveshananda. He was a disciple of Swami Nikhilananda at the New York Ramakrishna-Vivekananda Center on the Upper East Side, until they had an extreme falling-out, and Winslow left New York and settled in Chicago.

Winters, Shelley (1922–2006). American actress, from St. Louis. She worked on the New York stage and moved to Hollywood in 1943. She won Academy Awards for her supporting roles in *The Diary of Anne Frank* (1959) and *A Patch of Blue* (1965), and was nominated again for *The Poseidon Adventure* (1972). She played Natalia Landauer in the film of *I Am a Camera* in 1955. From the early 1990s, she played Roseanne's grandmother in the T.V. comedy series "Roseanne." Isherwood met her at the start of the 1950s, and she appears in *D.1*. She was married to actor Tony Franciosa from 1957 to 1960.

Wintle, Hector. Isherwood's contemporary at Repton, where his novel about life in the sixth form inspired Isherwood to start writing one, too. Later, Isherwood followed Wintle to medical school. Wintle became a doctor, and he published three novels, *The Final Victory* (1935), *Edgar Prothero* (1936), and *The Hodsall Wizard* (1938). He appears as Philip Linsley in *Lions and Shadows*, and the character Philip Lindsay in *All the Conspirators* is partly based on him.

Wonner, Paul (1920–2008). American painter, originally from Arizona. He settled on the West Coast, first in Los Angeles, and from the early 1980s in San Francisco, with his longtime partner, the painter Bill Brown. Wonner's work has been called American Realist or Californian Realist, and he has been grouped with other painters who came to maturity in the 1950s—Richard Diebenkorn, Wayne Thiebaud, David Park, Nathan Oliveira—as a Bay Area Figurative Artist or Figurative Abstractionist. He taught painting at UCLA and elsewhere, and, with Brown, had close friendships with a number of writers, for instance, May Sarton, Diane Johnson, and Janet Frame.

Wood, Christopher (Chris) (d. 1976). Isherwood met Chris Wood in September 1932 when Auden took him to meet Gerald Heard, then sharing Wood's luxurious London flat. Wood was about ten years younger than Heard, handsome and friendly but shy about his maverick talents. He played the piano well, but never professionally, wrote short stories, but not for publication, had a pilot's license, and rode a bicycle for transport. He was extremely rich (the family business made jams and other canned and bottled goods), sometimes extravagant, and always generous; he secretly funded many of Heard's projects and loaned or gave money to many other friends (including Isherwood). In 1937, he emigrated with Heard to Los Angeles and, in 1941, moved with him to Laguna. Their domestic commitment persisted for a time despite Heard's increasing asceticism and religious activities, but eventually they lived separately though they remained friends. From 1939, Wood was involved with Paul Sorel, also a member of their shared household for about five years. Wood appears throughout *D.1* and *Lost Years*.

Wood, Natalie (1938–1981). American actress, of French and Russian background; her mother was a ballet dancer. She appeared in her first movie when she was five years old and was a star by the time she was nine. Her films include: *Happy Land* (1943), *Tomorrow is Forever* (1946), *Miracle on 34th Street* (1947), *Rebel Without a Cause* (1955), *The Searchers* (1956), *Splendor in the Grass* (1961), *West Side Story* (1961), *Gypsy* (1962), *Inside Daisy Clover* (1966), *Penelope* (1966), and *Bob and Carol and Ted and Alice* (1969). She married twice to the same man, actor Robert Wagner, and in between married Richard Gregson, father of her daughter Natasha Gregson Wagner, also an actress. She drowned after falling overboard from a yacht.

Woodbury, Dana. A neighbor of Dorothy Parker and Alan Campbell; he lived across the street from them on Norma Place in West Hollywood, and they introduced him to Isherwood. Bachardy painted his portrait several times.

Woodcock, Patrick (1920–2002). British doctor. His patients included many London theater stars—including John Gielgud and Noel Coward—and also actors based in New York and Hollywood. Isherwood and Bachardy first met him with Gielgud and Hugh Wheeler in London in 1956, and a few days later Woodcock visited Bachardy at the Cavendish Hotel to treat him for stomach cramps. Woodcock also prescribed vitamins for Isherwood at the same time and became a friend. He appears in *D.1*.

Woolf, James (Jimmy) (1919–1966). British film producer. With his older brother, John Woolf, he produced *Room at the Top* (1958) starring Simone Signoret (who won an Academy Award) and Laurence Harvey, rumored to have been his lover. The brothers also produced *Moulin Rouge* (1952), *Of Human Bondage* (1964), *Life at the Top* (1965), and *Oliver!* (1968). Jimmy Woolf committed suicide.

Worsley, T.C. (Cuthbert) (1907–1977). English writer, theater critic, and schoolmaster. He was a friend of Stephen Spender and in 1937 accompanied him to the Spanish Civil War on an assignment for *The Daily Worker*. He returned to Spain to join an ambulance unit and wrote about this period for *The Left Review* and in *Behind the Battle* (1939), as well as in a later fictionalized memoir, *Fellow Travellers* (1971). During the 1950s, he was a theater critic for *The Financial Times*. John Luscombe was his much younger companion. Worsley appears in *D.1* and *Lost Years*.

Worton, Len (d. late 1990s). English-born monk of the Ramakrishna Order; he settled at Trabuco and became Swami Bhadrananda. He was a British army medic during World War II, stationed in Hong Kong, and was captured there by the Japanese when he remained behind tending the wounded. He was spared execution because of his medical skills and because he was wearing Red Cross insignia. He spent four years as a prisoner of war in Japan.

Wright, Thomas E. (Tom). American writer, from New Orleans. Isherwood met him while both were fellows-in-residence at the Huntington Hartford Foundation in 1951, and they had a casual affair which lasted about eight months. Wright was then about twenty-four years old and had taught at New York University. He was a childhood friend of Speed Lamkin and of Speed's sister, Marguerite, and he knew Tennessee Williams and Howard Griffin. He also became close to Edward James and lived at James's Hollywood house as a caretaker for a number of years. Bachardy drew the first of several portraits of Wright in 1960. Wright published novels, travel books—including one about Mexico which Isherwood mentions reading, *Into the Maya World* (1969)—and *Growing Up with Legends: A Literary Memoir*. He eventually settled in Guatemala. He appears in *D.1*.

Wyberslegh Hall. The fifteenth-century manor house where Isherwood was born and where his mother returned to live with his brother after the war; it was part of the Bradshaw Isherwood estate. During most of the nineteenth century, Wyberslegh was leased to a family called Cooper who farmed the surrounding fields, lived in the rear wing of the house, and rented out the front rooms. When Isherwood's parents married in 1903, his grandfather, John Isherwood, divided Wyberslegh Hall into two separate houses. Frank and Kathleen Isherwood moved into the front of the house, and, for roughly half the twentieth century, the Coopers continued to live in the back overlooking the farmyard. The Nazi incendiary bombs which Isherwood mentions were dropped on January 9, 1941, when a tenant was occupying the front half of Wyberslegh. Richard was then performing his wartime national service as a farmhand at Wyberslegh farm and helped Jack Smith and other farmhands put out the fires.

Wystan. See Auden, W.H.

Yatiswarananda, Swami. Hindu monk of the Ramakrishna Order. He was sent from India to live at the Vedanta centers in Switzerland and Philadelphia. In the early 1940s, he supervised the Hollywood Vedanta Society during an absence of Prabhavananda. The visit is described in *D.1*.

yoga. Union with the Godhead or one of the methods by which the individual soul may achieve such union. Isherwood and Prabhavananda translated the yoga sutras or aphorisms of the Hindu sage Patanjali (*circa* 4th C. BC to 4th C. AD) as *How to Know God* (1953); they describe the aphorisms as a compilation and reformulation of yoga philosophy and practices—spiritual disciplines and techniques of meditation—referred to even earlier in four of the Upanishads and in fact handed down from pre-historic times.

York, Michael (b. 1942) and Pat McCallum. British film star and his wife, a photographer and writer. He was educated at Oxford, where he acted for the Oxford University Dramatic Society. His real name is Michael York-Johnson. In 1965, he had a small role in Franco Zeffirelli's stage production of *The Taming of the Shrew*, for Olivier's National Theatre Company and afterwards appeared in the movie with Richard Burton and Elizabeth Taylor. Subsequent films include *Romeo and Juliet* (1968), *Justine* (1969), *Something for Everyone* (1970), *Cabaret* (1972), *England Made Me* (1973), *The Three Musketeers* (1974), *Murder on the Orient Express* (1974), *The Weather in the Streets* (1983), *The Wide Sargasso Sea* (1993), *Austin Powers: International Man of Mystery* (1997), *Austin Powers: The Spy Who Shagged Me* (1999), *Borstal Boy* (2000), *Megiddo* (2001), and *Austin Powers: Goldmember* (2002). He has often appeared on television, notably in "The Forsyte Saga" and "Curb Your Enthusiasm," and on Broadway, and he has a prize-winning career recording audio books. She was born in Jamaica to an English diplomat and his American wife, educated in England and France, married for the first time as a teenager, and in 1952 had a son, Rick McCallum, later a Hollywood producer. From the early 1960s, she worked as a fashion journalist at *Vogue*, then as a photographer and travel editor at *Glamor* before becoming a freelance photographer with a few lessons from David Bailey. Her portraits of celebrities have appeared in *Life*, *People*, *Town and Country*, *Playboy*, and *Newsweek*; she has published several books and had numerous gallery exhibitions, in particular of her later avant-garde work featurig dissected cadavers and nude portraits of professional people at work. She met Michael York when photographing him for a magazine profile, and they married a year later, in 1968. Isherwood was introduced to them in 1969, probably at the Vadims' lunch party described in the diary entry of January 15.

Yorke, Henry (1905–73). English novelist, educated at Eton and Oxford; his pen name was Henry Green. His mother, a daughter of Lord Leconfield, was brought up at Petworth, and his father, trained as a classicist, became wealthy manufacturing brewery and plumbing equipment. Yorke worked in a factory of his father's and rose through the firm to become a managing director. His novels draw on his experience of both working-class and upper-class life, and also on his time in the National Fire Service during World War II: *Blindness* (1926), *Living* (1929), *Party Going* (1939), *Caught* (1943), *Loving* (1945), *Back* (1946), *Concluding* (1948), *Nothing* (1950), *Doting* (1952). He also wrote a memoir, *Pack My Bag: A Self Portrait* (1940). He married Adelaide "Dig" Biddulph, daughter of 2nd Baron Biddulph, in 1929, and they had a son, Sebastian, but Yorke had a reputation as an adulterer and, during his last two decades, a heavy drinker. Isherwood first mentions him and his wife during his 1947 trip to England recorded in *Lost Years*; the Yorkes also appear in *D.1*.

Zeigel, John (Johnny, Jack) (b. 1934). American professor of literature; a doctor's son; educated in Sewanee, Tennessee, at Pomona College and Claremont Graduate University in California, and at Harvard. He majored in classics as an undergraduate and taught ancient Greek and Latin language and literature as well as modern literature. He also plays the violin. Zeigel was a Ph.D. candidate at Claremont and a teaching assistant at Pomona when Isherwood became friendly with him in 1961. Evidently they met in the late 1950s, probably through Zeigel's companion, Ed Halsey, who was briefly a monk at Trabuco. Zeigel met Gerald Heard who arranged counseling with Evelyn Hooker in 1957 or 1958. Hooker certified Zeigel was homosexual so he would not be drafted to fight in the Korean War. Isherwood gave Zeigel an introduction to the poet Witter Bynner, who had known Willa Cather, and after visiting Bynner in Santa Fe, Zeigel decided to write his dissertation about Cather. He moved with Halsey to Ajijic, Mexico, to begin working on the dissertation, then in 1962 took a part-time teaching job back in Pasadena at the California Institute of Technology. Halsey was killed in a car crash that autumn, a tragedy on which Isherwood possibly drew in *A Single Man*. Zeigel, instead of returning to Mexico as he had planned, became an assistant professor of English and Humanities at Cal. Tech. In 1975, after the death of his father, he left California to care for his mother and to manage family investments and a cattle ranch in Colbran, Colorado where his parents lived. He was made professor of English at Mesa State College in Grand Junction, Colorado, and spent the rest of his teaching career there, until he retired in 1998.

Index

Works by Isherwood appear directly under title; works by others appear under authors' names.

Page numbers in *italic* indicate entries in the Glossary (pages 601–722).